T0145103

# IFIP Advances in Information and Communication Technology

**631**

## Editor-in-Chief

*Kai Rannenberg, Goethe University Frankfurt, Germany*

## Editorial Board Members

TC 1 – Foundations of Computer Science
*Luís Soares Barbosa, University of Minho, Braga, Portugal*

TC 2 – Software: Theory and Practice
*Michael Goedicke, University of Duisburg-Essen, Germany*

TC 3 – Education
*Arthur Tatnall, Victoria University, Melbourne, Australia*

TC 5 – Information Technology Applications
*Erich J. Neuhold, University of Vienna, Austria*

TC 6 – Communication Systems
*Burkhard Stiller, University of Zurich, Zürich, Switzerland*

TC 7 – System Modeling and Optimization
*Fredi Tröltzsch, TU Berlin, Germany*

TC 8 – Information Systems
*Jan Pries-Heje, Roskilde University, Denmark*

TC 9 – ICT and Society
*David Kreps, National University of Ireland, Galway, Ireland*

TC 10 – Computer Systems Technology
*Ricardo Reis, Federal University of Rio Grande do Sul, Porto Alegre, Brazil*

TC 11 – Security and Privacy Protection in Information Processing Systems
*Steven Furnell, Plymouth University, UK*

TC 12 – Artificial Intelligence
*Eunika Mercier-Laurent, University of Reims Champagne-Ardenne, Reims, France*

TC 13 – Human-Computer Interaction
*Marco Winckler, University of Nice Sophia Antipolis, France*

TC 14 – Entertainment Computing
*Rainer Malaka, University of Bremen, Germany*

# IFIP – The International Federation for Information Processing

IFIP was founded in 1960 under the auspices of UNESCO, following the first World Computer Congress held in Paris the previous year. A federation for societies working in information processing, IFIP's aim is two-fold: to support information processing in the countries of its members and to encourage technology transfer to developing nations. As its mission statement clearly states:

> *IFIP is the global non-profit federation of societies of ICT professionals that aims at achieving a worldwide professional and socially responsible development and application of information and communication technologies.*

IFIP is a non-profit-making organization, run almost solely by 2500 volunteers. It operates through a number of technical committees and working groups, which organize events and publications. IFIP's events range from large international open conferences to working conferences and local seminars.

The flagship event is the IFIP World Computer Congress, at which both invited and contributed papers are presented. Contributed papers are rigorously refereed and the rejection rate is high.

As with the Congress, participation in the open conferences is open to all and papers may be invited or submitted. Again, submitted papers are stringently refereed.

The working conferences are structured differently. They are usually run by a working group and attendance is generally smaller and occasionally by invitation only. Their purpose is to create an atmosphere conducive to innovation and development. Refereeing is also rigorous and papers are subjected to extensive group discussion.

Publications arising from IFIP events vary. The papers presented at the IFIP World Computer Congress and at open conferences are published as conference proceedings, while the results of the working conferences are often published as collections of selected and edited papers.

IFIP distinguishes three types of institutional membership: Country Representative Members, Members at Large, and Associate Members. The type of organization that can apply for membership is a wide variety and includes national or international societies of individual computer scientists/ICT professionals, associations or federations of such societies, government institutions/government related organizations, national or international research institutes or consortia, universities, academies of sciences, companies, national or international associations or federations of companies.

More information about this series at http://www.springer.com/series/6102

Alexandre Dolgui · Alain Bernard ·
David Lemoine · Gregor von Cieminski ·
David Romero (Eds.)

# Advances in Production Management Systems

## Artificial Intelligence for Sustainable and Resilient Production Systems

IFIP WG 5.7 International Conference, APMS 2021
Nantes, France, September 5–9, 2021
Proceedings, Part II

 Springer

*Editors*
Alexandre Dolgui ⓘ
IMT Atlantique
Nantes, France

David Lemoine ⓘ
IMT Atlantique
Nantes, France

David Romero ⓘ
Tecnológico de Monterrey
Mexico City, Mexico

Alain Bernard ⓘ
Centrale Nantes
Nantes, France

Gregor von Cieminski ⓘ
ZF Friedrichshafen AG
Friedrichshafen, Germany

ISSN 1868-4238          ISSN 1868-422X   (electronic)
IFIP Advances in Information and Communication Technology
ISBN 978-3-030-85904-6          ISBN 978-3-030-85902-2   (eBook)
https://doi.org/10.1007/978-3-030-85902-2

© IFIP International Federation for Information Processing 2021
This work is subject to copyright. All rights are reserved by the Publisher, whether the whole or part of the material is concerned, specifically the rights of translation, reprinting, reuse of illustrations, recitation, broadcasting, reproduction on microfilms or in any other physical way, and transmission or information storage and retrieval, electronic adaptation, computer software, or by similar or dissimilar methodology now known or hereafter developed.
The use of general descriptive names, registered names, trademarks, service marks, etc. in this publication does not imply, even in the absence of a specific statement, that such names are exempt from the relevant protective laws and regulations and therefore free for general use.
The publisher, the authors and the editors are safe to assume that the advice and information in this book are believed to be true and accurate at the date of publication. Neither the publisher nor the authors or the editors give a warranty, expressed or implied, with respect to the material contained herein or for any errors or omissions that may have been made. The publisher remains neutral with regard to jurisdictional claims in published maps and institutional affiliations.

This Springer imprint is published by the registered company Springer Nature Switzerland AG
The registered company address is: Gewerbestrasse 11, 6330 Cham, Switzerland

# Preface

The scientific and industrial relevance of the development of sustainable and resilient production systems lies in ensuring future-proof manufacturing and service systems, including their supply chains and logistics networks. "Sustainability" and "Resilience" are essential requirements for competitive manufacturing and service provisioning now and in the future. Industry 4.0 technologies, such as artificial intelligence; decision aid models; additive and hybrid manufacturing; augmented, virtual, and mixed reality; industrial, collaborative, mobile, and software robots; advanced simulations and digital twins; and smart sensors and intelligent industrial networks, are key enablers for building new digital and smart capabilities in emerging cyber-physical production systems in support of more efficient and effective operations planning and control. These allow manufacturers and service providers to explore more sustainable and resilient business and operating models. By making innovative use of the aforementioned technologies and their enabled capabilities, they can pursue the triple bottom line of economic, environmental, and social sustainability. Furthermore, industrial companies will be able to withstand and quickly recover from disruptions that pose threats to their operational continuity. This is in the face of disrupted, complex, turbulent, and uncertain business environments, like the one triggered by the COVID-19 pandemic, or environmental pressures calling for decoupling economic growth from resource use and emissions.

The International Conference on Advances in Production Management Systems 2021 (APMS 2021) in Nantes, France, brought together leading international experts on manufacturing, service, supply, and logistics systems from academia, industry, and government to discuss pressing issues and research opportunities mostly in smart manufacturing and cyber-physical production systems; service systems design, engineering, and management; digital lean operations management; and resilient supply chain management in the Industry 4.0 era, with particular focus on artificial intelligence-enabled solutions.

Under the influence of the COVID-19 pandemic, the event was organised as online conference sessions. A large international panel of experts (497 from 50 countries) reviewed all the submissions (with an average of 3.2 reviews per paper) and selected the best 377 papers (70% of the submitted contributions) to be included in these international conference proceedings. The topics of interest at APMS 2021 included artificial intelligence techniques, decision aid, and new and renewed paradigms for sustainable and resilient production systems at four-wall factory and value chain levels, comprising their associated models, frameworks, methods, tools, and technologies for smart and sustainable manufacturing and service systems, as well as resilient digital supply chains. As usual for the APMS conference, the Program Committee was particularly attentive to the cutting-edge problems in production management and the quality of the papers, especially with regard to the applicability of the contributions to industry and services.

The APMS 2021 conference proceedings are organized into five volumes covering a large spectre of research concerning the global topic of the conference: "Artificial Intelligence for Sustainable and Resilient Production Systems".

The conference was supported by the International Federation of Information Processing (IFIP), which is celebrating its 60th Anniversary, and was co-organized by the IFIP Working Group 5.7 on Advances in Production Management Systems, IMT Atlantique (Campus Nantes) as well as the Centrale Nantes, University of Nantes, Rennes Business School, and Audecia Business School. It was also supported by three leading journals in the discipline: Production Planning & Control (PPC), the International Journal of Production Research (IJPR), and the International Journal of Product Lifecycle Management (IJPLM).

Special attention has been given to the International Journal of Production Research on the occasion of its 60th Anniversary. Since its foundation in 1961, IJPR has become one of the flagship journals of our profession. It was the first international journal to bring together papers on all aspects of production research: product/process engineering, production system design and management, operations management, and logistics. Many exceptional scientific results have been published in the journal.

We would like to thank all contributing authors for their high-quality work and for their willingness to share their research findings with the APMS community. We are also grateful to the members of the IFIP Working Group 5.7, the Program Committee, and the Scientific Committee, along with the Special Sessions organizers for their support in the organization of the conference program. Concerning the number of papers, special thanks must be given to the local colleagues who managed the reviewing process as well as the preparation of the conference program and proceedings, particularly Hicham Haddou Benderbal and Maria-Isabel Estrepo-Ruiz from IMT Atlantique.

September 2021

Alexandre Dolgui
Alain Bernard
David Lemoine
Gregor von Cieminski
David Romero

# Organization

## Conference Chair

Alexandre Dolgui        IMT Atlantique, Nantes, France

## Conference Co-chair

Gregor von Cieminski      ZF Friedrichshafen, Germany

## Conference Honorary Co-chairs

Dimitris Kiritsis         EPFL, Switzerland
Kathryn E. Stecke       University of Texas at Dallas, USA

## Program Chair

Alain Bernard          Centrale Nantes, France

## Program Co-chair

David Romero          Tecnológico de Monterrey, Mexico

## Program Committee

Alain Bernard          Centrale Nantes, France
Gregor von Cieminski      ZF Friedrichshafen, Germany
Alexandre Dolgui        IMT Atlantique, Nantes, France
Dimitris Kiritsis         EPFL, Switzerland
David Romero          Tecnológico de Monterrey, Mexico
Kathryn E. Stecke       University of Texas at Dallas, USA

## International Advisory Committee

Farhad Ameri          Texas State University, USA
Ugljesa Marjanovic     University of Novi Sad, Serbia
Ilkyeong Moon         Seoul National University, South Korea
Bojan Lalic           University of Novi Sad, Serbia
Hermann Lödding      Hamburg University of Technology, Germany

## Organizing Committee Chair

David Lemoine         IMT Atlantique, Nantes, France

## Organizing Committee Co-chair

| | |
|---|---|
| Hichem Haddou Benderbal | IMT Atlantique, Nantes, France |

## Doctoral Workshop Chairs

| | |
|---|---|
| Abdelkrim-Ramzi Yelles-Chaouche | IMT Atlantique, Nantes, France |
| Seyyed-Ehsan Hashemi-Petroodi | IMT Atlantique, Nantes, France |

## Award Committee Chairs

| | |
|---|---|
| Nadjib Brahimi | Rennes School of Business, France |
| Ramzi Hammami | Rennes School of Business, France |

## Organizing Committee

| | |
|---|---|
| Romain Billot | IMT Atlantique, Brest, France |
| Nadjib Brahimi | Rennes School of Business, France |
| Olivier Cardin | University of Nantes, France |
| Catherine Da Cunha | Centrale Nantes, France |
| Alexandre Dolgui | IMT Atlantique, Nantes, France |
| Giannakis Mihalis | Audencia, Nantes, France |
| Evgeny Gurevsky | University of Nantes, France |
| Hichem Haddou Benderbal | IMT Atlantique, Nantes, France |
| Ramzi Hammami | Rennes School of Business, France |
| Oncu Hazir | Rennes School of Business, France |
| Seyyed-Ehsan Hashemi-Petroodi | IMT Atlantique, Nantes, France |
| David Lemoine | IMT Atlantique, Nantes, France |
| Nasser Mebarki | University of Nantes, France |
| Patrick Meyer | IMT Atlantique, Brest, France |
| Merhdad Mohammadi | IMT Atlantique, Brest, France |
| Dominique Morel | IMT Atlantique, Nantes, France |
| Maroua Nouiri | University of Nantes, France |
| Maria-Isabel Restrepo-Ruiz | IMT Atlantique, Nantes, France |
| Naly Rakoto | IMT Atlantique, Nantes, France |
| Ilhem Slama | IMT Atlantique, Nantes, France |
| Simon Thevenin | IMT Atlantique, Nantes, France |
| Abdelkrim-Ramzi Yelles-Chaouche | IMT Atlantique, Nantes, France |

# Scientific Committee

| | |
|---|---|
| Erry Yulian Triblas Adesta | International Islamic University Malaysia, Malaysia |
| El-Houssaine Aghezzaf | Ghent University, Belgium |
| Erlend Alfnes | Norwegian University of Science and Technology, Norway |
| Hamid Allaoui | Université d'Artois, France |
| Thecle Alix | IUT Bordeaux Montesquieu, France |
| Farhad Ameri | Texas State University, USA |
| Bjørn Andersen | Norwegian University of Science and Technology, Norway |
| Eiji Arai | Osaka University, Japan |
| Jannicke Baalsrud Hauge | KTH Royal Institute of Technology, Sweden/BIBA, Germany |
| Zied Babai | Kedge Business School, France |
| Natalia Bakhtadze | Russian Academy of Sciences, Russia |
| Pierre Baptiste | Polytechnique de Montréal, Canada |
| Olga Battaïa | Kedge Business School, France |
| Farouk Belkadi | Centrale Nantes, France |
| Lyes Benyoucef | Aix-Marseille University, France |
| Bopaya Bidanda | University of Pittsburgh, USA |
| Frédérique Biennier | INSA Lyon, France |
| Jean-Charles Billaut | Université de Tours, France |
| Umit S. Bititci | Heriot-Watt University, UK |
| Magali Bosch-Mauchand | Université de Technologie de Compiègne, France |
| Xavier Boucher | Mines St Etienne, France |
| Abdelaziz Bouras | Qatar University, Qatar |
| Jim Browne | University College Dublin, Ireland |
| Luis Camarinha-Matos | Universidade Nova de Lisboa, Portugal |
| Olivier Cardin | University of Nantes, France |
| Sergio Cavalieri | University of Bergamo, Italy |
| Stephen Childe | Plymouth University, UK |
| Hyunbo Cho | Pohang University of Science and Technology, South Korea |
| Chengbin Chu | ESIEE Paris, France |
| Feng Chu | Paris-Saclay University, France |
| Byung Do Chung | Yonsei University, South Korea |
| Gregor von Cieminski | ZF Friedrichshafen, Germany |
| Catherine Da Cunha | Centrale Nantes, France |
| Yves Dallery | CentraleSupélec, France |
| Xavier Delorme | Mines St Etienne, France |
| Frédéric Demoly | Université de Technologie de Belfort-Montbéliard, France |
| Mélanie Despeisse | Chalmers University of Technology, Sweden |
| Alexandre Dolgui | IMT Atlantique, Nantes, France |
| Slavko Dolinšek | University of Ljubljana, Slovenia |

| Sang Do Noh | Sungkyunkwan University, South Korea |
| Heidi Carin Dreyer | Norwegian University of Science and Technology, Norway |
| Eero Eloranta | Aalto University, Finland |
| Soumaya El Kadiri | Texelia AG, Switzerland |
| Christos Emmanouilidis | University of Groningen, The Netherlands |
| Anton Eremeev | Siberian Branch of Russian Academy of Sciences, Russia |
| Åsa Fasth-Berglund | Chalmers University of Technology, Sweden |
| Rosanna Fornasiero | Consiglio Nazionale delle Ricerche, Italy |
| Xuehao Feng | Zhejiang University, China |
| Yannick Frein | INP Grenoble, France |
| Jan Frick | University of Stavanger, Norway |
| Klaas Gadeyne | Flanders Make, Belgium |
| Paolo Gaiardelli | University of Bergamo, Italy |
| Adriana Giret Boggino | Universidad Politécnica de Valencia, Spain |
| Samuel Gomes | Belfort-Montbéliard University of Technology, France |
| Bernard Grabot | INP-Toulouse, ENIT, France |
| Gerhard Gudergan | RWTH Aachen University, Germany |
| Thomas R. Gulledge Jr. | George Mason University, USA |
| Nikolai Guschinsky | National Academy of Sciences, Belarus |
| Slim Hammadi | Centrale Lille, France |
| Ahmedou Haouba | University of Nouakchott Al-Asriya, Mauritania |
| Soumaya Henchoz | Logitech AG, Switzerland |
| Hironori Hibino | Tokyo University of Science, Japan |
| Hans-Henrik Hvolby | Aalborg University, Denmark |
| Jan Holmström | Aalto University, Finland |
| Dmitry Ivanov | Berlin School of Economics and Law, Germany |
| Harinder Jagdev | National University of Ireland at Galway, Ireland |
| Jayanth Jayaram | University of South Carolina, USA |
| Zhibin Jiang | Shanghai Jiao Tong University, China |
| John Johansen | Aalborg University, Denmark |
| Hong-Bae Jun | Hongik University, South Korea |
| Toshiya Kaihara | Kobe University, Japan |
| Duck Young Kim | Pohang University of Science and Technology, South Korea |
| Dimitris Kiritsis | EPFL, Switzerland |
| Tomasz Koch | Wroclaw University of Science and Technology, Poland |
| Pisut Koomsap | Asian Institute of Technology, Thailand |
| Vladimir Kotov | Belarusian State University, Belarus |
| Mikhail Kovalyov | National Academy of Sciences, Belarus |
| Gül Kremer | Iowa State University, USA |
| Boonserm Kulvatunyou | National Institute of Standards and Technology, USA |
| Senthilkumaran Kumaraguru | Indian Institute of Information Technology Design and Manufacturing, India |

| | |
|---|---|
| Thomas R. Kurfess | Georgia Institute of Technology, USA |
| Andrew Kusiak | University of Iowa, USA |
| Bojan Lalić | University of Novi Sad, Serbia |
| Samir Lamouri | ENSAM Paris, France |
| Lenka Landryova | Technical University of Ostrava, Czech Republic |
| Alexander Lazarev | Russian Academy of Sciences, Moscow, Russia |
| Jan-Peter Lechner | First Global Liaison, Germany |
| Gyu M. Lee | Pusan National University, South Korea |
| Kangbok Lee | Pohang University of Science and Technology, South Korea |
| Genrikh Levin | National Academy of Sciences, Belarus |
| Jingshan Li | University of Wisconsin-Madison, USA |
| Ming K. Lim | Chongqing University, China |
| Hermann Lödding | Hamburg University of Technology, Germany |
| Pierre Lopez | LAAS-CNRS, France |
| Marco Macchi | Politecnico di Milano, Italy |
| Ugljesa Marjanovic | University of Novi Sad, Serbia |
| Muthu Mathirajan | Indian Institute of Science, India |
| Gökan May | University of North Florida, USA |
| Khaled Medini | Mines St Etienne, France |
| Jörn Mehnen | University of Strathclyde, UK |
| Vidosav D. Majstorovich | University of Belgrade, Serbia |
| Semyon M. Meerkov | University of Michigan, USA |
| Joao Gilberto Mendes dos Reis | UNIP Paulista University, Brazil |
| Hajime Mizuyama | Aoyama Gakuin University, Japan |
| Ilkyeong Moon | Seoul National University, South Korea |
| Eiji Morinaga | Osaka Prefecture University, Japan |
| Dimitris Mourtzis | University of Patras, Greece |
| Irenilza de Alencar Naas | UNIP Paulista University, Brazil |
| Masaru Nakano | Keio University, Japan |
| Torbjörn Netland | ETH Zürich, Switzerland |
| Gilles Neubert | EMLYON Business School, Saint-Etienne, France |
| Izabela Nielsen | Aalborg University, Denmark |
| Tomomi Nonaka | Ritsumeikan University, Japan |
| Jinwoo Park | Seoul National University, South Korea |
| François Pérès | INP-Toulouse, ENIT, France |
| Fredrik Persson | Linköping Institute of Technology, Sweden |
| Giuditta Pezzotta | University of Bergamo, Italy |
| Selwyn Piramuthu | University of Florida, USA |
| Alberto Portioli Staudacher | Politecnico di Milano, Italy |
| Daryl Powell | Norwegian University of Science and Technology, Norway |
| Vittaldas V. Prabhu | Pennsylvania State University, USA |
| Jean-Marie Proth | Inria, France |
| Ricardo José Rabelo | Federal University of Santa Catarina, Brazil |

| | |
|---|---|
| Rahul Rai | University at Buffalo, USA |
| Mario Rapaccini | Florence University, Italy |
| Nidhal Rezg | University of Lorraine, France |
| Ralph Riedel | Westsächsische Hochschule Zwickau, Germany |
| Irene Roda | Politecnico di Milano, Italy |
| Asbjörn Rolstadås | Norwegian University of Science and Technology, Norway |
| David Romero | Tecnológico de Monterrey, Mexico |
| Christoph Roser | Karlsruhe University of Applied Sciences, Germany |
| André Rossi | Université Paris-Dauphine, France |
| Martin Rudberg | Linköping University, Sweden |
| Thomas E. Ruppli | University of Basel, Switzerland |
| Krzysztof Santarek | Warsaw University of Technology, Poland |
| Subhash Sarin | VirginiaTech, USA |
| Suresh P. Sethi | The University of Texas at Dallas, USA |
| Fabio Sgarbossa | Norwegian University of Science and Technology, Norway |
| John P. Shewchuk | Virginia Polytechnic Institute and State University, USA |
| Dan L. Shunk | Arizona State University, USA |
| Ali Siadat | Arts et Métiers ParisTech, France |
| Riitta Smeds | Aalto University, Finland |
| Boris Sokolov | Russian Academy of Sciences, Russia |
| Vijay Srinivasan | National Institute of Standards and Technology, USA |
| Johan Stahre | Chalmers University of Technology, Sweden |
| Kathryn E. Stecke | The University of Texas at Dallas, USA |
| Kenn Steger-Jensen | Aalborg University, Denmark |
| Volker Stich | RWTH Aachen University, Germany |
| Richard Lee Storch | University of Washington, USA |
| Jan Ola Strandhagen | Norwegian University of Science and Technology, Norway |
| Stanislaw Strzelczak | Warsaw University of Technology, Poland |
| Nick Szirbik | University of Groningen, The Netherlands |
| Marco Taisch | Politecnico di Milano, Italy |
| Lixin Tang | Northeastern University, China |
| Kari Tanskanen | Aalto University School of Science, Finland |
| Ilias Tatsiopoulos | National Technical University of Athens, Greece |
| Sergio Terzi | Politecnico di Milano, Italy |
| Klaus-Dieter Thoben | Universität Bremen, Germany |
| Manoj Tiwari | Indian Institute of Technology, India |
| Matthias Thüre | Jinan University, China |
| Jacques H. Trienekens | Wageningen University, The Netherlands |
| Mario Tucci | Universitá degli Studi di Firenze, Italy |
| Shigeki Umeda | Musashi University, Japan |
| Bruno Vallespir | University of Bordeaux, France |
| François Vernadat | University of Lorraine, France |

Agostino Villa              Politecnico di Torino, Italy
Lihui Wang                 KTH Royal Institute of Technology, Sweden
Sabine Waschull            University of Groningen, The Netherlands
Hans-Hermann Wiendahl      University of Stuttgart, Germany
Frank Werner               University of Magdeburg, Germany
Shaun West                 Lucerne University of Applied Sciences and Arts,
                             Switzerland
Joakim Wikner              Jönköping University, Sweden
Hans Wortmann              University of Groningen, The Netherlands
Desheng Dash Wu            University of Chinese Academy of Sciences, China
Thorsten Wuest             West Virginia University, USA
Farouk Yalaoui             University of Technology of Troyes, France
Noureddine Zerhouni        Université Bourgogne Franche-Comte, France

## List of Reviewers

Abbou Rosa
Abdeljaouad Mohamed Amine
Absi Nabil
Acerbi Federica
Aghelinejad Mohsen
Aghezzaf El-Houssaine
Agrawal Rajeev
Agrawal Tarun Kumar
Alexopoulos Kosmas
Alix Thecle
Alkhudary Rami
Altekin F. Tevhide
Alves Anabela
Ameri Farhad
Andersen Ann-Louise
Andersen Bjorn
Anderson Marc
Anderson Matthew
Anholon Rosley
Antosz Katarzyna
Apostolou Dimitris
Arica Emrah
Arlinghaus Julia Christine
Aubry Alexis
Baalsrud Hauge Jannicke
Badulescu Yvonne Gabrielle
Bakhtadze Natalia
Barbosa Christiane Lima
Barni Andrea

Batocchio Antonio
Battaïa Olga
Battini Daria
Behrens Larissa
Ben-Ammar Oussama
Benatia Mohamed Amin
Bentaha M.-Lounes
Benyoucef Lyes
Beraldi Santos Alexandre
Bergmann Ulf
Bernus Peter
Berrah Lamia-Amel
Bertnum Aili Biriita
Bertoni Marco
Bettayeb Belgacem
Bevilacqua Maurizio
Biennier Frédérique
Bititci Umit Sezer
Bocanet Vlad
Bosch-Mauchand Magali
Boucher Xavier
Bourguignon Saulo Cabral
Bousdekis Alexandros
Brahimi Nadjib
Bresler Maggie
Brunoe Thomas Ditlev
Brusset Xavier
Burow Kay
Calado Robisom Damasceno

Calarge Felipe
Camarinha-Matos Luis Manuel
Cameron David
Cannas Violetta Giada
Cao Yifan
Castro Eduardo Lorenzo
Cattaruzza Diego
Cerqueus Audrey
Chang Tai-Woo
Chaves Sandra Maria do Amaral
Chavez Zuhara
Chen Jinwei
Cheng Yongxi
Chiacchio Ferdinando
Chiari da Silva Ethel Cristina
Childe Steve
Cho Hyunbo
Choi SangSu
Chou Shuo-Yan
Christensen Flemming Max Møller
Chung Byung Do
Ciarapica Filippo Emanuele
Cimini Chiara
Clivillé Vincent
Cohen Yuval
Converso Giuseppe
Cosenza Harvey
Costa Helder Gomes
Da Cunha Catherine
Daaboul Joanna
Dahane Mohammed
Dakic Dusanka
Das Dyutimoy Nirupam
Das Jyotirmoy Nirupam
Das Sayan
Davari Morteza
De Arruda Ignacio Paulo Sergio de
De Campos Renato
De Oliveira Costa Neto Pedro Luiz
Delorme Xavier
Deroussi Laurent
Despeisse Mélanie
Di Nardo Mario
Di Pasquale Valentina
Dillinger Fabian
Djedidi Oussama

Dolgui Alexandre
Dolinsek Slavko
Dou Runliang
Drei Samuel Martins
Dreyer Heidi
Dreyfus Paul-Arthur
Dubey Rameshwar
Dümmel Johannes
Eloranta Eero
Emmanouilidis Christos
Ermolova Maria
Eslami Yasamin
Fast-Berglund Åsa
Faveto Alberto
Federico Adrodegari
Feng Xuehao
Finco Serena
Flores-García Erik
Fontaine Pirmin
Fosso Wamba Samuel
Franciosi Chiara
Frank Jana
Franke Susanne
Freitag Mike
Frick Jan
Fruggiero Fabio
Fu Wenhan
Fujii Nobutada
Gahan Padmabati
Gaiardelli Paolo
Gallo Mosè
Ganesan Viswanath Kumar
Gaponov Igor
Gayialis Sotiris P.
Gebennini Elisa
Ghadge Abhijeet
Ghrairi Zied
Gianessi Paolo
Giret Boggino Adriana
Gloeckner Robert
Gogineni Sonika
Gola Arkadiusz
Goodarzian Fariba
Gosling Jon
Gouyon David
Grabot Bernard

Grangeon Nathalie
Grassi Andrea
Grenzfurtner Wolfgang
Guerpinar Tan
Guillaume Romain
Guimarães Neto Abelino Reis
Guizzi Guido
Gupta Sumit
Gurevsky Evgeny
Habibi Muhammad Khoirul Khakim
Haddou Benderbal Hichem
Halse Lise Lillebrygfjeld
Hammami Ramzi
Hani Yasmina
Hashemi-Petroodi S. Ehsan
Havzi Sara
Hazir Oncu
Hedayatinia Pooya
Hemmati Ahmad
Henchoz El Kadiri Soumaya
Heuss Lisa
Hibino Hironori
Himmiche Sara
Hnaien Faicel
Hofer Gernot
Holst Lennard Phillip
Hovelaque Vincent
Hrnjica Bahrudin
Huber Walter
Husniah Hennie
Hvolby Hans-Henrik
Hwang Gyusun
Irohara Takashi
Islam Md Hasibul
Iung Benoit
Ivanov Dmitry
Jacomino Mireille
Jagdev Harinder
Jahn Niklas
Jain Geetika
Jain Vipul
Jasiulewicz-Kaczmarek Małgorzata
Jebali Aida
Jelisic Elena
Jeong Yongkuk
Johansen John

Jones Al
Jun Chi-Hyuck
Jun Hong-Bae
Jun Sungbum
Juned Mohd
Jünge Gabriele
Kaasinen Eija
Kaihara Toshiya
Kalaboukas Kostas
Kang Yong-Shin
Karampatzakis Dimitris
Kayikci Yasanur
Kedad-Sidhoum Safia
Keepers Makenzie
Keivanpour Samira
Keshari Anupam
Kim Byung-In
Kim Duck Young
Kim Hwa-Joong
Kim Hyun-Jung
Kinra Aseem
Kiritsis Dimitris
Kitjacharoenchai Patchara
Kjeldgaard Stefan
Kjersem Kristina
Klimchik Alexandr
Klymenko Olena
Kollberg Thomassen Maria
Kolyubin Sergey
Koomsap Pisut
Kramer Kathrin
Kulvatunyou Boonserm (Serm)
Kumar Ramesh
Kurata Takeshi
Kvadsheim Nina Pereira
Lahaye Sébastien
Lalic Danijela
Lamouri Samir
Lamy Damien
Landryova Lenka
Lechner Jan-Peter
Lee Dong-Ho
Lee Eunji
Lee Kangbok
Lee Kyungsik
Lee Minchul

Lee Seokcheon
Lee Seokgi
Lee Young Hoon
Lehuédé Fabien
Leiber Daria
Lemoine David
Li Haijiao
Li Yuanfu
Lim Dae-Eun
Lim Ming
Lima Adalberto da
Lima Nilsa
Lin Chen-ju
Linares Jean-marc
Linnartz Maria
Listl Franz Georg
Liu Ming
Liu Xin
Liu Zhongzheng
Lödding Hermann
Lodgaard Eirin
Loger Benoit
Lorenz Rafael
Lu Jinzhi
Lu Xingwei
Lu Xuefei
Lucas Flavien
Lüftenegger Egon
Luo Dan
Ma Junhai
Macchi Marco
Machado Brunno Abner
Maier Janine Tatjana
Maihami Reza
Makboul Salma
Makris Sotiris
Malaguti Roney Camargo
Mandal Jasashwi
Mandel Alexander
Manier Hervé
Manier Marie-Ange
Marangé Pascale
Marchesano Maria Grazia
Marek Svenja
Marjanovic Ugljesa
Marmolejo Jose Antonio

Marques Melissa
Marrazzini Leonardo
Masone Adriano
Massonnet Guillaume
Matsuda Michiko
Maxwell Duncan William
Mazzuto Giovanni
Medić Nenad
Medini Khaled
Mehnen Jorn
Mendes dos Reis João Gilberto
Mentzas Gregoris
Metaxa Ifigeneia
Min Li Li
Minner Stefan
Mishra Ashutosh
Mitra Rony
Mizuyama Hajime
Mogale Dnyaneshwar
Mohammadi Mehrdad
Mollo Neto Mario
Montini Elias
Montoya-Torres Jairo R.
Moon Ilkyeong
Moraes Thais De Castro
Morinaga Eiji
Moser Benedikt
Moshref-Javadi Mohammad
Mourtzis Dimitris
Mundt Christopher
Muši Denis
Nääs Irenilza De Alencar
Naim Mohamed
Nakade Koichi
Nakano Masaru
Napoleone Alessia
Nayak Ashutosh
Neroni Mattia
Netland Torbjørn
Neubert Gilles
Nguyen Du Huu
Nguyen Duc-Canh
Nguyen Thi Hien
Nielsen Izabela
Nielsen Kjeld
Nishi Tatsushi

Nogueira Sara
Noh Sang Do
Nonaka Tomomi
Noran Ovidiu
Norre Sylvie
Ortmeier Frank
Ouazene Yassine
Ouzrout Yacine
Özcan Uğur
Paes Graciele Oroski
Pagnoncelli Bernardo
Panigrahi Sibarama
Panigrahi Swayam Sampurna
Papakostas Nikolaos
Papcun Peter
Pashkevich Anatol
Pattnaik Monalisha
Pels Henk Jan
Pérès François
Persson Fredrik
Pezzotta Giuditta
Phan Dinh Anh
Piétrac Laurent
Pinto Sergio Crespo Coelho da
Pirola Fabiana
Pissardini Paulo Eduardo
Polenghi Adalberto
Popolo Valentina
Portioli Staudacher Alberto
Powell Daryl
Prabhu Vittaldas
Psarommatis Foivos
Rabelo Ricardo
Rakic Slavko
Rapaccini Mario
Reis Milena Estanislau Diniz Dos
Resanovic Daniel
Rey David
Riedel Ralph
Rikalović Aleksandar
Rinaldi Marta
Roda Irene
Rodriguez Aguilar Roman
Romagnoli Giovanni
Romeo Bandinelli
Romero David

Roser Christoph
Rossit Daniel Alejandro
Rudberg Martin
Sabitov Rustem
Sachs Anna-Lena
Sahoo Rosalin
Sala Roberto
Santarek Kszysztof
Satolo Eduardo Guilherme
Satyro Walter
Savin Sergei
Schneider Daniel
Semolić Brane
Shafiq Muhammad
Sharma Rohit
Shin Jong-Ho
Shukla Mayank
Shunk Dan
Siadat Ali
Silva Cristovao
Singgih Ivan Kristianto
Singh Sube
Slama Ilhem
Smaglichenko Alexander
Smeds Riitta Johanna
Soares Paula Metzker
Softic Selver
Sokolov Boris V.
Soleilhac Gauthier
Song Byung Duk
Song Xiaoxiao
Souier Mehdi
Sørensen Daniel Grud Hellerup
Spagnol Gabriela
Srinivasan Vijay
Stavrou Vasileios P.
Steger-Jensen Kenn
Stich Volker
Stipp Marluci Andrade Conceição
Stoll Oliver
Strandhagen Jan Ola
Suh Eun Suk
Suleykin Alexander
Suzanne Elodie
Szirbik Nick B.
Taghvaeipour Afshin

Taisch Marco
Tanimizu Yoshitaka
Tanizaki Takashi
Tasić Nemanja
Tebaldi Letizia
Telles Renato
Thevenin Simon
Thoben Klaus-Dieter
Thurer Matthias
Tiedemann Fredrik
Tisi Massimo
Torres Luis Fernando
Tortorella Guilherme Luz
Troyanovsky Vladimir
Turcin Ioan
Turki Sadok
Ulrich Marco
Unip Solimar
Valdiviezo Viera Luis Enrique
Vallespir Bruno
Vasic Stana
Vaz Paulo
Vespoli Silvestro
Vicente da Silva Ivonaldo
Villeneuve Eric
Viviani Jean-Laurent
Vještica Marko
Vo Thi Le Hoa
Voisin Alexandre
von Cieminski Gregor
Von Stietencron Moritz
Wagner Sarah
Wang Congke
Wang Hongfeng
Wang Yin

Wang Yingli
Wang Yuling
Wang Zhaojie
Wang Zhixin
Wellsandt Stefan
West Shaun
Wiendahl Hans-Hermann
Wiesner Stefan Alexander
Wikner Joakim
Wiktorsson Magnus
Wimmer Manuel
Woo Young-Bin
Wortmann Andreas
Wortmann Johan Casper
Wuest Thorsten
Xu Tiantong
Yadegari Ehsan
Yalaoui Alice
Yang Danqin
Yang Guoqing
Yang Jie
Yang Zhaorui
Yelles Chaouche Abdelkrim Ramzi
Zaeh Michael Friedrich
Zaikin Oleg
Zambetti Michela
Zeba Gordana
Zhang Guoqing
Zhang Ruiyou
Zheng Feifeng
Zheng Xiaochen
Zoitl Alois
Zolotová Iveta
Zouggar Anne

# Contents – Part II

## Engineering of Smart-Product-Service-Systems of the Future

**Lean and Six Sigma in Services Healthcare**

## New Trends and Challenges in Reconfigurable, Flexible or Agile Production System

## Production Management in Food Supply Chains

## Sustainability in Production Planning and Lot-Sizing

# Digital Transformation of SME Manufacturers: The Crucial Role of Standard

# Strategic Roadmapping Towards Industry 4.0 for Manufacturing SMEs

Elli Verhulst$^{(\boxtimes)}$ and Stine Fridtun Brenden

Norwegian University of Science and Technology, Trondheim, Norway
{elli.verhulst,stine.f.brenden}@ntnu.no

**Abstract.** In recent years, there is a growing focus on the role of small and mediumsized enterprises (SMEs), and their development towards Industry 4.0. One way of supporting SMEs in this effort, is by utilizing the method of strategic roadmapping. This article presents a) a theoretical framework for the use of strategic roadmapping towards industry 4.0, and b) insights from a validation of the framework in four pilots - Norwegian manufacturing companies. The framework offers a systemic view of the company by focusing on five dimensions: business and strategy, product, customers and suppliers, production processes, and factory and infrastructure. Simultaneously, the framework offers a stepwise method to look at these five dimensions from a strategic perspective in a holistic way. The empirical data from the pilot companies offer insights on how the companies take up the strategic roadmapping method into their strategic operations, as well as which topics related to Industry 4.0 get integrated into their future vision, strategies and plans. The results indicate that the use of the strategic roadmapping method supports companies in seeing diverse routes towards Industry 4.0 and provides support in prioritizing relevant projects and activities.

**Keywords:** Industry 4.0 · Manufacturing SME · Strategic roadmapping

## 1    Introduction

The fourth industrial revolution, often referred to by the term Industry 4.0, is connected to the development and growing availability of Information and Communication Technologies, which allow for greater flexibility and customized mass production in manufacturing companies [15]. Many definitions of the term Industry 4.0 have been proposed, both from researchers and industry [3,5,14,17]. For this paper, with the purpose of strategic roadmapping towards industry 4.0 for SMEs, the definition presented by Qin, Liu et al. [14] will be applied. From all the definitions considered, the one from Qin, Liu et al. [14] is one of the more integrative definitions of Industry 4.0, incorporating technological

© IFIP International Federation for Information Processing 2021
Published by Springer Nature Switzerland AG 2021
A. Dolgui et al. (Eds.): APMS 2021, IFIP AICT 631, pp. 3–12, 2021.
https://doi.org/10.1007/978-3-030-85902-2_1

and organisational aspects of companies. Industry 4.0 is then said to include five dimensions: business and strategy, product, production processes, customers and suppliers, and factory and infrastructure.

A small and medium sized enterprise (SME) is in Europe defined as a firm employing fewer than 250 people, with a total turnover that does not exceed 50 million euro [4]. Up to recently, only a small part of the academic research on Industry 4.0 has focused on SMEs, while a larger amount of the research focuses on large enterprises [12]. This is in stark contrast with the number of SMEs actively creating value in industry. Nine out of ten enterprises in the European Union are SMEs, and they generate two out of every three jobs [4]. Several scholars therefore indicate a strong need to study implementation of Industry 4.0 initiatives in SMEs, specifically. Research indicates that, despite the growing number of new tools and technologies, most of them are underexploited, if not ignored by SMEs [11]. Barriers for Industry 4.0 are said to be more evident in such companies [19] and there are indications that a methodical approach for implementation is missing [13,19], next to a lack of resources. Several scholars [11,13,19] highlight a lack of a strategic focus: an operational modus hinders SMEs to go beyond a day-to-day focus, making it hard to think and plan in the long term. This in turn leads to a lack of vision for Industry 4.0 that fits the company.

In conclusion, a clear research gap comes forward from several recent studies, pointing at a lack of strategic perspective and holistic view on industry 4.0 integration in SMEs. With this article, we aim at closing some of this research gap by presenting a theoretical, holistic framework for Industry 4.0 integration in SMEs, in combination with a strategic roadmapping method. Strategic roadmapping as a method offers the opportunity to combine a focus on strategic aspects, with a holistic view on all the different aspects that influence Industry 4.0 integration in SMEs [9]. Next to that, we present insights from practice on the application of the framework and method in four Norwegian manufacturing companies. The work presented in this paper is part of a broader project called *DigiFab*, which focuses on supporting Norwegian SMEs to transition towards Industry 4.0.

## 2   Framework for Strategic Roadmapping

It is important for companies to understand what is crucial for the successful realisation of a digital transformation, and extensive academic literature on adequate drivers and barriers for Industry 4.0 is currently lacking [19]. However, a limited number of scholars have studied success factors and challenges for the implementation of digital transformation, adoption of IT systems and other relevant aspects of Industry 4.0 in SMEs [19–21]. These studies point out a number of relevant factors related to organisational issues, aside from technological factors that are crucial in the shift towards Industry 4.0. These include the importance of strategy as a success factor for a digital transformation in SMEs and the value of having a holistic view [15], whereby one can see the connections between the different organisational and technological factors that influence the integration process of Industry 4.0 in SMEs.

## 2.1 Strategic Roadmapping

A roadmap is a specialised type of strategic plan that outlines activities an organisation can undertake over specified periods to achieve stated goals and outcomes [8]. Roadmapping as a method combines strategic thinking with an integrative view on the organisation and a timeline perspective. Roadmapping has been used in the context of Industry 4.0 for some time, often on an overall level to present a research agenda, upcoming technologies, or trends towards a digital revolution in manufacturing [7,16,18]. Some scholars also present roadmapping as a valuable integrative approach for supporting the implementation of, and transition to Industry 4.0 in manufacturing companies [1,2,6,9,10]. Important elements that come forward in the presented methods are the adaptation to the company's maturity level, customisation to the roadmap based on the organisation's needs, alignment of business areas and processes, and identification and prioritisation of projects. Erol et al. [6] emphasize the need for a systematic integration of generic Industry 4.0 concepts with company specific vision and strategies. The scholars point out a strong interest in methodological support to adopt Industry 4.0 concepts, whereby it is important to bridge a broad vision to the mapping of concrete action points or projects.

Available methods do however not combine an integrative approach and a strategic perspective that goes beyond the operational day-to-day level, with an adaptation to the specific needs of SMEs. Moreover, several of the theoretical models ask for further validation and refinement.

## 2.2 The Developed Framework for Strategic Roadmapping Towards Industry 4.0

In the *Digifab project*, strategic roadmapping takes a central place, with the overall aim to offer a holistic approach whilst simultaneously making the complexity of Industry 4.0 accessible for manufacturing SMEs. Roadmapping thereby adds a strategic perspective. The theoretical framework takes into account main influencing factors SMEs face when implementing Industry 4.0 [13,15,21]. Furthermore, the framework applies a definition with a holistic view on Industry 4.0 that includes five dimensions [14]. These are business and strategy, product, production processes, customers and suppliers, and factory and infrastructure (Fig. 1). Each of the dimensions demands changes over time in order for manufacturing companies to be able to reach an Industry 4.0 vision.

The strategic roadmapping for Industry 4.0 framework, illustrated in Fig. 2, connects the five dimensions with the main activities of a roadmapping method. It includes a time perspective, starting from the company's status today, and working toward a company vision for Industry 4.0 by defining routes to attain that vision. Combining the five dimensions with a strategic timeline offers the company a systemic overview, visually showing how the different areas are connected, and how they influence each other. Moreover, it makes the company actively reflect on the advantages of working towards integrating Industry 4.0

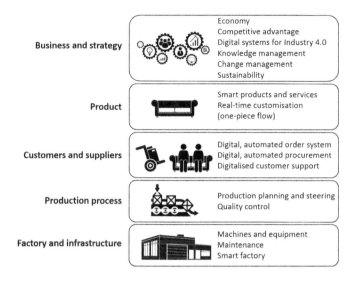

**Fig. 1.** Five dimensions within Industry 4.0, based on [14]

(Why - vision), what they want to achieve (What - goals) and how to make it happen (How – projects and activities).

Gathering insights on **today's status** is important in order for a company to know where they stand, before deciding where to go. This sounds logical and even almost banal. However, the importance of taking the time to stand still, look at how things are going and reflect on what one has reached – or not – is something that often is marginalised in companies. The insight for today's status is based on input from different sources within the company. It entails information gathered through questionnaires, a company visit and workshops with the management team and employees that focus on the Industry 4.0 maturity level of the company, main challenges and bottlenecks in the company related to the five dimensions, and insights on the company's values and culture.

**Company vision** builds on the insights from the mapping in today's status. Here the focus is on where the company sees itself in the longer run in relation to Industry 4.0, e.g. in five or 10 years from now. A generic Industry 4.0 vision for the five dimensions (the most right column in Fig. 2) serves as a template for a company to develop their own vision. It is important to keep in mind that this generic vision is the optimum maturity level on Industry 4.0 a company can reach, which can guide and inspire a company to develop its own vision and adapt it to its own current state and ambitions. This is important to ensure an organisational fit.

**Objectives and activities** that will make it possible to reach the vision are combined with a timeline perspective in the five areas. This is first performed in each dimension separately, building further on today's status and vision for that dimension, subsequently breaking it down into objectives for the coming years

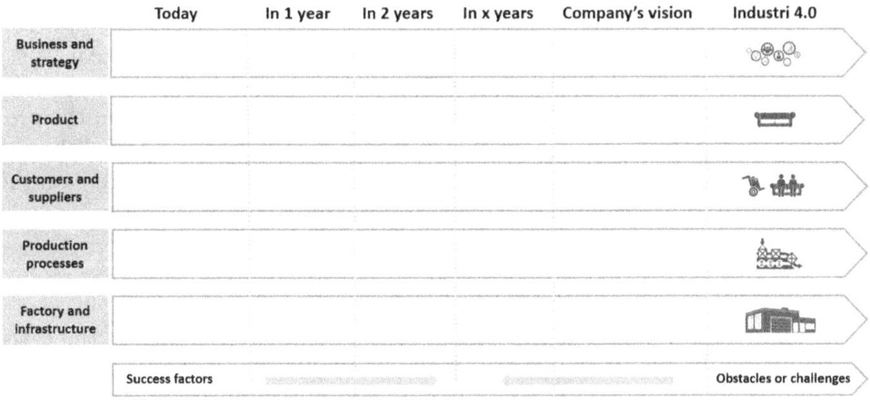

**Fig. 2.** Framework for strategic roadmapping towards Industry 4.0

as well as concrete projects and activities to reach those objectives. Afterwards, all objectives and activities are aligned in order to ensure that they support each other e.g. by looking at dependency, where some activities or objectives need to be reached before another project can be started. It also ensures a good distribution of projects and activities over time so that the roadmap is feasible to implement.

**Success factors and obstacles** give an overview of possible internal and external factors that can hinder or support a company to integrate Industry 4.0 into its processes and activities. Being conscious about these factors helps a company dealing with possible challenges when they occur.

The framework helps to get an overview of the routes a company can take to get closer to their Industry 4.0 vision on an overarching level. It can be used on different levels of detail, e.g. for detailed roadmaps or action plans in one or more dimensions, within a defined period (e.g. the coming year), or on specific, prioritised projects.

## 3    Validation and Discussion in Regard to the Pilot Cases

### 3.1    The Validation Process

During its development, the framework went through iterative cycles of development and testing in four pilot cases. The pilot cases are Norwegian manufacturing SMEs in the following branches: production of 1) metal doors and windows for boat industry, 2) wooden doors and windows for building industry, 3) furniture for offices and shops and 4) products in plastic.

The strategic roadmapping for Industry 4.0 framework has been used a minimum of two times in each pilot case: a) one workshop focusing on developing an initial Industry 4.0 roadmap in spring 2018 and b) a follow-up workshop in spring 2020 to evaluate the progress and to make adjustments to the roadmap.

The two workshops provided insights to the pilot cases as well as to the project team on how the framework is applied and understood in practice. All workshops and interviews were performed with the CEO and management team of each company, supplemented (when possible) with representatives from the production and sales/customer departments. At the end of the second workshop, a semi-structured interview was completed in each pilot, providing insights on the perception on and evaluation of the strategic roadmapping framework.

## 3.2   Example from Applying the Framework

In this paragraph, we present the final strategic roadmap from one of the cases in the project: The furniture producer Haugstad Møbel AS. Figure 3 gives an example of how they have utilized the framework. This roadmap, created in 2018 and updated in 2019 and 2020, visualizes in a structured and holistic way today's status, the company's vision for Industry 4.0 for 2028 and short to medium long term goals that make reaching this vision realistic. Feedback from the company indicates that the strategic roadmapping was strongly supporting the management at start for raising awareness within the company related to the need for digitalisation and Industry 4.0. On top of that, it helped the company to set concrete goals and to get an overall picture of how decisions influence other processes. The company connected the strategic roadmapping process to other strategic work, whereby the visual roadmap was made available and visible in meeting rooms. Despite that, the active use of the resulting roadmaps was mentioned to come on the background over time.

## 3.3   Insights from all Pilot Cases

All pilot cases emphasized that the framework gave them a clearer direction: several participants mentioned that it forces the company to think long term and strategically, and that it and supported them in prioritising activities and projects that focus on integrating Industry 4.0. A stronger focus on combining long-term goals with concrete, short-term actions was mentioned as a prerequisite for making the roadmapping method even more applicable for everyday use. The roadmap visualisation was said to give a good holistic overview that makes it more tangible and offers insights quickly in the progress made. Simultaneously, using the method was said to feel chaotic and foreign at the beginning. Participants mentioned they see the value of the framework, but that they need (to allocate) time and supervision to get to know the framework and methodology. Most pilots did not actively use the roadmapping framework between the two workshops; however, insights coming from the workshops have been used in other strategic work. One pilot made up to five versions of their Industry 4.0 roadmap on a regular basis. This pilot mentioned the importance of making a habit of using the methodology at least twice a year as part of strategic processes.

Empirical data from the pilot cases – i.e. the different versions of the roadmaps - offer insights on how topics related to Industry 4.0 get integrated into the pilot cases' future vision, strategies and plans. The results indicate that

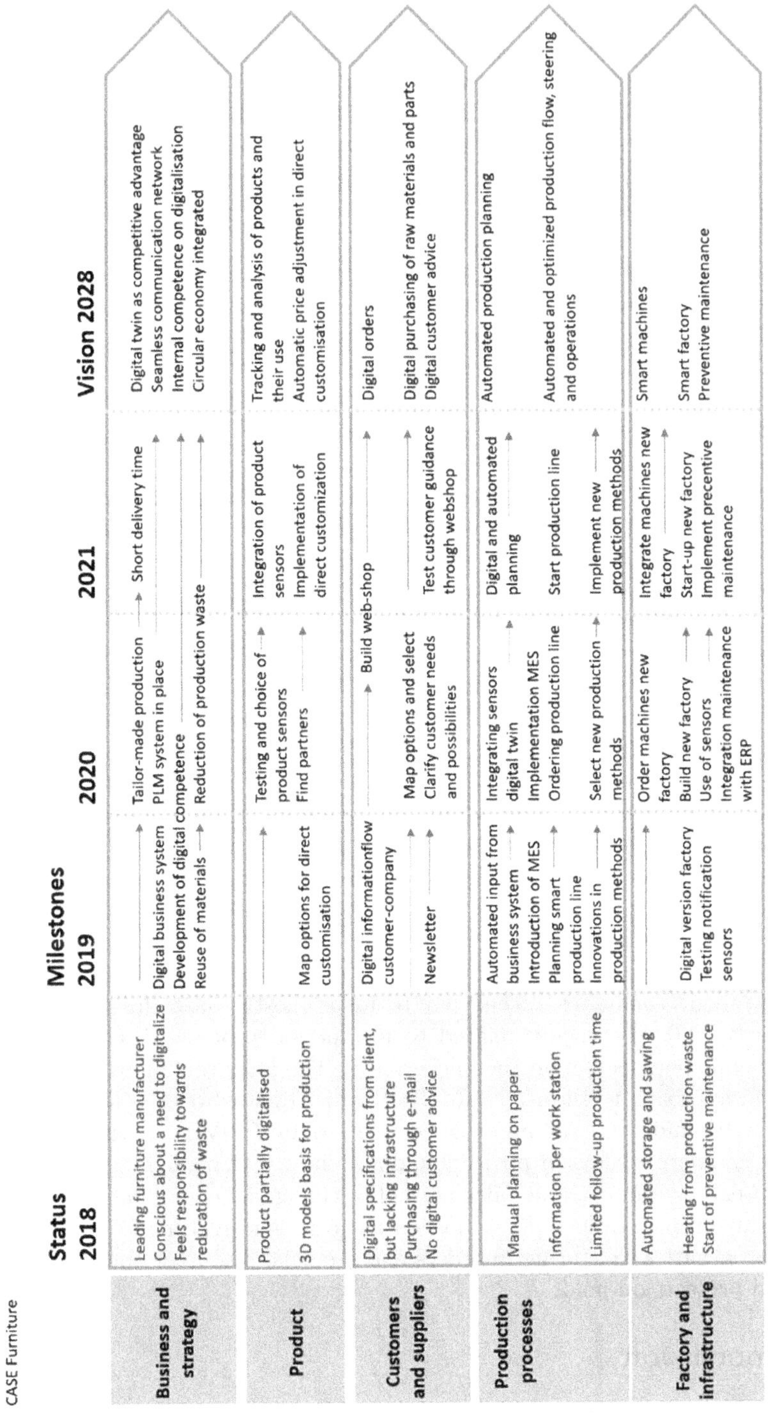

CASE Furniture

| | Status 2018 | Milestones 2019 | 2020 | 2021 | Vision 2028 |
|---|---|---|---|---|---|
| **Business and strategy** | Leading furniture manufacturer; Conscious about a need to digitalize; Feels responsibility towards reducation of waste | Digital business system; Development of digital competence; Reuse of materials | Tailor-made production; PLM system in place; Reduction of production waste | Short delivery time | Digital twin as competitive advantage; Seamless communication network; Internal competence on digitalisation; Circular economy integrated |
| **Product** | Product partially digitalised; 3D models basis for production | Map options for direct customisation | Testing and choice of product sensors; Find partners | Integration of product sensors; Implementation of direct customization | Tracking and analysis of products and their use; Automatic price adjustment in direct customisation |
| **Customers and suppliers** | Digital specifications from client, but lacking infrastructure; Purchasing through e-mail; No digital customer advice | Digital informationflow customer-company; Newsletter | Map options and select; Clarify customer needs and possibilities | Build web-shop; Test customer guidance through webshop | Digital orders; Digital purchasing of raw materials and parts; Digital customer advice |
| **Production processes** | Manual planning on paper; Information per work station; Limited follow-up production time | Automated input from business system; Introduction of MES; Planning smart production line; Innovations in production methods | Integrating sensors digital twin; Implementation MES; Ordering production line; Select new production methods | Digital and automated planning; Start production line; Implement new production methods | Automated production planning; Automated and optimized production flow, steering and operations |
| **Factory and infrastructure** | Automated storage and sawing; Heating from production waste; Start of preventive maintenance | Digital version factory; Testing notification sensors | Order machines new factory; Build new factory; Use of sensors; Integration maintenance with ERP | Integrate machines new factory; Start-up new factory; Implement precentive maintenance | Smart machines; Smart factory; Preventive maintenance |

**Fig. 3.** Strategic roadmap towards Industry 4.0 from Haugstad Møbel AS

the use of the roadmapping framework supported the companies in seeing more possibilities for moving towards Industry 4.0 from different perspectives, both on the short and long term. On a *business and strategic* level, mass customisation got mentioned in the vision of all four pilots as a competitive advantage that can be strengthened through Industry 4.0. Another common issue from all cases was the large need for defining and choosing a fitting IT system (more specifically ERP system) that fits SMEs and that connects different systems in the organisation in tread with Industry 4.0, incl. customer order systems, production planning, production processes, maintenance, and economics. Working with the framework raised awareness on the need for more knowledge and competences on Industry 4.0 amongst management and employees. This resulted in direct action in some pilots, e.g. through starting a hiring process with focus on the needed digital competences. In general, roadmapping for Industry 4.0 raised focus on *product* development within all pilots, whereby smart products and tracking of products through sensor technology came forward as main topics. Attention for *customers and suppliers* also raised, with topics including the improvement of customer services through automated digital ordering and customisation. Large gaps between the pilots became clear considering the maturity level on *production*: mapping of production planning and processes indicated a stringent need for improved efficiency in two of the pilots before any attention could go to the use of digital technologies that support Industry 4.0. However, three of the pilots envision a smart factory and Industry 4.0 infrastructure in a 3–5 years' timeframe. Other common topics on *factory and infrastructure* includes the vision towards automated maintenance systems.

When looking at the first versus last roadmap version, it shows that the overall vision on Industry 4.0 is stable in the four pilots. However, the last roadmap shows more clarity on how to get there: the early versions of the roadmaps include many projects and activities, which reflects the overwhelming feeling of the participants on starting with Industry 4.0. A better overview, prioritisation and spreading over time comes forward in the latest roadmap. Another change over time could be noticed on how the different dimensions were covered: the early strategic roadmaps turned out to focus mostly on technological aspects of Industry 4.0 - especially related to production processes - thereby putting the systemic perspective on the background. The later roadmaps reflect a more balanced approach to Industry 4.0. The last roadmaps were created during the COVID-19 pandemic. All pilots were affected by COVID-19 to some degree, mostly leading to delays of projects and activities, but no significant changes in the strategic roadmaps. One pilot needed to change its course radically due to a significant drop in orders. They needed to go in 'survival mode', shifting focus towards activities that bring in orders and financial resources and putting other planned projects on hold.

## 4   Conclusion

In this article we present a framework for strategic roadmapping towards Industry 4.0 for manufacturing SMEs. It aims to address strategic aspects in combination with the need for a systemic view of the company by focusing on five main dimensions. Development and validation of the framework in four pilot cases has offered insights on its application in practice: how it was used and how the users experienced it, main themes that emerged from the pilot cases' roadmaps for Industry 4.0 as well as changes in the roadmaps over time. The validation highlighted that the framework can support SMEs in seeing diverse routes towards Industry 4.0, and that it helps them to work more strategically on integrating Industry 4.0 in different dimensions of the company. As a future development, the use of the framework will be automated, which offers opportunities for further streamlining the data flow, a dynamic use of the roadmapping framework over a longer timeframe, as well as it will provide a digital interface with the companies and users. Further research will focus on how the application of the framework supports SMEs, thereby gathering data from a larger number of SMEs both in Norway and internationally.

**Acknowledgements.** This work is part of the project *Digifab – Automated digitalization and roadmap towards Industry 4.0 for manufacturing SMEs*, funded by the Research Council of Norway (NFR project nr. ES603431). The authors thankfully acknowledge the funding agency, Digifab project partners and the case companies.

# References

1. Anderl, R., et al.: Guideline Industrie 4.0 - guiding principles for the implementation of Industrie 4.0 in small and medium sized businesses, January 2015
2. Battistella, C., De Toni, A.F., Pillon, R.: The extended map methodology: technology roadmapping for SMEs clusters. J. Eng. Tech. Manage. **38**, 1–23 (2015). https://doi.org/10.1016/j.jengtecman.2015.05.006
3. Brettel, M., Friederichsen, N., Keller, M., Rosenberg, N.: How virtualization, decentralization and network building change the manufacturing landscape: an Industry 4.0 perspective. Int. J. Sci. Eng. Technol. **8**, 37–44 (2014)
4. European Commission: User guide to the SME definition. Report, European Union (2015). https://doi.org/10.2873/620234
5. Erol, S., Jäger, A., Hold, P., Ott, K., Sihn, W.: Tangible Industry 4.0: a scenario-based approach to learning for the future of production. Procedia CIRP **54**, 13–18 (2016). https://doi.org/10.1016/j.procir.2016.03.162. https://www.sciencedirect.com/science/article/pii/S2212827116301500. 6th CIRP Conference on Learning Factories
6. Erol, S., Schumacher, A., Sihn, W.: Strategic guidance towards Industry 4.0 - a three-stage process model. In: International Conference on Competitive Manufacturing (2016)
7. Ghobakhloo, M.: The future of manufacturing industry: a strategic roadmap toward Industry 4.0. J. Manuf. Technol. Manage. **29**(6), 910–936 (2018). https://doi.org/10.1108/jmtm-02-2018-0057
8. IEA: Energy technology roadmaps a guide to development and implementation. Report, OECD/International Energy Agency (2014)
9. Issa, A., Hatiboglu, B., Bildstein, A., Bauernhansl, T.: Industrie 4.0 roadmap: framework for digital transformation based on the concepts of capability maturity and alignment, vol. 72, pp. 973–978, January 2018. https://doi.org/10.1016/j.procir.2018.03.151

10. Leone, D., Barni, A.: Industry 4.0 on demand: a value driven methodology to imple-
    ment Industry 4.0. In: Lalic, B., Majstorovic, V., Marjanovic, U., von Cieminski,
    G., Romero, D. (eds.) APMS 2020. IAICT, vol. 591, pp. 99–106. Springer, Cham
    (2020). https://doi.org/10.1007/978-3-030-57993-7_12
11. Moeuf, A., Pellerin, R., Lamouri, S., Tamayo Giraldo, S., Barbaray, R.: The indus-
    trial management of SMEs in the era of Industry 4.0. Int. J. Prod. Res. **56**, 1–19
    (2017). https://doi.org/10.1080/00207543.2017.1372647
12. Müller, J.M., Buliga, O., Voigt, K.I.: Fortune favors the prepared: How SMEs
    approach business model innovations in Industry 4.0. Technol. Forecasting Soc.
    Change **132,** 2–17 (2018). https://doi.org/10.1016/j.techfore.2017.12.019. https://
    www.sciencedirect.com/science/article/pii/S0040162517312039
13. Orzes, G., Poklemba, R., Towner, W.T.: Implementing Industry 4.0 in SMEs: A
    Focus Group Study on Organizational Requirements, Book Section 9, pp. 251–278.
    Palgrave MacMillan (2020). https://doi.org/10.1007/978-3-030-25425-4
14. Qin, J., Liu, Y., Grosvenor, R.: A categorical framework of manufacturing for
    Industry 4.0 and beyond. Procedia CIRP **52**, 173–178 (2016). https://doi.org/10.
    1016/j.procir.2016.08.005
15. Rojko, A.: Industry 4.0 concept: background and overview. Int. J. Interact. Mobile
    Technol. (iJIM) **11**(5), 77–90 (2017). https://online-journals.org/index.php/i-jim/
    article/view/7072
16. Santosa, C., Mehrsaia, A., Barrosa, A.C., Araújob, M., Aresc, E.: Towards Industry
    4.0: an overview of European strategic roadmaps. In: Manufacturing Engineering
    Society International Conference 2017, MESIC 2017, Procedia Manufacturing, pp.
    972–979. Elsevier (2017). https://doi.org/10.1016/j.promfg.2017.09.093
17. Schumacher, A., Erol, S., Sihn, W.: A maturity model for assessing Industry 4.0
    readiness and maturity of manufacturing enterprises. Procedia CIRP **52**, 161–166
    (2016). https://doi.org/10.1016/j.procir.2016.07.040. The Sixth International Con-
    ference on Changeable, Agile, Reconfigurable and Virtual Production (CARV2016)
18. Sjøgren, J., Krogh, E., Christensen, L.C., Olsen-Skåre, K.H.: Digitalt veikart - for
    en heldigitalisert, konkurransedyktig og bærekraftig bae-næring. Report (2017)
19. Stentoft, J., Wickstrøm Jensen, K., Philipsen, K., Haug, A.: Drivers and barriers
    for Industry 4.0 readiness and practice: a SME perspective with empirical evidence.
    In: 52nd Hawaii International Conference on System Sciences (2019). https://hdl.
    handle.net/10125/59952
20. Vogelsang, K., Liere-Netheler, K., Packmohr, S., Hoppe, U.: Success factors for fos-
    tering a digital transformation in manufacturing companies. J. Enterp. Transform.
    **8**, 1–22 (2019). https://doi.org/10.1080/19488289.2019.1578839
21. Vrchota, J., Volek, T., Novotna, M.: Factors introducing Industry 4.0 to SMEs.
    Soc. Sci. **8**, 130 (2019). https://doi.org/10.3390/socsci8050130

# Developing Digital Supply Network's Visibility Towards Transparency and Predictability

Andreas M. Radke[1](✉) ⓘ, Thorsten Wuest[2] ⓘ, and David Romero[3] ⓘ

[1] mSE North America, Inc, Chicago, IL 60611, USA
aradke@mse-solutions.com
[2] West Virginia University, Morgantown, WV 26506, USA
thwuest@mail.wvu.edu
[3] Tecnológico de Monterrey, 14380 Mexico City, Mexico

**Abstract.** Despite advances in Industry 4.0 technologies, supply chains have fallen short in enabling agile supply chain responsiveness. Conceptually, the enablers of connectivity, visibility, and transparency are well defined, yet their operationalization remains a challenge. In this paper, we analyse a successful case study in the domestic electrical machinery industry and derive from it a proposal for data integration lifecycle phases and socio-technical domains to structure the challenges that need to be overcome as a prerequisite for digital supply networks' visibility towards transparency and predictability.

**Keywords:** Digital Supply Networks · Supply Chain Management · Visibility · Transparency · Predictability · Agility

## 1 Introduction

The COVID-19 black swan event has revealed that there is still a significant lack of connectivity and data exchange built into the fabric of global Supply Chains (SC) [1]. This puts Supply Chain Management (SCM) in the spotlight due to its limited, and also not sufficient, connectivity, visibility, and transparency capabilities [2, 3] for enabling agile supply chain responsiveness in the face of disruptive events towards resilient SC operations [4, 5]. A situation that in the light of Industry 4.0, and the post-COVID world, is no longer acceptable considering all available digital technologies such as the Internet of Things (IoT), End-to-End Digital Connectivity, Cloud Computing, Blockchain, and Predictive Analytics – capable of enabling superior "digital" SCM operations [6, 7].

In today's digital and global world, the traditional SC model continues to evolve not only from "analogue" to "digital" to facilitate electronic data interchange between all business partners [6], but also from a supply "chain" to a supply "network" model to create flexibility in the supplier base [7]. Together, these trends enable resilient Digital Supply Networks (DSNs), capable of creating "flexibility" and "responsiveness" in SC operations through digital technologies and reimagined business processes [6, 7].

© IFIP International Federation for Information Processing 2021
Published by Springer Nature Switzerland AG 2021
A. Dolgui et al. (Eds.): APMS 2021, IFIP AICT 631, pp. 13–21, 2021.
https://doi.org/10.1007/978-3-030-85902-2_2

In this research, we aim to address the question: How to enable three key digital capabilities of DSN: (i) visibility, (ii) transparency, and (iii) predictability to support agile practices in their SCM? Hence, we understand visibility – as the ability to trace, monitor, and obtain relevant data through a DSN to develop sustainable strategies and enable data-driven decisions that can lead to operational efficiency, increased customer service abilities, and improved on-time response to unexpected problems [2]; transparency – goes beyond "visibility", and entails the ability to gain insights through greater SC visibility by running data analytics to manage risks more effectively across SCs [3], and predictability – builds on "visibility" & "transparency" to create the ability to forecast possible SCM incidents before they happen to avoid or reduce their impact on SC operations [8]. Moreover, agility in SCM refers to – the ability of an SC to reconfigure dynamically and quickly as the market conditions change in response to the customer demand [9].

This paper presents a holistic analysis discussing the strategic, organisational, and technical challenges that need to be addressed as prerequisites for developing visibility, transparency, and predictability as digital capabilities in DSNs to enable agility in SCM.

This paper has been motivated out of lessons learned from consulting projects by the authors on "SCM (data) Integration", which call for more socio-technical approaches for the aforementioned digital capabilities building.

## 2  Literature Review

Lit review serves as background review.

According to the scientific and grey literature, developing "digital capabilities" such as visibility, transparency, and predictability in DSNs seems to be the most promising path towards the future of "agile" SCM [9].

As a starting point, from a strategic perspective, the literature highlights the urgency of evolving in the direction of data-driven SCM practices since these have proven to enhance the timeliness of decision-making and SC performance in terms of improved inventory levels, lower costs, and reduced risks while attempting to meet the changing customer demand [11–16]. As stated by [11], "[…] in the big data era, the decision-making of SCM must be increasingly-driven by data instead of experience". [12] proposes a data-driven SCM decision-making framework that focuses on the four operational flows of an SC (viz.: material, information, knowledge, and time flows) to support the data requirements for adopting data-driven SCM practices. Hence, in a data-driven SC, "information is shared across the entire SC to connect SC partners and provide end-to-end SC data access to SC managers" [13, 14]. Such data accessibility is being more and more understood as a critical source of value creation and competitive advantage [15] since data enables SC managers to gain greater "visibility" of their SC operations to identify trends in costs and performance, support process and planning control, capacity and inventory monitoring, and production optimization [14–16]. Therefore, data-driven SCs are leading to new SCM frontiers by focusing on the development of visibility, transparency, and predictability through digital capabilities in SC operations [14].

Next, from an organisational perspective, the literature remarks that the successful adoption of data-driven SCM practices needs to be supported by its respective data-driven decision-making culture [17–20]. In this context, a data-driven culture can be defined as the extent to which SC managers make decisions based on insights extracted from (end-to-end) SC data [18]. Developing a data-driven SCM culture requires SC partners to share their data to realize data-driven SCM capabilities and that SC managers base their decisions on data, rather than their instinct [19, 20]. This brings SC managers to the centre of the discussion when it comes to "transparency", in addition to "visibility". Transparent data-driven systems are crucial for building trust among SC partners for all to share their data and to enable the integration and the timely availability of end-to-end SC data. This will lead to more agile and resilient SC operations. SC transparency provides SC partners with new data-driven capabilities, allowing them to know what is happening upstream in the SC and communicate the insights opportunely across the SC to improve demand forecasting and process optimization [3, 21].

Lastly, from a technical perspective, the literature notes that SC partners' data sharing and data integration capabilities intended for offering end-to-end SC data availability supporting data-driven SCM decision-making practices are a prerequisite for developing "predictability" as part of the digital capabilities of SC operations [12, 22–24]. Real-time data sharing and data integration across SC partners allow for better data-driven decision-making providing all SC partners with the data they need when they needed for agile decision-making [12]. Addressing the "data sharing" and "data integration" challenge requires the adoption of interoperability standards and protocols by all SC partners aimed at the horizontal integration of their SCM systems [22, 23]. The timely electronic data interchange between SC partners increases "visibility" and "transparency" among them and promotes a high level of flexibility to respond quickly to problems or failures even before these have happened by predicting them [24].

## 3   Case Study

To answer our research question, we analyzed a novel case study of a manufacturer's supply network for domestic electrical machinery. The case study's supply network operates as follows: The highest value-adding sub-assemblies of the products, which are also the most sensitive from an intellectual property perspective, are assembled at its in-house plants. The distribution of finished products and spare parts is managed through the OEM's network of regional distribution centers and warehouses. The OEM company uses e-commerce to sell directly to the consumer (see Fig. 1).

The case company (OEM) has outsourced a large share of its lower value-adding activities to its SC partners, hereafter referred to as Contract Manufacturers (CMs). Taking advantage of their specific capabilities, a supply network has emerged. The resulting network needs to be digitally tied to the in-house nodes of the OEM's SC to achieve better "visibility" and "transparency" to enable the consideration of the CMs in the decision-making processes related to the demand and supply planning system.

The nature of the business is characterized by high-demand fluctuations driven by seasonality and promotional activities. Managing these fluctuations effectively requires developing "visibility" into the entire supply network. However, this is currently very limited as it is relying on the traditional means of information exchange. The usual approach of phone calls and e-mails with attachments creates latency in the information collection and delays the decision-making with the result that decisions are often outdated before they are communicated and confirmed by the CMs. Hence, iterative negotiations between the OEM and its CMs are the norm. Furthermore, to protect their profitability, the negotiations have imposed extended frozen horizons and penalties for changes in plans, limiting the supply network agility and flexibility.

The tentative solution to manage the supply network more efficiently and nimbly was to incorporate the CMs directly into the OEM's planning system with regular, automated data interfacing to enable matching demand with supply within the SC constraints. Achieving this integration presented not just a technical challenge but also an organisational and strategic one. The emerging problem and its accompanying challenges will be discussed in the following sections.

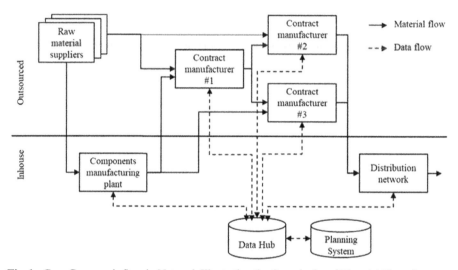

**Fig. 1.** Case Company's Supply Network Illustrating the Complexity of Material Flows between In-house and Outsourced Operations as well as the Solution Approach

### 3.1  Challenges in Digital Supply Network Visibility Across SC Partners

With the number of CMs (SC partners) that need to be integrated to achieve end-to-end "visibility" in the OEM's supply network, the stakeholders quickly noticed that the recurring challenges across their (data) integration were not only "technical" in nature but also "organisational" and "strategic"; these have been summarized in Table 1.

During the engineering of the data integration phase, reducing the effort to establish the integration with predefined data specifications is fundamental for both, reducing the barriers to onboarding and enabling "data integration" before changes to the data source and data destination systems require modifications. The ability to access and integrate data from multiple sources into the same target system was fundamental in achieving greater supply network "visibility". However, this, in turn, required the ability to manage less than perfect data quality, because each missing data source challenged the data integration efforts. In extension, the ability to identify whether data quality issues stem from a lack of business process compliance (i.e., data not collected) or from an error in the data extraction itself. Simplifying the unified data model to ease the harmonization of data deliveries is highly beneficial. Finally, the frequency of data exchange has to be appropriate with the often desired "real-time" data exchange, which may be appropriate for telemetric data from the shop floor but not for data-driven supply network decisions.

While pursuing the supply network (data) integration, organisational aspects played an important role too. The speed of defining or adjusting business processes and either developing or activating "user interfaces" in the target system was essential for data integration. The continuity of daily business operations had to be ensured for the switch to a supply network with increased "visibility".

Strategic aspects always play a role during this phase of the supply network (data) integration and can be considered risk factors. Outsourcing partners generally see the lack of "visibility" of their customers into their own operations as beneficial. This led to cases where motivating and onboarding CMs to the efforts was difficult. As a compromise, not all CMs could be integrated with the same degree of visibility, for example, inventories only or inventories and manufacturing capacity. Even they were motivated, the data model's technical specifications required negotiations around the degree of visibility into the CMs' lower-tier supply networks or granularity of inventory status. Likewise, the decision of whether data would flow only from the CM to OEM or in both directions was a critical discussion point, because the CM could benefit from better "visibility" into the OEM's demand signals and/or supply status to improve its operations too, promoting in this way greater supply network "transparency".

During the daily operations phase, the main technical challenge is the time that it takes to extract data. It is crucial to ensure "timely" extraction and reduce resource consumption on the data sources which are generally the systems of record. After extracting the data, continuous data quality assurance is required to prevent poor data from entering the data destination and disrupting daily operations and predictions. The data connection itself has to be secured to prevent unauthorized access and accidental data leaks. Organisationally, the challenges shift to the supply network coordination and timely execution of activities to use the gained "visibility" and achieve "agility" in decision-making and therefore in operations. A key challenge is in turn the completeness and seamlessness of the business processes which require "coordination" across SC partners because there is no longer an option to manually intervene and mitigate gaps. Particularly, new product introductions and product end-of-life activities turned out to be particularly challenging due to their "cross-functional" nature, which requires alignments between engineering, marketing, sales, sourcing, and production functions. Strategically, the "transparency" of data visibility within and across the supply network is challenging beyond data privacy as it faces the desire of some CMs to at times keep partners partially in the dark to avoid becoming easily replaceable.

Despite requiring clarification at the data integration phase, multiple challenges only occur at the end-of-life phase which can be either a change of product portfolio or the termination of an SC relationship. Ownership of the extracted data can become a point of contention; for example, who owns the raw data extract by the OEM, who owns the processed data, if any. Proof-of-deletion may become a challenge and in addition, as an organisational challenge, means a loss of historical data and the insights hidden within.

Even if the relationship ends, an SC partner may want to continue to signal its desire to regain the relationship and switch from providing capacity data to providing capability data. That, however, is hard to define for operational purposes. Hence, an organisational challenge is the continuity of the business process(es) when switching to a different CM or even returning the physical production process in-house. Strategically, the SC partner integration can mean a shift in the balance of power if the effort to switch is so high that the supply relationship becomes sticky and causes a supplier "lock-in". In this phase, we also need to consider changes to the business or the underlying information systems. Although there might be no change to the supply relationship or the intent on continuing with the data integration, changes to the business and business environment may require changes in the extraction data interfaces, as do changes to the data source like information system upgrades which may mean changes to the system's data model or a new system altogether.

**Table 1.** Challenges categorized by domain and data integration lifecycle phases

| Lifecycle phases | Strategic domain | Organisational domain | Technical domain |
|---|---|---|---|
| Engineering of data integration | • Motivate CM to share data<br>• Define the level of data sharing<br>• Define technical specifications<br>• Decide on mono- or bi-directional data exchange | • Define an agile business process<br>• Develop user interfaces development or activation<br>• Ensure continuity in business process execution (before & after integration) | • Develop data integration logic<br>• Harmonize data model<br>• Access data from multiple sources & merge in target system<br>• Manage data quality Identify missing data<br>• Define frequency data exchange |
| Daily Operations Management | • Ensure transparency of data visibility within and across organisations | • Execute (timely) business process steps within & across organisations<br>• Ensure complete and seamless to-be-process | • Extract, transfer and process data efficiently<br>• Monitor and control continuous data quality<br>• Secure data connection to authorized source(s) |
| End-of-Life Management | • Limit relationship stickiness due to the sunk cost of integration | • A switch contract manufacturer (supplier) or re-integration to in-house supply network<br>• Disconnect without loss of business (data) history | • Clarify data ownership and retention of data<br>• Switch exchange from production capacities to capabilities<br>• Re-adjust business process and interface |

## 3.2  Discussion on Possible Solution Directions

The objective of SC partners' data sharing and data integration efforts is to gain "visibility" across the company supply network for data-driven and agile decision-making. From initial assumptions, the only expected challenges are of a technical nature and can be easily overcome. To answer the research question of enabling "visibility" and "agility" of DSNs operationally, the case study showed a nuanced picture of various challenges occurring. These challenges vary across the lifecycle phase of data integration (see Table 1) and were rooted not just in the technical domain but also in the organisational and strategic ones, as would be expected by a socio-physical system.

These challenges were successfully addressed with an "orchestration" of multiple solution approaches across the domains they originated in. The solutions to the technical challenges revolved around the lowest possible "technical" complexity, which contradicts the increasingly demanding requirements of the "conceptual" view of full and comprehensive data integration (i.e., real-time, standing interfaces, etc.). Extraction and processing logics were decoupled without affecting the requirements on "timeliness" and "simultaneously" reduced the resource consumption on the data source system. Likewise, organisational solutions were simpler than would have been expected. Granular business process improvement [25] and employing checklists [26] proved highly effective. Although these methodologies are well-known, they apparently had not been applied consistently in this setting. The strategic challenges were solved by combinations of game theory and contract mechanism design. This was facilitated by clarity on the objectives and the existing business relationships.

# 4  Conclusions and Outlook

This paper provides an overview of the challenges faced by supply networks striving towards improved visibility across strategic, organisational, and technical domains in the proposed DSN data integration lifecycle phases of engineering of data integration, daily operations management, and end-of-life management. This aims to effectively operationalize "visibility" towards transparency and predictability with the overall objective to achieve agile DNSs. This is particularly relevant for SMEs and SCs where SMEs are involved as CMs and/or OEMs. The limitations of the case study are the use of a single case study and shifting investigations of solution approaches to future work. In future work solution approaches, cybersecurity, and the impact of increased "agility" on the organisational setup domain.

# References

1. Obaid AlMuhairi, M.: Why COVID-19 makes a compelling case for the wider integration of blockchain. WEF, Global Agenda Blog (2020). https://www.weforum.org/agenda/
2. Somapa, S., Cools, M., Dullaert, W.: Characterizing supply chain visibility – a literature review. The Int. J. Logist. Manag. **29**(1), 308–339 (2018)
3. Sodhi, M.S., Tang, C.S.: Research opportunities in supply chain transparency. Prod. Oper. Manag. **28**(12), 2946–2959 (2019)

4. Sharma, V., Raut, R.D., et al.: Ravindra Gokhale a systematic literature review to integrate lean, agile, resilient, green and sustainable paradigms in the supply chain management. Bus. Strateg. Environ. **30**(2), 1191–1212 (2021)
5. Kusiak, A.: Fundamentals of smart manufacturing: a multi-thread perspective. Annu. Rev. Control. **47**, 214–220 (2019)
6. Queiroz, M.M., Pereira, S.C.F., et al.: Industry 4.0 and digital supply chain capabilities: a framework for understanding digitalization challenges and opportunities. Benchmark. Int. J. **28**, 22.https://doi.org/10.1108/BIJ-12-2018-0435 (2019)
7. Sinha, A., Bernardes, E., Calderon, R., Wuest, T.: Digital Supply Networks: Transform Your Supply Chain and Gain Competitive Advantage with New Technology and Processes. McGraw-Hill Education, New York (2020)
8. Nguyen, T., Zhou, L., et al.: Big data analytics in supply chain management: a state-of-the-art literature review. Comput. Oper. Res. **98**, 254–264 (2018)
9. Shashi, C.P., Cerchione, R., Ertz, M.: Agile supply chain management: where did it come from and where will it go in the era of digital transformation?. Ind. Market. Manag. **90**, 324–345 (2020)
10. Büyüközkan, G., Göçer, F.: Digital supply chain: literature review and a proposed framework for future research. Comput. Ind. **97**, 157–177 (2018)
11. Li, Q., Liu, A.: Big data-driven supply chain management. Procedia CIRP **81**, 1089–1094 (2019)
12. Long, Q.: Data-driven decision making for supply chain networks with agent-based computational experiment. Knowl.-Based Syst. **141**, 55–66 (2018)
13. Sanders, N.R.: Big data-driven supply chain management: a framework for implementing analytics and turning information into intelligence. Pearson Financial Times (2014)
14. Yu, W., Chavez, R., et al.: Data-driven supply chain capabilities and performance: a resource-based view. Logist. Transp. Rev. **114**, 371–385 (2018)
15. Tan, K.H., Zhan, Y., et al.: Harvesting big data to enhance supply chain innovation capabilities: an analytic infrastructure based on deduction graph. Prod. Econ. **165**, 223–233 (2015)
16. Hazen, B.T., Boone, C.A., et al.: Data quality for data science, predictive analytics, and big data in supply chain management: an introduction to the problem and suggestions for research and applications. Prod. Econ. **154**, 72–80 (2014)
17. Lunde, T.Å., Sjusdal, A.P., Pappas, I.O.: Organizational culture challenges of adopting big data: a systematic literature review. In: Pappas, I.O., Mikalef, P., Dwivedi, Y.K., Jaccheri, L., Krogstie, J., Mäntymäki, M. (eds.) I3E 2019. LNCS, vol. 11701, pp. 164–176. Springer, Cham (2019). https://doi.org/10.1007/978-3-030-29374-1_14
18. Gupta, M., George, F.J.: Toward the development of a big data analytics capability. Inf. Manag. **53**, 1049–1064 (2016)
19. Sanders, N.R.: How to use big data to drive your supply chain. Calif. Manage. Rev. **58**(3), 26–48 (2016)
20. Davenport, R.T.H., Bean, R.: Big companies are embracing analytics, but most still don't have a data-driven culture. Harvard Business Review (2018)
21. Zhu, S., Song, J., Hazen, B.T., Lee, K., Cegielski, C.: How supply chain analytics enables operational supply chain transparency: an organizational information processing theory perspective. Int. J. Phys. Distrib. Logist. Manag. **48**(1), 47–68 (2018)
22. Vieira, A.A.C., Dias, L.M.S., et al.: Supply chain data integration: a literature review. Ind. Inf. Integr. **19**, 100161 (2020)
23. Brinch, M.: Big data and supply chain management: a content-based literature review. 23rd EurOMA Conference, pp. 1–13 (2016)
24. Gelper, S., Atan, Z., et al.: The Data Ambition Matrix: Awareness Andambition About Data Integration in Supply Chains. European Supply Chain Forum, Eindhoven (2019)

25. Hammer, M., Champy, C.: Reengineering the Corporation: A Manifesto for Business Revolution. Harper Business, New York (2004)
26. Gawande, A.: The Checklist Manifesto. Picador, London (2010)

# Proposing a Gamified Solution for SMEs' Use of Messaging Technology in Smart Manufacturing

Makenzie Keepers[1], Peter Denno[2], and Thorsten Wuest[1(✉)]

[1] West Virginia University, Morgantown, WV 26501, USA
mk0004@mix.wvu.edu, thwuest@mail.wvu.edu
[2] National Institute of Standards and Technology, Gaithersburg, MD 20899, USA
pdenno@nist.gov

**Abstract.** Small- and medium- sized enterprises (SMEs) face exceptional challenges in implementing smart manufacturing solutions. Specifically, SMEs often struggle with understanding advanced technologies well enough to implement them and reap the benefits. In this paper, we discuss one specific instance of this problem, namely implementation of data standards for effective business-to-business communications. We propose a possible solution to aid in lowering barriers for SMEs to access and apply technologies for data standardization, a vital part of effective business-to-business communications. Our solution takes a gamified approach by working to conceptualize the SMEs' data into a story with fill-in-the-blanks, similar to a Mad Lib™. We believe that the development and implementation of this tool would provide numerous benefits including, but not limited to, boosting morale, making new technology and standards more approachable, and improving the learning experience.

**Keywords:** SME · Smart Manufacturing · Data standards · Gamification · Mad Lib

## 1 Introduction

Data and the exchange thereof is at the very core of digital transformation and Smart Manufacturing (SM) [1]. Data and information are used in a variety of applications and use cases, from high-fidelity machine tool sensor data to qualitative inspection data to business process data such as invoices along the supply network. To ensure efficient and effective exchange of data and information, as well as semantically correct interpretation of the information, a variety of information models and data standards have been developed and widely adopted. For example, Open Platform Communications/Unified Architecture (OPC UA) is an information model and communication protocol for industrial automation, while the Open Application Group Integration Specification (OAGIS) focuses on data exchange in a supply chain setting.

© IFIP International Federation for Information Processing 2021
Published by Springer Nature Switzerland AG 2021
A. Dolgui et al. (Eds.): APMS 2021, IFIP AICT 631, pp. 22–30, 2021.
https://doi.org/10.1007/978-3-030-85902-2_3

The utility and benefits of standardized data formats are commonly accepted; however, the development, industry or organizational setting, and especially integration poses several challenges that hinder widespread adoption. Specifically, small- and medium-sized enterprises (SMEs) are struggling to match larger manufacturers in the use of advanced technologies and standards. In this paper, we propose a solution to one specific instance of this problem, namely barriers associated with SMEs adopting standardized data messaging within supply chains.

In the background section, the paper discusses messaging standards for communication, and the struggle of SMEs. Next, we describe gamification and how it relates to addressing the barriers faced by SMEs. Then, we propose an approach to effectively reduce these barriers faced by SMEs. Finally, we summarize our work, provide future work, and discuss limitations.

## 2 Background

### 2.1 Smart Manufacturing in SMEs

SMEs, representing a diverse set of organizations, are considered the backbone of US manufacturing and an integral part of most supply networks. The issues discussed herein do not apply to all SMEs and are generalized.

Despite the number and importance of manufacturing SMEs in the economy, their perspective is not always considered regarding the adoption of SM technologies and related business practices [2]. SM aims to help companies become more competitive by capitalizing on the three pillars: connectivity, virtualization, and data utilization [3]. The growth of SMEs, and their ability to contribute to society, may be hindered when their perspectives are overlooked. This may adversely impact the growth of the economy and the creation of globally competitive digital supply networks. A recent study on the state of SM adoption in manufacturing SMEs [4] found a distinct difference between how SMEs approach SM and how larger, international enterprises develop their SM strategies. These differences affect how effectively SM can be supported across different businesses. Table 1 highlights a selection of key differences between SMEs and large manufacturers that present barriers for the adoption of SM at SMEs and also for the efficient collaboration with mixed supply networks.

### 2.2 Communication, Interoperation, and the Plight of SMEs

In supply chains, communication supports joint work. Examples include alerting a supplier that more of what it supplies is needed, replying to the requester that what is requested is on its way, and asking for payment for doing so. The party making the request for supplies might be an original equipment manufacturer (OEM) of a complex product. These OEMs deal with many suppliers and likely have complex production plans. To orchestrate the work of its many suppliers and ensure smooth operation of its production lines, OEMs have typically sought uniform, efficient processes. Third parties, including logistics providers and government agencies involved in trade, also under the pressure of dealing with many customers, have followed suit [5]. Over several decades, a rough consensus regarding processes and terminology has emerged and has been encoded as the industry's best practice in standards. Among these are various messaging standards, including Electronic Data Interchange (EDI) emerging in the 1980s, and the Universal Business Language (UBL) [6] and OAGIS [7], in the early 2000s.

More recently, OEMs and upper-tier suppliers, most of whom have benefited by implementing messaging standards, have encouraged use of the standards among the smaller firms with which they do business. Standardization at the lower tiers promises to further improve the efficiency of everyone involved. Unfortunately, messaging standards involve complex information technology, and SMEs, owing to their size, are unlikely to possess the skills needed to implement them. Since the use of information technology in this use case does not concern revenue growth, it seems reasonable to conclude similar to Mithas et al. [8] that the cost of implementation may chip away at the SME's profitability. Further, a 2019 study noted that despite a booming economy, nearly two-thirds of small companies cannot cover their expenses and that SMEs are spending far less on research and development (R&D) than they have in the past [9].

**Table 1.** Selected key differences between SMEs and large manufacturers regarding adoption of SM (adapted from [2] – please see reference for complete list).

| No | Feature | SME | Large manufacturer |
|----|---------|-----|--------------------|
| 1 | Financial resources | Low | High |
| 2 | Software umbrella (incl. data analytics) | Basic | Comprehensive/integrated |
| 3 | Research & development | Low | High |
| 4 | Standards considerations | Low | High |
| 5 | Alliances w. universities/research institutions | Low | High |
| 6 | Important activities | Outsource | Internal |

## 2.3 Barriers to Adopting Standardized Messaging for SMEs

We can observe a mix of challenges that stem from i) difficulties within the technology adoption process itself, and ii) SMEs' internal factors as described in the previous subsection. There is no clean-cut line differentiating the two and they amplify each other.

SMEs have a general disadvantage compared to larger organizations when it comes to adoption of tools, methods, or standards that require substantial resources and/or specific domain knowledge [2]. While larger organizations have the overhead to provide such dedicated support, most manufacturing SMEs do not [10]. This is particularly troublesome in the IT space, where a dedicated, well-staffed IT department is common in larger organizations, while SMEs often rely on outsourced individuals to support their efforts [2]. Hence, SMEs are often left with limited options, such as to i) acquire/build the knowledge inhouse; ii) outsource implementation, integration, and maintenance of solutions to third parties; or iii) rely on solutions that are user friendly and do not require domain knowledge beyond what can be expected of a SME.

It is likely that SMEs find the messaging technology described in Sect. 2.2 to be arcane and daunting [4]. Conversely, SMEs are quite likely to maintain much of the essential information about their business processes in the form of spreadsheets [2]. To communicate with their large, IT-capable business partners, using a common messaging technology, SMEs must determine what part of their spreadsheets correspond to what parts of the standardized message form. This 'mapping' of information, the key task investigated in the paper, presents challenges for SMEs.

To-date, much of the support material and documentation of messaging standards are aimed at users with a specialized background in, and arcane knowledge of, messaging and mapping - mostly found in larger organizations. Hence it is no surprise that there is a disconnect between the adoption rate by SMEs and large organizations. Besides the IT resources and domain knowledge, further barriers that hinder SMEs from engaging in the efforts include production quantity (that might not 'require' automation of messages yet due to small quantities), diversity and number of customers and suppliers (diversity of messaging formats to deal with), state and integration of their IT systems, and language barriers and customs processes (for international business).

In essence, to fully address this problem and close the adoption gap between SMEs and large organizations, we need to understand and reflect on the stakeholders' requirements. Future methods will not succeed if they overwhelm SMEs with additional costs or complexity. Solutions to the mapping tasks need to be user-friendly with applications that reflect their reality, for instance by aligning with the use of their own spreadsheets, and thus leveraging tools familiar to SMEs. In the following sections, we propose the use of motivational affordances from the gamification domain to close the gap and make the goal of SMEs adopting advanced messaging standards more attainable.

## 3 Gamification as a Foundation for the Proposed Solution

### 3.1 Overview of Gamification

Gamification is "the use of design elements characteristic for games in non-game contexts" [11]. Common applications for gamification include education, training, and healthcare. Although little research is evident in manufacturing and business operations, gamification has expanded into numerous industries in recent years [12].

The game design elements depicted in Fig. 1 are occasionally referred to as motivational affordances. Often, by applying these motivational affordances in a non-game context, the situation may be considered gamified. However, with no clear lines drawn, this decision is ultimately left to the designer. It is common for a design to include multiple motivational affordances.

In Sects. 3.2 and 3.3, the opportunities, challenges, and limitations included are discussed specifically due to their relevance to the SMEs' challenges described above. Additional opportunities, challenges, and limitations may be prevalent with general implementations of gamification; however, we believe the following are the most applicable instances for the use-case discussed in this article.

**Fig. 1.** Key motivational affordances of gamification (based on: [13])

## 3.2 Opportunities with Gamification

We believe that gamification presents an opportunity to address the identified problem by making the performance of complex cognitive tasks more tractable. Gamification provides opportunities for both solution designers and users. For the solution designer, gamification provides guiding principles. The objective is to create a more intuitive experience for the user of the tool, aiding in its adoption and improving the user's ability to perform a complex cognitive task. For instance, "mapping" information from a SME's spreadsheets to a standards-based message. Further, an intuitive gamified experience which helps a user learn how to implement a new technology is likely to reduce the number of questions about its use, and thus requires less technical support.

*New technology* is used hereafter to refer to technologies, solutions, or systems that are new to the implementing entity and present an adoption challenge, with standard-based messaging as the focal instance in this paper.

For the user, gamification provides a morale-boosting approach to what is likely a daunting new technology. With the boost of morale, the learning and implementation of this new technology would seem less threatening, more attainable and 'fun', and makes the task worth doing. With majority of gamification affordances, they provide the user with a means to assess progress, and thus a sense of accomplishment. A key opportunity of gamification is that it makes challenging and complex tasks, such as adopting a new technology feel more attainable. Ultimately, the proposed solution in this work is meant to lower the barrier for SMEs to realistically implement data standardized messaging systems on their own. We believe that gamification significantly supports this transition for both the designer and user.

## 3.3 Challenges and Limitations of Gamification

Although gamification provides many opportunities to address the problem discussed in this article, it also presents some limitations. To implement gamification, significant time and effort is often necessary beyond that which is required for developing a conventional user interface. The development of the gamified tool likely requires multiple-disciplinary interaction for development and implementation. Additionally, designers must consider the balancing act between sufficiently developing the gamification so that it is user-friendly, while also not over engineering the solution which might lead to users spending time on tasks with little value-add. In a few instances, gamification can be seen as childish or not feel like a real solution to the problem.

# 4 Mad Lib: Gamification of Mapping Diverse Data to Standards

## 4.1 The Proposed Solution Explained

We propose the use of a Mad Lib, a word game where a story is completed by filling in blanks [14], to translate information from raw data (e.g., custom spreadsheet-based invoices), to a Mad Lib for validation, then to standardized data (e.g., OAGIS) (see Fig. 2). The process to use the Mad Lib will be as follows: (a) user (SME) will open their data source, likely in a spreadsheet, (b) the user will reference to a specific column of their Spreadsheet to complete the Mad Lib story, (c) after completing the Mad Lib, the user will read the Mad Lib like a story to validate that the information was referenced correctly, and (d) the tool will establish a connection between the initial spreadsheet through the Mad Lib to the standard's message form. We propose the combined use of human validation for conceptual dependencies and computer-aided validation for formatting such as numerical values, currencies, or text. By using this tool, SMEs will set up the connection for necessary linkages to appropriately implement the data standard without requiring the knowledge of the data standard.

Ideally, we anticipate this tool to be used once to set up the necessary connections for identical sets of information and/or each customer-supplier relationship; subsequent usage would reuse the knowledge about spreadsheet structure gained in the initial use. However, it is assumed that each different supply chain process (e.g. invoicing vs. advance ship notice) might require its own script, as would non-routine tasks. We also anticipate a multi-time use option for rare instances which do not require reoccurring connections. In general, the determination of this tool to be a one- or multi-time use tool will be made clear throughout the development process based on stakeholder feedback and programmability.

This proposed solution can be classified as gamification because it utilizes the story/theme, immediate feedback, and challenge affordances. These affordances are clearly involved in the Mad Lib solution because (a) a Mad Lib builds a story, (b) immediate feedback is provided to the user in the validation step wherein they are able to see if the story of their data is clear, and (c) the ability to conceptualize data from a data source to a story is likely to be challenging for new users. Additionally, we may consider adding extra affordances to enhance the user experience, such as badges, points, and leaderboards. We anticipate that these extra affordances may improve user enjoyment and excitement to use the tool, and thus the motivation to implement the standardized messaging technologies.

## 4.2 Efficacy of Mad Libs' Use by SMEs

Gamification provides opportunities to make new adoptions and implementations more widely accessible, by lowering the initial entry barriers of incorporating new and

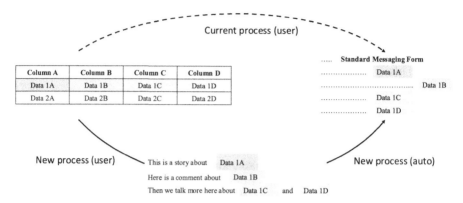

Fig. 2. Schematic of integrating gamification in mapping process via Mad Lib.

advanced technologies for SMEs. The Mad Lib provides a means for the SMEs to conceptualize data information into an easy-to-understand and intuitive format - a *coherent and sensible story*. Additionally, the tool encourages valid mapping, through a user-friendly validation step, which can be used, and in some cases is required, for B2B transactions. Most importantly, this tool will allow SMEs that use spreadsheets to use standard messaging techniques in communications with their business partners.

As with most proposed early-stage solutions, the potential challenges are difficult to anticipate and predict. From the viewpoint of initial implementation for a given SME, the formatting of the spreadsheets (or lack thereof) may impact the ability of the tool to be utilized as intended. In some cases, the spreadsheet may be at a point where it may be too difficult or time-consuming to reformat, that utilization of the Mad Lib can be considered unreasonable. We propose that initially the SME could validate the completed Mad Lib simply by verifying that the story produced makes sense. This however presupposes contextual expertise and knowledge and may introduce error and bias into the process. Another potential pitfall of the proposed Mad Lib solution is that the story line (script) may not apply in a given context. An example of this (outside our area of focus) is that scripts for fast-food and full-service restaurant scenarios differ; in fast-food restaurants you order before sitting down to eat. Additionally, most of today's typical messaging standards are built to accommodate a large amount of information. Including all of this information in a Mad Lib would be overwhelming and unnecessary. Thus, while developing the Mad Libs, we must assess the appropriate amount of information to include in each story line and consider the definition and intended ontological sense of each 'blank' in the story. We plan to address this by clearly defining each blank with a pop-up window, however this may still allow for human error in the process.

## 5   Conclusion, Outlook, and Limitations

In conclusion, we suggest the development and implementation of a Mad Lib-esque solution to lower the barrier for SMEs that are required by their business partners to use standards-based messaging. This Mad Lib will serve as a gamified method that allows SMEs to tell a story using data they typically manage in spreadsheets. The Mad Lib tool will then map that information into the standard message form which is compatible with technologies used in business communications.

Future work for this research includes expanding on a prototype for development of software for the proposed solution. We will begin this by investigating common business-to-business (B2B) communications, such as an invoice. Using the OAGIS messaging standard, we will develop Mad Libs for common B2B communications. After thorough development of an initial prototype and building out the initial idea presented in this paper, we will implement and test the prototype of the new gamified tool. Iterations and improvements of the tool will be completed throughout the process as necessary. Throughout the development, implementation, and testing phases, we will work closely with SMEs to receive feedback in order to develop a user-friendly and user-centric solution. The primary risk involved with this research is the scant exploration of the proposed solution. While the solution has been discussed amongst the authors in detail, it has not yet been proposed to end-users, SMEs. To address this limitation, we started a Delphi study to better assess the needs of SMEs in today's SM environment.

**Acknowledgements & Disclaimer.** This work was performed under the following financial assistance award 70NANB20H028 from U.S. Department of Commerce, National Institute of Standards and Technology and J. Wayne & Kathy Richards Faculty Fellowship at WVU.

Certain commercial software products are identified in this paper. These products were used only for demonstration purposes. This use does not imply approval or endorsement by NIST, nor does it imply these products are necessarily the best available for the purpose.

# References

1. Kusiak, A.: Smart manufacturing must embrace big data. Nat. News **544**, 7648 (2017)
2. Mittal, S., Khan, M.A., Romero, D., Wuest, T.: A critical review of smart manufacturing & Industry 4.0 maturity models: implications for small and medium-sized enterprises (SMEs). J. Manuf. Syst. **49**, 194–214 (2018)
3. Sinha, A., Bernardes, E., Calderon, R., Wuest, T.: Digital Supply Networks. McGraw-Hill Education, New York (2020)
4. Wuest, T., Schmid, P., Lego, B., Bowen, E.: Overview of Smart Mfg. in West Virginia. WVU Bureau of Business & Economic Research, Morgantown (2018)
5. Alicke, K., Rachor, J., Seyfert, A.: Supply Chain 4.0 – the Next-Generation Digital Supply Chain. McKinsey & Company Insights (2016)
6. UBL Version 2.2. http://docs.oasis-open.org/ubl/UBL-2.2.pdf. Accessed 17 June 2021
7. OAGI Homepage. https://oagi.org/. Accessed 17 June 2021
8. Mithas, S., Tafti, A., Bardhan, I., Mein Goh, J.: Information technology and firm profitability: mechanisms and empirical evidence. MIS Q. **36**(1), 205–224 (2012)
9. Govindarajan, V., Lev, B., Srivastava, A., Enache, L.: The gap between large and small companies is growing. why? Harv. Bus. Rev. (2019)
10. Mittal, S., Khan, M.A., Purohit, J.K., Menon, K., Romero, D., Wuest, T.: A smart manufacturing adoption framework for SMEs. IJPR **58**(5), 1555–1573 (2020)
11. Deterding, S., Dixon, D., Khaled, R., Nacke, L.: From game design elements to gameful-ness: defining "gamification". In: 15th MindTrek, pp. 9–15. Tampere (2011)

12. Keepers, M., Romero, D., Hauge, J.B., Wuest, T.: Gamification of operational tasks in man-ufacturing. In: Lalic, B., Majstorovic, V., Marjanovic, U., von Cieminski, G., Romero, D. (eds.) APMS 2020. IAICT, vol. 591, pp. 107–114. Springer, Cham (2020). https://doi.org/10.1007/978-3-030-57993-7_13
13. Hamari, J., Koivisto, J., Sarsa, H.: Does gamification work? - A literature review of empirical studies on gamification. In: 47th HICSS, Hawaii, pp. 3025–3034. IEEE (2014)
14. Mad Lib: The History of Mad Libs. https://www.madlibs.com/history/. Accessed 4 Sept 2021

# Lean First … then Digitalize: A Standard Approach for Industry 4.0 Implementation in SMEs

Daryl Powell[1,2(✉)], Richard Morgan[3], and Graham Howe[3]

[1] SINTEF Manufacturing AS, Raufoss, Norway
daryl.powell@sintef.no
[2] Norwegian University of Science and Technology, Trondheim, Norway
[3] University of Wales Trinity Saint David, Swansea, UK
{richard.morgan,graham.howe}@uwtsd.ac.uk

**Abstract.** The digitalization of manufacturing is the essence of Industry 4.0 realization. Many large manufacturers have developed ambitious digitalization strategies, and most have taken the first steps towards digital transformation. Unfortunately, the same cannot be said for small and medium-sized enterprises (SMEs). At the same time, SMEs contribute on average with more than 50% of the value to the economy in the European Union and with almost 100 million employees, represent approximately 70% of the European workforce. This makes the onset of Industry 4.0 and the accompanying digitalization of manufacturing a fundamental challenge for most SMEs, many of which already struggle to remain competitive in a rapidly evolving business climate. As such, in this paper, we aim to present an SME-friendly approach to Industry 4.0 implementation. We share practical insights from three SME case studies that enable us to propose the *lean first … then digitalize* approach to Industry 4.0 implementation in SMEs.

**Keywords:** Lean · Industry 4.0 · Digitalization · SMEs

## 1 Introduction

The digitalization of manufacturing has become a hot topic in recent years. The increased levels of connectivity provided by the industrial internet of things (IIoT) and big data analytics, combined with various methods of exploiting digital twins and artificial intelligence (AI), allows for the realization of so-called cyber-physical production systems (CPSs) that promise to advance the competitiveness of manufacturing organizations [1]. However, several analyses reveal that although most leading manufacturing firms have ambitious plans for digitalization, many of them are far from ready to benefit from it [2, 3]. This is particularly challenging for small- and medium-sized enterprises (SMEs), which are the backbone of the manufacturing sector in many European countries [4], yet typically lack the financial resources, advanced manufacturing technologies, research and development capabilities, and organizational culture required for a digital transformation [5]. [6] also suggests that there is little awareness of *Smart Manufacturing* (SM)

© IFIP International Federation for Information Processing 2021
Published by Springer Nature Switzerland AG 2021
A. Dolgui et al. (Eds.): APMS 2021, IFIP AICT 631, pp. 31–39, 2021.
https://doi.org/10.1007/978-3-030-85902-2_4

and its related topics among manufacturing SMEs. As such, many SMEs think they are ill-equipped to adopt the SM paradigm [7].

Many of the suppliers to the large manufacturing enterprises are SMEs, and as such, SMEs play a vital role in contemporary manufacturing value chains [8]. They contribute on average with more than 50% of the value to the economy in the European Union and with almost 100 million employees, represent approximately 70% of the European workforce. However, SMEs often lag behind when it comes to innovation – and this can impede industrial growth [9]. [10] presents several characteristics to help explain why SMEs struggle with change management in general, including a firefighting approach to solving problems, a lack of long-term planning, and a command-and-control culture. These are in direct contrast to the three Ls of lean described in [11]: *learning,* a *long-term perspective,* and *leadership.* For this reason, we position lean thinking and practice as a fundamental prerequisite for the successful digitalization of manufacturing, particularly in SMEs.

In this paper, we draw on insights from three industrial case studies to construct an SME-friendly Industry 4.0 implementation process. To guide the investigation, we adopt the following research question (RQ): *How can SMEs implement and benefit from Industry 4.0?*

## 2   Theoretical Background

### 2.1   Industry 4.0

There is much written about the benefits that industry 4.0 technologies can bring to manufacturing [12], and equally as much written about how the same technologies will lead to job losses [13]. Whilst this age-old argument remains, further analysis shows that new roles are, in fact, emerging for upskilled individuals. The challenge for SMEs of course, is to create environments where the skills and resources can be developed to capitalize on the potential economic impact that Industry 4.0 technologies promise. This involves boosting interdisciplinary thinking whilst appraising and deploying these technologies from a disciplined lean management perspective. [14] concludes that, whilst SMEs understand that the deployment of Industry 4.0 technologies cannot simply be ignored and that there was a clear desire for companies to begin (or in some cases, extend their levels of engagement), the same companies highlighted uncertainties in understanding both the full potential and the challenges of embracing these technologies.

### 2.2   Cyber-Physical Production Systems and Digital Lean Manufacturing

With the onset of digitalization, mechanical production systems evolve into cyber-physical production systems (CPPSs). As such, established lean methods gain a new, digitally enhanced edge. This phenomenon has been coined *digital lean manufacturing* [15]. However, such a paradigm assumes that the manufacturing firm has already adopted lean thinking and practice as an alternative means of managing the business. If not, as is the case of most SMEs, then the leap to a digital tomorrow simply risks automating the wasteful activities of today. As such, lean must also pave the way for effective digital transformation.

# 3   Research Design

Given the exploratory nature of our investigation and the how-based research question, we adopt case study research to explore how SMEs can implement and benefit from Industry 4.0. We specifically employ a multiple case research design, conducting semi-structured interviews with representatives from the industry 4.0 initiatives of three SMEs in Norway and Wales. The initiatives involve both the focal SMEs and the digitalization partners (contractors/consultants/researchers). Rich data that emerged from discussions with representatives of each of the three initiatives provided valuable insights that were subsequently used to construct the proposed *lean first ... then digitalize* approach to implementing Industry 4.0 in SMEs.

## 3.1   Case Study 1

Case study 1 is a medium-sized manufacturing and engineering support company located in Norway. The company is a contract producer of mechanical products with 21 employees and a turnover of 2MGBP (2019). The company initiated a project in 2020 to identify opportunities for digitalization, with a view to improving information flow within and across the firm's operations. The project was implemented over a three-month period, which included several workshops with the company.

Following an initial tour of the company's facilities, it became apparent that the company could benefit significantly from adopting basic lean thinking and practice – prior to any adoption of digital technology. For example, it was suggested that the company investigate lean production as a means of improving its operational performance, as well as lean leadership to help develop the continuous improvement culture required for successful lean- (and eventually digital-) transformation.

During further interventions with the SME, it was also observed that operators and employees relied heavily on paper-based work instructions and printed technical drawings and control plans. On further discussion, it became apparent that operators were often uncertain as to whether the documentation they were working from was the correct and current revision. As such, it was suggested that the company consider digitizing various analogue information, through the provision of tablets to the operators at the workstations, such that drawings and technical documentation could be stored online, readily available, and updated in near real-time.

The final stage in the digitalization project, and the first tangible effort for digitalization, was the suggestion of greater horizontal and vertical system integration. Here, the intended solution was to simply adopt a readily available digital collaboration platform (e.g., Microsoft Teams) to foster inter-organizational collaboration and improvement between the focal firm and its customers.

## 3.2   Case Study 2

The second case study draws upon work carried out at a South Wales based materials and casting company. The company produces high-value components for a range of sectors, including aerospace, automotive, and food and drink. With 211 employees and a turnover of 31MGBP, the organization predominantly utilizes foundry and machining processes to produce components for their wide-customer base. In 2019, the company began to explore opportunities for increased digitization and smart automation within their wax pattern production facility. Wax pattern production forms part of the company's investment casting operations. A collaborative research project was developed to implement a systematic approach to technology adoption.

Following initial scoping meetings and company visits, it was agreed that, although functional, the existing wax pattern production facility and procedures were not optimized for greatest operational performance. As such, a discrete event simulation model of the current-state system was developed and analyzed. In preparation for the development of the production model, digital data was recorded and, where this was not available, analogue data was collected and converted for digital input. Although understanding and deployment of lean principles was evident within the wider organization, the wax pattern production facility exhibited clear examples of waste activities which needed to be eliminated prior to the implementation of further digitalization measures.

Once lean principles had been adopted, the development team worked with automation integrators to adapt the process model, simulating a range of potential digitization and smart automation solutions. The simulation allowed a series of "what if" production scenarios to be evaluated. The simulations also allowed the impact various levels of investment would have on the production system as a whole. This helped to provide a solid business case for investment within the department. A Virtual Reality (VR) interface was used to engage production operators in the development process to evaluate how they would interact with proposed interventions and to obtain feedback and insights, from an operational standpoint. It was noted that during both the incorporation of lean principles and the integration of digital and automation technologies, the modelling and simulation work permitted a series of process improvement iterations to be evaluated and optimized prior to physical implementation.

### 3.3 Case Study 3

Case three is focused on work undertaken by a process and productivity improvement solutions provider based in Bridgend, Wales. Their business model focusses on supporting all areas of manufacturing, quality, and process engineering. This ranges from digital solutions integration to task-based activities and skills development through training. Established in 2005 by founding directors with over 50 years' experience in a broad range of manufacturing sectors, the organization has since developed a strong track record of supporting a wide range of SMEs across South Wales. The company has developed its own cloud-based interactive software solution, which is built upon a customizable and flexible framework, and can be used as a platform from which increased levels of digital data can be collected, recorded, analyzed and acted upon - with a view to increasing value adding activity and reducing operational waste. Through interviews with the customer-facing and development teams at the case company, it has been possible to identify recuring themes and patterns which typically emerge when SMEs advance towards the digitalization of their operations.

Prior to any digital intervention, it is often found that heuristics such as overall equipment effectiveness (OEE) are already being used to monitor performance. However, the implementation of the measurement, data collection and analysis tools tend to be labor intensive and inefficient. When embarking on a digitalization journey, many organizations are seen to be paralyzed by the size of the challenge and the range of technologies available. Value Stream Mapping (VSM) is frequently used as an evaluation technique. If they have not already been effectively embedded, lean principles need to be considered and implemented. This is especially relevant in medium- to high-volume production scenarios.

When implementing digital solutions, progressive SMEs can be tempted to move directly to autonomous measurement and data collection. However, this approach often bypasses production staff, resulting in little or no buy-in to the overarching goals associated with digitalization and continuous improvement. In terms of organizational culture, there are widespread, tangible benefits to involving production staff in the manual recording and inputting of data prior to introducing more autonomous methods of data collection. This approach encourages production staff to become involved in the transition and to recognize their ownership of both the data and resulting improvements.

Following the integration of automatic, digital production monitoring systems, control measures need to be implemented to ensure the data and analytics are being used to drive authentic productivity improvement interventions rather than merely serving as a manager's "finger pointing" function. Whilst digitalization has aided many manufacturers in realizing new and enhanced opportunities, it is important for SMEs to maintain focus on the greater purpose behind their adoption – the drive to increase value adding activity and to reduce operational waste. To fully realize its potential, digitalization in manufacturing and the insights which it can offer still needs to facilitate actionable improvement interventions or improved decision-making capabilities.

## 4 Discussion

Drawing on the practical insights from each of the three cases studies, it is clear to see that there is a greater need for SMEs to approach the digital manufacturing frontier first by adopting lean as an alternative business strategy. The level of organizational readiness observed in each of the three case studies also indicated the requirement of a *lean first ... then digitalize* approach. Additionally, we also observed the need to distinguish between digitization and digitalization [16]. Digitization simply refers to taking analog information and encoding it in a digital format so that computers can store, process, and transmit such information. Digitalization, on the other hand, is the process of employing digital technologies and information to transform business operations [17]. We consider these three phases as the foundations of an SME-friendly approach to Industry 4.0 implementation:

We also noticed that each of the lean principles described in [18] provide a suitable point of departure for digitization and digitalization initiatives, examples of which are presented as reference guidelines for Industry 4.0 implementation in SMEs in Table 1:

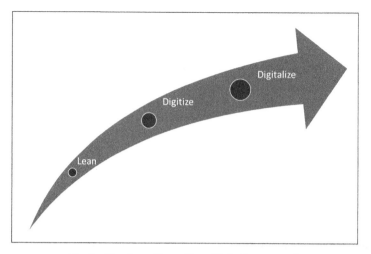

**Fig. 1.** The *Lean First…Then Digitalize* approach

**Table 1.** Reference guidelines for Industry 4.0 Implementation in SMEs.

| Lean (analogue) | Digitize | Digitalize |
|---|---|---|
| Customer value | CAD-CAM, e-documents, tablets | H & V system integration, (Big) data analytics |
| Value stream(s) | Basic modelling (e.g., Excel) | Advanced simulation |
| Create flow | Andon boards (stop, call, wait), Internet of things (IoT) | H & V system integration, Digital shadow/digital twin |
| Establish pull | Electronic Kanban | Artificial intelligence |
| Seek perfection | Digital team boards/SFM | Advanced simulation, VR, AR |

## 5  Conclusion

We set out to answer the research question: *How can SMEs implement and benefit from Industry 4.0?* As such, we carried out three exploratory case studies, first to gauge the organizational readiness levels of SMEs embarking on a digital transformation, and then second to develop an SME-friendly approach to the subsequent implementation efforts that ensue. We discovered that the digital transformations of SMEs must often begin far from adopting digital technologies – rather by first adopting lean thinking and practice as an alternative business strategy. In addition to the low levels of awareness of smart manufacturing, SMEs also lack awareness of what lean manufacturing really entails. As such, we developed a *Lean First … Then Digitalize* approach to implementing Industry 4.0 in SMEs. This is very much in line with the research of [19], which argues that *"lean is needed as a foundation for successful digitalization"* (p. 82), although the authors also suggest a positive correlation between the lean maturity and digital maturity of firms. A lean first approach was also presented by [20], which suggests lean management be applied first in service operations before such firms turn to automation and IT solutions. Automating (or indeed digitizing) wasteful operations will only lead to *digital waste* [15, 21].

There was no evidence from either of the case studies that suggested "lean improvements" are more effective than "smart improvements" to reach the companies' specific objectives. Rather (as illustrated in Fig. 1.), smart improvements become lean improvements when digital technologies are deployed to enhance fundamental lean practices. We suggest that these two "types" of improvement can be considered progressive, with organizational learning at the core.

Furthermore, we suggest that SMEs should begin with developing an awareness of the core lean principles – *customer value, value stream, flow, pull,* and *perfection* [18]; before selecting and adopting lean and/or digital tools. As such, the tools themselves should be seen as countermeasures to the real problems which the firms are struggling with, rather than simply implementing them on a nice-to-have basis. Discovering real problems demands a deep understanding of the gemba (lean jargon for "the real place", e.g., the shopfloor – where value is created), and as such, digitalization efforts begin there. Gemba, as they say, is the greatest teacher [22].

Though it was not a primary aim of the study, we also discovered that this approach also serves as a means of realizing *greener*, more sustainable production – where the resulting elimination of waste has impacts on both the economic, social, and environmental bottom-lines. We therefore conclude that both lean and digitalization have a core role in the sustainability efforts of organizations, including SMEs.

In terms of limitations, we understand that the insight gained from just three case studies cannot be assumed to be representative of all SMEs. However, we do suggest that many SMEs could benefit from adopting a *Lean First … Then Digitalize* approach to Industry 4.0 implementation. Further work should seek to apply this framework to a wider sample of SMEs, in different industries and in different countries.

**Acknowledgements.** The authors acknowledge support of the Research Council of Norway for the research project Lean Digital.

# References

1. Sjödin, D.R., Parida, V., Leksell, M., Petrovic, A.: Smart factory implementation and process innovation: a preliminary maturity model for leveraging digitalization in manufacturing moving to smart factories presents specific challenges that can be addressed through a structured approach focused on people, processes, and technologies. Res. Technol. Manag. **61**(5), 22–31 (2018)
2. Myklebust, O., Lodgaard, E., Sørumsbrenden, J., Torvatn, H.: Lær av de Beste - Hvordan skaffe seg konkurransekraft gjennom digitalisering (2020). https://www.sintef.no/contentassets/7f290c56456c4172a077ab7521a13e87/l-r_av_de_beste_-_rapport_til_nfd.pdf. Accessed 28 Mar 2021
3. Björkdahl, J.: Strategies for digitalization in manufacturing firms. Calif. Manage. Rev. **62**(4), 17–36 (2020)

4. Wiesner, S., Gaiardelli, P., Gritti, N., Oberti, G.: Maturity models for digitalization in manufacturing - applicability for SMEs. In: Moon, I., Lee, G.M., Park, J., Kiritsis, D., von Cieminski, G. (eds.) APMS 2018. IAICT, vol. 536, pp. 81–88. Springer, Cham (2018). https://doi.org/10.1007/978-3-319-99707-0_11

5. Mittal, S., Khan, M.A., Romero, D., Wuest, T.: A critical review of smart manufacturing & Industry 4.0 maturity models: implications for small and medium-sized enterprises (SMEs). J. Manuf. Syst. **49** 194–214 (2018)

6. Wuest, T., Schmid, P., Lego, B., Bowen, E.: Overview of Smart Manufacturing in West Virginia. Bureau of Business & Economic Research (2018)

7. Moeuf, A., Pellerin, R., Lamouri, S., Tamayo-Giraldo, S., Barbaray, R.: The industrial management of SMEs in the era of Industry 4.0. Int. J. Prod. Res. **56**(3), 1118–1136 (2018)

8. Müller, J.M., Buliga, O., Voigt, K.-I.: Fortune favors the prepared: how SMEs approach business model innovations in Industry 4.0. Technol. Forecast. Soc. Change **132**, 2–17 (2018)

9. Mittal, S., Khan, M.A., Purohit, J.K., Menon, K., Romero, D., Wuest, T.: A smart manufacturing adoption framework for SMEs. Int. J. Prod. Res. **58**(5), 1555–1573 (2020)

10. Ates, A., Bititci, U.: Change process: a key enabler for building resilient SMEs. Int. J. Prod. Res. **49**(18), 5601–5618 (2011)

11. Netland, T.H., Powell, D.J.: A lean world. In: Netland, T.H., Powell, D.J. (eds.) The Routledge Companion to Lean Management, pp. 465–473. Routledge, New York (2017)

12. Dalenogare, L.S., Benitez, G.B., Ayala, N.F., Frank, A.G.: The expected contribution of Industry 4.0 technologies for industrial performance. Int. J. Prod. Econ. **204**, 383–394 (2018)

13. Rajnai, Z., Kocsis, I.: Labor market risks of industry 4.0, digitization, robots and AI. In: 2017 IEEE 15th International Symposium on Intelligent Systems and Informatics (SISY), pp. 000343–000346. IEEE (2017)

14. UWTSD: Industry 4.0: Welsh Manufacturing Sector Technological Needs Scoping Study. University of Wales, Trinity St. David, Swansea (2018)

15. Romero, D., Gaiardelli, P., Powell, D., Wuest, T., Thürer, M.: Digital lean cyber-physical production systems: the emergence of digital lean manufacturing and the significance of digital waste. In: Moon, I., Lee, G., Park, J., Kiritsis, D., von Cieminski, G. (eds.) Advances in Production Management Systems. Production Management for Data-Driven, Intelligent, Collaborative, and Sustainable Manufacturing. APMS 2018. IFIP Advances in Information and Communication Technology, vol. 535, 11–20. Springer, Cham (2018). https://doi.org/10.1007/978-3-319-99704-9_2

16. Bloomberg, J. Digitization, Digitalization, and Digital Transformation: Confuse Them at Your Peril (2018). https://www.forbes.com/sites/jasonbloomberg/2018/04/29/digitization-digitalization-and-digital-transformation-confuse-them-at-your-peril/?sh=69188fab2f2c. Accessed 28 Mar 2021

17. Muro, M., Liu, S., Whiton, J., Kulkarni, S. Digitalization and the American Workforce (2017). https://www.brookings.edu/research/digitalization-and-the-american-workforce/. Accessed 28 Mar 2021

18. Womack, J.P., Jones, D.T.: Lean Thinking: Banish Waste and Create Wealth in Your Corporation. Simon and Schuster, New York (1996)

19. Lorenz, R., Buess, P., Macuvele, J., Friedli, T., Netland, T.H.: Lean and digitalization—contradictions or complements? In: IFIP International Conference on Advances in Production Management Systems, pp. 77–84. Springer (2019)

20. Bortolotti, T., Romano, P., Nicoletti, B.: Lean first, then automate: an integrated model for process improvement in pure service-providing companies. In: Vallespir, B., Alix, T. (eds.) Advances in Production Management Systems. New Challenges, New Approaches. APMS 2009. IFIP Advances in Information and Communication Technology, vol. 338. pp. 579–586. Springer, Berlin, Heidelberg (2009). https://doi.org/10.1007/978-3-642-16358-6_72

21. Alieva, J., von Haartman, R.: Digital muda-the new form of waste by Industry 4.0. In: Proceeding International Conference on Operations and Supply Chain Management (OSCM), pp. 269–278 (2020)
22. Ballé, M., Chartier, N., Coignet, P., Olivencia, S., Powell, D., Reke, E.: The Lean Sensei Go. See. Challenge. . Lean Enterprise Institute Inc., Boston (2019)

# Analyzing the Impact Level of SMEs Features Over Digital Transformation: A Case Study

Melissa Liborio Zapata[1,2]($\boxtimes$), Lamia Berrah[2], and Laurent Tabourot[1]

[1] Laboratoire Systèmes et Matériaux Pour la Mécatronique (SYMME), Université Savoie Mont Blanc, 74940 Annecy, France
melissa.liborio-zapata@univ-smb.fr

[2] Laboratoire d'Informatique, Systèmes, Traitement de l'Information et de la Connaissance (LISTIC), Université Savoie Mont Blanc, 74940 Annecy, France

**Abstract.** Digital Transformation (DT) represents a real challenge for companies worldwide, not only because of its complexity due to technology's fast evolution, but also because of the lack of appropriate guidance. Available approaches are judged generic as they do not take into account the specific context of companies. In this sense, this work explores the influence of context in DT success and introduces a performance indicator to measure the impact of the company features that represent its specific context on the dimensions involved in a DT. As the second phase in a research project aimed to build a quantitative model that explains this relationship, this paper focuses on the application of the Impact Level (IL) factor in a real case scenario. The goal is to validate a previous theoretical analysis and also to identify changes in the results with a different characterization of company features. Relevant findings confirm the critical importance of Culture ($f_3$) and R&D investment ($f_9$) for DT success, but many differences arise from the comparative analysis that reveals the DT process as highly contextual. Future work will be focused on translating the insights of both studies into a quantitative model that presents the IL as an aggregator but also with the possibility to provide enough detail for better decision-making during the DT process.

**Keywords:** Digital Transformation · Small and medium enterprises (SMEs) · Manufacturing · Impact analysis · Impact level performance indicator

## 1 Introduction

Digital Transformation (DT) represents a real challenge for companies worldwide and the complexity of its implementation takes a different magnitude for the manufacturing sector in particular [1]. Pressure is high for manufacturers in the digital era due to the fast evolution of the technologies that digitalize the means of production [2]. But just as the technological options and possibilities grow, so does the complexity of a DT for manufacturers [3]. Substantial research work has been produced related to the concept, its strategic options, the technologies to use, as well as the models and frameworks to guide its application [4]. Government programs in many countries are also numerous in promoting the DT of manufacturers and with that, the growth of their economies [5].

© IFIP International Federation for Information Processing 2021
Published by Springer Nature Switzerland AG 2021
A. Dolgui et al. (Eds.): APMS 2021, IFIP AICT 631, pp. 40–48, 2021.
https://doi.org/10.1007/978-3-030-85902-2_5

Despite such enthusiasm, the efforts are not helping to increase digitalization levels or the understanding surrounding the concept and the implications for organizations [6]. As the produced works provide only partial answers to manufacturers' needs, they have still not found the necessary guidance [7]. One of the reasons is that the approaches are considered generic as they often do not take into account the company's specific context [8]. In the DT *scenario*, however, the company's context is relevant because it represents its environment and the means it has to face the challenge [9]. This relevance of the context over the DT has not yet been established, but determining it could help companies to prepare better for this type of change. In order to do so, this research has previously introduced the Impact Level (IL) factor, an indicator that measures the level of impact of the company features over the dimensions involved in a DT [10]. A particular focus on SMEs is used due to their importance for economic growth as they represent around 95% of the businesses in countries like France [11], aim of this work.

Since that first analysis was based on experts' practical experience and provided theoretical conclusions about the IL indicator, the new stage of this research is focused on a practical application. An application through a case study allows not only to validate the findings, but also to find links and proportions between, respectively, features and dimensions in order to go deeper in the definition of the IL indicator. With this goal, this study performed an Impact Analysis on seven manufacturing companies that are fairly advanced in their DT process. In order to be concise, the results of one of them, particularly illustrative, are presented in this work. The insight provided by this study is highly relevant for the future definition of a quantitative model that helps decision-makers to take the appropriate measures before starting and during a DT.

Based on this, the aim of this paper is to present the results of the qualitative application of the IL indicator in a case study. The paper is organized as follows. Section 2 reviews the basis of the definition of the IL indicator and the results of the previous Impact Analysis. Section 3 presents the case study and the methodology followed to apply the qualitative approach of the Impact Analysis, as well as the obtained results. Finally, Sect. 4 presents the conclusions and perspectives of this research work towards the building of a quantitative model of the IL indicator.

## 2   The Basis of the Impact Level Factor

### 2.1   Digital Transformation

DT is generally considered as "*the profound and accelerating transformation of business activities, processes, competencies and models to fully leverage the changes and opportunities brought by digital technologies*" [12]. Even though research literature proposes a growing number of definitions for the concept, not one has been recognized as official [13]. Thus this definition has been chosen for its relevance to this work, as it considers the transformation of the business not as a consequence of the technology introduction, but as a preparation to get the most out of technology. In this sense, the dimensions that change during a DT are central to its success, and therefore this work explores how the features of companies, in particular those characterizing SMEs, impact the course of those changes in a positive way, facilitating them or in a negative way, making it more challenging than they already are.

## 2.2  DT Dimensions

The DT dimensions are the aspects of the company involved in the changes induced by digitalization. Research literature, however, has yet to propose an official set of DT dimensions [7]. For that reason, in this work, a collection of the most representative is included, based on the ones used by relevant proposals [10]. Hence, Fig. 1 presents the set D of the 12 dimensions $dj$, $j \in \{1,2,3,4,5,6,7,8,9,10,11,12\}$, with a brief description of the changes expected in each of them during a DT.

| ID | Dimension | Expected Changes |
|---|---|---|
| $d_1$ | Strategy | Digital Strategy definition and implementation |
| $d_2$ | Business Models | Innovation of the organization's value proposition |
| $d_3$ | Investment | Planning related to the realization of the Digital Strategy |
| $d_4$ | Customer | Digital Experience definition |
| $d_5$ | Products and Services | Creation of Smart and Connected Products and Services |
| $d_6$ | Business Process | Processes creation, redesign and automation |
| $d_7$ | Culture | Change towards Innovation and Collaboration |
| $d_8$ | Organizational Structure | Flexibility, Agility and Cross-functional Collaboration |
| $d_9$ | Leadership | Leaders aware and prepared for the Digital Era |
| $d_{10}$ | (Strategic) Partnerships | Collaboration with customers and competitors |
| $d_{11}$ | Employee Competences | Digital Competences |
| $d_{12}$ | Technology | Digital Technologies selection and implementation |

**Fig. 1.**  DT dimensions [10].

## 2.3  Company Features

The company features are the set of characteristics of companies that represent their context and that are relevant when facing a DT. In Fig. 2, the following set $F$ of the ten features $f_i$, $i \in \{1,2,3,4,5,6,7,8,9,10\}$, characterize the typical manufacturing SMEs, according to a previous analysis [10].

| ID | SME Features |
|---|---|
| $f_1$ | Limited resources (financial, technical, human) |
| $f_2$ | Organizational Structure less complex with informal strategy & decision making |
| $f_3$ | Culture with low flexibility for change and experimentation |
| $f_4$ | Personnel engaged in multiple domains of the organization |
| $f_5$ | Low regard for business processes and standards |
| $f_6$ | Product development with high levels of customization |
| $f_7$ | Industry Knowledge focused in a specific domain |
| $f_8$ | Strong Customer/Supplier Relationships |
| $f_9$ | Low investment in R&D and lack of alliances with Universities |
| $f_{10}$ | Low adoption of new technologies |

**Fig. 2.**  Manufacturing SMEs features [10].

## 2.4  Impact Factor and Impact Analysis

The Impact Level $ILij$ is conceptualized as a performance indicator that shows the level of the positive or negative effect that a given feature $f_i$ has over a given dimension $d_j$ and is represented by the following function [10]:

$$ImpLev : F \times D \rightarrow I$$

$$(f_i, d_j) \rightarrow ImpLev\,(f_i, d_j) \; = \; IL_{ij}$$

The previous work was focused on the analysis of the Impact Level $ILij$ by using the given set of SMEs features defined by their current stereotypical characterization. This theoretical analysis performed by the knowledge and experience of the researchers and validated by industry experts resulted in the following findings [10]:

- The features that make the strongest impact over the dimensions are Limited resources $f_1$, Culture with low flexibility for change and experimentation $f_3$ and Low investment in R&D and lack of alliances with Universities $f_9$.
- The dimensions that are more impacted by the features are Strategy $d_1$, Products and Services $d_5$ and Technology $d_{12}$.
- The industry experts that were consulted confirmed the general conclusions, as they recognize the importance of the limitation of resources and the critical role of the organizational culture as key components of DT success.

Industry experts also make a strong remark regarding the risks of generalizing the SMEs features. Therefore, in this new analysis, their stereotypical features will be substituted with those of a real case scenario of an SME in manufacturing. The goal is twofold, first to validate the researchers' assumptions in the first analysis and also, to identify how the differences in characteristics will affect the resulting IL.

## 3  Case Study

### 3.1  Methodology

The context of this research is defined by the French manufacturing sector, specifically companies in technologically advanced markets that are accelerating the DT of SMEs in the sector. The explorative nature of this research work allows the use of the case research method [14]. The choice of a single case study of a manufacturing SME obeys mainly the need to be concise and achieve representability at the same time [15]. In addition, the unique characteristics of the selected case, different from the SMEs stereotype, but typical of the described markets, are found relevant for the goal of this work.

The analysis performed in this study consists of the comparison of the theoretical values of the $IL_{ij}$ based on typical manufacturing SMEs features and those of our case study based on a real *scenario*. In order to prepare the values of the $IL_{ij}$ for the case study, data about the company was collected from primary (semi-structured interviews) and secondary sources (websites and news articles) [16], as a way to also achieve the triangulation needed to avoid any bias in the process [14]. The interviews were held with the relevant personnel to characterize the company features, to understand the digitalization of its production process and to identify the impact of each feature on the DT dimensions. The outcome was validated with the information available in the secondary sources. The data analysis started by defining the $IL_{ij}$ of the value of each feature $f_i$ on each dimension $d_j$, according to the information collected. This qualification was performed with a 4-level scale composed of 2 *criteria*, an intensity and a sense of this impact, to keep a practical approach. The resulting four levels are described as follows.

- **L+:** Low influence of the feature in support of the change in the dimension.
- **L−:** Low influence of the feature against the change in the dimension.
- **H+:** High influence of the feature in support of the change in the dimension.
- **H−:** High influence of the feature against the change in the dimension.

The results were entered in a matrix that displayed the individual qualification of the IL of all the possible combinations between features $f_i$ and dimensions $d_j$, for the case study, along with the results of the previous theoretical analysis. This display allows comparing both sets of values to understand the links between features and dimensions.

## 3.2 Case Description

The company selected for this study is a French manufacturing SME that produces a small variety of patented mechanical pieces as a supplier for the automobile and aeronautics industries. Founded in the 1990s, the company of around 100 employees has found success *via* its efforts of technological innovation. Searching for new ways to achieve efficiency and growth, the company started its DT a few years back, and now it has already digitalized its line of production. For confidentiality reasons, the company name is not provided and is labelled just as "Company X". In Fig. 3, the values of its features are displayed along with those of typical SMEs (differences marked in yellow).

| ID | Feature | SMEs | Company "X" |
|---|---|---|---|
| $f_1$ | Resource availability | Low | Low |
| $f_2$ | Organizational formality | Low | Low |
| $f_3$ | Culture flexibility | Low | High |
| $f_4$ | Personnel engagement | Multiple domains | Multiple domains |
| $f_5$ | Respect of processes and standards | Low | High |
| $f_6$ | Product customization | High | High |
| $f_7$ | Industry Knowledge | High inside industry/ Low outside industry | High inside industry/ Low outside industry |
| $f_8$ | Customer/Supplier Relationships | Strong | Strong |
| $f_9$ | R&D Investment and Alliances | Low | Medium |
| $f_{10}$ | Adoption of new technologies | Low | Low |

**Fig. 3.** Comparison of company features.

## 3.3 Findings and Discussion

Figure 4 presents the qualification of the $IL_{ij}$ of the company features $f_i$ over the DT dimensions $d_j$. Each combination contains the value obtained by the first analysis performed by the authors and the new analysis based on the case study. Different colours identify the sense of the impact, blue for " +" and red for "-". Additionally, the intensity of the colour is synchronized with the intensity of the impact for better visual interpretation. Finally, the three features in Company "X" that differ from typical SMEs (see Fig. 3) are also identified with a small triangle in blue (▲).

| Dimension/Feature | Analysis | $f_1$ | $f_2$ | $f_3$ ▲ | $f_4$ | $f_5$ ▲ | $f_6$ | $f_7$ | $f_8$ | $f_9$ ▲ | $f_{10}$ |
|---|---|---|---|---|---|---|---|---|---|---|---|
| $d_1$ | 1st analysis | H- | H- | H- | | H- | H+ | H- | H+ | H- | H- |
|  | New | L- | H- | H+ | | H+ | L+ | L- | H+ | L+ | |
| $d_2$ | 1st analysis | | | H- | H- | | | H- | L+ | | |
|  | New | | | H+ | | | L+ | L- | H+ | | |
| $d_3$ | 1st analysis | H- | H- | H- | | | | H- | | H- | H- |
|  | New | L- | L+ | H+ | | | L+ | L- | H+ | H+ | |
| $d_4$ | 1st analysis | H- | | H- | | H- | | | H+ | H- | H- |
|  | New | | | | | | | | | | |
| $d_5$ | 1st analysis | H- | | H- | H- | | H+ | H- | H+ | H- | H- |
|  | New | | | | | | | | | | |
| $d_6$ | 1st analysis | H- | H- | H- | H- | H- | | H- | H+ | | |
|  | New | L- | L- | H+ | L- | H+ | | L- | H+ | L+ | |
| $d_7$ | 1st analysis | H- | | H- | L+ | | | | H+ | H- | H- |
|  | New | | L+ | H+ | L+ | L- | | | | L+ | |
| $d_8$ | 1st analysis | H- | H- | H- | L- | | | | | | |
|  | New | L- | H+ | H+ | L+ | L+ | | L- | | L+ | |
| $d_9$ | 1st analysis | H- | H- | H- | | | | | | | |
|  | New | | L+ | H+ | | L+ | | L- | L+ | L+ | |
| $d_{10}$ | 1st analysis | | H- | H- | | | | | H+ | H- | |
|  | New | L+ | L+ | H+ | | | | L- | H+ | L+ | |
| $d_{11}$ | 1st analysis | H- | | H- | H- | | | H- | | H- | H- |
|  | New | L- | L+ | H+ | L- | H+ | L+ | L- | | L+ | |
| $d_{12}$ | 1st analysis | H- | H- | H- | | | H+ | H- | H+ | H- | H- |
|  | New | L- | H+ | H+ | | H+ | L+ | L- | H+ | H+ | |

**Fig. 4.** Impact Level ($IL_{ij}$) for the first analysis (theoretical) and the new analysis (case study).

Considering that the company has not yet redefined the Customer Experience ($d_4$) or implemented new Smart Products ($d_5$), these two dimensions were excluded from the analysis. For the rest, two types of findings are identified.

1. The differences in the values of $IL_{ij}$ in the theoretical analysis and the real case scenario, for example, between Resource availability ($f_1$) and Strategy ($d_1$).
2. The differences in the values of $IL_{ij}$ that resulted as a consequence of the company features that does not follow the typical characterization of an SME, for example, between Culture ($f_3$) and Strategy ($d_1$).

The main findings are organized and presented by the features perspective as follows.

**Culture and R&D are key.** Previously, the impact of the features Resources ($f_1$), Culture ($f_3$) and R&D investment ($f_9$) on DT success was clearly identified in the theoretical analysis using the typical characterization of SMEs. For the real case scenario, all three are confirmed as influential on DT success with certain differences. In the case of resources availability $(f_1)$, the case study confirmed its importance as it impacted many DT dimensions; however, the impact was not high as previously considered. Though the need for resources is real, the company covered it either by using financing or government aid designed to impulse the digitalization of the manufacturing sector.

On the other hand, as the value of the Culture $(f_3)$ feature changes for this company to "High", the intensity of the impact is confirmed, but in the opposite sense, as having a flexible culture strongly supports the DT. The company also deviates from the SME stereotype regarding R&D investment $(f_9)$, as it makes efforts to search for innovation through these activities. The result is a positive impact as it was anticipated in the first analysis, but in more dimensions than it was previously considered. On the contrary, the initial company's low adoption of new technologies $(f_{10})$ did not have the expected effect on its DT. Once they were convinced of the change, they went forward with the introduction of the required technologies.

**Flexibility in Structure and Processes.** Company X's high regard for processes $(f_5)$, another atypical feature for an SME derived from the highly regulated markets it serves, confirmed the expected effects on DT dimensions. However, this feature also showed other unexpected impacts, such as the fact that processes were seen as a barrier to becoming an agile company. Other interesting facts were identified in relation to the low complexity of the organizational structure and the informality of strategy and decision-making $(f_2)$. Considered a disadvantage in theory, in reality, they provided agility, well suited with a DT process that needs speed to advance further.

**Knowledge Is an Asset.** Regarding the features of high ability to customize products $(f_6)$ and strong customer/supplier relationships $(f_8)$, in the case study, they both translate into knowledge. In both cases, this knowledge helped identify what they need when defining technology investments. In consequence, the positive impact on the DT was higher than anticipated. On the contrary, their Industry Knowledge $(f_7)$, constrained to the industry they participate in, had a subtle negative influence as it limits the visibility of what is needed when expanding to other industries. Finally, the engagement of company personnel in multiple domains $(f_4)$ initially was considered to have minimal impact, mainly negative. In reality, this feature showed a lower effect, and in some cases, it switched into a positive one because the personnel welcomed some of the changes thanks to the positive organizational culture.

In summary, these results confirm the critical importance of Culture $(f_3)$ and R&D Investment $(f_9)$ for DT success. In addition, the Organizational Structure $(f_2)$ and the Industry Knowledge $(f_7)$ become more relevant, not in the same measure that the first two, but still with a significant effect. On the other hand, even when its importance is still high, the Resources availability $(f_1)$ become less determinant in the case study when there is access to resources other than their own and the will and creativity to get them. For companies considering a DT, a culture open to innovation, along with an investment in R&D activities, are requirements to support the change.

## 4   Conclusions and Perspectives

This paper focuses on the application of the Impact Level (IL) factor indicator in a real case to validate a previous theoretical analysis and also to identify changes in the results with a different characterization of company features. The main findings confirm the critical importance of Culture ($f_3$) and R&D investment ($f_9$) for DT success, but many differences arise from the comparative analysis that revealed the DT process as highly contextual. The perspectives of this research work are focused on three main objectives. First, the validation of the IL indicator through the analysis of the remaining six case studies in order to refine the understanding of its behavior. Secondly, the definition of more sophisticated numerical or symbolic scales by transforming the insight obtained from the real case scenarios. Finally, the building of an IL aggregate expression that could be defined either numerically or symbolically to determine the DT readiness of SMEs. Some frameworks such as the Visual Management principles and Fuzzy subsets theory could be used to achieve these objectives.

**Acknowledgements.** The financial support from CONACYT (Grant 707990) is gratefully acknowledged.

## References

1. Schwab, K.: The Fourth Industrial Revolution. Crown Publishing Group, New York (2017)
2. Mittal, S., Khan, M.A., Romero, D., Wuest, T.: Smart manufacturing: characteristics, technologies and enabling factors. Proc. Inst. Mech. Eng. Part B J. Eng. Manuf. **233**, 1342–1361 (2019).
3. Frank, A.G., Dalenogare, L.S., Ayala, N.F.: Industry 4.0 technologies: Implementation patterns in manufacturing companies. Int. J. Prod. Econ. **210**, 15–26 (2019).
4. Felch, V., Asdecker, B., Sucky, E.: Maturity models in the age of Industry 4.0 – do the available models correspond to the needs of business practice? In: Proceedings of the 52nd Hawaii International Conference Systematic Science, vol. 6, pp. 5165–5174 (2019)
5. European Commission: European SME Action Programme. https://ec.europa.eu/docsroom/documents/36142/attachments/1/translations/en/renditions/pdf
6. Colli, M., Berger, U., Bockholt, M., Madsen, O., Møller, C., Wæhrens, B.V.: A maturity assessment approach for conceiving context-specific roadmaps in the Industry 4.0 era. Annu. Rev. Control. **48**, 165–177 (2019)
7. Liborio Zapata, M., Berrah, L., Tabourot, L.: Is a digital transformation framework enough for manufacturing smart products? The case of small and medium enterprises. Procedia Manuf. **42**, 70–75 (2020).
8. Cimini, C., Pinto, R., Cavalieri, S.: The business transformation towards smart manufacturing: a literature overview about reference models and research agenda. IFAC-PapersOnLine. **50**, 14952–14957 (2017)
9. Nightingale, D.J.: Architecting the Future Enterprise. The MIT Press, Cambridge (2015)
10. Liborio Zapata, M., Berrah, L., Tabourot, L.: Towards the definition of an impact level factor of SME features over digital transformation. In: Lalic, B., Majstorovic, V., Marjanovic, U., von Cieminski, G., Romero, D. (eds.) APMS 2020. IAICT, vol. 591, pp. 123–130. Springer, Cham (2020). https://doi.org/10.1007/978-3-030-57993-7_15

11. European Commission: What is an SME. https://ec.europa.eu/growth/smes/business-fri endly-environment/sme-definition_en.
12. Demirkan, H., Spohrer, J.C., Welser, J.J.: Digital innovation and strategic transformation. IT Prof. **18**, 14–18 (2016)
13. Vial, G.: Understanding digital transformation: a review and a research agenda. J. Strateg. Inf. Syst. **28**, 118–144 (2019)
14. Voss, C., Tsikriktsis, N., Frohlich, M.: Case research in operations management. Int. J. Oper. Prod. Manag. **22**, 176–209 (2002)
15. Yin, R.K.: Case Study Research: Design and Methods. SAGE Publications, Thousand Oaks (2009)
16. Kothari, C.R.: Research Methodology: Methods and Techniques. New Age International (P) Limited, New Delhi (2004)

# Digital Transformations Towards Supply Chain Resiliency

# Information Distortion in a Fast Fashion Supply Network: The Impact of Digitalization

Maria Antonietta Turino$^{(\boxtimes)}$, Marta Rinaldi, Marcello Fera, and Roberto Macchiaroli

Department of Engineering, University of Campania "Luigi Vanvitelli", Aversa, Italy
mariaantonietta.turino@unicampania.it

**Abstract.** During the last decades, Information Sharing has gained a global attention among academic researchers. It has been widely demonstrated that such strategy improves the supply chain management and mitigates the bullwhip effect. In this research, the information distortion and its impact on a Fast Fashion Supply Chain has been modeled. The aim of this paper is to analyze how false data can affect the system performance. A simulation model has been developed in order to reproduce the behavior of the players. Then, different scenarios with different levels of digitalization and distortion have been tested. Results show that both upstream and downstream distortions have a disruptive impact on the system and a strong ripple effect. Moreover, negative effects result to be not linear, and small distortions already show great disruptive effects.

**Keywords:** Information sharing · Ripple effect · Fast fashion · Information distortion

## 1 Introduction

Information sharing (IS) is a well-known collaboration strategy widely analyzed in literature. Information flow and communication in the network allows people to exchange information and data, improving the robustness and the resilience of the system [1].

The aim is to connect players, in order to receive and use information to enhance flexibility and make decisions [2]. Information sharing is considered one of the best ways to solve the bullwhip effect (BWE) [3]. BWE is defined as the amplification of the order variability moving upstream the supply chain (SC) [4]. Its propagation depends on the structure of the network and the behavior of the supply chain players [5]. Such phenomenon well describes the effect of the distortion of information or the lack of communication in the supply chain. Lee et al. [6] stated how the information related to the order quantity falsifies the real needs increasing upstream the inventory level. In general, the benefits provided by the information sharing on BWE and SC performance have been widely debated and they continue to be discussed [7, 8]. Costantino et al. [9] tested different inventory policies to quantify the impact of the information sharing on both inventory level and variance. Tang et al. [10] have investigated the need of sharing information and its effect on the SC robustness. However, the accuracy of the information

© IFIP International Federation for Information Processing 2021
Published by Springer Nature Switzerland AG 2021
A. Dolgui et al. (Eds.): APMS 2021, IFIP AICT 631, pp. 51–60, 2021.
https://doi.org/10.1007/978-3-030-85902-2_6

was not enough discussed in literature, and many studies start from the assumption that transmitted data are true [11]. Nevertheless, in practice, distorted information is often shared among the supply chain partners, either willfully or unintentionally [12]. For such reasons, our research aims at evaluating the disruptive effect of distorted information on the SC. The focus of the study is on information distortion and the research question is: how does information distortion affect the supply chain success?

The purpose of this study is to quantify the impact of the false-data sharing on the SC performance. A simulation model has been developed in order to reproduce the behavior of the SC players and test different levels of distorted communication. Moreover, scenarios with distortions are compared with a system where digitalization assures the information truthfulness. The Fast Fashion sector has been chosen because of its variable environment and its need of flexibility. In fact, the fashion industry is characterized by rapid changes in trend and demand. For such reason a reactive SC is required. In such context, a good and real information sharing can strongly influence the Supply chain success. In the same way, a distorted data sharing can have a disruptive effect on the network.

Starting from the previous considerations, the blockchain has been chosen as suitable technology to exchange information in the network. The inalterable nature of the blockchain promotes collaborative relationships among players and it makes the supply chains more transparent and efficient [13]. Moreover, the quality and security of the information enhances the inventory control accuracy, reduces the errors, and improves the decision-making processes [14]. Finally, recent studies indicate that the blockchain has a good applicability in all industries, including the fashion sector [15, 16].

The rest of the paper is organized as follows: Sect. 2 defines the logics and the structure of the simulation model; Sect. 3 introduces the numerical example. Simulation results are presented in Sect. 4 and the last section deals with the main findings of the study.

## 2   Simulation Model

A three-level supply network has been modelled to test the impact of the information distortion on the performance of a Fast Fashion SC. The flow of both material and information have been designed starting from a framework developed by Martino et al. [17]. In particular, the authors have identified three SC levels and three different phases of production and distribution considering the characteristics of the sector. Moreover, other researchers have modeled the same network and phases including production, distribution, and sale for a generic Fast Fashion product life cycle [18]. Wang et al. [19] have also summarized the potential benefits provided by the adoption of the blockchain for each production phase in the Fast Fashion industry. Thus, starting from the analysis of the current literature, three phases that characterized the fashion supply chain have been considered: Pre-Season, In-season and Post-Season (Fig. 1). The flow of material starts with the Suppliers, which deal with the production and provide goods to a single Central Warehouse and ends with the Retail Stores that directly face the final customers. The main logic of the system and the information exchange has been developed considering the specific content.

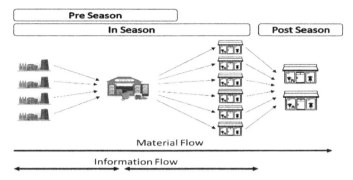

**Fig. 1.** Supply chain network

The Pre-Season phase defines the beginning of the whole process with the development of the New Collection and a first forecast of the demand. Then, orders are scheduled and assigned to suppliers. In fact, the first batch is produced before the sales season and delivered to the warehouse. During the In-Season phase the Central Warehouse manages orders to Suppliers and deliveries to the Retail Stores. A constant monitoring of the deviations between the expected orders and the real sales is carried out. The aim is to quickly respond to changes and adapt the production to the request. In such stage, the level of communication between partners and the accuracy of the information shared between them play a crucial role for the success of the whole system. For such reason, the model mainly focuses on such central phase. However, the pre-season phase has been considered since it defines the starting point of the system, while the outcomes of the simulation model represent the input data of the next phase. In fact, the unsold goods are moved from Stores to the Factory Outlet Stores, with the beginning of the Post-Season phase. Since each level of the SC deals with different activities, specific logics have been modelled to reproduce the behavior of the players, according to their role in the network. Moreover, a single product has been modelled.

## 2.1 First Level: Suppliers

As already asserted, each player plays a specific role, according to the level of the network. Suppliers handle the production process: they receive orders from the Central Warehouse, and they weekly update the production quantity. In fact, the model allows choosing between three different production batches, defined considering the production capacity and the level of flexibility of the player. Moreover, a double check is performed to define the right quantity to produce, matching the demand forecasting and the current stock position. This control is fundamental to continuously align the expected events with the real situation.

Concerning the delivery flows, if possible, the supplier prepares the order and sends it on time. In fact, if the real daily demand is lower than the inventory position, orders are sent the same day; otherwise, the orders will be joined to the orders of the following week and delivered late.

## 2.2  Second Level: Central Warehouse

The Central Warehouse is the central player of the network; he manages and supervises the whole process, choosing the right quantity to order to each supplier and the right quantity to deliver to each retailer. In fact, the warehouse handles two different flows of materials, both upstream and downstream. Moreover, he has to communicate with the partners in order to collect and update the real-time data, with the aim of making the supply chain flexible to the variable environment.

The player adopts a classical economic order interval (EOI) policy, based on periodic orders and variable order quantities. In fact, at fixed periodic interval (T), the inventory level is checked, and an order is placed. The quantity ordered allows the current stock to reach the order-up-to level (OUTL) which is the level of stock that should allow to satisfy the demand since the next order. The theoretical order ($TO_t$) placed at time T is computed as:

$$TO_t = OUTL - TI_t \tag{1}$$

Where, the theoretical inventory position ($TI_t$) is calculated considering the stock quantity (I) at $t - 1$, the order placed (O) at $t - LT$ and received at t and the expected demand (E) at t:

$$TI_t = I_{t-1} + O_{t-LT} - E_t \tag{2}$$

In fact, the reorder process depends on the demand forecasting, which is based on the study of the trend of sales of the previous seven days. Once the order quantity is defined, the warehouse assigns a production amount to the suppliers, according to their performance. In fact, he defines at the beginning of the pre-season phase the percentage of orders to assign to each supplier. Then, during the sales season, he starts from such defined percentage, but he also checks the stock level; if the supplier is not able to satisfy on time the order, he assigns to another player the quantity that would be send late. For this reason, suppliers could tend to alter the data relating to their stock, in order to receive the higher number of orders.

Looking to the second flow of material, the warehouse has to manage and decide the quantity to deliver to each retail store. If possible, the orders are sent and delivered on time. Otherwise, if the stock level is not high enough to satisfy all the retailers, a new distribution based on the order percentage is planned. In fact, the warehouse organizes the daily shipping to the i-th store ($S_i$) considering its order quantity at t ($Q_i$), all the orders received at time t and the available amount in stock:

$$S_{i,t} = \frac{Q_{i,t}}{\sum_i Q_{i,t}} \times I_t \tag{3}$$

For such reason, the shops could tend to falsify and increase the data relating to their order to obtain as many goods as possible.

## 2.3  Third Level: Retail Stores

The stores are the final players of the supply chain, and they directly face the customers. The inventory management is based on a particular inventory policy, which joins the EOI and the economic order quantity (EOQ) policy. In fact, when the inventory position is lower than the order point (OP), an order is placed. The order quantity is variable, and it allows to replenish the shelf considering its maximum capacity. In fact, in such case, it is important to respect the capacity restriction due to the available space in the shop. Moreover, a particular logic has been implemented in the model, which lead to satisfy the higher number of customers. In fact, if a retailer is not able to satisfy a customer, the model chooses randomly another store with enough capacity to provide the product to the customer the same day. Thus, a stock-out situation leads to a loss of sale.

## 2.4  Key Performance Indicators

A set of key performance indicators (KPIs) have been defined to analyze the performance of the Fashion Supply Chain, and to test the negative effect due to a distorted information. Four KPIs strictly related to the bullwhip effect have been chosen. In particular, the KPIs evaluate and quantify the main factors related to the distortion of demand and its propagation through the supply chain. Moreover, some indicators aim at highlighting the difference between the scenarios with and without digitalization. A brief description of the KPIs and their formulation are presented below:

- *Deviation between the orders placed by the Warehouse and the customer demand ($O_W D_C$)*: it is calculated as the deviation between the orders placed by the central warehouse ($O_w$) and the real requests of the end customers (d). It is computed at the end of the sales season. The KPI provides information on the ripple effect due to the distortion:

$$O_w D_C = \frac{\sum_t O_{w,t} - \sum_t d_t}{\sum_t O_{w,t}} \times 100 \ [\%] \tag{4}$$

- *On time deliveries ($OTD_W$)*: it is computed as the number of days (NOTD) in which the all the order placed by the retailers have been sent on time over the total days of the year (N):

$$OTD_w = \frac{N_{OTD}}{N} \times 100 \ [\%] \tag{5}$$

- *Orders recovered from the best Supplier ($OR_W$)*: it indicates the number of items that cannot be fulfilled on time by a supplier and are assigned to another one. The KPI sums all the items recovered and delivered on time because of such practice.
- *Supply chain Inventory level ($I_{SC}$)*: it reflects the average amount of items in stock considering all the players of the supply network. It is computed as the sum of the average inventory position of retail stores ($I_R$), suppliers ($I_S$) and the warehouse ($I_W$):

$$I_{SC} = \sum_R \bar{I}_R + \bar{I}_W + \sum_S \bar{I}_S \ [\text{item}] \tag{6}$$

## 3  Numerical Application

A discrete event simulation model was developed using Microsoft Excel. A supply chain composed of 3 suppliers, one central warehouse and 10 retail stores have been modelled. For each level, the logics described have been implemented. The main objective of the study is to evaluate how the negative effect varies increasing the data falsification. For this purpose, four-scenarios have been tested in order to study the impact of the false-information sharing on the system performance. The scenarios differ from each other in terms of percentage of distortion between the central warehouse and the other SC players. In particular, retail stores falsify the order quantity increasing their request, in order to reach a higher order percentage and prevent late deliveries. Suppliers alter the inventory position data in order to receive higher order quantities, trying to satisfy the request on time exploiting their flexibility. Following, the four scenarios are described:

- *Scenario 1:* retail stores alter data; the retail stores increase the order quantity by 1%, 5%, 10% respect to the reorder lot defined by the order policy, while the suppliers communicate the real inventory data.
- *Scenario 2:* suppliers alter data; suppliers communicate a higher stock amount, increasing the real data of 5%, 25%, 50%, while the retailers place an order in according to the reorder policy.
- *Scenario 3:* both retailers and suppliers alter data; suppliers share the data on the level of stock increasing the real inventory position by 25% and stores grow the order quantity of 5%.
- *Scenario 4:* information distortion is avoided. In such case, a digital technology (blockchain) is implemented and used to share information. Digitalization allows transferring real-time data. Moreover, the truthfulness of information is assured.

### 3.1  Input Data

Input data have been set considering the high variability of the sector. The data have been set considering a real context; however, the flexibility of the model allows changing and testing different data. The average expected customer demand has been fixed at 150 items/day, which consists of 50'400 items/year considering a simulation period of 12 months with time step of 1 day. The daily customer demand has been modelled as a normal distribution with N $(d, \sigma_d)$. Retail stores have been grouped in two groups (group one includes retailers from 1 to 5 and group 2 the other ones); groups differ in input parameters (Table 1).

**Table 1.** Input data – retail stores

|         | Demand mean (items) | Demand dev. std. (items) | Procurement lead time (days) | Shelf maximum capacity (items) |
|---------|---------------------|--------------------------|------------------------------|--------------------------------|
| Group 1 | 10                  | 10                       | 2                            | 147                            |
| Group 2 | 20                  | 20                       | 2                            | 207                            |

As already explained, two production periods are identified and the 65% of the total production is assigned to the In-season phase. Moreover, the warehouse shares the orders among the suppliers starting from a fixed percentage due to their reliability. Suppliers input data are shown in Table 2.

**Table 2.** Input data – suppliers

|  | Supplier 1 | Supplier 2 | Supplier 3 |
|---|---|---|---|
| Production flexibility (%) | 5% | 10% | 25% |
| Delivery lead time (days) | 7 | 6 | 5 |
| Order percentage (%) | 20% | 30% | 50% |
| Nominal production quantity (item/day) | 30 | 45 | 75 |

## 4 Results and Discussion

In this section, the outcomes of the four scenarios are discussed. In particular, scenarios 1, 2, and 3 are compared with the last one, which reflects a situation without distortion. Since the customer demand has been generated randomly, 20 replications have been run for each scenario. Results show the mean value derived from all the replications.

Looking to the first scenario, data falsification is relating to the orders placed by the retailers, which lead to a higher SC average stock (Table 3). Moreover, even though suppliers are flexible, they are unable to adequately respond to such request, whit a resulting decrease of on time deliveries and recovered orders. A strong ripple effect is shown by the increasing distance between the warehouse's orders and the customer demand. Moreover, the trend seems to be not linear, since the first small increases show a greater effect compared to the worst scenario.

**Table 3.** Results of first scenario

|  | 1% | 5% | 10% | Digitalization |
|---|---|---|---|---|
| $O_wD_C$ | 8,28% | 57,56% | 63,80% | 3,40% |
| $OTD_w$ | 90,96% | 52,60% | 47,12% | 96,44% |
| $OR_W$ | 970 | 779 | 612 | 1267 |
| $I_{SC}$ | 4537 | 4864 | 5342 | 4354 |

In the second scenario, suppliers falsify data (Table 4). Therefore, orders recovered by the central warehouse strongly decrease, with a consecutive effect on the on time deliveries. Overall, the SC average stock decreases, since suppliers are unable to satisfy requests and stock out situations occur. For such reason, the warehouse tends to increase the order quantity, increasing the ripple effect.

**Table 4.** Results of second scenario

|           | 5%      | 25%     | 50%     | Digitalization |
|-----------|---------|---------|---------|----------------|
| $O_wD_C$  | 14,17%  | 23,94%  | 29,54%  | 3,40%          |
| $OTD_w$   | 82,47%  | 74,25%  | 68,77%  | 96,44%         |
| $OR_W$    | 599     | 379     | 172     | 1267           |
| $I_{SC}$  | 4180    | 4076    | 4014    | 4354           |

In the last scenario, all the players falsify information (Table 5). Such double distortion causes a negative propagation and joins effects of the previous scenarios: the SC average stock reaches the worst result, but at the same time recovered orders and on time deliveries fall. The negative impact is confirmed by the amplification of the distance between orders and real customer demand.

**Table 5.** Results of third scenario

|           | 5%–25%  | Digitalization |
|-----------|---------|----------------|
| $O_wD_C$  | 62,40%  | 3,40%          |
| $OTD_w$   | 49,04%  | 96,44%         |
| $OR_W$    | 357     | 1267           |
| $I_{SC}$  | 4600    | 4354           |

## 5    Conclusions

Information sharing is a well-known topic of research. This study focuses on information distortion and proposes a simulation model to test the effect of altered data on a three-echelon Fast Fashion Supply Network. The sector is characterized by rapid changes in trend and demand, and short product life cycles. In such context, the supply chain needs rapid adaptations to the customer requests and the information sharing can play a crucial role providing strong benefits. Different levels of digitalization and distortion have been tested with the aim of understanding the impact of the information sharing on the supply chain performance. Thus, different scenarios have been modelled to quantify the impact of the false data sharing. Moreover, this study proposes the blockchain as suitable technology to exchange information in the network because it guarantees the truthfulness and security of the shared data.

Results show an evident impact of false information sharing on the network. Small distortions cause great ripple effects increasing the deviation between the orders placed by players moving upstream the supply chain. Such phenomenon strongly affects the performance of the system, causing two opposite effects on the inventory level of the whole SC. In fact, when retail stores alter data, the inventory level of the SC increases, since more quantities are requested. When suppliers alter data, the stock out situations increase, since suppliers are not able to satisfy the request. In both cases, the on time deliveries decrease.

The study shows that such negative effects could be eliminated using a digital technology, able to avoid false communications. In fact, the introduction of the blockchain leads great improvements if compared to the previous scenarios. Thus, this research offers an interesting starting point to the Fast Fashion practitioners and suggests the use of the blockchain to share information and increase the performance of the system.

Future researchers will focus on the impact of the technology on the economic and environmental performance of the supply chain to test its applicability.

**Acknowledgements.** This work was supported by the University of Campania Luigi Vanvitelli under SCISSOR project - V:alere 2019 program.

# References

1. Treber, S., Benfer, M., Häfner, B., Wang, L., Lanza, G.: Robust optimization of information flows in global production networks using multi-method simulation and surrogate modelling. CIRP J. Manuf. Sci. Technol. **32**, 491–506 (2021)
2. Datta, P.P., Christopher, M.G.: Information sharing and coordination mechanisms for managing uncertainty in supply chains: a simulation study. Int. J. Prod. Res. **49**(3), 765–803 (2011)
3. Jeong, K., Hong, J.-D.: The impact of information sharing on bullwhip effect reduction in a supply chain. J. Intell. Manuf. **30**(4), 1739–1751 (2017). https://doi.org/10.1007/s10845-017-1354-y
4. Cachon, G.P., Randall, T., Schmidt, G.M.: In search of the bullwhip effect. Manuf. Serv. Oper. Manag. **9**(4), 457–479 (2007)
5. Montanari, R., Ferretti, G., Rinaldi, M., Bottani, E.: Investigating the demand propagation in EOQ supply networks using a probabilistic model. Int. J. Prod. Res. **53**(5), 1307–1324 (2015)
6. Lee, H.L., Padmanabhan, V., Whang, S.: Information distortion in a supply chain: the bullwhip effect. Manag. Sci. **43**(4), 546–558 (1997)
7. Zhao, X., Xie, J.: Forecasting errors and the value of information sharing in a supply chain. Int. J. Prod. Res. **40**(2), 311–335 (2002)
8. Van Belle, J., Guns, T., Verbeke, W.: Using shared sell-through data to forecast wholesaler demand in multi-echelon supply chains. Eur. J. Oper. Res. **288**(2), 466–479 (2021)
9. Costantino, F., Di Gravio, G., Shaban, A., Tronci, M.: The impact of information sharing on ordering policies to improve supply chain performances. Comput. Ind. Eng. **82**, 127–142 (2015)
10. Tang, L., Yang, T., Tu, Y., Ma, Y.: Supply chain information sharing under consideration of bullwhip effect and system robustness. Flex. Serv. Manuf. J. **33**(2), 337–380 (2021). https://doi.org/10.1007/s10696-020-09384-6
11. Cannella, S., Dominguez, R., Framinan, J.M., Bruccoleri, M.: Demand sharing inaccuracies in supply chains: a simulation study. Complexity **2018** (2018)
12. Niranjan, T.T., Wagner, S.M., Aggarwal, V.: Measuring information distortion in real-world supply chains. Int. J. Prod. Res. **49**(11), 3343–3362 (2011)
13. Xue, X., Dou, J., Shang, Y.: Blockchain-driven supply chain decentralized operations–information sharing perspective. Bus. Process. Manag. J. **27**(1), 184–203 (2020)

14. Cole, R., Stevenson, M., Aitken, J.: Blockchain technology: implications for operations and supply chain management. Supply Chain Manag. Int. J. **24**(4), 469–483 (2019)
15. Chen, C., Gu, T., Cai, Y., Yang, Y.: Impact of supply chain information sharing on performance of fashion enterprises. J. Enterp. Inf. Manag. **32**(6), 913–935 (2019)
16. Lim, M.K., Li, Y., Wang, C., Tseng, M.L.: A literature review of blockchain technology applications in supply chains: a comprehensive analysis of themes, methodologies and industries. Comput. Ind. Eng. **154**, 107133 (2021)
17. Martino, G., Iannnone, R., Fera, M., Miranda, S., Riemma, S.: Fashion retailing: a framework for supply chain optimization. Uncertain Supply Chain Manag. **5**(3), 243–272 (2017)
18. Şen, A.: The US fashion industry: a supply chain review. Int. J. Prod. Econ. **114**(2), 571–593 (2008)
19. Wang, B., Luo, W., Zhang, A., Tian, Z., Li, Z.: Blockchain-enabled circular supply chain management: a system architecture for fast fashion. Comput. Ind. **123**, 103324 (2020)

# Smartwatch Integration in Digital Supply Chains

Ioan-Matei Sarivan$^{(\boxtimes)}$ ⓘ, Casper Schou ⓘ, Ole Madsen ⓘ,
and Brian Vejrum Wæhrens ⓘ

Department of Materials and Production, Aalborg University,
Fibigerstræde 16, 9220 Aalborg, Denmark
ioanms@mp.aau.dk
https://www.mp.aau.dk/

**Abstract.** As the maturity of digital integrated supply chains grows, the amount of operations which are tracked or which are directly controlled by a computerised system has seen a rapid increase with the emergence of the Industry 4.0 paradigm. This is notably desired in high-cost countries where having an overview on the supply chain is crucial in ensuring the delivery dependability for the customer. However, the manual tasks are overlooked to a certain extent. Most digitalisation initiatives have the worker interact manually with the manufacturing execution systems (MES) using terminals placed around the shop floor. Two scenarios in which the worker has to interact with a MES are given in this paper and a digital solution is proposed to solve the implied shortcomings concerning the interface between production planning and shop-floor production. A solution comes under the form of an open-source, freely available smartwatch app designed to be used by the workers for fast and easy interaction with the MES and enterprise resource planning system while at the same time serving as a task deployment method. The solution proposal is aligned with extant initiatives of obtaining end-to-end supply chain digitalisation while enabling the worker's fast responsiveness upon task deployment.

**Keywords:** Digital supply chain · Smartwatch · Smart manufacturing systems · Task deployment

## 1 Introduction

An end-to-end digitalized supply chain offers the ability to oversee the flow of data and materials, and the use of distributed resources in a centralised manner. The technical realisation of a digital supply chain is made through the successful integration of its business and technical systems (e.g. Enterprise Resource Planning System, Manufacturing Execution System etc.) [1]. However, the amount of captured details when attempting a systematic digitalisation can either be

Supported by MADE (Manufacturing Academy of Denmark).

© IFIP International Federation for Information Processing 2021
Published by Springer Nature Switzerland AG 2021
A. Dolgui et al. (Eds.): APMS 2021, IFIP AICT 631, pp. 61–67, 2021.
https://doi.org/10.1007/978-3-030-85902-2_7

overwhelming or insufficient [2]. As a consequence, it becomes a question of granularity, in how much detail a process can be broken down concerning if an activity can be completely, partially or not at all digitalized [3].

This paper presents a granular approach towards digitalisation of supply chains by targeting manual operations carried out by blue-collar workers. It is aligned with the efforts of combating factors that are hindering the implementation of digital technologies within small-medium enterprises (SME) like management awareness and required skill intensity [4]. A technical artefact is proposed, meant for small medium enterprises to experiment with the use of smartwatch devices in their attempt of optimising manual task deployment and coordination. The artefact is an open source, proof of concept on how wearable devices can be used as a digital integration tool in an industrial manufacturing environment.

The paper is structured as follows: the conducted literature study within the context of manufacturing supply chain digitalisation in high cost environments is presented in Sect. 2 which serves as motivation for the development of the artefact. Moreover, the perspective of two use-cases from industry is given, where digitalisation of manual labour is relevant. The third section presents related work regarding support of manual labour using smartwatches. The proposed digital solution is presented in Sect. 4. The paper is concluded with a discussion on how the presented artefact can fit in the digitalisation initiatives taken in high-cost environments, while placing it in relation with already existing similar technologies.

## 2    Motivation for Using Smartwatches in Industry

To motivate the development of the artefact presented in Sect. 4, a literature study was conducted regarding the perspective of supply chain digitalisation in high-cost environments. A study backed up by empirical data gathered from industry in these environments (e.g. Sweden) point out that the main challenge of the northern industry is to compete with companies from low-cost countries (e.g. Eastern Europe, Asia) on product cost [5]. Therefore, manufacturing companies which stay located in high-cost environments should choose to invest in the maturity of their digital supply chain in order for them to gain an edge on the quality, delivery dependability, flexibility, innovation, cost efficiency and sustainability dimensions of their business [5].

However, investing in the digital maturity of the supply chain also needs to make it possible for the existing methods to be combined with the new methods in clever ways to unlock performance [6]. Nygaard et al. [6] state this target can be achieved by frequent experimentation with new technologies in order to prevent taking costly and overwhelming decisions. It is further stated that there is a need for more efforts into the operational design of IoT based solutions.

Following the literature study in the context of high-cost environments, a strong motivation was found for supply chain digitalisation backed up by national plans in most of Europe [7]. Moreover, discussions with two industrial partners have further strengthened the motivation of looking into how the traditional human labour can benefit from digitalisation. The discussions were in line with the companies' initiative of developing their methods of adopting and utilising new enabling technologies that assist the workers in performing their daily tasks [8]. Two use-cases are hereby given:

## 2.1 Case 1

Automotive manufacturer in Germany: while assembling electronic components, the worker must confirm the successful assembly of the item within the MES by using a touchscreen interface available in various locations around the factory floor. The worker expressed the desire for a system which is highly portable, for it to be carried inside the body of the car when necessary and that it facilitates the interaction needed with the MES in order to complete the task.

## 2.2 Case 2

Metal structures manufacturer in Denmark: work force coordination of the human welders happens off-line without having the status of the weldments being tracked within the MES system. The MES system is only used to a limited extent by specialised workers within the factory. The management expressed the desire of having all the workers in the factory interact with the MES system, but without high training investments. Their main expectation is to facilitate task deployment and fast reaction regarding events triggered by various machines around the factory that need immediate attention, e.g. a milling machine needs a tool change to continue the programmed task.

# 3  State of the Art: Smartwatches in Industry

The usage of wearable technologies in the manufacturing industry for digital integration is documented in literature to a limited extent in terms of having an artefact specifically designed for integration of human labour in a digital supply chain. The use of smartwatches is however encouraged as it makes the interaction with the digital platforms easier when compared with traditional methods [9]. E.g., to use a bar-code scanner, the worker needs to stop what they are doing, lift the scanner, perform the scanning action, put the scanner back and continue whatever they were doing [9]. The interaction is made easier likely because of the available data input and output methods supported by the touchscreen.

Several examples from literature were selected by the authors to illustrate existing efforts in having smartwatches present in industrial environments. These however are specifically focused on exploiting the available sensors on the smartwatch to perform activity recognition in healthcare and sports applications [10], having energy efficiency as main research motivation. The available sensors on a smartwatch are exploited in a manufacturing application too, where artificial intelligence is used to classify the task performed by the worker as successful or unsuccessful based on audio and accelerometer data gathered by the sensors available on the smartwatch [11,12].

Solutions targeted to industrial manufacturing revolving around digital integration using smartwatches are available commercially, being provided by various start-up companies, in the spirit of Industry 4.0. Such a company exists in Germany and offers a proprietary smartwatch which is advertised as being able to be integrated in the MES of the customer company [13].

Following the state of the art analysis that use of smartwatches in industrial manufacturing environments, a research gap has been identified in what concerns the use of these devices in supply chain digitalisation attempts by providing the blue-collar worker a wearable and direct interface for the MES, which is also freely available in an open-source form.

# 4   Task Deployment Using Smartwatches

Even though, several solutions are already commercially available, the artefact presented in this section aligns with the initiative of providing small medium companies the possibility of experimenting with smartwatch like technologies before deciding on a large scale implementation, as supported by Nygaard et al. [6]. A prototype was developed under the form of an open source WearOS application [14]. The application can run on any smartwatch device that has WearOS as an operating system, thus taking advantage of the widely spread familiarity with the Android system which runs on smartphones. Therefore, little to no training is required for the user to get acquainted with the functionalities of the application once it is installed on the smartwatch. While in use, the app can receive tasks for the worker to perform automatically, from the MES.

When an event tracked by the MES occurs, e.g. a CNC machine needs to be unloaded, the task is sent to all the connected smartwatches which have the necessary roles (assigned in the MES) to fulfil the task. A task can have one of the following statuses:

- PENDING: the task is pending acquirement by an available worker
- ACQUIRED: the task was acquired by worker with smartwatch ID "x" and it is in progress
- COMPLETED: the task was completed by worker with smartwatch ID "x"

The process happens in real time upon user's interaction with the system using the graphical user interface (GUI) available on the smartwatch's screen, which can be observed in Fig. 1 and 2.

Once a new task is created by an event triggered in the MES, the task is sent to all the connected smartwatches of the workers who can perform the task, under several predefined criteria. The task is added to a stack through which the worker can navigate using the "PREV" and "NEXT" navigation buttons available on the app's GUI (Fig. 1). Depending on the priority of the task, it is displayed with a yellow background (NORMAL) or red background (URGENT). When a new task is received, a chime is played and haptic signal is performed

**Fig. 1.** Photo of the task coordinator app while showing an "urgent" task generated by "Robot2" with the description "Task Execution Error". The second task our of two tasks available in queue for the user is displayed. The "prev" and "next" buttons allows the user to navigate through the tasks queue

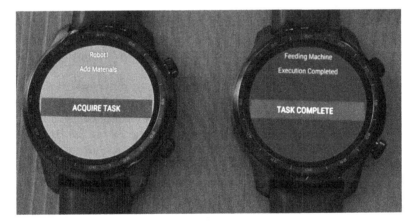

**Fig. 2.** Photo of the task coordinator app running on two smartwatches. The smartwatch on the right hand side displays a "normal" task. The smartwatch on the left hand side displays an "acquired" task which is currently being performed by the worker.

in order to alert the worker depending on the task's priority. The tasks are displayed in a first in - first out pattern, but urgent tasks take priority over a normal task by being displayed on screen as soon as it arrives. A task can be acquired by an available worker who has the necessary roles by tapping the "ACQUIRE TASK" button. Once acquired, the task disappears from all the other smartwatches connected to the MES. Upon completion of the task, the user must tap "TASK COMPLETE", as observed in Fig. 2. This will update the status of the task on the MES's side. The origin of the task and the description of the task are also displayed on the GUI, where the origin of the task is represented by an IoT device that requires the worker's attention or by manual tasks that is queued in the manufacturing execution system.

The perquisites of implementing this open source artefact for digital integration purposes are the availability of a MES and a sufficient number of WearOS smartwatches. The code for both the server and the client is available at https://github.com/MateiSarivan/Industrial-SmartWatch.

## 5   Discussion and Conclusion

The presented technical artefact is meant to provide manufacturing enterprises with a tool to experiment with while attempting to achieve end-to-end supply chain digitalisation. The focus is on the deployment and management of manual tasks. Such it is expected to overcome the usual obstacles that are in the way of implementing new technologies [4]. This comes as an auxiliary system which is meant to be implemented inside the MES platform and used by the shop floor workers for fast responsiveness upon task deployment. By being designed around an element with a high degree of familiarity of a smartwatch, no training investment is expected to be made by the companies for using the device. The artefact competes with current solutions commercially available, by having an open source design which anyone can use as-is or further develop to fit individual use-cases, without having to engage into long-term commitments with an external know-how supplier [6].

The prototype as it is in the current state, aligns with the use-cases presented in Sect. 2. The wearable nature of the device allows the worker to have it "at hand" in places where a traditional digital interface is impossible to be placed, like the body of the car. The software design allows for fast worker response upon task deployment as required in the second use-case. Further development of the artefact can take into account more complex interactions with the digital platforms available in the manufacturing environment, such as data collection using the equipped sensors [11,12].

Testing of the artefact was only performed in a controlled environment in the laboratory of the affiliated university, as access to real production environment Was limited at the time. Given the functionalities of the prototype, like the ability to send tasks immediately to the factory personnel as they emerge and the "at hand" interface of the app, an infrastructure is established to address the use cases and the identified research gap which motivated the development.

**Acknowledgements.** The authors would like to thank MADE FAST Association for providing financial support, and the MADE FAST associated companies for providing relevant information from the manufacturing industry. MADE FAST is financed by the member companies, the Innovation Fund Denmark under the Danish government, private funds and Universities.

# References

1. Groover, M.P.: Automation, Production Systems, and Computer-Integrated Manufacturing, 4th edn. Pearson, London (2016)
2. Forrester, J.W.: Industrial Dynamics. The M.I.T. Press and John Wiley & Sons Inc., New York, London (1961)
3. Fleischmann, A., Oppl, S., Schmidt, W., Stary, C.: Contextual Process Digitalization. Springer, Cham (2020). https://doi.org/10.1007/978-3-030-38300-8
4. Bell, T.: Supply chain integration systems by small engineering to order companies: the challenge of implementation. J. Manuf. Technol. Manag. **21**(1), 50–62 (2009)
5. Sansone, C., Hilletofth, P., Eriksson, D.: Evaluation of critical operations capabilities for competitive manufacturing in a high-cost environment. J. Glob. Oper. Strategic Sourcing **13**(3), 229–250 (2020). https://doi.org/10.1108/JGOSS-10-2019-0055
6. Nygaard, J., Colli, M., Wæhrens, B.V.: A self-assessment framework for supporting continuous improvement through IoT integration. In: 1st International Conference on Industry 4.0 and Smart Manufacturing (ISM 2019), Procedia Manufacturing, pp. 344–350. Elsevier, Italy (2019). 13(3), pp. 229–250 (2020)
7. Colli, M., Cavalieri, S., Cimini, C., Madsen O., Wæhrens, B.V.: Digital transformation strategies for achieving operational excellence: a cross-country evaluation. In: Proceedings of the 53rd Hawaii International Conference on System Sciences (HICSS), pp. 4581–4590. University of Hawaii, Hawaii (2020)
8. Saabye, H., Kristensen, T.B., Wæhrens, B.V.: Real-time data utilization barriers to improving production performance: an in-depth case study linking lean management and Industry 4.0 from a learning organization perspective. Sustainability 12(21), 8757 (2020)
9. Lindell, C.: Wearables on the manufacturing floor. Food Eng. **90**, 52–55 (2018)
10. Garcia-Ceja, E., Galván-Tejada, C.E., Brena, R.: Multi-view stacking for activity recognition with sound and accelerometer data. Inf. Fusion **40**, 45–56 (2018)
11. Sarivan, I.-M., et al.: Enabling real-time quality inspection in smart manufacturing through wearable smart devices and deep learning. In: 30th International Conference on Flexible Automation and Intelligent Manufacturing (FAIM2021). Procedia Manufacturing, pp. 373–380. Elsevier, Greece (2021)
12. WorkerBase SmartWatch, WorkerBase GmbH. https://workerbase.com/connected-worker-platform. Accessed 15 July 2021
13. Introduction to Wear OS, Google. https://developer.android.com/training/wearables. Accessed 14 July 2021
14. Sarivan, I.-M., et al.: Deep learning-enabled real time in-site quality inspection based on gesture classification. In: Weißgraeber, P., Heieck, F., Ackermann, C. (eds.) Advances in Automotive Production Technology – Theory and Application. ARENA2036, pp. 221–229. Springer, Heidelberg (2021). https://doi.org/10.1007/978-3-662-62962-8_26

# Cash-Flow Bullwhip Effect in the Semiconductor Industry: An Empirical Investigation

Chintan Patil[(✉)] and Vittaldas V. Prabhu

Harold and Inge Marcus Department of Industrial and Manufacturing Engineering, Pennsylvania State University, University Park, PA 16802, USA
chintan@psu.edu

**Abstract.** Cash flow bullwhip effect (CFB) is the amplification of working capital variance along a supply chain. High CFB is a sign of inefficient working capital (WC) management and can lead to a significant reduction in financial resilience. CFB can be used as a measure of a company's ability to manage operational risks and corresponding resilience. We investigate the existence of CFB and the traditional bullwhip effect (BWE) in a sample of 238 semiconductor companies over 2010-Q1 to 2020-Q4. These companies' average CFB and BWE are 3.95 and 2.77, respectively. We find that CFB and BWE of a semiconductor company are negatively associated with company size, degree of seasonality in demand, and company's payment policy conservativeness; and positively associated with procurement and payment lead times.

**Keywords:** Bullwhip effect · Working capital · Supply chain · Cash flow

## 1 Introduction

Working capital (WC) and cash conversion cycle (CCC) are important financial metrics for supply chains. Inefficient management of WC and CCC can lead to insolvency and can increase the risk of bankruptcy [1]. Numerous attempts have been made from a supply chain perspective on studying WC and CCC (e.g. [2–4]).

Hofmann and Kotzab [2] built a conceptual model of a supply chain to study the effect of inventory management, optimization of payment period, and collaboration between supply chain partners on the CCC of supply chain entities. Peng and Zhou [3] studied the effect of speed of cash turnover and discount rates on profitability in a two-echelon supply chain. Pirtila et al. [4] performed an empirical analysis of WC management in the Russian automotive industry. They showed that firms in the Russian automotive industry do not manage WC in a collaborative manner. None of these studies has in any manner focused on WC variance propagation in supply chains.

WC and CCC variance propagations in supply chains remain poorly studied by the scientific community [5]. Such studies could be especially critical for the semiconductor industry, where lead times are long, investments are huge, raw material supplies are tight, and working capital requirements are large.

© IFIP International Federation for Information Processing 2021
Published by Springer Nature Switzerland AG 2021
A. Dolgui et al. (Eds.): APMS 2021, IFIP AICT 631, pp. 68–77, 2021.
https://doi.org/10.1007/978-3-030-85902-2_8

Supply chains experience two kinds of risks: Operational and Disruption risks [6]. Bullwhip effect (BWE), the amplification of material flow variability in a supply chain, is a measure of how well a supply chain manages its operational risks. On the other hand, the ability to recover from disruptions is measured by resilience. Dominguez et al. [7] showed that BWE avoidance increases a supply chain's resilience.

BWE has received significant attention in the literature [8]. However, this is not the case for cash flow bullwhip in supply chains [5]. Cash flow bullwhip was originally characterized by Tangsucheeva and Prabhu [1] as the ratio of variance in a supply chain entity's CCC to the variance of its demand. Henceforth, we refer to Tangsucheeva and Prabhu's [1] definition of cashflow bullwhip as CFB-TP. Mathematical formulation of a supply chain entity's CFB-TP is given as follows.

$$\text{CFB - TP} = \frac{var(CCC)}{var(Demand)} \tag{1}$$

A caveat in the formulation of CFB-TP is that CCC is measured in number of days, whereas Demand is measured in dollars. Since CFB-TP represents the ratio of variance of 2 different units of measurement, it cannot be used to check if financial variances are either attenuating or amplifying across a supply chain. To overcome this shortcoming, Patil and Prabhu [9] redefined cashflow bullwhip (henceforth CFB) as the amplification of working capital variance across supply chain entities. CFB is characterized as the ratio of WC variance to Demand variance. Like BWE avoidance increases resilience, it is likely that CFB avoidance increases financial resilience. Mathematical formulations of CFB and BWE are given as follows.

$$\begin{aligned} CFB &= \frac{var(WC)}{var(Demand)} \\ &= \frac{var(Inventory + Cash + Trade\ Receivables - Trade\ Payables)}{var(Demand)} \end{aligned} \tag{2}$$

CFB larger than 1 is indicative of a company not managing its WC efficiently. Our objective in this paper is twofold: (i) measure the CFB and BWE experienced by companies in the semiconductor industry, and (ii) examine the association of company-level variables like company size, lead-time, payment policy, liquidity ratio, demand seasonality, demand autocorrelation with CFB and BWE. To the best of our knowledge, this is the first paper that empirically (i) examines the existence of CFB in a supply chain of any kind, and (ii) identifies the company-level variables that probably drive CFB.

## 2   Data

Public companies in the U.S. are obligated to release quarterly financial data. We access these data between 2010-Q1 and 2020-Q4 for manufacturers in the semiconductor industry (North American Industry Classification System (NAICS) codes 33242, 334413, 334111-334118, 334210-334290) from the Compustat database. Companies with missing values in cost of goods sold (COGS), cash, trade payables, trade receivables, and inventory; and with data series smaller than ten calendar quarters are dropped from the sample. We end up with 238 unique companies in our sample. Compustat database does not report companies' demand series. Therefore, we use each company's COGS series as a proxy for the demand series. As seen in Eqs. (2) and (3), calculations of a company's CFB and BWE require WC series and Order Quantity series. WC for company i in period t is given as follows.

$$WC_{i,t} = Cash_{i,t} + Invt_{i,t} + TR_{i,t} - TP_{i,t} \quad (3)$$

where $Casht_{i,t}$, $Invt_{i,t}$, $TR_{i,t}$, and $TP_{i,t}$ are the cash position, inventory, trade receivables, and trade payables of company $i$ in period $t$, respectively. Compustat does not report the order quantity series for companies. Therefore, we use a measure called Purchases as a proxy for order quantity. Purchases of company $i$ in period t is given as follows.

$$Purchases_{i,t} = COGS_{i,t} + Invt_{i,t} - Invt_{i,t-1} \quad (4)$$

where $COGS_{i,t}$ is the cost of goods sold by company $i$ in period $t$.

The WC, Purchases, and COGS series for each company are first-differenced. This operation detrends the data series. Detrending is critical because each company's data series span over $4-11$ years ($10-44$ quarters), and this is a long enough duration for data series to exhibit stochastic trends. Our objective is to capture WC, Purchases, and COGS series' variances about stochastic trends, not the overall data series variances.

## 3  Measures and Hypotheses

CFB for company i is determined by substituting its WC and COGS series in Eq. (2)'s numerator and denominator, respectively. Similarly, BWE for company i, is calculated by substituting its Purchases and COGS series in Eq. (3)'s numerator and denominator, respectively.

We develop several hypotheses to identify the drivers of bullwhip effects (CFB and BWE) in semiconductor industry. These hypotheses on the association of CFB and BWE with company-level variables are based on (i) the analytical model of CFB by Patil & Prabhu [9], and BWE by Chen et al. [10]; and (ii) intuition originating from our knowledge of supply chain literature.

CFB has been shown to be associated with supply chain entity's (or company's) procurement lead time, payment lead time, payment policy conservativeness, demand autocorrelation, and supply chain stage by Patil and Prabhu [9]. On the other hand, BWE has been shown to be associated with lead time, demand autocorrelation, and supply chain stage by Chen et al. [10]; and additionally, with supply chain entity (or company) size, and seasonality ratio by Cachon et al. [11]. Our intuition suggests that since liquidity is important in keeping semiconductor supply chain operations running, liquidity ratio also is associated with CFB and BWE experienced by semiconductor companies. Therefore, we study the association of CFB and BWE with company variables like: company size, liquidity ratio, demand seasonality, procurement and payment lead times, payment policy conservativeness, demand autocorrelation, and supply chain stage. The reader should note that we do not claim that the aforementioned company variables constitute an exhaustive set of variables associated with CFB and BWE. Next, we discuss our hypotheses on the nature of association of the abovementioned company-level factors on CFB and BWE.

Intuition suggests that large semiconductor companies might have smaller inventory variance than smaller companies because the former can aggregate inventories across multiple products and can leverage the concept of vendor-managed inventory (VMI). VMI can allow larger companies to push inventory fluctuations onto their suppliers. Large companies also tend to enforce short payment terms on customers and extended payment terms on their suppliers [2], thereby pushing cash flow risks onto their supply chain partners. Therefore, we hypothesize the following.

Hypothesis 1: CFB and BWE are negatively associated with semiconductor company size.

The mean value of a company's COGS over 2010-Q1 to 2020-Q4 is used as a proxy for a company's size.

Semiconductor supply chains are capital intensive [12, 13] because they need to upgrade processes continuously. Therefore, semiconductor companies require high liquidity to operate efficiently. The need to maintain high liquidity can motivate companies to push cash flow uncertainties onto supply chain partners. Therefore, we hypothesize the following.

Hypothesis 2: CFB is negatively associated with semiconductor company's liquidity ratios.

The liquidity ratio of company $i$ is given as follows.

$$LR_i = \frac{mean(Cash_i)}{mean(Invt_i)} \qquad (5)$$

where $Cash_i$ and $Invt_i$ are the cash and inventory series of company $i$.

COGS series of companies in the sample indicate that semiconductor companies experience seasonality in COGS. Given the capital-intensive nature of semiconductor industry, we believe companies have the motivation to smooth their purchases and WC, even more so when demand seasonality exists. Cachon et al. [11] found this to be the case across 74 industries in the U.S. economy. Therefore, we hypothesize the following.

Hypothesis 3: CFB and BWE are positively associated with demand seasonality.

To measure seasonality in demand experienced by a company, we use the seasonality ratio. Seasonality ratio for company $i$ is given as follows.

$$SR_i = \frac{Var(COGS_i) - Var(Seasonally\,adjusted\,COGS_i)}{Var(COGS_i)} \qquad (6)$$

where, $COGS_i$ is the COGS series of company $i$ and $Seasonallyadjusted COGS_i$ is the deseasonalized $COGS_i$ series. Method described in [14] is used to deseasonalize $COGS_i$ series.

Chen et al. [10] show that BWE is increasing in procurement lead time. Parallelly, Patil & Prabhu [9] show that CFB is increasing in procurement and payment lead times. Therefore, we hypothesis the following.

Hypothesis 4: CFB and BWE are positively associated with semiconductor company's procurement and payment lead times.

Procurement and payment lead times are not reported in the Compustat database. Therefore, we use Days Payable Outstanding (DPO) as a proxy for the sum of procurement and payment lead times, LT. DPO for company $i$ is given as follows.

$$LT_i = DPO_i = \frac{365}{4 \times \frac{mean(COGS_i)}{mean(TP_i)}} \tag{7}$$

where, $TP_i$ is the trade payables series of company $i$.

Patil and Prabhu [9] show that a supply chain entity's payment policy not only affects its own CFB but also entire upstream supply chain's CFB. Specifically, they show that the more conservative a company's payment policy, the higher the CFB it experiences. Therefore, we hypothesize the following.

Hypothesis 5: CFB is positively associated with a semiconductor company's payment policy's conservativeness.

Company $i$'s payment policy conservativeness is measured by its payment policy parameter which is given as follows.

$$\alpha_i = \frac{mean(TP_i)}{mean(TR_i)} \tag{8}$$

where $TP_i$ and $TR_i$ are the trade payables and trade receivables series of company $i$. Higher the $\alpha_i$, the more conservative company i's payment policy.

Chen et al. [10] show that a supply chain entity experiences BWE greater than 1 when the entity's demand autocorrelation ($\phi$) is greater than 0. We find that, out of the 238 companies in the sample, 234 have demand (COGS) autocorrelation greater than 0. Patil and Prabhu [9] show that for $\phi > 0$, CFB is increasing in $\phi$ as the latter approaches $\phi_{treshold}$. For $\phi > \phi_{treshold}$, CFB is decreasing in $\phi$. Patil and Prabhu [9] find $\phi_{treshold}$ to be greater than 0.75 in all the cases that they explore. Therefore, we hypothesize the following.

Hypothesis 6: CFB and BWE experienced by semiconductor companies are positively associated with demand autocorrelation.

Patil and Prabhu [9] and Chen et al. [10] analytically show that CFB and BWE experienced by upstream supply chain companies are larger than their downstream counterparts. Checking the prevalence of these effects in semiconductor supply chain would require complete knowledge of the upstream and downstream supply chain partners of all the companies in our sample. This is beyond the scope of the present paper. However, we can use the NAICS codes of companies in the sample to classify them into the following categories: (i) Semiconductor Machinery Manufacturing (SMM), (ii) Semiconductor and Related Devices Manufacturing (SRDM), and (iii) Computer and Communications Manufacturing (CCM). These three categories could be assumed to form a notional semiconductor supply chain shown in Fig. 1. Assuming these three categories relate to different stages of the semiconductor supply chain, we hypothesis the following.

Hypothesis 7: The differences in the mean CFBs and BWEs experienced by SMM, SRDM, and CCM companies are statistically significant.

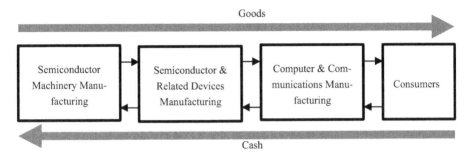

**Fig. 1.** Model of semiconductor supply chain structure studied in the paper

## 4   Analysis

In this section, we first quantify the CFB and BWE experienced by companies in the semiconductor industry and then investigate the association of company-level variables with CBF and BWE.

### 4.1   Existence of CFB and BWE in the Semiconductor Industry

Based on NAICS codes, the 238 semiconductor companies in the sample belong to eight different sub-industries. Table 1 reports the average and standard deviation of CFB and BWE experienced by these eight sub-industries. All semiconductor sub-industries but Electronic Computer Manufacturing, on average, experience CFB and BWE greater than 1. This shows that CFB and BWE exist in the semiconductor industry. The Telephone Apparatus Manufacturing industry, on average, experiences the largest CFB (5.27), and the Semiconductor & Related Device Manufacturing industry, on average, experiences the largest BWE (3.32).

To check if COVID-19 related supply chain disruptions had any effect on CFB and BWE experienced by semiconductor companies, we performed a t-test on the mean CFB (and BWE) experienced by semiconductor companies over the periods: 2010-Q1 to 2019-Q4 and 2010-Q1 to 2020-Q4. The test fails to reject the null hypothesis that the mean CFB (and BWE) of the 2 populations are the same. Therefore, based on the data we have, we infer that the COVID-19 related disruption did not have significant effect on the CFB and BWE experienced by the semiconductor supply chain. However, it should be noted that our analysis is based on quarterly data which does not provide the temporal granularity that monthly data would. Monthly data might be able to capture the real effect of COVID-19 related disruptions on the CFB and BWE experienced by the semiconductor industry.

**Table 1.** CFB and BWE experienced by sub-industries in the semiconductor industry.

| Sub-industry | CFB | | BWE | |
|---|---|---|---|---|
| | Mean | Std. dev | Mean | Std. dev |
| Semiconductor Machinery Manufacturing | 2.89 | 2.98 | 2.98 | 4.86 |
| Electronic Computer Manufacturing | 0.93 | 0.66 | 1.12 | 1.26 |
| Computer Storage Device Manufacturing | 3.00 | 3.81 | 1.98 | 1.33 |
| Computer Terminal & Other Computer Peripheral Equipment Manufacturing | 4.90 | 8.06 | 2.85 | 2.68 |
| Telephone Apparatus Manufacturing | 5.27 | 4.72 | 2.60 | 1.44 |
| Radio & Television Broadcasting & Wireless Communications Equipment Manufacturing | 2.73 | 3.46 | 2.02 | 1.86 |
| Other Communications Equipment Manufacturing | 2.67 | 2.35 | 1.41 | 0.94 |
| Semiconductor & Related Device Manufacturing | 4.27 | 4.71 | 3.32 | 3.90 |

## 4.2 Association of CFB and BWE with Company-Level Variables

To find the association of CFB and BWE with company variables, we test the hypothesses discussed in Sect. 3. To test these hypotheses, we build multiplicative regression models of the following types.

$$
\begin{aligned}
log(CFB_i) = a + b_1 \log(size_i) + b_2 \log(LR_i) + b_3 \log(SR_i) \\
+ b_4 \log(LT_i) + b_5 \log(\alpha_i) + b_6 \log(\phi_i) + c_1 SMM_i \\
+ c_2 SRDM_i + \epsilon_i
\end{aligned} \tag{9}
$$

$$
\begin{aligned}
log(BWE_i) = a + b_1 \log(size_i) + b_2 \log(LR_i) + b_3 \log(SR_i) \\
+ b_4 \log(LT_i) + b_5 \log(\alpha_i) + b_6 \log(\phi_i) + c_1 SMM_i \\
+ c_2 SRDM_i + \epsilon_i
\end{aligned} \tag{10}
$$

where, $b_j$ for $j \in \{1, 2, \ldots, 6\}$ are the regression coefficients of the continuous variables and $c_j$ for $j \in \{1, 2\}$ are the regression coefficients of dummy variables $SMM_i$ and $SRDM_i$. To avoid the dummy variable trap, we use $3 - 1 = 2$ dummy variables to represent the 3 supply chain stages (SMM, SRDM, and CCM). For a company that is an SMM, $c_1 = 1$ and $c_2 = 0$. For a company that is an SRDM, $c_1 = 0$ and $c_2 = 1$. Finally, for a company that is a CMM, $c_1 = 0$ and $c_2 = 0$. Only 2 dummy variables are sufficient to indicate if company $i$ is an SMM, SRDM, or a CCM company. $\in_i$ additively includes company-specific effects and random noise in the model.

We use multiplicative regression models instead of additive regression models for two reasons: (i) plotting CFB and BWE against $size, LR, SR, LT, \alpha$, and $\phi$ suggests a log-linear relationship between the variables, and (ii) multiplicative regression models have better explanatory power than additive regression models over our sample of semiconductor companies. The regression models are fit over winsorized data. Winsorizing the top and bottom 1% data dampens the effect of significant outliers. Table 2 reports the regression coefficients of models presented in Eqs. (9) and (10).

**Table 2.** Regression model summary

|  | log(CFB) | log(BWE) |
|---|---|---|
| SMM | −0.2 | −0.06 |
| SRDM | −3.65*** | 0.34*** |
| log(size) | −0.09** | −0.06** |
| log(LR) | 0.09 | −0.01 |
| log(SR) | −0.06** | −0.05** |
| log(LT) | 0.9*** | 0.35* |
| log($\alpha$) | −1.32*** | −0.38** |
| Log($\phi$) | 0.25 | 0.34** |
| Adjusted R-squared (%) | 35.2% | 13.6% |
| F-statistic | 17.11*** | 5.66*** |
| N | 238 | 238 |

***, **, and *denote statistical significance for p-value < 0.001, 0.01, and 0.05, respectively.

The R-squared values of the log(CFB) and log(BWE) models reported in Table 2 are low (35.2% and 13.6%, respectively), but it should be noted that this is an econometric study and therefore is not concerned with the models' predictive powers. The sole objective of these regression models is to make inferences about the association of company-level variables with CFB and BWE. Therefore, low R-squared values are not of much concern here. Results in Table 2 support Hypotheses 1, 3, and 4. The coefficients of company size, seasonality ratio (SR), and procurement + payment lead time (LT) are statistically significant for the models presented in Eqs. (9) and (10). Contrary to Hypothesis 5, results indicate that the payment policy parameter is negatively associated with CFB and BWE. The more conservative the company's payment policy, the smaller are its CFB and BWE. Hypothesis 6 is only supported for log(BWE). Hypothesis 7 is not entirely supported by the results. Dummy variable SRDM's coefficient is found to be statistically significant. However, the dummy variable SMM's coefficient is not statistically significant. It means that the differences in the mean values of bullwhip effects (CFB and BWE) experienced by SRDM and CMM are statistically significant.

In Table 2, we see that a 1% increase in company size is, on average, associated with a 0.09% and 0.06% drop in CFB and BWE, respectively. 1% increase in seasonality ratio is, on average, associated with a 0.06% and 0.05% drop in CFB and BWE, respectively. 1% increase in lead time is, on average, associated with a 0.9% and 0.35% rise in CFB and BWE, respectively. 1% increase in payment policy parameter is, on average, associated with a 1.32% and 0.38% drop in CFB and BWE, respectively. 1% rise in demand autocorrelation is, on average, associated with a 0.34% rise in BWE.

## 5 Conclusion

This paper is a step in the direction of empirically investigating the existence of CFB in supply chains. Our results indicate that semiconductor companies experience financial and operational risks in the form of CFB and BWE. These effects experienced by semiconductor companies are associated with company size, lead-time, payment policy, liquidity ratio, demand seasonality, and demand autocorrelation. We also observe the CFB experienced by SRDM companies to be smaller than that by CMM companies, and BWE experienced by SRDM companies to be larger than that by CMM companies.

Future research should focus on analyzing the relationship between CFB and financial resilience, and test if CFB reducing strategies increase financial resilience. It would be remiss of us not to list the limitations of this work. Our sample consists of only those companies which were public for more than ten quarters between 2010-Q1 and 2020-Q4. Therefore, our sample is vulnerable to size bias and survivorship bias. Another limitation of this paper is that it uses imperfect proxies for demand, order quantity, and lead times. Future explorations should investigate the prevalence of CFB across all major industries using better proxies for demand, order quantity, and lead times.

## References

1. Tangsucheeva, R., Prabhu, V.: Modeling and analysis of cash-flow bullwhip in supply chain. Int. J. Prod. Econ. **145**, 431–447 (2013). https://doi.org/10.1016/j.ijpe.2013.04.054
2. Hofmann, E., Kotzab, H.: A supply chain-oriented approach of working capital management. J. Bus. Logist. **31**, 305–330 (2010). https://doi.org/10.1002/j.2158-1592.2010.tb00154.x
3. Peng, J., Zhou, Z.: Working capital optimization in a supply chain perspective. Eur. J. Oper. Res. **277**, 846–856 (2019). https://doi.org/10.1016/j.ejor.2019.03.022
4. Pirttilä, M., Virolainen, V.M., Lind, L., Kärri, T.: Working capital management in the Russian automotive industry supply chain. Int. J. Prod. Econ. **221**, 107474 (2020). https://doi.org/10.1016/j.ijpe.2019.08.009
5. Lamzaouek, H., Drissi, H., El Haoud, N.: Cash flow bullwhip—literature review and research perspectives. In: Logist 2021, vol. 5, p. 8 5:8 (2021). https://doi.org/10.3390/LOGISTICS5010008
6. Ivanov, D.: Supply chain management and structural dynamics control. In: Structural Dynamics and Resilience in Supply Chain Risk Management. International Series in Operations Research & Management Science (ISOR), vol. 265. Springer, Cham (2018). https://doi.org/10.1007/978-3-319-69305-7_1
7. Dominguez, R., Cannella, S., Framinan, J.M.: On bullwhip-limiting strategies in divergent supply chain networks. Comput. Ind. Eng. **73**, 85–95 (2014). https://doi.org/10.1016/j.cie.2014.04.008

8. Wang, X., Disney, S.M.: The bullwhip effect: progress, trends and directions. Eur. J. Oper. Res. **250**, 691–701 (2016)
9. Patil, C., Prabhu, V.: Working capital distress caused by supply chain cash-flow bullwhip. Working paper (2021)
10. Chen, F., Drezner, Z., Ryan, J.K., Simchi-Levi, D.: Quantifying the bullwhip effect in a simple supply chain: the impact of forecasting, lead times, and information. Manage. Sci. **46**, 436–443 (2000). https://doi.org/10.1287/mnsc.46.3.436.12069
11. Cachon, G.P., Randall, T., Schmidt, G.M.: In search of the bullwhip effect. Manuf. Serv. Oper. Manag. **9**, 457–479 (2007). https://doi.org/10.1287/msom.1060.0149
12. Mönch, L., Chien, C.F., Dauzère-Pérès, S., et al.: Modelling and analysis of semiconductor supply chains. Int. J. Prod. Res. **56**, 4521–4523 (2018)
13. Chen, Y.J., Chien, C.F.: An empirical study of demand forecasting of non-volatile memory for smart production of semiconductor manufacturing. Int. J. Prod. Res. **56**, 4629–4643 (2018). https://doi.org/10.1080/00207543.2017.1421783
14. Vanek, F., Sun, Y.: Transportation versus perishability in life cycle energy consumption: a case study of the temperature-controlled food product supply chain. Transp. Res. Part D Transp. Environ. **13**, 383–391 (2008). https://doi.org/10.1016/j.trd.2008.07.001
15. Ravindran, A.R., Warsing, D., Jr.: Supply Chain Engineering: Models and Applications. CRC Press, Boca Raton (2016)

# Analytics with Stochastic Optimization: Experimental Results of Demand Uncertainty in a Process Industry

Narain Gupta[1], Goutam Dutta[2(✉)], Krishnendranath Mitra[3], and M. K. Tiwari[4]

[1] Management Development Institute Gurgaon, Gurugram, India
[2] Production and Quantitative Methods Area, Indian Institute of Management, Ahmedabad, India
goutam@iima.ac.in
[3] Department of Business Management, University of Calcutta, Calcutta, India
[4] National Institute of Training in Industrial Engineering, Mumbai, India

**Abstract.** The key objective of the research is to report the results of testing of a two stage stochastic linear programming (SLP) model with recourse using a multi scenario, multi period, menu driven user friendly DSS in a North American steel company. The SLP model and the DSS is generic which can be applied to any process industry. It is capable of configuring multiple materials, multiple facilities, multiple activities and multiple storage areas. The DSS is developed using 4th Dimension programming language, and the SLP model was solved using the IBM CPLEX solver. The value of the SLP solution derived from the experimentation of the DSS with a real-world instance of one steel mill is 1.61%, which is equivalent to a potential benefit of US$ 24.61 million.

A set of experiments were designed based on the potential joint probability scenarios, and the demand distributions expected skewness. The research reports a few interesting patterns emerged from optimization results when the volatility in demand of finished steel rises and the distribution of the demand skewness changes from left to right tail. The academic contribution of this research is two folds. Firstly, the depicting potential contribution to profit in a steel company using a SLP based DSS under probabilistic demand scenarios. Secondly, the optimization experiments confirm that the value of SLP solution increases with the increase in demand uncertainty. The research has applied implications that the practicing managers would be encouraged to look for more optimization based business solutions, and the prescriptive analytics discipline will fetch more scholarly and industry attention.

**Keywords:** Decision support system · Planning with optimization · Stochastic programming

© IFIP International Federation for Information Processing 2021
Published by Springer Nature Switzerland AG 2021
A. Dolgui et al. (Eds.): APMS 2021, IFIP AICT 631, pp. 78–88, 2021.
https://doi.org/10.1007/978-3-030-85902-2_9

# 1   Introduction and Motivation

In the field of decision support systems (DSS), analytics and digitization are the two new concepts. Stochastic Linear Programming (SLP) is one tool that synthesizes mathematical programming, Decision theory, and statistics. SLP is an important tool in managing an uncertain business environment and making an optimal managerial decision. In the two stage SLP with recourse there are two stages of decision making. The decisions of the first stage are taken before the realization of the outcomes from random parameters. The second stage is a rectification process where the parameters' randomness is considered with an objective function where long-term expected profits can be maximized.

The stochastic linear programming is a very important area of research from present businesses point of view. The industry 4.0 has taken place. Many new technologies are emerging to address the business environment uncertainties. The SLP is one of the prescriptive analytics modeling tool to address this uncertainty in the demand and the other critical parameters of the optimization models. Presently the corporate managers do the planning based on intuition, and the scientific methods of optimization planning have started to take place in a few founding industries such as airlines, hotels, and FMCG. A few manufacturing firms have also made attempts in this area by purchasing user friendly software produced by IBM, and other market leaders. The attempts have been made to integrate such planning and optimization software with the existing ERP systems of the organizations. These attempts have not been much successful so far. The scale of benefits from optimization are now very much known to the practicing managers and greatly appreciated. The INFORMS prestigious Franz Edelman award is given every year to the organizations practice and demonstrate the realization of the profits from large scale optimization implementation in their industries.

A series of earlier publications, namely [1–5] which deals with DSS in process industry using multi-period optimization is the motivation behind this present research. A study [6] described a customized DSS for a North-American integrated steel plant that helped achieve a 16% to 17% increment in its bottom line. Similarly, the DSS resulted in an impact of 6.72% and 12.46% when real-world data from an integrated aluminum and a pharmaceutical firm, respectively, were used. This study is motivated by this high impact potential.

The results of the applications in process industries were discussed using SLP's key performance parameters like Value of Stochastic Solution (VSS), Expectation of Expected Solution (EEV), Expected Value of Perfect Information (EVPI). We depict the practice implications of modeling under probabilistic scenarios. The research is an initiative to model production operations with probabilistic demand of finished steel using an SLP approach. It further demonstrates how managers of companies with little knowledge of OR/MS can impact their profitability using a generic and user-friendly SLP-based DSS [6].

The SLP based DSS is capable of addressing the following questions:

1. How does the VSS vary with demand variability?
2. How does modifying demand variability impact the EVPI?
3. How does the probability distribution of demand impact the VSS?
4. How does the discrete empirical probability distribution of demand impact the EVPI?

The rest of the paper is organized as review of literature in Sect. 2, the optimization model in Sect. 3, the design of database and model definition in Sect. 4, and DSS application in in Sect. 5 with Sect. 6 as results and conclusion.

## 2   Review of Related Literature

This work was also motivated by [2] on mathematical modeling for integrated steel plant. A linear program was presented [1] by describing the fundamental principles of relational database construction. The modeling evidences are a North American steel mill [2] to a Western Indian pharma firm [3] and an Eastern Indian aluminium company [5].

An SLP survey [7] argues that flexible and robust "near-optimal" resource allocation decisions can be made with SLP models in the face of uncertainty. Earlier studies [2, 4] developed design principles of a multi-period linear programming model and evaluated the optimization-based DSS with a real data set. Literature [3, 5] confirmed a potential substantial impact in financial terms with a DSS in a pharmaceutical company and an aluminum plant. It is observed that other than [3, 8] little has been documented in the literature on 2 stage SLP-based DSS in a process industry.

Reported studies [4–6] discussed the basic concepts of database structure and SLP-based DSS design. We reviewed the relevant literature and presented in the context of our research in Table 1.

There has been a series of attempts to design and develop the SLP based DSS ( Table 1). The focus of most of them have only been towards the mathematical model, algorithmic power, rigor of research. *It is unlikely that the practicing executive with little knowledge of OR/MS can implement such solutions. Our research bridge this gap of academic research versus the applied research.* The study primarily focuses on modeling demand uncertainty using a two stage SLP based user friendly menu driven DSS. Experiments set have been developed to test the multi-scenario optimization-based DSS with data from an integrated steel company from North America. The experiments are designed by varying volatility in the market demand with probability of the possible economic scenarios.

## 3   Optimization Model and Schema

There are six fundamental elements of the model's database structure, namely *Times, Scenarios,Materials, Facilities, Activities, Storage-Areas.* In line with the previous research by [4] Scenarios are added as an additional fundamental element [6]. They [6] discussed implementing a DSS in an SLP model, its detailed mathematical formulation, and the designed and developed database structure in their studies. The details of the model parameters, variables and the steps of optimization and model assumptions are referred to by [4, 17]. The model maximizes the expected contribution to a company's profit (nominal or discounted) subject to *Material Balance, Facility Input, Facility Output, Facility Capacity, Storage Inventory Balance, Storage Capacity, Set of Implementability (Non-Anticipativity) Constraints, and the Bounds on decision variables.* The model's objective and constraints are discussed in Appendix-A in brief to simplify understanding the context of the steel industry optimization application for this paper's audience.

**Table 1.** Review of recent literature on DSS and stochastic modeling

| Functional context | Mathematical model and solution platform | Source |
|---|---|---|
| *Decision support system* | | |
| Cost architecture analysis in assembly systems | Integer linear program solved using IBM CPLEX OPL 12.6, automation in projects | [9] |
| Asset liability management modeling for life insurance company in India | Two-stage stochastic programming and Big data analytics based DSS | [10] |
| Marine and liner vessel fuel and speed optimization using weather archived data | Fuel consumption function for speed optimization a metaheuristic optimization approach | [11] |
| *A two-stage stochastic programming approach* | | |
| Thermal power plant, and modeling electricity demand uncertainty, and wind speed | A two-stage stochastic program for optimal scheduling of energy storage system | [12] |
| Energy sector, uncertain energy prices, $CO_2$ emission reduction | Two-stage stochastic mixed-integer programming approach | [13] |
| Value-based closed-loop supply chain network tactical and strategic planning | Three echelon, multi-commodity, multi-period two-stage stochastic program | [14] |
| Production, inventory and backordering cost reduction in manufacturing organizations | Two-stage stochastic programming for scheduling and lot-sizing under random demand, Solved using GAMS | [15] |
| Retail staffing and employee scheduling in Switzerland | Two-Stage Stochastic Programming with simple recourse, MILP in uncertain demand environment simulation | [16] |

## 4  Defining Models and Designing Demand Scenarios

Demand scenarios are designed based on the volatility in demand and change in probability of demand scenario occurrence (Table 2). Stochastic Optimization Model Instances: The detailed definition of various model instances is referred to [7]. The algebraic codes of the model instances are as follows: Perfect Information Solution ($Z_{PI}$), Mean Value Solution ($Z_{MV}$), SLP Solution ($Z_{SLP}$), The expectation of Expected Value Solution ($Z_{EEV}$). Where:

$$VSS = (Z_{SLP}) - (Z_{EEV}); EVPI = (Z_{PI}) - (Z_{SLP}).$$

## 5  DSS Application in Steel Company

This study describes the application of the SLP model in a steel company. The steel company profile and the large-scale SLP optimization model are described in Table 3 and Fig. 1A. The optimal solution of SLP in steel company is shown in a user-friendly DSS layout in Fig. 2.

**Table 2.** Definition of probability scenarios and demand cases

|  | Expected economic situation | | |
|---|---|---|---|
| Probability scenario cases | Low | Reg | High |
| Case i (right skewed) | 0.75 | 0.15 | 0.10 |
| Case ii (equally likely) | 0.33 | 0.33 | 0.33 |
| Case iii (left skewed) | 0.10 | 0.15 | 0.75 |
| Demand cases | Low | Reg | High |
| Case 1: 20% demand volatility | 80% of D | D | 120% of D |
| Case 2: 30% demand volatility | 70% of D | D | 130% of D |
| Case 3: 40% demand volatility | 60% of D | D | 140% of D |

**Table 3.** Characteristics of the industry and variability of optimization

| Production parameters | | Model parameters | |
|---|---|---|---|
| Annual turnover (Mn US$) | 1,400 | No of variables | 44100 |
| Annual production (tons) | 860,000 | No of constraints | 40472 |
| Sell price ratio | 7.38 | No of coeff (non-zeros) | 168600 |
| Market demand ratio | 1,841.12 | (LP density – nonzeros) | 0.0094% |
| Buy price ratio | 147.99 | No of materials | 632 |
| Facility activity ratio (T/H) | 3,240.83 | No of facilities | 56 |
| Activity cost ratio (US$/ton) | 178.57 | No of activities | 1286 |
|  |  | No of planning periods | 3 |
|  |  | No of scenarios | 3 |

## Impact of SLP in the Steel Company

1. One would only solve an MV problem when the information about the future is not available and expect to achieve $Z_{MV}$ (US$ 1,640.9 million), while once the planning horizon is over, one would end up realizing only $Z_{EEV}$ (US$ 1,618.9 million). A loss of ($Z_{MV} - Z_{EEV}$) US$ 21.99 million is incurred due to the non-availability of any information about the future.
2. The $Z_{MV}$ is highest among all the solutions, namely $Z_{MV}$, $Z_{PI}$, $Z_{SLP}$, and $Z_{EEV}$. This indicates that practically $Z_{MV}$ (US$ 1640.9 million) can never be realized. In the long run, one can only realize the $Z_{PI}$ (US$ 1633.9 million) even with the availability of perfect information.

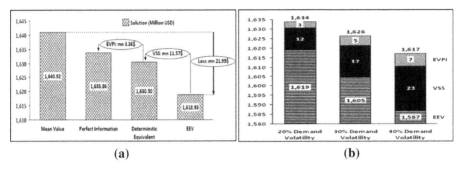

**Fig. 1.** A. Results of SLP model B. SLP optimization results

3. The impact of SLP is measured in terms of VSS (US\$ 11.6 million), the improvement in net (contribution to) profit by $Z_{SLP}$ concerning $Z_{EEV}$. The VSS as a percentage of $Z_{EEV}$ is 0.71%.

4. There are instances when it is possible to get perfect information about the future and know which scenario would occur with certainty. In these situations, it is expected to achieve a long-run solution as $Z_{PI}$ (US\$ 1,633.9 million). When the information about the future scenario occurrence is partially available, $Z_{SLP}$ (US\$ 1,630.5 million) can be attained (Table 4). The expected value of perfect information is given by the difference between $Z_{PI}$ and $Z_{SLP}$ (US\$ 3.4 million).

Birge (1997) established the principles that a decreasing trend is observed in solutions ($Z_{MV} \geq Z_{PI} \geq Z_{SLP} \geq Z_{EEV}$) for a maximization SLP. Our results were consistent with these principles (Table 5).

**Table 4.** Equally likely scenario, 20% demand volatility

| Planning periods | Total revenue | Cost of purchases | Cost of inv carrying | Cost of activities | Net profit | Total steel |
|---|---|---|---|---|---|---|
| Unit | Mn US\$ | Mn US\$ | Mn US\$ | Mn US\$ | Mn US\$ | Mn Tons |
| Grand total | 3,295 | 801.3 | 121.1 | 742.4 | 1,630.5 | 2.18 |
| Unit basis | 1510 | 367.2 | 55.5 | 340.2 | 747.2 | NA |
| Scenario: L | 1,080 | 266.5 | 40.4 | 244.2 | 529.3 | 2.19 |
| Scenario: R | 1,102 | 266.8 | 40.2 | 248.9 | 546.5 | 2.18 |
| Scenario: H | 1,112 | 268.0 | 40.5 | 249.3 | 554.6 | 2.18 |

*Note: Table should be replicated for a Cartesian product of demand variability of 20%, 30% and 40%, and the three probability distributions like left-skewed, equally likely, and right-skewed.*

***Trend Analysis with Volatility in Market Demand:*** This section addresses questions 1 and 2 raised in the introduction section. The important inferences from different solutions due to a change in demand volatility from 20% to 40% are listed below: see Fig. 1B.

**Table 5.** Experiment results from multi-scenario planning (Mn means million)

| Probability distribution | | Right skewed (R) | | Equally likely (E) | | Left skewed (L) | |
|---|---|---|---|---|---|---|---|
| | | Demand volatility cases | | | | | |
| | | 20% | 40% | 20% | 40% | 20% | 40% |
| | | M11 | M3 | M4 | M6 | M7 | M9 |
| PI | Mn $ | 1605 | 1551 | 1633 | 1617 | 1658 | 1669 |
| MV | Mn $ | 1610 | 1567 | 1640 | 1640 | 1661 | 1679 |
| SLP | Mn $ | 1603 | 1548 | 1630 | 1610 | 1654 | 1661 |
| EEV | Mn $ | 1593 | 1523 | 1618 | 1587 | 1646 | 1644 |
| VSS | Mn $ | 10.6 | 24.6 | 11.6 | 23.4 | 7.6 | 16.2 |
| VSS (% of EEV) | % | 0.67 | 1.61 | 0.71 | 1.47 | 0.46 | 0.98 |
| EVPI | Mn $ | 1.5 | 2.8 | 3.4 | 6.7 | 4.5 | 8.8 |

The experiments demonstrate that VSS, EVPI increases, and $Z_{EEV}$ and total contribution of SLP decreases with increased volatility in finished goods demand; see Fig. 3A, 3B, and 3C. The results are consistent in the remaining experiments of probability skewness, i.e., right-skewed and left-skewed.

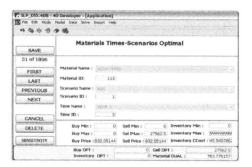

**Fig. 2.** Optimal solution layout

***Trend Analysis over Probability Distribution:*** This section addresses questions 3 and 4 raised in the introduction section. The VSS is the lowest with a left-skewed probability distribution. As the probability of a lower demand situation increases, the VSS increases (Fig. 3A). The impact of optimization on the VSS using SLP becomes more visible. Companies are more concerned about profits when the probability of low demand is high.

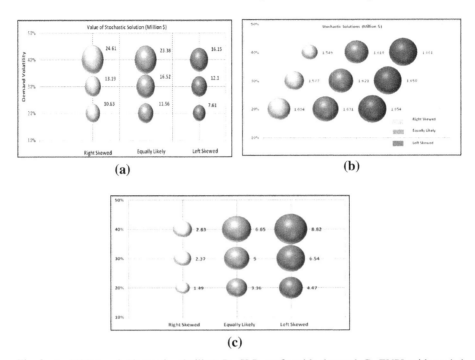

**Fig. 3.** A. VSS trend (demand volatility) B. SLP profit with demand C. EVPI with varied probability

EVPI consistently increases with a variation in the discrete probability distribution from a right-skewed to a left-skewed demand situation (Fig. 3C). Figure 3B clearly shows that the $Z_{SLP}$ increases with the increasing probability of high demand situations (Right skewed to Left skewed distribution). The $Z_{EEV}$ increases at a steeper rate than $Z_{SLP}$ with the increase in the probability of high demand situations (i.e., Right skewed to left-skewed distribution). One may notice that when the probability of a low-demand situation is high, one plans for a $Z_{MV}$, but realizes $Z_{EEV}$. The difference in $Z_{MV}$ and $Z_{EEV}$ happens to be more due to incorrect first-stage decisions. The SLP corrects the first stage decisions so that the losses due to $Z_{MV}$ expectations are minimized, and the VSS is maximized. When the high-demand situation occurs with a high probability of occurrence (left-skewed distribution), the $Z_{EEV}$ starts approaching $Z_{SLP}$, and so the VSS in this situation is lower than the VSS in the right-skewed probability distribution.

## 6    Conclusion, Limitations, and Future Directions

The necessity of stochastic optimization in the process industry is highlighted in this study for demand uncertainty of supply chain. We have witnessed unprecedented growth of Analytics that encompasses operations research, decision theory, statistics, computer science, and industry 4.0. Multi-stage optimization based on Stochastic Linear Programming (SLP) is a very useful concept in operations research. It has applications in the manufacturing industries like integrated iron and steel.

Applying the multi-scenario model in the DSS demonstrated a significant potential to improve the company's (contribution to) profit. A typical observation was that the profit from applying the SLP solution (in terms of VSS and EVPI) increases with the spread of demand distribution. Our results emphasize the need for SLP in an uncertain demand situation to improve profitability, especially when the occurrence probability of the low demand situation is very high. A consistent pattern in EVPI is observed; that is, EVPI is the highest when the occurrence probability of high demand is high (left-skewed) in each of the three demand cases, that is 20%, 30%, and 40%. Findings of the experiments obtained from this research, where we observed the pattern of VSS and EVPI with the change in the probability distribution of the demand, should be studied in various other industries. The DSS being generic, can be tested for modeling uncertainty in multiple input parameters simultaneously with various realistic probability distributions in various process industries. A Monte Carlo simulation in the SLP-based DSS for nonlinear or fuzzy LP applications may reveal interesting trends and inferences.

## Appendix A: Outline of the SLP Model from Gupta et al. (2018)

In order to avoid the repetition of the model formulation in the academic literature, but to improve the ready readability of the paper we present here a brief outline of the model published in Gupta et al., (2018) as follows.

### Equation 1: Nominal objective function

$$Z_N = \sum_{l \in L} p_l \sum_{t \in T} Z(l, t) \tag{1}$$

Where,

### Equation 2: Objective function as a function of time and scenario

$$Z(l, t) = \sum_{j \in M} c_{jlt}^{sell} x_{jlt}^{sell} - \sum_{j \in M} c_{jlt}^{buy} x_{jlt}^{buy} - \sum_{j \in M} c_{jlt}^{inv} x_{jlt}^{inv} - \sum_{(j,j') \in M^{conv}} c_{jj'lt}^{conv} x_{jj'lt}^{conv}$$
$$- \sum_{(i,k) \in F^{act}} c_{iklt}^{act} x_{iklt}^{act} - \sum_{i \in F} c_{ilt}^{cap} x_{ilt}^{cap} \tag{2}$$

The C and X represents the respective cost parameter, and decision variables. The subscript l, t, j, i, k indicates the set of scenarios, time horizons, materials, facilities, and activities respectively. Subject to the various material balance, facility input, facility output, facility capacity, storage inventory, storage capacity, set of implement ability (non anticipativity) constraints and bounds.

# References

1. Fourer, R.: Database structures for mathematical programming models. Decis. Support Syst. **20**(4), 317–344 (1997). https://doi.org/10.1016/S0167-9236(97)00007-9
2. Dutta, G., Fourer, R.: An optimization-based decision support system for strategic and operational planning in process industries. Optim. Eng. **5**(3), 295–314 (2004). https://doi.org/10.1023/B:OPTE.0000038888.65465.4e
3. Dutta, G., Fourer, R., Majumdar, A., Dutta, D.: An optimization-based decision support system for strategic planning in a process industry: the case of a pharmaceutical company in India. Int. J. Prod. Econ. **106**(1), 92–103 (2007). https://doi.org/10.1016/j.ijpe.2006.04.011
4. Dutta, G., Fourer, R.: Database structure for a class of multi-period mathematical programming models. Decis. Support Syst. **45**(4), 870–883 (2008). https://doi.org/10.1016/j.dss.2008.02.010
5. Dutta, G., Gupta, N., Fourer, R.: An optimization-based decision support system for strategic planning in a process industry: the case of aluminium company in India. J. Oper. Res. Soc. **62**, 616–626 (2011). https://doi.org/10.1057/jors.2010.8
6. Gupta, N., Dutta, G., Fourer, R.: An expanded database structure for a class of multi-period, stochastic mathematical programming models for process industries. Decis. Support Syst. **64**, 43–56 (2014). https://doi.org/10.1016/j.dss.2014.04.003
7. Birge, J.R.: State-of-the-Art-survey—stochastic programming: computation and applications. INFORMS J. Comput. **9**(2), 111–133 (1997). https://doi.org/10.1287/ijoc.9.2.111
8. Dutta, G., Gupta, N., Mandal, J., Tiwari, M.K.: New decision support system for strategic planning in process industries: computational results. Comput. Ind. Eng. **124**, 36–47 (2018). https://doi.org/10.1016/j.cie.2018.07.016
9. Salmi, A., David, P., Blanco, E., Summers, J.D.: A review of cost estimation models for determining assembly automation level. Comput. Ind. Eng. **98**, 246–259 (2016). https://doi.org/10.1016/j.cie.2016.06.007
10. Dutta, G., Rao, H.V., Basu, S., Tiwari, M.K.: Asset liability management model with decision support system for life insurance companies: computational results. Comput. Ind. Eng. **128**, 985–998 (2019). https://doi.org/10.1016/j.cie.2018.06.033
11. Lee, H., Aydin, N., Choi, Y., Lekhavat, S., Irani, Z.: A decision support system for vessel speed decision in maritime logistics using weather archive big data. Comput. Oper. Res. **98**, 330–342 (2018). https://doi.org/10.1016/j.cor.2017.06.005
12. Daneshvar, M., Mohammadi-Ivatloo, B., Zare, K., Asadi, S.: Two-stage stochastic programming model for optimal scheduling of the wind-thermal-hydropower-pumped storage system considering the flexibility assessment. Energy **193**, 116657 (2020). https://doi.org/10.1016/j.energy.2019.116657
13. Mavromatidis, G., Orehounig, K., Carmeliet, J.: Design of distributed energy systems under uncertainty: a two-stage stochastic programming approach. Appl. Energy **222**, 932–950 (2018). https://doi.org/10.1016/j.apenergy.2018.04.019
14. Badri, H., Fatemi Ghomi, S.M.T., Hejazi, T.H.: A two-stage stochastic programming approach for value-based closed-loop supply chain network design. Transp. Res. Part E Logist. Transp. Rev. **105**, 1–17 (2017). https://doi.org/10.1016/j.tre.2017.06.012
15. Ramaraj, G., Hu, Z., Hu, G.: A two-stage stochastic programming model for production lot-sizing and scheduling under demand and raw material quality uncertainties. Int. J. Plan. Sched. **3**(1), 1–27 (2019). https://doi.org/10.1504/ijps.2019.102993

16. Parisio, A., Neil Jones, C.: A two-stage stochastic programming approach to employee scheduling in retail outlets with uncertain demand. Omega **53**, 97–103 (2015). https://doi.org/10.1016/j.omega.2015.01.003

17. Gupta, N., Dutta, G., Tiwari, M.K.: An integrated decision support system for strategic supply chain optimisation in process industries: the case of a zinc company. Int. J. Prod. Res. **56**, 5866–5882 (2018). https://doi.org/10.1080/00207543.2018.1456698

# Modelling Critical Success Factors for the Implementation of Industry 4.0 in Indian Manufacturing MSMEs

Pulok Ranjan Mohanta$^{(\boxtimes)}$ ⓘ and Biswajit Mahanty ⓘ

Indian Institute of Technology Kharagpur, Kharagpur, India
`pulok@iitkgp.ac.in`

**Abstract.** This research is based on the recent endeavors of the manufacturers to achieve improvements in supply chain resiliency by implementing Industry 4.0 standards. The large enterprises, in today's scenario, are pursuing their supplier organizations to incorporate information and communication technology (ICT) based advancements to their production systems. But successful implementation of such technologies in the resource deficient supplier organizations depends on several factors that need due consideration of the decision makers. In this paper, we have identified 14 such critical success factors from extant literature for implementation of technologies conforming to Industry 4.0. Further, through an empirical investigation comprising of 222 micro, small and medium enterprises (MSMEs) from the automotive manufacturing sector in India, we have made use of exploratory factor analysis and structural equation modelling to identify the hidden constructs and examine the influencing relationships between them.

**Keywords:** Industry 4.0 · MSMEs · Supply chain · Information sharing · Structural equation modelling

## 1 Introduction

The manufacturing industry across the globe are going through the Industry 4.0 led transformations. An essential requirement for the large enterprises (LEs) in the manufacturing industry is to improve their supply chain resilience against the propensity of efficiency [1], particularly in the face of disruptions of different kinds. For the LEs to achieve higher level of resilience, digitalization by including advanced technologies into the production processes is believed to be a key enabler [2]. Thus, to attain Industry 4.0 standards, manufacturers are keen to integrate advanced digital technologies such as intelligent robotics, artificial intelligence (AI), Internet of Things (IoT), Cyber-physical Systems (CPS), and Cloud manufacturing [3, 4] into their manufacturing supply chains, which include their supplier organizations, mainly categorized as Micro, Small and Medium Enterprises (MSMEs). One key advantage of these technologies is the attainment of enhanced capability of sharing information. The importance of information sharing for attaining resilience in supply chain was adequately discussed in the literature [5, 6].

© IFIP International Federation for Information Processing 2021
Published by Springer Nature Switzerland AG 2021
A. Dolgui et al. (Eds.): APMS 2021, IFIP AICT 631, pp. 89–97, 2021.
https://doi.org/10.1007/978-3-030-85902-2_10

The transition of the existing manufacturing scenario to a complete digital scenario of Industry 4.0 involves a number of challenges related to the formulation of new strategies of implementation, modification of the physical infrastructure, up-gradation of the manufacturing technologies, bringing out improvements in the organizational practices, and to the formulation of new coordination and contracting policies. This transition requires the manufacturing organizations to initiate a number of projects for adoption of different enabling technologies [7]. The adoption of advanced technologies would be challenging considering technical issues, quality and availability of infrastructure, data handling capability, skill gaps, etc. [8]. Typically, in the automotive sector, LEs are having ample resources to invest in advanced-technology adoption endeavors such as R&D and technology acquisition. In contrast, their supplier enterprises, falling in the category of MSMEs, are known for low technology levels and scarce capital resources, implying that such endeavors are difficult for them to pursue [7–9]. Accordingly, external interventions are necessary to assist them to succeed in upgrading to Industry 4.0 [10]. Nevertheless, the technology implementation initiatives in the inbound supply chains of LEs depend on several factors and at the same time such implementation is expected to impact the MSMEs in several dimensions. A proper understanding of these factors and impacts would be helpful to the decision makers including business leaders, government agencies, investors and training providers [1].

An accepted definition of critical success factors (CSFs) describes them as the things that must go well to ensure success [11]. Adequate understanding of CSFs helps in reducing complexity and enabling the decision makers to concentrate on most important CSFs while making decisions at various levels of the decision-making process [7, 12, 13]. In this paper, we investigate the CSFs for implementation of Industry 4.0 technologies from a viewpoint of information sharing. The research objectives lie in the identification of the hidden constructs from these factors and then analyzing the influencing relations between them.

The rest of the paper is organized as follows. Section 2 reviews the related literature and presents the identified CSFs. Section 3 describes the method used for survey and initial analysis of the responses. Section 4 gives the details of the formulation of the hypotheses and structural equation modelling. Section 5 discusses the findings and Sect. 6 concludes the paper.

## 2    Related Literature

The literature in the manufacturing domain are found to include several recent publications related to the integration of advancements in information and communication technologies (ICT) into production systems. The transition to Industry 4.0 is possible for the manufacturing enterprises with the inclusion of advanced digital technologies such as intelligent robotics, Cloud manufacturing, artificial intelligence (AI), Internet of Things (IoT), and Cyber-physical Systems (CPS) into their production systems [14, 15]. Such transition is believed to be capable of transforming the entire value creation process and hence many countries have taken initiatives in developing ecosystems supporting this technological revolution [10, 16]. Many of the articles published are related to the CSF analysis for adoption of different technologies such as ERP, Cloud computing, Big Data,

RFID, 3D printing, and Augmented Reality [8, 17–19]. While the concept of Industry 4.0 encompasses all of these technologies, articles addressing the CSFs in the context of their inclusive implementation are few [20–22], more so in the context of MSMEs in India.

Since the response of manufacturing industries towards the ongoing technology revolution is influenced by several factors unique for each nation, such as the available ICT infrastructure and the economic and political scenarios [23, 24], we find the requirement of a study on the CSFs for implementation of Industry 4.0 technologies in the Indian manufacturing MSMEs. Accordingly, we explored the pertinent literature to identify 14 CSFs and ascertained their relevance by discussion with 5 experts from academia and 5 experts from industry. The CSFs identified are listed in Table 1.

**Table 1.** CSFs for implementation of Industry 4.0 technologies

| CSF no. | Description of the critical success factor |
|---------|--------------------------------------------|
| CSF 1 | Management commitment |
| CSF 2 | Adequate financial support |
| CSF 3 | Cross functional team formation |
| CSF 4 | Efficient information management system (ERP, MES, PLM etc.) |
| CSF 5 | High performance computing |
| CSF 6 | Automated electronic data capture mechanism |
| CSF 7 | Integration of wearable and mobile devices to cloud |
| CSF 8 | Cyber security measures |
| CSF 9 | Collaboration for information sharing |
| CSF 10 | Consistent network connectivity |
| CSF 11 | Semantic technologies |
| CSF 12 | Collaboration platforms availability |
| CSF 13 | Decentralization of decision making |
| CSF 14 | Traceability system implementation |

## 3 Materials and Method

A questionnaire-based survey was used to capture the perceptions of the MSMEs in the automotive sector of India on the enabling power of the CSFs (listed in Table 1) for advanced technology implementation, primarily focusing on information sharing on a 7-point Likert scale, where "1" means "very low" and "7" means "very high" enabling power. The questionnaire was administered to 280 manufactures from five major automotive manufacturing hubs in India located in and around Jamshedpur in the east, Chennai in the south, Pune in the west, Delhi in the north and Sanand in the northwest of India and obtained inputs from 222 enterprises. This accounted for a response rate of 79%. For any clarification on the questionnaire, the researcher was available either in person or on the telephone, which ensured the removal of ambiguity. All the 222 responses were complete and hence were considered fit for further analysis.

The survey responses were subjected to exploratory factor analysis (EFA) to reduce the 14 CSFs into a smaller number of highly inter-correlated sets of hidden constructs using SPSS 22.0. The data collected for this study met the requirements prescribed by Hair et al. [25], with case-to-variable ratio more than 5:1, KMO 0.780 (greater than 0.5), p-value less than 0.05 and the EFA-yielded-constructs together explaining 77.54% of the variance. The Cronbach's alpha values for all the four identified components were found to be more than the acceptable norm of 0.6. The extracted constructs and their associated variables are shown in Table 2.

# 4   Hypothesis Development and Structural Equation Modelling

## 4.1   Hypotheses Development

On the basis of the opinions gleaned through the questionairre survey and available literature (25–29), the following seven hypotheses are formulated:

*H1: Management initiatives for adoption of Industry 4.0 technologies (F1) positively influence the availability of advanced network and information management systems (F2)*
*H2: Management initiatives for adoption of Industry 4.0 technologies (F1) positively influence collaboration initiatives for information sharing (F4)*
*H3: Advanced information acquisition and processing technologies (F3) positively influence availability of advanced network and information management systems (F2)*
*H4: Advanced information acquisition and processing technologies (F3) positively influence collaboration initiatives for information sharing (F4)*
*H5: Availability of advanced network and information management systems (F2) positively influence the Implementation of Industry 4.0 (II4)*
*H6: Advanced information acquisition and processing technologies (F3) positively influence the Implementation of Industry 4.0 (II4)*
*H7: Collaboration initiatives for information sharing (F4) positively influence the Implementation of Industry 4.0 (II4)*

## 4.2   Structural Equation Modelling

We used structural equation modeling (SEM) using AMOS software package to test the validity of the hypotheses developed. The technique used was maximum likelihood estimation (MLE). In the path diagram produced using AMOS (see Fig. 1), the CSFs, extracted constructs and error terms are shown as rectangular boxes, ellipse and circles respectively. While arrows between a pair of constructs represents the hypothesized relations, numbers beside them are the corresponding standardized direct effects. To bring in clarity, Fig. 1 shows the standardized estimates for constructs and CSFs only. The quality of the model was evaluated using several goodness of fit indices usually

**Table 2.** Hidden constructs from exploratory factor analysis

| Components | Variables | Factor loadings | Cronbach's alpha |
|---|---|---|---|
| F1: Management initiatives for adoption of Industry 4.0 technologies | Management commitment (CSF 1) | .919 | .862 |
| | Adequate financial support (CSF 2) | .825 | |
| | Decentralization of decision making (CSF 13) | .816 | |
| | Cross functional team formation (CSF 3) | .802 | |
| F2: Advanced network and information management systems | Efficient information management system (ERP, MES, PLM, etc.) (CSF 4) | .926 | .919 |
| | Cyber security measures (CSF 8) | .910 | |
| | Consistent network connectivity (CSF 10) | .882 | |
| | Traceability system implementation (CSF 14) | .829 | |
| F3: Advanced information acquisition and processing technologies | High performance computing (CSF 5) | .890 | .893 |
| | Automated electronic data capture mechanism (CSF 6) | .867 | |
| | Semantic technologies (CSF 11) | .841 | |
| | Integration of wearable and mobile devices to cloud (CSF 7) | .822 | |
| F4: Collaboration initiatives for information sharing | Collaboration platforms availability (CSF 12) | .916 | .834 |
| | Collaborations for information sharing (CSF 9) | .808 | |

recommended. The values obtained are chi-square/degrees of freedom 3.515, root mean square error of approximation (RMSEA) 0.107, comparative fit index (CFI) 0.903 and Tucker-Lewis index (TLI) 0.878. These indices are found to be in the acceptable range suggesting that the model fits the data well.

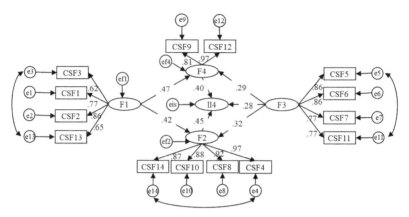

**Fig. 1.** Estimates for the constructs in SEM using AMOS

# 5   Results and Discussion

Figure 1 shows that all the hypotheses are supported by our survey responses. Figure 2 shows a conceptual model derived out of the structural equation model:

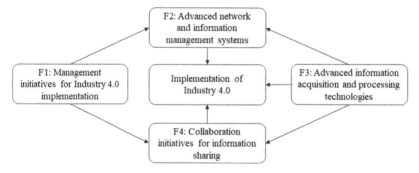

**Fig. 2.** Conceptual model derived out of the structural equation model

The following broad inferences could be drawn from the results of this analysis:

1) The role of the management for technology upgradation of MSMEs is indispensable. The role of the management would be that of a resource provider as well as that of a collaboration facilitator for the partner organizations. The management needs to be proactive in establishing collaborations for information exchange, developing standard norms and capabilities.

2) Information collaboration involving advanced technologies demands an efficient network and information management system consisting of good IT infrastructure, robust architecture, and well laid-out processes. Thus, the initiatives for implementing Industry 4.0 standards would improve the capability of data acquisition and the ability to improve the information management system of the MSMEs.

3) The technological upgradation will lead to the utilization of sensors and IoTs to generate big data in order to enhance a data centric production scenario. It would support the collaboration initiatives and information management by acquiring required data more effectively.
4) Adoption of Industry 4.0 technologies would also be beneficial for the organizations in their collaboration endeavors for effective exchange of information. For this, there is a need of a shared enterprise system comprising of information management systems for virtual integration of the collaborating enterprises.
5) The establishment of collaboration with other enterprises for sharing of information supported by advanced technology and information management systems would lead to the effective implementation of Industry 4.0.

The validity of the hypotheses also corroborates to the existing literature. To reap the benefits from the adoption of advanced technologies through information sharing, the MSMEs need to direct their efforts towards technology upgradation. This requires adequate patronage and active involvement of the top management [26, 27], typically in planning and management of the technology upgradation process [27, 28].

The technology upgradation process involves the inclusion of advanced technologies such as Internet of things, cloud computing, and robotics into the production systems. Analysis of Big data, traceability systems such as RFID and IoT, and the use of a shared information system would help managers to make more precise, practical, and reasonable decisions [29, 30]. Hence, an effective system for information collaboration can emerge from technology upgradation of MSMEs.

Since most of the manufacturing MSMEs are part of bigger supply chains, the intention of implementing new standards and technologies of Industry 4.0 would be to improve the supply chain. In addition, due to the recent pandemic-led disruptions, supply chain resiliency is the key focus [1, 3]. While the role of information sharing in a resilient supply chain is prominent [5, 6], an inter-organizational information system is the need of the hour [30].

## 6   Conclusion

In this research work, we have identified the critical success factors (CSFs) for the adoption of advanced technologies for Industry 4.0 implementation and derived the hidden constructs using EFA with the help of a questionnaire survey among Indian automotive manufacturing MSMEs. Subsequently, we have used structural equation modelling (SEM) to test hypotheses describing the influencing relations among the identified constructs.

The findings suggest the importance of management initiatives, advanced network and information management systems, advanced information acquisition and processing technologies, and collaboration initiatives for information sharing for the successful implementation of Industry 4.0 in the Indian manufacturing MSMEs.

There are some limitations that may be addressed in forthcoming research. Even though a reasonable sample size is considered, it could have been larger, so that the validity is improved. Moreover, our respondents were from automotive industries only and hence the study can be extended to other sectors as well with necessary precautions.

# References

1. Goffman, E.: In the wake of COVID-19, is glocalization our sustainability future? Sustain. Sci. Pract. Policy **16**(1), 48–52 (2020)
2. Autio, E., Mudambi, R., Yoo, Y.: Digitalization and globalization in a turbulent world: centrifugal and centripetal forces. Glob. Strateg. J. **11**, 3–16 (2021)
3. Van Hoek, R.: Research opportunities for a more resilient post-COVID-19 supply chain – closing the gap between research findings and industry practice. Int. J. Oper. Prod. Manag. **40**, 341–355 (2020)
4. Chowdhury, P., Paul, S.K., Kaisar, S., Moktadir, M.A.: COVID-19 pandemic related supply chain studies: a systematic review. Transp. Res. Part E Logist. Transp. Rev. **148**, 102271 (2021)
5. Thomas, A.V., Mahanty, B.: Dynamic assessment of control system designs of information shared supply chain network experiencing supplier disruption. Oper. Res. Int. J. **21**(1), 425–451 (2018). https://doi.org/10.1007/s12351-018-0435-9
6. Mehrjerdi, Y.Z., Shafiee, M.: A resilient and sustainable closed-loop supply chain using multiple sourcing and information sharing strategies. J. Clean. Prod. **289**, 125–141 (2021)
7. Ghobakhloo, M., Ching, N.T.: Adoption of digital technologies of smart manufacturing in SMEs. J. Ind. Inf. Integr. **16**, 100107 (2019)
8. Wielicki, T., Arendt, L.: A knowledge-driven shift in perception of ICT implementation barriers: comparative study of US and European SMEs. J. Inf. Sci. **36**(2), 162–174 (2010)
9. Subrahmanya, M.H.B.: External technology acquisition of SMEs in the engineering industry of Bangalore: what prompts them to move faster for acquisition? J. Manuf. Technol. Manag. **25**(8), 1174–1194 (2014)
10. Reischauer, G.: Industry 4.0 as policy-driven discourse to institutionalize innovation systems in manufacturing. Technol. Forecast. Soc. Change **132**, 26–33 (2018)
11. Dinter, B.: Success factors for information logistics strategy - an empirical investigation. Decis. Support Syst. **54**(3), 1207–1218 (2013)
12. Bai, C., Sarkis, J.: A grey-based DEMATEL model for evaluating business process management critical success factors. Int. J. Prod. Econ. **146**(1), 281–292 (2013)
13. Junior, C.H., Oliveira, T., Yanaze, M.: The adoption stages (Evaluation, Adoption, and Routinisation) of ERP systems with business analytics functionality in the context of farms. Comput. Electron. Agric. **156**, 334–348 (2019)
14. Ghobakhloo, M.: The future of manufacturing industry: a strategic roadmap toward Industry 4.0. J. Manuf. Technol. Manag. **29**(6), 910–936 (2018)
15. Oztemel, E., Gursev, S.: Literature review of Industry 4.0 and related technologies. J. Intell. Manuf. **31**(1), 127–182 (2018). https://doi.org/10.1007/s10845-018-1433-8
16. Singh, S., Mahanty, B., Tiwari, M.K.: Framework and modelling of inclusive manufacturing system. Int. J. Comput. Integr. Manuf. **32**(2), 105–123 (2019)
17. Loh, T.C., Koh, S.C.L.: Critical elements for a successful enterprise resource planning implementation in small-and medium-sized enterprises. Int. J. Prod. Res. **42**(17), 3433–3455 (2004)
18. Masood, T., Egger, J.: Augmented reality in support of Industry 4.0—implementation challenges and success factors. Robot. Comput.-Integr. Manuf. **58**, 181–195 (2019)
19. Yeh, C.C., Chen, Y.F.: Critical success factors for adoption of 3D printing. Technol. Forecast. Soc. Change **132**, 209–216 (2018)
20. Chiarini, A., Belvedere, V., Grando, A.: Industry 4.0 strategies and technological developments. An exploratory research from Italian manufacturing companies. Prod. Plann. Control **31**(16), 1385–1398 (2020)

21. Sony, M., Naik, S.: Critical factors for the successful implementation of Industry 4.0: a review and future research direction. Prod. Plann. Control **31**(10), 799–815 (2020)
22. Xu, L.D., Xu, E.L., Li, L.: Industry 4.0: state of the art and future trends. Int. J. Prod. Res. **56**(8), 2941–2962 (2018)
23. Castellacci, F.: Technological paradigms, regimes and trajectories: manufacturing and service industries in a new taxonomy of sectoral patterns of innovation. Res. Policy **37**(6–7), 978–994 (2008)
24. Dalenogare, L.S., Benitez, G.B., Ayala, N.F., Frank, A.G.: The expected contribution of Industry 4.0 technologies for industrial performance. Int. J. Prod. Econ. **204**, 383–394 (2018)
25. Hair, J.F., Black, W.C., Babin, B.J., Anderson, R.E.: Multivariate Data Analysis. Pearson Education Limited, Harlow (2013)
26. Dutta, D., Bose, I.: Managing a big data project: the case of Ramco cements limited. Int. J. Prod. Econ. **165**, 293–306 (2015)
27. Sony, M., Naik, S.: Critical factors for the successful implementation of Industry 4.0: a review and future research direction. Prod. Plann. Control **31**(10), 799–815 (2019)
28. Gunasekaran, A., et al.: Big data and predictive analytics for supply chain and organizational performance. J. Bus. Res. **70**, 308–317 (2017)
29. Zhong, R.Y., Xu, C., Chen, C., Huang, G.Q.: Big data analytics for physical internet-based intelligent manufacturing shop floors. Int. J. Prod. Res. **55**(9), 2610–2621 (2017)
30. Clegg, B., Little, P., Govette, S., Logue, J.: Transformation of a small-to-medium-sized enterprise to a multi-organisation product-service solution provider. Int. J. Prod. Econ. **192**, 81–91 (2017)

# A Framework Integrating Internet of Things and Blockchain in Clinical Trials Reverse Supply Chain

Yvonne Badulescu[1,2]([✉]) [iD] and Naoufel Cheikhrouhou[2] [iD]

[1] Faculty of Business and Economics, University of Lausanne, Lausanne, Switzerland
yvonne.badulescu@hesge.ch
[2] Geneva School of Business Administration, University of Applied Sciences Western Switzerland (HES-SO), Rue de la Tambourine 17, 1227 Carouge, Switzerland
naoufel.cheikhrouhou@hesge.ch

**Abstract.** Efficiency and resilience of the clinical trials supply chain are of particular prevalence in the current global context. The unique characteristic of the reverse logistics flow in the supply chains for clinical trials is the foundation for the digital transformation framework presented in this paper. This paper proposes a novel framework that integrates internet of things (IoT) and blockchain technology for the reverse logistics supply chain for clinical trials. The framework is implemented in a Contract Research Organisation operating clinical trials in Europe and North Africa and results are discussed. The main contribution of the proposed novel framework is the integration and interaction of both IoT and blockchain in a reverse logistics process.

**Keywords:** IoT · Blockchain · Reverse logistics · Clinical trials

## 1 Introduction

Efficient, effective, and secure logistics processes for clinical trials is essential for the advancement and development of improved medical treatments, pharmaceuticals, and vaccines. Clinical trials are organised and run by contract research organisations (CRO) and have a characteristic reverse logistics supply chain that requires a high level of confidentiality and security. The reverse logistics process of a clinical trials supply chain is a unique context for digital transformation. Reverse logistics using internet of things (IoT) devices and blockchain in the logistics process can generate new data, such as additional real-time tracking and traceability capabilities as well as temperature monitoring and the possibility to alert logistics providers of immediate issues. Putting this into a blockchain based system can create a full unchangeable record of the clinical trial supply chain transactions.

To date, blockchain and IoT devices have not been considered together in the reverse logistics process of CRO operating clinical trials. The current technological extent of blockchain in CRO is limited to document sharing ensuring the information security of

© IFIP International Federation for Information Processing 2021
Published by Springer Nature Switzerland AG 2021
A. Dolgui et al. (Eds.): APMS 2021, IFIP AICT 631, pp. 98–106, 2021.
https://doi.org/10.1007/978-3-030-85902-2_11

the patient recruitment process and medical records. This highlights the importance of tracking and tracing the location of the incoming shipments, as well as the condition of its contents, to optimise the complementary processes in CRO such as the arrival logistics area, re-distribution, and scheduling laboratory testing to support the resilience of the reverse logistics process.

This paper proposes a framework of a reverse logistics clinical trials supply chain that integrates digital technology, in the form of blockchain and IoT, into the logistics process. The paper is structured in the following way: the next section reviews the literature of the use of IoT and blockchain in CRO; Sect. 3 presents the proposed framework; Sect. 4 presents the results of the framework implementation in a real CRO; and finally, Sect. 5 concludes the paper including the limitations of the proposed solution and suggested future.

## 2   Literature Review

Reproducibility, data sharing, personal data privacy, traceability concerns and patient and investigators enrolment in clinical trials are big medical challenges for contemporary clinical research and have a direct impact on the supply chain [1]. Although technological integration is most prevalent in other industries, there have been several applications and research related to blockchain which is progressively edging toward the healthcare field [2–4].

IoT devices are already prevalent in the clinical trials industry. [5] highlighted the importance of IoT devices in both recruiting participants in clinical trials and patients' follow-up process. However, a previous study by [6] showed that smart wearables in many cases have limited capabilities for monitoring, signal processing, and communication due to operating in uncontrolled environments. Due to these constraints, applying standard security and privacy requirements is extremely challenging. [7] provide a proof-of-concept study of a digital health application enabling the recruitment process in clinical trials by using IoT data and blockchain. The results show that the combination of these technologies can be implemented to provide a robust digital clinical trials information system.

In terms of the use of blockchain in transportation systems, [8] study the use of blockchain in intelligent transportation systems (ITS) which and conclude that blockchain represents the future of autonomous transportation systems such as ridesharing and taxi services [9], and [10] develop an approach to securely share information in the transportation of dangerous goods using Smart Contracts based on blockchain technology. Blockchain aids with the secure storage of transportation documents, and facilitates the custom documents which increases efficiency of inspections and consequently reduces overall lead-time [11]. [12] explore the digital transformation of logistics documentation by smart contracts and blockchain technology via analysis of the literature and use cases from industry. They find that logistics providers are digitalising the documentation in the logistics process to increase transparency, to reduce the associate time and cost related to preparing paper-based documentation, which is highly prone to counterfeit and human errors, and to facilitate trade documents with customs authorities for the transportation and tracking of shipments across borders.

Logistics in a supply chain can be supported using Smart Contracts based on blockchain technology [13]. Suppliers search for transportation partners which are in the blockchain and linked by Smart Contracts that outline the terms and conditions as well as fees, payment terms and shipping times. Smart Contracts which are stored in blockchains can decrease administration and service costs as they can be automatically trigger the logistics activities in a decentralised way as well as improve efficiency in payment of suppliers to transportation partners based on the pre-defined conditions in the contract [13]. [14] develop a system architecture and a software prototype tested on the Ethereum test network for the use of Smart Contracts based on blockchain for the fair exchange of goods between suppliers and customers, and [15] develop a framework for using blockchain and smart contracts in the pharmaceutical supply chain which covers logistics processes, supplier transactions, tracking and tracing of shipments, quality monitoring of shipments, among others. They test the use of Smart Contracts for the real time tracing by storing the sequence of timestamps during the logistics process and quality monitoring of shipments. [16] propose a Smart Logistics solution based on Smart Contracts with a focus on transportation of dangerous goods which includes supplier recommendation, contract negotiation, condition monitoring via IoT devices and payment to suppliers. Their Smart Contract is on a cloud-based system which deploys the agreed upon Smart Contract on an Ethereum Virtual Machine which executes and monitors the details in the Smart Contract. They develop a prototype and find that the proposed architecture allows for quicker payments to suppliers and updated supplier ratings based on the details in the Smart Contracts.

Traceability accounts as one of the top issues to track the error back to its original source and is currently lacking in the traditional clinical trials database systems [17]. Blockchain technologies can enforce a high level of transparency, traceability and control over the interconnected processes in clinical trials' supply chains [18], however, in this regard, very little research exists. [19] propose a 4 components blockchain platform architecture for clinical trial and precision medicine built on top of the traditional blockchain to achieve reliable and safe transaction. [20] propose a private, permissioned Ethereum blockchain network maintained by regulators, the pharmaceutical industry and CRO to be used in parallel with traditional clinical data management systems in order to increase trust in the data they hold and the credibility of trials findings. [21] explore the role of blockchain in clinical trials data management and develop a proof-of-concept of a patient-facing and researcher-facing system based on Smart Contracts. The system uses a web-based interface to allow users to run trials-related Smart Contracts on an Ethereum network allowing patients to grant researchers access to their data and allowing researchers to submit queries for off-chain data. [22] develops a proof-of-concept study aiming to implement a process allowing the collection of patients' informed consent, which is bound to protocol revisions, storing and tracking the consent in a secure, unfalsifiable and publicly verifiable way, and enabling the sharing of this information in real time. [23] design workflows for using smart contracts in healthcare and clinical trials which focus on cost and healthcare optimisation.

Although IoT and blockchain are used independently for patient follow-up in the clinical trials industry, there is an opportunity to extend these technologies to the reverse logistics supply chain to improve the overall resilience of material and information flows. It has been highlighted in the literature that both IoT and blockchain are not only beneficial in many situations, but also complementary of each other. The following section proposes a novel framework that addresses the integration of both technologies to handle material and information flows which show to improve the digital activities in the reverse logistics clinical trials supply chain.

# 3 Proposal for a New Framework

Currently, the existing logistics processes in clinical trial companies use physical logistics-related documentation transfer. In addition, the visibility provided by digital documentation based on blockchain allow CRO to prioritise shipments based on information collection at the source, track and trace shipments and the environmental conditions via the installed IoT devices. This, in turn, allows them to better schedule arrivals, laboratory testing, redistribution of packaging, and decreased supply chain costs due to improved planning and prioritisation.

The specific nature of a clinical trials supply chain that works in a reverse logistics flow, requiring a high level of security, creates a unique context for the proposed framework development.

The proposed framework, shown in Fig. 1 integrating IoT and blockchain in clinical trials consists of 1) tracking and tracing of location and environmental conditions via IoT devices and a communication infrastructure; and 2) secure digital logistics document sharing between supply chain stakeholders based on blockchain technology and infrastructure.

## 3.1  IoT Integration

The implementation of real-time sensors can eliminate many manual tasks and allow rapid reaction in case of a problem. A sensor could quickly detect a problem during the journey, sending an alert to the logistician. After looking at the data transmitted by the sensor, the logistician could instruct the carrier to continue the delivery or to proceed with an intervention on the shipment. The delivery process can then be interrupted to investigate the problem and restart the delivery process if necessary. This eliminates the need to wait for the package to arrive and be examined after delivery to determine if the sample is usable.

The IoT devices are physically attached to the packages to monitor geo-localization and temperature and connected to relay communication nodes that connect to a base station which send the information to the clinical trials company which is monitoring the shipment.

## 3.2  Blockchain Implementation in Digital Documentation

The logistics documentation is accompanied by a digital version based on blockchain which is a transaction protocol that executes, controls, and documents each event in the reverse logistics flow of the clinical trials supply chain based on the terms of the Smart Contract with the transporters. The information is stored and shared on various servers globally based on the blockchain process. This information is then accessible by the clinical trials company logistics management team.

Blockchain implementation to track and trace logistics and transportation documents, which will allow those with the right to obtain the information about the various actions taken on the transportation documents, for example, who was the last person to view the document or the time when the document was opened. The documents concern the sending and receiving of the clinical trials products. An open-source project "HyperLedger", used as a framework for the private blockchain which allows the use of Smart Contracts, has consensus algorithms to validate transactions and maintains confidentiality due to access restrictions. However, the storage capacity is limited and the blockchain does not store the documents themselves, therefore a secure cloud storage is used which is compatible with the blockchain system.

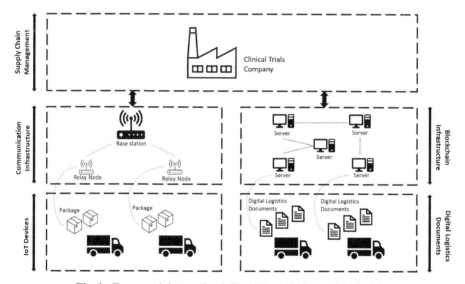

**Fig. 1.** Framework integrating IoT and blockchain in clinical trials.

The structure of the "HyperLedger" blockchain consists of several nodes (participants) which perform different roles: *Peers*: members of the organization participating in the blockchain network; *Endorser peer*: This type of node performs two important functions within the blockchain. These are the validation of transactions, thus verifying the certifications of the participant who sends a transaction request and the execution of the smart contract. Once the two previous tasks are completed, the endorser peer can approve or reject a transaction; *Ordered peer or ordering node*: this is the node that will receive the block and add it to the blockchain. In addition, this node will forward the approved blocks to the network participants. We define the components using the "HyperLedger" Business Network Definition model: 1) Network Participants: Carriers, Clients, CRO; 2) Assets: Transportation documents for goods; 3) Transactions: Transmission of transport documents when goods are sent and received as well as Smart Contracts. These documents digitally accompany the goods throughout the transport and are exchanged between the network participants.

The process begins by the clinical trials company connecting to the platform using the login information and obtain their certificate that they can use the blockchain. The system creates certificates for all new users. Then the user uploads the document to the platform which is automatically saved in the external cloud storage. A request will then be sent to the *endorser peer* or approved node, to check whether the user uploading the document to the network is certified and has the authorisation to perform this action. Once the transaction is validated, the approved node will call the *ordered peer* to add the new block to the blockchain and then distribute it to all the nodes in the network. After completion of the verification and validation steps, the document is indefinitely saved on the external cloud storage.

The framework integrating IoT and blockchain in a reverse logistics clinical trials supply chain is implemented in a real case and the results are presented and discussed in the next section.

## 4   Results of Implementation

The framework is applied to a CRO with headquarters in Switzerland that specializes in clinical trials and plans to integrate IoT devices onto the shipments and digital logistics documentation based on blockchain in their reverse logistics supply chain. Figure 2 illustrates the business processes related to the outgoing and incoming logistics of the reverse logistics supply chain for clinical trials implemented in the CRO. Contrary to the typical forward material and information flow of a supply chain, Fig. 2 starts with the outgoing logistics process first in which the CRO requests the medical samples for the clinical trials to be sent to the clients/investigators. They first request the shipment to be picked up from their depots by the transporter. Upon the reception of the order request, the transporter creates and shares a contract proposal to the clinical trials company, which accepts or rejects it. Each step of the digital contract process is based in the HyperLedger blockchain to record each action of the *peers*. If the contract is accepted, the shipment is picked up and delivered to the clients.

The incoming logistics process begins with the clients requesting the pickup of the clinical samples to the transporter. The packaging wears an IoT device monitoring the real-time temperature of the samples and their geo-localisation and sends an alert to the transporter in the occurrence of a problem, which then relays the information to the CRO. The CRO then makes a logistics decision based on the alert information and the geo-localisation, which can include reducing the transportation priority or changing the rest of the transportation journey to express and communicates this to the transporter which makes the necessary changes to the delivery as per the pre-agreements made in the Smart Contract.

The implementation of the new framework that integrates both IoT and blockchain in the reverse logistics process for clinical trials, impacts the entire CRO supply chain, from the outgoing logistics to the incoming logistics processes. The additional steps in the process shown in Fig. 2 are related to the agreement, or not, of the smart contract between the CRO and the transporter that not only outlines the conditions of transportation as

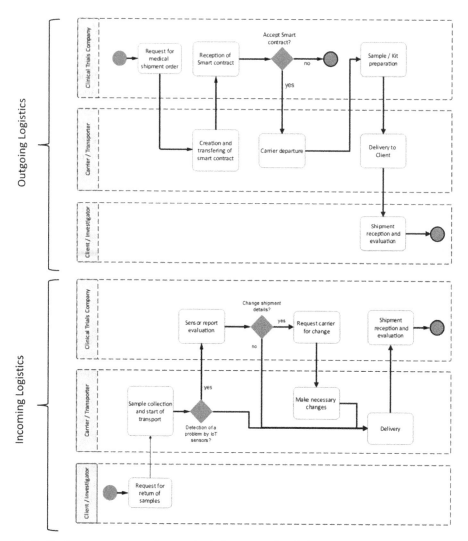

**Fig. 2.** Business process map of reverse logistics process for clinical trials integrating IoT devices and blockchain

in any contract but also covers and controls the relevant logistics activities related to the temperature control and geo-localisation from using IoT devices, as well as the management of the IoT devices themselves. Another notable impact is the decision-making opportunity in the incoming logistics process coming from the information from the IoT devices regarding the real-time temperature of the samples. This allows the CRO to reschedule the planned treatment and reception of the sample if it is considered expired, or to organise express shipment if the data indicates more rapid treatment is necessary. The terms of the Smart Contract also dictate and control the responsibilities and actions required by the parties in these instances. Additionally, all documentation and communication within the logistics process is recorded in the "HyperLedger" blockchain to ensure not only the security of the information but the validity of the data in case of any issues that occur during the transportation and logistics.

## 5   Conclusion

This paper proposes a new framework that integrates IoT and blockchain technology in the reverse logistics supply chain for clinical trials. The novelty lies in the combination of both IoT and blockchain in this unique context of contract research organisations (CRO) specialising in organising and running clinical trials. The framework is implemented in a CRO in Switzerland and the resulting dynamic processes are presented and discussed.

The most remarkable impacts on the CRO supply chain include the decision-making opportunity based on the data originating from the IoT devices in the incoming reverse logistics flow, as well as the integration of Smart Contracts which automatically execute the code in the blockchain ledger when the conditions regarding IoT data is triggered in the contract. The digitalisation of the information flow derived from the combination of the IoT with the blockchain allows CRO to more efficiently manage the material flows and internal processes pertaining to clinical trials.

On the other hand, clinical trials require a high level of confidentiality which is the reason for using a private blockchain, HyperLedger, however it generates high usage costs as each transaction is invoiced. The costly nature of using a private blockchain may deter other CRO to consider such a framework. Consequently, a direction for future research may be to design an alternative process that reduces the number of transactions and cost. Another direction for future research is to appropriate the framework in the pharmaceutical and medical devices industries which have comparable logistics requirements to CRO.

**Acknowledgements.** This work was supported by the Swiss National Science Foundation under project n° [176349].

## References

1. Tseng, J.-H., Liao, Y.-C., Chong, B., Liao, S.-W.: Governance on the drug supply chain via gcoin blockchain. Int. J. Environ. Res. Public Health **15**(6), 1055–1063 (2018)
2. HealthNautica. https://www.healthnautica.com/comppages/index.asp. Accessed 15 Mar 2021
3. FACTOM - Introducing Honesty to Record-Keeping. https://bitcointalk.org/index.php?topic=850070.0. Accessed 15 Mar 2021
4. Gem: The Best Crypto Portfolio Tracker Does the Work for You. https://gem.co/. Accessed 15 Mar 2021
5. Tehrani, N., Jin, Y.: How advances in the Internet of Things (Iot) devices and wearable technology will impact the pharmaceutical industry. RA J. Appl. Res. **4**(3), 1530–1533 (2018)
6. Li, W., Chai, Y., Khan, F., Jan, S.R.U., Verma, S., Menon, V.G., Kavita, X.L.: A comprehensive survey on machine learning-based big data analytics for IoT-enabled smart healthcare system. Mob. Netw. Appl. **26**(1), 234–252 (2021). https://doi.org/10.1007/s11036-020-01700-6

7. Singh, M., Katiyar, D., Singhal, S.: Blockchain technology in management of clinical trials: a review of its applications, regulatory concerns and challenges. Mater. Today Proc. (2021, in press). https://doi.org/10.1016/j.matpr.2021.04.095, https://www.sciencedirect.com/science/article/pii/S2214785321029515

8. Yuan, Y., Wang, F-Y.: Towards blockchain-based intelligent transportation systems. In: IEEE 19th International Conference on Intelligent Transportation Systems (ITSC), Rio de Janeiro, Brazil, pp. 2663–8, IEEE (2016)

9. Hewa, T., Ylianttila, M., Liyanage, M.: Survey on blockchain based smart contracts: applications, opportunities and challenges. J. Netw. Comput. Appl. **177**(1), 102857 (2021)

10. Imeri, A., Khadraoui, D.: The security and traceability of shared information in the process of transportation of dangerous goods. In: 2018 9th IFIP International Conference on New Technologies, Mobility and Security (NTMS), Paris, France, pp. 1–5. IEEE (2018)

11. Nasih, S., Arezki, S., Gadi, T.: Enhancement of supply chain management by integrating Blockchain technology. In: 2019 1st International Conference on Smart Systems and Data Science (ICSSD), Rabat, Morocco, pp. 1–2. IEEE (2019)

12. Merkaš, Z., Perkov, D., Bonin, V.: The significance of blockchain technology in digital transformation of logistics and transportation. Int. J. E-Serv. Mob. Appl. **12**(1), 1–20 (2020)

13. Zheng, Z., et al.: An overview on smart contracts: Challenges, advances and platforms. Future Gener. Comput. Syst. **105**(1), 475–491 (2020)

14. Alahmadi, A., Lin, X.: Towards secure and fair IIoT-enabled supply chain management via blockchain-based smart contracts. In: ICC 2019 - 2019 IEEE International Conference on Communications (ICC), Shanghai, China, pp. 1–7. IEEE (2019)

15. Jangir, S., Muzumdar, A., Jaiswal, A., Modi, C.N., Chandel, S., Vyjayanthi, C.: A novel framework for pharmaceutical supply chain management using distributed ledger and smart contracts. In: 2019 10th International Conference on Computing, Communication and Networking Technologies (ICCCNT), Kanpur, India, pp. 1–7. IEEE (2019)

16. Arumugam, S.S., et al.: IOT enabled smart logistics using smart contracts. In: 2018 8th International Conference on Logistics, Informatics and Service Sciences (LISS), Toronto, Canada, pp. 1–6. IEEE (2018)

17. Salah, K., Nizamuddin, N., Jayaraman, R., Omar, M.: Blockchain-based soybean traceability in agricultural supply chain. IEEE Access **7**(1), 73295–73305 (2019)

18. Benchoufi, M., Ravaud, P.: Blockchain technology for improving clinical research quality. Trials **18**(1), 335–340 (2017)

19. Shae, Z., Tsai, J.J.P.: On the design of a blockchain platform for clinical trial and precision medicine. In: 2017 IEEE 37th International Conference on Distributed Computing Systems (ICDCS), Atlanta, USA, pp. 1972–1980. IEEE (2017)

20. Angeletti, F., Chatzigiannakis, I., Vitaletti, A.: Towards an architecture to guarantee both data privacy and utility in the first phases of digital clinical trials. Sensors **18**(12), 4175–4202 (2018)

21. Maslove, D.M., Klein, J., Brohman, K., Martin, P.: Using blockchain technology to manage clinical trials data: a proof-of-concept study. JMIR Med. Inform. **6**(4), e11949 (2018)

22. Benchoufi, M., Porcher, R., Ravaud, P.: Blockchain protocols in clinical trials: transparency and traceability of consent. F1000Res **6**(1), 66 (2017)

23. Khatoon, A.: A blockchain-based smart contract system for healthcare management. Electronics **9**(1), 94–117 (2020)

# Smart Integration of Blockchain in Air Cargo Handling for Profit Maximization

Rosalin Sahoo[1]($\boxtimes$) (iD), Bhaskar Bhowmick[1], and Manoj Kumar Tiwari[2] (iD)

[1] Indian Institute of Technology, Kharagpur, Kharagpur 721302, India
[2] National Institute of Industrial Engineering (NITIE), Mumbai 400087, India

**Abstract.** Transportation and logistics management are critical to the economic growth of a country. Smart transportation is becoming possible thanks to the advancement of digital technology implementation. To decrease the vulnerability of digitized systems from cyber-attacks, Blockchain has recently become one of the most commonly adopted technologies for free, decentralized, and trustworthy intelligent transportation networks. By analyzing the potential application of Blockchain technologies in the smart transportation of cargo by air, this study hopes to contribute to the area of logistics management. For efficient and safe air cargo movement, we propose a distributed blockchain layered framework that helps in transforming air logistics operations in national and international trades and we validated our framework by mathematically comparing the cost with and without the implementation of Blockchain.

**Keywords:** Air cargo · Blockchain · Layered framework

## 1 Introduction

The air cargo industry has become an indispensable part of the global economy holding an important niche in the transport of high valued commodities. World logistics has turned upside down by the COVID-19 pandemic where countries have stopped their import/export and closed the doors of their Industries and Manufacturing sectors. A sudden disruption in supply and demand causes the need for establishing the vital connections of word trade and logistics together [1]. Maximum goods shipped by air cargo include perishable items, pharmaceuticals, medical protective equipments, and food supplies. Therefore, today the rapidly growing global trade requires cost-efficient, reliable, and highly secured air cargo operations and logistics management [2].

Blockchain is a decentralized transaction and data storage technology that helps individuals and businesses to store and share value without the use of conventional middlemen [3]. It's a cutting-edge technology with the ability to upend traditional economic and social structures and replace them with systems that are more accessible, reliable, and stable. Blockchain technology is based on a globally distributed ledger that verifies and approves transactions using the properties of a massive peer-to-peer network [4]. The Blockchain database is a permanent archive of transactions that occur between

© IFIP International Federation for Information Processing 2021
Published by Springer Nature Switzerland AG 2021
A. Dolgui et al. (Eds.): APMS 2021, IFIP AICT 631, pp. 107–114, 2021.
https://doi.org/10.1007/978-3-030-85902-2_12

individuals or between customers and businesses. There are a variety of Blockchain implementations that are still under development. A transaction value is available to any network user with access privileges, making the Blockchain-enabled transaction system extremely transparent [5]. It makes it very difficult to carry out fraudulent transactions. However, when making a transaction, a network user will select what information about their identity they want to share with the rest of the network, resulting in pseudonymity. Digital technologies have the potential to break down these barriers, allowing the supply chain to become a truly interconnected marketplace that is fully open to all stakeholders - from raw material, product, and portion manufacturers to trans-porters of those suppliers' finished products, and eventually to end-users [6].

The purpose of this research is to compare and analyze the traditional air cargo operations with digitized cargo management systems implementing blockchain. We have created a layered and structured framework to address the future role of air freight operations and services in transforming. This research would aid in identifying the latent opportunities presented by blockchain today, as well as the obstacles that come with these future opportunities. It will aid in comprehending the key factors of mainstream Blockchain acceptance in the logistics supply chain, as well as potential market models and competitive considerations.

The following sections are as follows: Sect. 2 explains an overview of the Concept of smart transportation in cargo logistics and challenges followed by the integration of blockchain in the air cargo logistics sector. Section 3 provides a comparison between the traditional model and our proposed layered framework and the applications that were derived from the literature and discussed potential business models and strategic considerations. Section 4 highlights the validation of the proposed model by considering the implementation of Blockchain. Section 5 summarizes the findings and conclusions drawn from the research.

## 2   Literature Review

The role of smart transportation and logistics will be described first, followed by a discussion of the issues and challenges that the logistics industry is facing. After that, we will study how Blockchain can help with transportation and logistics.

### 2.1   Concept of Smart Transportation in Air Cargo Logistics and Challenges

Air cargo is the most energy-intensive and inefficient mode of freight transport. As a result, the majority of freight shipped by air is both perishable and time-sensitive in the context of a broader manufacturing chain, or high value and low density. There has been very little contribution in air cargo-specific innovations using cutting-edge technologies to digitize the supply chain. A parallel transportation management system was developed by [3] for real-time ride-sharing services using blockchain. A methodology intelligent software agents can be used to monitor the activity of stakeholders in the blockchain networks to detect anomalies was proposed by [7]. Few researchers such as [8, 9], and [10] conducted a comprehensive survey on blockchain technology adoption by discussing its influences as well as the opportunities and challenges. Some of the challenges include

Traceability and visibility across the value chain, Lack of trust among the stakeholders, and Unsustainable and inefficient use of resources in cargo handling.

## 2.2  Integration of Digitization and Blockchain in Cargo Logistics

Blockchains are an immutable set of records that are cryptographically linked together for audit.The activities and decisions of stakeholders participating in air freight operations will have a direct impact on the whole transport chain management process. As a result, digitization technology will assist freight staff in developing new business logic and economic models, resulting in a variety of innovations and increased competitiveness by stimulating consumer demand. A container stacking and dispatching problem was studied by [11] using blockchain management. Implementation of blockchain using machine learning has been researched by many authors to design a decentralized, privacy-preserving, and secured network [12–14], and [15]. To highlight more on the application part of the blockchain and the cost-saving perspective, we developed a layer-based framework. The position of logistics is critical for the growth of civilization and the development of new frameworks, as these would open up new avenues for inquiry. As a result, the proposed model offers a simple and dependable architecture that is suitable for both transportation and logistics. Given the importance of logistics and transportation, it is essential to provide a comprehensive and consistent model/framework. To bridge this void, we have proposed a layered infrastructure that better addresses the need for smart transportation and logistics via blockchain.

# 3  Comparison of the Proposed Model with Traditional Model

## 3.1  Traditional Model

Traditional air freight includes the flow of physical goods, information, and financial transactions involving the key stakeholders such as shipper/buyer, Freight forwarders, and airlines. Often it is difficult to track the province of goods and the shipment status when the process relies on manual and paper-based transactions causing friction in global trade. Block chain can potentially help overcoming the frictions enabling data transparency and aces among the potential stakeholders, creating single source of truth as shown in Fig. 1.

In traditional model, the most common challenges airline freight forwarders face are usage of defunct technologies, unstructured data and no transparent information sharing. There exists difficulties in inability to track cargo consignments in real time once outside the carrier network and lack of integration throughout the supply chain leading to higher costs and lower efficiency.

## 3.2  Proposed Framework

To make the transportation and logistics system secure and efficient, the proposed framework combines Blockchain technology with IoT to make the system efficient and resilient against all the disruptions in freight handling (Fig. 2).

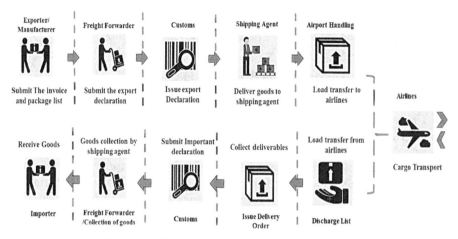

**Fig. 1.** Traditional air cargo supply chain frameworks

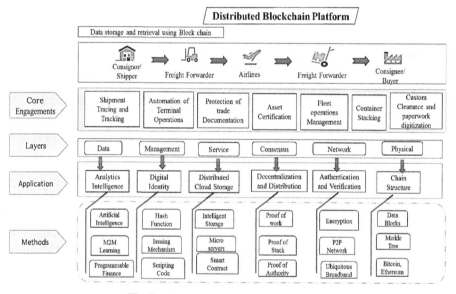

**Fig. 2.** Blockchain in air cargo framework

In this section, we propose a layered framework that signifies the prominent application of blockchain in handling global air freight logistics, supply chain transparency, and traceability. These frameworks useful to alleviate global trade managing procurement, transportation, trace and track, collaboration, customs, and trade finance depicted in Fig. 3. The layers of the blockchain manage the different architectures of the freight movement process.

*Data Layer:* This layer refers to the process of data masking providing the reproduction of data in a nonproduction environment to support the test and development process without exposing sensitive data. It eliminates all the identifiable features in sensitive data and enables data anonymization to remain available, thereby eliminating the risk of leaking sensitive information This layer provides the chained data blocks, together with the related techniques including asymmetric encryption, time-stamping, hash algorithms and merkle trees.

*Management Layer:* This layer refers to different functions of managing data using hash function, issuing code and cryptic framework. In Blockchain architecture, all the client nodes are connected with peer-to-peer networks. Once data is sent, it is received and encrypted using various hashing techniques and is broadcast. Only the legitimate person will receive the broadcast transaction, and this is only possible due to Blockchain enabled IoT sensors. In logistics, Blockchain technology not only maintains the integrity of transported products, but it keeps a complete record of the overall shipping process.

*Service Layer:* The service layer specifies the mechanisms of distributed networking, data forwarding and verification. Most models, systems and application scenarios are composed of large numbers of distributed, autonomous, dynamic decision-making devices or vehicles.

*Concensus Layer:* The consensus layer determines which the consensus algorithm is adopted to achieve consensus on some information among untrusted parties in decentralized systems. Currently, there are a lot of consensus protocols being applied in blockchain systems, which could be roughly divided into consensus protocols with proof of concept (e.g., Proof of Work (PoW), Proof of Stake (PoS), Delegated Proof of Stake (DPoS) and Proof of Authority (PoA). According to the different types of blockchains, consensus protocols are selected differently. For example, due to the loose control and poor synchronization, the consensus protocol in permissionless blockchain (public blockchain) widely selects the incentive- based consensus schemes, such as PoW and PoS.

*Network Layer:* This layer is responsible for transferring data. Blockchain uses a P2P network of devices instead of a middleman and transmits blocks through consensus. Once a block is created, it is broadcast on the P2P network and is verified by each node according to the predefined specifications. Each node discards invalid data and passes the valid data to its neighboring node. Thus, Blockchain data is verified by multiple nodes and then finally appended into a Blockchain. If a node forwards invalid data to its neighboring node, it is disconnected from the P2P network for a specific period of time. Therefore, only valid data are transmitted over the network.

*Physical Layer:* The physical layer encapsulates various kinds of physical entities (e.g., devices, vehicles, assets and other environmental objects) involved in ITSs, such as traffic lights, digital cameras, cars, and so on. The devices in this layer collect data and forward it to the upper layer for validation. IoT plays the main role in this layer by providing massive interconnectivity of devices. Sensors at this layer provide various functions, including temperature monitoring, traffic data collection, light, pressure monitoring, etc. Further, Blockchain-based sensor devices enhance data security and privacy. A large amount of data is collected through these sensors and is stored in decentralized blockchain ledgers.

## 4  Validation of Proposed model-Consideration of Blockchain Implementation

By enabling a more learner, streamlined, and error-free operation, blockchain will save money. It will accelerate the physical movement of commodities by increasing the visibility and predictability of logistics operations. The ability to trace the provenance of products will allow responsible and sustainable supply chains at scale, as well as aid in the fight against product counterfeiting. Our proposed framework is validated by providing an adequate mathematical explanation in this section. The model is tested by applying the blockchain concept in the profit maximization problem of a freight forwarding firm.

$$\underset{q^a, T_B}{Max} \ \pi_B = \underbrace{v.\min\left[D_B, q_B^a\right]}_{Revenue} - \underbrace{\left[r^a + (s^a + w)(1 - T_B) + (s_B^a + w_B)T_B\right]}_{Total \ cost \ of \ air \ shipping} - \underbrace{\left[I_B + c(q_B^a)\right]T_B}_{Cost \ savings \ due \ to \ Blockchain}$$

$$(1)$$

Where,

$$D_B \sim N\left(\mu_B, \sigma_B^2\right), \quad where \ \sigma_B > 0 \tag{2}$$

$$s_B^a = s^a - g\left(\tau^a\right)^a > 0, \quad where \ g\left(\tau^a\right)^a > 0 \tag{3}$$

$$w_B = w - g\left(\tau^a\right)^w > 0, \quad where \ g\left(\tau^a\right)^w > 0 \tag{4}$$

$$T_B \in \{0, 1\} \tag{5}$$

Notations:

$q^a$:  Quantity shipped over Air
$T_B$:  Binary decision variable
$\pi_B$:  Expected Profit
$v$:  Unit Revenue
$D_B$:  Demand
$q_B^a$:  Shipped quantity tracked and traced using Blockchain
$r^a$:  Revenue cost by air shipment
$s^a$:  Shipping cost
$w$:  Waiting time translated into monetary value
$s_B^a$:  Shipping cost traced using Blockchain
$w_B$:  Waiting time tracked via Blockchain
$I_B$:  Implementation of Blockchain
$c$:  Variable cost
$\mu_B$:  Std. Deviation of normally distributed products are being shipped
$\sigma_B$:  Variance of demand
$\tau$:  Saved time
$A$:  Shipments through Air
$B$:  Blockchain Integration

Equation (1) depicts the profit obtained through the implementation of Blockchain. The first term represents total revenue, the second term is the total cost of air shipping and the third term represents the cost savings due to the implementation of Blockchain. Equation (2) represents the demand function and Eq. (3)–(4) is the decrease storage cost of air shipping. Equation (5) represents the binary decision variable, if 1 then Blockchain is implemented, if 0, then otherwise. To the best of our knowledge, there is a scarcity of publicly available datasets of Blockchain cost implementation, therefore we assumed the objective function to be non-linear and we tried to solve the problem by random data generation of cost concerning demand variability. Figure 3 depicts the decrease in total cost with the implementation of blockchain technologies in handling air cargo system.

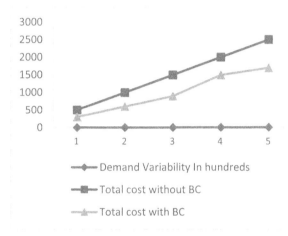

**Fig. 3.** Total air transport cost

Blockchain technology brings down the total cost improving efficiency in air cargo shipping and logistics operations preventing damage to cargo. It helps to automate the payments process that are repetitive most of the time so as to improve the customer service. Thus, with these contributions, the current study makes an effort to provide practical insights and also introduce a new channel for further study in fulfilling the voids regarding the adoption of blockchain technologies in the freight logistics industry.

## 5   Conclusion

Transportation and logistics management are critical for the growth of country. Smart transportation is now a possibility thanks to the advancement of IoT products. This study suggests a layered structure for integrating Blockchain into air freight shipping and logistics to make it more effective and resistant to security attacks. Blockchain, on the other hand, is a decentralized mechanism that operates on a peer-to-peer network. It is made up of chain blocks, each of which documents a series of transactions while still keeping track of previous blocks and their hashing. Smart pay, improved fault tolerance, real-time knowledge sharing, privacy, and protection are just a few of the benefits that convergence brings to transportation and logistics. This post has aided research in a variety of areas. First, it gives a thorough description of smart transportation and logistics, as well as their significance in today's growth age. Second, using Blockchain-enabled air freight movement offers a robust platform for smart shipping and logistics. Third, the proposed structure is further tested using statistical modeling by examining profit maximization by blockchain implementation. This study can be extended by developing a system-level model in the future to assess the scalability and interoperability of different platforms used in smart transportation and logistics using more real-world case studies.

# References

1. Sobieralski, J.B.: COVID-19 and airline employment: Insights from historical uncertainty shocks to the industry. Transp. Res. Interdiscip. Perspect. **5**, 100123 (2020)
2. Loske, D.: The impact of COVID-19 on transport volume and freight capacity dynamics: an empirical analysis in German food retail logistics. Transp. Res. Interdiscip. Perspect. **6**, 100165 (2020)
3. Yuan, Y., Wang, F.: Towards blockchain-based intelligent transportation systems, pp. 2663–2668 (2016)
4. Qian, Y., et al.: Towards decentralized IoT security enhancement: a blockchain approach. Comput. Electr. Eng. **72**, 266–273 (2018)
5. Pervez, H., Haq, I.U.: Blockchain and IoT based disruption in logistics. In: 2019 2nd International Conference on Communication, Computing and Digital systems, C-CODE 2019, pp. 276–281 (2019)
6. Helo, P., Shamsuzzoha, A.H.M.: Real-time supply chain—a blockchain architecture for project deliveries. Robot. Comput. Integr. Manuf. **63**, 101909 (2020)
7. Dey, S.: Securing majority-attack in blockchain using machine learning and algorithmic game theory: a proof of work, pp. 7–10. arXiv (2018)
8. Phan, T.D., Do, T.T.H., Do, H.H., Pham, V.H.: A survey on opportunities and challenges of Blockchain technology adoption for revolutionary innovation. In: ACM International Conference Proceedings Series, pp. 200–207 (2018)
9. Chen, F., Wan, H., Cai, H., Cheng, G.: Machine learning in/for blockchain: future and challenges, pp. 1–28. arXiv (2019)
10. Schmidt, C.G., Wagner, S.M.: Blockchain and supply chain relations: a transaction cost theory perspective. J. Purch. Supply Manag. **25**(4), 100552 (2019)
11. Lawrence, H., Yulia, L., Mahwish, A.: A multi-agent system with blockchain for container stacking and dispatching. In: 21st International Conference on Harbor, Maritime and Multimodal Logistics Modeling and Simulation, HMS 2019, pp. 79–87 (2019)
12. Chen, X., Ji, J., Luo, C., Liao, W., Li, P.: When machine learning meets blockchain: a decentralized, privacy-preserving and secure design. In: Proceedings of the 2018 IEEE International Conference on Big Data, Big Data 2018, pp. 1178–1187 (2019)
13. Yin, H.H.S., Langenheldt, K., Harlev, M., Mukkamala, R.R., Vatrapu, R.: Regulating cryptocurrencies: a supervised machine learning approach to de-anonymizing the bitcoin blockchain. J. Manag. Inf. Syst. **36**(1), 37–73 (2019)
14. Kim, H., Kim, S.H., Hwang, J.Y., Seo, C.: Efficient privacy-preserving machine learning for blockchain network. IEEE Access **7**, 136481–136495 (2019)
15. Zhou, S., Huang, H., Chen, W., Zheng, Z., Guo, S.: PIRATE: a blockchain-based secure framework of distributed machine learning in 5G networks, pp. 84–91. arXiv (2019)

# Requirements on Supply Chain Visibility: A Case on Inbound Logistics

Ravi Kalaiarasan[1]([envelope]) [iD], Tarun Kumar Agrawal[1] [iD], Magnus Wiktorsson[1] [iD], Jannicke Baalsrud Hauge[1] [iD], and Jan Olhager[2] [iD]

[1] KTH Royal Institute of Technology, 15136 Södertälje, Sweden
ravika@kth.se
[2] Lund University, 22100 Lund, Sweden

**Abstract.** Events such as Covid-19 have revealed the vulnerabilities that companies face due to low visibility. Consequently, companies experience impact on their supply chains in terms of disruptions of material supply, deliveries, productivity and revenue. Thus, the importance of Supply Chain Visibility (SCV) in global and competitive markets with increasing sustainability demands has received widespread recognition. Yet, the literature provides limited understanding of requirements to consider when developing a SCV system. Addressing the gap, this study presents the findings from a case study during the first months of 2021 at a global manufacturing company developing a SCV system to improve their inbound flow. Using a system engineering perspective, this study presents requirements highlighted during early stage for a SCV system. The results indicate the importance of ensuring SCV system requirements to enable data collection, handling and usage for decision making leading to both supply chain sustainability and resilience. This study contributes to the understanding of SCV by presenting and categorizing requirements considered for real-time case at a manufacturing company when developing a SCV system.

**Keywords:** Supply Chain Visibility · Digitalization · Real-time data · Logistics

## 1 Introduction

In recent years, the importance of Supply Chain Visibility (SCV) has received widespread recognition. The literature provides numerous suggestions for defining SCV. Key characteristics of visibility is the accuracy, accessibility, availability, timeliness and usefulness of information [1–4]. Visibility is also considered to play an important role in the transition towards industry 4.0 by enabling real-time data and support decision making [5]. Despite its relevance, the level of SCV perceived by supply chain practitioners within the manufacturing industry is rather low. The study conducted by Srinivasan and Swink [6] revealed that industrial leaders conveyed the need to increase SCV. Events such as Covid-19 pandemic have shown the vulnerabilities that companies face due to low SCV [7]. Companies experienced disruptions in their material supply affecting deliveries, productivity and revenue [8–10]. In addition to affecting business performances,

© IFIP International Federation for Information Processing 2021
Published by Springer Nature Switzerland AG 2021
A. Dolgui et al. (Eds.): APMS 2021, IFIP AICT 631, pp. 115–122, 2021.
https://doi.org/10.1007/978-3-030-85902-2_13

low SCV is mentioned restrict the ability to achieve supply chain resilience [11]. Thus, handling supply chain disruptions, a global supply base, operations and markets, have increased the attention on SCV in order to attain sustainable and resilient supply chains [12–14]. Although previous research accentuated the importance and expected benefits of SCV, the requirements to consider when developing and implementing a SCV system in practice are fuzzy [15, 16]. This is problematic since failing to develop understanding of requirements may result in inadequate implementation leading to financial and reputational loss [10]. Addressing the gap, the objective of this paper was formulated as investigating the following research question: how are requirements on SCV formulated within manufacturing industry? This study presents the findings from a case study at a manufacturing company developing a SCV system to improve their inbound flow. This study contributes to the understanding of SCV by analysing and categorizing requirements considered by a manufacturing company when developing a SCV system. The requirements can be considered as a support for further research on how to plan, develop and implement SCV systems.

## 2  Methodology

In order to address the objective, this research is based on a single case study within the manufacturing industry. Given the complexity of supply chains, adopting a case study method can be beneficial in order to understand specific problems [17]. This can in turn contribute to development of more generic solutions. In addition, case studies are stated to support in-depth understanding which in turn can contribute to development of theory in exploratory research [18, 19].

The selected case company is in the top-three in terms of revenue in its business, has a global manufacturing presence and is considered to be world-class in its industry. They have approximately 50 000 employees in 100 countries. In order to improve their supply chains in terms of performance and sustainability aspects, they have identified areas where they want to improve SCV. They also experienced SCV challenges in recent covid-19 pandemic through supply chain disruptions of material supply and uncertainties leading to production issues. Therefore, they can be considered as suitable representative of a manufacturing company on a digitalization journey, facing number of SCV challenges.

The study has followed a project, run from the company's headquarter in Northern Europe, which started at the case company in January 2021. Data collection occurred between January and March 2021. It included participation in six project meetings, three semi-structured interviews and analysis of project documents. The project team was cross-functional and consisted of members with responsibilities for supply chain development and data & mobility services. In total, three semi-structured interviews and participation in seven project meetings took place. The collected data was discussed and reviewed by the authors (consisting of three professors, one postdoc, and one PhD student) to establish a detailed picture.

In order to categorize requirements from the collected data, a system engineering perspective was adopted. System engineering, well recognized to enable the engineering of complex system, support to understand requirements needed to successfully implement a system [20, 21]. Here four categories of requirements were used. *Business requirements* can support to understand the general aim with the system describing what the system should do and desired features. *Functional requirements* describe what functions a system should deliver. *Nonfunctional requirements* refer to the behavior of and constraints placed upon the system. *Stakeholder's requirement* support to understand the expectations and needs from stakeholders. It also increases the possibility for project success and acceptance [22, 23]. It is important to ensure that the requirements are verifiable, realistic, consistent and understandable by stakeholders [22].

## 3   Results

The results from the data gathering and analysis is presented according to Current state, (detailing the process steps, flow of goods, IT-system as well as pain points) and Expressed requirements on the SCV system (mapped to the four categories of Business, Functional, Non-functional and Stakeholder requirements).

### 3.1   Current State

Semi-structured interviews were conducted with the project member with experience in supply chain development. This helped in understanding and visualizing the current state and related pain points in the inbound flow.

**Inbound Flow.** A general representation of the inbound flow at the case company is presented in Fig. 1, which depicts the main process steps, flow of goods & packaging and overview of the IT-systems. Orders to suppliers are predominately sent by EDI (Electronic Data Interchange) using the ERP (Enterprise Resource Planning) system at the case company. This is followed by order confirmations sent by the suppliers. Consecutively, suppliers will send an order for packaging, as they are normally required to use the packaging material produced by the case company. Once the parts are packed and ready for delivery, the suppliers send delivery confirmations to the company. Certain percentage of deliveries from suppliers are sent to cross-docks, called x-dock at the case company. The parts from suppliers arrive in pallet or box level as indicated in Fig. 1. Operations performed at the x-docs are mainly unloading, sorting and loading. No extensive verifications of parts delivered from suppliers are performed. The pallets are eventually loaded at the x-dock, picked-up and delivered to logistic centers. Once the deliveries to the logistic centers arrive, there is an arrival confirmation at gate unloading confirmation, storing confirmation, booking confirmation and stock status update in the WMS (Warehouse Management System). Eventually, the arrival confirmation can be seen by the production site in order to place an order to the logistic center.

**Current Pain Points.** Primarily the pain points (summarized in Table 1) consist of factors affecting operations but are also related to restricted capabilities to control, plan and analyze the inbound flow. 90–95% current deliveries are one time, so one challenge with current approach is that there is no-real time information from the supplier at the point of loading (Truck 1 in Fig. 1). Despite the fact that information exchange occurs using IT systems, this approach is in essence trust based. Consequently, the deliveries

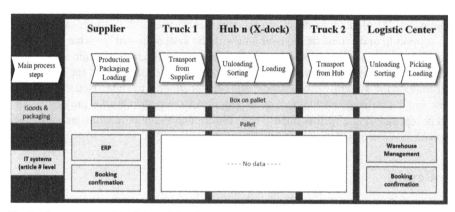

**Fig. 1.** Current state of the inbound flow in terms of main processes, flow of goods & packaging and IT system used.

cannot be checked and verified in real-time. This has in turn led to choosing speed transports to solve deviations leading to additional costs. Together with lack of ability to verify deliveries in real-time, the complexity of the supply base forces the company to keep additional stock to ensure material availability for production. The lack of visibility pointed out in Fig. 1 causes blind spots in the flow. Thefts have occurred in certain blind spots causing issues. As production rate is planned to increase, additional investments are required to increase the storage capacity if the current delivery situation remains. The study also revealed that there is no unified visualization of the connection between article number, pallet, truck and hub. This leads to reduced ability to monitor, plan and act proactively. The importance of these abilities for both current and future operations were further realized during the covid-19 pandemic when the company suffered from lack of SCV resulting in restricted capabilities to plan and be proactive.

**Table 1.** Summary of expressed pain points.

| Pain point | Description |
| --- | --- |
| Speed transports | No real-time information causes speed transports |
| Additional inventory levels | The complexity of the supply base forces the company to keep additional stock to ensure material availability for production |
| Blind-spots | Lack of SCV causes difficulties to identify spots of deviations in the flow |
| Additional investments | Additional investments in storage capacity due to the lack of SCV |
| Lack of unified view | No unified visualization of article number, pallets, truck and hub |
| Limited proactivity | Difficulties to identify deviations in time and take appropriate actions |

## 3.2   Requirements on SCV System

As explained in the previous chapter, the identified requirements were categorized into business, functional, non-functional and stakeholder requirements.

**Business Requirement.** Business requirement included both short term and mid-long-term goals describing both desired states and effects of implementing SCV system. Summary of the business requirements are presented in Table 2.

*Short Term.* Real-time monitoring was expected to result in improved resource utilization by reducing manual work spent on deviation handling. Flow optimization reduced inventory levels and unnecessary investments in logistics centers were stated to be anticipated effects. In addition, reducing the environmental impact by reducing speed transports and optimizing transports were other desired outcomes.

*Mid-Long Term.* Improved forecast and planning abilities are expected to be the outcome of implementing a SCV system. This was in turn expected to enable virtual stock across the supply chain. It was also clear that going forward, a SCV system should support the case company to become more autonomous in the supply chain. Having a SCV system is considered to support automation of processes and flows.

**Table 2.** Business requirements

| Business requirement | Description |
| --- | --- |
| Real-time monitoring | Enable real-time monitoring from point of loading at supplier till delivery required |
| Improve forecast and planning accuracy | Utilize real-time data to further improvement of forecasts |
| Virtual stock | Implement virtual across the supply chain |
| Autonomous flows | Attain more autonomous flow and processes |

**Functional Requirement.** The functional requirements identified describes the high level functions the SCV system is expected to perform. They were based on supporting operations and improve the ability to work proactively. The functional requirements are listed in Table 3.

**Non-functional Requirements.** The non-functional requirements indicated the expected characteristics of SCV system. They were based on the data collection, connectivity, analysis, handling and security. In addition, the importance of integration and interface with the current systems was identified. The non-functional requirements are listed in Table 4.

**Stakeholder Requirement.** No explicit goals describing stakeholder requirements on the SCV system could be identified in the project during the time period for data collection. One stakeholder requirement, which is related to business requirements, pointed out

**Table 3.** Functional requirements.

| Functional requirement | Description |
| --- | --- |
| Deliver digital view of article number, boxes and pallets | The ability to have unified view of deliveries in real-time for the inbound flow |
| Provide continuous updates regarding Estimated Time of Arrival (ETA) upon for deliveries | This function was crucial for operations and for instance support the reduction of speed transports |
| Deliver a track and trace function | The function was needed to support deviations |
| Deliver measurement of fill rate | This function will enable actual measurement of fill-rate in real-time |
| Deliver measurement of $CO_2$ consumption | This function will enable higher accuracy of the measurement and the sharing of information |

**Table 4.** Non-functional requirements

| Non-functional requirement | Description |
| --- | --- |
| Data collection and connectivity through registration of events in real-time | Important to capture key steps such as departure, vehicle on road position, arrival cross-docks and logistics centers. The data collection and connectivity should be used to visualize order number, supplier shipping confirmation, pallet ID, Vehicle ID, Vehicle position, Site information, Production unit need and stock status |
| Data analytics for the measurement's calculations | Data analytics should be applied to utilize the components of digital thread in order to provide performance indicators and reports |
| Set up data security to ensure data protection | To ensure data protection |
| Integration and interface with related systems | Critical to make sure integration and interfaces with related system which includes ERP, WMS, and TMS (Transport Management System) |

that the system should deliver alerts and notifications regarding deviations of deliveries in order to timely provide stakeholders with information to take appropriate actions.

## 4   Discussion and Conclusion

Supply Chain Visibility has received widespread attention from both researchers and practitioners. This paper strengthens the view that SCV is considered to be crucial for performance and decision making in both current and future operations in global manufacturing [6, 7].

This case study has identified and categorized the requirements expressed during early phases of a new SCV system for inbound logistics within a global manufacturing company, known to be a leader in their business. The current state of inbound flow confirms that the company suffers from lack of SCV [6] and the related pain points accentuates the consequences in terms of disruptions in material supplies [10]. Also, this study highlights that sustainability is a key driver for SCV [13].

From a systems engineering perspective, the requirements can be categorized into mainly business, functional and non-functional requirements. It can be noted that some of the requirements (both functional and non-functional) can be further detailed. For instance, the track and trace function might at this stage be insufficient, exposed for multiple interpretations and lack stakeholder input [22]. Even though the project is still in its early phase, the study indicated that stakeholders' requirements are not specified. Not including stakeholders' requirements and inputs might limit the utilization of the SCV system for future large-scale implementation.

Previous literature already highlighted the importance of gathering real-time data for SCV [5]. However, this study underlines additional points to consider. For manufacturing companies on a digitalization journey with complex supply chains facing SCV challenges, this study indicates the importance of developing SCV systems requirements for collecting, connecting, analyzing, securing and applying real-time data for decision making, which incorporates both the goals for the company as well as external supply chain challenges. This can in turn enable both sustainable and resilient supply chains.

The study contributed to the existing body of knowledge on SCV by presenting and analyzing early-stage requirements set within a large global manufacturing company, in the midst of external supply chain challenges and digitalization journey. Future studies will include analyzing further realization steps for the case, in relation to this initial requirement data. It can also include incorporating various stakeholders' requirements for specifying and testing a SCV system. Such a study will support to identify additional requirements to consider for a large-scale implementation of SCV systems.

**Acknowledgement.** We gratefully acknowledge the funding from Produktion2030 and Vinnova for this research, as part of the research project Production Logistic Visibility (LOVIS).

# References

1. Barratt, M., Oke, A.: Antecedents of supply chain visibility in retail supply chains: a resource-based theory perspective. J. Oper. Manag. **25**, 1217–1233 (2007)
2. Williams, B.D., Roh, J., Tokar, T., Swink, M.: Leveraging supply chain visibility for responsiveness: the moderating role of internal integration. J. Oper. Manag. **31**, 543–554 (2013)
3. Vernon, F.: Supply chain visibility: lost in translation? Supply Chain Manag. Int. J. **13**, 180–184 (2008)
4. Kalaiarasan, R., Olhager, J., Wiktorsson, M., Jeong, Y.: Production logistics visibility – perspectives, principles and prospects. In: Säfsten, K., Elgh, F. (eds.) Advances in Transdisciplinary Engineering. IOS Press (2020)

5. Zeller, V., Hocken, C., Stich, V.: Acatech Industrie 4.0 maturity index – a multidimensional maturity model. In: Moon, I., Lee, G.M., Park, J., Kiritsis, D., von Cieminski, G. (eds.) APMS 2018. IAICT, vol. 536, pp. 105–113. Springer, Cham (2018). https://doi.org/10.1007/978-3-319-99707-0_14
6. Srinivasan, R., Swink, M.: An investigation of visibility and flexibility as complements to supply chain analytics: an organizational information processing theory perspective. Prod. Oper. Manag. **27**, 1849–1867 (2018)
7. Sharma, A., Adhikary, A., Borah, A.B.: Covid-19's impact on supply chain decisions: strategic insights from NASDAQ 100 firms using Twitter data. J. Bus. Res. **117**, 443–449 (2020)
8. Caridi, M., Moretto, A., Perego, A., Tumino, A.: The benefits of supply chain visibility: a value assessment model. Int. J. Prod. Econ. **151**, 1–19 (2014)
9. Yu, M.-C., Goh, M.: A multi-objective approach to supply chain visibility and risk. Eur. J. Oper. Res. **233**, 125–130 (2014)
10. Swift, C., Guide, V.D.R., Muthulingam, S.: Does supply chain visibility affect operating performance? Evidence from conflict minerals disclosures. J. Oper. Manag. **65**, 406–429 (2019)
11. Bregman, R., Peng, D.X., Chin, W.: The effect of controversial global sourcing practices on the ethical judgments and intentions of U.S. consumers. J. Oper. Manag. **36**, 229–243 (2015)
12. Suh, C., Lee, I.: An empirical study on the manufacturing firm's strategic choice for sustainability in SMEs. Sustainability **10**, 572 (2018)
13. Kittipanya-ngam, P., Tan, K.H.: A framework for food supply chain digitalization: lessons from Thailand. Prod. Plan. Control **31**, 158–172 (2020)
14. Bai, C., Sarkis, J.: A supply chain transparency and sustainability technology appraisal model for blockchain technology. Int. J. Prod. Res. **58**, 2142–2162 (2020)
15. Somapa, S., Cools, M., Dullaert, W.: Characterizing supply chain visibility – a literature review. Int. J. Logist. Manag. **29**, 308–339 (2018)
16. Lee, Y., Rim, S.-C.: Quantitative model for supply chain visibility: process capability perspective. Math. Probl. Eng. **2016**, 1–11 (2016)
17. Yin, R.K.: Case Study Research and Applications: Design and Methods. SAGE, Los Angeles (2018)
18. Eisenhardt, K.M.: Building theories from case study research, p. 20 (2020)
19. Gehman, J., Glaser, V.L., Eisenhardt, K.M., Gioia, D., Langley, A., Corley, K.G.: Finding theory-method fit: a comparison of three qualitative approaches to theory building. J. Manag. Inq. **27**, 284–300 (2018)
20. Kossiakoff, A.: Systems Engineering: Principles and Practices, 2nd edn. Wiley, Hoboken (2011)
21. Defense Systems Management College: Systems Engineering Fundamentals: Supplementary Text. The Press, Fort Belvoir (2001)
22. Koelsch, G.: Requirements Writing for System Engineering. Apress, Berkeley (2016)
23. Hull, E., Jackson, K., Dick, J.: Requirements Engineering, 2nd edn. Springer, London (2005). https://doi.org/10.1007/b138335

# Liner Ship Freight Revenue and Fleet Deployment for Single Service

Jasashwi Mandal[1]([⊠]) [iD], Adrijit Goswami[1] [iD], Nishikant Mishra[2],
and Manoj Kumar Tiwari[1,3] [iD]

[1] Indian Institute of Technology Kharagpur, Kharagpur, India
[2] Hull University Business School, University of Hull, Hull, UK
[3] National Institute of Industrial Engineering, Mumbai, India

**Abstract.** For single-liner ship service, this study optimizes containerized cargo revenue minus ship operating costs. The decisions to optimize the ship fleet and shipment plan are included in this proposed model. This optimization problem is formulated as a mixed-integer non-linear programming model and solved it using LINGO. The proposed model is applied to a liner service route provided in the computational study and the results are analyzed in case of different scenarios of container shipment demand and different freight rates.

**Keywords:** Liner ship · Freight revenue · Fleet deployment

## 1 Introduction

Liner service is an efficient mode of marine freight transportation in which a ship owner sails the ship to a series of ports in sequential order. Each liner shipping company operates its own services along fixed routes with a set of ports of call. Containerized freight is mostly transported by liner ships on regularly scheduled service routes. Specific routes, specific ports, fixed ports of call and ship schedules, and fixed freight rates are among its four fixed characteristics. Liner shipping firms are mainly concerned with making a profit. To raise the shipping profit, the two most important aspects are reduction in operating cost and increasing freight revenue. Several fixed costs such as crew costs, repair and maintenance costs, marine insurance, stores, lubricant costs, and other major costs of fuel consumption, transshipment cost, berthing cost, loading and unloading cost altogether constitute the operating cost. The majority of freight revenue comes from the transportation of containerized cargo. Since freight rates vary between ports depending upon several types of cargo, the shipping plan selection has an impact on freight revenue. There is generally a high demand for cargo transportation, particularly during peak seasons, which makes determining the loading strategy more difficult.

Existing studies on freight revenue maximization focus on several topics such as optimal itinerary selection for individual customer, cargo slot allocation, speed optimization, and ship deployment. For example, [1] developed a bi-level optimization model to solve

© IFIP International Federation for Information Processing 2021
Published by Springer Nature Switzerland AG 2021
A. Dolgui et al. (Eds.): APMS 2021, IFIP AICT 631, pp. 123–131, 2021.
https://doi.org/10.1007/978-3-030-85902-2_14

the practical problem of deciding the optimal collection of itineraries to have for individual customers in order to optimize the expected profit. Next, [2] addressed a slot allocation optimization model for a liner shipping network to take the right decisions on which containers to transport to maximize the company's profit. A liner seasonal service and freight revenue management problem is proposed to determine the number of different type of containers to assign to all ship routes, number of vessels deployed on the routes, and also the voyage speed of ships so that the total liner seasonal profit is maximized [3]. Different Revenue Management problems are investigated for liner shipping by analyzing supply, demand, customer segments, capacity control, pricing and RM decisions [4]. A mixed-integer non-linear non-convex programming model to plan itineraries of ports of call, employ vessels to visit the itineraries, and find the container shipping strategy with the vessels deployed to optimize overall benefit is developed and solved using a column generation based heuristic method [5]. An joint modeling approach is presented for the interconnected problems of liner ship deployment, network design, and empty repositioning [6]. An integrated mathematical model and matheuristic is proposed for the liner ship fleet deployment and repositioning problem to simultaneously determine the ships for each route and their corresponding moving cost to the assigned shipments [7].

Liner shipping company fulfils container shipment demand from longterm contracts as well as spot market. The problem of determining the optimal freight rate of spot containers between each origin port and destination port of a network in order to maximize overall profit is presented as a two-stage stochastic programming model [8]. The problem of container assignment that involves finding the optimal freight rates, number of container shipments and the transporting strategy for containers in a liner shipping network is proposed to optimize overall shipping profit [9]. Furthermore, the problems of finding optimal speed, refueling strategies, and shipment planning are also addressed in [10] with the objective of maximizing the cargo revenue minus bunkering cost for a single route. The number of containers loaded for each OD pair has a significant impact on the total profit. In this study, the freight revenue management problem focuses on two major decisions: fleet optimization and flow of containerized cargo between each origin-destination pair. Based on these two decisions, the profit of the manager is maximized. The problem formulated is a nonlinear mixed integer programming model and solved using LINGO.

The remainder of this paper is organized as follows: Sect. 2 provides the problem description and mathematical formulation of the problem. Applicability of the model and results are discussed in Sect. 3. Finally, Sect. 4 provides the concluding remarks.

## 2  Problem Description

In most instances, a liner ship route's itinerary creates a loop, and a vessel deployed on the route can visit a port more than once. Let us take a liner route operated by a shipping company that covers a set of ports. For example, Fig. 1 presents a service route covering eight ports. The number mentioned on the voyage legs of the route is the voyage length or the distance between two ports. Colombo and Salalah are visited twice in this shipping route by a vessel during its voyage. In practice, a shipping company typically employs the same type of vessels on the route to ensure weekly service frequency. If each port of call is visited once in a week by the vessels deployed on the service route, then the service frequency is weekly. Let I be the set of ports of call for the service route that implies it has |I| number of ports of call and |I| legs. Any port along the circular liner route can be called the origin port. The vessels depart from one port and returns to it after completing a round trip. The round-trip time is set because the liner ship route has specific schedules. All the ports of call have a fixed arrival time window for the deployed vessels. After arriving at the port, the vessel needs a port time including container handling time. The round-trip time includes both port operation and voyage time.

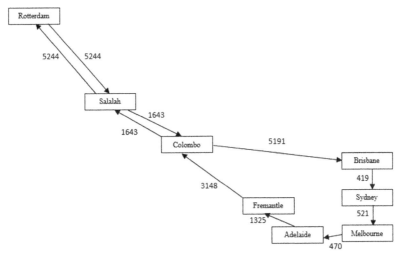

**Fig. 1.** Example of a liner service route with voyage length (nautical-miles).

We have considered the liner route with known voyage time for each leg, i.e., the vessel speed is fixed for all legs. On the liner ship route, every origin and destination port pair has a container shipping requirement weekly. Before the voyage, it is presumed that the number of containers to be shipped and the freight rate for each (O, D) port pair are given. It is also presumed that the shipping demand for each (O, D) port pair can be met in part. In most cases, the shipping company's ships are either chartered or their own ships. These ships are classified into various groups based on the loading capacity, berth occupancy fees, fixed operating costs, weight of the ships, and other ship-related characteristics. The aim of this study is to maximize the profit of the company considering the freight revenue and costs related to operating ships, fixed charges and berth occupancy charges for a liner ship service. The decisions to be taken here are to find the optimal fleet (type of ships and how many ships will be deployed) and the optimal quantity of containers to be loaded for each origin and destination pair.

## 2.1 Model Formulation

In this section, we present the mathematical model for the proposed profit maximization problem for a liner shipping company.

### Model Parameters

The interpretations of the notations for sets, indices and parameters used here are mentioned below.

*Sets*

$I$    Set of port of calls
$E$    Set of vessels

*Indices*

$\alpha$    Index of port of calls
$e$    Index of vessels

*Parameters*

$K_{\alpha\beta}$    Freight rate of transporting containers from $\alpha$th port of call to $\beta$th port of call of the route
$C_e^{opr}$    Operating cost for vessel type $e$
$C_e^{fix}$    Fixed cost for vessel type $e$
$C^{berth}$    Berth occupancy charge
$T$    Average handling time of one TEU container
$x_{min}$    Minimum number of vessels deployed on the route
$x_{max}$    Maximum number of vessels deployed on the route

### Decision Variables

$x$    Required number of vessels deployed on the route
$y_e$    $\begin{cases} 1, & \text{if vessel type } e \text{ is used} \\ 0, & \text{otherwise} \end{cases}$
$p_{\alpha\beta}$    Number of containers loaded at $\alpha$th port of call to be shipped to $\beta$th port of call of the route
$q_\alpha$    Number of containers on the vessel when it arrives at port of call $\alpha$ (TEUs).

### Mixed-Integer Non-linear Programming Model

The objective function (1) is designed to maximize the profit of the shipping company. The first term in the objective function shown in Eq. (1) denotes the freight revenue of the shipping company achieved by transporting the containers from the origin ports to the destination ports according to the shipment demand. It also includes the operating costs for ships (second component), fixed costs related to ships (third component) and berth occupancy charge (fourth component). The mathematical model for the mixed integer non-linear programming is formulated as follows:

$$\text{Max} = \sum_{\alpha \in I} \sum_{\beta \in I} K_{\alpha\beta} p_{\alpha\beta} - \sum_{e \in E} x C_e^{opr} y_e - \sum_{e \in E} C_e^{fix} y_e - \sum_{\alpha \in I} TC_e^{berth} q_\alpha \qquad (1)$$

Subject to

$$q_\alpha + \sum_{\beta \in I, \beta \neq \alpha} \left( p_{\alpha\beta} - p_{\beta\alpha} \right) = q_{\alpha+1} \quad \forall \alpha \in I \qquad (2)$$

$$q_1 = \sum_{\substack{\alpha \in I \\ 2 \leq \alpha \leq m}} \sum_{\substack{\beta \in I \\ \beta < \alpha \\ 1 \leq \beta \leq m-1}} p_{\alpha\beta} \qquad (3)$$

$$q_{m+1} = q_1 \qquad (4)$$

$$0 \leq p_{\alpha\beta} \leq D_{\alpha\beta} \qquad \forall \alpha, \beta \in I, \alpha \neq \beta \qquad (5)$$

$$p_{\alpha\beta} \in \mathbb{Z} \qquad \forall \alpha, \beta \in I, \alpha \neq \beta \qquad (6)$$

Constraint (2) presents the relation between the total amount of containers loaded on a ship when it visits a port and the amount of containers for each origin and destination pair. Constraints (3) and (4) determine the total amount of containers loaded on a ship when it visits the first port of call during the round trip. Next, constraint (5) enforces that the quantity of loaded containers must not exceed the corresponding origin-destination shipment demand. Constraint (6) ensures that the loading quantity is a non-negative integer.

$$\sum_{e \in E} y_e = 1 \qquad (7)$$

$$q_\alpha - \sum_{e \in E} Cap_e y_e \leq 0 \quad \forall \alpha \in I \qquad (8)$$

$$168x \geq \sum_{\alpha \in I} (t_\alpha + Tq_\alpha) \qquad (9)$$

$$x_{\min} \leq x \leq x_{\max} \qquad (10)$$

Constraint (7) states that only one type of vessel can be deployed on the liner service route. Constraint (8) restricts that the number of containers on each leg should not exceed the capacity of the vessel deployed on the leg. Constraint (9) determines the number of ships required to be deployed on the route which lies in a range specified by the constraint (10). The problem formulated is a mixed-integer non-linear programming model and solved using LINGO.

## 3    Computational Study and Results

In this section, the proposed modelling approach is applied to a liner service route containing four ports: Ho Chi Minh, Laem Chabang, Singapore, and Port Klang. The distance of the voyage lengths are 589, 755, 187, and 830 (in nautical miles) respectively for all the legs. Hence, the liner service route has four ports of call. Let us consider that the ports have similar characteristics except for their location and the demand for containerized freight. Three vessel types are considered in this study based on the loading capacity, voyage related fixed costs and their operating costs (e.g., repair and maintenance cost). The capacities of the three type of vessels are 3000, 4500 and 6000 TEU respectively. The operating costs for the ships are 51923, 76923 and 115384 USD/week respectively. Similarly, the voyage related fixed costs are 219080, 282840, and 340000 USD. The average port handling time for the ships is 0.012 TEU/hour. The weekly container shipping demand (in TEUs) and container freight rate (USD/TEUs) for each origin-destination pair are provided in Table 1.

**Table 1.** Container shipment demand and Freight rates for each (O, D) pair.

| Origin-destination ports | Demand (TEUs) | Freight rate (USD/TEUs) |
|---|---|---|
| Ho Chi Minh-Laem Chabang | 897 | 477 |
| Ho Chi Minh-Singapore | 333 | 117 |
| Ho Chi Minh-Port Klang | 546 | 203 |
| Laem Chabang-Ho Chi Minh | 1332 | 376 |
| Laem Chabang-Singapore | 1026 | 187 |
| Laem Chabang-Port Klang | 1365 | 213 |
| Singapore-Ho Chi Minh | 813 | 271 |
| Singapore-Laem Chabang | 456 | 112 |
| Singapore-Port Klang | 858 | 266 |
| Port Klang-Ho Chi Minh | 417 | 407 |
| Port Klang-Laem Chabang | 573 | 164 |
| Port Klang-Singapore | 1118 | 458 |

While optimizing the fleet, the constraints (7) and (8) determine the type of the vessel needed to be deployed. Using the above data, the vessel type obtained is of capacity 6000 TEUs. Next, the constraints (9) and (10) find the required number of vessels. Hence, in this case two vessels of capacity 6000 TEUs are employed to visit all the ports. Constraints (2), (3), (5), and (6) help to find the amount of loaded containers to satisfy the demand for each origin and destination pair. In Table 2, the optimal values of the number of containers loaded at $\alpha$th port of call to be shipped to $\beta$th port of call of the route ($p_{\alpha\beta}$) are shown. Therefore, the flow of containers on the legs of the route are 3290, 5543, 4914 and 4253 TEUs respectively. Next, the container shipment demand for the origin-destination pairs are changed. If the container shipment demand increases, the change in the freight revenue, the operating costs and the total profit of the manager are represented in Fig. 2. Similarly, Fig. 3 represents the freight revenue, the operating costs and the total profit of the manager with respect to the increasing freight rate.

**Table 2.** Number of containers loaded and Freight rates for each (O, D) pair.

| Origin-destination ports | Number of containers loaded (TEUs) | Freight rate (USD/TEUs) |
|---|---|---|
| Ho Chi Minh-Laem Chabang | 897 | 477 |
| Ho Chi Minh-Singapore | 156 | 117 |
| Ho Chi Minh-Port Klang | 546 | 203 |
| Laem Chabang-Ho Chi Minh | 1332 | 376 |
| Laem Chabang-Singapore | 1026 | 187 |
| Laem Chabang-Port Klang | 1365 | 213 |
| Singapore-Ho Chi Minh | 813 | 271 |
| Singapore-Laem Chabang | 0 | 112 |
| Singapore-Port Klang | 858 | 266 |
| Port Klang-Ho Chi Minh | 417 | 407 |
| Port Klang-Laem Chabang | 573 | 164 |
| Port Klang-Singapore | 1118 | 458 |

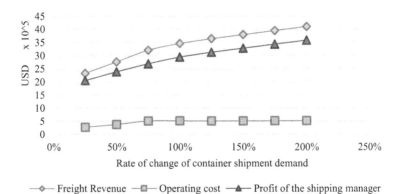

**Fig. 2.** Change in freight revenue, operating costs and the total profit of the manager with respect to the increase in container shipment demand.

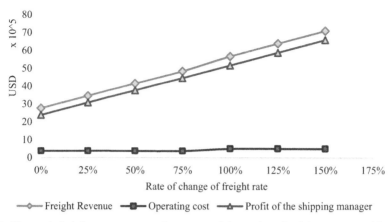

**Fig. 3.** Change in freight revenue, operating costs and the total profit of the manager with respect to the increase in freight rate.

## 4   Conclusion

This paper considers determining the ship fleet and container loading quantity jointly with the purpose of maximizing the freight shipping revenue while reducing the ship operating cost for one liner service. This optimization problem is formulated as a mixed-integer nonlinear programming model. For the computational study provided here, the proposed problem can be solved using LINGO. In conclusion, some discussions are made on the numerical results obtained for separate scenarios. We have shown if the container shipment demand increases, how the profit of the manager also increases. These observations may assist the shipping company in better understanding the consequences of various scenarios and making better liner service decisions.

## References

1. Wang, S., Wang, H., Meng, Q.: Itinerary provision and pricing in container liner shipping revenue management. Transp. Res. Part E Logist. Transp. Rev. **77**, 135–146 (2015). https://doi.org/10.1016/j.tre.2014.06.020
2. Zurheide, S., Fischer, K.: A revenue management slot allocation model for liner shipping networks. Marit. Econ. Logist. **14**, 334–361 (2012). https://doi.org/10.1057/mel.2012.11
3. Wang, Y., Meng, Q., Du, Y.: Liner container seasonal shipping revenue management. Transp. Res. Part B Methodol. **82**, 141–161 (2015). https://doi.org/10.1016/j.trb.2015.10.003
4. Meng, Q., Zhao, H., Wang, Y.: Revenue management for container liner shipping services: critical review and future research directions. Transp. Res. Part E Logist. Transp. Rev. **128**, 280–292 (2019). https://doi.org/10.1016/j.tre.2019.06.010
5. Wang, S., Meng, Q.: Liner shipping network design with deadlines. Comput. Oper. Res. **41**, 140–149 (2014). https://doi.org/10.1016/j.cor.2013.08.014
6. Monemi, R.N., Gelareh, S.: Network design, fleet deployment and empty repositioning in liner shipping. Transp. Res. Part E Logist. Transp. Rev. **108**, 60–79 (2017). https://doi.org/10.1016/j.tre.2017.07.005

7.  Wetzel, D., Tierney, K.: Integrating fleet deployment into liner shipping vessel repositioning. Transp. Res. Part E Logist. Transp. Rev. **143**, 102101 (2020). https://doi.org/10.1016/j.tre.2020.102101

8.  Wang, Y., Meng, Q.: Optimizing freight rate of spot market containers with uncertainties in shipping demand and available ship capacity. Transp. Res. Part B Methodol. **146**, 314–332 (2021). https://doi.org/10.1016/j.trb.2021.02.008

9.  Wang, S., Liu, Z., Bell, M.G.H.: Profit-based maritime container assignment models for liner shipping networks. Transp. Res. Part B Methodol. **72**, 59–76 (2015). https://doi.org/10.1016/j.trb.2014.11.006

10. Wang, S., Gao, S., Tan, T., Yang, W.: Bunker fuel cost and freight revenue optimization for a single liner shipping service. Comput. Oper. Res. **111**, 67–83 (2019). https://doi.org/10.1016/j.cor.2019.06.003

# Digitization of Real-Time Predictive Maintenance for High Speed Machine Equipment

Rony Mitra[1](✉) ⓘ, Mayank Shukla[1] ⓘ, Adrijit Goswami[1] ⓘ,
and Manoj Kumar Tiwari[1,2] ⓘ

[1] Indian Institute of Technology Kharagpur, Kharagpur, India
`ronymitra92@iitkgp.ac.in`
[2] National Institute of Industrial Engineering, Mumbai, India

**Abstract.** In the recent decade, state-of-the-art techniques of maintenance in manufacturing firms have evolved. Redefining itself to come up with a whole new perspective by including a regime of digitization. From inter-compatibility to intra-network communication between hardware to highly interactive user interfaces have made the managing of necessary procedures extremely transparent. Even complex inclusions are easy to monitor following the current trends and digital transformation. Data generated through sources is big and unmanageable with a lack of filtering technologies to identify useful processable content. The proposed framework helps notify end-users by monitoring and identifying certain user-based settings and business functions. Suggested findings used machine learning (ML) algorithms surpass any previous claimed results. The modeling approach ensures consistent and reliable performance. Inclusive integration of notifying tools into trending smart devices has been tested and validated in this study. The coupling of multidiscipline open-source web-based technologies with minimum expense has been in focus for designing such applications. The best-identified set of tools that help enable the management of workflow multitasks, and their semantic arrangement through the latest state-of-the-art and scientific tools for generic work environments is covered in this study.

**Keywords:** Predictive maintenance · Condition monitoring · Machine learning · Random forest regression · Applications

## 1 Introduction

Key management of core technical challenges and administrative issues during the action of the work life cycle in an organization to retain and restore original efficiency and functioning of the holistic system is termed Maintenance. Recently startups have shown up providing maintenance support and services in a more accessible way, therefore, making them among the best, and most coveted subjects in modern business ventures. One viable method to improve maintenance efficiency without compromising quality

© IFIP International Federation for Information Processing 2021
Published by Springer Nature Switzerland AG 2021
A. Dolgui et al. (Eds.): APMS 2021, IFIP AICT 631, pp. 132–140, 2021.
https://doi.org/10.1007/978-3-030-85902-2_15

is condition-based maintenance (CBM). In the past, this has mainly been achieved by utilizing the expertise and know-how of personnel gathered by training as well as long years of servicing the equipment. The downside of this approach is the dependence on a stable workforce with low fluctuation, which is often difficult to maintain in service projects. Furthermore, the results are often subjective depending on the experience level and interpretation of conditions by experts therefore mostly unreliable.

Vibration signals provide valuable information about the insights of the equipment on the operating condition. The presence of a fault can be detected by inspecting the vibration signal; also, it can localize the position of the fault and the health state of the equipment. Vibration signals can be captured through vibration sensors (e.g.; displacement sensors, velocity sensors, and accelerometers).

Sensitiveness of sensor for displacement measurement is highly sensitive in the low-frequency range ($\sim$<1 kHz), likewise, a flat amplitude response is more effective for velocity sensors within the range of 2 to 10 kHz. On the contrary, accelerometers come up with the best amplitude performance in a high range of frequency (i.e. Tens of kHz). This paper focuses on predictive maintenance by using vibration signals from the wireless vibration and temperature measuring sensors to reduce downtime and improve productivity and monitoring. The remainder of this paper is organized as follows. In Sect. 2, the related literature is reviewed. In Sect. 3, we described the challenges in predictive maintenance. In Sects. 4 and 5, solution methodologies and their results are explained. Finally, the conclusions and possible future research directions are exposed in Sect. 6.

## 2  Literature Review

Fault detection of the rotating machinery can be identifying by the fault diagnosis technique, which can also be applied to get the details regarding about equipment's operating condition [1]. The rudimentary aim of the diagnosis for fault are mentioned as: (i) measuring and analyzing real time health of the equipment's (ii) determining reason behind breakdown or equipment failure and (iii) anticipating fault by developed in the modeled line and its trend [2]. Vibration signals of the machine can offer an early warning to the operator to make a crucial decision before any major failure or break down and reduce the unscheduled downtime. Vibration signal amplitude portrays basic image of the challenging problem, whilst frequency can give us some information about the source of the defect [3]. Condition Monitoring (CM) is used to provide alarms and actions to prevent production of out-of-specification components and avoid to machine breakdowns. Also, it is defined as a process of monitoring a parameter, which is indicative for a significant change of a developing failure of condition in machinery [4]. To avoid break down and preventing unexpected system downtime [5–7], CM can estimate the remaining useful life of the equipment. Predictive Maintenance (PdM) is aimed to determine the equipment working condition. Other than replacing the equipment at frequently intervals, PdM allows real-time evaluation and close monitoring of the equipment's, to alert the system administrator with replacement and servicing suggestions when the equipment is about to malfunction or fail [8]. PdM brings cost saving in the industries, since the requirement of PdM is based on the actuality, rather than estimating the condition of the machine's equipment and its performance [9]. In the existing architectures GE digital Discrete Manufacturing Software, Manufacturing Execution Systems (MES) (such as ABB discrete manufacturing operations management software, SAP MES, Siemens SIMATIC IT for Discrete Manufacturing) demands new processing and analysis techniques for PdM with vibration sensor data.

Jemielniak summarized the signal processing techniques and collated advanced signal analysis techniques, the filtering (low-pass, high-pass and band-pass), averaging and RMS are the most effective [10]. Chiementinl et al. worked on the early detection methods for fatigue measurement of ball bearings using adaptive wavelet analysis [11]. Computational intelligence methods include evolutionary method like Genetic Algorithm (GA) [12], Artificial Neural Network (NN) [13], support vector machine (SVM) [14] and one-class SVM [15] have been effectively in machines for automatic detection and identification of faults.

# 3 Problem Statement

Global challenges and intense competitiveness in the markets recently rising steadily to attain peak. The firms are under constant pressure and in regular need to adapt recent advances for increasing efficiency, delivering better solutions, reduction in costs, and finally accepting digital transformation to capture inherited benefits to leverage technology to enhance management.

Real-time condition monitoring of machinery is a pivotal technique for guaranteeing the efficiency and quality of any production process. It may cost a big capital for the company in case of system failure. So, future prediction of the machine failure in advance and choosing desired alternatives, before the failure happens, is an essential precaution. At the operating time, most of the machines will vibrate and vibration has a strong relationship with the condition of the equipment or spare parts of the machine. Our main focus is predictive maintenance on high-speed rotating machinery such as wind turbines, baggage handling systems in the airport terminal. Experiencing digital transformation in real the data generated drives applications of artificial intelligence (AI) and big data shape our daily lives and decrease the intricacies of working processes in all sectors of industry and other important fields.

In this era, disruptive technologies play a crucial role in innovation, sensors harness invaluable analogous signals present from all around and all forms of physicality in nature. Mobile phones, wearable, smart devices, etc. are an essential part of daily life. Ever since the inception of these gadgets have formed an integral part of the human lifestyle and the average daily time humans remain engaged dealing with this electronical hardware is remarkably high. The purpose of this approach highlighting user-friendliness and inbuilt features of such tangible assets. Features such as touch interfacing, high-definition sensitive screens, high-quality compact processor and ram, etc. are all packed in with wireless connectivity. High-end interfacing is yet not been able to drive heavy machines and power electronics easily. This makes it an engineering problem of interdisciplinary approach as it remains untouched due to unfilled research gaps. While data is being collected more rapidly, the identification and inter-compatibility of already evolved technologies and software are still not known. The skills needed in the development of merging technologies and decipher inter-compatibility are tried to be resolved in this paper using machine learning framework, application open-source platform, and recent advances in interfacing.

The core objective of this paper is to create a dashboard that helps managers to predict machine equipment failure in advance and according to that schedule maintenance activity. Also, we can achieve maximum productivity by minimizing downtime of the machinery equipment, and parallel it will reduce maintenance cost as well as labor cost.

# 4  Solution Approach

A generic framework comprising of IoT integrated auxiliaries with external systems is the demand of the times in the current domain. With an integrated and real-time data cascascading, monitoring, streaming anticipatory model. The system would able to diagnose breakdowns, trigger alerts using simple logical rules (like ladder logic), and alarm authorized officials to execute intelligent decisions manually or automatically in real-time with all necessary safety guidelines and improvised compulsory regulations.

In this study, the infrastructure with embedded sensors and gateways, the buffering of all information flow from the production lines constantly integrating system and allowing accessibility to the components prior to developing the predictive maintenance system is assumed available. To enable communication between machines, collect data from devices, monitor live data, and manage historical data in enhanced view with suitable use of machine learning and AI techniques using the suitable platform is provided in this study. Collected data by devices from locations, with the least data dissipation, is offered, and the data from all sensors are processed in lesser time than usual to evaluate essential decisive actions within time boundaries.

## 4.1  Proposed Model

The proposed approach is a case study in the industrial environment to introduce signal processing techniques (time domain, frequency domain) that provides useful information to discriminant faults according to their installed position. The novelty of the work is to condition monitoring of spare parts of the equipment in high-speed machinery and estimate the remaining useful life of the equipment by using raw vibration data collected from the sensor installed in the machine equipment (Fig. 1). Sensors are communicating with the data acquisition system (DAQ) to the base station and through edge computer raw data was stored in the cloud by using python programming. Then feature extraction and data filtering is applied to get processed data. There after compile processed data with historical maintenance data and train by using random forest algorithm. Finally, for the new data we predict machine equipment conditions in advance. This prediction can be monitored by the managers in computer system, smart phone, and smart watch.

## 4.2  Data Collection

Within this research we monitored machine equipment by using 18 wireless, temperature, velocity and vibration capturing sensors. Vibration data from several month of recording

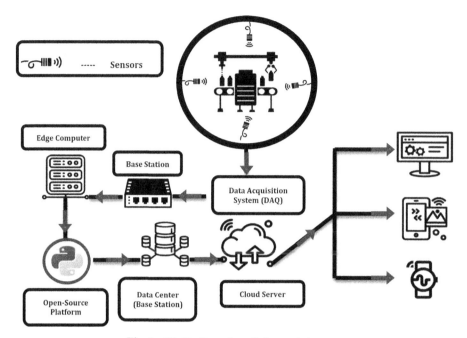

**Fig. 1.** Digitization of predictive maintenance

were collected from the sensors by using python language through edge computer and store them in the cloud so that in future space complexity would not be an issue. Raw data are stored in the form of temperature, acceleration (with respect to X-axis, Y-axis, Z-axis), velocity (with respect to X-axis, Y-axis, Z-axis).

### 4.3 Feature Extraction

Feature extraction can be divided into time domain analysis and frequency domain analysis. We extract mean (Eq. 1), RMS (Eq. 2), peak-to-peak (Fig. 2), standard deviation (Eq. 3), kurtosis (Eq. 4), and Skewness (Eq. 5) to analyze time domain features. Also, apply FFT and IFFT filter out raw data.

$$Mean, \mu = \frac{1}{N} \sum_{i=1}^{N} x_i \tag{1}$$

$$RMS = \sqrt{\frac{1}{N} \sum_{i=1}^{N} x_i^2} \tag{2}$$

$$\sigma = \frac{1}{N} \sum_{i=1}^{N} (x_i - \mu)^2 \tag{3}$$

$$x_{kurt} = \frac{1}{N\sigma^4} \sum_{i=1}^{N} (x_i - \mu)^4 \tag{4}$$

$$\widetilde{\mu}_3 = E\left[\left(\frac{x-\mu}{\sigma}\right)^3\right] = \frac{E[X^3] - 3\mu\sigma^2 - \mu^3}{\sigma^3} \tag{5}$$

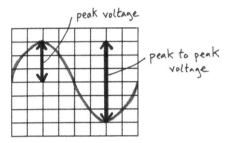

**Fig. 2.** Peak-to-peak

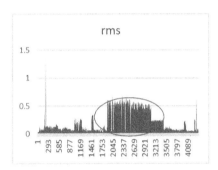

**Fig. 3.** High rms value in idle time

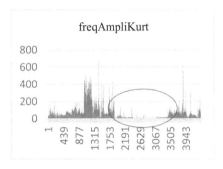

**Fig. 4.** Low kurtosis value in idle time

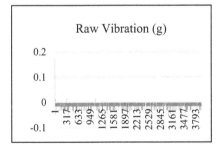

**Fig. 5.** Low rms with low kurtosis in idle time

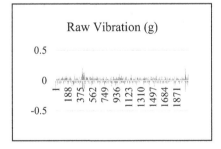

**Fig. 6.** Low rms value in working condition

## 4.4 Data Filtering

Based on the experimental work, the vibration input can be divided into idle condition and working condition. Since idle time data has no impact in fault prediction, it is better to remove idle time data. We observed that idle time data has high rms (Fig. 3) and low kurtosis (Fig. 4) value. Therefore, it is better to remove vibration data with rms value greater than 0.4 g and as well as low kurtosis value less than 25 (since it is noted that sometime low rms can be recorded in idle machine condition (Fig. 5)). Therefore, we use only data generated in working condition of a machine i.e.; with low rms and high kurtosis (Fig. 6). Also noticed that there are some gravitational effects (Fig. 7) on the sensors and it will impact on the dataset. To overcome this issue, we use FFT and IFFT for further filtering of the data.

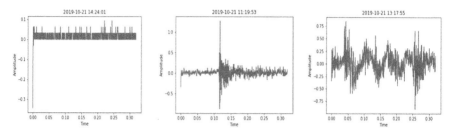

**Fig. 7.** Gravitational effects and some external effects in reading

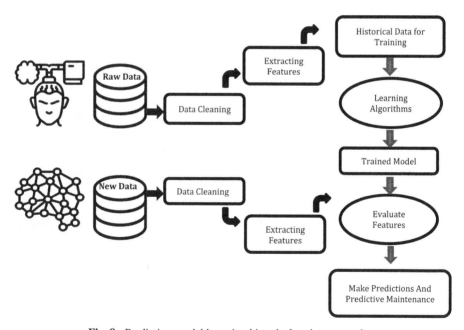

**Fig. 8.** Prediction model by using historical maintenance data

### 4.5  ML Algorithm and Predictions

To predict machine conditions in advance, we arrange historical maintenance data and processed data collected from sensors and use Random Forest (RF) as a supervised machine learning algorithm to train and validate our model.

Predictive models are generated using RF. Multiple randomized decision trees (DT generates "forest" (ensemble) while simple average predictions get aggregated with RF. With an increase in the count of variables than the observations and sample performance are enhanced. RF is both classification and regression performing supervised machine learning. This work proposes an improvement where monitoring of machine equipment conditions is achieved. To attain this, data of status (activated and deactivated alarms) and data of operations (about historical maintenance) are employed to articulate the model. Work contributions include speed in processing of collected inputs, automation scalability, expansion, and economics. Results show predictive accuracy improvement of 94.7%, compared to the previous historical data. For real-time monitoring, currently, processed data is used for prediction by using the same trained algorithm.

## 5   Results and Discussion

Opensource interfacing libraries for sensing analog signals and prediction using ML libraries when coupled have shown potential for the development of highly sensitive, inter-compatible, customizable, real-time, web-based tools. Nevertheless, intranet practices do not have security concerns. The analysis and post-processing of ML algorithms performed over edge cloud reflected highly efficient graphs, metrics, with substantial user control and full web ergonomics. Ergonomics for web development has also shown the potential of learning using ML based on customer preference. The suggested framework was tested twice using simulations and once a perturbation was mimicked to create a need disruption to invoke resilience procedure. The outcome has substantial deliverables as prediction graph as bad percentage showed in Fig. 8 by using python in the cloud and can display in desktop, smartphone as well as a smartwatch.

## 6   Conclusion and Future Scope

A real time vibration-based fault detection using ML techniques was introduced in this study. This method used the vibration sensors attached to the machine parts to obtain the vibration data, which was processed, complied with historical maintenance data and then fed to the ML decision-making models. The ML techniques used is RF algorithm and this will make a real-time decision whether the machine or its equipment's are safe to operate or not, which can be monitored using a smartphone. However, these findings are only based on indoor testing and for that it is not required to go infield for engineers to check machine conditions. Though sensor installation is costly, but by predicting machine failure in advance company can minimize machine downtime, labor hour and can increase productivity. In future sensor fusion can be used to minimize the overall cost of the entire system by minimizing the number of sensor and cost of the down time of the system.

## References

1. Jardine, A.K.S., Lin, D., Banjevic, D.: A review on machinery diagnostics and prognostics implementing condition-based maintenance. Mech. Syst. Signal Process. **20**, 1483–1510 (2006). https://doi.org/10.1016/j.ymssp.2005.09.012

2. Caesarendra, W., Tjahjowidodo, T.: A review of feature extraction methods in vibration-based condition monitoring and its application for degradation trend estimation of low-speed slew bearing. Machines **5** (2017). https://doi.org/10.3390/machines5040021
3. Peng, Z., Kessissoglou, N.: An integrated approach to fault diagnosis of machinery using wear debris and vibration analysis. Wear **255**, 1221–1232 (2003). https://doi.org/10.1016/S0043-1648(03)00098-X
4. Hu, Q., Si, X.S., Zhang, Q.H., Qin, A.S.: A rotating machinery fault diagnosis method based on multi-scale dimensionless indicators and random forests (2020)
5. Xiahou, T., Zeng, Z., Liu, Y.: Remaining useful life prediction by fusing expert knowledge and condition monitoring information. IEEE Trans. Ind. Inform. **17**, 2653–2663 (2021). https://doi.org/10.1109/TII.2020.2998102
6. Xia, T., Song, Y., Zheng, Y., Pan, E., Xi, L.: An ensemble framework based on convolutional bi-directional LSTM with multiple time windows for remaining useful life estimation (2020)
7. Motahari-Nezhad, M., Jafari, S.M.: Bearing remaining useful life prediction under starved lubricating condition using time domain acoustic emission signal processing (2021)
8. Jung, D., Winslett, M.: Vibration analysis for IoT enabled predictive maintenance (2017). https://doi.org/10.1109/ICDE.2017.170
9. Tickoo, O., Iyer, R., Illikkal, R., Newell, D.: Modeling virtual machine performance: challenges and approaches. Perform. Eval. Rev. **37**, 55–60 (2010). https://doi.org/10.1145/1710115.1710126
10. Jemielniak, K.: Commercial tool condition monitoring systems, pp. 711–721 (1999)
11. Chiementin, X., Bolaers, F., Dron, J.P.: Early detection of fatigue damage on rolling element bearings using adapted wavelet. J. Vib. Acoust. Trans. ASME. **129**, 495–506 (2007). https://doi.org/10.1115/1.2748475
12. Huang, J., Hu, X., Yang, F.: Support vector machine with genetic algorithm for machinery fault diagnosis of high voltage circuit breaker. Meas. J. Int. Meas. Confed. **44**, 1018–1027 (2011). https://doi.org/10.1016/j.measurement.2011.02.017
13. Fernando, H., Surgenor, B.: An unsupervised artificial neural network versus a rule-based approach for fault detection and identification in an automated assembly machine. Robot. Comput. Integr. Manuf. **43**, 79–88 (2017). https://doi.org/10.1016/j.rcim.2015.11.006
14. Yan, X., Jia, M.: A novel optimized SVM classification algorithm with multi-domain feature and its application to fault diagnosis of rolling bearing. Neurocomputing **313**, 47–64 (2018). https://doi.org/10.1016/j.neucom.2018.05.002
15. Saari, J., Strömbergsson, D., Lundberg, J., Thomson, A.: Detection and identification of windmill bearing faults using a one-class support vector machine (SVM). Meas. J. Int. Meas. Confed. **137**, 287–301 (2019). https://doi.org/10.1016/j.measurement.2019.01.020

# Engineering of Smart-Product-Service-Systems of the Future

# Integrating Failure Mode, Effect and Criticality Analysis in the Overall Equipment Effectiveness Framework to Set a Digital Servitized Machinery: An Application Case

Claudio Sassanelli(✉) ⓘ, Anna de Carolis ⓘ, and Sergio Terzi ⓘ

Department of Management, Economics and Industrial Engineering, Politecnico di Milano, Piazza Leonardo da Vinci 32, 20133 Milan, Italy
{claudio.sassanelli,anna.decarolis,sergio.terzi}@polimi.it

**Abstract.** Digital transformation and servitization have been merging in a coalescing paradigm called Digital Servitization, changing not only companies' business model but also their portfolio and thus their business. This hybrid paradigm is increasingly overwhelming manufacturing companies, compelling them to change their business model and provide more complex solutions to survive. Indeed, first a business model shift is needed (bringing to cope with organizational/managerial aspects), and then a suitable new technology stack has to be implemented. In the extant literature it is not clear how companies can define which are the improvements to implement on Smart Connected PSSs. These modifications on the physical products, if flanked by a concurrent definition and structuring of data requirement on the database on the cloud, would also lead to a better comprehension of the solution functioning, enabling to know which are the causes leading to breakdowns and performance and quality losses during the use phase. To address this, the paper proposes a method combining the Failure Mode and Effect Analysis with the Overall Equipment Effectiveness framework.

**Keywords:** Digital servitization · Machinery sector · Industry 4.0 · FMECA · OEE

## 1 Introduction

Digital servitization is increasingly overwhelming manufacturing companies, forcing them to change their business model and provide more complex solutions [1, 2]. Indeed, traditional physical products need to be enriched with digital technologies to enable the provision of the sol called Product-Service Systems (PSSs). Often shifting the possession of the product from the user to the provider, PSSs generate profit through the sale either of their use or of the result obtained with their use by the customer [3]. The benefits triggered by PSS are multiple and have been unveiled by the academics as valuable both for the provider and the customer (being able to enhance providers' competitiveness on

© IFIP International Federation for Information Processing 2021
Published by Springer Nature Switzerland AG 2021
A. Dolgui et al. (Eds.): APMS 2021, IFIP AICT 631, pp. 143–152, 2021.
https://doi.org/10.1007/978-3-030-85902-2_16

the market, to better meet customer needs and to decrease environmental impact than traditional business models) [4, 5].

In addition, digital technologies [6–8], recognized under the Industry 4.0 (I4.0) paradigm [9], can further increase and empower PSSs' functionalities. Their embedding on traditional physical products triggers the possibility of new knowledge-based services capable to address not only the monitoring functionality, but also to allow to control, optimize and even automatize the behaviour of the solution. By the way, the provision of Smart, Connected PSSs would require first of all a business model shift, bringing with itself organizational and managerial (human resources-related, and customer-related) challenges and in a second step also technical/technological [10]. Indeed, as suggested by [11], once the company manages to realize its need of business model shift and copes with the organizational/managerial aspects, a suitable new technology stack is needed to be implemented to actually develop and deliver the Smart, Connected PSS. Here, a new infrastructure made up of multiple levels is composed by product hardware, embedded software, a connectivity part, a product cloud remotely running on servers, a security tools suite, a gateway for managing external information sources and finally also an integration with enterprise business systems (e.g., ERP, CRM, PLM).

As a result, digital transformation and servitization have been merging in a unique coalescing paradigm called Digital Servitization [6–8], changing not only companies' business model but also their portfolio and thus their business [12].

The way products change into new integrated and more complex solutions is strongly driven and affected by the data that can be potentially generated, gathered and analysed along their entire lifecycle and somehow shared among customers and providers. Often also the relationship provider-customer becomes stronger (lightening the user from the commitment of managing the solution during its use phase).

To be able to deliver such data-driven solutions and address a sheltered digital servitization, manufacturers need to enact multiple modifications: cultural [13], customer-related [14, 15], and also in the way the company approaches the PSS design [16, 17]. However, most of the times, in the extant literature it is not clear how companies can define which are the improvements to implement on their product to enrich them with smart Connected functionalities and provide data-driven services.

The modifications on the physical products, if flanked by a concurrent definition and structuring of the data requirements to be stored on the database on the cloud, would also lead to a better comprehension of the functioning model of the solution, enabling to know which are the causes leading to availability, performance and quality losses during the use phase. For this reason, this paper has the aim of proposing a method combining the Failure Mode, Effect and Criticality Analysis (FMECA) [18] with the Overall Equipment Effectiveness (OEE) framework [19] and apply it in a pilot case. OEE is the standard metric for measuring internal performance of manufacturing productivity, effective for detecting losses, comparing progress, and enhancing the productivity of manufacturing equipment. It measures how well a manufacturing operation is utilized compared to its full potential, keeping in consideration the system quality (good pieces delivered), performance (speed) and availability (interruptions). Company A, developing and delivering machineries for plastic objects and containers decoration, has been chosen. In particular, from an analysis of the product portfolio, Machine 1 (Screen Printing) was selected as pilot machine to apply the method proposed. First of all, after analysing the bill of materials of the machine, detecting the main constructing groups and decomposing them at a detailed level, the FMECA analysis is implemented to detect the main failure causes on the machine and to find the most critical components. Once prioritized through the FMECA method, each critical component is evaluated to understand which data are required to be monitored, which type of loss (availability, performance, or quality) the critical components cause, which sensors can be embedded on the product to improve its smart functionalities and to which part of the OEE (Availability, Performance and Quality) each loss of time due to the critical components' failures analysed is related.

The paper is structured as follows. Section 2 presents the research methodology and describes the Company A and the pilot machine chosen. Section 3 show the results, then discussed in Sect. 4. Finally, Sect. 5 concludes the paper and envisages future research.

## 2 Research Methodology

To achieve the research objectives, an application case was conducted.

### 2.1 The Pilot Case: Company a and Machine 1

From '60ies Company A delivers dry-offset and silk-screen printing machines for the decoration of plastic objects and containers. Along the time, the product portfolio was enriched with hot foil printers, multipurpose machines for flexible tubes and pails and heat transfer machines for digital printing. Today, Company A's portfolio includes over 40 basic models characterized by a good flexibility (due to a modular structure and a range of accessories) to face with the changing market's needs.

Company A is a convenient sample for this research, being involved in a funded project. Its choice for this research is justified by the fact that it represents a purposive case, since it is suitable to the particular problem or representative of a special population. In addition, this case is also an idiographic study, i.e. the intensive study of an individual case (20). Indeed, the method embedding the FMECA in the OEE framework, proposed and presented in this paper, has been applied in Company A in an mixed interpretative/interactive way [21] on the basis of a previous strategic analysis which brought to the need of improving the company's product portfolio with smart functionalities. Company A represents the traditional small manufacturing company, developing, producing and delivering industrial machineries according to the typical product-based business model. At the same time, Company A has also realized that it needs to go towards the digital servitization transformation to survive in the market. For this reason, a strategic analysis was performed in the company through a series of interviews with employees from all the functions involved in the order development process, detecting several hurdles throughout its digital servitization path. Indeed, among the others, Company A requires effort mainly in the Technical Department (with the introduction of smart components enabling to both define the machine operating models and provide data-driven services), in the R&D (to study innovative solutions) and IT (with the development of a database). In addition, the analysis previously performed led also to the choice of a pilot machine to be used in this research. Machine 1 (Screen printing) is one of the latest realizations added to the Company A's portfolio. It is characterized by several electronic components on board and it has been already subject of incremental technical improvements (direct engine transmission and updating of the loading system). All the knowledge of its design and development, as for example the structure of the machine and its mechanisms, is completely known and easy retrievable. The main customers and thus users of this machines are European manufacturers of small tubes, bottles and mascara. Finally, Machine 1 has been chosen for this research by Company A since, despite its recent introduction on the market, it has already medium-high sales volumes (representing more than the 25% of the company's total sales).

## 2.2  The Research Process

Several workshops were performed in the company to conduct the research. First, the company involved in the research had been analysed from a strategic perspective (through a series of interviews with all the functions involved in the order development process), leading to the detection of its main issues related to the digital servitization (this preliminary phase is fully described in [22]). In parallel, also an analysis of the company's product portfolio was performed, providing as a result the choice of Machine 1 (Screen Printing) as pilot case for this research. In addition, to enhance its awareness about this paradigm, the company was also gradually introduced to all the concepts related to both servitization and digital transition. Indeed, to provide them further practical evidence, a demonstration of how a database could be implemented on their solutions was provided by a digital provider belonging to the Politecnico di Milano's ecosystem. Therefore, two theoretical workshops were organized to introduce the company first to the FMECA method and then to the performance measurement of production systems (with a special focus on OEE). Finally, two workshops to conduct the research were organized. In total, seven workshops were conducted, with a total duration of about 66 h. In all the workshops after the strategic and product portfolio analysis, several employees of Company A Technical Department were involved in a pervasive way (involving the technical director, the Electronic Department Manager, one electronic department engineer and one mechanical department engineer). A wrap-up of the workshops conducted is also provided in Table 1.

## 3  Results

Through the research process described in the previous section, it was possible to apply the method embedding the FMECA results in the OEE framework. First, in Subsect. 3.2, the results from the FMECA are presented: after analysing the bill of materials of the machine, Machine 1's main constructing groups have been detected and then decomposed at a more detailed levels, the main failure causes on the machine are also declined and components prioritized in terms of risk priority. Then, in Subsect. 3.2, once prioritized through the FMECA, each critical component has been evaluated to understand which data are required to be monitored, which type of loss (availability, performance, or quality) the critical components cause and which sensors can be embedded on the product to improve its smart functionalities.

**Table 1.** The research process

| Workshop | Duration | Aim |
|---|---|---|
| 1. Strategic and product portfolio analysis (interviews) [23] | 14 h | Gaps detection in the company from a digital servitization perspective and choice of Machine 1 as pilot case |
| 2. Servitization and digital transformation | 20 h | Introduction to the main concepts on PSS, Servitization transition, Digital transformation, Smart Connected Products and Technology Stack |
| 3. Digital provider demonstration | 6 h | Application of a technology provider solution for industrial machineries monitoring on Machine 1 and dashboarding configuration |
| 4. Presentation of strategic analysis results and FMECA theoretical session | 3 h | Presentation of the gaps in the company in terms of digital servitization. Focus on the need for the Technical Department to define the machine operating models through the introduction of sensors and the analysis of the data generated |
| 5. FMECA practical session | 6 h | Decomposition of Machine 1 and detection of the failure modes, causes and effects. Risk Prioritization Number (RPN) definition |
| 6. OEE theoretical session | 2 h | Introduction to performance measurement of production systems with a focus on OEE |
| 7. Combined FMECA- OEE practical session | 5 h | Possible activity to be done and sensors to be embedded, variables to be monitored, frequency of detection, unit of measure, standard interface, reference KPI, warning and fault rules |

### 3.1 FMECA on Machine 1

First of all, Machine 1 was decomposed in twenty main constructing groups (i.e. Pneumatic system, Flame treatment air and gas system, Electrical system, control actuators and signalling devices, Central and peripheral structure of the machine, Plateau, spindles and rotating air distributor/intake, Drying group (LED/Mercury), Deionizer group, Flame treatment group, Quality control group, Paddle group, Screen printing head unit, Tailstock (printing and vision system), Discharge belt, Loading conveyor, Good parts unloading, Unloading of scrap pieces, Piece loading and handling unit (pick and place), Electronic rear positioning, Front electronic positioning, Automatic mechanical search). Then, the machine constructing groups were exploded in more detailed levels of components, leading to obtain 229 single parts. Therefore, the FMECA was performed to detect the most critical components to be monitored to better understand and define the machine operating model and provide more effective product-related data-driven services. Three categories of failure modes were considered in the analysis (clogging, opacification, breakdown/malfunction) led by 18 failure causes (e.g., damage during format change, accidental damage or breakage, transmission hardening, clogging, component mortality, reduction of reflection or emission, dirt, overheating of the motor, usury) and bringing to five failure effects, and thus status of the machine (1. Discontinuity in power circuits/Disturbances on the signal circuit; 2. Transmission hardening, games, movement inaccuracy; 3. Downtime for suction not adequate on the spindles; 4. Downtime; 5. No Downtime). Based on these failure modes, causes and effects, the two indexes Probability (P, in which time frame the failure occurs) and Severity (S, duration and effects of the fault) were set (defining their scales reported in Table 2) and used to calculate the Risk Prioritization Number (RPN), i.e. the numeric assessment of risk assigned to a failure.

**Table 2.** FMECA: Probability and Severity scale for RPN calculation

| Probability | | Severity | |
|---|---|---|---|
| Scale | Description: in which time frame the failure occurs) | Scale | Description: duration and effects of the fault |
| 1 | >3 years | 1 | No downtime, no safety problems |
| 2 | 1 year < x <= 3 years | 2 | Downtime <= 1 h, no safety problems |
| 3 | 6 months < x <= 1 year | 3 | 1 h < downtime <= 8 h, no safety problems |
| 4 | 3 months < x <= 6 months | 4 | 8 h < downtime <= 16 h, no safety problems |
| 5 | 1 month < x <= 3 months | 5 | 16 h < downtime <= 1 week, no safety problems |
| 6 | <= 1 month | 6 | 1 week < downtime <= 1 month |
| – | – | 7 | Downtime > 1 month, no safety problems |

The matrix of the critical components for Machine 1 was obtained (Table 3) neglecting 57 components considered not relevant by the Technical Department for the purpose of the analysis. As a result, the matrix assessed 172 components (Fig. 1), evidencing that 1 component is classified as very important (bolditalic part in the bottom right of the matrix), 24 as moderately important (italic part at the centre of the matrix), 147 as ordinary (bold part at the top left of the matrix).

In particular, the aim of these tasks is to understand on which components it is better to act first to be able to improve the machine so that the data generated during its use phase could be better exploited. Indeed, the analysis started on the 16 most critical components, i.e., those in the italic and bolditalic zones of the matrix in Table 3.

**Table 3.** Criticality priority matrix of Machine 1 based on RPN

| P/S | 1 | 2 | 3 | 4 | 5 | 6 | 7 |
|-----|---|---|---|---|---|---|---|
| 1 | 9 | 12 | 42 | 54 | 0 | 15 | 8 |
| 2 | 1 | 13 | 1 | 5 | 2 | 4 | 0 |
| 3 | 0 | 0 | 3 | 0 | 0 | 1 | 0 |
| 4 | 0 | 0 | 1 | 0 | 0 | 1 | 0 |
| 5 | 0 | 0 | 0 | 0 | 0 | 0 | 0 |
| 6 | 0 | 0 | 0 | 0 | 0 | 0 | 0 |

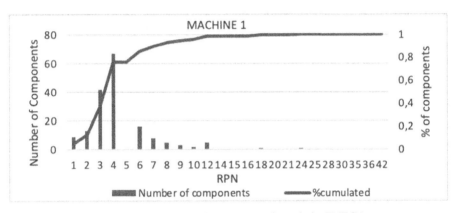

**Fig. 1.** Classification of components through the FMECA

## 3.2 Embedding the FMECA in the OEE Framework for Machine 1

Once prioritized the components, the concept of performance measurement of production systems was introduced to Company A, with a focus on the OEE framework. Therefore, the failures related to the Machine 1's critical components were declined by the Technical Department in terms of: which corrective actions are ongoing, which corrective actions could be introduced, which are the variables to be monitored (with which frequency of detection and with which unit of measure), which sensors can be embedded and which standard interface to use. Based on this analysis, each of these components was also declined in terms of OEE. That is to say that it was understood to which part of the OEE (Availability, Performance and Quality) each loss of time due to the critical components' failures analysed is related. Once defined the reference KPI, warning and fault rules have been envisioned to monitor the failures.

For example, the rotary table of the turret intermittent group is characterized by a dynamic mechanism that is controlled by the engine. The monitoring of this mechanism can provide important information to the machinery provider. Its break-down/malfunctioning is due to the mechanical wear of internal components and causes the transmission hardening, with the creation of play and movement inaccuracy. The company is already measuring the motor winding temperature and the motor absorption to try to control this component but applying the method also realizes that other actions can be done, as e.g., the refinement of the measurement of the brushless motor torque and the monitoring of vibrations. In addition, also test cycles on all brushless motors of the machine should be implemented to gather enough knowledge on its behaviour and fully understand its functioning model in relation to the different output to be produced. Indeed, the dynamic behaviour of an axis managed by the brushless motor has to be controlled. To do this, a temperature and torque detection should be revealed every 2 ms. The Technical Department operators also added that dealing with this component it is necessary to contextualize the results of torque/current/lag error/vibration within one production cycle. In addition, they also planned to generate template cycles to be used as reference to detect the status of the components monitored. Thus, vibration warning and fault ranges should be defined based on the studies to be performed in the future on the machine so that, based on the amount of $mm/ms^2$ observed, a warning and fault status could be detected by the machine and communicated to both the machinery user and provider. In terms of sensors to be embedded on the system, the electric current/torque are already detected by the drives of the motion control system. Using a real time field bus, new sensors should detect vibrations or temperatures in different points of the windings of the brushless motors. The intervention on such component would affect both the Availability and Performance of the machine's OEE. Finally, also the other critical components were analysed following this level of detail but are not reported here due to the page limit.

## 4   Discussion

The method embedding the FMECA in the OEE framework revealed to be effective and brought to relevant results in Company A. Indeed, it supported the company in exploring new solutions to avoid failures, in understanding how each component contributes to the machine's OEE reduction (and to its single indexes, Availability, Performance and Quality) and in enabling the delivery of new services. However, it also raised the need to perform further analysis to better define functioning models of certain components (that can beat the same time be supported by the method). In addition, the last step of this method should consist in the evaluation of the feasibility and convenience of both the physical (hardware/component) improvements to be done on the product and the data analysis to be implemented. The Machine A (Screen Printing) needs to be used only as a pilot case to discover how to apply the method and to understand how to start to enrich a traditional machinery with smart connected functions, making it more compliant with the customers' needs and expectations. Therefore, the results obtained on this machine can also be extended to the entire company's portfolio. Nevertheless, to implement the corrective actions on the machine defined with the application of the method, a database and a cloud platform (hardware and software architecture) need to be built by Company A, enabling the collection of the data envisioned as those useful to understand the functioning model of the machine and to provide more effective data-driven services to the customer during its usage. Finally, after the improvement of the product portfolio with smart functionalities and the creation of a database, a service department (able to deliver the data-driven services added to the machines) should be created in the company.

# 5   Conclusions

This paper presents a method embedding the FMECA in the OEE framework to initialize the development of a smart servitized solution. In particular, its application in Company A (on Machine 1 (Screen Printing)) was triggered by a previous strategic assessment of the company's criticalities in the digital servitization domain. First, the machine selected has been decomposed in its constructive groups and single components based on its bill of material. Then, the FMECA has been implemented, first declining each component in terms of failure modes, failure causes and failure effect and then prioritizing the components based on the RPN calculated through the P and S indexes. Therefore, components were prioritized based on the RPN and the most critical ones were assessed. Using the OEE framework, each loss of time due to the component failure has been analysed to evaluate which sensors could be embedded to improve the monitoring and control functionalities of Machine 1. Per each of them, several factors have been explored: which corrective actions are ongoing, and which could be introduced, which are the variables to be monitored (with which frequency of detection and with which unit of measure), which sensors can be embedded and which standard interface to use. Based on this analysis, each of these components was also declined in terms of OEE. That is to say that it was understood to which part of the OEE (Availability, Performance and Quality) each loss of time, due to the critical components' failures analysed, is related. Then, warning and fault rules have been defined to monitor components' failures. It must be said that the adoption of this method requires the creation of the entire technology stack (with a database and a cloud, that so far are missing in Company A) and the development of a service function able to deliver data-driven services. Finally, this research is not free from limitations. The method, applied to one single case, can be integrated in a unique methodology with the strategic analysis previously conducted in the company to detect the digital servitization hurdles. Then, the integrated method should be applied in a systematized way in other companies willing to pursue the digital servitization path.

# References

1. Porter, M.E., Heppelmann, J.E.: How smart, connected products are transforming competition. Harv Bus Rev. **92**, 64–89 (2014)
2. Bilgeri, D., Wortmann, F., Fleisch, E.: How digital transformation affects large manufacturing companies' organization. In: ICIS 2017 Transforming Society with Digital Innovation, pp. 91–99 (2018)
3. Baines, T., et al.: State-of-the-art in product-service systems. Proc. Inst. Mech. Eng. Part B J. Eng. Manuf. **221**(10), 1543–1552 (2007). [Internet]. Accessed 2014 Jul 11
4. Mont, O.: Introducing and developing a Product-Service System (PSS) concept in Sweden, IIIEE Reports 2001, 6, Lund, Sweden (2001)
5. Sassanelli, C., Rossi, M., Pezzotta, G., de Pacheco, D.A.J., Terzi, S.: Defining Lean Product Service Systems (PSS) features and research trends through a systematic literature review. Int. J. Prod. Lifecycle Manag. **12**(1), 37–61 (2019)
6. Paschou, T., Rapaccini, M., Adrodegari, F., Saccani, N.: Digital servitization in manufacturing: a systematic literature review and research agenda. Ind Mark Manag. **89**, 278–292 (2020)
7. Pirola, F., Boucher, X., Wiesner, S., Pezzotta, G.: Digital technologies in product-service systems: a literature review and a research agenda. Comput. Ind. **123**, 103301 (2020)

8. Gaiardelli, P., et al.: Product-service systems evolution in the era of Industry 4.0. Serv. Bus. **15**(1), 177–207 (2021). https://doi.org/10.1007/s11628-021-00438-9

9. Rüßmann, M., et al.: Industry 4.0: The Future of Productivity and Growth in Manufacturing Industries [Internet] (2015)

10. Klein, M., Biehl, S., Friedli, T.: Barriers to smart services for manufacturing companies – an exploratory study in the capital goods industry. J. Bus. Ind. Market. **33**, 846–856 (2018)

11. Porter, M.E., Heppelmann, J.E.: How smart, connected products are transforming companies. Harv Bus Rev. **93**(10), 96–114 (2015)

12. Adrodegari, F., Pashou, T., Saccani, N.: Business model innovation: process and tools for service transformation of industrial firms. Procedia CIRP **64**(June), 103–108 (2017)

13. Bustinza, O.F., Gomes, E., Vendrell-Herrero, F., Tarba, S.Y.: An organizational change framework for digital servitization: evidence from the Veneto region. Strateg. Change **27**, 111–119 (2018)

14. Raddats, C., Baines, T., Burton, J., Story, V.M., Zolkiewski, J.: Motivations for servitization: the impact of product complexity. Int. J. Oper. Prod. Manag. **36**(5), 572–591 (2016)

15. de Senzi Zancul, E., et al.: Business process support for IoT based product-service systems (PSS). Bus. Process. Manag. J. **22**(2), 305–23 (2016). [Internet]. Accessed 2017 Jan 31

16. Sassanelli, C., Fernandes, S.D.C., Rozenfeld, H., Da Costa, J.M.H., Terzi, S.: Enhancing knowledge management in the PSS detailed design: a case study in a food and bakery machinery company. Concurr. Eng. Res. Appl. 1–14 (2021)

17. Pezzotta, G., Sassanelli, C., Pirola, F., Sala, R., Rossi, M., Fotia, S., et al.: The Product Service System Lean Design Methodology (PSSLDM): integrating product and service components along the whole PSS Lifecycle. J. Manuf. Technol. Manag. **48**(2), 1270–1295 (2018)

18. Department of Defense Washington DC: Military Standard Procedures for Performing a Failure Mode, Effects and Criticality Analysis (1980)

19. Nakajima, S.: Introduction to TPM: Total Productive Maintenance [Internet]. Productivity Press, Inc. (1988)

20. Williamson, K.: Research Methods for Students, Academics and Professionals. Quick Print, Wagga Wagga (2002)

21. Ellström, P.-E.: Knowledge creation through interactive research: a learning perspective. In: HSS 2007 Conference, pp. 1–12 (2007)

22. Sassanelli, C., De Carolis, A., Terzi, S.: Initiating an industrial machinery producer to digital servitization: a case study. In: 18h IFIP WG 51 International Conference on Product Lifecycle Management, PLM 2021 (2021)

# Transformation of Manufacturing Firms: Towards Digital Servitization

Slavko Rakic[1]([✉]) [ID], Ivanka Visnjic[2] [ID], Paolo Gaiardelli[3] [ID], David Romero[4] [ID], and Ugljesa Marjanovic[1] [ID]

[1] Faculty of Technical Sciences, University of Novi Sad, Novi Sad, Serbia
slavkorakic@uns.ac.rs
[2] ESADE Business School, Ramon Llull University, Barcelona, Spain
[3] University of Bergamo, Bergamo, Italy
[4] Tecnológico de Monterrey, Mexico City, Mexico

**Abstract.** Digital technologies are disrupting servitization in manufacturing firms. In the last decade, manufacturing firms transform their business models from traditional offers of physical goods to digital solutions for their customers. In this paper, we investigate the transformation of digital servitization in manufacturing firms. We challenge relations between traditional and digital service portfolio offered by applying linear regression on the data obtained from 690 manufacturing firms from the Republic of Serbia from 2015 to 2020. The results show that firms significantly increase the offer of traditional services from 2015 to 2018. Moreover, results demonstrate a rapid growth of digital services in the period from 2018 to 2020. The application of traditional and digital services in manufacturing firms increased by 30% in five years.

**Keywords:** Servitization · Digital transformation · Product-related services · Digital services

## 1 Introduction

The growing *Service Economy* has changed the way manufacturing firms do their business [1], driving the implementation of *service business models*, as clearly evidenced by a growth of about 20% in the last decade [2, 3]. Also, the implementation of *digital technologies* in manufacturing firms has rapidly expanded due to the beginning of the Industry 4.0 era [4]. The combination of these two paradigms has transformed the traditional service business models into the *digitalized service business models* [5, 6]. Traditional product-related services such as maintenance, installation, revamping are often used by firms with low technology levels [7] while manufacturing firms with more research activities have more opportunity to get involved in the development of new digital services [8]. From this perspective, the literature provides an overview that *digital servitization* represents the implementation of digital technologies in the offer of product-related services [9]. In such a context, digital technologies such as Big Data

© IFIP International Federation for Information Processing 2021
Published by Springer Nature Switzerland AG 2021
A. Dolgui et al. (Eds.): APMS 2021, IFIP AICT 631, pp. 153–161, 2021.
https://doi.org/10.1007/978-3-030-85902-2_17

Analysis, Virtual Reality, the Internet of Things, influence the characteristics of new "smart services" [10]. However, the influence of traditional product-related services on the implementation of digital services is still neglected [8]. This gap in the literature makes a misunderstanding in the transformation process from "traditional" to "digital service" [10]. Although the literature presents digital technologies as triggers for digital servitization, the role of traditional services needs to be illuminated [10]. Accordingly, this paper aims to shed the light on how manufacturing firms transform their business from the offer of traditional services to the offer of digital services, and which traditional services influence the use of digital services. Hence, this led to the following research questions:

RQ1: *What is the implementation trend of traditional and digital services in manufacturing firms?*
RQ2: *To what extend traditional services influence the use of digital services?*

In line with the proposed research questions, Fig. 1 depicts the research framework.

**Fig. 1.** Research framework

To answer the two research questions, we used data from the European Manufacturing Survey (EMS) obtained from 690 firms from the Republic of Serbia. The results show that firms significantly increase the offer of traditional services from 2015 to 2018. Moreover, results demonstrate the rapid growth of digital services in the period from 2018 to 2020.

## 2 Literature Review

Through time, the application of traditional services has included a variety of technologies, which have contributed to a more comprehensive range of services [11]. The increased application of digital technologies in manufacturing firms influenced the implementation of digital services [12]. For instance, *product-oriented firms* are introducing digital technologies to increase product-service efficiency and value while changing processes and business models [13]. Industrial cases such as IBM, Piaggio, Canon, and Kone have shown how new business models and smart services can be delivered with high efficiency and effectiveness [14]. A precondition for the use of digital services is the existence of a *service ecosystem* in the firms, which plan to deliver these types of services [10]. The ecosystem must strive for the firms to provide systemic, dynamic, and contextual interaction between firms that offer digital services and their customer [15]. A *service ecosystem* includes both internal and external resources of the firm. *Internal organisational aspects* possess the ability to use internal resources to achieve a delivery strategy of traditional and digital services [10]. For instance, *internal resources* for digital servitization are digital technologies, organisation structures, or knowledge [1]. Previous studies show that along with digital technologies traditional product-related services could be drivers for digital servitization [8]. According to the previous research which investigates the role of product-related services in manufacturing firms, this research involves eight product-related services [7, 16]. The product-related services presented in this study could be divided into two groups. The first group consists of services that are closely related to product characteristics such as installation, maintenance and repair, design, consultation and planning, and take-back services [7, 16]. The second group includes services that are not closely related to the product such as training, remote support for clients, software development, and revamping and modernization [7, 16]. Gebauer [18] presented the *innovation potential* of product-related services, which could be drivers for the transformation of manufacturing firms. The first type of transformation is from product to service-oriented firms, and the second is from product-service systems to digitalized-service systems [13]. According to the digitalized service business models, this research involves digital services which are proposed from the Cambridge Service Alliance, EMS, and previous research that investigates digital servitization [18, 19]. *Digital services,* which are shown in this research, are examples of the application of digital technologies into traditional services [21]. Digital services are different from traditional since the marginal cost of digital services is smaller than that of traditional, and they are substitutes for traditional products [9]. The relationship between traditional and digital services is only presented in a way how they affect the financial performance of firms [20, 21]. However, the relationship of the transformation process from product-related to digital services remains neglected [8]. In such a context, this study examines the influence of product-related traditional services on the use of digital services.

## 3   Methodology

Data for this empirical study derive from the European Manufacturing Survey [24]. The objective of this regular, triennial questionnaire is to systematically monitor the innovation behaviour of European manufacturers at the firm level. The final sample comprises 690 manufacturing firms from the Republic of Serbia, operating during the period 2015–2020. Concerning the descriptive statistics, the sampled firms report, on average, a company size of 230 employees (SD = 90.2). In total, 266 companies are small firms (less than 50 employees), 318 companies are medium-sized (between 100 and 249 employees), and 106 firms are large enterprises (more than 250 employees). The largest industry in the sample is the Manufacture of Food Production (NACE 10) with 17.1%, followed by the Manufacture of Electrical Equipment (NACE 27) and Manufacture of Machinery and Equipment n.e.c. (NACE 27) with 8.2%. In third place are the Manufacture of Rubber and Plastic products (NACE 22), Manufacture of Basic Metals (NACE 24), Manufacture of Motor Vehicles, Trailers and Semi-trailers (NACE 29), and

Manufacture of Furniture (NACE 31) with 5.3%. The remaining manufacturing sectors together have 45.3% of the total sample. According to the classification of UNIDO for developing countries [25], the sample can be divided into three groups of technology intensity: low technology intensity firms constitute 47.1%% of the total sample, while 20.6%% and 32.4%% of the sample consist of medium- and high-technology-intensive companies respectively. To analyse the impact of product-related services on the use of digital services, the authors adopt linear regression. The dependent variable is the use of digital services, and the independent variables are product-related services shown in the research framework depicted in Fig. 1. The dependent variable is presented as a latent variable of the average values from digital services used per firm. According to the previous research on digital servitization, the industry sector is used as the control variable [26]. To prevent the effect of the external factors on the relationships between traditional and digital services The NACE Rev 2.2 classification was used to define the firm's sector.

## 4  Results and Discussion

Figure 2 and Fig. 3 present the usage trend of traditional and digital services in the manufacturing sector of the Republic of Serbia in the period from 2015 to 2020. In particular, in Fig. 2, eight product-related services are presented: Installation (PR1), Maintenance and Repair (PR2), Training (PR3), Remote Support for Clients (PR4), Design, Consulting, Project Planning (PR5), Software Development (PR6), Revamping (PR7), and Take-back Services (PR8).

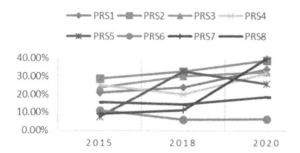

**Fig. 2.**  Use of product-related services in manufacturing firms

Figure 3 depicts the trends of the application of digital services – Digital Services for Product Utilization (DS1), Digital Services for Customized Product Configuration or Product Design (DS2), Digital Monitoring of Operating Status (DS3), Mobile Devices for Diagnosis, Repair or Consultancy (DS4), and Data-based Services based on Big Data Analytics (DS5).

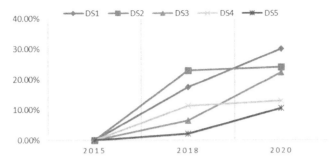

**Fig. 3.** Use of digital services in manufacturing firms

Table 1 depicts the use of traditional and digital services in the manufacturing firms according to the technology level of the firm.

**Table 1.** The use of services according to the technology level of the firm

| Services | Technology intensity | | |
|---|---|---|---|
| Tradition/Digital | Low-tech firms (%) | Med-tech firms (%) | High-tech firms (%) |
| Installation, start-up | 31 | 71 | 47 |
| Maintenance and repair | 33 | 74 | 51 |
| Training | 26 | 60 | 40 |
| Remote support for clients | 15 | 34 | 24 |
| Design, consulting, project planning | 20 | 46 | 31 |
| Software development | 6 | 14 | 9 |
| Revamping | 25 | 57 | 36 |
| Take-back services | 1 | 3 | 2 |
| Digital services for product utilization | 20 | 46 | 29 |
| Digital services for customized product configuration or product design | 13 | 29 | 22 |
| Digital monitoring of operating status | 20 | 46 | 29 |
| Mobile devices for diagnosis, repair or consultancy | 6 | 14 | 9 |
| Data-based services based on big data analytics | 5 | 11 | 7 |

Result analysis reveals that manufacturing firms increase the use of traditional services over the years (Fig. 2). In particular, Installation (PR1), Maintenance and Repair (PR2), Training (PR3), and Revamping (PR7) progressive increase their use in all three research rounds. Furthermore, Remote Support for Clients (PR4), Design, Consulting, Project Planning (PR5), and Take-back Services (PR8) have progressed with some oscillation in 2018, while Software Development (PR6) is the only service that has declined over the years. Moreover, the results show that manufacturing firms increase the use of digital services over the years, especially in the period from 2018 to 2020: DS1, DS3, and DS5 have the highest level of usage increase in the manufacturing sector, while DS2 and DS4 are increased in the period from 2015 to 2018 but stays on a similar level in 2020. Finally, as shown in Table 1, the same trend of the use of services is observed in the firms with low, medium, and high technological intensity, despite their percentage difference. In other words, the achieved results underline that the same types of services have the largest share regardless of the percentage of use and type of technological intensity.

In conclusion, referring to RQ1 (*What is the implementation trend of the traditional and digital services in manufacturing firms?*) it can be stated that the process of servitisation had two significant phases. After a first phase during which firms have significantly increased their offer of traditional product-related services (from 2015 to 2018), an expansion of digital services began (from 2018 to 2020). Moreover, regardless of their origin, the application of traditional and digital services in manufacturing firms increased by 30% in the period from 2015 to 2020.

Table 2 reports the main effects of the linear regression model, used to test RQ2 and where the regression coefficients for the independent variables reflect the influence on the dependent.

**Table 2.** Results of the linear regression

| Product-related services | Model parameters |
|---|---|
| Industry sector | .258 |
| Installation, start-up | .087 |
| Maintenance and repair | −.029 |
| Training | .304*** |
| Remote support for clients | .121* |
| Design, consulting, project planning | .057 |
| Software development | .278** |
| Revamping | .158* |
| Take-back services | .070 |
| R | .610 |
| $R^2$ | .370 |
| Sig. | .000 |

In the regression model, the overall model is significant, $R^2 = .370$, $p < .001$. Four predictors had significant coefficients – training (B = .304, p < 0.001), software development (B = .278, p < 0.01), revamping (B = .158, p < 0.1), and remote support for clients (B = .121, p < 0.1), thus supporting the idea to include these product-related services in the service portfolio to increase the use of digital services. Nevertheless, installation, maintenance and repair, design, consulting and project planning, and take-back services show no statistically significant effect on the use of digital services. Moreover, results show that the control variable does not make a significant effect on model construction.

Outcomes of the linear regression depict the results for the RQ2: *To what extend traditional services influence the use of digital services.* The four from eight product-related services show a positive effect on the use of digital services, providing support for RQ2.

## 5  Conclusions

This research investigates what is the trend in the implementation of *digital services* and how traditional services influence the use of digital services in manufacturing firms. Therefore, this study provides theoretical and practical implications for how manufacturing firms could employ digital services. The empirical results show that traditional services, which are not closely related to product characteristics significantly, influence the use of digital services. These findings fill the gaps in the literature in the transformation process from traditional product-service systems to digitalized product-service systems. Moreover, these results show that product-related services are drivers of digital servitization along with digital technologies. Additionally, results show what is a trend in the implementation of traditional and digital services and how managers of manufacturing firms could combine traditional with digital services. Findings show that from the practical side manufacturing firms have two ways to involve servitization in their firms. One side is like a traditional servitization with services that are closely related to products (e.g., installation, maintenance). On the other side, they could involve digital technologies with traditional services that are not related to the product to involve digital servitization in manufacturing firms.

This study adds to the understanding of the implementation of service business models and transformation towards digital servitization. We take an initial step toward formulating a model that simultaneously analyses the impact of product-related services on a business model based on digital services. As such, we extend the empirical work of Sklyar et al. [10] and find that product-related services that are not closely related to product characteristics influence the use of digital services. Our findings advise manufacturing firms to provide training, software development, revamping or modernization, and remote support for clients to increase the use of digital services and catch up with the current trend [2, 3]. Additionally, research results show that services that are closely related to products, such as installation, maintenance and repair, design, consulting and project planning, and take-back services are inhibitors of the implementation of digital services in the processing sector. The results of the research complement the existing research that examines the drivers of digital servitization [1, 8].

This study is limited only to datasets considers in the Republic of Serbia. In such a context, further research could consider datasets from the other members of the EMS consortium to show the wider framework of digital servitization. For further research, authors could involve interviews with experts from practice to finds more drivers for digital servitization, which are not based on digital technologies or product-related services. Additionally, further research is necessary to estimate the challenges of firms over the years. With this information production managers could find how firms change their offer of digital services. The development of these ideas could be especially useful for manufacturing firms facing the challenges of "digital servitization".

# References

1. Paschou, T., Rapaccini, M., Peters, C., Adrodegari, F., Saccani, N.: Developing a maturity model for digital servitization in manufacturing firms. In: Anisic, Z., Lalic, B., Gracanin, D. (eds.) IJCIEOM 2019. LNMIE, pp. 413–425. Springer, Cham (2020). https://doi.org/10.1007/978-3-030-43616-2_44
2. Neely, A., Benedetinni, O., Visnjic, I.: The servitization of manufacturing: further evidence. In: 18th European Operations Management Association Conference, Cambridge, pp. 1–9 (2011)
3. Mastrogiacomo, L., Barravecchia, F., Franceschini, F.: A worldwide survey on manufacturing servitization. Int. J. Adv. Manuf. Technol. **103**(9–12), 3927–3942 (2019)
4. Lalic, B., Rakic, S., Marjanovic, U.: Use of Industry 4.0 and organisational innovation concepts in the Serbian textile and apparel industry. Fibres Text. Eastern Eur. **27**(3), 10–18 (2019)
5. Gaiardelli, P., Songini, L.: Successful business models for service centres: an empirical analysis. Int. J. Prod. Perform. Manage. 1–26 (2021)
6. Romero, D., Gaiardelli, P., Pezzotta, G., Cavalieri, S.: The impact of digital technologies on services characteristics: towards digital servitization. In: Ameri, F., Stecke, K.E., von Cieminski, G., Kiritsis, D. (eds.) APMS 2019. IAICT, vol. 566, pp. 493–501. Springer, Cham (2019). https://doi.org/10.1007/978-3-030-30000-5_61
7. Bikfalvi, A., Lay, G., Maloca, S., Waser, B.R.: Servitization and networking: large-scale survey findings on product-related services. Serv. Bus. **7**(1), 61–82 (2013)
8. Marjanovic, U., Lalic, B., Medic, N., Prester, J., Palcic, I.: Servitization in manufacturing: role of antecedents and firm characteristics. Int. J. Eng. Manage. **2**, 133–144 (2020)
9. Marjanovic, U., Rakic, S., Lalic, B.: Digital servitization: the next "big thing" in manufacturing industries. In: Ameri, F., Stecke, K.E., von Cieminski, G., Kiritsis, D. (eds.) APMS 2019. IAICT, vol. 566, pp. 510–517. Springer, Cham (2019). https://doi.org/10.1007/978-3-030-30000-5_63
10. Sklyar, A., Kowalkowski, C., Tronvoll, B., Sörhammar, D.: Organizing for digital servitization: a service ecosystem perspective. Bus. Res. **104**, 450–460 (2019)
11. Rabetino, R., Harmsen, W., Kohtamäki, M., Sihvonen, J.: Structuring servitization-related research. Int. J. Project Org. Manage. **38**(2), 350–371 (2018)
12. Paschou, T., Rapaccini, M., Adrodegari, F., Saccani, N.: Digital servitization in manufacturing: a systematic literature review and research agenda. Ind. Mark. Manage. **89**, 278–292 (2020)
13. Lerch, C., Gotsch, M.: Digitalized product-service systems in manufacturing firms. Res. Technol. Manag. **58**(5), 45–52 (2015)
14. Ardolino, M., Rapaccini, M., Saccani, N., Gaiardelli, P., Crespi, G., Ruggeri, C.: The role of digital technologies for the service transformation of industrial companies. Int. J. Prod. Res. **56**(6), 2116–2132 (2018)
15. Edvardsson, B., Tronvoll, B., Gruber, T.: Expanding understanding of service exchange and value co-creation: a social construction approach. J. Acad. Market. Sci. **39**(2), 327–339 (2011)
16. Marjanovic, U., Lalic, B., Majstorovic, V., Medic, N., Prester, J., Palcic, I.: How to increase share of product-related services in revenue? Strategy towards servitization. In: Moon, I., Lee, G.M., Park, J., Kiritsis, D., von Cieminski, G. (eds.) APMS 2018. IAICT, vol. 536, pp. 57–64. Springer, Cham (2018). https://doi.org/10.1007/978-3-319-99707-0_8
17. Rakic, S., Simeunovic, N., Medic, N., Pavlovic, M., Marjanovic, U.: The role of service business models in the manufacturing of transition economies. In: Lalic, B., Majstorovic, V., Marjanovic, U., von Cieminski, G., Romero, D. (eds.) APMS 2020. IAICT, vol. 592, pp. 299–306. Springer, Cham (2020). https://doi.org/10.1007/978-3-030-57997-5_35

18. Gebauer, H., Krempl, R., Fleisch, E., Friedli, T.: Innovation of product-related services. Manag. Serv. Qual. **18**(4), 387–404 (2008)
19. Zaki, M.: Digital transformation: harnessing digital technologies for the next generation of services. J. Serv. Mark. **33**(4), 429–435 (2019)
20. Rakic, S., Pavlovic, M., Marjanovic, U.: A precondition of sustainability: Industry 4.0 readiness. Sustainability **13**(12), 6641 (2021). https://doi.org/10.3390/su13126641
21. Zivlak, N., Rakic, S., Marjanovic, U., Ciric, D., Bogojevic, B.: The role of digital servitization in transition economy: an SNA approach. Tehnicki vjesnik Tech. Gazette **26**(8), 10 (2021)
22. Visnjic, I., Van Looy, B.: Servitization: disentangling the impact of service business model innovation on the performance of manufacturing firms. SSRN J. (2012). 10.2139/ssrn.2117038
23. Kohtamäki, M.: The relationship between digitalization and servitization – the role of servitization in capturing the financial potential of digitalization. In: Technological Forecasting, p. 9 (2020)
24. Jäger, A.: European Manufacturing Survey 2021. https://www.isi.fraunhofer.de/en/themen/industrielle-wettbewerbsfaehigkeit/fems.html
25. United Nations Industrial Development Organization: Classification of Manufacturing Sectors by Technological intensity (ISIC Revision 4). https://stat.unido.org/content/focus/classification-of-manufacturing-sectors-by-technological-intensity-%2528isic-revision-4%2529;jsessionid=4DB1A3A5812144CACC956F4B8137C1CF
26. Martín-Peña, M.L., Sánchez-López, J.M., Díaz-Garrido, E.: Servitization and digitalization in manufacturing: the influence on firm performance. J. Bus. Ind. Market. **35**(3), 564–574 (2019)

# Service Shop Performance Insights from ERP Data

Shaun West[1]([✉]) [ID], Daryl Powell[2,3] [ID], and Ille Fabian[1] [ID]

[1] School of Technology and Architecture, Lucerne University of Applied Sciences and Arts,
Lucerne, Switzerland
{shaun.west,fabian.ille}@hslu.ch
[2] Department of Industrial Ecosystems, SINTEF Manufacturing AS, Horten, Norway
daryl.powell@sintef.no
[3] Department of Economics and Technology Management, Norwegian University of Science
and Technology, Trondheim, Norway

**Abstract.** Enterprise Resource Planning (ERP) systems offer firms a wealth of readily available transactional data. However, deriving insights from such data often demands the examination of multiple issues simultaneously. In this paper we use simple data mining to analyze ERP data from 27 service shops over a period of 35 months. The data has been used to provide valuable business performance insights to the service shop managers. Though the granular ERP data needed to be supplemented by further data in some instances, we found it has the potential to provide real insights into a firm's performance. Such simple data mining approaches can be standardized and automated across service centers for insights that can be used to drive continuous improvement activities within and across sites. We also suggest that this initial, exploratory study opens exciting avenues for further research into business analytics and, business intelligence pipelines.

**Keywords:** Data mining · Enterprise Resource Planning · Business performance · Service center

## 1 Introduction

This exploratory study examines the improvement of service workshop operations management using existing Enterprise Resource Planning (ERP) data, and what insights could support decision-making processes. The authors were given a set of data from the ERP system of a service provider who repairs a range of industrial equipment. The company has provided these services for over 60 years, yet the service shop managers only had the ERP data in the form of a monthly cash flow statement. The system was set up to provide monthly financial statements based on the transactions logged in the ERP system, and workshop managers and the operations lead confirmed that they were using it to help with planning and control. This was outside the ERP system's initial requirements and outside the finance department's direct control, apparently being a grassroots initiative.

© IFIP International Federation for Information Processing 2021
Published by Springer Nature Switzerland AG 2021
A. Dolgui et al. (Eds.): APMS 2021, IFIP AICT 631, pp. 162–171, 2021.
https://doi.org/10.1007/978-3-030-85902-2_18

Prior studies [1–4] confirmed that ERP data provides a log of business transactions as a basis for financial reporting, and firms such as SAP have invested in tools like HANA to extract insights from data. Studies have considered the data for decision support within businesses, in the area of customer service support [5], data mining for business analytics [6], as well as knowledge assimilation and more advanced decision support [7, 8]. Others have considered manufacturing firms or pure service businesses, but not an industrial repair and overhaul business that provides both workshop and field-based services. For these reasons, the research question for this paper is: *"what service center insights can be learnt from data mining of ERP systems?"*.

## 2  Literature Review

The research question examines benchmarking and decision-making to support management planning, control, and performance in the context of operational excellence [9]. Using ERP data for operations management has been identified [10] as a strategical and a tactical approach in manufacturing firms and made-to-order businesses [11, 12].

Data mining can [13] reveal insights within a firm, and has been used in many different business areas, within the sales function, customer support and manufacturing [1, 3–5]. Increasingly, it is now being used for forward-looking business analytics [6], coupled with machine learning and other techniques. Studies [2, 14–16] describe using ERP systems to support lean or continuous improvements within firms in different manufacturing contexts, including in SMEs similar in size to typical service workshops. The maturity model [16] provides a five-level model with examples showing how and where ERP systems can support pull production in firms.

Benchmarking can improve organization performance by comparison with others, as described in the literature [17]. Often based on external benchmarks, it can be used internally to consider cross-business performance and used to support knowledge management and continuous improvement [18]. With the use of business analytics systems based on ERP data [19–21] further insights can be gleaned [7].

Decision-making is enhanced when data is presented in a form such as clear visual representation [17], that allows a team to assimilate it into knowledge and deliver actionable decisions and forecasts [23, 24]. ERP data can and should be used to support decision-making [8].

## 3  Methodology

Statistical analysis of an ERP data set of monthly cash flow and invoicing data from 27 service centers in one region. The data were provided as flat CSV files to be imported into MS Excel, SPSS, or Microsoft Power BI. Additional contextual management data to support the analysis was collected separately and additional tables were created. The steps applied to the raw ERP data were: data cleaning; collecting contextual information for the ERP data set; structuring the flat data to the structure of the business and its sub-business units; exploring the invoicing data and monthly cashflows.

# 4   Results

Around 3MB of raw data from 27 service shops were collected in a CSV file (later moved into Excel) over a 35-month period. Billing captured 40,000 lines of data whereas monthly cashflows provided 39 lines of data per month. Tables 1 and 2 provide an overview of the data fields.

**Table 1.**  Billing data fields

| | |
|---|---|
| Location code | Workshop location and sub-business unit |
| Customer code | The code for the customer |
| Zip code | The delivery Zip code |
| Net sales value | The value on the invoice |
| Net cost | The costs associated with the work |

**Table 2.**  Monthly cashflow data fields

| Data field | Data field | Data field |
|---|---|---|
| Jobs sold | CM/sales | Amortization |
| Hours sold | Overhead costs | EBIT |
| External sales | Overhead recovery | Interest charges |
| Internal sales | Debt provisions | Profit before tax (PBT) |
| Total sales | Gross profit | BPT/Sales |
| Cost of goods sold | Gross profit/sales | WIP provisions |
| Labor costs | Admin costs | Bad debt provisions |
| Under absorption cost | Sales costs | Pre interest PBT |
| Employee social costs | Distribution costs | Internal rechanges |
| Material costs | Mgt overheads | Pre-exceptional PBT |
| Job expenses | Rent (nominal) | Exceptional costs |
| Provisions | EBITDA | PBT contribution |
| Contribution margin | Depreciation | No of Employees |

## 4.1   Initial Analysis in Excel for Data Cleaning

The data was assessed in Excel for correction and validation and was generally consistent, though there were issues with the invoicing data, where the calculated margins could be very large (both negative and positive outliers). With no apparent reason for this, the outliers were removed for the analysis in this study. Discussions with the local individual workshop managers confirmed that there were quality issues with the data as often invoices were issued without them being linked directly to a repair project. This came about mostly due to additional repairs needed or to invoicing separately for out-of-scope items. At times invoicing was also used as a price adjustment mechanism, breaking a clear link between the cost of goods sold and the invoice value and creating a quality issue with the data. When analyzing this type of data is necessary to understand its contextual meaning.

The monthly cashflow data was used in the management control processes and every branch manager was sent a month end cash flow (Fig. 1). The sheet had evolved over time, allowing branch managers to talk knowledgably with the head of the operations, the business units and other branch managers, as well as with the finance departments. All branch managers were used to using the figures and could take meaningful actions based on the sheets.

| | Jul-07 Actual | Aug-07 Actual | Sep-07 Actual | Oct-07 Actual | Nov-07 Actual | Dec-07 Actual | Jan-08 Actual | Feb-08 Actual | Mar-08 Actual | Apr-08 Actual | May-08 Actual | Jun-08 Actual | 2007/ 2008 |
|---|---|---|---|---|---|---|---|---|---|---|---|---|---|
| JOBS SOLD | 46 | 39 | 41 | 49 | 39 | 31 | 54 | 51 | 34 | 57 | 41 | 34 | 516 |
| HOURS SOLD | 2,331 | 1,901 | 1,682 | 2,719 | 2,758 | 1,442 | 1,934 | 1,939 | 1,782 | 2,279 | 1,908 | 1,333 | 24,008 |
| EXTERNAL SALES | 73 | 83 | 80 | 109 | 101 | 118 | 78 | 80 | 69 | 146 | 63 | 59 | 1,059 |
| INTERNAL SALES | 20 | 6 | 5 | 2 | 4 | 4 | 4 | 5 | 2 | 9 | 6 | 5 | 72 |
| TOTAL SALES | 93 | 89 | 86 | 110 | 105 | 122 | 82 | 85 | 71 | 155 | 69 | 65 | 1,132 |
| COST OF SALES: | | | | | | | | | | | | | |
| LABOUR COSTS | 27 | 21 | 20 | 32 | 34 | 17 | 22 | 22 | 20 | 25 | 21 | 15 | 277 |
| NON-PROD LABOUR | (3) | (1) | (1) | (4) | (3) | 1 | 0 | (0) | 3 | (0) | 2 | 1 | (6) |
| WORKS NI & SUPN | 3 | 3 | 2 | 3 | 3 | 3 | 2 | 3 | 3 | 3 | 3 | 3 | 31 |
| MATERIAL COSTS | 15 | 24 | 7 | 18 | 18 | 57 | 27 | 11 | 13 | 56 | 11 | 9 | 266 |
| JOB EXPENSES | 0 | 0 | 0 | 0 | 0 | 0 | 0 | 0 | 0 | 0 | 0 | 0 | 0 |
| ADD BACK WIP PROV. MOVE. | 6 | 0 | 0 | 0 | 0 | 0 | 0 | 0 | 0 | 0 | 0 | 0 | 0 |
| MAN. OP | 51 | 42 | 57 | 61 | 53 | 45 | 30 | 49 | 52 | 71 | 33 | 38 | 563 |

**Fig. 1.** Extract of the data provided to the managers from the monthly cashflows

## 4.2   Collection and Integration of Contextual Information

Initial analysis confirmed contextual information was missing: the floor space for each workshop and the hierarchical relationship with the business (e.g., its sub-business unit, based on business model, not location). Contextual information supports the wider interpretation of the data and helps to frame the data within the larger picture. The information was stored on a separate system within the finance department.

## 4.3   Structuring of the Flat Data

The business structure was used to create a synthetic consolidation of the invoice data and the cashflows. This created business level, and business unit perspectives of the data sets and individual branches. The data collected did not provide the consolidation as a business unit or business level on the same basis as the data from the individual branches. For this reason, a synthetic consolidation was needed, although this may have not fully considered the intra-business trading fully. This confirms that data cleansing to remove artifacts is an important activity where more focus should be given.

## 4.4  Exploration of the Invoicing Data

The invoicing data was explored using a set of scatter plots before moving to box plots, which offer more insights than the final scatter plots. Here, benchmarking shows high variance in both the size of individual jobs and the margins per job. Figure 2 shows the breakdowns from the business to the BU and finally to location 1 (LO01), in effect allowing consolation that was created as part of the structing of the flat data. Location 17 (LO17) has a Gross Profit (GP) margin that is slightly lower than the average for Business Unit (BU) B it has long tails, whereas LO19 with lower variance provided a more reliable project Gross Profit. Improvement in individual service shops' performance (e.g., no invoices with less than zero GP) could improve the firm's margin overall. The boxplots for the individual locations from on BU allow a comparison for each individual location and provides the comparisons with the other BUs as well as the business in aggregate.

Noise in the data from the scatter plots, required additional contextual information to understand the challenges with the data quality. Discussions confirmed that there were many invoices created where they were used for billings without matching the costs directly with them. Interestingly, anomaly data appears to be associated with the smaller invoice values and a filter applied at ±80% GP margin may provide a pragmatic approach to data cleansing.

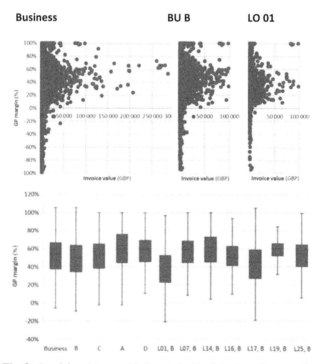

**Fig. 2.** Invoicing data provided statistical insights into the operations

Analysis of the customer base confirmed that the top ten customers on the consolidated business or at the business-unit level contributed less than 10% to the total sales volume (Fig. 3) but, when drilling down to individual service centers, the top ten customers often contributed 30% or more of the total sales, so the loss of two key accounts could create challenges for a service shop. This reliance on large orders is an insight into the underlying business model, as well as the risk of missing targets due to failing to win a large order. The firm focused on the overall business reliance on single customers and large orders rather than undertaking a detailed customer analysis at the business unit level providing new insights for the managers.

The use of customer post codes helped to visualize the geographic distribution of the customer base. However, we assumed that the invoice address was the ship-to address, but the data showed this was not always the case. Discussions within the firm confirmed that there could be problems with confusions between ship-to and bill-to addresses. This suggests a problem with both data quality and data structure.

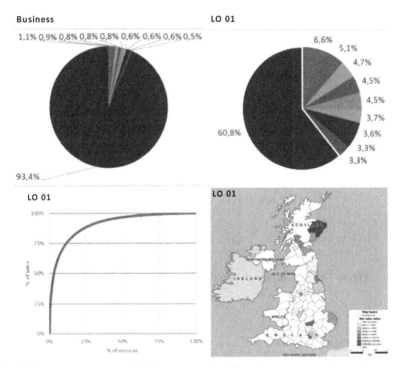

**Fig. 3.** Different perspectives of business sustainability (customers, project size, customer locations)

## 4.5   Exploration of the Monthly Cashflows

Trend data were plotted again at the three different levels of the firm (Fig. 4). In the past, only the spreadsheet had been used, but visuals presented the data in a more actionable form. The upper charts show data on a rolling 18-month basis, rather than a financial year basis, and lean approaches were used to convert the charts into control charts with a regression trend line to provide a form of forecast. The basic regression model was used. There was only a basic understanding of statistics, and the approach is a early approach with time-series data and commonly applied within firms with a lean/six-sigma improvement program. The upper and lower control lines were based on the ±2 sigma around the mean. The middle charts show the number of booked working hours in the month with LO 01 showing significant variance month-on-month. Compared with monthly under/over absorption, the weakness in management control was clear, as shown in the lower chart. Sales/hours show increases in value added and reflect the knowledge intensity, whereas costs/hours show the change in productivity. Other combinations of data (e.g., ROS%, Jobs Sold, Jobs sold/hours sold) were tested to understand the dynamics of the business. In several workshops there was under-absorption of labor one month and over-absorption the next, to the detriment of the ROS (in percentage and in absolute terms). These data could provide the basis for forecasting.

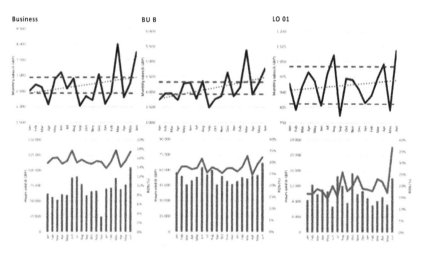

**Fig. 4.** Different perspectives providing different insights into the data

## 4.6  Benchmarking Metrics

To understand if the business is based on spares (materials), sales or labor, the materials cost/COGS was plotted on a monthly basis. Significant differences could be found between locations and business units, reflecting different business models. For example, one location was an outlier with high sales per square meter. Discussions with the business confirmed that the branch's sales contained a large volume of field service sales significantly greater than most locations.

# 5  Discussion

Our research question was *"what service center insights can be learnt from data mining of ERP systems?"* Drawing on ERP data from 27 service centers, we combined supplementary data and applied data mining to derive useful and useable insights. Using ERP data as a source was found to agree with the literature [10–12]. However, the use of visualizations to support decision-making, as seen in this study, was missing from the firm's approach. Data mining gave new insights into the firm's performance and again agrees with the literature [3–5]. What was missing was translating the financial data into actionable visual insights that could be used on an operational and tactical basis in the service shops. Also missing was an attempt to draw forecasts from the insights, or to use the insights as a benchmarking tool to support understanding of business performance [14, 16, 17].

## 5.1  Managerial Relevance

The approach moves away from traditional financial reporting and into the domain of business intelligence and analytics. Using ERP data integrated with contextual insights and additional information (ie., floor areas, employees etc.), lessons can be learnt and shared within the service network. This should improve performance, becoming the basis of continual improvement, and decision support (operational, tactical and strategic). Implementing such a tool needs careful consideration to ensure its usability and management use. This study confirmed that data collection (transactional data), derived parameters (hypercube dimensions) and insights should be structured around the individual operational locations and business units, as well as the overall business, so relevant information reaches different managerial levels to support business decisions. Doing so it became possible to gain new insights into the business.

## 5.2  Academic Implications

There are too few published examples of performance management of service business. Much of the available data is either at too high a level to allow detailed analysis or has critical data missing that prevents analysis. The data here confirms that granular ERP data, with some supplementary data, can provide real insights into a firm's performance. The lessons learnt from the analyses demand further investigation and could provide the basis for sharing lessons and experiences within a service business. Statistical-based models could be developed and integrated with business intelligence and analytics solutions to provide forecasting capabilities, based on the business structures that the data has described. The models could support decision-making on operational, tactical, and strategic levels.

# 6  Conclusions and Future Research

The analysis of the ERP data gave the firm new insights into the business's performance, and more insights could be gleaned when additional data was added to provide new KPIs.

The reports could be generated automatedly and shared monthly to support the branch managers, the business units, and the whole firm. The process could be represented as OLAP cubes to provide real-time business intelligence. Transforming insights into a visual form can offer actionable information that can support tactical and strategic decisions. For instance, a wide difference between GP% and the ROS% could identify a service center location where prices were maintained, albeit with a low total sales volume. Additional market data outside of the ERP would be required to confirm this.

This is an initial study and has not taken full account of the literature, the potential for operational improvements, or improved decision support. Advanced analytics has not been applied to the datasets. The possibility remains to create forecasts from the data and business simulations. The usability of the original uncleaned data set and the traditional management reports were not investigated in this study, nor were the use and useability of the new insights in terms of their support to the decision-making process. This should be investigated further, as should their integration into existing processes to ensure the impact within the business. There are therefore limitations to the study that demand further investigations in these areas.

# References

1. Cohen, M.C.: Big data and service operations. Prod. Oper. Manage. (2018). https://doi.org/10.1111/poms.12832
2. Powell, D., Alfnes, E., Strandhagen, J.O., Dreyer, H.: The concurrent application of lean production and ERP: towards an ERP-based lean implementation process. Comput. Ind. (2013). https://doi.org/10.1016/j.compind.2012.12.002
3. Kusiak, A.: Data mining: Manufacturing and service applications. Int. J. Prod. Res. (2006). https://doi.org/10.1080/00207540600632216
4. Berry, M.J.A., Linoff, G.S.: Data mining techniques: for marketing, sales, and customer relationship management. Portal.Acm.Org (2004)
5. Hui, S.C., Jha, G.: Data mining for customer service support. Inf. Manage. (2000). https://doi.org/10.1016/S0378-7206(00)00051-3
6. Ledolter, J.: Data mining and business analytics with R. Data Min. Bus. Anal. R (2013). https://doi.org/10.1002/9781118596289
7. Elbashir, M.Z., Collier, P.A., Sutton, S.G., Davern, M.J., Leech, S.A.: Enhancing the business value of business intelligence: the role of shared knowledge and assimilation. J. Inf. Syst. (2013). https://doi.org/10.2308/isys-50563
8. Holsapple, C.W., Sena, M.P.: EP plans and decision-support benefits. Decis. Support Syst. (2005). https://doi.org/10.1016/j.dss.2003.07.001
9. Ivanov, D., Tang, C. S., Dolgui, A., Battini, D., Das, A.: Researchers' perspectives on Industry 4.0: multi-disciplinary analysis and opportunities for operations management. Int. J. Prod. Res. (2020). https://doi.org/10.1080/00207543.2020.1798035
10. Chiarini, A., Vagnoni, E.: Strategies for modern operations management. Benchmarking Int. J. (2017). https://doi.org/10.1108/bij-11-2015-0115
11. Aslan, B., Stevenson, M., Hendry, L.C.: Enterprise resource planning systems: an assessment of applicability to make-to-order companies. Comput. Ind. (2012). https://doi.org/10.1016/j.compind.2012.05.003
12. Kakouris, A.P., Polychronopoulos, G.: Enterprise resource planning (ERP) system: an effective tool for production management. Manage. Res. News (2005). https://doi.org/10.1108/01409170510784878

13. Larose, D.T.: Discovering Knowledge in Data: An Introduction to Data Mining (2005). https://doi.org/10.1002/0471687545
14. Powell, D., Alfnes, E., Strandhagen, J.O., Dreyer, H.: ERP support for lean production. In: Frick, J., Laugen, B.T. (eds.) APMS 2011. IAICT, vol. 384, pp. 115–122. Springer, Heidelberg (2012). https://doi.org/10.1007/978-3-642-33980-6_14
15. Powell, D.: ERP systems in lean production: new insights from a review of lean and ERP literature. Int. J. Oper. Prod. Manage. (2013). https://doi.org/10.1108/IJOPM-07-2010-0195
16. Powell, D., Riezebos, J., Strandhagen, J.O.: Lean production and ERP systems in small- and medium-sized enterprises: ERP support for pull production. Int. J. Prod. Res. (2013). https://doi.org/10.1080/00207543.2011.645954
17. Agrahari, A., Srivastava, S.K.: A data visualization tool to benchmark government tendering process: insights from two public enterprises. Benchmarking (2019). https://doi.org/10.1108/BIJ-06-2017-0148
18. McGinnis, T.C., Huang, Z.: Rethinking ERP success: a new perspective from knowledge management and continuous improvement. Inf. Manage. (2007). https://doi.org/10.1016/j.im.2007.05.006
19. Shi, Z., Wang, G.: Integration of big-data ERP and business analytics (BA). J. High Technol. Manage. Res. (2018). https://doi.org/10.1016/j.hitech.2018.09.004
20. Gupta, S., Qian, X., Bhushan, B., Luo, Z.: Role of cloud ERP and big data on firm performance: a dynamic capability view theory perspective. Manag. Decis. (2019). https://doi.org/10.1108/MD-06-2018-0633
21. Chou, D.C., Bindu , H., Chou, A.Y.: BI and ERP integration. Inf. Manage. Comput. Secur. (2005). https://doi.org/10.1108/09685220510627241
22. Erasmus, P., Daneva, M.: An experience report on ERP effort estimation driven by quality requirements. In: CEUR Workshop Proceedings (2015)
23. Karsak, E.E., Özogul, C.O.: An integrated decision making approach for ERP system selection. Expert Syst. Appl. (2009). https://doi.org/10.1016/j.eswa.2007.09.016
24. Grover, P., Kar, A.K., Dwivedi, Y.K.: Understanding artificial intelligence adoption in operations management: insights from the review of academic literature and social media discussions. Ann. Oper. Res. (2020). https://doi.org/10.1007/s10479-020-03683-9

# From Qualitative to Quantitative Data Valuation in Manufacturing Companies

Hannah Stein[1,2(✉)], Lennard Holst[3], Volker Stich[3], and Wolfgang Maass[1,2]

[1] Saarland University, Campus A5 4, 66123 Saarbrücken, Germany
hannah.stein@iss.uni-saarland.de
[2] German Research Center for Artificial Intelligence (DFKI), Stuhlsatzenhausweg 3, 66123 Saarbrücken, Germany
[3] Institute for Industrial Management (FIR) at RWTH Aachen University, Campus-Boulevard 55, 52074 Aachen, Germany

**Abstract.** Since data becomes more and more important in industrial context, the question arises on how data-driven added value can be measured consistently and comprehensively by manufacturing companies. Currently, attempts on data valuation are primarily taking place on internal company level and qualitative scale. This leads to inconclusive results and unused opportunities in data monetization. Existing approaches in theory to determine quantitative data value are seldom used and less sophisticated. Although quantitative valuation frameworks could enable entities to transfer data valuation from an internal to an external level to take account of progress in digital transformation into external reporting. This paper contributes to data value assessment by presenting a four-part valuation framework that specifies how to transfer internal, qualitative to external, quantitative data valuation. The proposed framework builds on insights derived from practice-oriented action research. The framework is finally tested with a machine tool manufacturer using a single case study approach. Placing value on data will contribute to management's capability to manage data as well as to realize data-driven benefits and revenue.

**Keywords:** Data value · Data valuation framework · Industry 4.0 · Intangible assets · Case study research

## 1 Introduction

Through transformation by Industry 4.0, manufacturing companies become data-intense environments, supporting transmission, sharing, and analysis of data [1]. Using and analyzing these data becomes a necessity for economic survival [2]. Manufacturing companies improve processes and decision making, offer new services and business models based on data, e.g., through selling machine process data on top of the machine itself [3–5]. Although data seems a valuable asset, numerous challenges are related to discovering data value. Existing metrics and measurement tools struggle to value data consistent and comprehensible [6]. Therefore, data are seldom reported in balance sheets

© IFIP International Federation for Information Processing 2021
Published by Springer Nature Switzerland AG 2021
A. Dolgui et al. (Eds.): APMS 2021, IFIP AICT 631, pp. 172–180, 2021.
https://doi.org/10.1007/978-3-030-85902-2_19

or management reports [7]. Furthermore, most data valuation approaches are of qualitative nature, i.e., they do not result in a quantitative, monetary value. Transferring external accounting valuation approaches to data, i.e., cost- income-, or market-based, failed until today. Due to absence of markets, data are rather valued company-internally based on quality, usage or costs. Internal valuation means that company-internal stored data are valued without exchanging results with third parties, i.e., no cash flows are generated. External data valuation implies that valuations are exchanged with third parties or value and cash flows are generated through selling data themselves. In this paper, we present a framework that specifies the process from internal, qualitative to external, quantitative valuation in order to strengthen potential data monetization. Based on extensive literature review and action research, we develop a theoretical four-step valuation framework. It includes criteria-, cost-, reporting- and transaction-based data valuation methods. Framework evaluation is performed through a theory-testing single case study together with a large machine tool manufacturer [8]. We wrap up the paper with a conclusion and future work description. Thus, we provide manufacturing companies with tools and knowledge for valuing and monetizing data as an asset.

## 2   State of the Art

Data value can basically be determined by deriving either a non-financial, i.e., qualitative, or financial, i.e., quantitative, value. An initial overview of the distinction between qualitative and quantitative approaches and their different objective dimensions is provided by the Gartner framework [9].

### 2.1   Qualitative Data Valuation Approaches

The result of case study research by Otto implies qualitative measurements to be suitable in drawing more attention to data as a resource and to uncover cause-effect relationships between data and business value [10]. Laney takes up this point and defines three qualitative value measures: intrinsic, business and performance. Intrinsic value evaluates data sets in terms of predefined data quality dimensions. Business value describes the fulfillment of usage requirements of the considered dataset that arise from relevant business processes. Finally, performance value provides information about contributions of considered datasets to achievement of business goals [9].

### 2.2   Quantitative Data Valuation Approaches

In order to determine monetary value of data in business context, Moody and Walsh examined the three major asset paradigms in accounting theory as approaches for quantitative data valuation methods: Cost (Historical Cost), Market (Current Cash Equivalent) and Utility (Present Value) approaches [11]. These were also included in the Gartner framework [9]. While describing pros and cons of the paradigms in context of data valuation and developing general principles and ideas for valuing data, they do not provide a complete methodology [11]. More sophisticated approaches, e.g., by Zechmann (Usage based) or Heckman (Market based), build upon the qualitative analysis of data sets with regard to impact correlations and data attributes [12, 13]. They incorporate costs, quality criteria or specific usage aims for data sets within their approaches. Next, we present a framework to sensitize potential users to the topic of data evaluation and the dependence of qualitative and quantitative evaluation methods. The framework illustrates the procedure of a data valuation process.

As result of the literature review, we compare existing approaches with our proposed data valuation framework (cf. Sect. 3). Criteria include (1) deployment of method in manufacturing context, (2) integration of several valuation methods into one framework and (3) whether the methods are applied practically. Thereby we rate, whether a criterion is fulfilled (x), partly fulfilled (o) or not fulfilled (-). We summarize this within Table 1.

**Table 1.** Comparison of Literature to Envisioned Framework.

| Publication | Manufacturing context | Integration of valuation methods | Method application |
|---|---|---|---|
| Otto (2015) [10] | o | - | - |
| Laney (2017) [9] | - | x | o |
| Moody & Walsh (1999) [11] | - | - | - |
| Zechmann (2018) [12] | o | o | x |
| Heckman (2015) [13] | o | - | - |
| This study | x | x | x |

## 3    Proposed Framework

We propose a four-step data valuation framework for manufacturing industry based on extensive literature research and practice-oriented action research [14]. The data valuation process is described as follows: First, data are valued by ranking criteria of usage and quality for datasets. The second step values data quantitatively by deriving a monetary value based on costs, incorporating usage and quality from step one. Furthermore, sharing an enhanced criteria-based data valuation within a management report enables a reporting-based, external valuation of data. The fourth step represents transaction-based data valuation, i.e., a quantitative data value is derived externally through selling data, e.g., solely or as part of a new business model. We categorize the four steps in Table 2, assigning them to internal or external, respectively qualitative or quantitative valuation context. In general, we recommend to apply the full integrated valuation framework iteratively from (1)–(4). Still, independent application of the single approaches is possible. Within the next sections, we describe the detailed framework steps for data valuation in manufacturing.

**Table 2.** Categorization of data valuation approaches

|  | Qualitative valuation | Quantitative valuation |
|---|---|---|
| Internal valuation | (1) Criteria-based | (2) Cost-based |
| External valuation | (3) Reporting-based | (4) Transaction-based |

**Criteria-based Data Valuation.** Use of data is one of its essential value drivers [11]. If data remain unused, they are valueless to the company, but cause costs that arise in the course of data value chain [15], e.g., for collection, processing and storage. Aim is to identify data-driven business processes that generate highest added-value through data use. Related work in manufacturing context addresses the issue of qualitative data ranking when starting data valuation activities [16]. Especially steps one to three of the six-phase assessment are of particular importance for a qualitative ranking of relevant data sets. Step one serves to initially determine most valuable data sets for the considered system or company segment in the form of a data catalog. To accomplish step two, a catalog of relevant use cases for the data sets based on step one is defined. Subsequently data attributes (e.g., data quality, data sourcing, data processing and analysis) and threshold values are established in step three as every use case has different intrinsic requirements regarding aforementioned attributes [16]. Data quality represents one of the most important data attributes. Defining quality level of data gives an impression on the fit of data to usage targets. Assessing data quality is broadly covered in literature [17–19]. For valuing data internally on a specific valuation date in manufacturing context, the following quality criteria for acquisition and storage of data could be included: completeness, conciseness, relevance, correctness, reliability, accuracy, precision, granularity, currency and timeliness [20].

**Cost-based Data Valuation.** After criteria-based valuation, internal accounting approaches for cost-based valuation are adapted. This enhances understanding of data cost structure, enables better data planning, controlling, coordinating and decision making. Furthermore, it prepares the determination of minimum data value for future data monetization (cf. step (4)). Within cost accounting, costs for material, production and maintenance are summarized. Regarding data, costs arise for collecting, storing, pre-processing and maintaining [21]. In manufacturing context, collection costs can be captured by sensor costs. Assuming 0.42 € per sensor attached to a machine, collection costs for resulting data sets equal 0.42 €. Storage costs depend on companies' infrastructure. While Amazon Web Services offer 1GB storage for 0.019€ per month, a local storage infrastructure might be more costly. Pre-processing costs are linked to data structure and quality. Neatly collected, structured data need no pre-processing before using them, whereas semi- or unstructured data with lacking quality will probably produce more costs. Maintenance costs include overhead costs per data set for hardware (e.g. PCs), software (e.g. ERP-system), personnel (e.g. Chief Data Officer) and IT-security (e.g. security system). By applying adapted cost accounting techniques for data, interrelated costs should be collected, summed up and documented. They result in a quantitative, financial data value [21]. Furthermore, level of usage and data quality are considered and yield to value impairment or enhancement. E.g., a quality level of 98% enhances data value, while quality levels below 50% decrease data value by a percentage of 5%. An application example will be presented in Sect. 4.

**Reporting-based Data Valuation.** Building on a broader understanding for potential usage and value of existing data, developing qualitative data reporting for external data valuation is the next step. Within annual financial statements, companies report on their financial situation and success in a complete, reliable and concise manner. Aim of these statements is documentation, profit determination and information. It should represent all value-relevant information. For example, if revenue streams and business models build on data, they should be included in the statement. Still, data are seldom reported in annual statements [7]. Therefore, we propose to report on data in terms of an integrated data reporting within the annual financial statement, in the style of sustainability reporting. CSR is also added to the management report, which is part of the financial statement [22]. The data report includes for example descriptions of data strategy, i.e., data use, and quality (cf. step 1), data governance, data finance (cf. step (2)) as well as technical aspects and data security. Within this report, potentials and risks of data are presented. Through publication of the annual statement, including management and data reporting, the significance of data for the company is noticed by third parties, e.g., competitors or potential collaborators. This can lead to cooperation with third parties, e.g., by developing new business models together or through monetizing data by enhancing data trading, as presented in the next step of our framework.

**Transaction-based Data Valuation.** The framework concludes with the transaction-based data valuation through direct data monetization. In its simplest form, data can be sold directly to a customer or possibly be traded on marketplaces. Existing business models, such as those of SCHUFA [23] or IOTA market places [24], are based on this form of data monetization. Nevertheless, many companies, especially in the manufacturing sector, hesitate to sell data directly or offer it on marketplaces as they do not know how much of their own existential know-how is represented in the data and what the value of their data is [25]. However, new data-based business models, such as the as-a-Service or subscription-based business model, enable direct data monetization through a participatory business approach for manufacturers [26]. Due to the fact that these new business models are not viable without data from the customer operation phase of products [27, 28] such as the rotation of a spindle for usage-based billing, it is reasonable to assume that the value of this specific data corresponds to the value of the recurring revenues from this business relationship [29]. With the customer lifetime value (CLV), there is an existing approach to calculate the recurring revenues from subscription business models over the entire lifetime of a customer relationship [30, 31]. With a total of all calculated CLVs within the customer or subscriber base, the data within these new business models can be endowed with an exact monetary value.

# 4    Case Study

## 4.1    Methodology

We derived the valuation framework based on existing literature and the practice-oriented action research approach. Quantitative data valuation for manufacturing companies is explorative in its nature, calling for an open, explorative research approach. Therefore, we adopted deductive application of case study research [32, 33]. We applied the proposed valuation framework together with a large German machine tool manufacturer from Nov 2020–Feb 2021. Within the company, we conducted interviews and performed the four parts of the valuation framework with Head of Process Management, Head of Corporate Master Data Management and Head of Subscription. During the case study, further departments, such as controlling, accounting, information technology department or higher management level have been included. The following section describes our results.

## 4.2    Case Study Results

Applying the criteria-based valuation method, we set up a data catalog together with the case study partner. The data catalog consists of 11 customer-related data sets, e.g., historic purchase data, machine runtimes or historic service errors. In a first step, the data sets were ranked regarding certain criteria, such as their frequency of use or their timeliness. Together with an interdisciplinary team of experts within the company, a use case catalog was developed as well, consisting of 17 use cases, e.g., potential for sales of new machinery, potential for additional service business or potential for increasing service efficiency. Afterwards, the data sets were linked to the use cases to identify and rank the data sets with the highest potential of further monetary relevance. In a last step, the findings from the criteria-based rankings were compared, indicating that the machine-related data sets, which have a high potential value, are qualitatively not sufficient for broad use, especially in terms of timeliness. In this context, specific requirements for the real-time capability of data acquisition have already been formulated.

Criteria-based valuation indicates high value for machine-related data. In a next step, cost-based valuation is applied to machine usage data, stored in the CRM system of the company. More accurately, we valued a sample of 100 customers in automotive context. Collection, storage, pre-processing and maintenance costs are evaluated. Machines delivered to customers are already equipped with measurement sensors and settled via customer payments, i.e., they do not produce collection costs in the first place. For the storage costs, controlling calculated 0.01 € of overhead per customer and month. Pre-processing is unnecessary, as data delivered by customers are already structured. Hardware, network, and security costs that are necessary for data maintenance within the company equal 0.34 € per customer per month. Our partners state that three sales employees, responsible for these customers, need 2% of their working hours to maintain the data. This equals personnel costs of 480 € for all customers per month. Due to current sufficient quality and usage, (cf. criteria-based valuation), there is no value impairment due to these factors. On the valuation date, the cost-based value of the data set equals 6.180 €. Setting up this calculation is uncomplicated from the partner's point of view, as controlling could easily calculate the necessary overheads, and the sales department could give required information on the workload for maintenance.

Conducting integrated data reporting together with the company required answering several questions in context of general data use, quality, corporate digitalization, data governance and finance, data security and legal aspects, as well as technical data management. Answering these questions required discussions with several company-internal departments such as IT-departments, management, accounting. Furthermore, questions on whether data cooperation with third parties take place were not answered due to security concerns. Questions on the overall importance of data for business activities were difficult to answer, as the company does not yet possess a unified data strategy. In general, the machine tool manufacturer representatives appreciated the idea of data reporting, as (1) they needed to deal more intensely with the topic of data, which could lead to a better awareness of data relevance, and (2) sharing the report within the management report could lead to strategic data alliances with third parties in the future.

Starting the transaction-based valuation approach, the case study partner expressed concerns about whether data from the production context would ever be sold. Currently no data is sold directly to customers or traded on marketplaces. However, the company is in the middle of setting up a new subscription business to offer customers a flexible business model tailored to their needs. In this context, three machine-related data sets were defined, all of which are critical to the success of monetization in the subscription business. Since the subscription business cannot function without these data sets, they are compared with the calculated CLV of the potential subscription customers. Therefore, we calculated the value of each individual data set equivalent to the overall CLV of the prospect yearly customer base. In general, the company representatives stated to have achieved a better understanding which data sets provide potential data value, which usage possibilities are there and that profit opportunities based on data exist. Furthermore, they got a better idea of how to initially estimate and calculate data value.

## 5  Conclusion and Future Research

Our work presents a data valuation framework that considers internal and external as well as qualitative and quantitative valuation perspectives. The framework was applied and validated through a single case study approach together with a large machine tool manufacturer. Results show that the framework provides support for enhancing the understanding of existing corporate data quality and usage, value calculations and potential revenues for future data transactions or data-based business models.

Due to the exploratory nature of this work, several limitations need to be overcome by future research. First, we evaluated the framework with a single company and focused on CRM data in the manufacturing context. Additional studies with multiple companies in different domains could enhance the gathered knowledge and could support generalization of the results. Further, more data sets could be considered in order to develop data valuation of the entire data collection of companies. Second, the company's feedback on reporting-based valuation revealed that adaptations are necessary so they are better able to answer the data reporting questions widely. Therefore, we will further develop data reporting together with a management consultancy and iteratively validate the report structure with companies. The framework will furthermore be transferred to a technical prototype, leading companies through the application process, based on future study results.

**Acknowledgement.** This work is part of the research project "Future Data Assets" (grant number: 01MD19010C), funded by the German Federal Ministry for Economic Affairs and Energy (BMWi) within the scope of the "Smart Data Economy" technology program, managed by the DLR project management agency. The authors are responsible for the content of this publication.

# References

1. Davis, J., Edgar, T., Porter, J., Bernaden, J., Sarli, M.: Smart manufacturing, manufacturing intelligence and demand-dynamic performance. Comput. Chem. Eng. **47**, 145–156 (2012)
2. Schüritz, R.M., Seebacher, S., Satzger, G., Schwarz, L.: Datatization as the next frontier of servitization-understanding the challenges for transforming organizations. In: International Conference on Information System (2017)
3. Hartmann, P.M., Zaki, M., Feldmann, N., Neely, A.: Big data for big business? A taxonomy of data-driven business models used by start-up firms. Cambridge Service Alliance, 1–29 (2014)
4. Wixom, B.H., Ross, J.W.: How to monetize your data. MIT Sloan Manage. Rev. **58**(13) (2017)
5. Pavlović, M., Marjanović, U., Rakić, S., Tasić, N., Lalić, B.: The big potential of big data in manufacturing: evidence from emerging economies. In: Lalic, B., Majstorovic, V., Marjanovic, U., von Cieminski, G., Romero, D. (eds.) Advances in Production Management Systems. Towards Smart and Digital Manufacturing, vol. 592, pp. 100–107. Springer, Cham (2020). https://doi.org/10.1007/978-3-030-57997-5_12
6. OECD: Measuring the Digital Transformation: A Roadmap for the Future. OECD Publishing Paris (2019)
7. Kanodia, C., Sapra, H.: A real effects perspective to accounting measurement and disclosure: implications and insights for future research. J. Account. Res. **54**, 623–676 (2016)
8. Cavaye, A.L.: Case study research: a multi-faceted research approach for IS. Inf. Syst. J. **6**, 227–242 (1996)
9. Laney, D.B.: Infonomics: How to Monetize, Manage, and Measure Information as an Asset For Competitive Advantage. Routledge, New York (2017)
10. Otto, B.: Quality and value of the data resource in large enterprises. Inf. Syst. Manag. **32**, 234–251 (2015)
11. Moody, D., Walsh, P.: Measuring the value of information-an asset valuation approach. ECIS 496–512 (1999)
12. Zechmann, A.: Nutzungsbasierte Datenbewertung: Entwicklung und Anwendung eines Konzepts zur finanziellen Bewertung von Datenvermögenswerten auf Basis des AHP. Epubli, Berlin (2018)
13. Heckman, J.R., Boehmer, E., Peters, E.H., Davaloo, M., Kurup, N.G.: A pricing model for data markets. In: Conference Proceedings (2015)
14. Baskerville, R.L.: Investigating information systems with action research. Commun. Assoc. Inf. Syst. **2**, 19 (1999)
15. Fadler, M., Legner, C.: Managing data as an asset with the help of artificial intelligence. Competence center corporate data quality, Lausanne (2019)

16. Holst, L., Stich, V., Frank, J.: Towards a comparative data value assessment framework for smart product service systems. In: Lalic, B., Majstorovic, V., Marjanovic, U., Cieminski, G.V., Romero, D. (eds.) Advances in Production Management Systems. Towards Smart and Digital Manufacturing: IFIP WG 5.7 International Conference, APMS 2020, Novi Sad, Serbia, August 30–September 3, 2020, Proceedings, Part II, pp. 330–337. Springer International Publishing, Cham (2020). https://doi.org/10.1007/978-3-030-57997-5_39
17. Pipino, L.L., Lee, Y.W., Wang, R.Y.: Data quality assessment. Commun. ACM **45**, 211–218 (2002)
18. Wang, R.Y., Strong, D.M.: Beyond accuracy: what data quality means to data consumers. J. Manag. Inf. Syst. **12**, 5–33 (1996)
19. Batini, C., Cappiello, C., Francalanci, C., Maurino, A.: Methodologies for data quality assessment and improvement. ACM Comput. Surv. (CSUR) **41**, 1–52 (2009)
20. Despeisse, M., Bekar, E.T.: Challenges in data life cycle management for sustainable cyber-physical production systems. In: Lalic, B., Majstorovic, V., Marjanovic, U., von Cieminski, G., Romero, D. (eds.) Advances in Production Management Systems. Towards Smart and Digital Manufacturing, vol. 592, pp. 57–65. Springer, Cham (2020). https://doi.org/10.1007/978-3-030-57997-5_7
21. Stein, H., Maass, W.: Monetäre Bewertung von Daten im Kontext der Rechnungslegung. In: Trauth, D., Bergs, T., Prinz, W. (eds.) Monetarisierung Von Technischen Daten: Innovationen Aus Industrie Und Forschung,. Springer, Berlin (2021)
22. The Sustainability Code. https://www.deutscher-nachhaltigkeitskodex.de/en-gb/, Accessed 14 June 2021
23. How Schufa Works. https://www.schufa.de/en/about-us/company/schufa-works/how_schufa_works.jsp, Accessed 14 June 2021
24. IOTA Industry Marketplace: Executive Summary. https://industrymarketplace.net/executive_summary, Accessed 14 June 2021
25. Mayer, J., Niemietz, P., Trauth, D., Bergs, T.: A concept for low-emission production using distributed ledger technology. Procedia CIRP **98**, 619–624 (2021)
26. Schuh, G., Frank, J., Jussen, P., Rix, C., Harland, T.: Monetizing Industry 4.0: design principles for subscription business in the manufacturing industry. In: IEEE International Conference on Engineering, Technology and Innovation (ICE/ITMC), pp. 1–9 (2019)
27. Schuh, G., Wenger, L., Stich, V., Hicking, J., Gailus, J.: Outcome economy: subscription business models in machinery and plant engineering. Procedia CIRP **93**, 599–604 (2020)
28. Abramovici, M., Gobel, J.C., Neges, M.: Smart engineering as enabler for the 4th industrial revolution. In: Fathi, M. (ed.) Integrated systems: Innovations and applications, pp. 163–170. Springer, Cham (2015). https://doi.org/10.1007/978-3-319-15898-3_10
29. Tzuo, T., Weisert, G.: Subscribed: Why the subscription model will be your company's future- and what to do about it. Penguin (2018)
30. Gupta, S., et al.: Modeling customer lifetime value. J. Serv. Res. **9**, 139–155 (2006)
31. Borle, S., Singh, S.S., Jain, D.C.: Customer lifetime value measurement. Manage. Sci. **54**, 100–112 (2008)
32. Lee, A.S.: A scientific methodology for MIS case studies. MIS Q. **13**(1), 33 (1989)
33. Benbasat, I., Goldstein, D.K., Sead, M.: The case research strategy in studies of information systems. MIS Q. **11**(3), 369–386 (1987)

# A Methodology to Build a Framework for Collaboration Performance Assessment in PSS Delivery

Mourad Harrat[(⊠)], Farouk Belkadi, and Alain Bernard

Ecole Centrale de Nantes, Laboratory of Digital Sciences of Nantes,
LS2N UMR 6004, Nantes, France
{mourad.harrat,farouk.belkadi,alain.bernard}@ls2n.fr

**Abstract.** Companies are more and more seeking for external partners in order to manage new solutions at their development and use phases, especially when the type of these solutions is Product-Service Systems (PSS). PSS have some organizational particularities which increase the complexity of collaboration processes. In this context, collaborating efficiently with the different partners is a key aspect to reduce the risk of failure of PSS projects, and is influenced by various organizational factors and practices. This paper proposes a methodology in four steps to build a decision-aid framework supporting collaboration assessment and management in the presented context. Important factors and performance indicators are identified based on literature review and industrial practices. Then, Fuzzy techniques as well as decision trees are used to build the assessment systems. Three case studies are conducted to explore industrial practices and to confront the different elements of the proposed framework, and finally to validate the assessment framework.

**Keywords:** Collaboration · Product-Service Systems (PSS) · Assessment

## 1 Introduction

Product-Service Systems (PSS) can be defined as a combination of tangible products and intangible services which jointly responds to customer's needs [1]. Compared to pure products, these business models have an increased complexity and require multi-disciplinary domains and competences along its whole lifecycle [2], which can be hardly held without the involvement of external partners. Thus, PSS contexts necessitate closer relationships with the different stakeholders, entailing a shift from a transactional to relational orientated partnerships [3]. This is due to the organizational differences induced by servitized environments, such as in contract management, life-cycle management, degree of data collection from customer, etc. [2]. Consequently, collaborations become a central aspect to manage in order to ensure the success of PSS development and use phases. To achieve effective collaborations in PSS context, two different aspects need to

© IFIP International Federation for Information Processing 2021
Published by Springer Nature Switzerland AG 2021
A. Dolgui et al. (Eds.): APMS 2021, IFIP AICT 631, pp. 181–191, 2021.
https://doi.org/10.1007/978-3-030-85902-2_20

be explored and timely operationalized: (1) the factors influencing on inter-firm collaboration performance, and (2) the organizational practices suggested in PSS situations. The operationalization of these aspects can be performed by defining some relevant Key Performance Indicators (KPIs) and enabling an assessment process. Most of existing research works which established assessment frameworks for PSS context, focused on aspects such as design, functionalities, sustainability, customer satisfaction, maintenance, etc. [4]. In difference with classical New Product Development (NPD) projects, studies addressing collaboration performance assessment in PSS are still limited. For example, some authors dealt with the evaluation of collaborative networks readiness [5]. However, the literature is missing an extensive framework to assess collaboration performance for PSS contexts. To deal with this gap, this paper introduces a methodology for building such framework. The collaboration problem and context are clarified in Sect. 2. Then, the methodology is explained progressively in Sect. 3. An ongoing validation process is detailed in the conclusion.

## 2   Inter-firm Collaboration Within PSS Context

Before any partnership, and particularly to carry on PSS development and use phases, a recurring problem for firms is how to choose the relevant partners and how to ensure that these partners are able to participate in a collaborative PSS delivery. This ability translates into a set of capabilities and competences related to PSS which are required for each collaborating party. However, assessing a partner based solely on his competences is not sufficient, because the causes of an occurring problem during the collaboration can be beyond the control of the partner and being more related to the relationship [6] (Fig. 1).

For example, some important criteria of success as efficient tasks coordination, an appropriate communication frequency, well-established routines of knowledge sharing, and an atmosphere of trust are needed to avoid risks of project failure. This requires an assessment at each step of the collaboration process.

Consequently, an assessment during collaboration is necessary. The assessment framework in this step is not only based on collaboration factors as mentioned above, but also on PSS organizational practices. Indeed, PSS business models have some particularities that we need to consider in this assessment process. One of these particularities is the necessity of a whole life-cycle consideration with long-term relationships with partners [7, 8], as the collaboration process is extended beyond the sale of the product. This implies the necessity to involve all supply chain partners in early design, which may include maintenance and after sales services [2, 3]. Another particularity is that contractual mechanisms of PSS show higher complexity regarding terms and agreements, with higher risk levels which is shared throughout the life-cycle [9, 10]. The permanent evolution of the PSS is also another characteristic of PSS, requiring closer interactions with customers [2, 9], and higher levels of adapting common standards and processes [10].

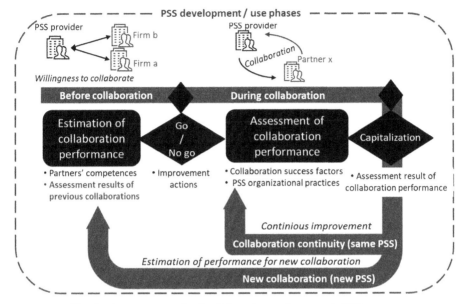

**Fig. 1.** Assessment cycle of collaboration performance in PSS context

# 3  A Methodology for Building a Framework to Assess Collaboration Performance

The proposed methodology takes place in four main steps, which are presented in the following sub-sections. A fifth step which concerns the validation of the framework in is explained in conclusion as a work in progress.

## 3.1  Identification of Factors Impacting Collaboration Performance

The first step was to identify factors leading to successful collaboration. Here, a focus should be done on the performance of relationships between partners rather than the particular partners' performance only [6, 11]. Thus, we performed a literature review from a selection of 60 articles dealing with collaboration factors, chosen according to three criteria: (1) the addressed subject is about collaboration success factors; (2) the addressed context is collaboration between industrial organizations; and (3) the dependent variables or addressed concepts have a close meaning or similar characteristics to "collaboration performance". Thus, 41 factors were identified and then grouped into main categories. After that we filtered them according to three criteria: (1) the frequency of citation; (2) the adaptation to the context of the framework (inter-organizational collaboration, without including inter-personal factors); and finally (3) the significance of the impact of these factors by checking the consistency of the different studies. The references and more details of this literature review are mentioned by Harrat et al. (2020) [12], and cannot be cited in this paper due to its size limitation. This review resulted on

10 retained factors which are: trust, commitment, shared vision and values, shared language, knowledge sharing, shared goals and interests, social network ties, coordination, communication quality, and interdependence.

## 3.2　Case Studies Presentation and Validation Process

In order to validate the list of factors and to consolidate the whole framework, three case studies were conducted by means of 5 different interviews (Table 1).

**Table 1.** Details of the exploratory case study conducted

| Case | PSS name | PSS type | Interview with | Business area | Main activity |
|------|----------|----------|----------------|---------------|---------------|
| Case A | Connected shoe | Product-oriented | 1-OEM | Leather and shoes | Shoe manufacturer |
| | | | 2-Consulting Partner | Electronics | IoT assistance |
| Case B | Drilling robot | Use-oriented | 1- OEM | Construction and public works | Public works |
| | | | 2-Supplier | Robotics | Robot provider |
| Case C | Surgery robot | Product-oriented | Supplier | Robotics | Robot provider |

Case A concerns the development project of a connected shoe, with the ability to detect falls and alert the emergency services. Since the latter services are integrated, the solution is considered as a product oriented-oriented PSS.

Case B is about the development project of a robot that ensures the drilling activity in construction sites. This robot is intended to be rented to construction sites according to their drilling needs. It is then considered as a use-oriented PSS.

Case C concerns the development project of a mobile robot providing support for surgical and imagery activities in an operating theater. This robot is sold by the OEM with assistance services, so it is considered as a product-oriented PSS.

Through semi structural interviews, the industrials were first questioned about different aspects, and especially the factors that are in relation with their collaboration performance. Except case C where there were globally no collaborative issues, the other cases present some considerable challenges.

In Case A, some trust issues were mentioned by the interviewee, between the OEM and the electronic design partner. The latter was considering his own constraints and was not able to adapt the solution to the OEM needs. Thus, the relationship was not balanced as described by the interviewee, which leaded the OEM to search for new partners in the project. On the other hand, Case B had a similar problem as Case A, as the specifications of the OEM did not meet the standards of his robot supplier. The ground clearance of the robot had to be high enough to be able to work in varied terrain, which was not in accordance with the standards of the supplier. The latter required a minimum height to ensure the safety of the operators. The OEM ended up accepting this constraint and the design was pursued and a final prototype was developed. At the end, the interviewee mentioned that the prototype was finally "a failure". This lets them to consider preparing a new project in future with a new supplier and a different design.

During the interviews, the industrials were asked to assess the importance of the different factors arose from our literature review. As a result of this assessment, we performed a second filtering of the factors, by keeping the ones considered most important according to the industrials, which are: trust, commitment, communication quality, project coordination and shared language.

### 3.3   KPIs Building and Validation

After validating the list of factors, the next step is to build the corresponding KPIs, by defining quantitative measures as percentages. This form of measure allows us to counter the subjectivity biases of human factors. The KPIs we define are adapted from indicators that are already validated by the literature. For each factor, we use references from both a PSS and a generic perspective to express the KPIs.

**Table 2.** KPIs adapted from authors corresponding to collaboration performance factors

| Factors | Corresponding KPIs | References with a generic perspective | References with a PSS perspective |
|---|---|---|---|
| Trust | Pre-transaction costs | [13] | Same KPIs for social capital (e.g. trust, commitment) are applied in PSS perspective, according to Zhang (2017) [7] |
| | Post-transaction costs | [13] | |
| | History of collaboration | [14] | |
| | Knowledge protection | [15] | |
| Commitment | Participation | [16, 17] | |
| | Commitment to schedule | [18, 19] | |
| | Responsiveness | [20] | |
| | Percentage of committed risk | | [10, 21] |
| Communication quality | Effectiveness of meetings | [22] | |
| | Openness of communication channels | [23] | |
| | Frequency of knowledge sharing | | [10, 24, 25] |

*(continued)*

**Table 2.** (*continued*)

| Factors | Corresponding KPIs | References with a generic perspective | References with a PSS perspective |
|---|---|---|---|
| | Reciprocity in knowledge sharing | | [10, 24, 25] |
| | Percentage of customer data collection | | [2] |
| Project coordination | Planning adjustment | [18] | |
| | Conflict solving | [18, 26] | |
| | Percentage of involvement of use phase departments in early design | | [2, 3] |
| Shared language | Percentage of mutual adaptations | | [10, 27] |
| | Percentage of adequacy of standards | | [10] |

We argue that the KPIs adapted from references with a generic perspective can be applied to PSS context. Indeed, the activity of the partners in a PSS development project generally come from conventional business fields (e.g. an OEM in automotive industry who wants to add specific services needs complementary skills) [12]. In addition, Zhang (2017) [7] proved that the effect of social capital (i.e. trust) on operational performance is higher in the PSS context.

On the other hand, the KPIs adapted from references with a PSS perspective highlight some organizational characteristics of PSS that need to be considered.

Yet, the suitability of the KPIs was assessed by interviewees from two previous case studies (Case B and C). The result of this process is presented in Table 2.

### 3.4 Building Membership Functions Using Fuzzy Techniques

The next step of the proposed methodology is the use of fuzzy techniques for the construction of the assessment system. The choice of this tool is leaded by the capacities of fuzzy techniques to counter ambiguity and imprecision problems in calculations, and the possibility to use linguistic terms instead of precise numerical values [28], especially in solving socioeconomic problems. This choice is also inspired from Ayadi et al. (2013) [28], who used fuzzy techniques to assess trust level between partners in supply chain. Thus, we start by defining the membership functions, by considering the KPI as inputs, and the different success factors of collaboration performance as outputs. A mix of triangular and trapezoidal functions is chosen for the representation of the membership function. This choice is leaded by the suitability of these functions to represent human factors [28]. They are also widely used in practice [28, 29], which is explained by Barua et al. (2014) [29] using an interval-based theory. The different values of membership

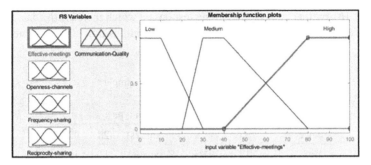

**Fig. 2.** Illustrative membership functions "Effectiveness of meetings"

functions are chosen with an intuition method based on the experience of the authors and will be validated in a further step through in-depth case studies.

We illustrate this step with the example of the input "Effectiveness of meetings" which is one of the KPIs of the output "Communication Quality" (Fig. 2). The input variable is represented by the following formula:

$$\text{Effectiveness of meetings} = (\text{Number of meetings conducting to ideation or decision making})/(\text{Total number of meetings}).$$

Three MFs are defined to describe the value of transaction costs: Low (0, 0, 10, 30), medium (20, 30, 40, 75) and high (40, 80, 100, 100). In this way, all the presented KPIs as inputs and the factors as outputs have their defined membership functions.

### 3.5 Inference Rules Definition

The next step of the methodology is to generate fuzzy inference rules, by using Mamdani's fuzzy inference method which is considered suitable for human inputs [28]. To improve the reliability of the assessment, the definition of these rules is performed by a combination of two iterations, which correspond to two methods presented in the literature. In the first iteration, we conduct a survey from which we extract data by using decision trees (C4.5 learning system). Decision trees provide a well-understood mechanism for inducing classification rules from data, as suggested by Hall and Lande (1998) [30]. Secondly, we involve experts for rules definition and adjustment, by using semi-structured interviews, which is a widely presented method and recommended by Ayadi et al. (2013) [28].

In the first iteration, we performed a survey through a structured questionnaire from where we extract data by using the data mining process.

**Data Collection.** For the purpose of data collection, we performed an online survey targeting people who already worked in a multi-partner project for product and/or service delivery. The questionnaire was sent to 353 people from three different networks of engineering schools. Thus, 67 responses were received, which constitutes a response rate of 19%. Respondents were asked to consider 1 to 3 partners in their answers. Among 67, 54 respondents considered 2 partners, while 40 considered 3 partners. This results to a total of 161 partners considered in the responses, which increase the data size to be used in the data mining process. Tables 3 show the composition of the sample.

**Table 3.** Composition of the sample

| Geographical position | | | Solution type | | |
|---|---|---|---|---|---|
| France | 33 | 49% | Product(s) | 28 | 42% |
| Africa | 12 | 18% | Service(s) | 11 | 16% |
| Europe (other) | 7 | 10% | Product oriented PSS | 14 | 21% |
| USA | 6 | 9% | Use oriented PSS | 3 | 4% |
| Asia/Oceania | 4 | 6% | Result oriented PSS | 11 | 16% |
| Germany | 4 | 6% | | | |
| America (other) | 1 | 1% | | | |

**Extraction of Inference Rules.** First, from the continuous input and output variables, classes were created to express intervals, which are necessary to build decision trees [30]. These classes correspond to the membership functions (e.g., low, medium, high) as shown in the third step of the methodology (Fig. 2).

Decision trees are then generated based on our survey data. The output factors (trust, commitment, communication quality, coordination, shared language) are considered as target variables, and their corresponding KPIs are put as attributes. After that, fuzzy rules are extracted by performing a depth search in the decision tree. The created rules correspond to each time a path reaches a leaf [30, 31]. This process is illustrated by the following example in the decision tree generated for trust factor.

Based on the selected path from the C4.5 in Fig. 3, we formulate the following rule: **IF** Restriction of Marketing plans = severe **AND** History of Collaboration = 1 year **AND** Pre-transaction costs = (medium **OR** high **OR** very high) **AND** Restriction of technical information = (partial **OR** severe) **THEN** trust = Medium.

The percentages shown in the figure represent the population responses' which match the rule, and should be considered to improve reliability of the rules.

In the second iteration, the ongoing work is to conduct semi-structured interviews with experts from industry to adjust the inference rules. The results allow the generation of rules describing the relation between the KPIs and the factors in a first time, and between the factors and collaboration performance in a second time.

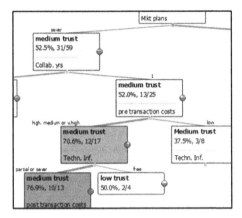

**Fig. 3.** Example of decision tree used for fuzzy rules generation

# 4   Conclusion and Perspectives

This paper proposes a methodology in four steps to build an assessment model for supporting inter-firm collaborations in PSS development and use phases. After ensuring the ability of partners to collaborate, the contribution of this framework is to assess the inter-firm relationships during the collaborations for PSS development and use phases. To do so, we first defined measures based on literature, and performed three case studies to validate them while analyzing industrial practices in terms of managing collaborations. The assessment system is built using fuzzy logic techniques. Furthermore, inference rules are created using decision trees by collecting data from a survey. The final step is the validation of the introduced framework through in-depth case studies. Indeed, the assessment model is currently being tested with industrials in order to make some adjustments. They are asked to enter values of the inputs to get the output value which is the performance of their collaboration in their PSS project. This assessment will finally allow the capitalization from their collaborations and the estimation of performance for future collaborations. This approach should help industrials to decide which strategy and improvement actions should be executed in order to collaborate effectively and to ensure the success of PSS delivery.

# References

1. Tukker, A.: Eight types of product-service system: eight ways to sustainability? Exp. Suspronet. Bus. Strateg. Environ. **13**, 246–260 (2004)
2. Wallin, J., Parida, V., Isaksson, O.: Understanding product-service system innovation capabilities development for manufacturing companies. J. Manuf. Technol. Manag. **26**, 763–787 (2015)
3. Resta, B., Gaiardelli, P., Cavalieri, S., Dotti, S.: Enhancing the design and management of the product-service system supply chain: an application to the automotive sector. Serv. Sci. **9**, 302–314 (2017)

4. Mourtzis, D., et al.: Performance indicators for the evaluation of product-service systems design : a review. In: Umeda, S., Nakano, M., Mizuyama, H., Hibino, H., Kiritsis, D., von Cieminski, G. (eds.) Advances in Production Management Systems: Innovative Production Management Towards Sustainable Growth, pp. 592–601. Springer, Cham (2017). https://doi.org/10.1007/978-3-319-22759-7_68
5. Durugbo, C., Riedel, J.C.K.H.: Readiness assessment of collaborative networked organisations for integrated product and service delivery. Int. J. Prod. Res. **51**, 598–613 (2013)
6. Johnsen, T.E., Johnsen, R.E., Lamming, R.C.: Supply relationship evaluation: the relationship assessment process (RAP) and beyond. Eur. Manag. J. **26**, 274–287 (2008)
7. Zhang, M., Guo, H., Zhao, X.: Effects of social capital on operational performance: impacts of servitisation. Int. J. Prod. Res. **55**, 4304–4318 (2017)
8. Baines, T., Shi, V.G.: A Delphi study to explore the adoption of servitization in UK companies. Prod. Plan. Control. **26**, 1171–1187 (2015)
9. Reim, W., Parida, V., Örtqvist, D.: Product-Service Systems (PSS) business models and tactics - a systematic literature review. J. Clean. Prod. **97**, 61–75 (2015)
10. Bastl, M., Johnson, M., Lightfoot, H., Evans, S.: Buyer-supplier relationships in a servitized environment: an examination with Cannon and Perreault's framework. Int. J. Oper. Prod. Manag. **32**, 650–675 (2012)
11. Varoutsa, E., Scapens, R.W.: The governance of inter-organisational relationships during different supply chain maturity phases. Ind. Mark. Manag. **46**, 68–82 (2015)
12. Harrat, M., Belkadi, F., Bernard, A.: Towards a modeling framework of collaboration in PSS development project: a review of key factors. Procedia CIRP. **90**, 736–741 (2020)
13. Dyer, J.H., Chu, W.: The role of trustworthiness in reducing transaction costs and improving performance: empirical evidence from the United States, Japan, and Korea. Organ. Sci. **14**, 57–68 (2003)
14. Belkadi, F., Messaadia, M., Bernard, A., Baudry, D.: Collaboration management framework for OEM – suppliers relationships: a trust-based conceptual approach. Enterp. Inf. Syst. **11**, 1–25 (2016)
15. Jean, R.J.B., Sinkovics, R.R., Hiebaum, T.P.: The effects of supplier involvement and knowledge protection on product innovation in customer-supplier relationships: a study of global automotive suppliers in China. J. Prod. Innov. Manag. **31**, 98–113 (2014)
16. Servajean-Hilst, R.: Necessary governing practices for the success (and failure) of client-supplier innovation cooperation. In: Moreira, A.C., Ferreira, L.M.D.F., Zimmermann, R.A. (eds.) Innovation and Supply Chain Management. CMS, pp. 79–100. Springer, Cham (2018). https://doi.org/10.1007/978-3-319-74304-2_4
17. Wibisono, Y.Y., Govindaraju, R., Irianto, D., Sudirman, I.: Interaction quality and the influence on offshore IT outsourcing success. In: Proceedings of 2017 International Conference on Data and Software Engineering. ICoDSE 2017, January 2018, pp. 1–6 (2018)
18. Westphal, I., Thoben, K.-D., Seifert, M.: Measuring collaboration performance in virtual organizations. In: Camarinha-Matos, L.M., Afsarmanesh, H., Novais, P., Analide, C. (eds.) PRO-VE 2007. ITIFIP, vol. 243, pp. 33–42. Springer, Boston, MA (2007). https://doi.org/10.1007/978-0-387-73798-0_4
19. Pemartín, M., Rodríguez-Escudero, A.I.: Is the formalization of NPD collaboration productive or counterproductive? Contingent effects of trust between partners. BRQ Bus. Res. Q. **24**, 2–18 (2020)
20. Davis-Sramek, B., Omar, A., Germain, R.: Leveraging supply chain orientation for global supplier responsiveness: the impact of institutional distance. Int. J. Logist. Manag. **30**, 39–56 (2019)
21. Zou, W., Brax, S.A., Rajala, R.: Complexity in product-service systems: review and framework. Procedia CIRP **73**, 3–8 (2018)

22. Chirumalla, K., Bertoni, A., Parida, A., Johansson, C., Bertoni, M.: Performance measurement framework for product-service systems development: a balanced scorecard approach. Int. J. Technol. Intell. Plan. **9**, 146–164 (2013)

23. Norman, P.M.: Protecting knowledge in strategic alliances resource and relational characteristics. J. High Technol. Manag. Res. **13**, 177–202 (2002)

24. Kohtamäki, M., Partanen, J.: Co-creating value from knowledge-intensive business services in manufacturing firms: the moderating role of relationship learning in supplier – customer interactions. J. Bus. Res. **69**, 2498–2506 (2016)

25. Weigel, S., Hadwich, K.: Success factors of service networks in the context of servitization – development and verification of an impact model. Ind. Mark. Manag. **74**, 254–275 (2018)

26. Wu, G., Liu, C., Zhao, X., Zuo, J.: Investigating the relationship between communication-conflict interaction and project success among construction project teams. Int. J. Proj. Manag. **35**, 1466–1482 (2017)

27. Li, H., Yang, Y., Singh, P., Sun, H., Tian, Y.: Servitization and performance: the moderating effect of supply chain integration. Prod. Plan. Control. 0, 1–18 (2021).

28. Ayadi, O., Cheikhrouhou, N., Masmoudi, F.: A decision support system assessing the trust level in supply chains based on information sharing dimensions. Comput. Ind. Eng. **66**, 242–257 (2013)

29. Barua, A., Mudunuri, L.S., Kosheleva, O.: Why trapezoidal and triangular membership functions work so well: towards a theoretical explanation. J. Uncertain Syst. **8**, 164–168 (2014)

30. Hall, L.O., Lande, P.: Generation of fuzzy rules from decision trees. J. Adv. Comput. Intell. Intell. Inform. **2**, 128–133 (1998)

31. Das, A., Desarkar, A.: Decision tree-based analytics for reducing air pollution. J. Inf. Knowl. Manag. **17**, 1–20 (2018)

# Digital Servitization and Smart Services for the New Normal

Giuditta Pezzotta[1]([⊠]), Nicola Saccani[2], Federico Adrodegari[2], and Mario Rapaccini[3]

[1] Department of Management, Information and Industrial Engineering, University of Bergamo, Bergamo, Italy
`giuditta.pezzotta@unibg.it`
[2] Department of Mechanical and Industrial Engineering, University of Brescia, Brescia, Italy
`{nicola.saccani,federico.adrodegari}@unibs.it`
[3] Department of Industrial Engineering, University of Florence, Firenze, Italy
`mario.rapaccini@unifi.it`

**Abstract.** The COVID-19 Pandemic has caused an economic breakdown, especially in the manufacturing industry. Manufacturing companies have used Digital service to stay in contact with their customers and as a source of revenue even during the general lockdown An exploratory focus group has been carried out to understand the problems and opportunities manufacturing companies went and are going through and to suggest open research questions that both research and industry should explore further. To deeply analyses the content of the discussion, Latent Dirichlet Allocation (LDA) was used to identify the main research topics. 4 topics were identified as the most relevant to be investigated: digitalization and collaboration emerged as the most interesting trends that will characterize the new normal. For each of the topics, insights, critical points and research questions are presented and discussed, reporting the main evidence of the focus group discussion.

**Keywords:** Digital servitization · Smart service · COVID-19 · Latent Dirichlet Allocation (LDA) · Focus group

## 1 Introduction

The Covid-19 pandemic has caused vast economic breakdown across the world. In this period, business activity has collapsed. The manufacturing industry was one of the most severely affected sectors regarding the negative impact of product sales. In this context, servitization, the shift from a product-centric to a service-centric business model [1, 2], has played a key role as a countercyclical stabilizer and has helped manufacturers stabilize their businesses. Even if the (hard and soft) lockdowns deeply impacted both product and service businesses, the latter showed higher resilience [3]. In particular, advanced services [4] and smart product-service system [5, 6] have been less impacted than basic (reactive) maintenance, repair, and training services. Therefore, it is not surpassing that,

© IFIP International Federation for Information Processing 2021
Published by Springer Nature Switzerland AG 2021
A. Dolgui et al. (Eds.): APMS 2021, IFIP AICT 631, pp. 192–201, 2021.
https://doi.org/10.1007/978-3-030-85902-2_21

during the pandemic, innovation initiatives related to introducing new technologies and developing new smart services have been accelerated in manufacturing companies.

Smart Services are defined as "*digital-enabled business solution supplied within an ecosystem which provides economic and sustainable value to the customer by integrating into a unique offer intelligent products with data-enabled services allowed by physical and digital infrastructures*" [6]. In this perspective, smart services make it possible to re moteise services traditionally delivered in the presence.

Digitalization and servitization, i.e. digital servitization [7], have been recognized as a "*proactive weapon for acceleration and implementation to respond to the crisis*" [3] since they can improve organizational resilience [8]. In this sense, it represents an excellent strategy to respond to the unthinkable changes that the COVID-19 pandemic requires the manufacturing company to face and support the definition of the so-called "new normal" [9].

Given the innovativeness of the topic and the scarce amount of literature regarding the holistic evolution of the unexpected changes in digital servitization research concerning the COVID-19 effects and the new normal definition, this study relies on the exploratory findings of an international focus group. The aim is to identify the main research topics that the scientific community will have to investigate to support companies in facing the new normal in the best possible way through the digital and smart service offering. To this end, problems and opportunities that both research and industry should explore further have been extracted from the discussion using topic modelling technique. The paper is structured as follow: Sect. 2 explain the methodology adopted to identify the main topics, Sect. 3 provides insight into the results, and Sect. 4 deeply investigates each topic by reporting the main insights and open questions. Finally, Sect. 5 concludes the paper.

## 2  Methodology

'*A focus group is, at its simplest, an informal discussion among selected individuals about specific topics*' [10]. In the literature, it has been used in multiple domains to explore a subject or a phenomenon. The reasons behind its selection lay in its main feature, which is the interaction of participants [10] to constitute a common understanding of a complex new phenomenon. The focus groups' scope is to provide insights about how experts perceive a new phenomenon [11].

The focus group is theoretically characterized by homogeneity of the participants, but it requires a sufficient variation among participants interest, knowledge and background to allow for contrasting opinions, the following selection criteria have been used to identify the experts:

- Research topics: All the participants selected are academic with a background in digital servitization with a different focus: strategical, tactical, and operational to guarantee homogeneity and variation among participants.
- Geographical distribution: For a global perspective on the topics, the experts must come from various countries.
- Experience in industry: To evaluate both theoretical and practical perspectives, experts deeply involved in industry-oriented projects are required.

The current study relied on the participation of 8 different experts, coordinated by two moderators to keep the discussion. The work was organized into a session group meeting anticipated by a preliminary individual evaluation of 4 statements the moderators have identified from the literature [3] and have used to stimulate the discussion. During the debate, the experts confronted each other starting from four statements but also diverging reporting their insights on the main topic. The discussion lasted half a day, and a detailed transcript created.

Since many different topics have been discussed in a divergent way, topic modelling has been applied to provide insights. Topic modelling refers to a set of machine learning algorithms providing insights into the 'latent' semantic topics in a collection of documents [12]. Latent Dirichlet Allocation (LDA) has been selected among all the algorithms available since it represents the simplest and most popular statistical topic modelling. LDA is an unsupervised, non-parameterized and generative probabilistic topic modelling [13], where each sentence is a probability distribution of topics and each topic is a probability distribution of words representing the main ideas of the focus group participants. Then, the LDA algorithm aims to model a comprehensive representation of the text (namely corpus) by inferring topics from recurring patterns of word occurrence in the documents. Topics are heuristically located on an intermediate level between the corpus and the documents and can be imagined as content-related clusters. To carry out the analysis Text Analytics Toolbox of Matlab was used.

Grounded on the methodology proposed by [6, 14], the main steps to perform the analysis are represented in Fig. 1.

Based on the LDA modelling results, four clusters were obtained and represented through word clouds for a preliminary interpretation. The sentences belonging to the four clusters were analyzed and the insights achieved are discussed in the following.

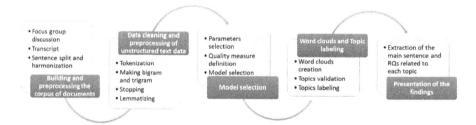

**Fig. 1.** Research methodology

## 3    Definition of the Main Topics

### 3.1    General Statements and Agreement Level

Four statements, based on [3], were used to support the discussion during the roundtable:

- Statement #1: The provision of digital and remote services has helped (and will help) the manufacturing value chain to be more resilient

- Statement #2: The adoption of digital technologies to provide services has been enormously accelerated due to the pandemic
- Statement #3: Use and outcome-based services are suffering due to the risk perceived during the pandemic
- Statement #4: Customer acceptance of digital services, their willingness to pay and share data have accelerated.

An instant poll, results reported in Fig. 2, among the panellists assessed the level of agreement about the statements on a 1 (strongly disagree) - 4 (strongly agree) scale.

The panel had a high agreement on statements 1 and 2, a moderate agreement on statement 4. At the same time, we found conflicting opinions on statement 3.

The discussion started with the four topics, but the panellists discussed several aspects of digital servitization and pandemic in a divergent way. Some topics were touched upon multiple times in support of the different statements. As previously mentioned, to really extract the main topics characterizing the impact of the pandemic on the digital servitization and the new normal trends, LDA analysis on the transcript was performed. A summary of the main findings is described in the following.

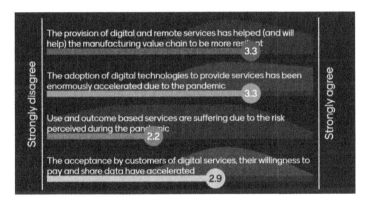

**Fig. 2.** Statement instant poll results

## 3.2 Word Cloud on the General Discussion

A word cloud of the transcript text was created to highlight the most used terms in the discussion. Wordcloud creates a chart from the unique elements of the text with sizes corresponding to their frequency counts (Fig. 3).

Some of the most used words are directly linked to the 4 statements under discussion, while other words show additional themes that emerged during the discussion: customer, platform, IoT, remote, chain, risk, and AI. This underlines how digitalization and collaboration are the most interesting topics and will characterize the new research trends.

**Fig. 3.** Genaral wordcloud

## 3.3 LDA Topics Identification

LDA analysis aims to extract and cluster the main themes discussed. We identified 4 LDA Topics that maximize the coherence of the groups, and their main content is summarised in the word cloud reported in Fig. 4.

In particular, the moderators punctually analyzed the sentences allocated to each group by the LDA algorithm to investigate and provide an in-depth interpretation of the topics covered. The word cloud allows a synthetic representation of the most repeated words and thus facilitated the interpretation and characterization of each group. The aim was to understand the contents and insights of each topic. In addition, several critical points and research questions that both research and industry should explore were extracted from the discussion.

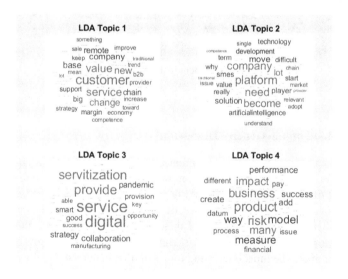

**Fig. 4.** LDA topics word cloud

The four topics obtained from the LDA are:

- TOPIC 1: provision of new forms of value through remote services in a B2B environment, involving different value chain actors
- TOPIC 2: Platform, IoT, and Artificial Intelligence as new technologies to answer customer needs
- TOPIC 3: Digital Servitization Emerged During the Pandemic as a New Strategy to Provide Smart Service Thanks to Collaborations
- TOPIC 4: Understand and Measure Business and Financial Risk and the Impact of New Business Models (Pay-Per-X, Outcome-Based...).

# 4 Results and Discussion

In the following, each topic is explained. Then, starting from the insights and the critical points that emerged along with the focus group, open research questions to be addressed by future research and discussions within the managerial and scientific communities are reported.

**TOPIC 1: Provision of New Forms of Value Through Remote Services in a B2B Environment, Involving Different Value Chain Actors**
During the pandemic and general lockdowns worldwide, digital and remote services helped all the value chain actors be more resilient [8]. In fact, due to the already well-known and strong convergence of digitalization and servitization [15], the growth of digital services accelerated due to customer demands that occurred during the pandemic. To date, it remains unclear whether the adoption of these technologies will also drive value chain actors to servitize their business in the future, and to what extent their role in the supply chain will favour or obstacle their servitization path (e.g., technology provider, component manufacturer, OEM,...). However, this rapid transition highlighted the risk of having a concentration of profits in the hands of a few value chain orchestrators and the inability of other players to achieve the expected benefits and margins.

Along with this transformation, it emerged the disruptive and relevant role of "pure" software players. This role has been seen often as a threat, but who in most cases have complementary capabilities and need to team up with manufacturers and vice versa.

Open questions for further research are:

- What about pre-sales services? Remote has been extensively used also in this perspective. How to consider remote services along the lifecycle of the solution?
- Remote has great potential, but why should customers pay the service as much as before when done remotely? Remote service must be packaged in a way that can keep the margins. How to deal with co-creation issues, value for the customer and pricing for remote services or "assisted self-service" by the customers?
- How to increase the capabilities of different actors and roles in the value chain? For instance, technicians should become "data-savvy" technicians, and SMEs should grow their digital capabilities to be part of such value chains.

**TOPIC 2: Platform, IoT, and Artificial Intelligence as New Technologies to Answer Customer Needs**

Digital technologies played a key role in service delivery during the pandemic. In particular, IoT, Platforms, and AI emerged as the most widely adopted technologies [16].

As previously mentioned, the pandemic fostered the growth of remote services. Also, it enabled manufacturing companies to deliver solutions with a high level of personalization obtained from insights based on data collected via the platform. However, to ensure a high level of personalization, this data's processing turns out to be "human-based". The role of humans is still critical to add value for the customer during data processing. This is not the only direction taken by manufacturing companies, in fact in many are also trying to mass-customize their services through the adoption of AI, which can support standardization and automation of some activities of the service delivery process [17].

It must be emphasized that although it emerged how the role of digital technologies is undisputed to provide both customized and standard services, the technology-push approach adopted by many companies during the pandemic is limiting. Therefore, rather than focusing on the best "technologies", it is essential to identify the digital functionalities to enhance customer involvement, which can be enabled by (one or more) technologies.

Open questions for further research emerged are:

- Companies need to focus on the functionalities the technologies can provide more than on the technologies themselves. So how to change their approach from technology-push to solution-driven?
- The notion of "platform economy" is more and more agreed upon. How to support companies, especially SMEs, in this transition?
- How will Artificial Intelligence support companies to achieve mass customization within the provision of services? Through which paths and mechanisms?

**TOPIC 3: Digital Servitization Emerged During the Pandemic as a New Strategy to Provide Smart Service Thanks to Collaboration**

Everyone agrees that the accelerated adoption of digital services experienced during the pandemic can also be successful in the new normal. A critical success factor is collaboration to develop and deliver services among different value chain partners, which is even more relevant than in the "traditional" servitized value chain [18]. However, how to improve this close collaboration between all actors is still an open question. Indeed, data sharing is an issue that involves trust, technical capabilities, system interoperability, etc. These aspects are particularly critical in complex systems, where different parts (e.g. of a production line) are produced by various OEMs collaborating with different IT providers: data belonging to one part of the systems may not be disclosed to suppliers of other parts of the same system. In this context, creating close collaboration means moving from a single-actor perspective to a multi-stakeholder perspective, involving manufacturing companies and IT providers from the early definition of servitization business models [15].

Open questions for further research are:

- Are value chain actors collaborating to provide digital services to build up a common business model for their servitized offerings?
- Close collaboration between manufacturing and IT players is needed to develop valuable digital services. How can such collaboration be defined, supported and measured?
- It is hard for a company to have all the competence and skills regarding digital servitization. How to identify proper competence and collaboration mechanisms to answer customer needs? How to orchestrate all the actors in the value chain?

**TOPIC 4: Understand and Measure Business and Financial Risk and the Impact of "New" Business Models (i.e. Pay-Per-X, Outcome-Based...)**
The panel discussed the role of restrictions, the evolution of risk perception, and the phenomenon of digital transformation related to offering and adopting pay-per-x and outcomes-based business models. In this period, these business models are seen as an opportunity by customers to reduce their risk [19]. Currently, many manufacturing companies are hesitating to invest in Capex, in this perspective Opex based business models can help a lot. On the other hand, the producers/suppliers of these business models may perceive that they bear too much risk [20]. This will make them withdraw from offering these business models or reduce their scope. This issue is very contextual, with significant differences between B2C and B2B and across industries, so a contextual approach to studying these business models should be taken.
     Open questions for further research emerged are:

- How has the pandemic impacted the customer's and supplier's perception and adoption of these "new" business models (their willingness to adopt these new business models, risk perception)?
- How to forecast and measure the risk of these services?
- How can we improve/facilitate the measurement of performance indicators? How can we measure the margins and support companies to improve the margins?

# 5   Conclusions

In an environment of high uncertainty and uneven impacts, it remains too early to talk of a real recovery. However, research and industry should understand the problems and opportunities of the effects of COVID-19 and explore further the impact on the ongoing and future business. Smart services have been used by manufacturers to stabilize their businesses. A focus group with experts and the use of the topic modelling Latent Dirichlet Allocation (LDA) allowed identifying four main topics that research and industry should focus on. Providing new forms of value, collaborating with the value chain actors, developing digital technologies, and answering customer needs are some of the main concepts that characterise the main four topics.

However, this work presents some limitations, in particular linked to the methodology adopted. Although the composition of the focus group reflects the topic under investigation in terms of skills, knowledge and number of experts involved, the final contribution is undoubtedly influenced by the specific interest of each expert and the impact of COVID in individual countries. Moreover, the reported perspective is exclusively academic, in order to address the issue, future research should consider refining the results obtained by enlarging the focus group by involving experts from the industrial world.

**Acknowledgements.** This paper was inspired by the activities of the Scientific advisory board of ASAP Service Management Forum, an industry–academia community aimed at developing knowledge and innovation in product-service systems and service management (www.asapsm f.org).

# References

1. Kwak, K., Kim, W.: Effect of service integration strategy on industrial firm performance. J. Serv. Manage. **27**, 391–430 (2016)
2. Raddats, C., Baines, T., Burton, J., Story, V.M., Zolkiewski, J.: Motivations for servitization: the impact of product complexity. Int. J. Oper. Prod. Manage. **36**, 572–591 (2016)
3. Rapaccini, M., Saccani, N., Kowalkowski, C., Paiola, M., Adrodegari, F.: Navigating disruptive crises through service-led growth: the impact of COVID-19 on Italian manufacturing firms. Ind. Mark. Manage. **88**, 225–237 (2020)
4. Baines, T., Lightfoot, H., Smart, P.: Servitization within manufacturing: exploring the provision of advanced services and their impact on vertical integration. J. Manuf. Technol. Manage. **22**, 947–954 (2011)
5. Lim, C., Maglio, P.P.: Clarifying the concept of smart service system. In: Maglio, P.P., Kieliszewski, C.A., Spohrer, J.C., Lyons, K., Patrício, L., Sawatani, Y. (eds.) Handbook of Service Science, Volume II. SSRISE, pp. 349–376. Springer, Cham (2019). https://doi.org/10.1007/978-3-319-98512-1_16
6. Pirola, F., Boucher, X., Wiesner, S., Pezzotta, G.: Digital technologies in product-service systems: a literature review and a research agenda. Comput. Ind. **123**, 103301 (2020)
7. Paschou, T., Rapaccini, M., Adrodegari, F., Saccani, N.: Digital servitization in manufacturing: a systematic literature review and research agenda. Ind. Mark. Manage. **89**, 278–292 (2020)
8. Zhang, J., Qi, L.: Crisis preparedness of healthcare manufacturing firms during the COVID-19 outbreak: digitalization and servitization. Int. J. Env. Res. Public Health **18**, 5456 (2021)
9. The new normal definition and meaning | Collins English Dictionary. https://www.collinsdi ctionary.com/dictionary/english/the-new-normal. Accessed 17 June 2021
10. Wilkinson, S.: Focus group methodology: a review. Int. J. Soc. Res. Methodol. **1**, 181–203 (1998)
11. Krueger, R.A., Casey, M.A.: Focus groups: a practical guide for applied research (2014)
12. Mahmood, A.A.: Literature survey on topic modeling. Technical report, Department of CIS, University of Delaware Newark, Delaware (2009)
13. Blei, D.M.: Latent dirichlet allocation. J. Mach. Learn. Res. **3**, 30 (2003)
14. Maier, D., et al.: Applying LDA topic modeling in communication research: toward a valid and reliable methodology. Commun. Methods Meas. **12**, 93–118 (2018)
15. Gaiardelli, P., et al.: Product-service systems evolution in the era of Industry 4.0. Serv. Bus. **15**(1), 177–207 (2021). https://doi.org/10.1007/s11628-021-00438-9

16. Lepore, D., Micozzi, A., Spigarelli, F.: Industry 4.0 accelerating sustainable manufacturing in the COVID-19 era: assessing the readiness and responsiveness of Italian regions. Sustainability **13**, 2670 (2021)
17. Pallant, J.L., Sands, S., Karpen, I.O.: The 4Cs of mass customization in service industries: a customer lens. J. Serv. Mark. **34**, 499–511 (2020)
18. Ayala, N., Gaiardelli, P., Pezzotta, G., Le Dain, M.A., Frank, A.G.: Adopting service suppliers for servitisation: which type of supplier involvement is more effective? J. Manuf. Technol. Manage. (2021, ahead-of-print)
19. Sjödin, D., Parida, V., Jovanovic, M., Visnjic, I.: Value creation and value capture alignment in business model innovation: a process view on outcome-based business models. J. Prod. Innov. Manag. **37**, 158–183 (2020)
20. Kohtamäki, M., Parida, V., Oghazi, P., Gebauer, H., Baines, T.: Digital servitization business models in ecosystems: a theory of the firm. J. Bus. Res. **104**, 380–392 (2019)

# Improving Maintenance Service Delivery Through Data and Skill-Based Task Allocation

Roberto Sala[1]($^{(\boxtimes)}$) ⓘ, Fabiana Pirola[1] ⓘ, Giuditta Pezzotta[1] ⓘ,
and Mariangela Vernieri[2]

[1] Department of Management, Information and Production Engineering,
University of Bergamo, Viale Marconi, 5, 24044 Dalmine, BG, Italy
roberto.sala@unibg.it
[2] Balance Systems, Via Ruffilli, 8/10, 20060 Pessano con Bornago, MI, Italy

**Abstract.** Maintenance service delivery constitutes one of the most problematic tasks for companies offering such service. Besides dealing with customers expecting to be served as soon as possible, companies must consider the penalties they are incurring if the service is delivered later than the deadline, especially if the service suppliers want to establish long and lasting relationships with customers. Despite being advisable to use appropriate tools to schedule such activity, in many companies, planners rely only on simple tools (e.g., Excel sheets) to schedule maintenance interventions. Frequently, this results in a suboptimal allocation of the interventions, which causes customer satisfaction problems. This paper, contextualised in the Balance Systems case study, proposes an optimisation model that can be used by planners to perform the intervention allocation. The optimisation model has been developed in the context of the Dual-perspective, Data-based, Decision-making process for Maintenance service delivery (D3M) framework, which aims to improve the maintenance service delivery by making a proper use of real-time and historical data related to the asset status and the service resources available. The proposed model tries to cope with the current problems present in the company's service delivery process by proposing the introduction of a mathematical instrument in support of the planner. Being strongly influenced by the contextual setting, the model discussed in this paper originates from the D3M framework logic and is adapted to the company necessities.

**Keywords:** Maintenance · Product-service systems · Decision-making · Task allocation

## 1 Introduction

The productivity of the industrial assets depends on how production and maintenance schedules complement themselves [1]. Production constraints, established maintenance policies, and unexpected failures significantly impact the company schedule [2] and, in turn, the definition of the maintenance calendars. The definition of effective maintenance schedules, or the capacity to intervene as soon as the failure presents itself, becomes even

© IFIP International Federation for Information Processing 2021
Published by Springer Nature Switzerland AG 2021
A. Dolgui et al. (Eds.): APMS 2021, IFIP AICT 631, pp. 202–211, 2021.
https://doi.org/10.1007/978-3-030-85902-2_22

more important in the Product-Service System (PSS) context [3], where maintenance is offered as a service by a supplier to a customer. From the maintenance supplier perspective (the one assumed in this work), knowledge and experience guide not only the definition of the maintenance policies (in agreement with the customer) but also the operational decisions (e.g., the allocation of the maintenance intervention to the technician), which are rarely executed using supporting tools (e.g., software) as guidance [4] despite the availability of data collectable from the field [5]. Even in an Industry 4.0 context, data gathering and analysis are not fully exploited if decision-makers rely on their experience instead of field data to make decisions [6]. Despite the availability of maintenance and scheduling software on the market, their adoption is not obvious for various reasons. For instance, companies may be used to carry out these activities without their support or may consider too expensive the costs for the adoption of a software. On the other hand, companies may not be so aware of the potentialities and benefits that the introduction of such software in their processes may lead to and, thus, may decide to avoid such an investment.

Maintenance delivery is a complex process encompassing a series of activities and relations between decision-makers and should guarantee a satisfactory result for the customer and the supplier [7]. Especially in the PSS context, where the access to the asset operational data depends upon the PSS typology, the contractual and data sharing agreements between the stakeholders, it is necessary to be reactive and make suitable decisions to tackle in a short time the assets problems. To do so, it is necessary to make use of all the asset and service data available and evaluate the intervention alternatives matching the requests with the available solutions, figuring out the resolution scenario that maximises the utility for the customer and the provider [8]. Despite this, some authors identified a lack in terms of approaches able to manage, process and match data and information, especially as far as the service characteristics are considered [9]. A possible solution consists of adopting optimisation models that can improve the intervention allocation according to various objective functions and constraints. A literature review on the topic showed how the adoption of optimisation is becoming increasingly important [10] and requires additional research based on case studies. This paper is settled into this research stream, where an optimisation model (part of a wider framework for the improvement of the maintenance service delivery) is proposed in response to the willingness of a company to improve the maintenance delivery process in terms of management and decision-making. Considering a literature perspective, researchers proposed various resolution approaches focused on different objectives such as costs minimization [11], profit maximisation [12], optimisation of the resolution teams [13] and others. This paper proposes an approach that has as objective the minimization of the number of late (i.e., tardy) interventions justified by the idea that the maintenance supplier has to pay a penalty for every tardy intervention.

The authors [14] developed a Dual-perspective, Data-based, Decision-making process for Maintenance service delivery (D3M) framework to exploit at best the possibilities offered by data-based decision-making that rely on the asset and service data. The D3M framework uses the asset and service data to support the service planner in organising the service delivery process by allocating the intervention requests to service technicians. Maintenance execution decisions should be adapted to the assets requiring it as well as to other factors such as supplier constraints and customer requirements and characteristics (e.g., location, availability of maintenance workforce, contractual constraints, SLA). The development of the optimisation model discussed in the following sections of the paper should be seen as a part of the development of the D3M framework, in the scope of ameliorating the interventions allocation based on the data collected and processed during the D3M framework application. From a practical perspective, the model presented in this paper is strongly influenced by the collaboration with an Italian manufacturing company and, thus, it is also shaped based on the company necessities.

This paper deals with the proposal of an optimisation model, part of the D3M framework, able to handle asset and service data to support decision-making related to the maintenance service delivery. Section 2 describes the model formulation, while Sect. 3 focuses on its application. Eventually, Sect. 4 and 5 discuss the analysis results and concludes the paper delineating future research.

## 2    Problem Description and Model Formulation

### 2.1    Problem Description

Maintenance requests are generated by customers and shared with the maintenance supplier. Each maintenance request is evaluated by the provider who selects the resolution approach. Each request can be fulfilled in one or more manners, each one requiring different skills, execution times, and costs. The maintenance supplier can satisfy the customers' requests in different ways depending on the context: remote support, send spare parts to customer premise, return the failed part to the provider premise, send technician to customer premise.

Every request $R$ has its characteristics (e.g., failure typology, gravity, skills required to solve the problem), which determines if it can be solved in one or more ways. Not all the resolution modes are similar in terms of competencies required (from now on referred to as skills), execution times (varying depending on the executor's experience), and costs. Therefore, the planner must match the request with the proper resolution mode considering all these factors trying to maximise customer satisfaction and minimise the resolution's times and costs. In the model, expenditures and costs are linked to the execution of interventions happening after a specific due date, which can be defined by the forecasted failure of a component or defined by Service Level Agreements (SLA) contracts. The assumptions are: (i) the other costs related to the intervention are covered by the service contract established between the stakeholders, and (ii) the maintenance supplier wants to minimise the costs arising when the suppliers fail in executing the intervention before the due date.

The model aims to be the bridge between asset- and service-related information. Thus, the model wants to use all the available information source (e.g., maintenance requests, resolution modes, technicians' skills, and calendar) to optimise maintenance service delivery while minimising the number of tardy interventions. The originality of the model is in merging real-time (e.g., RUL, technicians' calendar, position, and availability) and historical data (e.g., customer information, skills of the technicians) coming from different sources (asset and service-related) and guided in their collection and use by the D3M framework structure, to identify the best resolution modes and executors considering the customer and suppliers constraints and interests. The inclusion of real-time data such as the technicians' and resources availability, as well as the contextual conditions (e.g., failure typology, customer information and history) would allow for an improved allocation of the interventions as well as for the selection of the solution the best fits the failure characteristics, the resolution strategies, and the companies' constraints.

The notation in the following is used to model the problem discussed hereabove:

- $R$: the set of intervention requests.
- $M_r$: the set of modes that can be used to fulfil the intervention request $r \in R$.
- $T$: the set of available technicians.
- $S$: the set of skills required by mode $m \in M_r$
- $W$: the set of windows available for each technician. Each window delimitates the period where the technician is available to execute the intervention.

Each intervention request $r \in R$ defines a set of modes $M_r$ that can be used to execute the intervention and fulfil the request. Each technician $t \in T$ has a set of intervention request $r \in R$ already assigned before the next planning. Thus, before the planning, technicians already have some free and busy spots in their calendar. The availability windows can be described through a start date $s_w$, and an end date $e_w$, which can be used to define the length $\theta_w = e_w - s_w$ of the window $w \in W$. There is an infinity window available for each technician, which is the one just after the last busy block in the calendar.

The problem that wants to be modelled consists in assigning an intervention request $r \in R$ to be executed in a specific mode $m \in M_r$ to a specific window $w \in W$, which is associated with a single technician $t \in T$ simultaneously minimising the number of tardy jobs executed. Other assumptions characterising the model are:

- Each technician $t \in T$ owns a set of skills that define its competencies and the intervention modes they can execute. Skillsets influence the technician's ability to execute deal with certain requests, resolution typologies (e.g., on-field vs remote) and execution length.
- At the moment of the intervention allocation, the schedule of the technicians is not blank. There are windows of availability and unavailability for all the technicians.
- Each time an on-field intervention is performed, the technician leaves from (and returns to) the headquarter before executing the following one.

The parameters used in the model are the following:

- $\delta_{rms}$: binary, representing the requirements of each mode in terms of skills. In particular:

$$\delta_{rms} = \begin{cases} 1 \ \textit{if the mode } m \in M_r \textit{ requires the skill } s \in S \\ \qquad\qquad 0 \ \textit{otherwise} \end{cases}$$

- $\omega_{mts}$: binary, associating the skills of each technician with the mode of resolution. In particular:

$$\omega_{mts} = \begin{cases} 1 \ \textit{if the technician } t \in T \textit{ has the skill } s \in S \\ \quad \textit{required for the mode } m \in M_r \\ \quad\quad 0 \ \textit{otherwise} \end{cases}$$

- $M_{rm}$: number of skills required to perform the intervention that satisfied the request $r \in R$ with the mode $m \in M_r$ so that $M_{rm} = \sum_{s \in S} \delta_{rms}$.
- $DD_r$: due date of intervention request $r$, defined as $\min\{SLA_r; RUL_r\}$, where $SLA_r$ is the date before which the request $r$ has to be executed according to the SLA stipulated between the stakeholders and $RUL_r$ is the residual life before the breakdown of the component associated with the request $r \in R$.
- $t_{rm}^{TOP}$: travelling time for the technician to reach (and come back from) the location of the intervention request $r$ addressed with the mode $m \in M_r$. This value is not dependent on the single technician $t \in T$.
- $t_r^{SS}$: time to get the spare parts in place for the execution of the intervention in mode $m \in M_r$ fulfilling the request $r \in R$. This time is dependent upon the intervention request $r$, which determines the necessity of spare parts.
- $t_{rmt}^{INT}$: the time required to perform the intervention in mode $m \in M_r$ by the technician $t \in T$;
- $M$: a constant, large number for modelling purpose.

Finally, the following variables:

- $C_r =$ completion time of intervention $r \in R$;
- $U_r = \begin{cases} 1 \ \textit{if the request } r \in R \textit{ is satisfied after the due data } DD_R \\ \quad\quad 0 \ \textit{otherwise} \end{cases} =$ tardiness
- $x_{rmtw} = \begin{cases} 1 \ \textit{if request } r \in R \textit{ is satisfied in mode } m \in M_r \\ \quad \textit{by operator } t \in T \textit{ in windows } w \in W \\ \quad\quad 0 \ \textit{otherwise} \end{cases}$

Each request must be allocated to a single technician, and all the tasks have to be allocated.

## 2.2 Model Formulation

The model formulated is described by equations from (1) to (11). In particular, the objective function (1) minimises the number of tardy interventions. Such an objective is relevant in the considered case because it minimises the number of requests that should be re-negotiated with the customers (i.e., if the planning returns a tardy job, it may be possible to renegotiate the terms of that job with the customer). Constraint set (2) stipulates that each intervention request is allocated precisely once, whereas constraint set (3) guarantees that each window contains at most one task. Constraint sets (4), (5), and (6) define the completion time of the intervention executed through the mode $m \in Mr$, assuring that the intervention starts after the beginning of the availability window and concludes before its end, leaving the technician the time to travel back to the headquarter. Constraint set (7) defines the match between technician and intervention mode based on the skills required (by the mode) and owned (by the technician). Constraint set (8) introduces the decision variable $U_r$ that assumes the value $U_r = 1$ only if the intervention is tardy, which means that completion time $C_r \geq DD_r$, otherwise $U_r = 0$, which means that the intervention satisfied the condition of being completed before the due date ($C_r \leq DD_r$). Constraint sets from (9) and (11) define the domains of the variables.

$$\min Z = \sum_{r \in R} U_r \tag{1}$$

$$\sum_{m \in M_r, t \in T, w \in W : \theta_w \geq \min\left(t_{rmt}^{INT}\right)} x_{rmtw} = 1 \; \forall r \; in \; R \tag{2}$$

$$\sum_{r \in R, m \in M_r, t \in T} x_{rmtw} \leq 1 \; \forall w \in W \tag{3}$$

$$C_r \geq \sum_{m \in M_r, t \in T, w \in W} \left(s_w + t_{rmt}^{INT} + \max\left(t_{rm}^{TOP}; t^{SS}\right)\right) \forall r \in R \tag{4}$$

$$\theta_w \geq \left(t_{rmt}^{INT} + t_{rm}^{TOP} * 2\right) * x_{rmtw} \forall r \in R, m \in M_r, t \in T, w \in W \tag{5}$$

$$C_r \leq \sum_{m \in M_R, t \in T, w \in W} \left(e_w - t_{rm}^{TOP}\right) \forall r \in R \tag{6}$$

$$x_{rmtw} * M_{rm} \leq \sum_{s \in S} \delta_{rms} * \omega_{mts} \forall r \in R, m \in M_r, t \in T, w \in W \tag{7}$$

$$C_r \leq DD_r + M * U_r \; \forall r \in R \tag{8}$$

$$U_r \in \{0, 1\} \; \forall r \in R \tag{9}$$

$$x_{rmtw} \in \{0, 1\} \; \forall r \in R, m \in M_r, t \in T, w \in W \tag{10}$$

$$C_r \geq 0 \; \forall r \in R \tag{11}$$

## 3  Model Application

### 3.1  Balance Systems

Due to space constraints, a complete description of the model application cannot be reported. For this reason, in the following, a summary of three applications run using historical data of Balance Systems (BS) is reported. BS is a manufacturing company headquartered in Italy that sells balancing machines and offers maintenance services to customers who buy the company products. The company manages the scheduling of the maintenance requests through a Microsoft Excel® sheet that is filled by the planner with the support of service technicians. Due to how this process is managed, the scheduling activity is time-consuming. The company has five service technicians, two of them execute only remote support, while the remaining carry out on-field activities.

An analysis of more than 150 maintenance reports allowed to summarise the information useful to run the model (e.g., request, technician, customer, travel time, execution time, spare parts used). Data has been used to create scenarios that differentiate themselves for the number of requests. The data used in the three scenarios have been randomly generated based on the historical data available from BS. Each scenario aimed to verify that the model could guarantee the minimisation of the number of tardy interventions while demonstrating that it would be possible to introduce such an instrument in the planning process to support the planner in his activity.

This information has been used to feed the optimisation model that has been modelled in Cplex12 and solved using an Intel® Core™ i5-7200 CPU @ 2.50 GHz, 2 cores. Table 1 reports the results of the application in three different scenarios. The number of requests in each scenario has been established following a discussion with the company, interested in verifying the applicability of the model under different circumstances. The number of requests in each scenario has been calibrated based on the average number of requests that the company is usually responding to in a normal situation. Variations to the number of requests allowed to test the behaviour of the model in various scenarios. Following, the applicability of the model in different (more complex) scenario is foreseen in the future also to investigate the model limitations in the of variables' handling capabilities.

Each of the requests has been associated with one or more resolution modes, depending on its characteristics. The resolution mode "Send spare parts" has not been modelled, since it does not occupy the technicians' time. Once defined all the inputs, these have been used to run the model and verify its performance. The model has been evaluated according to running time and the capability of minimising the number of tardy interventions.

**Table 1.** Summary of model application in three scenarios

| Scenario | Request | Available technicians | Resolution time | Tardy interventions |
|----------|---------|-----------------------|-----------------|---------------------|
| 1 | 7 | 5 | 5–10 s | 0 |
| 2 | 10 | 5 | 20–30 s | 1 |
| 3 | 12 | 5 | 25–35 s | 1 |

## 4  Discussion

The optimisation model was tested in three scenarios with an increasing number of intervention requests. The results show that the model can optimise the intervention allocation and select the approach to maintenance delivery to minimise the number of tardy interventions. Scenario 2 and Scenario 3 see the presence of one tardy intervention each caused the impossibility, for the model in its current version, to allocate more than one request in the same window. Thus, this forces the model to search and allocate new requests in the successive windows, limiting the possibility to execute some interventions in the shortest possible time.

The model is not free form limitations and opportunities for improvement:

- *The model allocates only one intervention request per window.* Currently, even though a customer sends multiple requests, each one is treated singularly due to modelling purposes. Relaxing this constraint would favour a faster resolution of the intervention requests, especially when multiple intervention requests are received from the same customer. In this case, the travel time would reduce (the technician would need to travel only once to solve multiple problems). In other words, this would disclose the opportunity to execute opportunistic maintenance, anticipating future interventions.

- *The model allocates an intervention to a single technician.* Based on the context in which the model has been developed (i.e., the Balance Systems context), the model currently does not include the possibility to create teams of technicians for the resolution of interventions requests. At the moment, this limitation may prevent the application of this model to other, more complex, contexts.

- *The model cannot postpone, re-schedule or re-allocate interventions already confirmed.* Currently, the model is not able to change the size of the availability windows by moving the interventions or re-allocating them to technicians that may be free and that can execute them before the moment initially established. Of course, anticipating the intervention also requires confirmation from the customer.

- *The model does not consider costs.* This assumption is part of the model boundaries, which assume the execution costs as part of the contract signed with the customer and already covered. Something that could help in optimising the intervention allocation is considering the penalty cost associated with each customer in case of tardy intervention.

- *The model does not consider weight for the requests.* All the requests are characterised by the same weight, which means that simple and complex problems have the same importance for the model and are differentiated only by the execution and travel times. Something that could help in prioritising the intervention requests is considering weights for the interventions that would allow distinguishing between them facilitating their prioritisation.

- *The model works in a deterministic fashion.* The model considers data and inputs as deterministic, neglecting the possibility that there could be variations in the time required for reaching the customer or the time required to perform the maintenance (e.g., because new problems emerge). Using a stochastics approach would increase the realism and the efficiency of the model in terms of allocation.

## 5  Conclusions

This paper presented an optimisation model developed in the context of the D3M framework [14]. The model aims to optimise the maintenance delivery by improving the intervention allocation and identifying the optimal resolution strategy considering the objective function (i.e., minimisation of the number of tardy interventions).

The model, adapted to fit Balance Systems necessities, was applied in three different scenarios created using the data collected from the company. The three scenarios were dimensioned based on the average workload of the company in terms of service requests. The application of the model allowed to show the benefits and limitations connected to the model development status. The model presented in this paper should be considered as a first version of a model able to overcome the limitations listed in Sect. 4 and, thus, able to be applied in different a various scenario. In addition, it should be clarified that the baseline used for the model development was based on to the D3M framework logical structure. Being still in a development phase, the model can, at the moment, handle simple situations and requires to be integrated with the human knowledge for the consideration of external factors (e.g., costs).

Following the model application and the company's discussion, it has been possible to identify a set of limitations (Sect. 4) that will guide future development and research.

**Acknowledgements.** This research is supported by MADE Competence Center in the project "PRocessi, strumEnti e dAti a supporto delle deciSiOni di MaNutenzione 4.0 (REASON4.0)".

# References

1. Rahman, A.R.A., Husen, C.V., Pallot, M., Richir, S.: Innovation by service prototyping design dimensions & attributes, key design aspects, & toolbox Abdul. In: 23rd ICE/IEEE International Technology Management Conference, pp. 587–592 (2017)
2. Potes Ruiz, P.A., Kamsu-Foguem, B., Noyes, D.: Knowledge reuse integrating the collaboration from experts in industrial maintenance management. Knowl. Based Syst. **50**, 171–186 (2013)
3. Ardolino, M., Rapaccini, M., Saccani, N., Gaiardelli, P., Crespi, G., Ruggeri, C.: The role of digital technologies for the service transformation of industrial companies. Int. J. Prod. Res. 1–17 (2017)
4. Gopalakrishnan, M., Bokrantz, J., Ylipää, T., Skoogh, A.: Planning of maintenance activities - a current state mapping in industry. Procedia CIRP **30**, 480–485 (2015)
5. Bumblauskas, D., Gemmill, D., Igou, A., Anzengruber, J.: Smart maintenance decision support systems (SMDSS) based on corporate big data analytics. Expert Syst. Appl. **90**, 303–317 (2017)
6. Karim, R., Westerberg, J., Galar, D., Kumar, U., Karim, R.: Maintenance analytics – the new know in maintenance. IFAC-PapersOnLine. **49**, 214–219 (2016)
7. Mathieu, V.: Service strategies within the manufacturing sector: benefits, costs and partnership. Int. J. Serv. Ind. Manage. **12**, 451–475 (2001)
8. Kuo, T.C., Wang, M.L.: The optimisation of maintenance service levels to support the product service system. Int. J. Prod. Res. **50**, 6691–6708 (2012)
9. Rondini, A., Tornese, F., Gnoni, M.G., Pezzotta, G., Pinto, R.: Hybrid simulation modelling as a supporting tool for sustainable product service systems: a critical analysis. Int. J. Prod. Res. **55**, 6932–6945 (2017)
10. Afshar-Nadjafi, B.: Multi-skilling in scheduling problems: a review on models, methods and applications. Comput. Ind. Eng. 107004 (2020)
11. Agnihothri, S.R., Mishra, A.K.: Cross-training decisions in field services with three job types and server-job mismatch*. Decis. Sci. **35**, 239–257 (2004)

12. Pal, D., Vain, J., Srinivasan, S., Ramaswamy, S.: Model-based maintenance scheduling in flexible modular automation systems. In: IEEE International Conference on Emerging Technologies and Factory Automation, ETFA, pp. 1–6. Institute of Electrical and Electronics Engineers Inc. (2017)

13. Xu, Z., Ming, X.G., Zheng, M., Li, M., He, L., Song, W.: Cross-trained workers scheduling for field service using improved NSGA-II cross-trained workers scheduling for field service using improved NSGA-II. Int. J. Prod. Res. **53**, 1255–1272 (2014)

14. Sala, R., Bertoni, M., Pirola, F., Pezzotta, G.: Data-based decision-making in maintenance service delivery: the D3M framework. J. Manuf. Technol. Manage. **32**, 122–141 (2021)

# Setting the Stage for Research on Aftermarket Production Systems in Operations Management

Clemens Gróf[(✉)] and Torbjørn H. Netland

ETH Zurich, Zurich, Switzerland
`clgrof@ethz.ch`

**Abstract.** Reduction of resource consumption is essential to the Sustainable Development Goals. One key strategy for achieving this is product life-extension. Products can be maintained or remanufactured to a condition that is better than new. A fragmented literature uses terms such as remanufacturing, maintenance, repair, and overhaul (MRO), or simply service, to refer to various forms of product life-extension. Even though different terms are used, these operations share common characteristics and challenges that are distinctive compared to traditional manufacturing. Addressing the ambiguous use of terms for product life-extending operations and the fragmented body of knowledge, we introduce Aftermarket Production Systems (AmPS) as an umbrella term for industrialized product life-extending operations. The purpose of this paper is to provide an overview of AmPS-related literature and to discuss future directions for research on industrialized product life-extending operations.

**Keywords:** Closed-loop supply chain · Remanufacturing · Product-service systems (PSS) · MRO · Circular economy

## 1 Introduction

One way to reduce global resource consumption is to establish a closed material loop [1]. This can be done by reusing used products or recycling materials [2], often referred to as the circular economy [3]. Closing the material loop and decoupling resource use from economic growth contributes to sustainable development [3].

In reaction to growing public pressure and for economic reasons [4], manufacturing organizations have developed new and more sustainable technologies and business models. One such business model is servitization [5]. Pivotal elements of servitized offerings, such as product-service systems (PSS) [6], are product life-extending operations such as maintenance, refurbishment, and repair [7, 8]. Another example of a business model that supports product life-extending operations is the one of independent third-party remanufacturers. These companies acquire used products (cores), return them through industrial operations to a condition like new, and distribute these products on the aftermarket [9, 10]. Operations management scholars study product life-extending operations most commonly as remanufacturing. Unfortunately, there is no universal definition for

© IFIP International Federation for Information Processing 2021
Published by Springer Nature Switzerland AG 2021
A. Dolgui et al. (Eds.): APMS 2021, IFIP AICT 631, pp. 212–219, 2021.
https://doi.org/10.1007/978-3-030-85902-2_23

remanufacturing. Furthermore, other terms, such as maintenance, repair, and the abbreviation MRO (maintenance, repair, and overhaul), are often used synonymously when referring to remanufacturing [9, 11]. The related literature focuses on specific functional activities and neglects discussing operational interdependencies in the closed-loop supply chain system [12–14], as well as variation in operational characteristics depending on the industrial setting. The ambiguous use of terms and the focus on isolated operational challenges have led to a fragmented body of research.

To address the ambiguity and fragmentation of the literature, we provide an overview of the scientific literature that discusses product life-extending operations. We introduce the term "Aftermarket Production Systems" (AmPS) to refer to all industrialized product life-extending operations. We position AmPS as a sub-system in the circular economy and closed-loop supply chains. We provide an overview of terms for product life-extending operations, illustrate how the concept of AmPS is linked to the literature, and discuss future directions for research.

## 2 Background

### 2.1 Circular Economy

The circular economy is discussed in the literature as an alternative to the currently dominating linear economy [15]. The key element of the circular economy is a reduction in resource consumption by means of a closed material loop [2]. The circular economy is seen as an operationalization of sustainable development with the goals of environmental quality, economic prosperity, and social equity [3]. Industrial implementation of the circular economy idea involves the R-principles, or waste hierarchies (ranking based on value retention) [1–3]. In the literature, the R-principles are most often discussed as reduce, reuse, and recycle (3R framework) [2, 3]. However, Reike et al. identified up to 38 different R-principles—remanufacturing being one of them—with differences in the interpretation of their meaning and their ranking in regards to value retention [1].

### 2.2 Servitization and Product-Service Systems

Operationalization of the circular economy and closed-loop supply chains requires new business models [4]. One business model that fosters industrial sustainability is servitization [5]. Servitization refers to the strategic transition of manufacturers to not only offer a physical product but to provide integrated bundles of products and services [16]. A special manifestation of a servitized offering are PSS. A PSS is defined as a combination of product and service to deliver functionality while reducing impact on the environment [6]. The literature typically distinguishes between three main categories of PSS: product-oriented, use-oriented, and result-oriented [17]. The importance of the service increases from the first to the latter, and the relationship between the provider and the customer shifts from a transaction-based to a relationship-based interaction [8, 17]. As part of this move from product transactions toward the provision of functionality, the ownership structure changes such that the PSS provider remains the owner of the product [17]. Therefore, the provider is motivated to keep requirements for spare parts low and, as a result, reduce waste [18]. In order to provide services, the provider of PSS often has to establish and manage a network of partnering organizations [19].

Pivotal elements of PSS are refurbishment, repair, and maintenance, either as product-related services for customers owning the product, or as means to keep the provider's assets operational in order to fulfill the promised functionality [7, 8]. There are several good reasons for manufacturing companies to shift their business towards service provision. Competitive advantage is achieved by the higher margins of services and the stable cash flow throughout the product's life [8]. A comprehensive service offering can serve as a barrier to competition if the customer is tied to the organization by a service contract or is simply dependent on the offered services [16]. Manufacturers have certain advantages when offering product-related services, such as existing customer contacts, knowledge about the product, and having the required technology to provide the services [8]. However, the provision of product-related services is not limited to the original equipment manufacturer (OEM). These services are also provided by component manufacturers, independent or subcontracted companies, and end-users (the operator) [8, 13], which are in competition or in a partnership with the manufacturer. Additional challenges faced by manufacturers when providing product-related services and PSS include the increasing integration of products, services, and PSS provider operations with customer operations, the management of reverse logistics, and related data exchange [20–22]. Industry 4.0 technologies, such as the Internet of Things, cloud computing, and big data analytics are considered solutions to overcome these challenges since they facilitate integration of organizations and enable information-based service and value provision [21].

## 2.3   Remanufacturing and Product Life-Extending Operations

Product life-extending operations keep a product in usable condition or return it to a condition that is even better than new. These operations establish a closed material loop by retaining a product's value in the form of the material, energy, and labor involved in the product's creation [12]. These operations are most often studied using the term remanufacturing, which currently has no universal definition [1]. The definition in the automobile industry requires the product to be as new or even better and fully warranted [23]. Other terms for product life-extending operations include repairing, refurbishing, repurposing, reconditioning, reprocessing, restoration, or upgrading [1, 3, 24]. The aviation industry refers to them as MRO [11]. In the operations management literature, the definition of Lund [25] is often used, which considers remanufacturing as an industrial process whereby a product is transformed at its end of life to a condition like new. In this literature, remanufacturing is closely linked with closed-loop supply chain research [9, 12, 14, 26]. The servitization literature refers to product life-extending operations as product-related services or simply as service [7, 8, 22]. The service operations literature considers product life-extending operations a special type of service [27–30]. As addressed in Sect. 2.2, product life-extending operations are run by different types of organizations. These operations can be part of a PSS-integrated service, offered as a stand-alone aftermarket service, or they can support other business models, such as the one of an independent third-party remanufacturer.

Product life-extending operations have been studied from different perspectives. These operations include traditional production operations, such as disassembly, inspection, machining, and assembly, but they are also more closely connected with the market and customer [31, 32]. Customer interaction and the processing of a used product lead to uncertainties in the return rate of the products to be processed, the condition of the product, the work scope, and spare part requirements [12, 13, 33–35]. These uncertainties complicate operations and make them very labor-intensive [12, 13]. Most studies focus on these uncertainties and related specific functional activities, such as planning, scheduling, and decision making [12, 13, 32], and investigate rather isolated cases. This research has not considered important operational interdependencies [12, 13]. Guide [12] identified seven operational characteristics of remanufacturing in the literature. These seven characteristics include the condition of the returned products and the disassembly of the products. Taking these two challenges as an example, the disassembly process is dependent on the product quality. Guide stressed the importance of understanding operational interdependencies when developing a planning and control system [12]. Furthermore, product life-extending operations are exposed to the challenges of closed-loop supply chains. By focusing on specific functional activities, the interplay on higher system levels in the closed-loop supply chain is often not considered [14].

On the one hand, product life-extending operations share many characteristics. On the other hand, these characteristics vary depending on the industrial setting. This can be illustrated with another remanufacturing challenge identified by Guide [12]: the need to balance core returns with demand. A company that is remanufacturing consumer goods with high variation in the timing of the product return experiences this challenge differently than a company that remanufactures expensive industrial goods whose return can be forecasted or even scheduled [26]. Categorizations of remanufacturing, and therefore implicitly categorizations of industrial product life-extending operations, have been provided based on customer relations [10], product design and the strategic focus of the manufacturer [9], and production strategy [12]. Despite awareness of these differences, these categorizations are barely used in the available literature.

## 2.4 Proposing a Definition of Aftermarket Production Systems

The fragmented research on product life-extending operations and insufficient categorization of studied cases hampers the advancement of this research field. We introduce the term "Aftermarket Production System" (AmPS) as a sub-system within the circular economy and closed-loop supply chains. AmPS is defined as industrialized product life-extending operations taking place at a dedicated facility, and include technology and organizational principles within the system borders of this facility. The position of AmPS in the circular economy and closed-loop supply chains is illustrated in Fig. 1.

AmPS consider all product life-extending operations from major repair to remanufacturing as defined by the automotive industry. The introduction of AmPS as an umbrella term builds on the fact that product life-extending operations share commonalities. However, the term is introduced with an awareness of variations in their characteristics (see Sect. 2.3). AmPS process a product (core) returning from use, which may have already reached its end of life. After being processed, the product is either returned to the operator for use or sold on the aftermarket. In some cases, parts go partially through the same production process as new products do. The prefix "aftermarket" is used to emphasize that AmPS are part of a product recovery system that provides supply for subsequent secondary market transactions or provides standalone after-sales services. Product recovery and after-sales service both take place in the aftermarket [36]. AmPS can also be part of a PSS delivery system. PSS can include monitoring the product in use and applying Industry 4.0 solutions to provide access to information about its condition [21, 37]. AmPS are part of the circular economy at the micro level [3] and enable a higher value retention than recycling.

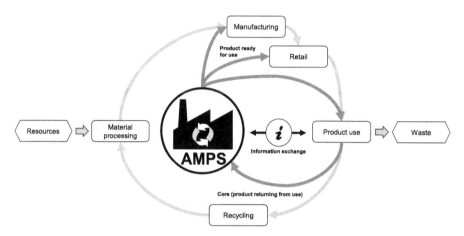

**Fig. 1.** Aftermarket Production Systems (AmPS) within the circular economy

## 3   Directions for Future Research

Remanufacturing is considered the backbone of the operationalization of the circular economy [14]. This is applicable to product life-extending operations in general. Even though remanufacturing has received increasing attention among operations management scholars, industry reports show low efficiency and long throughput times compared to manufacturing [13, 32, 35]. This is an indication of the need for more research on industrial product life-extending operations and their improvement.

The literature stresses the importance of considering the operational interdependencies of industrialized product life-extending operations [12] and the importance of understanding them within the context of the supply chain [14] (see Sect. 2.3). This must be considered when studying AmPS. Different categorization criteria for remanufacturing and industrial product life-extending operations are provided in Sect. 2.3. As mentioned in that section, these categorizations are barely used. Studying the commonalities between AmPS in different industry settings can lead to synergies in research and advancement of the field, which are goals of the introduction of the umbrella term AmPS. However, the distinction is essential when researching AmPS in order to know whether findings are applicable to AmPS in general or only to specific settings. One crucial distinction for future research is the type of company that is running the AmPS, since AmPS are run by different types of companies (see Sect. 2.2 and Sect. 2.3). An independent remanufacturer of car parts, for instance, faces a different mix of challenges in its operations than a provider of PSS for aircraft engines, who overhauls them [9, 10]. Much of the literature on servitization, PSS, and remanufacturing approaches operations from the perspective of an OEM [7, 8, 22, 27, 28, 34] or it does not indicate the type of company that was studied.

AmPS can be part of a PSS delivery system. One of the concerns of the community, who does research on servitization, is the PSS delivery system [38]. Related literature examines operations for delivering PSS, which implicitly considers maintenance and repair operations [7, 22]. The integration in the network for PSS delivery [19, 39] must be taken into account in research on AmPS. Additionally, the integration of AmPS in IT infrastructure and the application of Industry 4.0 solutions that support PSS delivery must be considered [21, 37, 40, 41]. However, to advance industrial product life-extending operations, research on operations at the plant level is required. A well designed AmPS can support a business model, as manufacturing can support the corporate strategy [42]. Furthermore, most related research does not take service operations into consideration [22]. The service operations literature studies how variables such as customer interaction, variation in workflow, and fluctuation of processed volumes affect operations [29]. These are characteristics of AmPS. Hence, establishing a link with the service operations literature can contribute to the advancement of AmPS.

## 4  Conclusion

Product life-extending operations play a pivotal role in the operationalization of a circular economy and the support of service-centered business models. They face distinctive challenges compared to conventional manufacturing. Related research has not considered operational interdependencies and studied rather isolated cases. The term remanufacturing is used broadly in the field of operations management to refer to product life-extending operations. Many other terms are used synonymously. Furthermore, little attention is paid to the respective industrial setting of the operations, such as the type of organization performing the work. Research is fragmented, and the operations that have been studied are insufficiently categorized. Taking a systems perspective, we defined AmPS and positioned them within the circular economy and closed-loop supply chains. AmPS collectively refers to all industrialized product life-extending operations, including applied technology and organizational principles at the dedicated facility. The definition of AmPS serves as a basis for future research. It aims to reduce ambiguity when referring to product life-extending operations, and it emphasizes the importance of operations at the plant level that support the respective business model in use. We linked AmPS to related research streams in operations management—which include remanufacturing, closed-loop supply chain, servitization, PSS, and service operations—and highlighted important considerations for future research. The future research we outlined is particularly relevant for an OEM on its journey toward a servitized product offering, independent organizations that are running product life-extending operations, and operators who are maintaining their assets.

# References

1. Reike, D., Vermeulen, W.J.V., Witjes, S.: The circular economy: new or refurbished as CE 3.0? — exploring controversies in the conceptualization of the circular economy through a focus on history and resource value retention options. Resour. Conserv. Recycl. **135**, 246–264 (2018)
2. Ghisellini, P., Cialani, C., Ulgiati, S.: A review on circular economy: the expected transition to a balanced interplay of environmental and economic systems. J. Clean. Prod. **114**, 11–32 (2016)
3. Kirchherr, J., Reike, D., Hekkert, M.: Conceptualizing the circular economy: an analysis of 114 definitions. Resour. Conserv. Recycl. **127**, 221–232 (2017)
4. Kleindorfer, P.R., Singhal, K., van Wassenhove, L.N.: Sustainable Operations Management. Prod. Oper. Manag. **14**(4), 482–492 (2005)
5. Smart, P., Hemel, S., Lettice, F., Adams, R., Evans, S.: Pre-paradigmatic status of industrial sustainability: a systematic review. Int. J. Opt. Prod. Manage. **37**(10), 1425–1450 (2017)
6. Baines, T.S., et al.: State-of-the-art in product-service systems. Proc. Inst. Mech. Eng. B J. Eng. Manuf. **221**(10), 1543–1552 (2007)
7. Baines, T., et al.: Towards an operations strategy for product-centric servitization. Int. J. Opt. Prod. Manage. **29**(5), 494–519 (2009)
8. Oliva, R., Kallenberg, R.: Managing the transition from products to services. Int. J. Serv. Ind. Manage. **14**(2), 160–172 (2003)
9. Abbey, J.D., Guide, V.D.R.: A typology of remanufacturing in closed-loop supply chains. Int. J. Prod. Res. **56**(1–2), 374–384 (2018)
10. Östlin, J., Sundin, E., Björkman, M.: Importance of closed-loop supply chain relationships for product remanufacturing. Int. J. Prod. Econ. **115**(2), 336–348 (2008)
11. Ayeni, P., Ball, P., Baines, T.: Towards the strategic adoption of Lean in aviation Maintenance Repair and Overhaul (MRO) industry. J. Man. Tech. Manage. **27**(1), 38–61 (2016)
12. Guide, V.D.R.: Production planning and control for remanufacturing: industry practice and research needs. J. Oper. Manag. **18**, 467–483 (2000)
13. Seitz, M.A., Peattie, K.: Meeting the closed-loop challenge: the case of remanufacturing. Calif. Manage. Rev. **46**(2), 74–89 (2004)
14. Goltsos, T.E., Ponte, B., Wang, S., Liu, Y., Naim, M.M., Syntetos, A.A.: The boomerang returns? Accounting for the impact of uncertainties on the dynamics of remanufacturing systems. Int. J. Prod. Res. **57**(23), 7361–7394 (2019)
15. Stahel, W.R.: The circular economy. Nature **531**(7595), 435–438 (2016)
16. Vandermerwe, S., Rada, J.: Servitization of business: adding value by adding services. Eur. Manag. J. **6**(4), 314–324 (1988)
17. Tukker, A.: Eight types of product–service system: eight ways to sustainability? Exp. SusProNet. Bus. Strat. Env. **13**(4), 246–260 (2004)
18. Isaksson, O., Larsson, T.C., Rönnbäck, A.Ö.: Development of product-service systems: challenges and opportunities for the manufacturing firm. J. Eng. Des. **20**(4), 329–348 (2009)
19. Pawar, K.S., Beltagui, A., Riedel, J.C.K.H.: The PSO triangle: designing product, service and organisation to create value. Int. J. Opt. Prod. Manage. **29**(5), 468–493 (2009)
20. Moro, S.R., Cauchick-Miguel, P.A., Mendes, G.H.S.: Product-service systems benefits and barriers: an overview of literature review papers. Int. J. Ind. Eng. Manag. **11**(1), 61–70 (2020)
21. Gaiardelli, P., et al.: Product-service systems evolution in the era of Industry 4.0. Serv. Bus. **15**(1), 177–207 (2021). https://doi.org/10.1007/s11628-021-00438-9
22. Baines, T., Lightfoot, H.W.: Servitization of the manufacturing firm. Int. J. Opt. Prod. Manage. **34**(1), 2–35 (2013)

23. CLEPA, MERA, APRA, ANRAP, FIRM, CPRA: A Definition of Remanufacturing. Remanufacturing Associations Agree on International Industry Definition. https://cdn.ymaws.com/apra.org/resource/resmgr/european/reman_definition.pdf. Accessed 4 Mar 2021
24. Khan, M.A., Mittal, S., West, S., Wuest, T.: Review on upgradability – a product lifetime extension strategy in the context of product service systems. J. Clean. Prod. **204**, 1154–1168 (2018)
25. Lund, R.T.: Integrated Resource Recovery. Remanufactuing: The Experience of the United States and Implications for Developing Countries. The World Bank, Washington DC (1984)
26. Guide, V.D.R., Jayaraman, V., Linton, J.D.: Building contingency planning for closed-loop supply chains with product recovery. J. Oper. Manag. **21**(3), 259–279 (2003)
27. Johansson, P., Olhager, J.: Linking product–process matrices for manufacturing and industrial service operations. Int. J. Prod. Econ. **104**(2), 615–624 (2006)
28. Johansson, P., Olhager, J.: Industrial service profiling: matching service offerings and processes. Int. J. Prod. Econ. **89**(3), 309–320 (2004)
29. Wemmerlöv, U.: A taxonomy for service processes and its implications for system design. Int. J. Serv. Ind. Manage. **1**(3), 20–40 (1990)
30. Buzacott, J.A.: Service system structure. Int. J. Prod. Econ. **68**, 15–27 (2000)
31. Chase, R.B., Ravi Kumar, K., Youngdahl, W.E.: Service-based manufacturing: the service factory. Prod. Oper. Manag. **1**(2), 175–184 (1992)
32. Vasanthakumar, C., Vinodh, S., Ramesh, K.: Application of interpretive structural modelling for analysis of factors influencing lean remanufacturing practices. Int. J. Prod. Res. **54**(24), 7439–7452 (2016)
33. Srinivisan, M., Gilbert, K., Bowers, M.: Lean Maintenance Repair and Overhaul. Changing the Way You Do Business. McGraw Hill, New York (2014)
34. Vogt Duberg, J., Johansson, G., Sundin, E., Kurilova-Palisaitiene, J.: Prerequisite factors for original equipment manufacturer remanufacturing. J. Clean. Prod. **270**, 122309 (2020)
35. Kurilova-Palisaitiene, J., Sundin, E., Poksinska, B.: Remanufacturing challenges and possible lean improvements. J. Clean. Prod. **172**, 3225–3236 (2018)
36. Durugbo, C.M.: After-sales services and aftermarket support: a systematic review, theory and future research directions. Int. J. Prod. Res. **58**(6), 1857–1892 (2020)
37. Sala, R., Pirola, F., Dovere, E., Cavalieri, S.: A dual perspective workflow to improve data collection for maintenance delivery: an industrial case study. In: Ameri, F., Stecke, K.E., von Cieminski, G., Kiritsis, D. (eds.) APMS 2019. IAICT, vol. 566, pp. 485–492. Springer, Cham (2019). https://doi.org/10.1007/978-3-030-30000-5_60
38. Lightfoot, H., Baines, T., Smart, P.: The servitization of manufacturing. Int. J. Opt. Prod. Manage. **33**(11/12), 1408–1434 (2013)
39. Cavalieri, S., Romero, D., Strandhagen, J.O., Schönsleben, P.: Interactive business models to deliver product-services to global markets. In: Prabhu, V., Taisch, M., Kiritsis, D. (eds.) APMS 2013. IAICT, vol. 415, pp. 186–193. Springer, Heidelberg (2013). https://doi.org/10.1007/978-3-642-41263-9_23
40. Stoll, O., West, S., Gaiardelli, P., Harrison, D., Corcoran, F.J.: The successful commercialization of a digital twin in an industrial product service system. In: Lalic, B., Majstorovic, V., Marjanovic, U., von Cieminski, G., Romero, D. (eds.) APMS 2020. IAICT, vol. 592, pp. 275–282. Springer, Cham (2020). https://doi.org/10.1007/978-3-030-57997-5_32
41. Pirola, F., Boucher, X., Wiesner, S., Pezzotta, G.: Digital technologies in product-service systems: a literature review and a research agenda. Comput. Ind. **123**, 103301 (2020)
42. Hill, T.: Manufacturing Strategy. Text and Cases, 2nd edn. Palgrave, Basingstoke (2000)

# Smart Landscaping Services

Kai-Wen Tien[✉], William E. Sitzabee, Phillip Melnick, and Vittaldas V. Prabhu

The Pennsylvania State University, University Park, PA 16802, USA
kut147@psu.edu

**Abstract.** Landscaping services industry is estimated to be about $100 billion in the US. These services tend to be labor-intensive and are varied in scales ranging from single-family homes to large hospitality and leisure enterprises such as resorts and golf courses. From a management perspective the three main objectives of landscaping services are maintaining aesthetics, pest control, and lowering cost. Some of the major activities in landscaping include mowing lawns, pruning shrubs, clearing leaves, trimming hedges, and mulching. Operating cost depends on staffing level, frequency of activities, and associated fuel consumption, which have been investigated in several studies. The focus of this paper is to make landscaping services smarter by using decision-support models for managing them. Specifically, this paper proposes a two-stage optimization model for lawn mowing. The first-stage model assigns appropriate pieces of equipment and staff to various areas to minimize both operating costs and labor costs. The second-stage model optimizes the schedule of activities based on the desired due times for various areas. A numerical study is used for demonstrating the application of the decision-support model. Future direction for smart landscaping through better decision-making based on data from IoT sensors for monitoring growth, soil conditions, and weather data is also proposed.

**Keywords:** Landscaping · Workforce planning · Mowing · Mixed integer programming

## 1 Introduction

For many institutions such as universities, schools, resorts, and municipalities, it is important to maintain green areas' aesthetically pleasing and pest-free. Landscaping is labor-intensive and can require a sizable budget in these institutions. For instance, Central Park in New York City spends over $10 million for landscaping every year [1]. Some of the major activities in landscaping include mowing lawns, seasonal planting, pruning shrubs, clearing leaves, trimming hedges, fertilizing, irrigating, and mulching. Typically, lawn mowing tends to require significantly more resources than other activities and hence is the focus of this paper.

Mowing lawns at a high frequency ensures aesthetics are maintained and weeds controlled but increases associated labor and fuel cost and adverse environmental impact [2]. It is estimated that a typical gasoline-powered lawnmower generates as much pollution

© IFIP International Federation for Information Processing 2021
Published by Springer Nature Switzerland AG 2021
A. Dolgui et al. (Eds.): APMS 2021, IFIP AICT 631, pp. 220–227, 2021.
https://doi.org/10.1007/978-3-030-85902-2_24

as 43 cars, resulting in lawn care producing 13 billion pounds of toxic pollutants per year [3]. Moreover, for some turfgrass species, high mowing frequency kills it because it cannot produce enough leaf area for photosynthesis between mows. On the other hand, low mowing frequency can reduce the cost, but turfgrass may grow so much between mows that mowing removes too much leaf surface, leading to scalping [4, 5]. Additionally, low mowing frequency may cause a greater buildup of thatch, which in turn can cause slower microbial decomposition [2].

Some of the recent developments in mowing have mainly focused on robotic mowers, which could help in lowering costs in high-wage locations [6, 7]. However, at present and for the foreseeable future, much of lawn mowing can be expected to be largely labor-intense [8]. Thus, there is still a need for decision-support tools for lawn mowing services, especially for institutions having large green areas. Le et al. (2010) is the only paper we found that discusses a decision-making model for landscaping [1]. They developed an ant-colony heuristic algorithm to schedule maintenance activities for green areas in a university in Taiwan.

There is a need for generic mathematical models that can lead to smart landscaping services that managers use to plan and schedule activities based on resource constraints. As a first step towards such smart landscaping services, this paper proposes a two-stage optimization model for lawn mowing and its practical applicability is demonstrated through a case study. The paper is organized as follows. In Sect. 2, the proposed mathematical model is presented. Section 3 shows the case study and the results. Section 4 proposes a futuristic smart landscape service through better decision-making based on data from IoT sensors for monitoring growth, soil conditions, and weather data. Section 5 presents conclusions and directions for future work.

## 2 Optimization Model for Planning and Scheduling

We decompose the lawn mowing optimization problem into two stages with the objective of minimizing operating cost and labor cost. In the first stage we assign the set of available mowing equipment and staff to various areas that need to be mowed to minimize the total cost. In the second stage the assigned tasks are scheduled to reduce tardiness.

### 2.1 Equipment-Technician Assignment Problem

One of the model's inputs is $A$ areas, indexed by $i$, that need to be mowed. Another input is the planning span denoted by $D$, in days, indexed by $l$, $l = 1, \ldots, D$. There are $E$ types of mowing equipment and $\alpha_j$ pieces for type $e$ equipment, $j = 1, \ldots, E$. There are $W$ technicians, indexed by $k = 1, \ldots, W$, who have different salary costs. The optimization model is as follows.

$$Min \quad z_1 = \sum_{i=1}^{A}\sum_{j=1}^{E}\sum_{k=1}^{W}\sum_{l=1}^{D} C_j^e P_{ij} x_{ijkl} + \sum_{j=1}^{E}\sum_{k=1}^{W}\sum_{l=1}^{D} C_k^r t_{jkl}^r + C_k^o t_{jkl}^o \qquad (1\text{-}1)$$

$$s.t. \quad \sum_{j=1}^{E}\sum_{k=1}^{W}\sum_{l=1}^{D} x_{ijkl} = 1, \qquad\qquad \forall i \qquad (1\text{-}2)$$

$$\sum_{i=1}^{A}\sum_{l=1}^{D}x_{ijkl} \le M\,\Delta_{jk}, \qquad\qquad \forall k \qquad (1\text{-}3)$$

$$\sum_{l=1}^{D}\delta_{il} = 1, \qquad\qquad \forall i \qquad (1\text{-}4)$$

$$\sum_{j=1}^{E}\sum_{k=1}^{W}x_{ijkl} = \delta_{il}, \qquad\qquad \forall i, l \qquad (1\text{-}5)$$

$$\sum_{i=1}^{A}P_{ij}x_{ijkl} = t_{jkl}^{r} + t_{jkl}^{o}, \qquad\qquad \forall j, k, l \qquad (1\text{-}6)$$

$$t_{jkl}^{r} \le T^{r}n_{jkl}, \qquad\qquad \forall j, k, l \qquad (1\text{-}7)$$

$$t_{jkl}^{o} \le T^{o}n_{jkl}, \qquad\qquad \forall j, k, l \qquad (1\text{-}8)$$

$$\sum_{k=1}^{W}n_{jkl} \le \alpha_{jl}, \qquad\qquad \forall j, l \qquad (1\text{-}9)$$

$$\sum_{j=1}^{E}n_{jkl} \le 1, \qquad\qquad \forall k, l \qquad (1\text{-}10)$$

$$x_{ijlk}, \delta_{il} \in \{0, 1\}; t_{jkl}^{r}, t_{jkl}^{o} \ge 0; n_{jkl} \in \{0\} \cup \mathbb{N}$$

Equation (1-1) is the objective function to minimize equipment operation costs and labor costs. $C_{j}^{e}$ is the hourly cost of equipment $j$ including fuel cost; or equipment rental and fuel cost. $P_{ij}$ is the hours to mow area $i$ by using equipment $j$. $C_{k}^{r}$ and $C_{k}^{o}$ are the unit labor cost in regular hours and overtime, respectively. The decision variable $x_{ijkl}$ is a binary variable which indicates equipment-technician pair $(j/k)$ is used for mowing area $i$ on day $l$. Other two decision variables, $t_{jkl}^{r}$ and $t_{jkl}^{o}$, are the number of regular hours and overtime hours, respectively, used in day $l$ of worker type $k$ using equipment $j$.

Equation (1-2) ensures that all areas are mowed by assigning one equipment-technician pair to every area. Equation (1-3) describes technican capability for operating a specicific type of equipment; $\Delta_{jk}$ is equal to one if technician $j$ is able to operation equipment type $k$; $M$ is a sufficient large contant. Equation (1-4) and Eq. (1-5) decides the day for mowing an area where variable $\delta_{il}$ is set to 1 if area $i$ is cleared on day $l$. Equation (1-6) constrains the total time that a piece of equipment is used to match the available hours of the assigned technician. Equation (1-7) and (1-8) constrain the time assigned for a technician is within allowable limits of regular and overtime hours denoted by $T^{r}$ and $T^{o}$, respectively. The number of available pairs of equipment-technician $(j/k)$ on day $l$ is denoted by $n_{jkl}$. Equation (1-9) restricts the pieces of type $k$ equipment assigned on day $l$ should not exceed the number available. Equation (1-10) constrains that one technician can be only assigned to at most one piece of equipment in a day.

It should be noted that the model assumes the technicians are available all the time during the planning period, and this assumption holds for some organizations. The assumption can be relaxed by substituting $T^r$ with $T^o$ to $T_{kl}^r$ and $T_{kl}^o$, which allows some technicians have a tolerance for working on other tasks.

## 2.2 Workforce Scheduling Problem

Based on the optimal assignment, the second stage optimization model tries to schedule the tasks to be completed before the due time of each area. It can be especially essential when there are periodic pedestrian traffics that would making mowing inconvenient and inefficient. Here we assume that the initial schedule of each technicians is blank. Define $m$ and $n$ are index aliases of $i$. $C_m^d$ is the tardiness cost for area $m$. $P_m$ is the optimal process time for area $m$ from the result of stage one. The scheduling model for an equipment-technician pair $(j/k)$ on day $l$ is shown below.

$$Min \quad z_2 = \sum_{m=1}^{A} C_m^d d_m^+ \tag{2-1}$$

$$s.t \quad v_m = s_m + P_m \qquad\qquad m = 1, \ldots, A \tag{2-2}$$

$$d_m^+ - d_m^- = v_m - DT_m, \qquad\qquad m = 1, \ldots, A \tag{2-3}$$

$$s_m \geq v_n - M\delta_{mn}, \qquad\qquad m = 1, \ldots, A; n = 1, \ldots, A; m \neq n \tag{2-4}$$

$$s_n \geq v_m + M(\delta_{mn} - 1), \qquad\qquad m = 1, \ldots, A; n = 1, \ldots, A; m \neq n \tag{2-5}$$

$$\delta_{mn} + \delta_{nm} = 1, \qquad\qquad m = 1, \ldots, A; n = 1, \ldots, A; m \neq n \tag{2-6}$$

$$\delta_{mn} = \{0, 1\}, \qquad\qquad m = 1, \ldots, A; n = 1, \ldots, A; m \neq n$$

$$v_m, s_m, d_m^+, d_m^- \geq 0, \qquad\qquad m = 1, \ldots, A$$

The objective function (2-1) is to minimize the costs of tardiness. Equation (2-2) calculates the completion time of each task. Equation (2-3) is the tardiness constraint, where $d_m^+$ is zero if the task is scheduled earlier than due time, otherwise, $d_m^-$ is zero. $DT_i$ is the due time of the task in area $i$. Equation (2-4) and (2-5) are the schedule constraints that determine the order of the tasks, where $M$ is a sufficiently large number. If the technician mows area $m$ before area $n$ ($\delta_{mn} = 1$), it will activate Eq. (2-5) and release Eq. (2-4), and vice versa. Equation (2-6) makes sure the consistency of the schedule.

## 3  Case Study

A large institution (name withheld) has 30 areas that need to be mowed. The operation team has three technicians and four pieces of equipment available to perform the tasks every day. Based on the records, the operating time $p_{ij}$ of each type of equipment for each area can be obtained. The planning span is from Monday to Wednesday ($L = 3$), and the cost of labor and equipment are shown in Table 1. In the second stage, the due time for each task is set to be eight hours, and the tardiness cost $c_m^d$ is one in order to minimize the overtime labor hours. In this study, techincans can operate all types of euqipments, which means $\Delta_{jk} = 1$ for all $j$.

**Table 1.** Hourly rate of technician and equipment

| Technician ($k$) | Regular hour cost ($c_k^r$) | Overtime cost ($c_k^o$) |
|---|---|---|
| Tech1 | $25.00 | $37.50 |
| Tech2 | $28.00 | $42.00 |
| Tech3 | $31.00 | $46.50 |

| Equipment ($j$) | Hourly cost ($c_j^e$) | Availability ($\alpha_j$) |
|---|---|---|
| Equip1 | $74.00 | 1 |
| Equip2 | $51.00 | 2 |
| Equip3 | $80.00 | 1 |

The problem is solved by CPLEX/GAMS solver on NEOS [9], and the relative optimality criterion is set to be 0.01. This particular formulation for the application is solved in less than one minute, which is adequate for the purpose of daily or weekly use. The resulting solution is illustrated in Fig. 1. The optimal cost, consisting of operating costs and labor costs, is 5669.31. The corresponding total time required is 62.66 h, and the total tardiness is 3.82 hours.

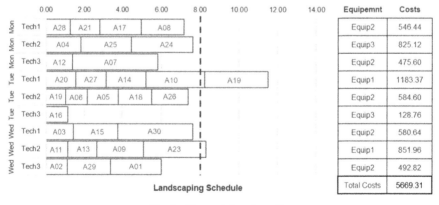

**Fig. 1.** Computational results

## 4   IoT-Driven Models for Smart Landscaping

Now there is a slew of affordable and easy-to-use networked sensors as a part of the IoT (Internet-of-Things) megatrend. These IoT sensors can be networked wirelessly, which makes them very attractive for innovating landscaping services, and enabling their digital transformation towards smart landscaping systems. For instance, a first step would be use IoT GPS trackers mounted on mowing equipment to accurately estimate the time required to mow a specific area. Such IoT GPS trackers would be an early step in digitizing the process to acquire key performance data ($P_{ij}$ in the models above) instead of relying on experience-based guesstimates or expensive manual time studies. As in Fig. 2, using GPS trackers, the measurements from sensors, types of equipment used, and actual operation time of each area will be gathered and stored in a database. These records coupled with weather data can establish a regression model to predict operation times ($P_{ij}$). Moreover, such a digitized smart system will provide new insights in terms of how $P_{ij}$ is influenced by factors such as time-of-day, specific technician, weather, and seasonality, enabling higher fidelity models and more effective decisions. Similarly, low-cost soil sensors coupled with weather data can be used to predict grass growth and optimize lawn mowing frequency that takes into turfgrass health and environmental considerations. The resulting models and decision-support systems will help make landscaping services smarter than in the current manual, labor-intense operations or for a fleet of robotic lawn mowers in the future.

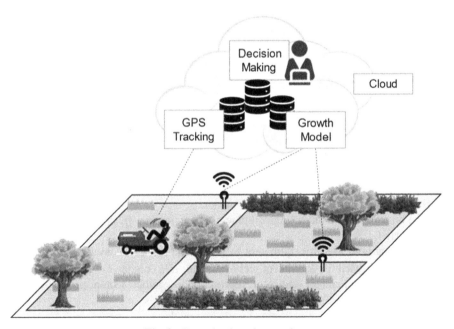

**Fig. 2.**  Smart landscaping services

# 5  Conclusion

Maintaining green areas is important for many organizations since the green area provides water filtration, temperature regulation, carbon dioxide absorption, oxygen re-lease into the atmosphere, and aesthetics. However, only a few studies have focused on models for decision-making in landscaping services. This paper proposes a two-stage optimization model. In the first stage, the model assigns appropriate equipment-techni-cian pairs to various areas considering their availability and efficiency to minimize op-erating and labor costs. In the second stage, the activities are scheduled optimally so that any tardiness is minimized and provide the start time for each activity. Scheduling of mowing activities may be necessary in practice when there are considerable pedes-trian traffics at different times of the day or days of the week.

A case study is used to demonstrate the usage of the model in a relatively small scale to illustrate the opportunity for making such labor-intense services smarter. It should be noted that in this model we assume that every technician only operates one equipment on any given day (constraint (1-9)). In practice, a technician may operate multiple equipment if they have time available, and in such situations the optimal cost generated by the model should be treated as an upper bound because the real cost will be lower. This issue can be addressed in the future by increasing the granularity of the time period from one day to hours but this may take longer to compute the solution for large-scale problems. This issue could also be addressed by reformulating the constraints in the model.

Future works will expand the model to include other significant activities in landscaping services such as pruning shrubs and leaf cleaning to provide a more comprehensive solution. IoT-driven models for smart landscaping services (Fig. 2) present a fertile opportunity to fully engineer these processes and leverage technological advances to improve the productivity of these services.

**Notation**

$C_j^e$ The hourly cost of equipment $j$ including fuel cost; or equipment and fuel cost.

$C_k^r$ The regular hourly cost of technician $k$.

$C_k^o$ The overtime hourly cost of technician $k$.

$P_{ij}$ The hours to mow area $i$ by using equipment $j$.

$T^r$ The allowable regular time hours of a technician.

$T^o$ The allowable overtime hours of a technician.

$x_{ijkl}$ The binary variable indicating equipment-technician pair $(j/k)$ is used for mowing area $i$ on day $l$.

$t_{jkl}^r$ The regular hour used for equipment-technician pair $(j/k)$ on day $l$.

$t_{jkl}^o$ The overtime used for equipment-technician pair $(j/k)$ on day $l$.

$\Delta_{jk}$ The binary parameter indicator representing whether techincan can operate equipment type $j$ or not.

$\delta_{il}$ A binary variable indicating if area $i$ is cleared on day $l$.

$n_{jkl}$ The number of pairs of equipment-technician $(j/k)$ on day $l$.

$\alpha_{jl}$ The total available equipment $j$ on day $l$.

$C_m^d$ The tardiness cost of area $m$.

$d_m^+$ The tardiness variable of area $m$.

$d_m^-$ The earliness variable of area $m$.
$v_m$ The completion time of area $m$.
$s_m$ The start time of area $m$.
$P_m$ The process time of area $m$ (from model 1).
$DT_m$ Due-time of area $m$.
$\delta_{mn}$ The binary variable indicating if area $m$ is cleared before area $n$.

*Note I , m, n are alias indices of area; j is the index of equipment types; k is the index of technicians; l is the index of planning days.*

# References

1. Lee, H.Y., Tseng, H.H., Zheng, M.C., Li, P.Y.: Decision support for the maintenance management of green areas. Expert Syst. Appl. **37**, 4479–4487 (2010). https://doi.org/10.1016/j.eswa.2009.12.063
2. Busey, P., Parker, J.H.: Energy conservation and efficient turfgrass maintenance. In: Turfgrass, pp. 473–500. Wiley (2015)
3. Hitchler, L.: Grass lawns are an ecological catastrophe. In: One Only Nat. Energy (2018). https://www.onlynaturalenergy.com/grass-lawns-are-an-ecological-catastrophe/. Accessed 9 Apr 2021
4. Happer, J.K., Happer, J.C.: Mowing Turfgrasses: grass cutting is the major time-consuming operation in the maintenance of any turfgrass area (2016). https://extension.psu.edu/mowing-turfgrasses. Accessed 1 Apr 2021
5. Christians, N.E., Patton, A.J., Law, Q.D.: Mowing, rolling, and plant growth regulators. In: Fundamentals of Turfgrass Management, 5th edn., pp. 209–224. Wiley, Hoboken (2016)
6. Mechsy, L.S.R., Dias, M.U.B., Pragithmukar, W., Kulasekera, A.L.: A mobile robot based watering system for smart lawn maintenance. In: International Conference on Control, Automation and Systems, pp. 1537–1542. IEEE Computer Society (2017)
7. Weiss-Cohen, M., Sirotin, I., Rave, E.: Lawn mowing system for known areas. In: 2008 International Conference on Computational Intelligence for Modelling Control and Automation, CIMCA 2008, pp. 539–544 (2008)
8. Wu, B., Wu, Y., Aoki, Y., et al.: A study on the reduction of mowing work burden for maintaining landscapes in rural areas: experiment design for mowing behaviors analyze. In: 2019 IEEE International Conference on Dependable, Autonomic and Secure Computing, International Conference on Pervasive Intelligence and Computing, Intl Conf on Cloud and Big Data Computing, International Conference on Cyber Science and Technology Congress (DASC/PiCom/CBDCom/CyberSciTech), pp. 533–536. IEEE, Fukuoka (2019)
9. Dolan, E.D.: NEOS Server 4.0 administrative guide. arXiv Prepr cs/0107034 1–38 (2001)

# The Contracting of Advanced Services Based on Digitally-Enabled Product-Service Systems

Shaun West[1](✉) ⓘ, Zou Wenting[2], and Oliver Stoll[1] ⓘ

[1] School of Technology and Architecture, Lucerne University of Applied Sciences and Arts, Lucerne, Switzerland
{shaun.west,oliver.stoll}@hslu.ch
[2] Aalto University, Otakaari 1B, AALTO, 00076 Espoo, Finland
wenting.zou@aalto.fi

**Abstract.** This paper uses an integrative literature review to explore the contracting of advanced services based on digitally-enabled product-service systems (PSS). The need for this study was derived from studies that have highlighted the difficulty of selling and buying advanced services with a digital element. Other studies confirmed that firms had challenges with obtaining the value expected from such advanced service contracts. The integration of digital into a PSS value proposition increases the complexity and the potential application of advanced services. The emergence of digitally-enabled PSS (based on "Smart Products" in many cases) suggests a need to understand the sales and contracting process better. An integrative literature review was chosen, as the literature was fragmented between different fields. Forty-eight papers were selected as relevant, and 14 were then considered key to creating an initial model for the contracting process. It was identified that there were limited examples of contracts with high degrees of value-co-creation and that the ability of manufacturing firms to translate a value proposition successfully into a binding contract was weak. The contract negotiation process was found to be well defined, yet the governance of such contacts over their duration was again weak. For these reasons, a model based on the lifecycle was proposed. The model should be further integrated into the contracting process for services. This is an initial study, and it is recommended that further research should test the model.

**Keywords:** Value propositions · Offers · Contracts · Product-service system · Digital

## 1 Introduction

Reflecting the increasing importance of services in all industries, recent studies report a significant rise of interest in servitization of manufacturing firms over the past two decades [1, 2]. This interest continues to grow rapidly and in line with the progress of research on service-dominant logic [3]. Servitization is considered a competitive strategy

© IFIP International Federation for Information Processing 2021
Published by Springer Nature Switzerland AG 2021
A. Dolgui et al. (Eds.): APMS 2021, IFIP AICT 631, pp. 228–237, 2021.
https://doi.org/10.1007/978-3-030-85902-2_25

for manufacturing to survive in competitive markets to address diverse customer needs and then gain revenue growth [4].

New research interests have arisen in the area of advanced services, where value is co-created through in-depth interactions and extensive capability integration between the manufacturing firm and the customer [1]. These interactions are based on the exchange of data, information, and knowledge. With digital transformation, the importance of advanced services has increased substantially in line with the introduction of digital technologies such as big data and the Internet of Things [5]. This inevitably increases risks and complexities for manufacturing firms during a collaborative process with their customers, which complicates the design of service contracts [6].

The previous studies [7, 8] have identified that some buyers are not used (or are unable) to buy advanced services, whereas some suppliers cannot develop a suitable contract for advanced services. A study in the Harvard Business Review shows that firms lose between five to forty percent of value on a deal because of inefficient contract management [9]. Furthermore, inefficient contracts could also lead to legal problems due to the usage, storage, and ownership of data in addition to more traditional contractual disputes. Because of these challenges, designing and managing contracts for advanced services is difficult and complex. However, very little work has been done to provide guidance on the contract design for the effective delivery of advanced services.

The integration of digital into PSS, or in some cases Smart Products, means that many more firms are developing PSS with advanced service attributes. Yet, without a contract that reflects the offer, the likelihood of successful commercialization is limited. Therefore, there is an urgent need to identify a contracting model for advanced services contracts in B2B contexts to improve today's research-based knowledge of advanced services contracts. The purpose of this paper is to provide initial input into such a model, which can then be further tested and further developed.

## 2    Research Framework and Methodology

PSS is a mature research field, yet the commercializing of digitally-enabled PSS remains poorly investigated, particularly the development of advanced service contracts. There is a need for a model to describe the development of advanced service contracts within industrial contexts. Given the degree of variety in the research knowledge relating to the contracting of advanced services, an integrative literature review has been chosen for this study. This approach allows integrating knowledge from other research fields with a critical focus on how value is co-created through in-depth interactions and extensive capability integration between the manufacturing firm and the customer. The outcome of this review will be a theoretical model which can be applied for future studies. This approach requires that the literature and the main ideas and relationships of the issue are critically analyzed.

The methodology based on the integrative literature review follows the conceptual structure of the contracting process for advanced services within an industrial PSS context [10, 11]. Web of Science was chosen as the database for the literature search. The analysis of the literature is based on the premise that digitally-enabled PSS provides the opportunities for outcome- or performance-based contracts. The data set was built by searching the following keywords (within the titles, abstract and keywords fields): TS = (servitization AND "outcome based contracts") OR TS = (servitization AND "performance based contact") OR TS = ("contract" AND "service-dominant logic") OR TS = (Servitization AND procurement). The search was limited to literature within B2B environments that was relevant to long-life industrial PSS assets.

## 2.1 Dimension Selection for the Literature Review

Advanced service within the context of servitization is focused on value creation [1]; there has been significant discussion about value-in-exchange, value-in-use, and value-in-context within the literature [3]. Therefore, based on the association with advanced services within a PSS environment, it was considered important to focus on four aspects of value: drivers; creation (or co-creation); delivery; and capture. Marketing theory and service science, particularly SD logic, support the approach and the appropriate selection of the four value dimensions. Additionally, from SD logic [3, 4], value creation and co-creation demand a multi-actor perspective, making this an important consideration for the literature review. Considering the PSS focus of the review, the lifecycle perspectives are essential, as there are phases in the lifecycle where different activities are necessary for contracting [5]. Returning to the contracting focus of the study, papers that describe contracting within the context of advanced services should be assessed.

# 3  Integrative Literature Review

New digitalized technologies are being integrated into traditional PSS offers and value propositions within the context of servitization [12, 13]. Given the servitization focus and the importance of value co-creation, PSS and, in particular, advanced services literature have been assessed according to the value drivers; value creation; value co-creation; and value capture. Short insights from the process will provided. 48 papers were identified and considered to provide useful insights for this integrative literature review, 14 of those were considered to provide significant value in building the model (Table 1).

## 3.1 Assessment of Literature

The literature gap in advanced service contracting was identified in 2016 [15] and continues to exist. The conflict between the value proposition and the contract is described naïvely [26]. This demonstrates that some firms find it challenging to transfer the value proposition's intent into a legally binding agreement between the parties and are unable to, in effect, create a fully formed offer (or solution) [27]. There exists a poor understanding of the transformation of the value proposition into a boilerplate (or template) contract and the ongoing re-application of boilerplate contracts as the basis for an advanced service contract [17].

One study [19] described contract structures, including the form of typical advanced service agreements and the actor responsibilities for service delivery. Other papers tend to take a more marketing-led view of contracts, rather than using the fundamentals of contracting. The actors included in the contracting process have been well described [17] and the model proposed gives a solid foundation for converting a value proposition into a contract, considering the actors who are involved and when they are included in the contracting process. The use of the lifecycle supports the application of service-dominant logic during the pre-negotiation and negotiation process. However, it lacks governance and contract management during the delivery phase [20]. Linking the pre- and post-contract phases, including the importance of multi-actor engagement, is essential and confirms the importance of contract governance in the delivery phase [14].

**Table 1.** Key references from the literature study ordered based on the dimension

| Reference | Value drivers | Value creation | Value delivery | Value capture | Multi-actor perspective | Lifecycle perspectives | Contract process |
|---|---|---|---|---|---|---|---|
| [14] Broekhuis, M., & Scholten, K. (2018) 12 | x | x | x | X | X | X | X |
| [15] Essig, M., Glas, A. H., Selviaridis, K., & Roehrich, J. K. (2016) | x | x | x | x | X | X | X |
| [16] Hou, J., & Neely, A. (2018) | x | x | X | x | X | x | x |
| [17] Liinamaa, J., Viljanen, M., Hurmerinta, A., Ivanova-Gongne, M., Luotola, H., & Gustafsson, M. (2016) | X | X | X | X | X | x | X |
| [18] Rialland, A., Nesheim, D. A., Norbeck, J. A., & Rødseth, Ø. J. (2014) | x | x | x | X | X | X | X |
| [19] Kleemann, F. C., & Essig, M. (2013) | x | x | X | x | X | | X |
| [20] Nystén-Haarala, S., Lee, N., & Lehto, J. (2010) | x | x | x | | X | X | X |
| [21] Raja, J. Z., Frandsen, T., Kowalkowski, C., & Jarmatz, M. (2020) | X | X | X | x | x | x | |
| [22] Razmdoost, K., & Mills, G. (2016) | X | X | x | x | | X | X |
| [23] Smith, L., Maull, R., & Ng, I. C. L. (2014) | x | X | X | | X | x | x |

(*continued*)

**Table 1.** (*continued*)

| Reference | Value drivers | Value creation | Value delivery | Value capture | Multi-actor perspective | Lifecycle perspectives | Contract process |
|---|---|---|---|---|---|---|---|
| [24] Tjendani, H. T., Anwar, N., & Putu Artama Wiguna, I. (2018) | x | x | X | x | X | | X |
| [25] Zou, W., Brax, S. A., Vuori, M., & Rajala, R. (2019) | x | x | x | | X | X | x |
| [26] Ng, I. C. L., Maull, R., & Yip, N. (2009) | X | X | X | | X | | x |
| [27] Töytäri, P. (2018) | X | X | x | | | | |

Notes: X – major relevance; x – minor relevance

The application of service-dominant logic in terms of value identification and co-creation within an advanced service contract is essential and not in any way easy to deliver [22, 23, 26]. Risk is an alternative form of value, and the transfer of risk [16] can be viewed as an essential part of the value proposition that needs to be included in the contract [16]. The importance of a shared understanding of the customer's value creation process within the context of advanced service is described in the literature [21]. Missing within this approach is risk perception; the approach may be combined with others [24] to quantify the value of risk transfer in an advanced service contract.

A few of the papers reviewed [18, 25] consider the buyer's (or beneficiary's) perspective. There are several ways to achieve this; during the contracting process or during the delivery phase, the application of standard KPIs appears to be common (e.g., the measurement of availability or reliability).

In summary, there is a gap in the literature on the design of suitable forms of contract based on the underlying value proposition. This is partially described through the value identifying processes, but the details of integrating this into a form of contract are missing. The contracting process is well described in the literature, yet the governance of the contract in the delivery phase is poorly defined. The aspect of digital PSS is absent within the literature reviewed with respect to the translation of the value proposition into a contract, during the contracting and delivery phases.

## 4 Discussion

Based on the literature assessed in this study, a model for digitally-enabled PSS contracts is proposed to support the timely and effective contract development of an advanced service. The issue of asset ownership has challenged the traditional buyer-supplier contractual relationships and requires a multi-actor perspective. This perspective may be guided via value co-creation aspects and becomes important [3]. In an industrial environment, the value co-creation process may be a triadic (or more) relationship and here, contracts can describe roles and obligations between parties.

Customer segmentation, expressed as preferences, supports an approach to value propositions, offers, and the associated contract where multiple options exist. This adds a layer of complexity to the contracting process; nevertheless, when developing Smart Products, firms should know that their customers may expect advanced services. Forms of contract (based on the patterns described) should be developed in parallel with the development of the digitally-enabled PSS during the beginning-of-life phase. There are boilerplate forms of contract for advanced services that exist, yet with the advent of digitally-enabled PSS, they may not reflect fully the value co-creation that takes place. A contract that does not reflect the value proposition's promise may explain why performance has been below the expected value.

A model for the development of contracts that match the value proposition and is applicable for digitally-enabled PSS is needed. Figure 1, based on the synthesis of the findings, provides such a model; it describes the generic tasks necessary to take a value proposition to a contract. The figure also connects to the middle of the product's life, where the installed base may call for renegotiations or where the supplier may wish to offer new value propositions to support the customer. The end-of-life provides the opportunity for upgrades, which can provide a route to contract extensions. It also provides a potential for "retirement services". A contracting process for a single contract, covering the four phases: 'pre-sales' where the value proposition is "sold" to the customer, the "detailed sales" where the offer is jointly detailed, the "negotiation phase" where the contract is settled and finally a "contract delivery and governance phase" where the long-term successful outcome of the contract is managed. The four steps represent a modification to the model [17] through the addition of the delivery and governance phase. The process is predicated on having in place a clearly defined value proposition, the underlying hybrid offer definition, and appropriate boilerplate contracts. Value identification is based on the negotiation phase and the preference for a practical value proposition during the first phase of the contracting process. The model assumes high-value contracts, and it may be more difficult without some automation to use it for lower-value contracts.

### 4.1 Managerial Implication

Contracts standardize and regulate the value co-creation and value capture of advanced services associated with PSS [7, 8]. Without contracts, we organizations cannot achieve this outcome successfully, and furthermore, contracts should reflect the promised value proposition that was sold. When creating contracts, firms should consider flexibility within the type of contracts as the value propositions can change over the equipment's lifecycle, and actor risk-tolerance and resources may also evolve [14, 17]. Specific contracts have been identified along the lifecycle, and the different types are in no way exhaustive [12, 16, 18]. There will be other forms of contract for different solutions that have not been identified in this integrative literature review. To succeed in value co-creation and value capture with of advanced services, management needs to focus on the specific form of contract in a digitally-enabled PSS environment.

## 4.2 Theoretical Implications

SD logic is compelling for servitization strategies and PSS-based business models [1, 3, 23]. Yet without contracts that standardize the value co-creation and value capture, it is difficult to imagine how they can be successfully implemented. A conceptual and practical gap exists between the value co-creation and value capture and the contracting of advanced services coming from (digital-enabled). Different disciplines (including law and finance specialists) will be required to fill this research gap successfully [8]; only by doing so will the value be obtained that is promised by digitally-enabled PSS. Without

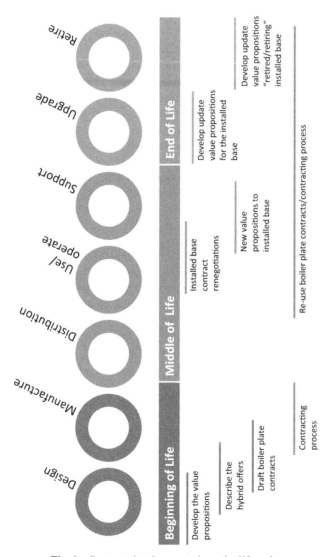

**Fig. 1.** Contract development along the lifecycle

contracts that can support value co-creation and value capture from advanced services based on digitally-enabled product-service systems, there will remain a detachment between value contracted and value delivered. This is important for SD logic research, as firms move to advanced services and move to value proposition based on value-in-use or value-in-context with multi-actor value co-creation processes [22, 24, 25].

## 5   Conclusions and Recommendations

Commercialization of innovation requires firms to sell the value proposition and convert it into a contract to deliver value to the different parties. Without a contract it is not possible to convert innovation into commercial reality. Digital brings to PSS a new level of possibilities and increases complexity, often for manufacturing firms who are more used to selling products rather than advanced services. Drafting advanced service contacts that support multi-actor value creation is a multi-disciplinary process drawing from diverse research areas, such as servitization, PSS, project management, general management, marketing, finance, legal and risk management. The research gap identified in this study on one side was rather surprising in that some firms have successfully created advanced services and have contracted them. However, the multi-actor approach to value creation and how value capture takes place through contract is poorly understood and to understand it fully needs a wider team of researchers than is normally the case. It also requires firms to understand more fully the middle-of-life phase and the opportunities it offers, and how to design value propositions and appropriate contracts that allow their delivery.

A model has been proposed that is integrated with the lifecycle and linked to a generic contracting process. Three recommendations are made from this study:

1. The model proposed should be tested and further adapted;
2. More research into how firms design contracts for advanced services by using digital technologies within a PSS context is required;
3. The influence of contract design on service (or business) performance should be studied.

## References

1. Baines, T., Ziaee Bigdeli, A., Bustinza, O.F., Shi, V.G., Baldwin, J., Ridgway, K.: Servitization: revisiting the state-of-the-art and research priorities. Int. J. Oper. Prod. Manage. **37**(2), 256–278 (2017). https://doi.org/10.1108/IJOPM-06-2015-0312
2. Kreye, M.E., Roehrich, J.K., Lewis, M.A.: Servitising manufacturers: the impact of service complexity and contractual and relational capabilities. Prod. Plan. Control **26**(14–15), 1233–1246 (2015). https://doi.org/10.1080/09537287.2015.1033489
3. Brodie, R.J., Löbler, H., Fehrer, J.: Evolution of service-dominant logic: towards a paradigm and metatheory of the market and value cocreation? Ind. Mark. Manage. **79**, 3–12 (2019)
4. Smith, L., Maull, R., Ng, I.C.L.: Servitization and operations management: a service dominant-logic approach. Int. J. Oper. Prod. Manage. **34**(2), 242–269 (2014). https://doi.org/10.1108/IJOPM-02-2011-0053

5. Latonen, T., Akpinar, M.: Designing an advanced services contract in servitization strategy: a study from the B2B sector in Finland
6. Raddats, C., Kowalkowski, C., Benedettini, O., Burton, J., Gebauer, H.: Servitization: a contemporary thematic review of four major research streams. Ind. Mark. Manage. **83**, 207–223 (2019)
7. Stoll, O., West, S., Hennecke, L.: Procurement of advanced services within the domain of Servitization: preliminary results of a systematic literature review. In: West, S., Meierhofer, J., Ganz, C. (eds.) Smart Services Summit. PI, pp. 73–81. Springer, Cham (2021). https://doi.org/10.1007/978-3-030-72090-2_7
8. West, S., Gaiardelli, P., Ozbek, D., Züst, S.: Exploring the alignment between servitization based value propositions and contracts. In: Baines, T., Bigdeli, A., Rapaccini, M, Saccani, N., Adrodegari, F. (eds.) Servitization: A Pathway Towards a Resilient, Productive and Sustainable Future. Spring Servitization Conference May 2021, Aston (2021)
9. Stouthuysen, K., Slabbinck, H., Roodhooft, F.: Controls, service type and perceived supplier performance in interfirm service exchanges. J. Oper. Manage. **30**(5), 423–435 (2012). https://doi.org/10.1016/j.jom.2012.01.002
10. Torraco, R.J.: Writing integrative literature reviews: guidelines and examples. Hum. Resource Dev. Rev. **4**(3) (2005). https://doi.org/10.1177/1534484305278283
11. Snyder, H.: Literature review as a research methodology: an overview and guidelines. J. Bus. Res. **104** (2019). https://doi.org/10.1016/j.jbusres.2019.07.039
12. Grubic, T., Jennions, I.: Do outcome-based contracts exist? The investigation of power-by-the-hour and similar result-oriented cases. Int. J. Prod. Econ. **206** (2018). https://doi.org/10.1016/j.ijpe.2018.10.004
13. Kohtamäki, M., Parida, V., Oghazi, P., Gebauer, H., Baines, T.: Digital servitization business models in ecosystems: a theory of the firm. J. Bus. Res. **104**, 380–392 (2019)
14. Broekhuis, M., Scholten, K.: Purchasing in service triads: the influence of contracting on contract management. Int. J. Oper. Prod. Manage. **38**(5) (2018). https://doi.org/10.1108/IJOPM-12-2015-0754
15. Essig, M., Glas, A.H., Selviaridis, K., Roehrich, J.K.: Performance-based contracting in business markets. Ind. Market. Manage. **59** (2016). https://doi.org/10.1016/j.indmarman.2016.10.007
16. Hou, J., Neely, A.: Investigating risks of outcome-based service contracts from a provider's perspective. Int. J. Prod. Res. **56**(6) (2018). https://doi.org/10.1080/00207543.2017.1319089
17. Liinamaa, J., Viljanen, M., Hurmerinta, A., Ivanova-Gongne, M., Luotola, H., Gustafsson, M.: Performance-based and functional contracting in value-based solution selling. Ind. Market. Manage. **59** (2016). https://doi.org/10.1016/j.indmarman.2016.05.032
18. Rialland, A., Nesheim, D.A., Norbeck, J.A., Rødseth, Ø.J.: Performance-based ship management contracts using the Shipping KPI standard. WMU J. Marit. Affairs **13**(2), 191–206 (2014). https://doi.org/10.1007/s13437-014-0058-9
19. Kleemann, F.C., Essig, M.: A providers' perspective on supplier relationships in performance-based contracting. J. Purchas. Supply Manage. **19**(3) (2013). https://doi.org/10.1016/j.pursup.2013.03.001
20. Nystén-Haarala, S., Lee, N., Lehto, J.: Flexibility in contract terms and contracting processes. Int. J. Manag. Projects Bus. **3**(3) (2010). https://doi.org/10.1108/17538371011056084
21. Raja, J.Z., Frandsen, T., Kowalkowski, C., Jarmatz, M.: Learning to discover value: value-based pricing and selling capabilities for services and solutions. J. Bus. Res. **114** (2020). https://doi.org/10.1016/j.jbusres.2020.03.026
22. Razmdoost, K., Mills, G.: Towards a service-led relationship in project-based firms. Constr. Manage. Econ. **34**(4–5) (2016). https://doi.org/10.1080/01446193.2016.1200106

23. Smith, L., Maull, R., Ng, I.C.L.: Servitization and operations management: a service dominant-logic approach. Int. J. Oper. Prod. Manage. (2014). https://doi.org/10.1108/IJOPM-02-2011-0053
24. Tjendani, H.T., Anwar, N., Putu Artama Wiguna, I.: Two stage simulation to optimize risk sharing in performance-based contract on national road a system dynamic and game theory approach. ARPN J. Eng. Appl. Sci. **13**(15), 4432–4439 (2018)
25. Zou, W., Brax, S.A., Vuori, M., Rajala, R.: The influences of contract structure, contracting process, and service complexity on supplier performance. Int. J. Oper. Prod. Manage. **39**(4), 525–549 (2019). https://doi.org/10.1108/IJOPM-12-2016-0756
26. Ng, I.C.L., Maull, R., Yip, N.: Outcome-based contracts as a driver for systems thinking and service-dominant logic in service science: evidence from the defence industry. Eur. Manage. J. **27**(6) (2009). https://doi.org/10.1016/j.emj.2009.05.002
27. Töytäri, P.: Selling solutions by selling value. In: Practices and Tools for Servitization: Managing Service Transition (2018). https://doi.org/10.1007/978-3-319-76517-4_15

# Lean and Six Sigma in Services Healthcare

# Fast Track in Emergency Services an Integrative Review

Sandra Maria do Amaral Chaves[1(✉)] ⓘ, Robisom Damasceno Calado[1(✉)] ⓘ,
Sara Avelar Coelho[1(✉)] ⓘ, Olavo Braga Neto[2(✉)] ⓘ,
Alexandre Beraldi Santos[1(✉)] ⓘ, and Saulo Cabral Bourguignon[3(✉)] ⓘ

[1] Universidade Federal Fluminense, Rio das Ostras, Brazil
{sandrachaves,robisomcalado,saraavelar,alexandreberaldisantos}@id.uff.br
[2] Ministério da Saúde, DAHU-SAES-MS, Distrito Federal, Brazil
olavo.neto@saude.gov.br
[3] Universidade Federal Fluminense, Niterói, Brazil
saulocb@id.uff.br

**Abstract.** The objective of this study was to investigate the methodological theoretical framework of the Lean approach in overcrowded emergency services with the implementation of Fast Track. With the integrative review method, it was defined the guiding question: Which Lean principle is related to the application of Fast Track in overcrowded emergency services? The following database were used: Web of Science, PubMed, and Scopus in which 268 records were collected and 14 articles were selected. It was verified that Fast Track is a fast flow that favors the application of the principle of Continuous Flow, essential in emergency services with overcrowding, as it aims to reduce the average length of stay of patients with lower acuity. It was identified in the methodological theoretical framework the demand for more in-depth studies; and that Fast Track is an approach to reducing overcrowding in emergency services associated with the Continuous Flow principle. It was concluded that Fast Track proved to be an effective management strategy in reducing the average length of stay and waiting time of patients, which resulted in improvements in the flow of care and increased patient satisfaction, consequently providing improved performance in the emergency services.

**Keywords:** Emergency departments · Fast track · Lean healthcare

## 1 Introduction

Emergency overcrowding occurs when the demand for emergency services exceeds the capacity to provide care. The problem of overcrowding in emergency services is global in scope [16]. Lean thinking has been applied in emergency services to reduce waste, particularly in relation to the average length of stay of

© IFIP International Federation for Information Processing 2021
Published by Springer Nature Switzerland AG 2021
A. Dolgui et al. (Eds.): APMS 2021, IFIP AICT 631, pp. 241–249, 2021.
https://doi.org/10.1007/978-3-030-85902-2_26

low-acuity patients, being value stream mapping the main tool [9,10]. Thus, the implementation of Fast Track for management of hospital systems has allowed to improve the flow of patients, speeding up the fulfillment of relevant activities. This study finds its main justification in the need to understand the theoretical and methodological aspects presented in the literature regarding the concept of Fast Track related to Lean Healthcare or lean management. Considering that there are still gaps in the literature regarding Fast Track applications as a robust intervention in emergency services, in this perspective, the guiding question of this study was built: What is the Lean Principle related to the application of Fast Track in overcrowded emergency services? The objective of the study was to investigate the theoretical and methodological framework of the Lean approach in overcrowded emergency services with the implementation of Fast Track.

## 2    Method

Integrative Review - it was carried out in 6 steps as shown in Fig. 1.

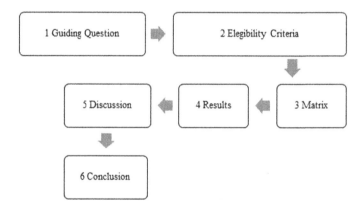

**Fig. 1.** Demonstrative scheme of the 6 steps performed in the integrative review of this article in the period from January the 1st, 2016 to December 31st, 2020.

1. In order to select the guiding questions, the terms Medical Subject Headings (MeSH) were verified, after which search strategies were constructed using only the English language, in the Web of Science, PubMed and Scopus databases, in the period January the 1st, 2016 to December 31st, 2020.
2. The selected articles addressed the Lean principle and their interventions in emergency services in overcrowded health sectors. However, records of systematic review articles, records of scientific events, letters to the editor, book chapters, theses and dissertations, and studies on overcrowding and others without the Lean approach were excluded. The PRISMA flowchart [12] was applied for this purpose.

3. Creation of a matrix with analysis units divided into three categories: Category I - Lean Principles; Category II - Intervention approach; III - Indicators.
4. Presentation of results according to the distribution of articles in the units of analysis after categorization.
5. Interpretation and discussion of the results obtained.
6. Presentation of the review and conclusions.

# 3   Results

In the Scopus database, the following descriptors were used: "emergency department", "emergency room", "emergency units", "emergency ward", "lean", "lean six sigma", "sigma metrics", "six sigma", filters were applied: pub year 2015 and pub year 2021 and limited to English language and records of article types as well as limited to subarea: medicine and nursing - 72 records were collected. In the PubMed database, the following descriptor was added: "crowding", with filters: full text pub year 2016 and pub year 2020 and limited to English language as well as articles PUBMED – 31 records were collected. And in the Web of Science database, the same strategy used for the PubMed database was applied, with the filters: types of documents: article language: English, which resulted in the collection of 165 records.

The collection of records took place from January 11th to 28th, 2020, which were selected in three phases, according to the PRISMA 2020 Flowchart [12], (Fig. 2). First phase of the 268 records, 06 articles were excluded using the duplicate identification strategy in the Mendley® reference manager. Second phase, 190 articles from the remaining 262 were excluded, after reading the titles and abstracts, for not meeting the scope of this research. The remaining 72 articles were fully read, 51 articles were excluded, as they dealt with the Lean approach in relation to overcrowding in order health services that did not directly involve the emergency services. Third phase, the full reading of the 21 selected articles was carried out, in order to identify the units of analysis according to the categories described in the method of this study, 07 articles were excluded for dealing with evaluations on the Lean approach without information on the interventions performed, 14 articles were selected and submitted to the analysis matrix, shown in Table 1.

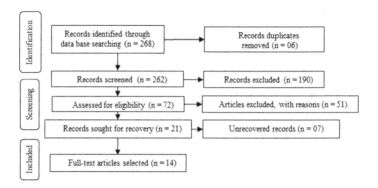

**Fig. 2.** PRISMA 2020 Flowchart - it demonstrates the steps of selected published articles from January 11th to 28th, 2020.

**Table 1.** Distribution of analysis units by categories in the articles

| Reference | Category I Lean Principles | | | | | Category II Intervention Approaches | | | | | | Category III Indicators | | |
|---|---|---|---|---|---|---|---|---|---|---|---|---|---|---|
| | UNITS OF ANALYSIS | | | | | | | | | | | | | |
| | AI | BI | CI | DI | EI | AII | BII | CII | DII | EII | FII | AIII | BIII | CIII |
| [1] | * | * | | | * | * | * | | * | | | * | | |
| [2] | | * | | | | * | | | | | | * | | |
| [3] | * | * | | | * | * | | | | | | * | | |
| [4] | | * | * | * | * | | | | | | | * | * | * |
| [5] | * | * | | | | * | | * | | | * | * | * | |
| [6] | * | * | | | | * | | | | | | | * | |
| [7] | | * | | | * | * | | | * | | * | | | |
| [8] | * | | | | | | | | | | | | * | * |
| [10] | | | | | | | | | | | | * | * | * |
| [11] | | | | * | | | | | | | | * | | * |
| [13] | * | | | | | * | | * | * | | | | | |
| [14] | | | * | | | * | | | | | | * | * | |
| [15] | * | * | | * | | * | | | * | | * | | | * |
| [16] | | * | * | * | | * | | | | | | | * | * |

LEGEND: **AI** (Added Value) **BI** (Flow of Value) **CI** (Continuous Flow) **DI** (Pull Production) **EI** (Perfection) **AII** (Value Stream Mapping) **BII** (Report A3) **CII** (5S) **DII** (Kaizen) **EII** (Plan Do Check and Act) **FII** (Define Measure Analyze Improve and Control) **AIII** (LOS = Length of Stay) **BIII** (WT = Waiting Time) **CIII** (LWBS = Leave Without Being Served).

## 3.1   Category I – Lean Principle

**Added Value Principle:** In the Lean methodology, the use for any purpose other than the creation of value is a waste and must be eliminated [8]. An important focus of Lean is waste reduction, which is defined as anything that "does not add value" to the customer, such as waiting time [3]. Identifying and reducing non-value-added activities is the key to improving patient flow [1].

**Value Stream Principle:** In the methodology, they used direct observation to collect flow data from patients in the emergency room [4]. They aimed to increase the flow to the client/patient, where the value is effectively measured by the time spent by the patient before receiving care within the emergency service [6].

**Principle of Continuous Flow:** The improvement of process contributes to the enhancement of clinical services and facilitates the transitions (flow) of patients in the various stages of medical treatment [5]. Lean focuses on achieving continuous flow through a process, identifying value and eliminating waste [1].

**Principle of Pull Production:** The "pull as far" strategy that was employed so that, hen rooms were available, patients were brought back immediately and screened by the bedside by the head of nursing [15]. Intervention with a series of processes, improvement steps based on Lean methodologies, and aimed at reorganizing the radiology process flow, resulted in a shift to a "pull" system instead of a "push" system, in which patients were actively moved to the subsequent step in their tests by radiology technicians [16].

**Perfection Principle:** This principle is one of the fundamentals of Lean, in which improvement efforts must be fast and interactive, as methods change quickly and frequently and, in its study, it involved improvements in the emergency sector personnel model, resulting in a dramatic reduction in waiting time [4]. In a comprehensive project that involved all visits to the emergency department, including less severe cases that need Fast Track and high-acuity cases that need observation and longer treatment time, and incorporated lean principles as an approach to improvement [4]. One of the steps for improvement consisted of adding a physician or triage team to the system [11].

## 3.2   Category II – Intervention Approaches

**Value Stream Mapping (VSM):** The VSM is a key tool to identify opportunities to reduce waste and further integrate process steps, thus improving processing efficiency [1]. The VSM in the Lean approach serves to identify (and potentially remove) any "steps that do not add value" in the process [5]. Process mapping is a component of Lean management, which was applied in the psychiatric service linked to the emergency department where there was a significant reduction in the length of stay of psychiatric patients in the emergency department [2]. After identifying the barriers to efficiency, these were grouped by similar ideas or concepts and the VSM was then executed [7]. A3 Report: analyzed activities using process mapping and problem solving spreadsheet or A3 report as a visual tool to identify sources of waste [1].

**Five Sense (5S):** They applied the 5S as a visual technique for the reorganization of the nursing station, as they followed the theoretical path of the Lean methodology, choosing the techniques considered the most important in health management, such as VSM and 5S [5]. With the 5S technique, a consensus was reached among employees about what materials and equipment should be available and how and where they should be placed in each zone [13].

**Kaizen:** The Kaizen team's first activity was to develop a high-level process map to understand the main steps in the process. The use of Six Sigma Lean strategies served to streamline the computerized tomography processing of the emergency department in a tertiary care unit, created a Kaizen team, whose first activity was to develop a high-level process map to understand the main steps of the process [7]. The rapid improvement workshop technique corresponds to the application of Kaizen in [13,15].

**Plan, Do, Check and Act (PDCA):** In the construction of the MFV, the teams' responses were structured based on the PDCA Process to initiate some strategies to improve the teams' performance [1].

**Define, Measure, Analyze, Improve and Control (DMAIC):** Some studies point to the use of Lean Six Sigma and additional management or mathematical tools in their bibliography referenced one of their studies on DMAIC [5].

### 3.3   Category III – Indicators

**Length of Stay (LOS):** They used a color code to process the time from the patient's arrival to the emergency room until the patient leaves the emergency room [5]. There was a statistically significant improvement in the mean length of stay between the pre-mapping group and the post-mapping group [2]. The intervention with Lean approach reduced the overall length of stay in the emergency service and in the subgroup analysis the LOS rate reduced for non-critical cases, but no significant improvement for critical and semi-critical cases [11]. Waste in the emergency patient flow process is considered not only as a problem, but also as the basis for measuring and improving process outcomes (e.g., LOS) [1].

**Waiting Time:** To study the issue of overcrowding an interesting new case study was based on the implementation of the Lean methodology to reduce emergency waiting times and improve the flow of patients between the emergency room and recovery areas [4]. They reported having reduced waiting times during peak hours in diagnostic imaging and particularly in X-ray with Lean Healthcare [6]. In a cross-sectional observational study, comparing crowding, radiology response times, patient waiting times in the emergency department, and the proportion of patients who leave unattended (LWBS) report their reductions [16]. Through systems engineering approaches such as Lean methodologies, emergency room leaders can potentially reduce patient wait times [16].

**Leave Without Being Served (LBWS):** In an external review, Lean training, Fast Track, patient centered approach, door-to-doctor approach, performance reporting and an action-based peak capacity protocol the LWBS indicator decreased, and in Fast Track areas with mid-level providers, improved patient flow occurs and reduces crowding, wait times, LOS and LWBS rates in the emergency room without affecting the quality of care [10]. Waiting times and services to identify the main causes of overcrowding were calculated, and Fast Track was applied for urgent cases that do not require observation or long stay, this allows to manage more patients with fewer staff [4]. Lean principles in the redesign of frontline operations in the emergency department and regression-adjusted difference estimates for LWBS rates showed that it did not change significantly [15]. In their study, they obtained a result of improvement in the rate of patients leaving a health service without being treated after their arrival record (LWBS) [11].

## 4 Discussion

Lean principles are fundamental in the management process, considering that they are intrinsically related and sequential. These principles were referenced in the studies, especially in the theoretical foundation, not making it clear how they guided their interventions.

As for the relationship between Lean principles and interventions, it was found that the principle of Pull Production was described in only three of the fourteen studies analyzed, contrary to the Perfection principle, which was frequently cited, which leads us to the understanding that they are related to the application of strategies for continuous improvement. The Value Stream principle was also mentioned, in this sense, the added value represents the patient's satisfaction when being assisted in a shorter time than expected.

The principle of Continuous Flow is related to the application of fast track strategies, which are essential in overcrowded emergency services, as it aims to reduce the average length of stay of patients with lower acuity, and which represent one of the main factors that contribute to overcrowding in emergency services, creating bottlenecks.

The time factor has been extensively studied in relation to the stay of patients in emergency services and we can relate it to the Added value principle, as this factor was considered as a value attributed to patients who had their time clocked and recorded in their care for analysis of results of the LOS and LWBS indicators.

A large part of the studies showed positive results, highlighting the importance of the Lean approach in relation to improving the quality of emergency services.

## 5 Conclusion

In this study, it was found that the Lean Continuous Flow principle is related to the application of Fast Track, and consists of a fast flow, which in the context of emergency services contributes to reducing overcrowding. It was found that for the healthcare area, the flow of care with Fast Track represents the application in an order or sequencing of the planning and control of resources, of the type "Shortest Operating Time first" (SOT), where small tasks are processed quickly, aimed at patients with lower risk or complexity. Thus, with the performance of care with Fast Track, improvements in the flow of care are achieved, resulting in patient satisfaction through the shorter waiting time and shorter average length of stay for patients in the emergency room.

Fast Track proved to be an effective management strategy in reducing the average length of stay of the patient and consequently, it would improve the performance of these services. Public and private sector managers devote resources and efforts to mitigating the effects of overcrowding. The improvement in the quality of care for patients in emergency services with the Lean approach was demonstrated in the articles, however, from a theoretical perspective, we saw that future research is still needed to consolidate a theoretical and methodological framework that can serve as a reference for application Fast Track. The methodological theoretical framework in the application of the flow of care with Fast Track in the Lean Healthcare approach in overcrowded emergency services was not clearly presented in the articles, further studies using the method intervention (DMAIC) are suggested, according to the Lean Six Sigma approach.

**Limitations:** The terminology on the different techniques, tools and methods used in the Lean approach is vast, it was observed that they are presented with different names in the articles, which led us to adopt Category II Intervention Approach, to be included regardless of nomenclature adopted in the articles.

**Acknowledgements.** The authors would like to thank the Brazilian Ministry of Health, Universidade Federal Fluminense and Euclides da Cunha Foundation. This Research is part of a "Lean Project in UPAs 24 h" that has been funded by the Brazilian Ministry of Health (TED 125/2019, number: 25000191682201908).

# References

1. Al Owad, A., Samaranayake, P., Karim, A., Ahsan, K.B.: An integrated lean methodology for improving patient flow in an emergency department-case study of a Saudi Arabian hospital. Prod. Plan. Control **29**(13), 1058–1081 (2018)
2. Alexander, L., Moore, S., Salter, N., Douglas, L.: Lean management in a liaison psychiatry department: implementation, benefits and pitfalls. BJPsych Bull. **44**(1), 18–25 (2020)
3. Balfour, M.E., Tanner, K., Jurica, P.J., Llewellyn, D., Williamson, R.G., Carson, C.A.: Using lean to rapidly and sustainably transform a behavioral health crisis program: impact on throughput and safety. Joint Comm. J. Qual. Patient Saf. **43**(6), 275–283 (2017)
4. Elamir, H.: Improving patient flow through applying lean concepts to emergency department. Leadership in Health Services (2018)
5. Improta, G., et al.: Lean thinking to improve emergency department throughput at Aorn Cardarelli hospital. BMC Health Serv. Res. **18**(1), 1–9 (2018)

6. Jessome, R.: Improving patient flow in diagnostic imaging: a case report. J. Med. Imag. Radiation Sci. (51), 1–11 (2020)
7. Klein, D., Khan, V.: Utilizing six sigma lean strategies to expedite emergency department CT scan throughput in a tertiary care facility. J. Am. Coll. Radiol. **14**(1), 78–81 (2017)
8. van der Linden, M.C., van der Linden, N.N., et al.: The impact of a multimodal intervention on emergency department crowding and patient flow. Int. J. Emerg. Med. **12**(1), 1–11 (2019)
9. Lobo, C.V.F., da Conceição, R.D.P., Calado, R.D.: Evaluation of value stream mapping (VSM) applicability to the oil and gas chain processes. Int. J. Lean Six Sigma **1**, 1–23 (2018)
10. Patey, C., et al.: Surgecon: priming a community emergency department for patient flow management. West. J. Emerg. Med. **20**(4), 654 (2019)
11. Peng, L.S., Rasid, M.F., Salim, W.I.: Using modified triage system to improve emergency department efficacy: a successful lean implementation. Int. J. Healthcare Manage. **14**(2), 419–423 (2021)
12. Rethlefsen, M.L., et al.: Prisma-s: an extension to the prisma statement for reporting literature searches in systematic reviews. Syst. Rev. **10**(1), 1–19 (2021)
13. Sánchez, M., Suarez, M., Asenjo, M., Bragulat, E.: Improvement of emergency department patient flow using lean thinking. Int. J. Qual. Health Care **30**(4), 250–256 (2018)
14. Swancutt, D., et al.: Not all waits are equal: an exploratory investigation of emergency care patient pathways. BMC Health Serv. Res. **17**(1), 1–10 (2017)
15. Vashi, A.A., Sheikhi, F.H., Nashton, L.A., Ellman, J., Rajagopal, P., Asch, S.M.: Applying lean principles to reduce wait times in a VA emergency department. Mil. Med. **184**(1–2), e169–e178 (2019)
16. White, B.A., Yun, B.J., Lev, M.H., Raja, A.S.: Applying systems engineering reduces radiology transport cycle times in the emergency department. West. J. Emerg. Med. **18**(3), 410 (2017)

# Information, Communication and Knowledge for Lean Healthcare Management Guidelines, a Literature Revision

Christiane Barbosa[1](✉) ⓘ, Adalberto Lima[2] ⓘ, Alberto Sobrinho[1] ⓘ,
Robisom Calado[3] ⓘ, and Sandro Lordelo[3] ⓘ

[1] Universidade Federal do Pará, Belém, Brazil
cllima@ufpa.br
[2] Universidade Federal do Pará, Abaetetuba, Brazil
[3] Instituto de Ciência e Tecnologia, Universidade Federal Fluminense,
Rio das Ostras, Brazil
sandrolordelo@id.uff.br

**Abstract.** The objective is to identify the management of information, communication and knowledge in Emergency Care Units (ECU). The method was a lexical and semantic analysis using VOSViewer, categorization of the results for a qualitative and cross analysis of the approaches on the subject. With the results of the practical application of the Lean Project in the ECU in 50 units it was possible to establish criteria and a parameter for knowledge management in this health area. The results categorize the publications in four central themes in healthcare: leadership and governance; quality in health; technology and information; Lean. The information flow is presented as a means, not an end activity and this is confirmed when the articles limit themselves to highlight the means of communication, applications or devices used for the dissemination of information, without further deepening of steps such as data collection, validation and valuation of information in health environments. The human factor is the main agent of change, because it needs clarity that the use of information and communication technologies do not make its final activity impossible, but it is a facilitating, agile and effective tool in the conduct of their work activities.

**Keywords:** Lean Healthcare · Information management · Communication management

## 1 Introduction

The healthcare environment is dynamic and its complexity of services, processes and application areas, both public and private is understood by society as chaos. The Unified Health System (SHS) is composed of several units, with different

ⓒ IFIP International Federation for Information Processing 2021
Published by Springer Nature Switzerland AG 2021
A. Dolgui et al. (Eds.): APMS 2021, IFIP AICT 631, pp. 250–257, 2021.
https://doi.org/10.1007/978-3-030-85902-2_27

purposes each. They complement each other in order to meet the demands generated by the population's needs.

Because it operates on a large scale and in a country with territorial extensions such as Brazil, the operational noncompliance of SUS either due to decentralized management or inconsistency of data or lack of communication or incompatibility of software used in the units may influence the level of service. Information and communication are used in national statistics and provide some transparency and access to official information, generating an overview of public health in Brazil.

Thus, this research aims to identify the existence of practices focused on information and communication management in urgency and emergency hospital environments. This study seeks to answer the three following base questions: i) How is information and communication management applied in the health area? ii) How does knowledge management occur in the health environment?

## 2 Basis of Concepts

Information needs a flow to drive improvements and for communication to be disseminated. The information flow requires constant evaluation, definition of collection instruments, transparency, discipline, continuous improvement to assist the decision making process and produce useful information [18]. The fragility of the information flow is presented: in redundancy, lack of clarity, reliability or diversity of sources; generate costs for the dissemination or creation in/of communication vehicles; fragmentation between sectors within the same organization; creation of communication barriers between the generating pole and the information consumer; rigidity to innovations; lack of objectivity; disorganization and inefficiency of systems and/or people.

A targeted and transparent workplace simplifies management when the work environment speaks to employees and managers. For Lean Healthcare to succeed, communication needs to be frequent in transmitting information, facilitating employee understanding and commitment [11] creates clear, measurable and bidirectional goals [16], and the results must be contemplated, demonstrated and transmitted to the right people [6].

The greatest practices in public sector are Information and Communication Technologies (ICTs) in government, the existing physical and network infrastructure and the provision of better services to citizens [9].

## 3 Method

The method was divided into two phases: 1. lexical and semantic analysis using VOSViewer with published articles on information and communication management in Lean Healthcare in Emerald Insights, Scopus/Elsevier, Medline/PubMed, Web of Science; 2. application of knowledge management in urgency and emergency of the Lean Project in Emergency Care Units (ECU).

# 4   Results

The authors citation network is formed by two clusters with two authors in each overlapping in pairs. The analysis of the impact factor of publications presents Tzortzopoulos, P.'s articles with the higher impact factor and the most cited work is from the year 2018 (Table 1). The highlight is for four authors with the highest number of publications being Improta, G. the most cited with five published works. The authors with the greatest number of works are little cited and were not listed. Therefore, the authorship and co-citation do not show an author as a reference in the subject studied.

**Table 1.** Relationship between author and quantity of quotations

| Author | Documents | Year | Citations |
|---|---|---|---|
| Improta, G. | 5 | 2020 (1) | 0 |
| | | 2019 (1) | 2 |
| | | 2018 (2) | 9 and 19 |
| | | 2017 (1) | 18 |
| Triassi, M. | 4 | 2020 (1) | 0 |
| | | 2019 (1) | 2 |
| | | 2018 (2) | 9 and 19 |
| Tzortzopoulos, P. | 4 | 2020 (2) | 0 and 1 |
| | | 2018 (2) | 1 and 6 |
| Junior, J. S. | 4 | 2020 (2) | 0 and 1 |
| | | 2019 (1) | 0 |
| | | 2018 (1) | 1 |

Triassi, M.'s publications have the second best impact factor and the most cited work is from 2018 in Mathematical Biosciences co-authored with Improta, G. and others. 65.8% are papers affiliated with Harvard Medical School (10) and the predominance is publication in English, in the form of scientific articles, and in countries such as the United States (77), including Brazil (12) and in three areas, Medicine (23.9%), Engineering (14%) and Computer Science (11%) and others.

In an analysis of the occurrence of words among the publications it was found that "Human" is the central key in a series of six clusters (Fig. 1). Specificity addressing Information and Communication is on the fringe in the form of Internet of Things, big data, information management, machine learning and automation indicating the least amount of work or even indirect approaches in some publications, corroborating the justification and need for research on the topic.

Lean Production has a direct link with Sustainability and indirect Healthcare. However it is not related to information system or management. It is possible to highlight the research gap on communication, since it is treated as a direct

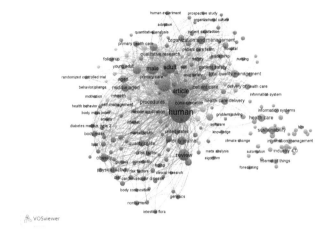

**Fig. 1.** The occurrence of words among publications

consequence and restricted to means of transmission of information, without connection or specificities related to the subject. Only one article deals with terminologies and weaves discussions related to the research object of this study, however, this same article makes a superficial mention about Lean approaches.

The health environment is dynamic and there has been a concern to search for more recent works with open content, which has resulted in five works published in the last three years [2,5,14,17,20]. One article described applicability in Brazil [17], the others presented cases in the UK, Tanzania, Sweden and Canada. All articles presented key words that were linked to the health area, eight with managerial approaches [1,2,4,5,13,17,20,22], five studies explicitly highlighted Lean or its minutiae/particularities [8,12–14,24] and five referred to information or communication or technology aspects [5,17,19,20,23], thus confirming the need to be read in their entirety.

In the phase of reading and analyzing the objectives of the selected articles, it was possible to categorize them into four central themes: leadership or governance; health quality; technology and information; Lean. In an analysis of the context of application and scope of the study, a prevalence was identified in the area of management, leadership [1,4], and governance [17].

Communication is present in all Lean leadership principles and information gains accompany the leader's behavior and improve the value of information to the customer. The way of communicating happens verbal and non-verbal [1]. In a chaotic and urgent approach, Caro [4] and Aij and Teunissen [1] argue that strategic leadership improves the efficiency and effectiveness of emergency services when there is horizontal and vertical systemic integration, requires information exchange and a communication system.

In health sector organizations in Brazil identified formal and informal governance instruments and quality aspects. Quality in the healthcare sector requires collective projects to create coordination and collaboration actions. The findings highlighted the lack of coordination and collaboration in the healthcare system as weaknesses that pose major risks to quality and safety in this environment [2].

Collaboration is the key to success and in dynamic and complex environments like hospitals highlights the need to know the essential elements of patients, to define people (technical staff), specialties, materials, technologies and information, working them in an aligned way [12, 14, 17].

Information and communication management is essential for understanding and referring actions. Communication is linked to how information is transmitted, uses technological resources to expedite when the patient's need is known [17] and any restrictions can affect the emergency service. Information technology links these services, ensures clear and effective communication between the parties [4] and should be applicable in all sectors and in a single area.

The health system uses communication technologies to transmit information and speed delivery to providers such as e-mail, telephone and messaging applications or online meetings. The combination of visual and textual information generates significant advances and efforts to create a comprehensive electronic record. An example of this is the operational efficiency of the organ center in Brazil.

The introduction of new technology makes information management more complex and requires systems that facilitate and promote information sharing in decentralized networks [10]. Information is linked to organizational image, requires standardization, unification of systems and simplified infrastructures [20] to reduce costs, increase reliability and implement operational efficiency.

With the use of Information and Communication Technology (ICT) and a shared platform there have been improvements in communication, creation of a revised and optimized workflow, information analysis, and real-time communication, qualitative feedback [19, 24] and management of their processes [21] especially in public sector management [7] in order to create value for government and society, making it smarter, more efficient, more effective, better communicated, more receptive and closer to citizens.

Finally the need to use Lean techniques and tools in healthcare highlights the importance of providing better quality and safety to patients [10, 13], managing costs, minimizing waiting times [1, 3, 12] and generating employee satisfaction [14]. Faced with increasing external pressure and challenges to overcome, hospitals and their branches of care that conduct their activities based on Lean knowledge have obtained good results in terms of humanizing patients and employees, eliminating barriers, centralizing efforts in service delivery and promoting integration between sectors even the sustainability of Lean [8, 15].

This study proposes a generalized flow Table 2 that aims at knowledge management, bringing people and processes together and defining a periodicity.

The purpose is to subsidize the activities of health professionals. However, it is necessary to distinguish two points on information and communication: i) their management for decision making and internal organization of working procedures in a health unit; ii) and, transparency, efficiency and use by clients/patients.

Thinking about a flow of Lean information and communication is to prioritize clarity, organization, trust, own identity and intelligibility.

**Table 2.** Proposed generalized information flow to the health environment.

| Data | Information management | Communication management | Knowlegde managementt | ECU's aplication |
|---|---|---|---|---|
| Source, types | Classification | Dissemination | Acquisition | Database creation in each unit |
| Collect, storage | Formating | Tecnologies | Storage | Data collection by sampling, recording of times per service and patient |
| Selection, sampling | Representation | Agility | Distribution | Enter the data in the form and/or system by patient |
| Filtration | Atualization | Reability | Use | Compile the cycle times per patient in spreadsheet |
| Accuracy | Applicability | Decision making | People: ability and competence | Check and understand the outliers in service cycle times per patient |
| Padronization | Padronization | Image | Process: resource and investiments | Visual management of average time (LOS-Length of stay), graph or Dashboard (Power BI) |

The Lean project in ECU was developed in three phases: planning (Dec/2019 to Mar/2020); implementation (Mar/2020 to Oct/2020) with on-site technical visits to the units with weekly follow-up and reporting; and remote monitoring of the results achieved (Nov/2020 to May/2021). The results obtained with information, communication and knowledge management are: production and distribution of 10 technical scientific journals on Lean tools applied to public health registered with DOI; Development and online distribution of handouts to train the continuous improvement teams; 10 meetings of the Lean Project in the UPAs with a total of 3,530 participants from 8 states and the Federal District, and with themes related to the tools and methods of the Lean Project; Production and dissemination of 3 videos to integrate the teams; Presentation and communication of the project through social media (Facebook and Youtube, website and Whatsapp group); Monitoring of the technical visits in the UPAs and percentage control of the execution through Dashboard - Power BI with weekly feedback; Production of 450 technical visit reports; Achievement of 678 improvement actions evidenced and accessible through the application; Self-assessment of the Lean Healthcare maturity level of 3 Project ECU, after project implementation; Delivery of 926 certificates from PROEX-UFF for training and qualification of the Lean Project members in the UPAs; 6 books or book chapters; Development of 3 softwares (improvement practices, self-assessment and overcrowding).

The methodology benefited about 13 million users of the public network in the country, more than 28,300 h of training for the 548 employees of the UPAs between the months of April and November 2020.

## 5 Conclusion

The use of information and communication technology is present in healthcare and this research confirmed that ICT is used to connect the system and help leaders in decision making, communication and management strategy. The least attention corresponds to functionality, technical specifications and cultural change issues as a means to be adopted in healthcare facilities.

Health units do not practice information management as a whole because there was no report on the origin and methods used to generate this information, allowing the conclusion that there is negligence on this issue. The human factor is resistant to the use of ICT in daily life, not the technology itself. The greatest benefit is in the activities of the professionals, but there is the neglect of information and communication for the patient, leaving them out of the whole process.

All information has its value, goes through a flow, is disseminated and reaches people. It should be the basis for decision making by managers. The usefulness and the audience to be served by this information varies. Knowledge must be transmitted in a clear, direct and accurate manner. The dynamics of the hospital environment offer opportunities for Lean Thinking to intervene in adjusting processes, streamlining patient care, organizing the work environment, and motivating employees. The result is an improvement in the quality of the service provided.

**Acknowledgment.** The authors would like to thank the Brazilian Ministry of Health, Fluminense Federal University and Euclides da Cunha Foundation. This Research is part of a "Lean Project in UPAs 24h" that has been funded by the Brazilian Ministry of Health (TED 125/2019, number: 25000191682201908).

# References

1. Aij, K.H., Teunissen, M.: Lean leadership attributes: a systematic review of the literature. J. Health Organ. Manage. **31**(7–8), 713–729 (2017)
2. Allen, D.: Translational mobilisation theory: a new paradigm for understanding the organisational elements of nursing work. Int. J. Nurs. Stud. **79**, 36–42 (2018)
3. Bhat, S., Jnanesh, N.: Enhancing performance of the health information department of a hospital using lean six sigma methodology. Int. J. Six Sigma Competitive Advantage **8**(1), 34–50 (2013)
4. Caro, D.H.: Code red: towards transformational leadership of emergency management systems **12**(2), 113–135 (2015)
5. Comes, T., Sandvik, K.B., Van de Walle, B.: Cold chains, interrupted: the use of technology and information for decisions that keep humanitarian vaccines cool. J. Human. Logist. Supply Chain Manage. **5**(1), 49–56 (2018)
6. Costa, L.B.M., Filho, M.G., Rentes, A.F., Bertani, T.M., Mardegan, R.: Lean healthcare in developing countries: evidence from Brazilian hospitals. Int. J. Health Plann. Manage. **32**(1), e99–e120 (2017)
7. Criado, J.I., Gil-Garcia, J.R.: Creating public value through smart technologies and strategies. Int. J. Public Sector Manage. **32**(5), 438–450 (2019)
8. Flynn, R., Rotter, T., Hartfield, D., Newton, A.S., Scott, S.D.: A realist evaluation to identify contexts and mechanisms that enabled and hindered implementation and had an effect on sustainability of a lean intervention in pediatric healthcare. BMC Health Serv. Res. **19**(1), 1–12 (2019)

9. Gil-Garcia, J.R., Pardo, T.A., Nam, T.: What makes a city smart? Identifying core components and proposing an integrative and comprehensive conceptualization. Inf. Polity **20**(1), 61–87 (2015)
10. Harrison, M.I., Paez, K., Carman, K.L., Stephens, J., Smeeding, L., Devers, K.J., Garfinkel, S.: Effects of organizational context on lean implementation in five hospital systems. Health Care Manage. Rev . **41**(2), 127–144 (2016)
11. Hwang, P., Hwang, D., Hong, P.: Lean practices for quality results: a case illustration. Int. J. Health Care Qual. Assur. **27**(8), 729–741 (2014)
12. Ishijima, H., Eliakimu, E., Mshana, J.M.: The "5s" approach to improve a working environment can reduce waiting time. TQM J. **28**(4), 664–680 (2016)
13. Kahm, T., Ingelsson, P.: Lean from the first-line managers' perspective-assuredness about the effects of lean as a driving force for sustainable change. Manage. Prod. Eng. Rev. **8**(12), 49–56 (2017)
14. Kaltenbrunner, M., Mathiassen, S.E., Bengtsson, L., Engström, M.: Lean maturity and quality in primary care. J. Health Organ. Manage. **33**(2), 141–154 (2019)
15. Leite, H., Bateman, N., Radnor, Z.: Beyond the ostensible: an exploration of barriers to lean implementation and sustainability in healthcare. Prod. Plann. Control **31**(1), 1–18 (2020)
16. Lorden, A.L., Zhang, Y., Lin, S.H., Côté, M.J.: Measures of success: the role of human factors in lean implementation in healthcare. Qual. Manage. J. **21**(3), 26–37 (2014)
17. Martins, S., Machado, M., Queiroz, M., Telles, R.: The relationship between quality and governance mechanisms: a qualitative investigation in healthcare supply-chain networks. Benchmarking Int. J. **27**(3), 1085–1104 (2020)
18. Massuqueto, K., Duarte, M.d.C.F.: Gerenciamento do fluxo da informação: Estratégia convergindo com a prática do lean office. Revista Intersaberes **10**(21), 676–687 (2015)
19. Murphy, B.P., O'Raghallaigh, P., Carr, M.: Nurturing the digital baby: open innovation for development and optimization. Health Inform. J. **26**(4), 2407–2421 (2020)
20. Petersilge, C.A.: The enterprise imaging value proposition. J. Digit. Imaging **33**(1), 37–48 (2020)
21. Sabet, E., Yazdani, N., De Leeuw, S.: Supply chain integration strategies in fast evolving industries. Int. J. Logist. Manage. **28**(1), 29–46 (2017)
22. Shaw, J.A., Kontos, P., Martin, W., Victor, C.: The institutional logic of integrated care: an ethnography of patient transitions. J. Health Organ. Manage. **31**(1), 82–95 (2017)
23. Sweeney, A., Clement, S., Filson, B., Kennedy, A.: Trauma-informed mental healthcare in the UK: what is it and how can we further its development? Mental Health Rev. J. **21**(3), 174–192 (2016)
24. Xu, D., et al.: Lay health supporters aided by mobile text messaging to improve adherence, symptoms, and functioning among people with schizophrenia in a resource-poor community in rural china (LEAN): a randomized controlled trial. PLoS Med. **16**(4), e1002785 (2019)

# HFMEA-Fuzzy Model for Lean Waste Assessment in Health Care Units: Proposal and Utilization Cases

Harvey Cosenza[1]($\boxtimes$) (iD), Nilra Silva[1]($\boxtimes$) (iD), Olavo Neto[2]($\boxtimes$) (iD), Luis Torres[3]($\boxtimes$) (iD), and Robisom Calado[1]($\boxtimes$) (iD)

[1] Federal Fluminense University, Niterói, RJ, Brazil
robisomcalado@id.uff.br
[2] Ministry of Health – Secretariat of Specialized Health Care – Department of Hospital, Home and Emergency Care, Brasilia, DF, Brazil
olavo.neto@saude.gov.br
[3] Campinas State University, Campinas, SP, Brazil

**Abstract.** The management of risks and failures in health systems is considered one of the most efficient solutions for providing care to patients. In practice, this activity also makes it possible to join new corporate opportunities, because it allows preventive measures resulting from social, scientific, and technological developments to be constantly evolving throughout the organization. The aim of the article is to present an application of the failure modes and their health effects analysis model (HFMEA) integrated with Fuzzy logic in three Emergency Care Units (UPAs) in Brazil. This mapping model analyzes and analyzes how failures occurred in the supply chain of UPAs. The methodology used included dividing the study into 4 stages, which were used to prepare the proposed model through the evaluation of failure modes integrated with Fuzzy logic. This work reports as one of the roots of the problems in the studied UPAs: the predominance of processing and creativity waste. These, associated with the factor Management and Leadership for a Lean environment, demonstrate that failures in leadership are linked to the dissatisfaction of management support professionals. In addition, a hierarchical analysis of the failure probabilities and evaluation of the eight Lean wastes found in the UPAs was presented.

**Keywords:** HFMEA-Fuzzy · Wastes · Lean healthcare

## 1 Introduction

Concern for patient safety has become a priority issue in recent decades. The methods and tools to carry out the exploration of risks that currently exist have been used extensively by industries, for example, aeronautics and nuclear. These have a complexity like that of health, where people and processes are exposed to human and technological failures, etc. The Failure Modes Analysis and Health Effects - HFMEA is one of the most used methods, being the best approach currently available, as mentioned by The Joint

© IFIP International Federation for Information Processing 2021
Published by Springer Nature Switzerland AG 2021
A. Dolgui et al. (Eds.): APMS 2021, IFIP AICT 631, pp. 258–268, 2021.
https://doi.org/10.1007/978-3-030-85902-2_28

Commission [1] to map the steps of a process, identify possible failures and their causes, and recommending measures to prevent failures and/or eliminate unnecessary risks and damages associated with health care. The Fuzzy linguistic model, on the other hand, consists of a tool capable of capturing vague and imprecise information, usually described in a natural language, and converting it to a numerical format, such as "high", "low" or "medium" enabling a more accurate and reliable interpretation. Its main objective is the computational modeling of human reasoning, imprecise, ambiguous, and vague. Therefore, it has been used as a strategy to support decision making for problems in the health area, using linguistic variables and the degrees of relevance that assume, respectively, the attributes and weights for the evaluation of each attributed factor, in order of importance.

Despite the advances already achieved, the following gap has not yet been filled: how to quantify a more accurate and reliable assessment of the probability of failures in emergency care units? As this problem involves failure mode analysis in the presence of qualitative variables that cannot be measured by binary evaluation mechanisms, this work proposes an unprecedented model.

The objective of this article is to present an integration model of HFMEA and Fuzzy logic to evaluate the eight lean wastes in three Emergency Care Units (UPAs) in Brazil, located in the municipality of Parque Vitória in Maranhão, Franco da Rocha in São Paulo and Trindade in Goiás. Therefore, the information used was collected through the internal reports of each UPA. This model proposed for failure analysis was specifically developed for this study, allowing the organization and assessment of the types of losses that do not add value to activities or processes considered important in UPAs. Thus, the data from this research show that the quality improvement processes are at different levels of maturity in the studied UPAs.

Therefore, this model is formulated to achieve the following secondary objectives:

- Classification of waste probabilities.
- Evaluation of the eight wastes found in the UPAs.
- Presentation of the Predominant Waste in the 3 UPAs.

## 2   Theoretical Framework

### 2.1   The 8 Lean Wastes

Lean mindset consists in continually reducing waste from activities that add value [2]. The waste or loss is capable of affecting what truly matters: the safety and quality of patient care. Therefore, a brief description the waste that can be found in the health care system: 1. Overproduction: Excessive monitoring of low-urgency patients, large number of patient visits, etc.; 2. Defects: Wrong handling of medications, carrying out tests inappropriately and mistaken discharge of patients; 3. Waiting: Delays in attending in-coming patients, delays in collection and test results, delay in patient discharge, unavailability of resources, delay in accessing systems, etc.; 4. Transportation: Medicines and materials moved unnecessarily; 5. Creativity/Unused talent: Misuse of professionals' skills; 6. Overprocessing: Doublechecking the same process and asking the same questions consecutively; 7. Inventory: Ineffective distribution/allocation of materials, medicines and downtime information and 8. Motion: Long journey to be made by patients, employees and medical staff. Therefore, this visualization allows health organizations to observe the specific failures that generate waste.

## 2.2  HFMEA

The Healthcare Failure Mode Effect Analysis (HFMEA) is a prospective risk assessment tool that adopts measures aimed to prevent or avoid failures. Therefore, it is used in healthcare environments to map, evaluate, control and review the information of risk-generating processes in further detail. The steps for implementing the HFMEA are defining the topic to be analyzed; assembling a multidisciplinary team; describing the process and performing a risk assessment so as to identify possible failure modes, root causes and effects of failures. Thus, estimating the probability of occurrence, severity and failure detection determine the priorities and define action plans towards achieving the expected results aimed to reduce risks [3].

## 2.3  Fuzzy Logic

Fuzzy sets consist of a theory in which everything is graduated, that is, between the certainty of being and the certainty of not being, there are infinite degrees of uncertainty [4]. In 1965, Zadeh [5] published an article on Fuzzy Sets that led to a series of studies on applica-tions of fuzzy systems. Therefore, more objectively, Fuzzy Sets are, if neces-sary, better suited to dealing with information imperfections than probability theory. Therefore, the transition between belonging or not to a set is gradual, that is, it is not abrupt. An element can belong to more than one fuzzy set with different degrees of associa-tion. Association functions can be defined from the user's experience and perspective, but it is common to use standard association functions such as triangular, trapezoi-dal, and Gaussian functions.

The membership functions $\mu_{Rij}$ ($r_{ij}$) and $\mu_{wj}$ ($w_j$) are assumed to be in the range of $[0, 1]$. All fuzzy sets are considered normalized, i.e., they have finite supports and assume the value of 1 at least once. For the evaluation of an $Xj$ environment, a fuzzy set is assumed which is computed on the bases of $r_{ij}$ and $w_j$, where vector $z = (w_1, w_2, \ldots, w_m, r_1, r_2, \ldots, r_n)$.

$$f(z) = \frac{\sum_{j=1}^{m} w_j r_j}{\sum_{j=1}^{m} w_j}$$

The membership function for weighted degrees is given by: $\mu_{\tilde{R}_i}(\overline{r}_i) = \sup \mu_{Zi}(z) \, \overline{r}_i \in R$; $Z$: $(z) = \overline{r}_i$. The normalized value $\overline{r}_i$ for alternative i is given by:

$$\overline{r}_i = \frac{\sum_{j=1}^{m} w_j r_{ij}}{\sum_{j=1}^{m} w_j}$$

If the final degrees are *crisp values*, $r_1, r_2, \ldots, r_n$, then:

$$p_i = \bar{r}_i - \frac{1}{n-1} \sum_{j=1}^{m} \bar{r}_j.$$

Once $r_{ij}$'s and $w_j$'s are fuzzy variables and $p_i$ is also a *fuzzy* variable with membership function $\mu_{pi}(p) = \sup \mu_{\bar{R}}(\bar{r}_1, \bar{r}_2, \ldots, \bar{r}_n) \, p_i \in R; \bar{r}_1, \bar{r}_2, \ldots, \bar{r}_n; \mu_i (\bar{r}_1, \bar{r}_2, \ldots, \bar{r}_n) = p_i.$

# 3 Methods and Data

The stages comprise: 1. Theoretical Framework: mapping the research topics in the context of available literature and providing theoretical support; 2. Data analysis: document analysis of reports received from the Emergency Care Units; 3. Data Treatment: considerations used for elaborating the proposed modeling through an analysis of failure modes HFMEA and fuzzy logic; 4. Results: the effects of aligning the proposed objectives. It is worth mentioning that the quantitative data offered herein have been collected from the internal diagnoses of each UPA, which have used a maturity scorecard to find what was the probability of occurrence of good management practices, important points that should be addressed so that there was quality management in throughout processes.

Table 1.  Characterization of the degree of importance.

| Xi | Degree of importance |
|-----|----------------------|
| 1 | Very important |
| 0,8 | Important |
| 0,6 | Relatively important |
| 0,4 | Slightly important |
| 0,2 | No importance |

The first stage consists in characterizing the importance of factor in linguistic terms. The second stage was to recruit trained and experienced Lean professionals, out of which 5 specialists were invited to assess the degree of importance of the 8 wastes that were listed on a scale of 0.2 to 1 (Table 1), i.e., those that the specialists believed to be more important.

## 3.1 Supply Matrix Compilation

A Supply Matrix was compiled for each of the three UPAs based on the analysis performed as regards occurrence, severity and detection following the linguistic terms of the HFMEA. Probability of Occurrence: What is the probability of this type of failure to occur? a) Very High: up to one missed score; b) High: between 1 and 3 missed scores; c) Average: between 4 and 5 missed scores; d) Low: between 6 and 7 missed scores; and e) Very Low: all scores were missed. Severity: if the failure occurs, what damage can be inflicted? a) Very Little: failure will have little noticeable effect on discontent patients and/or professionals; b) Little: slight deterioration in performance with slight discontent from patients and/or professionals; c) Average: significant deterioration in performance of a system with patient and/or professional discontent; d) High: system stops working and there is great discontent from patients and/or professionals and e) Very High: same as the above, however. It affects/compromises the safety of patients and/or professionals. Detection: If failure occurs, what is the chance of being detected? a) Almost Certain; b) Alta High; c) Moderate; d) Remote; and e) Nearly impossible.

## 3.2  Demand Matrix Compilation

A Demand Matrix was also compiled based on the opinion of specialists. Grades were assigned to the degrees of membership by considering each of the 5 factors (Table 2) in order of importance. The result consisted in relating the factors of diagnoses received from each UPA to the Lean wastes associated according to their characteristics.

**Table 2.** Factors and degrees of membership of the demand matrix.

| Demand matrix | Degrees of membership | | | | |
|---|---|---|---|---|---|
| Factors | VI | I | RI | SI | NI |
| 1. Management and leadership in a lean environment | 0.67 | 1 | 0 | 0 | 0 |
| 2. Adherence to communication and interaction among staff members | 1 | 1 | 0.50 | 0 | 0 |
| 3. Management through indicators | 0.67 | 1 | 0 | 0 | 0 |
| 4. Physical and technological infrastructure | 1 | 1 | 0 | 0.50 | 0 |
| 5. Clear and well-defined processes | 1 | 0.67 | 0 | 0 | 0 |

## 3.3  Cross-checking of Data from Supply and Demand Matrices

Table 3 represents the cross-checking of data from the demand (Table 2) and supply matrices which allows observing the HFMEA failure analyzes according to occurrence, severity and detection criteria. The cross-evaluation form divides the set into 5 subsets, respecting proportionality.

**Table 3.** HFMEA of supply and demand matrices

Demand × Supply (Occurrence, Severity and Detection)

| Demand | Supply | | | | |
|---|---|---|---|---|---|
| | Very low | Low | Average | High | Very high |
| Very important | 1 | $1 - 1/N$ | $1 - 2/N$ | $1 - 3/N$ | $1 - 4/N$ |
| Important | $1 + 1/N$ | 1 | $1 - 1/N$ | $1 - 2/N$ | $1 - 3/N$ |
| Relatively important | $1 + 2/N$ | $1 + 1/N$ | 1 | $1 - 1/N$ | $1 - 2/N$ |
| Slightly important | $1 + 3/N$ | $1 + 2/N$ | $1 + 1/N$ | 1 | $1 - 1/N$ |
| No importance | $1 + 4/N$ | $1 + 3/N$ | $1 + 2/N$ | $1 + 1/N$ | 1 |

$N = 5$

Demand × Supply – Occurrence

| Demand | Supply | | | | |
|---|---|---|---|---|---|
| | Very low | Low | Average | High | Very high |
| Very important | 1 | 0,80 | 0,60 | 0,40 | 0,20 |
| Important | 1,20 | 1 | 0,80 | 0,60 | 0,40 |
| Relatively important | 1,40 | 1,20 | 1 | 0,80 | 0,60 |
| Slightly important | 1,60 | 1,40 | 1,20 | 1 | 0,80 |
| No importance | 1,80 | 1,60 | 1,40 | 1,20 | 1 |

$N = 5$

Demand × Supply – Severity

| Demand | Supply | | | | |
|---|---|---|---|---|---|
| | Very low | Low | Average | High | Very high |
| Very important | 1 | 0,80 | 0,60 | 0,40 | 0,20 |
| Important | 1,20 | 1 | 0,80 | 0,60 | 0,40 |
| Relatively important | 1,40 | 1,20 | 1 | 0,80 | 0,60 |
| Slightly important | 1,60 | 1,40 | 1,20 | 1 | 0,80 |
| No importance | 1,80 | 1,60 | 1,40 | 1,20 | 1 |

$N = 5$

Demand × Supply – Detecion

| Demand | Supply | | | | |
|---|---|---|---|---|---|
| | Almost certain | High | Moderate | Remote | Nearly impossible |
| Very important | 1 | 0,80 | 0,60 | 0,40 | 0,20 |

(*continued*)

**Table 3.** (*continued*)

| Demand × Supply – Detecion | | | | | |
|---|---|---|---|---|---|
| Demand | Supply | | | | |
| | Almost certain | High | Moderate | Remote | Nearly impossible |
| Important | 1,20 | 1 | 0,80 | 0,60 | 0,40 |
| Relatively important | 1,40 | 1,20 | 1 | 0,80 | 0,60 |
| Slightly important | 1,60 | 1,40 | 1,20 | 1 | 0,80 |
| No importance | 1,80 | 1,60 | 1,40 | 1,20 | 1 |
| **N = 5** | | | | | |

# 4 Result and Discussion

## 4.1 Ranking Waste Probabilities

The results shown throughout this section are based on Table 4 - Evaluation of expectations from data summarized in Table 3. It is worth mentioning that the present arguments are based on an Average - Fuzzy Index. Where, 1 - fully meets the expectation; close to 1 - slightly meets expectations and above 1 - exceeds expectations.

**Table 4.** Evaluation of expectations.

| Demand × Supply matrices results | | | | | | | | | |
|---|---|---|---|---|---|---|---|---|---|
| Factors | Parque Vitória – MA | | | Franco da Rocha – SP | | | Trindade – GO | | |
| | O | S | D | O | S | D | O | S | D |
| Management and leadership in a lean environment | 0,8 | 0,8 | 1,2 | 0,8 | 0,8 | 1 | 0,6 | 0,8 | 1,2 |
| Adherence to communication and interaction among staff members | 0,4 | 0,6 | 1 | 0,4 | 0,8 | 1 | 0,4 | 0,6 | 0,8 |
| Management through indicators | 0,4 | 0,8 | 1 | 0,6 | 0,8 | 1,2 | 0,4 | 0,8 | 1,2 |
| Physical and technological structure | 0,6 | 0,4 | 1 | 0,6 | 0,6 | 0,8 | 0,6 | 0,6 | 0,8 |
| Clear and well-defined processes | 0,4 | 0,6 | 1 | 0,4 | 0,6 | 0,8 | 0,8 | 0,6 | 1 |
| Total - fuzzy index | 2,6 | 3,2 | 5,2 | 2,8 | 3,6 | 4,8 | 2,8 | 3,4 | 5 |
| Average - fuzzy index | 0,52 | 0,64 | 1,04 | 0,56 | 0,72 | 0,96 | 0,56 | 0,68 | 1 |

It is found that UPAs of Trindade - GO and Franco da Rocha - SP are the most sensitive in terms of occurrence, i.e., 56%, to failures concerning the 5 factors and/or precursors of waste; as for severity, the Franco da Rocha - SP UPA has a 72% chance of the generated damage, thus disrupting and impacting the whole system of the unit and, finally, the detection criterion in the Parque Vitória - MA UPA has over 100% chances of being detected. In short, the rankings of HFMEA criteria as to occurrence, severity and detection rates provided the possibility of failures in relation to how often, how, and when they can occur, so that there are preventive measures to combat waste.

## 4.2   Assessment of 8 Lean Wastes Found in the UPAs

The assessment of the eight Lean wastes was correlated with situations found in the UPAs. This analysis allows assessing whether failures described in UPAs are the causes of the wastes found. Table 5 reports the causes, modes and effects of presented failures. Caption: Causes of failure: Why would the failure occur? Failure mode: What can go wrong? Effects of failure: What would be the consequences of failure?

**Table 5.**  Cause, mode and effect of wastes.

| Cause of failure | Mode of failure (Waste) | Effect of failure |
|---|---|---|
| Lean management/Leadership | Over-processing/Creativity | Dissatisfaction with leadership support |
| Adherence to communication and interaction among staff members | Defects | Failure of staff communication and interaction |
| Management through indicators | Waiting/Inventory | Lack of management through indicators |
| Physical and technological infrastructure | Overproduction | Failure in physical and technological structure |
| Clear and well-defined processes | Transport/Motion | Lack of clear and well-defined processes |

Therefore, relating Lean Management/leadership to the waste of over-processing and creativity is affected by presenting redundant or unnecessary steps in processes from the perspective of professionals and/or the patient, and it does not involve the perceptions of professionals about the identification and resolution of the problems which they deal with daily; relating adherence to communication and interaction among staff members to the waste of defects demonstrates the need to redo something due to incorrect or incomplete information, i.e. poor communication due to the fact that meetings for staff alignment do not exist or are inefficient. In addition, ranking Management through indicators of the wastes of waiting and inventory refers to demonstrating that waiting time is mostly associated with failures by some of the teams responsible for a part of the process, information that does not flow, and being unnecessarily stuck in the middle of the process; ranking the physical and technological structure with the waste of overproduction represents the losses derived from waiting for the process due to the lack of organization, quality and efficiency of the health environment. It is also due to loss of medication in addition to what was requested and/or surplus of products due to waiting for their moment of use in certain stages of the production process. In addition, relating clear and well-defined processes to the wastes of transport and motion represents demonstrating the loss of transporting products, medications and patients from one place to another unnecessarily. Therefore, this mapping allowed describing and analyzing the failures occurring in the supply chain of the UPAs, which are the causes of the wastes found. Observing these wastes is going to make it possible to eliminate what does not add value to the user/patient and all professionals involved from the processes.

## 4.3    Presentation of the Predominant Waste at the 3 UPAs

The HFMEA allows calculating the Risk Priority Number – RPN, indicator which is a product (O × S × D) to the three items: occurrence, severity and detection, which allows prioritizing the worst failures. Table 6 shows a quantitative visualization of the highest RPN values for the UPAs in Parque Vitória – MA, 0.77; Franco da Rocha – SP, 0.64, and Trindade – GO, 0.58, in which it was commonly found the element of Management and Leadership in a Lean environment, which is related to the waste of processing and creativity. The effect exerted by these wastes is dissatisfaction with leadership support. For this reason, although UPAs have different difficulties, however, they have the predominance of the same waste in common.

**Table 6.** Fuzzy HFMEA of UPAs waste.

Results of demand × Supply matrices

| Factors | Parque Vitória – MA | | | | Franco da Rocha – SP | | | | Trindade – GO | | | |
|---|---|---|---|---|---|---|---|---|---|---|---|---|
| | O | S | D | CPR | O | S | D | CPR | O | S | D | CPR |
| Lean management/Leadership | 0,8 | 0,8 | 1,2 | 0,77 | 0,8 | 0,8 | 1 | 0,64 | 0,6 | 0,8 | 1,2 | 0,58 |
| Adherence to communication and interaction among staff members | 0,4 | 0,6 | 1 | 0,24 | 0,4 | 0,8 | 1 | 0,32 | 0,4 | 0,6 | 0,8 | 0,19 |
| Management through indicators | 0,4 | 0,8 | 1 | 0,32 | 0,6 | 0,8 | 1,2 | 0,58 | 0,4 | 0,8 | 1,2 | 0,38 |
| Physical and technological infrastructure | 0,6 | 0,4 | 1 | 0,24 | 0,6 | 0,6 | 0,8 | 0,29 | 0,6 | 0,6 | 0,8 | 0,29 |
| Clear and well-defined processes | 0,4 | 0,6 | 1 | 0,24 | 0,4 | 0,6 | 0,8 | 0,19 | 0,8 | 0,6 | 1 | 0,48 |

The waste of creativity corresponds to a misuse of human talent. Not fully using people's capacity is associated with the current culture of Emergency Care Units (UPAs), therefore, it is necessary to improve health services seeking to better learn management and leadership tools and practices so that there is a reduction in the chances of error in management, information, activities, people, materials, among other actions.

## 5 Conclusion

The purpose of this article was to present an integration model of HFMEA and Fuzzy logic to evaluate the eight Lean waste in three Emergency Care Units (UPAs) in Brazil, located in the municipality of Parque Vitória in Maranhão, Franco da Rocha in São Paulo and Trindade in Goiás. Thus, three secondary objectives were achieved: classification of waste probabilities; evaluation of the eight waste found in the UPAs; and presentation of the predominant waste in the UPAs under analysis. This mapping allowed us to describe and analyze the failures that occurred in the supply chain of the UPAs, capable of eliminating what does not add value to the user/patient and to all professionals involved in the processes. The HFMEA/Fuzzy analysis quantitatively showed that among the eight wastes, the main ones among the UPAs were processing and creativity, which was associated with the generators factor of Lean Management and Leadership. The spread of the effect of these wastes generates dissatisfaction with leadership support. Therefore, although UPAs experience different difficulties, however, they have the predominance of the same waste in common. In short, a new lean culture must be fostered in the three UPAs used as the object of study so as to achieve efficient management. The application of HFMEA/Fuzzy model allowed identifying the failure modes belonging to the system in question through the eight wastes presented. However, the entire management team must be committed to experiencing a collaborative environment with the other professionals in their work units in such a way as to target continuous improvement. This work provides knowledge that points pout the need of developing new studies and research focusing on the Health Safety Leadership and Management errors, as the main gap to be filled by future work. Thus, measures can be taken on time so as to ensure continuity of improvement within the organization, which is going to allow the implementation and improvement of patient safety practices based on strategies to prevent failures for maximum reduction and/or elimination of unnecessary risks and damages associated with health care.

**Acknowledgements.** The authors would like to thank the Federal Fluminense University and the Fundação Euclides da Cunha, which made the research project funded by the Ministry of Health of Brazil viable; TED 125/2019, Number: 25000191682201908.

## References

1. DeRosier, J., Stalhandske, E., Bagian, J.P., Nudell, T.: Using health care failure mode and effect analysisTM: the VA national centre for patient safety's prospective risk analysis system. J. Qual. Improv. **28**(5), 248–267 (2002)
2. Lobo, C.V.F., Calado, R.C., Conceição, D.P.R.: Evaluation of value stream mapping (VSM) applicability to the oil and gas chain processes. Int. J. Lean Six Sigma (2018). https://doi.org/10.1108/ijlss-05-2018-0049

3. Helman, H., Andery, P.R.P.: Análise de falhas (aplicação dos métodos de FMEA – FTA). Belo Horizonte: Fundação Christiano Ottoni (1995)
4. Boole, G.: An Investigation of the Laws of Thought on Which are Founded the Mathematical Theories of Logic and Probabilities. Dover Publications, New York (1951)
5. Zadeh, L.A.: Fuzzy sets. In: Information and Control, vol. 8, pp. 338–353 (1965)

# Motivators to Application of DMAIC in Patient Care Processes

Milena Reis[1]([✉]) [iD], Luis Viera[2] [iD], Laryssa Amaral[2] [iD], José Farias Filho[2] [iD], Adriana Teixeira[3] [iD], and Robisom Calado[2] [iD]

[1] Universidade Federal do Rio de Janeiro, Rio de Janeiro, Brazil
milenaestanislau@macae.ufrj.br
[2] Universidade Federal Fluminense, Rio das Ostras, Brazil
{luisvaldiviezo,laryssaamaral,joserodrigues,robisomcalado}@id.uff.br
[3] Ministério da Saúde, DAHU-SAES-MS, Distrito Federal, Brazil

**Abstract.** Six Sigma comprises DMAIC which is a methodology that has attracted the attention of researchers in the field of Healthcare. This study aims to understand the motivators to apply DMAIC in process in the context of Six Sigma and Lean Six Sigma to be implemented in processes focused on direct patient care. The article covers literature on DMAIC from 1997 to October 2020. For such a purpose, a systematic literature review has been carried out based on empirical studies in Hospitals and Emergency units. A total of 21 articles were reviewed, and the topic Motivators were addressed in this article with the aim of performing DMAIC through Six Sigma/Lean Six Sigma. Based on the analyzed articles, the most frequent motivator to implement Six Sigma/Lean Six Sigma DMAIC in organizations lies in reducing process time (such as length of stay) and improving the quality of service. It was also noticed that DMAIC has been used in several health services, such as patient discharge and surgeries (femur surgeries, for instance. An application of DMAIC in 50 emergency care units (UPAs) corroborated the main motivators indicated in the literature.

**Keywords:** Lean Six Sigma · DMAIC · Motivators · Healthcare

## 1 Introduction

Healthcare is a field which has made remarkable advances in technology and treatment, but it is often overwhelmed by inefficiencies, errors, resource constraints and other problems that hinder accessibility and safe patient care [35]. Emergency departments have become increasingly overcrowded due to growing demand, among other factors, which adversely affects the performance of health services [1].

This article analyzes the use of the DMAIC cycle aimed to implement Lean Six Sigma Improvement Projects grounded in the Motivators. The present study seeks to understand the motivations and objectives of implementing the Lean

© IFIP International Federation for Information Processing 2021
Published by Springer Nature Switzerland AG 2021
A. Dolgui et al. (Eds.): APMS 2021, IFIP AICT 631, pp. 269–279, 2021.
https://doi.org/10.1007/978-3-030-85902-2_29

Six Sigma DMAIC cycle in Hospitals and Emergency Units. The Research question of the present study is "What were the motivations for implementing the Improvement Projects?

## 2    Lean, Six Sigma, Lean Six Sigma

The Six Sigma methodology is grounded in the work of Motorola and General Eletric [13]. It is a improvement methodology [18,33]. It is an approach that aims "to improve an organization's products, services and processes by continually reducing defects in organizations" [24]. By defect it is understood "anything that does not meet customer requirements" [27]. It aims to "make the process 99.99996% defect free" [13].

The purpose of using Six Sigma lies in "process improvement"; while a Lean approach "focuses on flow and speed up of processing by reducing or eliminating" waste [28]. Lean focuses on reducing inefficiency, cost [10], waste [10,25]. Lean manufacturing Lean Manufacturing comes from Toyota Production System [25]. Six Sigma and Lean methodologies are complementary, as both use "data collection and analysis to improve performance" and "focus on eliminating waste and redundancy in operational processes" [15]. When used together, they are called Lean Six Sigma (LSS). In literature, it is also referred to as Lean Sigma, Lean Six Sigma DMAIC and LSS DMAIC.

DMAIC refers to the 5 stages of the Lean Six Sigma methodology and it is similar in many to the steps to the PDCA (Plan-Do-Check-Act) cycle [22]. The term DMAIC is an acronym for "Define, Measure, Analyze, Improve and Control" (Define, Measure, Analyze, Improve and Control). DMAIC incorporates a clear problem-solving framework into a Six Sigma approach [17] to be used in order to reduce process variability based on statistics [15,36]. DMAIC can be considered as a continuous improvement method [12], and "a core methodology used in Six Sigma projects", used for improving, optimizing and reorganizing processes" [19].

## 3    Motivators

Motivators are crucial for designing the implementation and achieving results. The motivating factors for implementing Lean Six sigma in Indian SMEs are, namely, reducing cycle time, increasing customer satisfaction, improving process efficiency, increasing production capacity, reducing costs and continuous improvement [34]. It should be noted that there seems to be a gap in studies on motivational factors for Lean Six Sigma applications in Healthcare units.

As regards Lean process implementation, Antony et al. [6] conducted a systematic literature review on healthcare services and stated the following motivational factors: 1) improving consumer services; 2) increasing patient satisfaction; 3) surpassing others and gaining competitive advantage; 4) improving the process and operational efficiency; 5) improving the quality of service; 6) transforming organizational culture; 7) standardizing and streamlining the process; 8) reducing delays and operational time; 9) eliminating waste; 10) eliminating tasks that do not add value; 11) minimizing staff and administrative inefficiencies.

## 4   Materials and Methods

This article sought to answer research questions by selecting articles and performing their subsequent analysis. The analysis of articles corresponded to a systematic literature review.

### 4.1   Systematic Review

In this article, a systematic review was carried out so as to achieve its objective. Therefore, the sequence of stages presented by Moher et al. [26] was selected for its research design Fig. 1 shows a flowchart of the systematic review.

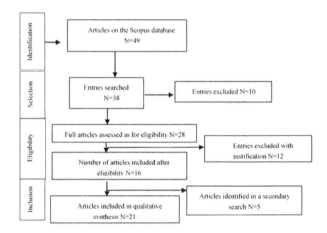

**Fig. 1.** Systematic review: manuscript selection flowchart

The present research was carried out between August and October 2020. Scopus was the database selected for the present research due to the fact that it contains the largest database of peer-reviewed abstracts and citations reviewed.

This article is focused on DMAIC implementations in hospital and emergency health environments. For such a purpose, in the stage of Identification, a simulation of keywords with variations of terms linked to DMAIC was carried out. The term "DMAIC" and its variants were combined and as well as the terms "hospital" or "emergency". For this article, a combination of the following keywords will be used: dmaic OR "define measure analyze improve control" AND hospital OR emergency AND application OR implementation AND patient. As a result, 49 entries were found.

Table 1 shows details about the search carried out on the Scopus Platform regarding search filters. In the Screening Stage, as a result of using search filters through the entries found in the previous stage, only articles published in journals in English were considered. The "year" search filter was not considered; therefore the search covered an analysis of all years, which resulted in an analysis ranging from 1997 to 2020 (10/12/2020). By inserting the search filters, 38 articles remained, of which 10 were excluded (unavailability, poster and abstract reading). Thus, 28 articles were selected for the stage of eligibility.

**Table 1.** Search filters

| Search filters | Search entries | Documents |
|---|---|---|
| Document Type, Type source and language | TITLE-ABS-KEY (dmaic OR "define measure analyze improve control") AND TITLE-ABS-KEY (hospital OR emergency) AND TITLE-ABS-KEY (application OR implementation) AND TITLE-ABS-KEY (patient) AND DOCTYPE ("ar") AND SRCTYPE(j) AND LANGUAGE (english) | 38 |

In the stage of eligibility, full articles were considered with detailed intervention through the DMAIC stages based on an empirical research method carried out in a clinical environment that focused on direct patient care. The 28 articles were examined based on the criteria shown in Table 2. Twelve out of the 28 articles were excluded, and 16 articles remained. In addition, in a secondary search based on the criteria shown in Table 2, 5 articles were identified in the reference sections of selected articles (secondary search) that started to comprise the list of articles that would be used. Thus, the research to be carried out in the next stages was based on 21 articles.

**Table 2.** Inclusion criteria

| Database: Scopus | Focus on patient care in a clinical setting |
|---|---|
| Published in peer-reviewed journals | Clear focus on implementing DMAIC Six Sigma |
| Language: English | Detailed in-process intervention through DMAIC |
| Research method: empirical - case study (single/multiple)/ action research | Exposure of pre-implementation and post-implementation measures |
| Articles published before October 2020 | Articles with an empirical approach |

# 5   Improvement Project Motivators

In this section, the results of the systematic literature review will be presented. The analyses related to the following aspects will be presented: motivators.

## 5.1   Problems and Project Motivators

The analyzed articles have DMAIC implementations through Sigma/Lean Six Sigma applications in hospital wards to solve/minimize a large variety of problems, such as patient dissatisfaction with the service, long waiting time in the process and a negative impact on the patient safety. Table 3 shows the problems to be solved and/or minimized for carrying out the Improvement Project. Regarding the need to reduce the length of stay, in addition to the patient's risks, Singh et al. [32] consider that an increase in length of stay generates "higher treatment costs for patients and a reduction in bed turnover rate".

**Table 3.** Problems

| Authors | Problems |
| --- | --- |
| [5] | Patient dissatisfaction with the discharge process |
| [3] | Multiple patient falls leading to high dissatisfaction rate |
| [4] | Failure to use the PHQ-2/9 depression screening tool in primary care |
| [7] | Long patient discharge time (on average 215.7 min) |
| [9] | Need to reduce average cycle time of an outpatient department service process (Outpatient Department), waiting time and average queue time. Decreased productivity in the organization affecting quality and patient care |
| [8] | The fact of not charging for the service increased the number of patients with a consequent longer length of service and waiting time, which compromised functioning of the services, in addition to crowding within and outside the Hospital |
| [11] | Long patient length of stay has led to long patient waits and delays and, thus staff was unable to meet growing demands for performing patients musculoskeletal procedures, especially in outpatient procedures |
| [14] | Patient safety |
| [16] | Inadequate patient waiting time |
| [21] | Unduly prolonged hospital stays of patients undergoing hip prosthesis surgery |
| [22] | Inappropriate length of stay at the Hospital |
| [20] | Inappropriate length of stay of patients undergoing hip replacement surgery |
| [23] | Inappropriate waiting time of patients needing femur fracture operation |

(*continued*)

**Table 3.** (*continued*)

| Authors | Problems |
|---------|----------|
| [2] | Frequent occurrence of medication errors in the pharmacy which leads to rework and increased waiting time for patients |
| [30] | Need for a path that streamlines the process of femur surgery |
| [29] | Inappropriate length of stay of patients undergoing knee replacement surgery |
| [31] | Inadequate patient transfer time |
| [32] | Delays in administering medications that could create problems such as increased length of stay, recovery time as a result of compromising patient safety |
| [37] | High incidence of human errors impacting patient safety in hospitals and the potential for errors with the use of insulins in the hospital setting |
| [38] | In addition to the hospital's mission to be patient-centered, the results of the voice of the customer (complaints from previous patient periods, feedback form and patient satisfaction survey record) which indicated the time of patient discharge as factoring in dissatisfaction |
| [39] | Impact on patient safety from possible errors in the hospital environment with regard to insulin dispensing and administration process (improper use of medication can cause serious problems to the patient) |

A total of 15 motivators for implementing Six Sigma/Lean Six Sigma using the structure of the DMAIC methodology could be obtained (Table 4). A compilation was carried out by considering the problem at times and the objectives of works shown in Table 3 and at other times based on the term "limitation" or the terms "limitations" or "challenges" in analyzed articles. It is worth mentioning that Al Kuwaiti and Subbarayalu [3] were the only ones to highlight the safety and satisfaction of employees as important factors for carrying out the improvement project. The motivator "reducing time" can be cited as the most widely considered towards the improvement projects under analysis, followed by "improving the quality of service" and "improving the process", "reducing errors; while "reducing rework, improving employee satisfaction" were less mentioned.

**Table 4.** Improvement projects motivators

| Motivators | Authors |
|---|---|
| Improving patient satisfaction | [3,5,32] |
| Improving employee satisfaction | [3] |
| Reducing errors | [2,4,5,37,39] |
| Improving the quality of service | [5,8,9,14,21,22] |
| Reducing risk | [3,14] |
| Improving patient safety | [3,14,32,39] |
| Improving staff safety | [3] |
| Process improvement (simplification/efficiency/ effectiveness/productivity/reliability) | [4,9,11,21,39] |
| Reducing time (waiting time, length of stay) | [2,7–9,11,16,20–23,29–32,38] |
| Improving patient services | [9,14,23] |
| Reducing delays | [11,32] |
| Increasing service capacity | [11,23] |
| Standardizing and streamlining the process | [20,21,23] |
| Reducing costs | [20,23,29,39] |
| Reducing rework | [2] |

As regards the 11 motivators cited by Antony et al. [6], 7 were found in the present work: improving services for consumers, increasing patient satisfaction, improving the process operational efficiency, improving the quality of services, standardizing, and streamlining the process, reducing delays and operational time and eliminating tasks that do not add value. As for those cited by Sodhi et al. [34], among others, the factors "increasing production capacity" and "reducing costs" were also considered.

## 6   Application of DMAIC Methodology in Patient Care Processes: The Lean Project Emergency Care Units (UPAs) 24 h

The Lean project in UPAs 24 h is a partnership between the Ministry of Health and the Fluminense Federal University with the objective of implementing the restructuring project - Humanization of Patient Care Flow in the UPAs – 24 h urgency and emergency network. The project team is composed of 30 professors and students and 20 specialists, working in management, research, training and advising the 50 UPAs distributed in 8 states and the Federal District. The methodology benefited about 13 million users of the country's public health network. More than 28,300 h of training were carried out for the 548 employees of the UPAs between the months of April and November/2020.

The methodology of the Lean Project in UPAs 24 h developed is integrated and oriented to the principles of the Lean Healthcare approach, to improve patient care processes. The implementation is guided by action research, using the DMAIC method (Define, Measure, Analyze, Improve and Control) to improve communication and the quality of the service offered. The use of continuous improvement and quality tools with servers and employees of the 50 UPAs is highlighted. The main motivators for the development of the project were improved service to SUS users in UPAS, reducing the length of stay of users as well as overcrowding in UPAS, in addition to a focus on patient safety and an increase in user and team satisfaction. In this controlled project, the appropriate indicators are monitored to assess and reduce overcrowding in UPAs. Indicators such as average length of stay were evaluated and monitored.

The project had assistance (in person) to identify potential opportunities for flow and process improvement, aiming at a more efficient management of resources. The project has the following main features: a) support for the development of organizational skills by planning face-to-face technical visits to the 50 UPAs in different states of Brazil; initial assessment of services and environment; involvement of the entire UPA network to impact culture change and reorganization; b) support in the use of DMAIC tools and in leading teams to change and reorganize processes; review of instruments, tools and construction of driving diagrams according to the DMAIC method; elaboration of educational and support materials related to intervention/change processes; monitoring activities and technical visits, as well as controlling the actions of each improvement implementation project; c) Monitoring and Presentation of Results: Develop the control methodology and perform remote monitoring of UPAs indicators; presentation of the results of the services and the project as a whole.

The methodology of the Lean project in UPAs generated qualitative and quantitative results. Even with difficulties especially related to the fact that the project takes place during the Covid-19 pandemic, the project had the benefits of greater humanization of treatment from patient reception to discharge, improvement in the organization of the environment and safety, greater flexibility and service capacity to patients during and after the Covid-19 period, increased user and team satisfaction and reduced waiting time for care. In 43 of the 50 UPAs – 24 h there was a 39.5% reduction in the patients' average length of stay in the establishments and at the end of the 6 months of remote monitoring for monitoring the indicators. The main motivators for the development of the project were improved service to SUS users in UPAS, reducing the length of stay of users as well as overcrowding in UPAS, in addition to a focus on patient safety and an increase in user and team satisfaction.

## 7   Discussion and Conclusions

This section outlines the main results of the systematic literature review and indicates directions for further research. The present work comprised the analysis of 21 articles through the Scopus database. Firstly, the review results have been discussed, which are: Motivators. Most factors cited by Antony et al. [6] as motivators could be found in the present review and some motivators cited by Sodhi et al. [34] have also been highlighted. "Reducing process time" was the most commonly cited motivator in literature review articles, followed by "improving the quality of service". Discharge of femur, knee and hip prosthesis surgeries were the most recurrent processes in the analysis of the articles.

This article was limited to revising articles that contained the terms "dmaic" and "define measure analyze improve control" as keywords in their title and abstract, but it is known that there are several articles that contain these terms in other parts of the manuscript with applications or implementations that have not been included. In addition, the research that supported the present research was limited to the Scopus database. These limitations can be further explored in later studies. In line with Antony et al. [6], the Lean Journey in the articles of the systematic literature review does not usually contemplate a long period. This translates into a research opportunity for further studies. The study made it possible to present, through articles selected from literature, results of the development of Six Sigma/Lean Six Sigma improvement projects based on the DMAIC cycle in different healthcare processes. Different motivators mentioned in the literature were considered in the action research presented regarding the 24-hour UPAs, such as improving user satisfaction, improving employee satisfaction, improving the quality of service, improving patient safety, reducing length of stay. It can be seen from the action research presented that among the various benefits obtained, the reduction in permanence time was a substantial reduction in most of the UPAs analyzed, which indicates that the DMAIC methodology can be successfully used in emergency units with a view to the continuous improvement.

**Acknowledgements.** The authors would like to thank the Brazilian Ministry of Health, Fluminense Federal University and Euclides da Cunha Foundation. This Research is part of a "Lean Project in UPAs 24 h" that has been funded by the Brazilian Ministry of Health (TED 125/2019, number: 25000191682201908).

# References

1. Ahsan, K.B., Alam, M., Morel, D.G., Karim, M.: Emergency department resource optimisation for improved performance: a review. J. Ind. Eng. Int. **15**(1), 253–266 (2019)
2. Al Kuwaiti, A.: Application of six sigma methodology to reduce medication errors in the outpatient pharmacy unit: a case study from the king Fahd university hospital. Saudi Arabia. Int. J. Qual. Res. **10**(2), 267–278 (2016)
3. Al Kuwaiti, A., Subbarayalu, A.V.: Reducing patients' falls rate in an academic medical center (AMC) using six sigma "DMAIC" approach. Int. J. Health Care Qual. Assur. **30**(4), 373–384 (2017)
4. Aleem, S., Torrey, W.C., Duncan, M.S., Hort, S.J., Mecchella, J.N.: Depression screening optimization in an academic rural setting. Int. J. Health Care Qual. Assur. **28**(7), 709–725 (2015)
5. Allen, T.T., Tseng, S.H., Swanson, K., McClay, M.A.: Improving the hospital discharge process with six sigma methods. Qual. Eng. **22**(1), 13–20 (2009)

6. Antony, J., Sreedharan, R., Chakraborty, A., Gunasekaran, A., et al.: A systematic review of lean in healthcare: a global prospective. Int. J. Qual. Reliab. Manage. **36**(8), 1370–1391 (2019)
7. Arafeh, M., et al.: Using six sigma DMAIC methodology and discrete event simulation to reduce patient discharge time in king Hussein cancer center. J. Healthcare Eng. **2018** (2018)
8. Bhat, S., Gijo, E., Jnanesh, N.: Application of lean six sigma methodology in the registration process of a hospital. Int. J. Prod. Perform. Manage. **63**(5), 613–643 (2014)
9. Bhat, S., Jnanesh, N.: Application of lean six sigma methodology to reduce the cycle time of out-patient department service in a rural hospital. Int. J. Healthcare Technol. Manage. **14**(3), 222–237 (2014)
10. Black, J.: Transforming the patient care environment with lean six sigma and realistic evaluation. J. Healthcare Qual. **31**(3), 29–35 (2009)
11. Cheung, Y.Y., Goodman, E.M., Osunkoya, T.O.: No more waits and delays: streamlining workflow to decrease patient time of stay for image-guided musculoskeletal procedures. RadioGraphics **36**(3), 856–871 (2016)
12. Chiarini, A.: From total quality control to lean six sigma: evolution of the most important management systems for the excellence (2012)
13. Coughlin, K., Posencheg, M.A.: Quality improvement methods-part II. J. Perinatol. **39**(7), 1000–1007 (2019)
14. Dang, A., Gjolaj, L.N., Whitman, M., Fernandez, G.: Using process improvement tools to improve the care of patients with neutropenic fever in the emergency room. J. Oncol. Pract. **14**(1), e73–e81 (2018)
15. DelliFraine, J.L., Langabeer, J.R., Nembhard, I.M., et al.: Assessing the evidence of six sigma and lean in the health care industry. Qual. Manage. Healthcare **19**(3), 211–225 (2010)
16. Gijo, E., Antony, J., Hernandez, J., Scaria, J.: Reducing patient waiting time in a pathology department using the six sigma methodology. In: Leadership in Health Services (2013)
17. Goh, T.N., Tang, L.C., Lam, S.W., Gao, Y.F.: Six sigma: a SWOT analysis. Int. J. Six Sigma Competitive Adv. **2**(3), 233–242 (2006)
18. Goh, T., Xie, M.: Statistical control of a six sigma process. Qual. Eng. **15**(4), 587–592 (2003)
19. Henrique, D.B., Godinho Filho, M.: A systematic literature review of empirical research in lean and six sigma in healthcare. Total Qual. Manage. Bus. Excellence **31**(3–4), 429–449 (2020)
20. Improta, G., et al.: Lean six sigma in healthcare: fast track surgery for patients undergoing prosthetic hip replacement surgery. TQM J. **31**(4), 526–540 (2019)
21. Improta, G., et al.: Lean six sigma: a new approach to the management of patients undergoing prosthetic hip replacement surgery. J. Eval. Clin. Pract. **21**(4), 662–672 (2015)
22. Improta, G., et al.: Improving performances of the knee replacement surgery process by applying DMAIC principles. J. Eval. Clin. Pract. **23**(6), 1401–1407 (2017)
23. Improta, G., Ricciardi, C., Borrelli, A., D'Alessandro, A., Verdoliva, C., Cesarelli, M.: The application of six sigma to reduce the pre-operative length of hospital stay at the hospital Antonio Cardarelli. Int. J. Lean Six Sigma **11**(3), 555–576 (2020)
24. Kwak, Y.H., Anbari, F.T.: Benefits, obstacles, and future of six sigma approach. Technovation **26**(5–6), 708–715 (2006)

25. Lobo, C.V.F., Calado, R.D., da Conceicao, R.D.P.: Evaluation of value stream mapping (VSM) applicability to the oil and gas chain processes. Int. J. Lean Six Sigma **11**(2), 309–330 (2018)
26. Moher, D., Liberati, A., Tetzlaff, J., Altman, D.G., Group, P.: Preferred reporting items for systematic reviews and meta-analyses: the PRISMA statement. PLoS Med. **6**(7), e1000097 (2009)
27. Pande, P., Neuman, R., Cavanagh, R.: The Six Sigma Way Team Fieldbook: An Implementation Guide for Process Improvement Teams. McGraw Hill Professional, New York (2001)
28. Raval, S.J., Kant, R., Shankar, R.: Revealing research trends and themes in lean six sigma: from 2000 to 2016. Int. J. Lean Six Sigma **9**(3), 399–443 (2018)
29. Ricciardi, C., Balato, G., Romano, M., Santalucia, I., Cesarelli, M., Improta, G.: Fast track surgery for knee replacement surgery: a lean six sigma approach. TQM J. **32**(3), 461–474 (2020)
30. Ricciardi, C., Fiorillo, A., Valente, A.S., Borrelli, A., Verdoliva, C., Triassi, M., Improta, G.: Lean six sigma approach to reduce LOS through a diagnostic-therapeutic-assistance path at Aorna Cardarelli. TQM J. **31**(5), 657–672 (2019)
31. Silich, S.J., et al.: Using six sigma methodology to reduce patient transfer times from floor to critical-care beds. J. Healthcare Qual. **34**(1), 44–54 (2012)
32. Singh, A., Pradhan, S., Ravi, P., Dhale, S.: Application of six sigma and 5 s to improve medication turnaround time. Int. J. Healthcare Manage. 1–9 (2020)
33. Snee, R.D.: Lean six sigma-getting better all the time. Int. J. Lean Six Sigma **1**(1), 9–29 (2010)
34. Sodhi, H.S., Singh, D., Singh, B.J.: An empirical analysis of critical success factors of lean six sigma in Indian SMES. Int. J. Six Sigma Competitive Adv. **11**(4), 227–252 (2019)
35. Taner, M.T., Sezen, B., Antony, J.: An overview of six sigma applications in healthcare industry. Int. J. Health Care Qual. Assur. **20**(4), 329–340 (2007)
36. Tlapa, D., et al.: Effects of lean healthcare on patient flow: a systematic review. Value Health **23**(2), 260–273 (2020)
37. Trakulsunti, Y., Antony, J., Dempsey, M., Brennan, A.: Reducing medication errors using lean six sigma methodology in a Thai hospital: an action research study. Int. J. Qual. Reliab. Manage. **38**(1), 339–362 (2021)
38. Vijay, S.A.: Reducing and optimizing the cycle time of patients discharge process in a hospital using six sigma DMAIC approach. Int. J. Qual. Res. **8**(2), 169–182 (2014)
39. Yamamoto, J., Abraham, D., Malatestinic, B.: Improving insulin distribution and administration safety using lean six sigma methodologies. Hosp. Pharm. **45**(3), 212–224 (2010)

# Evaluation of Fast-Track Implementation on Emergency Department: A Literature Review

Luis Valdiviezo Viera[1]([⊠]) [iD], Milena Reis[2]([⊠]) [iD], Sandra Chaves[1] [iD],
Robisom Calado[1] [iD], Saulo Bourguignon[1] [iD], and Sandro Lordelo[1] [iD]

[1] Federal Fluminense University, Niterói, RJ, Brazil
luisvaldiviezo@id.uff.br
[2] Federal University of Rio de Janeiro, Rio de Janeiro, RJ, Brazil
milenaestanislau@macae.ufrj.br

**Abstract.** The fast-track is an important strategy for the management and control of overcrowding, which is increasingly used in hospital emergency services. This review article aims to identify the main approaches and criteria for analyzing and deciding on the implementation of the fast-track strategy in the Emergency Department. The bibliographic review considered articles published in the Scopus and MedLine databases during the period 2010 and 2020. In general, the bibliographic research showed: the relative scarcity of scientific communication, regarding the implementation of fast-track strategies in hospital management systems; the diversity of the implantation purposes; the preference for modelling and simulation techniques; the difficulty of *ex-ante* and *ex-post* evaluation from a social perspective. The evaluation of the effectiveness of the implementation of fast-track strategies, when operational criteria are used, is based on quantitative models and techniques. In Brazil, the efficient management of Emergency Care Units, UPAs 24-h, is essential because the UPAs are a fundamental part of the Brazilian health system.

**Keywords:** Fast track · Emergency department · Lean healthcare · Assessment methods

## 1 Introduction

Overcrowding in hospitals is the result of a lower capacity to meet the demand for hospital services. The increase in demand for health services is a phenomenon observed in hospitals in different cities around the world [1–4].

The increase in demand is justified by the growth of population groups or by the expansion of the supply of services [5]. In Brazil, [6] consider that overcrowding is a performance problem in the organization of Brazilian hospitals.

In general, overcrowding management seeks improving operational, economic-financial and quality performance indices, etc. Consequently, efficient management of overcrowding benefits patients and their families, medical professionals, hospital staff and society.

© IFIP International Federation for Information Processing 2021
Published by Springer Nature Switzerland AG 2021
A. Dolgui et al. (Eds.): APMS 2021, IFIP AICT 631, pp. 280–288, 2021.
https://doi.org/10.1007/978-3-030-85902-2_30

The implementation of fast-track in hospital systems is conceived as a strategy for the control and reduction of overcrowding of emergency and emergency services [7]. Thus, the implementation of fast-track by the management of hospital systems has allowed improving the flow of patients, accelerating the fulfilment of the relevant activities. Consequently, how to evaluate the effectiveness or contribution of fast-track in reducing overcrowding in the areas of emergency care in hospitals? The bibliographic review of the present work has as main objective to identify the main practices in the evaluation of the implementation of the fast-track strategy in health services in hospital urgencies and emergencies.

In Brazil, the Emergency Care Units, UPAs24-h, are integrated into the Brazilian emergency care network, showing their strategic relevance by enabling the reduction of demand pressure in the emergency services of hospitals. The University Federal Fluminense, UFF, in partnership with the Ministry of Health of Brazil, developed the Project for Restructuring Implantation of Humanization in the Flow of Care of Patients at UPAS-24 h. Brazil, with the implementation of the National Humanization Policy (PNH) promotes the humanization of health and integration of the hospital environment into the social environment [8–10].

## 2    Research Methodology and Data Collection

A bibliographic review was developed based on the guidelines proposed by Villas [11], and along with the PRISMA 2020 flowchart for new systematic reviews. The steps covered by this study are detailed below.

### 2.1    Search Strategy

The survey of bibliographic references of scientific literature was carried out on the CAPES Journal Portal, made available through the CAFe platform (Federal Academic Community) to researchers at Federal Fluminense University.

The bibliographic research was based on articles published in the last ten years found on Bases Scopus and MedLine. The parameters used in the bibliographic search were:

- Search period: LAST 10 years;
- Search material: ARTICLES;
- Language: ANY.

The CAFe Platform was accessed from April 9 to 11, 2020. The results generated are summarized in Table 1. In the total results, shown in the table, articles common to the databases under consideration are excluded.

### 2.2    Selection of Articles

The articles considered for the elaboration of a database correspond from search # 4 in Table 1, which uses Fast-Track and ED as search descriptor (ED refers to Emergency Department). Then, potentially relevant articles have been identified as a result of excluding articles that:

- Were listed as removed from the database;
- Were published on an untrusted site identified by the antivirus;
- Had a link redirecting to other sources;
- Were not made available to their full extent;
- Reported a certification error while being downloaded.

**Table 1.** Results generated by platform CAFe

| Search | Search descriptors | Data base | Outcome |
|---|---|---|---|
| 1 | Fast track | Total | 2.760 |
| | | Scopus | 1.304 |
| | | MedLine | 780 |
| 2 | Fast track and hospital | Total | 837 |
| | | Scopus | 539 |
| | | MedLine | 505 |
| 3 | Fast track and emergency room | Total | 42 |
| | | Scopus | 39 |
| | | MedLine | |
| 4 | Fast track and ED | Total | 204 |
| | | Scopus | 152 |
| | | MedLine | 72 |
| 5 | Fast track and lean-healthcare | Total | 10 |
| | | Scopus | 8 |
| | | MedLine | 7 |

## 2.3   Creating a Database

Potentially relevant articles were subjected to a selection criterion based on the relevance demonstrated by their abstracts. After analyzing the abstracts of articles, those who have not adequately meet at least one of this inclusion and/or exclusion selection criteria were excluded:

- Identification of used methodology and/or methods used: case study, experimental;
- Identification of use tools: documentary research, questionnaires, focus groups;
- Identification of data analysis techniques: statistical tests, regression analysis, descriptive statistics;
- Fast-track approach from a managerial perspective.

Articles that adequately met the inclusion criteria started to constitute the database for this purpose in this study. This database consists of twenty articles. Figure 1 illustrates this process.

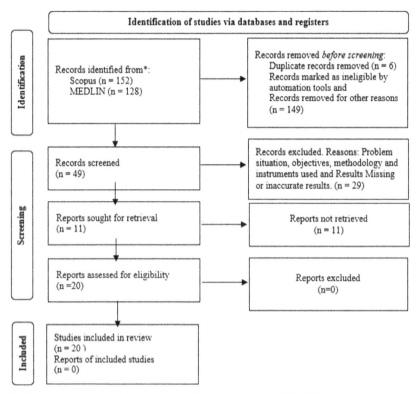

**Fig. 1.** Process of study and database establishment based on PRISMA 2020 flow diagram for systematic reviews

### 2.4 Database Analysis

The database created for this study was organized in a Microsoft Excel spreadsheet containing the following: Title; Authors; Geographic origin; Theoretical framework; Objectives; Problem situation; Participants; Methodology and Methods; Contribution; Strong points; Weak points.

The database analysis was carried out by considering the following aspects:

- Problem situation;
- Objectives;
- Methodology, methods and instruments used;
- Results.

## 3  Analysis of the Results

The main objectives, stated by the analyzed authors with fast-track strategies are organized in Table 2. Most authors propose to assess the implementation of fast-track in DE, while others seek to compare the fast-track with other strategies. The "comparison" objectives refer to a specific form of evaluation, which is why the comparison objective was broken down into more specific categories such as costs, medical resources, revenues, satisfaction, productivity, etc. It is observed that the formulation of objectives related to the identification of factors and/or variables that influence the success of the fast-track implementation, seek to identify the most important factors and measure their degree of influence.

**Table 2.** Objectives formulated of analyzed articles

| Objectives | Authors | Outcome |
|---|---|---|
| Evaluate the implementation of the fast-track ED | [1, 2, 12, 13] | 4 |
| Analyze the length of stay | [1, 4, 14] | 3 |
| Analyze service quality indicators | [1] | 1 |
| Analyze the variability times | [5] | 1 |
| Comparison of time reduction strategies | [4, 7, 15] | 3 |
| Comparison of cost reduction strategies | [7, 16, 17] | 3 |
| Comparison of revenue growth strategies | [18] | 1 |
| Comparison of resource use strategies | [4, 19] | 2 |
| Comparison of satisfaction strategies | [18, 20, 21] | 3 |
| Comparison of strategies to increase productivity | [18, 22, 23] | 3 |
| Identify success factors or variables that influence the times | [3, 24, 25] | 3 |

Table 3 lists the main methodologies described by the authors that would make it possible to achieve the objectives, to explain how the scientific study was developed. The use of observational methodologies, called before and after, which ex-ate an ex-post evaluation, predominates. A prospective approach suggests an ex-ante evaluation. Both types of assessment used statistical techniques to identify the statistical relevance of differences in patient groups in fast-track care, or without fast-track care.

Another important technique used is mathematical modelling, where the behaviour of the flow of queues in urgent and emergency services in hospitals is modeled. The results generated by the simulation are statistically validated. Statistical analysis is the method widely used both to analyze and validate indicators of observed patient groups as to analyze and validate indicators generated by simulating the behaviour of the patient flow.

The information and data obtained through the application of questionnaires are also subjected to statistical methods of analysis to grant scientific support to the conclusions obtained. Statistical analysis is aided by a range of commercial software. Modelling and simulation is also favoured by the availability of high performance, friendly software.

Table 4 describes the main results and findings identified by the analyzed authors. There is a wide dispersion of these results. In general, the fast-track, as expected, allows for a reduction in waiting times, length of stay and the number of patients who have dropped out of care. However, it is important to emphasize certain constraints. Some

**Table 3.** Methodologies used by the authors of the analyzed articles.

| Methodologies | Authors | Outcome |
|---|---|---|
| Observational study: before and after | [1, 7, 12, 21] | 4 |
| Retrospective and prospective review | [18, 22] | 2 |
| Prospective study | [17] | 1 |
| Economic analysis | [7] | 1 |
| Statistical analysis | [3, 5, 13] | 3 |
| Top-down and bottom-up approach | [24] | 1 |
| Cross-sectional study of questionnaire surveys | [20, 21, 25] | 3 |
| Mathematical modelling | [2, 4, 15, 19, 23] | 4 |

articles show that the reduction of times is achieved through the greater use of resources or that the increase in customer satisfaction compensates for an increase in costs.

Other alternatives for managing overcrowding were considered and compared with the fast track, such as the Dynamic Priority Queue, DPQ, [19] and Inter-professional Teamwork [7]. The results obtained indicate that these alternative strategies to the fast-track proved to be effective in reducing the times and overcoming costs.

**Table 4.** Results obtained by the authors of the analyzed articles.

| Results | Authors | Outcome |
|---|---|---|
| Increase in patients attended | [1, 12] | 2 |
| Reduced waiting time for low-risk patients | [1, 4, 12] | 3 |
| Reduction in waiting times for high-risk patients | [1] | 1 |
| Reduction in the number of patients who dropped out | [1, 12–14] | 4 |
| Reduction of the patient's length of stay | [2, 22] | 2 |
| Variation in treatment times | [5] | 1 |
| Shorter treatment time and wait time in alternative strategies to FT | [7, 19] | 2 |
| Lower costs in alternative strategies to FT | [7] | 1 |
| Lower costs per QALY in alternative strategies to FT | [7] | 1 |
| Additional medical staff to decrease FT strategy times | [12, 18, 21, 23] | 4 |
| Identification of the factors that influence waiting times in FT strategies | [3, 14] | 2 |
| Improving customer satisfaction with FT strategies | [20, 21] | 2 |
| Improving the productivity of ED resources | [15] | 1 |

## 4  Conclusions

Overcrowding of hospital emergency services is currently a worldwide phenomenon. Public and private sector managers devote resources and efforts to mitigate the effects of overcrowding. The fast-track proved to be an effective management strategy in reducing the patient's stay and, consequently, it would improve the performance of hospital services.

In the analysis of decisions to implement fast-track, the prevalence of the formulation of operational objectives, mainly related to the reduction of time, remained. On the other hand, economic and financial objectives, such as cost reduction; and quality objectives, such as patient and professional satisfaction, are less prevalent.

The main method of evaluating the decision to implement fast-track in hospital emergency services is statistical analysis. However, qualitative assessment methods are also used, through the application of questionnaires to users. These evaluation practices are of the *ex-post* type because they consider the results obtained by the normal operation of the fast-track, when evaluating the decision already executed. However, mathematical modelling allows for a prospective, or *ex-ante* type assessment, when assessing the implementation decision, with the results of the fast-track being generated or estimated and statistically validated.

In general, the use of mathematical models, simulation techniques, use of queues are the main methods and techniques for the analysis and evaluation of the decision to operationalize the fast-track. And, the evaluation of the effectiveness or benefits of implementing the fast-track predominates the use of models and techniques of univariate and multivariate statistical analysis. Other criteria used are the reduction of costs and the increase in revenues, the results of which are statistically valid.

**Acknowledgment.** The authors would like to thank the Brazilian Ministry of Health, Fluminense Federal University and Euclides da Cunha Foundation. This Research is part of a "Lean Project in UPAs 24 h" that has been funded by the Brazilian Ministry of Health (TED 125/2019, number: 25000191682201908).

## References

1. Chrusciel, J., et al.: Impact of the implementation of a fast-track on emergency department length of stay and quality of care indicators in the Champagne-Ardenne region: a before-after study. BMJ Open **9**, 1–8 (2019)
2. Hajjarsaraei, H., Shirazi, B., Rezaeian, J.: Scenario-based analysis of fast track strategy optimization on emergency department using integrated safety simulation. Saf. Sci. **107**, 9–21 (2018)
3. Gill, S.D., et al.: Investigation of factors that predict length of stay. Obj. Meth. Setting **30**, 1–11 (2020)
4. Fitzgerald, K., Pelletier, L., Reznek, M.A.: A queue-based monte carlo analysis to support decision making for implementation of an emergency department fast track. J. Healthc. Eng. **2017** (2017)

5. McCarthy, M.L., et al.: Provider variation in fast track treatment time. Med. Care **50**, 43–49 (2012)
6. Bittencourt, R.J., Hortale, V.A.: Intervenções para solucionar a superlotação nos serviços de emergência hospitalar: Uma revisão sistemática. Cad. Saude Publica **25**, 1439–1454 (2009)
7. Liu, J., Masiello, I., Ponzer, S., Farrokhnia, N.: Interprofessional teamwork versus fast track streaming in an emergency department—an observational cohort study of two strategies for enhancing the throughput of orthopedic patients presenting limb injuries or back pain. PLoS ONE **14**, 1–17 (2019)
8. Souza, L.A.deP., Mendes, V.L.F.: O conceito de humanização na Política Nacional de Humanização (PNH). Interface Commun. Heal. Educ. **13**, 681–688 (2009)
9. Nunes, L.C.: Implantação da política nacional de humanização (pnh): conquistas e desafios para a assistência em saúde, 2–7 (2018)
10. Brasil. Política Nacional de Humanização, 368 (2013)
11. Villas, M.V., Macedo-Soares, T.D.L.vanA.de, Russo, G.M.: Bibliographical research method for business administration studies: a model based on scientific journal ranking. BAR Brazilian Adm. Rev. **5**, 139–159 (2008)
12. Saidi, K., et al.: Effets de la création d'un circuit court au sein d'un service d'urgence adulte. Annales françaises de médecine d'urgence **5**(6), 283–289 (2015). https://doi.org/10.1007/s13 341-015-0593-9
13. Gardner, R.M., Friedman, N.A., Carlson, M., Bradham, T.S., Barrett, T.W.: Impact of revised triage to improve throughput in an ED with limited traditional fast track population. Am. J. Emerg. Med. **36**, 124–127 (2018)
14. Copeland, J., Gray, A.: A daytime fast track improves throughput in a single physician coverage emergency department. Can. J. Emerg. Med. **17**, 648–655 (2015)
15. Kaushal, A., et al.: Evaluation of fast track strategies using agent-based simulation modeling to reduce waiting time in a hospital emergency department. Socioecon. Plann. Sci. **50**, 18–31 (2015)
16. Wijnen, B.F.M., et al.: Cost-effectiveness of an integrated'fast track' rehabilitation service for multi-trauma patients: a non-randomized clinical trial in the Netherlands. PLoS ONE **14**, 1–18 (2019)
17. Reurings, J.C., et al.: A prospective cohort study to investigate cost-minimisation, of Traditional open, open fAst track recovery and laParoscopic fASt track multimodal management, for surgical patients with colon carcinomas (TAPAS study). BMC Surg. **10** (2010)
18. Jeanmonod, R., DelCollo, J., Jeanmonod, D., Dombchewsky, O., Reiter, M.: Comparison of resident and mid-level provider productivity and patient satisfaction in an emergency department fast track. Emerg. Med. J. **30**, e12 (2013)
19. Ferrand, Y.B., Magazine, M.J., Rao, U.S., Glass, T.F.: Managing responsiveness in the emergency department: comparing dynamic priority queue with fast track. J. Oper. Manag. **58–59**, 15–26 (2018)
20. Hwang, C.E., Lipman, G.S., Kane, M.: Effect of an emergency department fast track on press-ganey patient satisfaction scores. West. J. Emerg. Med. **16**, 34–38 (2015)
21. Lutze, M., Ross, M., Chu, M., Green, T., Dinh, M.: Patient perceptions of emergency department fast track: a prospective pilot study comparing two models of care. Australas. Emerg. Nurs. J. **17**, 112–118 (2014)
22. Kosy, J.D., Blackshaw, R., Swart, M., Fordyce, A., Lofthouse, R.A.: Fractured neck of femur patient care improved by simulated fasttrack system. J. Orthop. Traumatol. **14**, 165–170 (2013)

23. La, J., Jewkes, E.M.: Defining an optimal ED fast track strategy using simulation. J. Enterp. Inf. Manag. **26**, 109–118 (2013)
24. Basta, Y.L., et al.: Organizing and implementing a multidisciplinary fast track oncology clinic. Int. J. Qual. Heal. Care **29**, 966–971 (2017)
25. Dinh, M.M., Enright, N., Walker, A., Parameswaran, A., Chu, M.: Determinants of patient satisfaction in an Australian emergency department fast-track setting. Emerg. Med. J. **30**, 824–827 (2013)

# An Adaptive Large Neighborhood Search Method to Plan Patient's Journey in Healthcare

Gérard Olivier[1,2][✉], Lucet Corinne[2], Brisoux Devendeville Laure[2], and Darras Sylvain[1]

[1] MIS Laboratory (EA 4290), University Picardie Jules Verne, Amiens, France
{olivier.gerard,corinne.lucet,laure.devendeville}@u-picardie.fr
[2] Evolucare Technologies, Fouilloy, France
{o.gerard,s.darras}@evolucare.com

**Abstract.** In this paper an adaptation of the Adaptive Large Neighborhood Search (ALNS) to a patient's care planning problem is proposed. We formalize it as an RCPSP problem that consists of assigning a start date and medical resources to a set of medical appointments. Different intensification and diversification movements for the ALNS are presented. We test this approach on real-life problems and compare the results of ALNS to a version without the adaptive layer, called ($\neg$A)LNS. We also compare our results with the ones obtained with a 0–1 linear programming model. On small instances, ALNS obtains results close to optimality, with an average difference of 1.39 of solution quality. ALNS outperforms ($\neg$A)LNS with a gain of up to 18.34% for some scenarios.

**Keywords:** ALNS · Planning · Healthcare · Optimization

## 1 Introduction

Improving the health care system is one of the biggest challenges many countries will have to face over the coming years. The complexity of scheduling problems in the healthcare domain is an issue that is increasingly being highlighted by healthcare facilities. This kind of problem belongs to the Resource Constrained Project Scheduling Problem (RCPSP) family that is NP-Hard [2,7]. The RCPSP problem consists of finding the best assignment of resources and start times to a set of activities. In healthcare it usually involves finding a starting date and medical resources (medical staff, rooms and medical equipment) for an appointment with a patient.

Nowadays, planning the patient's care is mostly done by hand, a difficult and time-consuming task due to the number of appointments and resources to consider, that can be challenged by various kinds of unexpected events. Scheduling problems have been the subject of many studies for decades in various fields [1,3], and they are of increasing interest in the healthcare domain [6,12].

© IFIP International Federation for Information Processing 2021
Published by Springer Nature Switzerland AG 2021

A. Dolgui et al. (Eds.): APMS 2021, IFIP AICT 631, pp. 289–297, 2021.
https://doi.org/10.1007/978-3-030-85902-2_31

The structure of considered problems differs according to the institutions, their size and the number of resources taken into account. This article is focused on scenarios designed with various planners from different health care facilities in France who deal with real-life problems every day. These scenarios focus on the planning of patient's care in different kinds of institutions. In this work we propose an Adaptive Large Neighborhood Search (ALNS) metaheuristic able to deal with large instances derived from these various real-world healthcare scenarios.

This article is structured as follows. In Sect. 2 we describe our problem. In Sect. 3 we present our ALNS algorithm, with the different movements used. In Sect. 4 we give further details on the scenarios and the corresponding results obtained by the ALNS metaheuristic. We compare the results of this method with the results obtained by the 0–1 linear programming model presented in [5]. In Sect. 5, we conclude with some remarks and perspectives.

## 2    Problem Definition

Our problem can be stated as follows. The horizon $H$ is decomposed into time-slots. We have a finite set of resources $R$ and each resource $r \in R$ is characterized by a set of properties $\Pi_r$ that determines which roles a resource will be able to hold in an appointment. Availability of each resource is known. $A$ is a set of appointments to be planned, such that each appointment $a \in A$ is characterized by its duration $duration_a$, a feasibility interval of time $[ES_a, LS_a]$, $qtreq_a^\pi$ the amount of resources with property $\pi$ required by $a$. $Essential_a$ and $Emergency_a$ are two coefficients used to respectively quantify the importance and the urgency of an appointment $a$. Both are used by planners to define priorities on some appointments and to specify which ones should be set as soon as possible within their feasibility interval, therefore they both occur as penalties in the objective function. $PreAssigned_a$ is a set of couples $(resource, property)$ pre-assigned to $a$. To each appointment $a$ could be associated a set of appointments $pred_a$ that must be planned before $a$.

We define the triplet $(a, t_a, R_a)$, where $a$ is an appointment, $t_a$ the starting date for $a$ and $R_a$ the set of resources assigned to $a$. A valid solution $Sol$ is a set of triplets that respect the hard time and resource constraints. We denote $A_{Sol}$ the set of scheduled appointments and $A_{\overline{Sol}}$ the set of unscheduled appointments. The quality of a solution is evaluated by the objective function $f$ that is the sum of weights $Essential_a$ of unplanned appointments $a \in A_{\overline{Sol}}$ plus the sum of delay impacts of planned urgent appointments. The aim is to find a valid solution while minimizing objective function $f$ defined in Eq. 1.

$$f(Sol) = \sum_{a \in A_{\overline{Sol}}} Essential_a + \sum_{a \in A_{Sol}} \frac{t_a - ES_a}{LS_a - ES_a} \times Emergency_a \qquad (1)$$

## 3    Adaptive Large Neighborhood Search

Adaptive Large Neighborhood Search is based on the Large Neighborhood Search framework defined in [10]. ALNS was first introduced in [9] and was applied on various problems, such as scheduling problems [4,8]. Different movements are iteratively applied to a solution in order to explore its neighborhood. Basically, a solution is partially destroyed then reconstructed using destruction and construction heuristics. In a scheduling problem, it usually consists of removing a number of appointments from a solution and then trying to reinsert them.

---

**Algorithm 1:** ALNS

---

1 **begin**

2     $Sol \leftarrow greedy()$ ;

3     $Sol_{best} \leftarrow Sol$ ;

4     $\forall m \in M, w_m^0 \leftarrow 1/|M|$ ;

5     **while** *stopping criterion are not met* **do**

6        **for** *segment s* **do**

7           $m \leftarrow RouletteWheel(M, w^s)$ ;

8           $Sol' \leftarrow m(Sol)$ ;

9           $\pi_m$ is updated according to $f(Sol')$ ;

10           $Sol' \leftarrow diversification(Sol')$ ;

11           $Sol \leftarrow Sol'$ ;

12           **if** $f(Sol) < f(Sol_{best})$ **then**

13              $Sol_{best} \leftarrow Sol$ ;

14           **end**

15        **end**

16        update weights $w_m^{s+1}$ for each movement $m$ ;

17     **end**

18     **return** $Sol_{best}$ ;

19 **end**

---

The general algorithm of our ALNS based on [9] is given in Algorithm 1. An initial solution is computed using a greedy algorithm described in Sect. 3.1. All movements $m$ in the set of movements $M$ are equally weighted at the beginning of the algorithm. As suggested in [9], the search is divided into blocks of consecutive iterations, called *segments*. For each iteration in a segment $s$, a movement $m$ is chosen with a roulette wheel selection according to its weight $w_m^s$ and is applied to the current solution $Sol$. During the course of a segment, a score $\pi_m$ is associated with each movement $m$. $\pi_m$ represents the total reward of the movement, relative to the results of its use. At the end of each segment $s$, weights $w_m^s$ are updated according to scores $\pi_m$. If the search stagnates, diversification movements such as *Swap* or *Left Shift* can be applied to the solution. In an extended period without improvement, a *reset* can also be performed on the current solution $Sol$. This process is iterated until stop conditions are met. The various movements that ALNS uses for diversification are described below.

## 3.1    Greedy Algorithm

To build the initial solution, a simple greedy algorithm is used. The order that the greedy algorithm uses to schedule each unscheduled appointment is random. For an appointment $a$ the greedy searches for the first timeslot $t_a$ on which a set of resources $R_a$ that exactly matches its resource requirement is available. This search starts either at $ES_a$ or at a random timeslot $t \in [ES_a, LS_a]$.

## 3.2    Weight Adjustment

As stated above, the search is divided into segments. The score $\pi_m$ of all movements $m$ is set to zero at the start of each segment. At each iteration within a segment, the score $\pi_m$ of selected movement $m$ is increased by adding a reward $\sigma$ according to the following conditions:

$$\sigma = \begin{cases} \sigma_1, & \text{if the produced solution is a new best overall solution} \\ \sigma_2, & \text{if the produced solution is better than the current one} \\ \sigma_3, & \text{if the produced solution is worse than the current one} \end{cases} \quad (2)$$

At the end of each segment $s$, weight $w_m^{s+1}$ is updated using formula 3:

$$w_m^{s+1} = (1 - r)\, w_m^s + r\, \left(\frac{\pi_m}{\theta_m}\right) \quad (3)$$

where $\theta_m$ is the number of times movement $m$ has been used during segment $s$. The reaction factor $r$ controls the strength of the adjustment from one segment $s$ to the following segment $s + 1$. If $r = 0$, there is no change and if $r = 1$, the weights for the new segment $s + 1$ only depends on the performance during the previous segment. Movements are selected using a roulette wheel selection method. The number of iterations of a *segment*, the different scores $\sigma_1$, $\sigma_2$ and $\sigma_3$ and the reaction factor $r$ are parameters of the ALNS algorithm.

## 3.3    Intensification Movements

We propose different intensification movements. These movements are assessed by the ALNS and used according to their performance during execution.

**Random Destruction and Greedy Reconstruction.** This movement removes a random number of appointments from the current solution. Then it tries to plan all unscheduled appointments using the greedy algorithm described above. Time complexity of this movement: $O(|A_{\overline{Sol}}| \times |H| \times |R|)$.

**Random Destruction and Optimal Reconstruction.** This movement also removes a random number of appointments and uses an optimal reconstruction to try to plan as many unscheduled appointments as possible. The order in which the appointments will be considered is obtained by a roulette wheel selection, its probabilities being calculated from the importance coefficients $Essential_a$ of unscheduled appointments $a$. For each $a$ of them and for each timeslot $t \in [ES_a, LS_a]$, a heuristic searches for the best possible set of resources $R_a$. This choice is based on a stress score $\alpha_r$ for each resource $r \in R_a$. $\alpha_r$ increases each time a movement fails to assign $r$ to an appointment. Conversely, it decreases each time resource $r$ is successfully assigned to an appointment. The most requested resources will fail more often, and therefore have a higher stress score. This allows the ALNS to keep track of the most stressed resources. The heuristic favors for an appointment $a$ the sets of resources $R_a$ with the lowest cumulated stress score. $t_a$ is chosen according to the stress score of $R_a$ and to $Emergency_a$: if appointment $a$ is urgent, earlier timeslots close to $ES_a$ are preferred. Appointments that cannot be rescheduled remain in $A_{\overline{Sol}}$. Time complexity of this movement: $O(|A_{\overline{Sol}}| \times |H| \times |R|)$.

**Difficulty Based Destruction and Optimal Reconstruction.** To each appointment $a$ is associated a difficulty score $\delta_a$. Whenever a movement fails to schedule $a$, its difficulty score $\delta_a$ increases. Conversely, each time an appointment $a$ is scheduled, its difficulty score $\delta_a$ decreases. This allows the ALNS to keep track of the most difficult appointments to schedule. Among all scheduled appointments, this movement uses a roulette wheel selection based on their difficulty score to select one appointment to remove from $A_{Sol}$, the other are randomly chosen in $A_{Sol}$. Then it tries to plan all unscheduled appointments ordered by their decreasing difficulty score (the most difficult being scheduled first). Time complexity of this movement: $O(|A_{\overline{Sol}}| \times |H| \times |R|)$.

**Targeted Destruction.** This movement extracts one unscheduled appointment $a$ according to its difficulty score $\delta_a$. Next it tries to find the best timeslot $t_a$ to plan $a$ by removing some scheduled appointments in order to satisfy all resources needed by $a$. First we randomly select a set $T \subseteq [ES_a, LS_a]$ of timeslots where $a$ could be scheduled by relaxing resource constraints. For each timeslot $t \in T$ we compute all combinations $C_t^i$ of appointments whose removal frees up resources that allow $a$ to be planned. Next we select a timeslot $t$ with a combination $C_t^*$ by probabilistic rules based on $Essential$ and $Emergency$ factors of contributive appointments. All appointments in $C_t^*$ are unplanned and $a$ is planned on $t$. Finally, the greedy algorithm is used to place as many unscheduled appointments as possible. Time complexity of this movement: $O(\max(A, H) \times |R| \times |T|)$.

### 3.4 Diversification Movements

Two diversification movements *Swap* and *Left Shift* are used during the execution of the ALNS. They allow the current solution to be perturbed, lowering the odds that the search remains trapped in a local minimum. They are applied to the current solution when a preset number of iterations without improvement is reached.

**Swap.** The *Swap* randomly chooses two triplets $(a_1, t_{a_1}, R_{a_1})$ and $(a_2, t_{a_2}, R_{a_2})$ of *Sol* and exchanges their start date $t_{a_1}$ and $t_{a_2}$, if the availability of resources allows it. Otherwise the exchange is aborted. Time complexity of this movement: $O((duration_{a_1} + duration_{a_2}) \times |R|)$.

**Left Shift.** This movement tries to shift the appointments of a subset of randomly chosen resources so that their respective schedules are as compact as possible. The start date $t_a$ of each affected appointment $a$ is brought as close as possible to its $ES_a$ date. Time complexity of this movement: $O(|A| \times |H| \times |R|)$.

### 3.5    Restart

If the search stagnates for too long, the solution will undergo a complete restart. All appointments $a \in A_{Sol}$ are unscheduled, then the greedy algorithm described above is used. However the greedy will consider appointments ordered by their decreasing difficulty score $\delta$. The most difficult appointments will be processed first. This way we expect to schedule difficult appointments in priority. Time complexity of this movement: $O(|A| \times |H| \times |R|)$.

## 4    Experimentations and Results

### 4.1    Experimentations

We generated 80 instances from four scenarios presented in [5] to test our ALNS algorithm. Instances are generated by varying some parameters: size of the instance, essential and emergency coefficients, precedence relationship and resource availabilities. As suggested in [11], we compare the results of ALNS to a version without the adaptive layer, called $(\neg A)$LNS to assess the effect of the adaptive layer. Both ALNS and $(\neg A)$LNS were implemented in C# and tests were run on an Intel i5-8350U processor. We also compare ALNS results to the optimal solutions reached by the linear programming model 0–1 presented in [5] and implemented under CPLEX 12.8.0.0.

We set the parameters of ALNS as follows. The maximum number of iterations in a *segment* was set to 100. The rewards $\sigma_1$, $\sigma_2$ and $\sigma_3$ were set respectively to 75, 20, and 0. For the ALNS, the reaction factor $r$ is set to 0.08. For the $(\neg A)$LNS, $r$ is set to 0. The number of destroyed appointments mentioned in Sect. 3.3 is randomly picked between 5 and 15. For the ALNS and $(\neg A)$LNS the computation time was limited to two minutes when for CPLEX it was limited to two hours.

**Table 1.** Comparisons of ALNS results with optimal solutions

| Instance | ALNS | | CPLEX | | $\Delta_f$ | $Gap_f$ |
|---|---|---|---|---|---|---|
| | $f(Sol_{best})$ | $|A_{\overline{Sol}_{best}}|$ | $f(Sol_{best})$ | $|A_{\overline{Sol}_{best}}|$ | | |
| SurgDep 1 | 0 | 0 | 0 | 0 | 0 | 0% |
| SurgDep 2 | 0 | 0 | 0 | 0 | 0 | 0% |
| SurgDep 3 | 1 | 1 | 0 | 0 | 1 | 100% |
| SurgDep 4 | 3.75 | 2 | 1.48 | 0 | 2.27 | 153.17% |
| SurgDep 5 | 14.54 | 3 | 9.54 | 0 | 5 | 52.4% |
| SurgDep 6 | 0 | 0 | 0 | 0 | 0 | 0% |
| SurgDep 7 | 4.31 | 2 | 1.48 | 0 | 2.83 | 196.13% |
| SurgDep 8 | 0 | 0 | 0 | 0 | 0 | 0% |

**Table 2.** Comparisons of ALNS results with (¬A)LNS

| Scenario | $|A|$ | $|H|$ | ALNS Average $Gap_f$ | ALNS Average $\Delta_f$ | Average $|A_{Sol_{best}}|/|A|$ |
|---|---|---|---|---|---|
| SurgDep | 16 | 52 | −1.76% | −0.14 | 93.06% |
| | 48 | 52 | −12.61% | 0.56 | 90.51% |
| RehabCenter | 96 | 48 | −5.26% | −1.04 | 99.45% |
| | 288 | 48 | −7.16% | −11.02 | 99.38% |
| Admission | 136 | 240 | −18.34% | −8.80 | 94.22% |
| | 408 | 240 | −2.93% | 24.41 | 95.66% |
| CardioRehab | 160 | 240 | −7.52% | −2.61 | 98.62% |
| | 480 | 240 | −17.92% | 25.88 | 98.78% |

## 4.2 Results

Our different approaches have been applied to all generated instances. Comparisons between the optimal results obtained by CPLEX and those obtained by ALNS are reported in Table 1. For CPLEX and ALNS we give $f(Sol_{best})$ and the number of unplanned appointments $|A_{\overline{Sol}_{best}}|$, the difference $\Delta_f$ between CPLEX and ALNS results and the $Gap_f$ from CPLEX ($Gap_f = (ALNS - CPLEX)/CPLEX$). Not surprisingly, CPLEX cannot conclude on large instances, and we can only obtain optimality on small instances (SurgDep scenario, the smallest of our scenarios with 16 appointments to schedule on one day). We note that ALNS obtains results close to optimality, with an average difference of 1.39 of solution quality computed by Eq. 1.

We next compare the results obtained by ALNS and (¬A)LNS. Results are reported in Table 2. We give for each scenario the number of appointments $|A|$, the number of timeslots $|H|$, the average $Gap_f$ ($Gap_f = (ALNS - (¬A)LNS)/(¬A)LNS$) from (¬A)LNS, the average difference $\Delta_f$ between (¬A)LNS and ALNS and the average percentage of scheduled appointments

$|Sol_{best}|$ / $|A|$. We see that ALNS outperforms $(\neg A)$LNS on all scenarios, especially on the largest ones with an improvement of 17.92% on the scenario with the highest number of appointments to schedule. The average gain obtained with ALNS exceeds the average gain reported in [11]. These results suggest that some combinations of movements may be especially efficient depending on the scenario. They are therefore favored by ALNS and thus lead to better solutions.

## 5    Conclusion and Perspectives

In this paper we presented an adaptation of the ALNS framework to a patient's healthcare planning problem. Such a method, whose execution times are very short and provide solutions quite close to the optimum, is very interesting for applications in the real world. Indeed, the first results of this approach are promising especially on large instances. Further tests should be conducted to study the impact of parameters on the performance of ALNS. We would also investigate how the weights of movements evolve according to the scenarios in order to understand why some of them seem more efficient than others from one scenario to another.

**Acknowledgements.** This project is supported by LORH project (CIFRE $N^o$ 2018/0425 between Evolucare and MIS Laboratory).

## References

1. Anthony, R.: Planning and control systems: a framework for analysis. Graduate School of Business Administration, Harvard University, Division of Research (1965)
2. Baptiste, P., Laborie, P., LePape, C., Nuijten, W.: Constraint-based scheduling and planning. In: Foundations of Artificial Intelligence, vol. 2, pp. 761–799. Elsevier (2006)
3. Blazewicz, J., Ecker, K., Pesch, E., Schmidt, G., Sterna, M., Weglarz, J.: Handbook on Scheduling. INFOSYS, Springer, Cham (2019). https://doi.org/10.1007/978-3-319-99849-7
4. Bueno, E.: Mathematical modeling and optimization approaches for scheduling the regular-season games of the National Hockey League. Ph.D. thesis, École Polytechnique de Montréal (2014)
5. Gérard, O., Devendeville, L., Lucet, C.: Planning problem in Healthcare domain. In: 17th International Workshop on Project Management and Scheduling PMS 2020 (2021)
6. Hall, R. (ed.): Handbook of Healthcare System Scheduling. ISOR, Springer, Cham (2012). https://doi.org/10.1007/978-1-4614-1734-7
7. Johnson, D., Garey, M.: Computers and Intractability: A Guide to the Theory of NP-completeness. WH Freeman, New York (1979)
8. Muller, L.: An adaptive large neighborhood search algorithm for the multi-mode RCPSP. DTU Manage. Eng. **3**, 25 (2011)
9. Ropke, S., Pisinger, D.: An adaptive large neighborhood search heuristic for the pickup and delivery problem with time windows. Transp. Sci. **40**(4), 455–472 (2006)

10. Shaw, P.: Using constraint programming and local search methods to solve vehicle routing problems. In: Maher, M., Puget, J.-F. (eds.) CP 1998. LNCS, vol. 1520, pp. 417–431. Springer, Heidelberg (1998). https://doi.org/10.1007/3-540-49481-2_30
11. Turkeš, R., Sörensen, K., Hvattum, L.: Meta-analysis of metaheuristics: quantifying the effect of adaptiveness in adaptive large neighborhood search. Eur. J. Oper. Res. **292**(2), 423–442 (2020)
12. Zhang, X., Ma, S., Chen, S.: Healthcare service configuration based on project scheduling. Adv. Eng. Inform. **43**, 101039 (2020)

# Capacity Management as a Tool for Improving Infrastructure in the Lean Healthcare: A Systematic Review

Adalberto Lima[1] , Christiane Barbosa[2(✉)] , Alberto Sobrinho[2] ,
Robisom Calado[3] , and Ana Paula Sobral[3]

[1] Universidade Federal do Pará, Abaetetuba, Brazil
[2] Universidade Federal do Pará, Belém, Brazil
`cllima@ufpa.br`
[3] Instituto de Ciência e Tecnologia, Universidade Federal Fluminense,
Rio das Ostras, Brazil
`ana_sobral@id.uff.br`

**Abstract.** The objective of this paper is to identify ways of managing capacity in healthcare settings and apply them to Emergency Care Units. The methodology used is based on three pillars: 1. lexical and content analysis of existing scientific articles in the Scopus database; 2. quantify and qualify the existence of publications by identifying the correlations between authors using VOSviewer; 3. proposed application of capacity management in 50 Emergency Care Units. The results showed that with Capacity Management it is possible to generate activities that add more value to the infrastructure and control service delivery, eliminating queues and bottlenecks in the healthcare environment, thus contributing to effective service through continuous flow in the urgency and emergency of 24 h ECUs.

**Keywords:** Capacity management · Capacity control · Lean healthcare

## 1 Introduction

The production of goods and services requires the dimensioning of resources that include machinery, equipment, human resources, space required for installations, and the definition of flows (material, information, and financial). The intention is to generate productivity to obtain quality and sustainability in the organizational environment.

In healthcare service delivery, the measurement of performance and the determination of capacity is unclear or no viable methods for such measurement are indicated in the literature. In hospitals where customer service saves lives or decrees deaths it is necessary to establish a set of measures or indicators to assess the operational efficiency of processes, the degree of capacity utilization, the availability of productive resources, and the quality of operations.

© IFIP International Federation for Information Processing 2021
Published by Springer Nature Switzerland AG 2021
A. Dolgui et al. (Eds.): APMS 2021, IFIP AICT 631, pp. 298–304, 2021.
https://doi.org/10.1007/978-3-030-85902-2_32

The Unified Health System is composed of several units, with different care objectives and that complement each other. This was the way to assign responsibilities to the Union, States, Federal District, and Municipalities through the management of health actions and services in a solidary and participatory manner.

In Brazil, the SUS brings together a network structure of basic care, hospital, home care, mobile emergency and emergency care units (ECU 24h) - whose goal is to meet the needs of the population. Otherwise, the scenario of uncertainty will be detrimental to the efficiency of health service delivery.

Therefore, it is essential to measure, evaluate performance and capacity to direct management and decision-making actions. Thus, this study aims to identify the measurement of capacity in the healthcare environment and propose an application in Emergency Care Units (ECU)

## 2 Theoretical Reference

### 2.1 Lean Healthcare

The dynamics that exist in the hospital environment offer opportunities for intervention that currently focuses on applying the principles and tools of Lean Thinking to adjust processes, streamline patient care, organize the work environment, and motivate employees.

Production capacity and operational efficiency are two improvements sought by healthcare executives and managers. They are concepts that require care when measured, as they vary with the local economy, over time due to resource scarcity [3,12] and decentralized management [5,10]. Therefore, it is necessary to plan, verify the availability of resources, define performance indicators and service level [3].

For hospital capacity planning, Ettelt et al. [5] stress investment in facilities, equipment, and technology. The number of hospitals varies with level of care and most countries define only the number and location of facilities for an existing infrastructure within a coverage area. Each differentiation of resources defines a capacity and the sum of all will determine the capacity of the system as a whole. Inadequate measurements or comparisons between the various capacities can result in inefficiency and non-accuracy of the system.

The hospital dynamic has variable demand, different pathologies and time periods. The service structure must offer an adequate infrastructure, personal and material resources, and a level of service to the patient.

In emergency areas it is important to have a specialized team to assist patients, ensure effective management of bed flow and capacity 24/7 [3,7,10,12]. Capacity management in hospitals allows for the correction of errors such as lost revenue, delays, operational inefficiency and patient dissatisfaction. The lack of an adequate model and reliable database can make it difficult to plan service delivery or to make resources available for internal activities.

# 3  Method

The research strategy was action research with a qualitative and quantitative approach, data collection with interviews, questionnaires, and official documents. The method is divided into phases: 1. lexical and content analysis of existing scientific articles in the Scopus database; 2. quantify and qualify the existence of publications by identifying the correlations between authors using VOSviewer; 3. proposed application of capacity management in 50 Emergency Care Units.

The purpose of this proposal is to provide employees with practical tools to use in their daily lives and to improve their activities, eliminating waste of time and improving the quality of patient service. The study presents a systematic application of the Lean Healthcare system in the 50 ECUs participating in the project.

# 4  Development

## 4.1  Study of the Process

The Scopus base (Elsevier) aggregated 473 articles (35.5%), three authors with three papers each, 240 in the medical field (29.9%), 126 in management, business and accounting (15.7%) and 73 in engineering (9.1%). The largest number of publications came from the United States of America and were affiliated with Harvard Medical School. In Brazil, 58.8% of the papers were published as articles in 2018 and 2019, two authors and two publications each - Augusto, B.P. and Tortella, G.L.

Using VOSviewer in the Scopus base (Elsevier) was observed that the word network is composed of seven clusters differentiated by colors (Fig. 1), each of them has a central word that connects the others through links, interrelating them, being "Human" the central link.

One of the branches in red ends in "total quality management" from which other terminologies emerge, loosely, without direct connection to the others, however, with some degree of importance. Implying that although there is no direct link such concepts are pertinent or have been studied in some context, which illuminates the originality and generates the need to deepen and fill the research gaps, gains of this article. The terms Lean and Lean Production stand out in isolation and surrounded by terminology such as "performance measurement", "operations management" and "six sigma", which allows us to conclude that there is a concern to measure and determine through metrics, the variables that can interfere or improve the capacity of the entire system, especially in healthcare.

Authors' analysis, applying the VOSviewer to at least one publication per author identified the existence of 38 authors divided into three clusters that are related to each other by means of two central ones, Hooper, L. and Cederholm, T.. There is a little variation in relation to the number of publications in this area of interest, as there is not one author who has published more than three papers. However, Cederholm, T. stands out in the number of citations from his

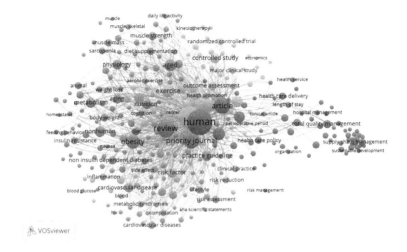

**Fig. 1.** Occurrence of key words in publications.

research in years prior to 2016, while Hooper, L. obtained more citations than works published from 2014 onwards, the most relevant succeeding 2018.

The analysis of the authors with the largest number of papers identified three clusters, one author in each. Individually, they have three publications, the most important and the most cited is Radnor, Z.

There is no consensus in the understanding and establishment of the relationship between Capacity Control, Capacity Measure, Effective Capacity, Lean Healthcare either by keyword or authorship or co-authorship. The publications do not treat the subject in a direct way or establish a concrete way of evaluation of capacity in health environments, especially among the authors on the largest number of published works.

Publications address the topic related to queuing concepts [8], discrete event simulation and operational research [13] applicable to healthcare, and with expressive results when combined with Lean to manage patient flow in hospital settings, especially with Value Stream Map (VSM), the implementation of 5S [4] and dynamic capacity measurement instruments (DC) [2].

The study by Leite et al. [9] identified 21 barriers to lean implementation and among them, the measurement and performance management system. This barrier shows that professionals and the SUS are limited to the service provided, i.e., non-urgent cases that reach the system create an unexpected demand, can make the process difficult and affect the capacity of that health unit. Similarly, the unavailability of resources for correct care can limit SUS processes, affect the performance of the technical team, and imply in its capacity.

During the application of the lean project in the ECUs, the monitoring of activities and the definition of indicators such as the average length of stay started in three units (Fig. 2). The capacity of the units needs control, quality, and statistics. The u-chart, applied in the ECUs, was used to monitor indicators such as the length of stay of patients as well as the average number of non-conformities per unit.

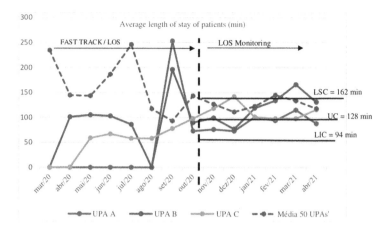

**Fig. 2.** Average length of stay patients (min).

The project brought to light the need to establish some measurement and representation parameters for the internal control of the activities, namely: Data collection by sampling, recording the times by service and patient; Point the data in the form and or system by patient; Compile the cycle times by patient in spreadsheet; Check and understand the outliers of the cycle times of services by patient; Visual management of the average time (LOS -Length of stay), present graph or Dashboard (Power BI).

In healthcare, Fast Track is a form of resource planning and control sequencing, such as Shortest operation time first (SOT), where small tasks are processed quickly, and thus a green rating, for example, is accomplished more quickly. The performance of care in the shortest time can be improved with Fast Track, achieving more productivity and a shorter Average Dwell Time for patients in the Urgent and Emergency care.

The concept of capability helps managers and policy makers identify the capabilities needed to increase change: organizational culture; data and performance; employee engagement; leadership engagement; process improvement and learning; service user focus; stakeholder and supplier focus; strategy and governance [6]. However, volatility, uncertainty, complexity and ambiguity in the organizational environment are barriers to performance effectiveness. Therefore, there must be a rapid change at the strategic, tactical and operational levels [1].

Decision making is based on the collection of information and the functioning of the chain and its environment. Thus it is possible to align, adapt and streamline the chain for environments of high uncertainty and risk to promove agility and responsiveness [11].

To calculate the patient journey, the following procedures were determined: random sample of 80 patients per month, 20 per week (preferably 10 in the morning shift and 10 in the afternoon shift) and in all the stages in the ECU - Reception, Triage, Consultation, Medications and Exams. The data should be entered into the standardized spreadsheet. The data collection worksheet has 4 tabs, which are "Data", "Dashboard", "Graphical Dashboard" and "Action

Plan". An example is shown in Table 1 of part of the standard spreadsheet available on ECUs.

Table 1. Example of a log sheet used in ECU's

| (5) Check-in time at reception | (6) Check-out time at reception | (7) Screening start time | (8) End time of screening |
|---|---|---|---|
| 1:00 | 1:02 | 1:03 | 1:05 |
| 3:10 | 3:13 | 3:16 | 3:18 |
| 5:52 | 5:53 | 5:54 | 5:55 |
| 6:05 | 6:10 | 6:13 | 6:15 |

Capacity measurement requires prior monitoring, the provision of standard information via remote systems and the creation of a big data to analyze the data received and establish service priorities according to the emergency. Therefore, it is essential to establish monitoring and control mechanisms linked to an information and communication system integrating the entire Unified Health System (SUS).

The definition of indicators, the use of methods for monitoring data such as DMAIC and A3 reporting in the ECUs have benefited patient care by more than 1Mi in reception, 5Mi in risk classification, 22Mi in orthopedics, and 5Mi in physicians. Each unit is able to identify needs and establish its indicators. However it is necessary to have standardized parameters that can guarantee a service level standard in the ECUs, the great benefit of the Lean application.

## 5   Conclusions

In the literature there are several management models, but not specific to the service in healthcare. The research points out some solutions to be applied in capacity management, but leaves gaps that need to be filled in order to improve the process of care in urgency and emergency in healthcare.

Another point that makes capacity management difficult is the human factor and its uncoordinated actions, generating overload or idleness. A method is needed to adjust capacity through overtime allocation or value chain balancing to eliminate idleness, varying, if possible, the size of the workforce, adjusting the infrastructure to new demand, or even subcontracting is an alternative.

The main barriers to Lean implementation are: a performance measurement and management system that is closely linked to capacity management; the unavailability of resources that affects the performance of the entire technical team (doctors and nurses).

Thus, the application of Lean Six Sigma in Emergency Departments is recommended, as well as the definition of indicators and their monitoring. Orienting the organizational culture and the management for data collection was one of the benefits of the Lean implementation in the 50 UPAs and the next step is the definition of a method for capacity management in this universe .

**Acknowledgment.** The authors would like to thank the Brazilian Ministry of Health, Fluminense Federal University and Euclides da Cunha Foundation. This Research is part of a "Lean Project in UPAs 24h" that has been funded by the Brazilian Ministry of Health (TED 125/2019, number: 25000191682201908).

# References

1. Alexander, A., Kumar, M., Walker, H.: A decision theory perspective on complexity in performance measurement and management. Int. J. Oper. Prod. Manag. **38**(11), 2214–2244 (2018)
2. de Araújo, C.C.S., Pedron, C.D., Bitencourt, C.: Identifying and assessing the scales of dynamic capabilities: a systematic literature review. Revista de Gestão **25**, 390–412 (2018)
3. Bonomi Savignon, A., Costumato, L., Marchese, B.: Performance budgeting in context: an analysis of Italian central administrations. Admin. Sci. **9**(4), 79 (2019)
4. Chadha, R., Singh, A., Kalra, J.: Integração enxuta e de filas para a transformação dos processos de assistência à saúde: um modelo de assistência à saúde enxuta **17**(3), 191–199 (2012)
5. Ettelt, S., et al.: Capacity Planning in Health Care: A Review of the International Experience, WHO, Geneva (2008)
6. Furnival, J., Boaden, R., Walshe, K.: A dynamic capabilities view of improvement capability. J. Health Organ. Manag. **33**(7-8), 821–834 (2019)
7. Gabutti, I., Mascia, D., Cicchetti, A.: Exploring "patient-centered" hospitals: a systematic review to understand change. BMC Health Serv. Res. **17**(1), 1–16 (2017)
8. Lantz, B., Rosén, P.: Measuring effective capacity in an emergency department. J. Health Organ. Manag. **30**(1), 73–84 (2016)
9. Leite, H., Bateman, N., Radnor, Z.: Beyond the ostensible: an exploration of barriers to lean implementation and sustainability in healthcare. Prod. Plan. Control **31**(1), 1–18 (2020)
10. Naslund, D., Norrman, A.: A performance measurement system for change initiatives. Bus. Process Manag. J. **25**, 1647–1672 (2019)
11. Sabet, E., Yazdani, N., De Leeuw, S.: Supply chain integration strategies in fast evolving industries. Int. J. Logist. Manag., 29–46 (2017)
12. Sharifi, S., Saberi, K.: Capacity planning in hospital management: an overview. Ind. J. Fund. Appl. Life Sci. **14**(2), 515–521 (2014)
13. Yip, K., Leung, L., Yeung, D.: Levelling bed occupancy: reconfiguring surgery schedules via simulation. Int. J. Health Care Qual. Assur. **31**(7), 864–876 (2018)

# MDE-S: A Case Study of the Health Company Diagnostic Method Applied in Three Health Units

Alexandre Beraldi Santos[1]([⊠]) , Robisom Damasceno Calado[1] ,
Sandra Maria do Amaral Chaves[1] , Stephanie D'Amato Nascimento[1] ,
Messias Borges Silva[2] , and Saulo Cabral Bourguignon[1]

[1] Universidade Federal Fluminense, Rio das Ostras, Brazil
{alexandreberaldisantos,robisomcalado,sandrachaves,sd_amato,
saulocb}@id.uff.br
[2] Universidade Estadual Paulista, Guaratingueta, Brazil
messias.silva@unesp.br

**Abstract.** The objective of this study was to evaluate the degree of organizational maturity in three Emergency Care Units (a.k.a. UPA in Brazilian acronym) in a city in the interior of the state of São Paulo - Brazil, with the following fundamental criteria observed: the utility, feasibility, and specialized technical knowledge of the health unit contextualized in this study. The method used was the Health Company Diagnostic Method (MDE-S), which is an approach to study and qualify the level and organizational maturity, that has the fundamental characteristic of providing a cycle of continuous improvement in the company. Among the most relevant results, the study demonstrated, in comparison with a study conducted in Basic Health Units in the interior of the state of Rio de Janeiro, an existent lack of knowledge in health management tools and basic administration fundamentals of the evaluated units. Finally, the study recommends that a strategic system be created for the management and deployment of goals and guidelines, based on Hoshin Kanri. It is considered that, despite the challenges that will certainly occur, this small-scale and well-developed diagnostic system will certainly be useful and may become an important step towards a continuous practice of efficient management in the investigated health unit.

**Keywords:** Maturity · MDE-S · Lean healthcare · Lean hospital

## 1 Introduction

The current scenario of organizations requires that they remain competitive throughout their existence. For Bhagat et al. [1], the reduction of barriers to negotiation between countries imposed by globalization, the use of labor from different countries, administration of a large amount of unstructured data, and the focus on the development of skills for customer satisfaction are imperative

© IFIP International Federation for Information Processing 2021
Published by Springer Nature Switzerland AG 2021
A. Dolgui et al. (Eds.): APMS 2021, IFIP AICT 631, pp. 305–313, 2021.
https://doi.org/10.1007/978-3-030-85902-2_33

issues for companies to continue to exist. According to Salomão [4], a survey carried out by the Datafolha institute at the request of the Brazilian Federal Council of Medicine (CFM) points out that 93% of Brazilians evaluate public and private health services as terrible, bad, or regular. Among the Public Health System (PHS) which in Portuguese means Unified Health System (SUS), the users, 87% declared themselves dissatisfied with the services offered, according to IBGE [3], in 2015, the final consumption of health goods and services in Brazil were R\$ 546 billion (9.1% of Gross Domestic Product - GDP). Of this total, R\$ 231 billion (3.9% of GDP) corresponded to government consumption expenditures and R\$ 315 billion (5.2% of GDP), expenses of families, and non-profit institutions in service of families. According to Calado et al. [2], companies are at different levels of maturity. The application of lean methods and tools is structured according to the principles and adequacy of the reality of each company. Some actions and programs are not very sustainable because they are out of alignment with strategic decisions. Therefore, with a diagnosis and alignment of goals and guidelines and the proper choice of Lean tools and methods application, a transformation of the production system can be achieved with considerable improvements.

## 2   Method

The method used in this work is based on the MDE- (from Portuguese acronym for Health Company Diagnostic Method), which is a method for studying and qualifying the level and maturity of the company, where the results are observed and evaluated, through the application of a set of methods and tools according to the Lean approach. The MDE-S implies gathering and systematizing information related to different aspects of business management that are not always known to everyone and normally a lot of information is not expressed and used in decision making [2]. The MDE-S, Fig. 1, consists of three main axes: Application of the Value Stream Mapping, the Capacity Management Survey, and the Company Diagnostic Survey, emphasizing that self-assessment and the gathering of evidence generate the identification of strengths and weaknesses that translate into possible improvements.

To achieve this study objective, the scope of the assessment was delimited, based on steps 7 to 14 of the MDE-S, except for step 11, in which information was obtained from the health professionals. These data were then categorized and calculated and, after being analyzed, produced the necessary inputs for the formulation of value judgments, presented, finally, in the final considerations and recommendations.

The MDE-S has already been applied to measure the degree of maturity on different occasions. In the central region of Portugal, the MDE-S was applied in a hospital using the Baldrige and LESAT methodology [7] and in three Basic Health Units - (UBS, in the Portuguese acronym) in the state of Rio de Janeiro, by Santos [5].

For MDE-S application in the studied units, was developed software for use in web and mobile platforms (tablets and smartphones) called MDE-S 1.0.

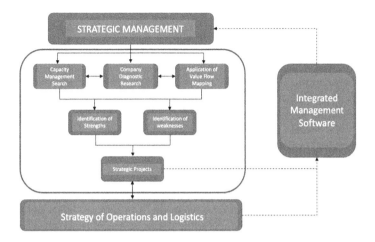

**Fig. 1.** Health company diagnostic method (MDE-S).

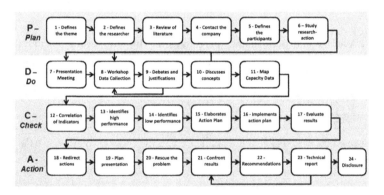

**Fig. 2.** Structure of the company diagnostic method.

The software was based on the combination of two tools: LESAT and the Baldrige Award, which, when combined, make it possible to measure results in terms of the institution's positioning in practice and performance. In this context, it is relevant to define the concept of practices and performance:

(a) Practices are the processes implemented in the company/institution to improve business management - that is, they are the management and technological tools implemented in the organization. These include organizational aspects, such as employee involvement and teamwork;

(b) Performance refers to the measurable results of the processes previously implemented in the organization, such as the volume of material in process, cycle times, and impacts on the business result, such as market share, customer satisfaction level, and employee morale.

# 3   Results and Discussion

This chapter aims to present the results obtained in the study, presenting, simultaneously, the analyzes corresponding to each aspect addressed in it. At this point, it is worth note that all the tables and figures in this chapter are authored by the authors of this study. In this stage of the evaluation, the indicators were verified in eight areas related to the criterion framework presented in the method. These diagnostic areas were analyzed in three Emergency Care Units (UPA, in brazilian acronym) in the city of Guarulhos (São Paulo, Brazil).

To determine a value for practice and performance it was necessary to differentiate a priori the survey topics related to both concepts. After this analysis, it is necessary to calculate the average of both concepts described. The results are presented in percentage numbers for later framing of practices versus performances. To obtain the General Practice and Performance Index of the Unit all PR values were calculated, adding up all the data obtained by the respondent regarding PR and then dividing it by the sum of possible PR scores. The Table 1 below shows a comparison of Practice and Performance of the three units evaluated.

**Table 1.** Average and performance percentage of the units evaluated in São Paulo

| Indicators | UPA 1 | | UPA 2 | | UPA 3 | |
|---|---|---|---|---|---|---|
| | Average (1 a 5) | Performance | Average (1 a 5) | Performance | Average (1 a 5) | Performance |
| Practices | 2,1 | 41% | 3,3 | 67% | 3,3 | 66% |
| Performance | 2,2 | 44% | 2,7 | 55% | 3,1 | 61% |

The results obtained by each UPA are also presented in the form of a radar graph that allows a more interactive interpretation of the results (Fig. 3).

Topics 1 to 7, in Fig. 3, concerns the methodology of the Baldrige award. Topic 8 is composed of questions that result from the LESAT methodology. In Fig. 3, topics 1 to 6 are composed of questions that intend to measure practices, and topics 7 and 8 relate to performance issues, which allows defining the organization's positioning.

To demonstrate the parameters that the method can compare, it follows the same set of data extracted from the MDE-S in three Basic Health Units – (UBS) located in the state of Rio de Janeiro (Brazil) present in the dissertation "Adoption of Lean practices in the service of public health: an evaluative study", 2019, shown in Table 2, that unlike the three UPAs in São Paulo, the UBS present in Table 2 were not the object of actions to implement lean management methods in health (Lean Healthcare).

The results obtained by each UBS are also presented in the form of a radar graph that allows a more interactive interpretation of the results (Fig. 4).

Figure 4 shows some strengths and weaknesses of the units evaluated. In the Graph, it is explicit that point 8 (regarding the LESAT tool) was not answered

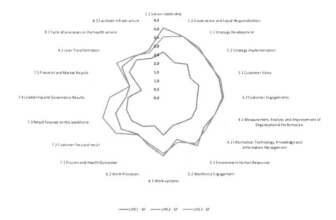

**Fig. 3.** Results of the eight dimensions analysis by UPA.

**Table 2.** Average and performance percentage of the units evaluated in Rio de Janeiro

| Indicators | UBS 1 | | UBS 2 | | UBS 3 | |
|---|---|---|---|---|---|---|
| | Average (1 a 5) | Performance | Average (1 a 5) | Performance | Average (1 a 5) | Performance |
| Practices | 2,3 | 46% | 2,4 | 47% | 2,9 | 59% |
| Performance | 1,7 | 34% | 1,7 | 33% | 2,2 | 44% |

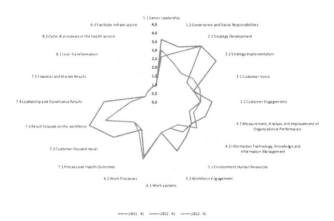

**Fig. 4.** Results of the eight analysis dimensions by UBS.

by the respondents, with classification note "1" assigned for this topic. It occurs because the units analyzed were not targeted at any Lean transformation before.

For a better perception of the relationship between the organization's position and its ability to answer to the challenges of competitiveness in the market, an analogy with the skill and performance of the boxers was used.

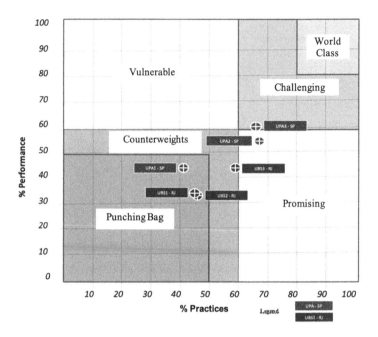

**Fig. 5.** Boxers' analogy between UPA and UBS.

For the definition of the organizational degree of maturity of the health units contextualized in this study, the classifications defined by Seibel [6] were considered, taking into account the General Indices of Practice and Performance obtained by applying the LESAT × BALDRIGE model. These indices are classified as: (i) Punching Bag; (ii) Counterweights; (iii) Vulnerable; (iv) Promising; (v) Challengers; and (vi) World Class.

In that case, the organizations classified as 'Challenging' or 'World Class' have a High Level of Maturity. Those classified as 'Vulnerable' or 'Promising' have Medium Organizational Maturity levels. Finally, those classified as 'Bag of Punch' or 'Counterweights' have a Low Organizational Maturity index. Figure 5 shows that the level of maturity is higher in UPAs than in UBS. It is worth mentioning that the UPAs surveyed have already undergone Lean Transformation training and assistance.

Figure 6 shows a boxplot that compares the UBS with the UPAS evaluated in the study. The graph shows a characteristic related to the values assigned as scores in the self-diagnosis, showing that the UPAs' results have a smaller range of values once these units have passed by a Lean transformation process.

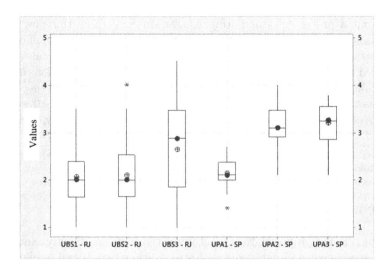

**Fig. 6.** Comparative boxplot on practices and performance of UPAs and UBS.

# 4   Conclusion

The analysis of the MDE-S allowed elaborating the outline of the entire evaluation. It facilitated a lot in the development of this study showing step-by-step how it should be carried out. The planning phase took the longest, then the data collection phase was delayed a few days due to the collection taking place during the COVID 19 pandemic period, which still did not delay the study progress. In the analysis of the Practices and Performances data, considering the evaluated units, it was found that the Units that have not yet undergone Lean transformation processes (training and on-site assistance in Lean Healthcare) have values and a low level of maturity, besides percentages below 60% for practice and below 60% for performance.

The analysis of the boxers' analogy (Fig. 5) helped to bring a quick visualization in which point the UPA was classified in the percentage of practices and performances. It also made it possible to visualize the positioning of the evaluated Unit in a graph. The study showed a lack of knowledge in Lean Healthcare methods and tools, by the managers and employees in Units valuated below 60% in the value of Practice and Performance. At this point, it was found that none of these evaluated units has a world-class degree of maturity, that is, 'Elevated'. Two units were located in the maturity classification zone as 'Punching Bag' (UBS1 -RJ, UBS2 - RJ, and UPA1 - SP), and a unit located in the maturity classification zone as 'Counterweight' (UBS3 - RJ), that is, all with maturity degrees classified as 'Low', those that obtained a "Medium" degree were (UPA2 - SP, and UPA3 - SP).

To adopt a new form of management for higher levels of maturity based on Lean principles, it is necessary for the organization, through its workforce, to accept the change. For this, it is considered the fundamental importance to carry out a process of stimulating motivation concerning this new form of organizational management. It is noteworthy that the entire process of changing the organizational culture needs substantive actions to clarify new goals and that it is desired to achieve and carry out processes to stimulate motivation that leads to the achievement desired. The concept of organizational culture encompasses all practices, habits, principles, and values within an organization. All workforce must understand this culture and share its values. If done properly, the process of understanding and absorbing a relevant change in an organization's culture can positively boost the behavior of its employees about the impacts caused by its implementation.

## 5   Recommendations

A recommendation proposed due this study is the creation of a strategic system for the management and deployment of organizational goals and guidelines. At this point, it is suggested to use a management system based on the so-called Hoshin Kanri because it is considered adequate for the studied context and, also, for having already been used in other contexts, presenting a high-efficiency index.

The second recommendation is the use of the MDE-S before and after training and assistance in Lean Healthcare implementation, that is, during the Lean transformation in Health.

Despite the challenges, the MDE-S is a well-developed diagnostic system and, it can be useful and is the first step to continue good management practices in health facilities. However, in order to adopt this new form of management it is necessary that the organization, through its servers, accept it.

It is noteworthy that the entire process of organizational culture change requires substantive actions to clarify new goals and objectives that one wants to achieve and carrying out processes to encourage motivation that lead to the achievement of what is desired.

The points covered in this article are fully applicable in the units surveyed, where you can comment and/or recommend the following analysis of improvements:

1. Implement a cost reduction strategy combined with the growth strategy through workshops with the management team.
2. Implement the strategic deployment, define goals and objectives for all areas in an integrated and transparent way, if possible with software support.
3. Implement the day-to-day routine management, starting with the work pattern of critical activities and treatment of bottlenecks.
4. Systematize a company's problem-solving method.
5. Implement 5S project and visual management in processes.
6. Develop current and future value stream mapping to guide improvements.
7. Implement autonomous maintenance, leverage TPM, to reduce downtime and involve operational personnel.

8. Develop the multifunctionality matrix.
9. Leverage an ideas plan with direct rewards and leverage the career plan.
10. Develop a training plan for managers and employees to implement Lean Enterprise.

Finally, it should be added that some of the problems founded about performance and operations show the importance of public administration managers becoming more relevant to the development of Lean projects in the health system to promote a culture of continuous improvement centered on care. to the patient, which, in itself, is an important and positive aspect of this work.

**Acknowledgements.** The authors would like to thank the Brazilian Ministry of Health, Fluminense Federal University and Euclides da Cunha Foundation. This Research is part of a "Lean Project in UPAs 24h" that has been funded by the Brazilian Ministry of Health (TED 125/2019, number: 25000191682201908).

# References

1. Bhagat, R.S., McDevitt, A.S., Baliga, B.R.: Global Organizations: Challenges, Opportunities, and the Future. Oxford University Press, New York (2017)
2. Calado, R.D., Batocchio, A., Calarge, F.A., Silva, M.B.: Enterprise Diagnostic Method: Organization Performance Improvement. GlobalSouth Press, USA (2015)
3. IBGE: Health satellite account - brazil: 2010–2015. https://www.ibge.gov.br/estatisticas/economicas/contas-nacionais/9056-conta-satelite-de-saude.html?=&t=o-que-e (2017)
4. Salomão, L.: Pesquisa diz que 93% estão insatisfeitos com sus e saúde privada (2014). http://g1.globo.com/bemestar/noticia/2014/08/pesquisa-diz-que-93-estao-insatisfeitos-com-sus-e-saude-privada.html
5. Santos, A.B.: Adoption of lean practices in the public health service: an evaluative study (2019). http://mestrado.cesgranrio.org.br/mestrado/principal.aspx
6. Seibel, S.: Um modelo de benchmarking baseado no sistema produtivo classe mundial para avaliação de práticas e performances da indústria exportadora brasileira (2004)
7. Valente, R.M.F.: Avaliação de unidades de saúde recorrendo à metodologia Baldrige e LESAT: Um caso de estudo de um hospital da região centro de Portugal. Ph.D. thesis, Universidade de Coimbra (2014)

# Kaizen and Healthcare: A Bibliometric Analysis

Sandro Alberto Vianna Lordelo[1]([⊠])(iD), Sara Monaliza Sousa Nogueira[1](iD),
José Rodrigues de Farias Filho[1](iD), Helder Gomes Costa[1](iD),
Christiane Lima Barbosa[2](iD), and Robisom Damasceno Calado[3](iD)

[1] Universidade Federal Fluminense, Niterói, Brazil
sandrolordelo@id.uff.br
[2] Universidade Federal do Pará, Belém, Brazil
cllima@ufpa.br
[3] Universidade Federal Fluminense, Rio das Ostras, Brazil
robisomcalado@id.uff.br

**Abstract.** The objective of this article was to map the state of the art, by means of bibliometric research, of how the scientific evolution of Kaizen has been occurring in theory, in practice and, especially, in the health area around the world, in order to guide future research in the same, as, for example, kaizen has been contributing to humanization within health centers. A broad structured search was performed using keywords and Boolean logic, in the PubMed, ISI and Scopus data bases. The identified publications were cataloged in EndNote and NVIVO 10 software for bibliometric analysis. Through these, 1,467 publications were identified, in the three databases, excluding the repeated publications and those without relevant information. The main key terms identified and in common in all publications were: "management", "lean", "quality" and "kaizen"; and the authors who publish the most were Van Aken, E.M., Farris, J. A., Suárez-Barraza, M.F. and Glover, W. J; and when kaizen applied to healthcare was evaluated, the most cited article among the 1,467 selected was the "Lean in healthcare: The unfilled promise?", by Radnor, ZJ, Holweg, M. and Waring, J., which has 316 citations in Scopus. It was evidenced that kaizen applied to health still has much potential to expand.

**Keywords:** Continuous improvement · Lean · Healthcare

## 1 Introduction

Lean is a management philosophy based on concepts from the Toyota Production System [25]. The Lean process evaluates operations step by step in order to identify waste and inefficiencies, create solutions and improve operations, adding value and continuous quality improvement [1,7,10].

© IFIP International Federation for Information Processing 2021
Published by Springer Nature Switzerland AG 2021
A. Dolgui et al. (Eds.): APMS 2021, IFIP AICT 631, pp. 314–322, 2021.
https://doi.org/10.1007/978-3-030-85902-2_34

As a lean tool, Kaizen is used to achieve continuous improvement [3,6], reducing waste and improving the quality and efficiency [16]. Kaizen seeks incremental improvements, small or large, to eliminate non-value-added actions, determining who are the customer and the current status, developing the vision of the future and implementing, measuring and evaluating performance [13], being already used in several areas [16], including health [7,24].

The general objective of this research was to map the state of the art, by means of bibliometric research, of how the scientific evolution of Kaizen has been occurring in theory, in practice and, especially, in the health area around the world.

## 2 Theoretical Framework

Kaizen is part of Lean philosophy and therefore embodies this lean thinking. Kaizen is a Japanese word formed by two terms, 'zen' meaning 'good', and 'kai', 'change'; that is, a term that constitutes "good change" or "improvement" [11].

At first, this lean thinking was focused on manufacturing companies, hence comes the denominations lean manufacturing or lean production. However, this philosophy quickly shifted to new areas, such as healthcare [16]. Hence the concept of "Lean Healthcare" emerged with a focus on adding value to the patient, eliminating wasteful activities from healthcare processes, and respecting all people [17].

The Kaizen philosophy has been successfully applied to quality improvement and safety in the medical field [12]. Physicians and healthcare managers work together to study and apply the principles of continuous improvement to achieve better efficiency, greater effectiveness, lower costs, and more satisfied patients [2].

Nowadays Kaizen is used in many different areas of medicine: radiology [12]; ophthalmology [20]; surgical units [24]; telemedicine [5]; exams [21]; among others.

And in the scientific literature there are already documents about Lean and Kaizen approaches in healthcare. Some, for example, in review formats, such as those presented by D'Andreamatteo et al. [8], Daultani et al. [4] and Mazzocato et al. [15], which bring dozens of other articles, with theories and practical cases, in general, about such applications.

## 3 Methodology

This research is based on the literature review of the state of the art of continuous improvement, especially kaizen and its use in healthcare. The first steps of this are the definition of the research keywords, the search for data in bibliographic information bases, and the construction of the library [22,23].

The keywords were organized according to the focus of the research, into three axes: a) theoretical axis; b) application axis; c) health axis. The main terms selected were kaizen, lean, improvement, concept, case, application, practice, implementation and health; all in English, to obtain international publications.

Then, these terms were entered using the Boolean logic with the connecting terms "AND" and "OR" in the PubMed, Web of Science and Scopus. As reported by Rodriguez et al. [19] the choice of such bases reduces the probability of using predatory sources or even no blinded reviewed ones.

In PubMed, the search was performed by selecting the option "Title/ Abstract" option, so the search for the desired words was performed over the title and abstract of the papers. In Web of Science, the option "TOPIC" was selected, which searches the title, abstract, author's keywords and Keywords Plus® (words or phrases that often appear in the titles of article's references, but do not appear in the title of the article). And in Scopus, the "TITLE-ABS-KEY", which searches the title, abstract and keywords of the documents in this database.

All research in the databases were performed on July 1, 2020, and these considered the documents already published in the first half of 2020. These results were exported to the EndNote X7 software for library construction and management. In such software, articles repeated ("duplicates") in the databases or with incomplete relevant information, such as title, authors, abstract and keywords, were excluded.

The remaining records were then exported to the NVivo 10 and VantagePoint software to perform the bibliometric analysis: year of publications; authors with the most published scientific articles and with the most citations; and journals with the most publications in the study areas.

# 4    Results and Discussions

After building the keyword tree and the definition of the Boolean logic for each axis, the queries were inserted in the three search bases and 1,795 documents were reached in axis 1 (kaizen in theory); 1,395 documents in axis 2 (kaizen in practice); 187 documents in axis 3 (kaizen in health). After assembling these 3,377 documents in EndNote software and deleting 1,753 duplicate documents, 1,624 documents remained. The Boolean sequences with the key terms and these results can be seen better in Table 1.

It is interesting to point out that about 90% of the documents raised in the third axis, had already been raised in the other two axes. And, about 2/3 of the documents in the second axis, had also already been lifted in the first axis. Also the repetition of documents between the bases; about 90% of them were in Scopus. Thus, relating the third axis with the second and the first axis, respectively, one notices that only 12.6% of the articles with Kaizen practices in healthcare, and 9.5% of theoretical documents about Kaizen and healthcare. This reveals the large field still to be explored.

Latter, the data from the documents were viewed one by one by the authors of this research in order to exclude files with excessive missing information (no title, abstract and authors' name) or even duplicate documents that passed the first software filter, leaving a total of 1,467 diverse documents in the library, among articles (53.3%), conference papers (34.3%), reviews (5.4%), book chapters (3.1%), books (1.3%), and others (2.6%).

**Table 1.** Number of selected articles per axis of interest

| AXIS | Base | No. of articles | No. of articles (excluding repeats) |
|---|---|---|---|
| 1. THEORETICAL ((kaizen) AND (improvement)) OR ((kaizen) AND (concept)) OR ((kaizen) AND (lean)) | Pub Med | 81 | 1.314 |
| | Web of Science | 546 | |
| | Scopus | 1.168 | |
| | Sub-Total | 1.795 | |
| 2. PRACTICAL ((kaizen) AND (application)) OR ((kaizen) AND (case)) OR ((kaizen) AND (practice)) OR ((kaizen) AND (implementation)) | Pub Med | 57 | 998 |
| | Web of Science | 464 | |
| | Scopus | 874 | |
| | Sub-Total | 1.395 | |
| 3. HEALTH ((((kaizen) AND (application)) OR ((kaizen) AND (case)) OR ((kaizen) AND (practice)) OR ((kaizen) AND (implementation))) AND (health*)) | Pub Med | 26 | 126 |
| | Web of Science | 48 | |
| | Scopus | 113 | |
| | Sub-Total | 187 | |
| | **Total** | **3.377** | **1.624** |

The main areas of study for these papers identified were: Engineering (31.2%); Business, Management and Accounting (23.3%); Decision Sciences (9.4%); Computer Science (9.3%); and Medicine and Health (9.1%).

Next, this data was exported from EndNote to the NVivo and VantagePoint software for bibliometric analysis. At this step, it was possible to obtain an overview of Kaizen in the three contexts: theoretical, practical and in healthcare. That is, from the 1,467 documents. The first result generated by VantagePoint was the historical line with the number of publications per year. The oldest document found is from 1948. But the recurrence of publications on these themes occurred since the late 1980s.

The highest peak of publications occurred in 2018, with a total of 125 articles. It is worth noting that this research was conducted considering only the 1st half of 2020, and therefore indicates a 'drop' in publications this year, which cannot yet be stated as true.

Next, all authors with at least one publication on the themes addressed here were identified. In all, 3112 different authors were surveyed, who published alone or in co-authorship with other researchers. The list of authors with the largest number of publications and the respective total number of articles published by them are Van Aken, E.M. (28), Farris, J.A. (26), Suarez-Barraza, M.F. (22), Glover, W.J. (19), Doolen, T.L. (18), Aleu, F.G. (11), Katayama, H. (11), Wor,ey, J.M. (11), Murata, K. (10) and Nahmens, K. (10). In this list one can highlight the work done by the first three researchers who published more than two dozen materials on the themes explored here, often as co-authorship.

In order to evalue whether there is a relationship between quantity and quality, Table 2 was prepared, with the first 10 names of authors who have accumulated more citations received for their scientific articles that were selected in this research. And to this, the h-index that these authors have in the Scopus database was added.

**Table 2.** Ranking of authors with the highest number of citations received

| Rank | Author | Number of citations | H index |
|------|--------|---------------------|---------|
| 1 | Bessant, Jonh | 625 | 38 |
| 2 | Van Aken, E. M. | 478 | 17 |
| 3 | Farris, Jennifer A. | 440 | 15 |
| 4 | Doolen, T. L. | 358 | 14 |
| 5 | Caffyn, Sarah | 331 | 8 |
| 6 | Radnor, Zoe Jane | 318 | 28 |
| 7 | Holweg, Matthias | 316 | 28 |
| 8 | Waring, Justin | 316 | 28 |
| 9 | Samson, Danny | 286 | 23 |
| 10 | Emiliani, M. L. | 270 | 23 |

Remembering that this index quantifies the productivity and impact of a scientist, evaluating their most cited articles [9]. And here, it will consider the citations and publications that the author has in the entire Scopus database, and not only the articles selected here.

These results were surprising because the most cited author, among the more than three thousand researchers, was J. Bessant who was not even on the previous list because he only had three articles selected in this research; but, who received 246, 193 and 186 citations in them, totaling 625 and an average of 208 citations per article.

The second most cited author was Van Aken, E. M., with 475 total citations and an average of 17 citations per article only. A relatively low value because of the large number of publications that this researcher has. And because only two of them have more than 100 citations and 20 have less than 10 citations, some with none. The same occurs with Farris J. A. and Doolen, T. L.

These results of Bessant, J., Van Aken, EM, Farris JA and Doolen, TL could already indicate that there is not necessarily a relationship between quantity and quality in publications, since the author who has the highest number of citations does not have as many papers of his authorship in the areas of this research. And the authors who have many publications in the areas of interest here did not received numerous citations per article. Therefore, it is also interesting to evaluate an author's h-index.

Other authors who were noteworthy were Radnor, Z. J., Holweg, M. and Wa-ring, J. Together they wrote the most cited article among the 1467 selected, "Lean in healthcare: The unfilled promise?" [18], which has 316 citations in Scopus as of the date of this research (07/2020), as can be seen in Table 3.

**Table 3.** Articles with highest citations and respective authors, journal and year of publication.

| Document title | Authors | Journal | Year | Citations |
|---|---|---|---|---|
| *Lean in healthcare: The unfilled promise?* | Radnor, Z.J., Holweg, M., Waring, J. | Social Science and Medicine | 2012 | 316 |
| *Evolutionary model of continuous improvement behaviour* | Bessant, J., Caffyn, S., Gallagher, M. | Technovation | 2001 | 245 |
| *Implementation of total productive maintenance: A case study* | Chan, F.T.S., Lau, H.C.W., Ip, R.W.L., Chan, H.K., Kong, S. | International Journal of Production Economics | 2005 | 215 |
| *An international study of quality improvement approach and firm performance* | Adam, E.E., Corbett, L.M., Flores, B.E., (...), Samson, D., Westbrook, R. | International Journal of Operations and Production Management | 1997 | 213 |
| *A review of data mining applications for quality improvement in manufacturing industry* | Köksal, G., Batmaz, I., Testik, M.C. | Expert Systems with Applications | 2011 | 212 |

It is interesting to note in Table 3 that the most cited article addresses Lean in healthcare (axis 3), which apparently quite aligned with the interests of this research, the others are from industry and engineering (axes 1 and 2).

It is also curious to note that all the authors surveyed in Table 3 are also in Table 2 and that most of them reached such a position in this one with only a single article. This reinforces that there is not necessarily a relationship between quantity and quality, since a single article may contain enormous quality; and vice versa, many articles may not. Thus, it is important to read these and evaluate whether they are aligned with the research objectives.

Another interesting example is that of Suárez-Barraza, M. F. with 23 articles, 262 citations and an h index of only 9 at Scopus, where there are several other works by him and in partnership with other authors. Thus, if only the h-index were considered as a selection criterion for further reading of the selected articles, it is likely that this author would not be selected. Therefore, several criteria are used when selecting the articles that will be read [22,23].

Lima et al. [14] in their literature review listed 19 areas of healthcare where kaizen can be applied. If at least one article per year were published in a specially index journal in each of these areas, there would be an exponential increase in this margin 10%.

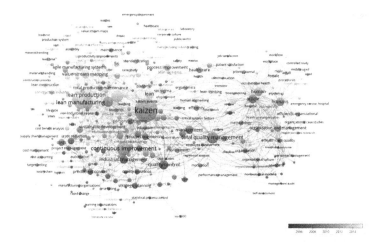

**Fig. 1.** Co-occurence of keywords

Another important analysis to be mentioned is the co-occurede of keywords that appear in the articles. Figure 1 was created using the VOSviewer software.

The word kaizen was the most common, followed by continuous improvement and lean manufacturing. It is also observed that the most recent terms (after 2012) are works related to health care and lean production. The health care part is focused on managing people and organizations, such as hospitals, while the lean production part, for process improvements, using VSM, Six Sigma and agile manufacturing systems.

## 5    Conclusion

Overall, the application of Kaizen in any area identifies problems and provides changes in the organization, with a focus on improving results. In healthcare, kaizen can bring continuous improvement to processes, improving management, impacting efficiency and productivity in hospitals, in improving customer satisfaction.

The research groups for developing articles together demonstrate that universities can achieve greater strength by publishing a larger number of articles per year, and also that they possibly become more multidisciplinary research centers that are able to apply the knowledge produced by academy to everyday life in society.

**Acknowledgements.** The authors would like to thank the Brazilian Ministry of Health, Fluminense Federal University and Euclides da Cunha Foundation. This Research is part of a "Lean Project in UPAs 24h" that has been funded by the Brazilian Ministry of Health (TED 125/2019, number: 25000191682201908).

# References

1. Álvarez-García, J., Durán-Sánchez, A., del Río, M.d.l.C., et al.: Systematic bibliometric analysis on kaizen in scientific journals. TQM J. (2018)
2. Berwick, D.M.: Continuous improvement as an ideal in health care (1989)
3. Brunet, A.P., New, S.: Kaizen in Japan: an empirical study. Int. J. Oper. Prod. Manage. (2003)
4. Daultani, Y., Chaudhuri, A., Kumar, S.: A decade of lean in healthcare: current state and future directions. Glob. Bus. Rev. **16**(6), 1082–1099 (2015)
5. van Dellen, J.R.: The philosophy of kaizen and telemedicine. World Neurosurgery **91**, 600–602 (2016)
6. Deming, W.E.: Out of the Crisis. MIT Press, Cambridge (2018)
7. Dickson, E.W., Singh, S., Cheung, D.S., Wyatt, C.C., Nugent, A.S.: Application of lean manufacturing techniques in the emergency department. J. Emerg. Med. **37**(2), 177–182 (2009)
8. D'Andreamatteo, A., Ianni, L., Lega, F., Sargiacomo, M.: Lean in healthcare: a comprehensive review. Health Policy **119**(9), 1197–1209 (2015)
9. Elsevier: Scopus: Access and use support center (2020). https://service.elsevier.com/app/answers/detail/a_id/14194/supporthub/scopus/. Accessed 20 July 2020
10. Farris, J.A., Van Aken, E.M., Doolen, T.L., Worley, J.: Critical success factors for human resource outcomes in kaizen events: an empirical study. Int. J. Prod. Econ. **117**(1), 42–65 (2009)
11. Goyal, S., Law, E.: An introduction to kaizen in health care. Br. J. Hosp. Med. **80**(3), 168–169 (2019)
12. Knechtges, P., Decker, M.C.: Application of kaizen methodology to foster departmental engagement in quality improvement. J. Am. Coll. Radiol. **11**(12), 1126–1130 (2014)
13. Liker, J.K., Meier, D.: Toyota Way Fieldbook. McGraw-Hill Education, New York (2006)
14. Lima, R.M., Dinis-Carvalho, J., Souza, T.A., Vieira, E., Gonçalves, B.: Implementation of lean in health care environments: an update of systematic reviews. Int. J. Lean Six Sigma (2020)
15. Mazzocato, P., Savage, C., Brommels, M., Aronsson, H., Thor, J.: Lean thinking in healthcare: a realist review of the literature. BMJ Qual. Safety **19**(5), 376–382 (2010)
16. Mazzocato, P., Stenfors-Hayes, T., von Thiele Schwarz, U., Hasson, H., Nyström, M.E.: Kaizen practice in healthcare: a qualitative analysis of hospital employees' suggestions for improvement. BMJ Open **6**(7), e012256 (2016)
17. Naidoo, L., Mahomed, O.H.: Impact of lean on patient cycle and waiting times at a rural district hospital in Kwazulu-natal. Afr. J. Prim. Health Care Family Med. **8**(1), 1–9 (2016)
18. Radnor, Z.J., Holweg, M., Waring, J.: Lean in healthcare: the unfilled promise? Soc. Sci. Med. **74**(3), 364–371 (2012)
19. Rodriguez, D.S.S., Costa, H.G., Carmo, L.F.R.R.S.d.: Multicriteria decision aid methods applied to PPC problems: a mapping of papers published in Brazilian journals. Gestão & Produção **20**(1), 134–146 (2013)
20. Sommer, A.C., Blumenthal, E.Z.: Implementation of lean and six sigma principles in ophthalmology for improving quality of care and patient flow. Surv. Ophthalmol. **64**(5), 720–728 (2019)

21. Sugianto, J.Z., et al.: Applying the principles of lean production to gastrointestinal biopsy handling: from the factory floor to the anatomic pathology laboratory. Lab. Med. **46**(3), 259–264 (2015)
22. de Toledo, R.F., Junior, H.M., Farias Filho, J., Costa, H.G.: A scientometric review of global research on sustainability and project management dataset. Data Brief **25**, 104312 (2019)
23. Treinta, F.T., Farias Filho, J.R., Sant'Anna, A.P., Rabelo, L.M.: Metodologia de pesquisa bibliográfica com a utilização de método multicritério de apoio à decisão. Production **24**, 508–520 (2014)
24. Ullah, M.F., Fleming, C., Fox, C., Tewary, T., Tormey, S.: Patient experience in a surgical assessment unit following a closed-loop audit using a kaizen lean system. Irish J. Med. Sci. (1971-) **189**(2), 641–647 (2020)
25. Womack, J.P., Jones, D.T., Roos, D.: The machine that changed the world: the story of lean production-Toyota's secret weapon in the global car wars that is now revolutionizing world industry. Simon and Schuster (2007)

# The Benefits of Deploying the Toyota Kata

Gislayne Vieira Borges[1(✉)] iD, Alexandre Beraldi Santos[1(✉)] iD,
Luis Fernando Torres[1(✉)] iD, Messias Borges Silva[2(✉)] iD,
Gabriel Nascimento Santos[1(✉)] iD, and Robisom Damasceno Calado[1(✉)] iD

[1] Universidade Federal Fluminense, Rio das Ostras, Brazil
{gislaynevb,alexandreberaldisantos,
gabrielnascimentosantos,robisomcalado}@id.uff.br
[2] Universidade de São Paulo, São Paulo, Brazil

**Abstract.** This research aims the identification of the theoretical benefits of Toyota Kata implementation. This analyze was done through a theoretical, combined, descriptive and bibliographic research, carried out through a literature review. To clarify these benefits, improvement kata and coaching kata was described according to the literature review. As this is a literature review of a subject of general implementation, the macro results of this research suggest as benefits, the creation of knowledge and organizational culture having as secondary results the efficiency improving and waste reduction. The results obtained will culminate in operational results such as the increase in the quality of services, growth in production volume, besides the tasks optimization and success in solving problems, which are typical characteristics of Toyota Kata, for involving people and routines, promoting a better engagement. This research showed also, when applied in its essence, Toyota Kata is an effective continuous improvement mechanism within organizations.

**Keywords:** Toyota Kata · Benefits · Improvement · Waste

## 1   Introduction

In many organizations, there is an implicit frustration due to the gap between the desired results and what actually happens. Goals are defined, but they are not achieved. The change does not occur [17].

Most companies are led, managed and occupied by caring and hardworking people who wants their organization, their team, to be successful. The conclusion is clear: it is not the people, but the current management system in which we work that is to blame. One problem is how we are managing our organizations, and there is a growing consensus that a new approach is needed. But we have not yet seen what this change should be [17].

Researches that have focus on individual importance show that in large, small and medium enterprises systematic strategy deployment is the most important

© IFIP International Federation for Information Processing 2021
Published by Springer Nature Switzerland AG 2021
A. Dolgui et al. (Eds.): APMS 2021, IFIP AICT 631, pp. 323–332, 2021.
https://doi.org/10.1007/978-3-030-85902-2_35

routine in terms of increasing performance. Specifically in SME's (small and medium enterprises), management support and leadership are the most important improvement routines for implementing lean practices and increasing operational performance [6].

Besides most large companies are indeed hierarchically organized in terms of business units, sites, product lines and teams, they don't necessarily grant each level the autonomy needed to effectively support performance improvement and adaptation. Such autonomy would demand the implementation (and the exercise) of the policy, intelligence and cohesion functions at all levels, as well endowing each organizational unit with the needed authority over the resources that are needs in order to exercise this autonomy [19].

In Japan, these patterns or routines are called kata (noun). The word derives from basic forms of movement in the martial arts, which are transmitted from master to student over generations [17].

In a more minimalist way, kata can be understood as the execution of a structured daily routine [20].

## 1.1   Toyota Kata

Usual lean manufacturing tools and also quality management tools are predominantly exploratory methods that allow optimization of previously existing assets and capabilities [1].

Toyota Kata is a proven and highly successful method for continuous improvement at the whole organization level. The Toyota Kata concept of a clear vision to strive towards is very appealing and powerful [21].

The idea behind the Toyota Kata is that it provides routines (katas) to practice that, over time and repetitions, will reconnect our brains in a way that systematic improvement and management become second nature to us. When this is achieved at an organizational level, it can allow the organization to adapt to changing circumstances faster and better than other organizations. The practice of the Toyota Kata is done through real problems and challenges related to work (increasingly difficult), which makes the practice economically viable as the Toyota Kata becomes second nature. In fact, it is a central idea of the Toyota Kata that everyone's work consists of two parts: real work and improving work [21].

According to Rother [17], a key concept underlying kata is that, although we often cannot exercise much control over the realities around us, we can exercise control over managing how we deal with them. The Toyota Kata is based on two concepts: the improvement kata and the coaching kata. According to Rother [17], the improvement Kata is the repetition of the management routine aimed at improving, adapting and evolving. The coaching kata is a repetition of the routine where leaders and managers teach the organization's employees the improvement kata. The technique called kata is passed on to all Toyota employees and this approach is a big part of what moves this company as an adaptive organization, towards continuous improvement. The objective is to develop systematic and scientific ways to develop solutions in dynamic and uncertain situations.

Furthermore, research in neuroscience reveals that, due to the plasticity of the human brain, it is possible to reconnect it to new ways of thinking and new habits, through repeated practice of a routine (kata). The focus of the 'Toyota Kata' is to provide routines (katas) for practice, which, over time, through a significant number of repetitions, allow the human brain to be rewired in such a way that systematic improvement and management of this improvement become secondary to nature [2].

**The Improvement Kata.** According to Rother [17], the improvement kata describes a routine of continuous improvement. This kata is also part of Toyota's way of managing people every day. The psychology of improvement kata is universal and, at Toyota, everyone is taught to operate according to this systematic approach. This applies to many different situations, not just in manufacturing. The content varies, but the approach is the same.

The improvement kata is a model of how to develop a capacity for continuous systematic improvement. This model must be learned by the leadership so that it can train and teach the entire organization [3].

In preparing for the use of the improvement kata, the organization must ensure that the improvements to be made are aligned with each other. To achieve this, an analysis of the value flow must be carried out. The future state value stream map is used to focus the efforts of the various improvement kata on the prioritized processes [20].

In summary, Toyota's improvement kata continuous repetition routine looks like this: (1) considering a vision, direction or target, and (2) with a first hand understanding of the current condition, (3) a next condition is defined target in the visual path. When (4) we strive to advance step by step towards this target condition, we encounter obstacles that define what we need to work with and with which we will learn [17].

While a high level of competence is not achieved, it is important don't deviate from the improvement kata routine because this will harm the learning of the improvement kata mental model and lower the probability of achieving good results in the long term [21].

**The Coaching Kata.** One of the main characteristic of Toyota Kata is that it offers an existing pattern to teaching, coaching and managing the improvement kata the coaching kata [21].

The goal of the coaching kata is to teach the improvement kata and bring it to the organization [17].

Mastery is the goal of any kata, and even people at the highest levels of Toyota are honing their skills and working to achieve that goal. Like the improvement kata, the training kata standard is also practiced at all levels at Toyota. Each employee is assigned to a more experienced employee a mentor who provides active guidance during the process of making real improvements or dealing with work related situations. This mentor, in turn, has his own mentor who is doing the same [17].

A person's need for coaching never goes away. Regardless of how much experience the person has acquired, it is unlikely that anyone can become so good at discerning the reality of a situation and applying the improvement kata that coaching will no longer be needed. The intention is that both the improvement kata and the coaching kata become increasingly second nature (automatic and reflective) as the person moves up in the organization [17].

Coaching Kata can help an organization to realize the possibilities inherent in other kata. In the case of Improvement Kata, it can help make improvement part of the company's "day to day" [16].

In a business environment, training a kata facilitates improved performance by improving a process [16].

### 1.2 Toyota Kata Opportunities and Barriers in Organizations

The development process is full of uncertainties that must be controlled in order to achieve the goals of improvement within the time, costs and results. However, there is little literature on how to manage these uncertainties, integrate developers, maximize knowledge sharing and minimize their losses in a development environment [15].

According to Michels et al. [10], their results suggest that the main barriers encountered in various contexts are: 1 - lack of direction and environ mental preparation (awareness), 2 - lack of value flow map, and 3 - lack of Additional coaching Kata meetings. The main opportunities were: 1 - short cycles, 2 - knowledge sharing and 3 - adaptability. Table 1 presents a brief summary of the main opportunities and barriers, in accord to the cited authors.

## 2    Method

According to Torres et al. [22], scientific research is characterized by the possibility of multiple classifications, however, its classification according to academic criteria is of fundamental importance for a better understanding. This methodology can be divided in 4 groups: research methods, objectives, nature and used techniques [4,7–9,18,24]. Based on these groups, it is possible to classify the present re-search. Thus, in agreement with the aforementioned authors, in terms of Research Methods, it is possible to say that this research presents deductive characteristics (because it allows conclusions based on true premises), with the focus strategy based on bibliographic research (through sampling, using technological resources of online search as an instrument to perform the research). About the Objectives, the research is classified as descriptive, as it seeks to understand the relationship between variables (aspects of the coach and improvement kata, with leadership). As for the Nature of the research, it is classified as applied, as it seeks to generate knowledge to enhance the use of kata philosophy. Finally, as a last classification criteria, a research can be differentiated according to the used technique for data collection, which in this specific case, was carried out through the review of documents.

**Table 1.** Summary of bibliographies (Source: Michels et al. [10])

| Authors/Year | Implementation model | Opportunities | Barriers |
|---|---|---|---|
| Soltero, C./ 2011 | None | Adaptability; TWI & Toyota Kata integration | Lack of a value screening flow |
| Soltero, C./ 2012a | None | Foster a culture of innovation throughout the organization; Address continuous improvement, favoring innovation for the external customer | Lack of experimentation with the client; lack of value sorting flow |
| Soltero, C./ 2012b | None | Creative inspiration; Standard problem solving process; Improvement in behavior patterns; Improvement impact; Short cycles in the use of Improvement Kata | Lack of value sorting flow |
| Reverol, J./ 2012 | None | Implementation of Coaching Kata that can facilitate the implementation of other Katas; Pre-work can mean the successful implementation of Toyota Kata routines; Develop more adaptable and valuable employees for the organization | Prepare the company's strategy and identify the value chain before applying the Toyota Kata; Appropriate time for the work of the coach and apprentice Coaching Kata; not following the routine steps of the TK, as proposed in the literature |
| Casten, et al./2013 | None | Knowledge sharing; Routine of learning and constant improvement; Leadership development; Adaptability | The preparation of the environment in the construction industry is different and even predictable; compared to the manufacturer's environment; Need for more experimentation with Toyota Kata routines |
| Tillmann, P; Ballard, G; Tommelein, I./ 2014 | None | Collaborative environment; Leadership development; Creating an environment with psychological security | Lack of consensus in the team in the use of routines and methods; Conflict of interest between those who support and those who do not support the initiative |
| Toivonen, T./ 2015 | None | Oppose cognitive prejudice; Creates alignment in the organization from the strategic level to the operational level; A greater number of more creative ideas for solving problems; A systemic approach to understand efforts to improve intangible aspects, which are nowadays most important in industries | It creates pressure on Coaches, who must be very competent in the teaching method to make it accessible to everyone in the organization. The amount of work to create a sustainable level of competence and support culture |
| Merguerian, et al./2015 | None | Apply the Toyota Kata first in the value-adding process; Automate the collection and process to improve exposed indexes | Collect data manually; Manual processes; Not all the collaborators were included for the cost estimate |
| Iberle, K./2015 | None | Short steps in the PDCA cycles are better adapted to people's workload: Coaching meeting 2 to 3 times a week; The questions in the training Kata push the trainer and the apprentice in the right direction | Low quality in the formation of hypotheses; Lack of constant entertainer and apprentice encounters; Lack of verification in the gemba; Coaching meeting only once a week |
| Ehni, M; Kersten, W. | Rother [17] & Bleicher (2011) | Definition of target condition; Using Hoshin Kanri; Using the value stream map | |
| Villalba-Diez et al. | None | | Lack of systemic use of PDCA |

In this informational scenario, the literature reviews, due to their summarizing aspect, mainly assume an important organic function, together with the indexes (here in the biblioteconomic sense), abstracts and specialized bibliographies. Often times more time is spent trying to identify whether a particular study has been carried out before than actually carrying it out [11].

## 3  Results

This step explains the knowledge of the addressed issues which were found during the application of the bibliographic review. The bibliographic research development was focused on the theoretical and practical research of Toyota Kata routines.

The authors emphasize that the researches for the literature review happened through the keywords "Toyota + system", "lean + manufacturing", "Toyota + Kata", "benefits", "improvement", "waste", "lean + management", "coaching + Kata" and "improvement + Kata". A fair amount of articles associated with the themes "Toyota system" and "lean + manufacturing" was found. Once the term "Kata" was added to the filter, the number of articles found (which would be useful for this study) decreased significantly.

As a result of the results, Table 2 shows in an objective way a summary of the bibliographic research regarding the aspects desired in this study.

**Table 2.** Summary of Bibliographic Review (Source: Prepared by the Authors).

| Authors/Year | Segment | Benefits |
| --- | --- | --- |
| De Oliveira et al./ 2018 | Product development | • The TK approach provided a knowledge management framework that made development management easier and simpler <br> • TK's cycle structure record provided an organized source of information, as they could access it and easily find what they needed. The coaching sessions increased the integration between the developers of the subsystem <br> • Development risks were reduced, as with the TK approach, it was certain that all the activities that each developer performed were in line with the development objectives. If they went out of the way, it was easy to see and correct <br> • Developers started to think about how what they were doing would lead them to achieve their Challenges |

(*continued*)

**Table 2.** (*continued*)

| Authors/Year | Segment | Benefits |
|---|---|---|
| Carvalho et al./ 2016 | Manufacturing firms | • Reduction of lead time<br>• Dramatic improvement in the quality of service perceived by customers and<br>• Increase in production, maintaining the same number of employees in the cell |
| Ferenhof et al./ 2017 | Service companies | • Alignment between company objectives and behavior in the workplace and the development of human skills related to work<br>• Building knowledge and organizational culture through routines involving all stakeholders in the development of consistent and stable processes |
| Merguerian et al./ 2015 | Health area | • Improvement of efficiency and reduction of waste in a multidisciplinary clinic<br>• 69% reduction in costs of clinical preparation and improvements in outcome metrics, such as clinical volume, family experience and, to a lesser extent, value-added time |
| Martins et al./ 2016 | Occupational Health & Safety | Problem solving through teamwork with a deep understanding of the PDCA |
| Bonamigo et al./ 2015 | Logistics area | Loss prevention of retail logistics |

# 4  Discussion

Toyota Kata is a proven and highly successful method of continuous improvement across the organization. Toyota Kata was discovered by Mike Rother while he researched Toyota's quality improvement methods. It is a holistic system method for improvement efforts that contains processes and behavior patterns for strategically aligned goal setting, problem solving, coaching, management and training. It is a simple and teachable approach that also covers the management of improvement efforts. The disadvantage of the approach is its focus on incremental improvement instead of revolutionary innovation [21].

However, a kata is more than just a methodology. It is a philosophy, perhaps even a state of mind. It impresses on the worker's mind the idea that there is no limitation when making improvements, as long as the worker knows where he is and where he wants to go [16].

Companies that create knowledgement use problem-solving routines to extend their corporate knowledge and to deal with complex challenges [12]. The Toyota Kata approach enables the establishment of management routines that can lead the development process, deal with uncertainties, stimulate organizational learning and integrate and align the entire development team with the development objectives [15].

In according to Sagalovsky [19], the team's responsibility is not limited to the mechanical execution of defined work tasks, but also to the ongoing improvement of their work process by, for instance, structuring their workplace and keeping it organized using tools like 5S, routines such as PDCA, A3 and the Improvement Kata in order to meet target cycle time, level workloads, among other gains, through operator's Standard Work. The team and team leader are also responsible for cross-training its members and setting up work rotation and for the replacement of any absentee.

To some authors like Michels et al. [10] and Noviyanti [13] it is clear that opportunities and barriers are faced and vary according to the area in which the routines are applied, besides the importance of a motivation system to the employees, because their discipline affect the competitiveness of the company.

In the healthcare sector, these approaches are seen as innovative, bringing about a radical change in the way things have been done to date. In fact, Walley [23] points out that when the service sector is compared to the industrial sector, it is widely felt that the service sector, and healthcare in particular, is lagging in terms of adopting new management innovations and improvements.

As commented by Hellstrom et al. [5], when the service sector is compared to the manufacturing or industrial sector, it is widely felt that the service sector, and healthcare in particular, is lagging in terms of adopting new management approaches [14]. For many years it has been considered sufficient for there to be only professional knowledge to ensure quality and safety in the delivery of healthcare services. Today's healthcare delivery systems are complex; however, calling for further organizational awareness in order to provide the appropriate medical care along the entire patient pathway, generating savings without incurring costs but also improving, for example, the standards of quality, flexibility and safety. Consequently, the problem with healthcare today is largely organizational and not only clinical.

According to Toivonen [21] there are several benefits of the Toyota Kata:

- Creates alignment in the organization from the strategic to the operational level
- It involves employees through common challenges and frequent successes.
- Common approach to improvement and management facilitates effective collaboration across the organization the implementation of Fractal allows the use of the collective intelligence of the organization for continuous improvement activities.

We conclude that the Toyota Kata approach provides a management structure that allows the alignment of developers, integration, reduction of development risks and knowledge management [15].

# 5   Conclusion

The challenge faced is not to direct executives and managers to implement new production or management techniques or to adopt new principles, but to achieve continuous systematic evolution and improvement across the organization, developing behavioral routines applied repeatedly and consistently: kata [17].

According to the result presented, the work can be seen as successful, since it reached its main objective, which was to identify, through a theoretical, mixed, descriptive and bibliographic research, carried out through a literature review, the benefits Toyota Kata deployment and routine. The authors consider as a scientific contribution the fact that this article presents a recent and updated bibliographic review carried out on the subject, which follows the steps of a research classification. As for perspectives, this article can encourage researchers as well as lean healthcare implementers to adopt the Toyota kata as a methodology for the multipliers of their institutions.

However, in addition to this study, the theme allows other studies to complement the subject, such as: case studies that show the practical application of the elaborated standard process, analyzing the filling of the chart in the kata cycles.

**Acknowledgements.** The authors would like to thank the Federal Fluminense University and the Fundação Euclides da Cunha, which made the research project funded by the Ministry of Health of Brazil viable; TED 125/2019, Number: 25000191682201908.

# References

1. Brandl, F.J., Ridolfi, K.S., Reinhart, G.: Can we adopt the Toyota Kata for the (re-)design of business processes in the complex environment of a manufacturing company? Procedia CIRP **93**, 838–843 (2020)
2. Dinis-Carvalho, J., Ratnayake, R., Stadnicka, D., Sousa, R.M., Isoherranen, J., Kumar, M.: Performance enhancing in the manufacturing industry: an improvement kata application. In: 2016 IEEE International Conference on Industrial Engineering and Engineering Management (IEEM), pp. 1250–1254. IEEE (2016)
3. Ferenhof, H.A., Da Cunha, A.H., Bonamigo, A., Forcellini, F.A.: Toyota Kata as a km solution to the inhibitors of implementing lean service in service companies. VINE J. Inf. Knowl. Manage. Syst. (2018)
4. Gil, A.C.: Métodos e técnicas de pesquisa social. atlas. São Paulo (1994)
5. Hellström, A., Lifvergren, S., Gustavsson, S.: Transforming a healthcare organization so that it is capable of continual improvement-the integration of improvement knowledge. In: 18th International Annual EurOMA Conference, Cambridge UK (2011)
6. Knol, W.H., Slomp, J., Schouteten, R.L., Lauche, K.: The relative importance of improvement routines for implementing lean practices. Int. J. Oper. Prod. Manage. (2019)
7. Lakatos, E.M., Marconi, M.d.A.: Metodologia científica. 2 edição. São Paulo: Atlas (1991)
8. Malhotra, N.K.: Pesquisa de marketing: foco na decisão. Pearson Prentice Hall, São Paulo (2011)

9. Martins, R.A.: Sistemas de medição de desempenho: um modelo para estruturação do uso. 258f. Ph.D. thesis, Tese (Doutorado em Engenharia de Produção)-Escola Politécnica da ... (1999)
10. Michels, E., Forcellini, F.A., Fumagali, A.E.C.: Opportunities and barriers in the use of toyota kata: a bibliographic analysis. Gepros: Gestão da Produção, Operações e Sistemas **14**(5), 262 (2019)
11. Moreira, W.: Sistemas de organização do conhecimento: aspectos teóricos, conceituais e metodológicos. Universidade Estadual Paulista (UNESP) (2018)
12. Nonaka, I., Takeuchi, H.: The Knowledge-Creating Company: How Japanese Companies Create the Dynamics of Innovation. Oxford University Press, Oxford (1995)
13. Noviyanti, N.P.A.: The analysis of employee discipline towards competitiveness of local bakery industry in Manado: case study of Kartini bakery and dolphin bakery. Jurnal EMBA: Jurnal Riset Ekonomi, Manajemen, Bisnis dan Akuntansi **6**(2) (2018)
14. Ohno, T.: Toyota Production System: Beyond Large-Scale Production. CRC Press, Boca Raton (1988)
15. Oliveira, M.S.D., Lozano, J.A., Barbosa, J.R.: The Toyota kata approach for lean product development. In: Transdisciplinary Engineering Methods for Social Innovation of Industry 4.0: Proceedings of the 25th ISPE Inc., International Conference on Transdisciplinary Engineering, 3–6 July 2018, vol. 7, p. 361. IOS Press (2018)
16. Reverol, J.: Creating an adaptable workforce: using the coaching kata for enhanced environmental performance. Environ. Qual. Manage. **22**(2), 19–31 (2012)
17. Rother, M.: Toyota Kata: Managing People for Improvement, Adaptiveness and Superior Results. McGraw Hill, New York (2009)
18. Ruy, M.: Aprendizagem organizacional no processo de desenvolvimento de produtos: estudo exploratório em três empresas manufatureiras. Universidade Federal de São Carlos, São Carlos (2002)
19. Sagalovsky, B.: Organizing for lean: autonomy, recursion and cohesion. Kybernetes (2015)
20. Soltero, C., Boutier, P.: The 7 Kata: Toyota Kata, TWI, and Lean Training. Productivity Press (2019)
21. Toivonen, T.: Continuous innovation-combining Toyota kata and TRIZ for sustained innovation. Procedia Eng. **131**, 963–974 (2015)
22. Torres, L.F.: Proposição de um método de treinamento da filosofia lean voltado ao nível operacional de empresas de autopeças. Universidade Estadual de Campinas, Campinas (2020)
23. Walley, P.: Designing the accident and emergency system: lessons from manufacturing. Emerg. Med. J. **20**(2), 126–130 (2003)
24. Yin, R.K.: Estudo de caso: planejamento e métodos. 4ª edição. Bookman, Porto Alegre (2001)

# An ACO Algorithm for a Scheduling Problem in Health Simulation Center

Simon Caillard$^{(\boxtimes)}$, Corinne Lucet, and Laure Brisoux-Devendeville

Laboratoire MIS (UR 4290), Université de Picardie Jules Verne,
33 rue Saint-Leu, 80039 Amiens Cedex 1, France
{simon.caillard,corinne.lucet,laure.devendeville}@u-picardie.fr

**Abstract.** SimUSanté is one of the biggest European simulating and training centers, proposing training sessions for all involed in healthcare: professionals, students, patients. This paper presents the timetabling problem encountered by SimUSanté with regard to the quality objectives and the time and resource constraints. To solve it, $SimUACO\text{-}LS$ which is the hybridization of the Min-Max Ants Colony Optimization algorithm $SimUACO$ with the variable neighborhood search $SimULS$ [3], is presented. $SimULS$, $SimUACO$ and $SimUACO\text{-}LS$ are compared in a set of representative instances [2], newly generated and derived from those of the Curriculum-Based Course Timetabling problem [1]. $SimUACO\text{-}LS$ always improves both results of $SimULS$ and $SimUACO$ by respectively 3.84% and 2.97%.

**Keywords:** Scheduling · Healthcare training · Timetabling · Operational research · Optimization

## 1 Introduction

SimUSanté, located in Amiens, France, is the biggest european healthcare simulation and training center that provides more than 400 different training sessions in a wide range of healthcare areas. Everyone involved in healthcare can meet up in training sessions and learn together. A training session corresponds to a set of activities followed by a group of learners. Such teaching requires equipment and classrooms that are very specific to hospital activities and building a timetable that take into account all these resource constraints is essential for the smooth-running of the center. The aim of SimuSanté is then to schedule a maximum number of activities while maximizing the timetable compactness of each session to schedule. So, in this paper, we focus on the Curriculum-Based Course Timetabling problem (CB-CTT) [6] which is NP-Hard [5] and has similarities with the SimuSanté problem. Nevertheless, the SimUSanté problem differs in a major point: the precedence relationships between activities. In addition, 10 out of the 13 of the CB-CTT constraints are hard in the SimUSanté problem. Moreover in the literature, the solvers of CB-CTT instances, allow to violate hard constraints, whereas this is prohibited in the SimUSanté problem. Nevertheless, CB-CTT remains the academic problem closest to that of SimUSanté.

© IFIP International Federation for Information Processing 2021
Published by Springer Nature Switzerland AG 2021
A. Dolgui et al. (Eds.): APMS 2021, IFIP AICT 631, pp. 333–341, 2021.
https://doi.org/10.1007/978-3-030-85902-2_36

Over the last ten years, CB-CTT has been widely studied, as can be seen in many surveys [10,13]. The best known methods for solving it are the local search methods and population-based meta-heuristics. The local search methods are based on neighborhood operators that start from an initial solution and modify a part of it to switch to another one. In addition, there are a few ways to avoid being trapped in a local minimum. The population-based methods use many solutions at the same time and combine them to obtain better ones. The best known population based methods to solve timetabling are the Ants Colony Optimization (ACO) [12] and the Genetic Algorithms (GA) [8]. Although genetic algorithms are efficient, most of the solutions obtained require an important repairing phase after crossover operator which potentially leads to remove the activities that violate hard constraints. In regards with the objectives of SimUSanté, ACO algorithms that always produce feasible solutions seem to be better suited.

Nevertheless, such methods have the drawback of converging too early in the search process towards a local optimum. For these reasons, hybrid methods which combine both approaches are often used [9,11]. So, in this paper, we propose *SimUACO-LS* algorithm that is a combination of *SimUACO*, a Max-Min Ant Colony Optimization (MMACO) system [14], with a Variable Neighborhood Search (VNS) *SimULS* [3]. ACO algorithms are efficient in exploring the search space solutions, but the quality of this search relies on the quality of the heuristic information used to guide it. We have therefore developed several specialized heuristics, which are not redundant with those used in *SimULS*, to lead ants in building solutions. Although these heuristics fit the SimuSanté problem, they can be applied to similar problems, with the same hard constraints by adapting the heuristics according to the desired qualitative characteristic. *SimUACO-LS* have been tested on various instances with results close to optimum solutions.

This paper is organized as follows: in Sect. 2, we formalize both the SimUsanté problem and the structure of the solutions. Section 3 gives the graph representation related to the formalization of the problem as an ACO problem and details our algorithm *SimUACO-LS*. In Sect. 4, the results provided by *SimUACO-LS* are compared to optimal results given by a mathematical model implemented with CPLEX [4]. Also, the relevance of the combination of a MMACO system with a VNS algorithm is assessed by comparing *SimUACO-LS* with *SimULS* and *SimUACO*. Finally, Sect. 5 concludes this paper and presents some perspectives.

# 2   The SimUSanté Problem

## 2.1   Data Definitions

An instance is composed of a set $S$ of training sessions to schedule over horizon $H$, defined by a set of working days $D$. Each day $d \in D$ is composed of a set $T_d$ of 9 time slots of one hour and has a starting time slot $start_d$, an ending one $end_d$ and a set $break_d \subset T_d$ of potential time slots for the lunch breaks.

Each session $s \in S$ is a subset of activities that must be scheduled in serie. Any scheduled session must have one lunch break per day. A session $s$ can be partially scheduled, i.e. some of its activities are unplanned in the final solution. The makespan of a session $s$ is the number of time slots from the start of the first activity of $s$ to the end of the last activity of $s$, including lunch break.

$A$ represents the set of all instance activities and $s_a$ is the session to which activity $a$ belongs. $\forall a \in A$, $duration_a$ is the number of consecutive time slots required to execute $a$. Some precedence constraints exist between activities and $pred_a$ denotes the subset of activities that must be completed before $a$ starts.

The set of resources $R$ represents both employees and rooms and $\Lambda$ is the set of resource types. A resource type corresponds to a skill for employees and to a specific equipment or characteristic for rooms. $\forall \lambda \in \Lambda$, we denote $R^\lambda$ the set of resources of type $\lambda$, and $qtav_\lambda^t$ its associated quantity of resources available at time slot $t$. To each resource $r \in R$, $\Lambda_r \subseteq \Lambda$ denotes the set of types associated to $r$, because employees may have several skills and rooms may have different equipments. The availability of resource $r$ is given by $isavailable_r^t$ which is equal to 1 if $r$ is available at time slot $t$ and 0 otherwise. If resource $r$ is assigned to activity $a$, $r$ must be available over $duration_a$. Moreover, all employee timetables must have one lunch break per day.

In order to be scheduled, activity $a$ needs specific resources over its entire duration. $qtreq_\lambda^a$ is the quantity of resources of type $\lambda \in \Lambda$ required by $a$ and $\Lambda_a = \{\lambda \in \Lambda | qtreq_\lambda^a \neq 0\}$ is the set of resource types required by $a$. The eligibility of activity $a$ to time slot $t$ is given by $iseligible_a^t$ which is equal to 1 if there are enough resources for scheduling $a$ at $t$, 0 otherwise.

Solving an instance of the SimuSanté problem consists in scheduling as many activities as possible of $A$, while minimizing the sum of the makespans of the sessions of $S$.

## 2.2  Solution and Evaluation

To build a solution, it is necessary to assign a start date $t_a$ (a time slot) and a set of resources $c_a \subseteq R$ to each scheduled activity $a$. Then, a solution $Sol$ is a set of triplets $\{(a_1, t_1, c_1), \ldots (a_i, t_i, c_i)\}$, where $a_i \in A$, $t_i \in H$ and $c_i \subseteq R$. A triplet $(a, t, c)$ can be added to a solution if it satisfies all constraints of resources ($c$ perfectly matches the resource requirements of $a$), of precedence and of operating rules. We denote $UA = A \backslash \{a | (a, t, c) \in Sol\}$, the set of unscheduled activities and $Sol_s = \{(a, t, c) \in Sol | a \in s\}$, the set of triplets associated to $s$.

Once a solution is built, its evaluation relies on two criteria: its compactness and the number of scheduled activities. The makespan $mk_s = t_{end_s} - t_{start_s}$ represents the total duration of scheduled activities of the session $s$, with respectively $t_{start_s} = \min_{(a,t,c) \in Sol_s} \{t\}$ and $t_{end_s} = \max_{(a,t,c) \in Sol_s} \{t + duration_a\}$, the starting time slot and the ending one of session $s$. We note that if $Sol_s = \emptyset$, then $mk_s = 0$.

The evaluation of solution $Sol$ denoted $Eval(Sol)$, is established with the sum of the $mk_s$ and the sum of penalties $\alpha$ given to each unplanned activity (see

Eq. 1). The SimUSanté problem aims to construct a valid solution $Sol^*$ such that $Eval(Sol^*)$ is minimum.

$$Eval(Sol) = \sum_{s \in S} mk_s + |UA| \times \alpha \tag{1}$$

# 3    An Ant Colony Optimization Algorithm: $SimUACO\text{-}LS$

An ant colony optimization algorithm (ACO) [7] is a bio-inspired algorithm which uses ant behavior to find food and return back to the nest. Each ant leaves pheromones on the trail from their nest to the food source. An ant moves randomly but when it detects pheromones, it follows the trail and reinforces it by leaving additional pheromones. The more ants follow a trail, the more attractive that trail becomes. Pheromones evaporate over time and therefore the least used or slower paths become the least attractive. Ants can then find the fastest trail from the nest to a source of food. A Max-Min Ant Colony Optimization (MMACO) system is derive from the ACO system. Its principle, as described in [14], is to maintain pheromones between a minimum and a maximum value in order to keep accessible a maximum of relevant paths.

We propose to solve SimUSanté problem by combining a MMACO system with a VNS algorithm [3]. Thus, the first step is to turn the problem into a path search in a graph. Because of the complexity of the SimUSanté problem, a solution will be defined as a collection of elementary paths rather than a simple path as detailed below.

## 3.1    Graph Representation

The SimUSanté problem can be modeled by a graph $\mathscr{G} = ((A, H, R), E)$. $A$ is the set of activities, $H$ and $R$ are respectively those of time slots and resources. $E = \{A \times H\} \cup \{(t, r)|t \in H, r \in R$ and $isavailable_r^t = 1\} \cup \{R \times R\}$. A triplet $(a, t, c)$ of a valid solution described in Sect. 2.2 is represented by the path $(a, t, r_1, r_2, \ldots, r_k)$ with $c = \{r_1, r_2, \ldots, r_k\} \subseteq R$. For more convenience, we also denote it $(a, t, c)$. There are different ways of scheduling a given activity $a$, so there are different valid paths $(a, t^i, c^i)$. Each ant $k$ builds a collection $\Pi^k = ((a_1, t_1, c_1), \ldots, (a_n, t_n, c_n))$, with $n \leq |A|$, of paths corresponding to a valid solution $Sol^k$ of the problem. Once, path $(a_x, t_x, c_x)$ is constructed, ant $k$ is guided by a heuristic, as explained in the Sect. 3.2, to jump from $R$ to $A$ in order to choose the next $a_{x+1} \in A$.

## 3.2    $SimUACO\text{-}LS$ Principle

In $SimUACO\text{-}LS$, each ant builds a valid solution by constructing its collection of paths. During this construction process, $selectActivities()$ chooses the next activity $a$ to plan, while $selectPath()$ schedules it at time slot $t$ with resources $c$. Both rely on random proportionality rules which is based only on heuristic information for $selectActivities$ and based on heuristic and pheromone information for *selectPath*. Once a valid solution is built, the variable neighborhood search *SimULS* (see Sect. 3.5) is applied to it with a probability of $\beta\%$.

The pheromone information is stored in $\tau$, a tridimensional matrix of size $|A| \times |H| \times |R|$ such that $\tau_{a,t,r}$ is the pheromone assigned to the allocation of the resource $r$ to activity $a$, at time slot $t$. To retrieve the pheromone information of a resource combination $c$ assigned to an activity $a$ at a time slot $t$, we simply sum the amount of pheromones of each resource involved in this combination, i.e.: $\tau(a,t,c) = \sum_{r \in c} \tau_{a,t,r}$.

The update pheromones phase occurs at the end of each iteration, when all ants have constructed their own solution. All the pheromones of the matrix $\tau$ are first evaporated with coefficient $\gamma$ $(0 < \gamma \leq 1)$: $\tau_{a,t,r} = \tau_{a,t,r} \times (1 - \gamma)$. Next all components $(a,t,c) \in Sol_{best}$ are rewarded: $\forall r \in c, \tau_{a,t,r} = \tau_{a,t,r} + 1/Eval(Sol_{best})$.

The pheromones represent the ant learning to find better solutions by exploring the search space around the best known solution while heuristic information is an essential knowledge to guide ants to promising areas of the search space. In *SimUACO* different heuristics are used to select both activities and resources. The following section describes these heuristics and the overall process used to guide an ant to schedule activities. An ant tries to schedule each activity only once. Let $UA^k$ be the set of not visited activities by the ant $k$, i.e. the unscheduled activities in solution of ant $k$.

### 3.3   Choose the Next Activity: *selectActivities*()

The aim of *selectActivities*() is to choose in $UA^k$ the next activity to schedule for ant $k$ according to a random probability rule $next_a^k$, described in Eq. 2. It is based on the heuristic information $poss_a$ which represents an estimation of the remaining scheduling possibilities of $a$ over $H$. This estimation considers the number of times there is $duration_a$ consecutive time slots with enough resources to schedule $a$.

$$next_a^k = \frac{\frac{1}{poss_a}}{\sum_{a' \in UA^k} \frac{1}{poss_a'}} \qquad (2)$$

The objective of this random proportionality rule is twofold: lead ants in selecting one of the more promising activities to schedule and allow a diversification of the search.

### 3.4   Choose Time Slot and Resources: *selectPath*()

From a given vertex activity $a$, *selectPath*() chooses a path from $a$, through a time slot vertex $t$ to a resource combination vertex $c$. This selection is made according to the random proportionality rule $p(a,t,c)$, described at Eq. 3, which is based on the heuristic information $\eta(a,t,c)$ and the pheromone $\tau(a,t,c)$.

$$p(a,t,c) = \frac{\tau(a,t,c) \times \eta(a,t,c)}{\sum\limits_{(a,t^i,c^i)} \tau(a,t^i,c^i) \times \eta(a,t^i,c^i)} \qquad (3)$$

The higher the value of the heuristic information and the related pheromone, the more profitable a path becomes. Note that we only consider possible paths which respect all the constraints of the SimUSanté problem with regard to the current state of the solution construction process.

The heuristic information $\eta(a,t,c)$, see Eq. 4, is based on three indicators: the makespan variation $\Delta(a,t,c)$, the number of idle time slots $Idle(a,t,c)$ and the remaining resources from $t$ to $t + duration_a$ and $Usage(a,t,c)$, if the path $(a,t,c)$ is chosen.

$$\eta(a,t,c) = \frac{1}{(1 + \Delta(a,t,c)) \times (1 + Idle(a,t,c)) \times (1 + Usage(a,t,c))} \qquad (4)$$

$\Delta(a,t,c)$ computes $Nmk_{s_a} - mk_{s_a}$, the difference between the new makespan $Nmk_{s_a} = \max_{t' \in \{t+duration_a, t_{end_{s_a}}\}}\{t'\} - \min_{t'' \in \{t, t_{start_{s_a}}\}}\{t''\}$ of session $s_a$, if $a$ is scheduled at time slot $t$ with the set of resources $c$, and its current makespan $mk_{s_a}$.

$Idle(a,t,c)$ counts the number of free time slots unusable if path $(a,t,c)$ is selected. A set of contiguous free time slots is considered as unusable if it is not possible to plan any unscheduled activities of $s_a$ over it, because the duration of each of them exceeds the quantity of free time slots in the set.

$Usage(a,t,c)$ is computed by Eq. 5. Over the interval $[t; t + duration_a[$, it adds up the quantities of resources remaining in $\Lambda_c = \bigcup_{r \in c} \Lambda_r$, if the path $(a,t,c)$ is chosen, with the aim of minimizing the use of more than one type resources. Let us note that $\Lambda_a \subseteq \Lambda_c$.

$$Usage(a,t,c) = \sum_{t' \in [t;t+duration_a[} \left( \sum_{\lambda \in \Lambda_c} \left( \sum_{r \in R^\lambda} isavailable_r^{t'} \right) - qtreq_\lambda^a \right) \qquad (5)$$

### 3.5    A Variable Neighborhood Search: *SimULS*

The VNS algorithm *SimULS* [3] is randomly applied with a rate of $\beta\%$ to the solutions constructed by ants, with the aim of improving them, and also to diversify the population. To proceed, it relies on the operators *saturator*, *intra*, *extra*, and *extra*[+]. These operators build a set of movements which individually schedules an activity by removing one or more scheduled activities of the current solution. The difference between these operators is the target session from which activities are removed. Thus, they define different neighborhoods.

Moreover, to escape from a local minimum, *SimULS* regularly uses *diversificator* to destroy a part of the current solution. The condition to use *diversificator* is reached when all activities are scheduled or when there is no improvement during a preset number of iterations. *diversificator* removes scheduled activities according to two criteria. The first one is the position of the activity in its session. The second one is the number of resources with multiple type assigned to the activity.

**Table 1.** Result averages between $SimULS$, $SimUACO$ and $SimUACO$-$LS$

| Instance family | $SimULS$ | | $SimUACO$ | | $SimUACO$-$LS$ | | CPLEX |
|---|---|---|---|---|---|---|---|
| | Eval | SD | Eval | SD | Eval | SD | |
| $Brazil1$ | 87.10 | 1.68 | 86.14 | 1.75 | 85.01 | 1.79 | 83.6 |
| $Italy1$ | 109.54 | 1.77 | 108.15 | 1.86 | 106.03 | 1.88 | 102.8 |
| $Brazil2$ | 179.10 | 1.79 | 179.04 | 4.20 | 172.74 | 4.09 | 165.6 |
| $Finland1$ | 364.30 | 1.34 | 354.36 | 4.95 | 342.55 | 4.86 | na |
| $Brazil6$ | 430.09 | 1.01 | 429.14 | 4.94 | 409.49 | 4.81 | na |
| $Finland2$ | 382.08 | 1.42 | 380.5 | 3.91 | 366.58 | 3.94 | na |
| $StPaul$ | 1427.48 | 1.03 | 1427.18 | 5.42 | 1402.14 | 5.20 | na |

## 4    Experimental Results

In this section, we test $SimUACO$-$LS$, the combination of $SimUACO$ and $SimU$-$LS$ on instances close to the SimUSanté problem. These instances are based on those of the CB-CTT problem [1] and have been modified to include characteristics inherent in the SimUSanté problem. They vary according to the following characteristics: $D_1$, the availabilities of employees, $C_1$, the skills of employees, $T_1$ and $T_2$, the types of the rooms, $A_1$ and $A_2$ the requirements of activities. For each of the followings CB-CTT instances: $Brazil1$, $Italy1$, $Brazil2$, $Brazil6$, $FinlandHighSchool$ ($Finland1$), $FinlandSecondarySchool$ ($Finland2$) and $StPaul$, we generated an instance family which represents a set of 16 instances varying the previously presented criteria [2].

$SimUACO$, $SimULS$ and $SimUACO$-$LS$ were written in Java and tested on an Intel i7 8700K processor with a running time limit of two hours. We implemented the mathematical model [4] with CPLEX solver (version 12.6) and set a time limit of 7200 s. First three columns $SimUACO$, $SimULS$ and $SimUACO$-$LS$ in Table 1 are split into sub-columns $Eval$ and $SD$. $Eval$ corresponds to the average of the objective function (Eq. 1, with $\alpha = |H|$) computed over all instances of family and over 20 runs for each of them. $SD$ corresponds to average of the standard deviation. On the CPLEX column are averages on optimal results when available, or $na$ otherwise. Figure 1 is another view of these results (without CPLEX ones) where the score of $SimUACO$-$LS$ is set to 100%. $SimUACO$-$LS$ has been used with the following empirically determined parameters: $\beta = 6$ and $\gamma = 0.1$. For both algorithms $SimUACO$ and $SimUACO$-$LS$, the quantity of ants is set to the number of activities.

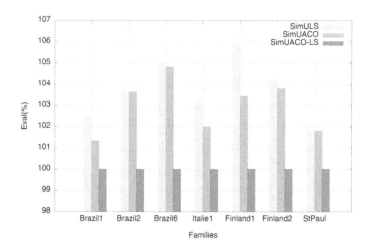

**Fig. 1.** Comparison between *SimuACO-LS*, *SimuACO* and *SimULS*

Our experiments show that for all methods here presented, all activities are scheduled. *SimUACO-LS* improves the results of *SimUACO* and *SimULS* by respectively 2.97% and 3.84%. Some optimal results can be reached on small instance families (*Brazil1*, *Brazil2* and *Italy1*) and *SimUACO-LS* diverges on average by 3.05% from these optimal results. The standard deviation for *SimUACO-LS* remains less than 5.20 for all the tested instances with an average of 3.80. This standard deviation is lower than that of *SimUACO* (3.86) but higher than that of *SimULS* (1.43). Although the standard deviation of *SimUACO-LS* is higher than *SimULS* one, its average *Eval()* is better which means that it better explores the solution space.

## 5  Conclusion and Perspectives

This paper presents the planning problem of the health training and simulation center SimUSanté. The proposed method *SimUACO-LS* is a hybridization of the Max Min Ant Colony Optimization algorithm *SimUACO* with the Variable Neighborhood Search *SimULS*. The combination of both metaheuristics allows the weaknesses of each to be overcome. Build a feasible solution is a difficult problem, mainly on the largest instances. Because *SimUACO-LS* strictly respects all hard constraints, it always produces feasible solutions for all instances. Moreover, all activities are scheduled. *SimUACO-LS* always improves the solution quality of *SimUACO* and *SimULS* by respectively 2.97% and 3.84%. The results of *SimUACO-LS* have an average gap of 3.05% with optimal when known. In addition, the schedules obtained are compact and relevant for SimUSanté. The next step will be to improve the search strategy for *SimUACO-LS* in the solution space.

**Acknowledgements.** This project is supported by the Hauts-de-France region of France and the SimUSanté center.

# References

1. Benchmarking Project for (High) School Timetabling. EEMCS Faculty, DMMP Group - High School Timetabling Project (HSTT). https://www.utwente.nl/en/eemcs/dmmp/hstt/
2. Caillard, S., Brisoux-Devendeville, L., Lucet, C.: Health Simulation Center Simusanté®'s Problem Benchmarks. https://mis.u-picardie.fr/en/Benchmarks-GOC/
3. Caillard, S., Brisoux Devendeville, L., Lucet, C.: Local search algorithm to solve a scheduling problem in healthcare training center. In: 17th International Workshop on Project Management and Scheduling (PMS 2020), Toulouse, France, April 2021
4. Caillard, S., Devendeville, L.B., Lucet, C.: A planning problem with resource constraints in health simulation center. In: Le Thi, H.A., Le, H.M., Pham Dinh, T. (eds.) WCGO 2019. AISC, vol. 991, pp. 1033–1042. Springer, Cham (2020). https://doi.org/10.1007/978-3-030-21803-4_102
5. Cooper, T.B., Kingston, J.H.: The complexity of timetable construction problems. In: Burke, E., Ross, P. (eds.) PATAT 1995. LNCS, vol. 1153, pp. 281–295. Springer, Heidelberg (1996). https://doi.org/10.1007/3-540-61794-9_66
6. Di Gaspero, L., Mccollum, B., Schaerf, A.: The second international timetabling competition (ITC-2007): curriculum-based course timetabling (track 3), February 2007
7. Dorigo, M., Maniezzo, V., Colorni, A.: Ant system: optimization by a colony of cooperating agents. IEEE Trans. Syst. Man Cybern. Part B (Cybern.) **26**(1), 29–41 (1996)
8. Gozali, A., Kurniawan, B., Weng, W., Fujimura, S.: Solving university course timetabling problem using localized Island model genetic algorithm with dual dynamic migration policy. IEEJ Trans. Electr. Electron. Eng. **15**, 389–400 (2020)
9. Kenekayoro, P., Zipamone, G.: Greedy ants colony optimization strategy for solving the curriculum based university course timetabling problem. Br. J. Math. Comput. Sci. **14**(2), 1–10 (2016)
10. Lewis, R.: A survey of metaheuristic-based techniques for university timetabling problems. OR Spectr. **30**(1), 167–190 (2008). https://doi.org/10.1007/s00291-007-0097-0
11. Matias, J., Fajardo, A., Medina, R.: Examining genetic algorithm with guided search and self-adaptive neighborhood strategies for curriculum-based course timetable problem. In: 2018 Fourth International Conference on Advances in Computing, Communication Automation (ICACCA), pp. 1–6 (2018)
12. Mazlan, M., Makhtar, M., Khairi, A., Mohamed, M.A.: University course timetabling model using ant colony optimization algorithm approach. Indones. J. Electr. Eng. Comput. Sci. **13**, 72–76 (2019)
13. Mazlan, M., Makhtar, M., Khairi, A., Mohamed, M.A., Rahman, M.: A study on optimization methods for solving course timetabling problem in university. Int. J. Eng. Technol. (UAE) **7**, 196–200 (2018)
14. Stützle, T., Hoos, H.H.: MAX-MIN ant system. Future Gener. Comput. Syst. **16**(8), 889–914 (2000)

# Proposed Method for Identifying Emergency Unit Profiles from the Monthly Service Number

Ana Paula B. Sobral⬤, Aline R. de Oliveira$^{(\boxtimes)}$ ⬤, Hevelyn dos S. da Rocha⬤,
Harvey José S. R. Cosenza⬤, and Robisom D. Calado⬤

Federal Fluminense University, Niterói, RJ, Brazil
alinerangel@id.uf.br

**Abstract.** The indicators are used to quantify how the results of the public emergency departments (ED) can be classified, with regard to the efficiency and quality of the services provided to its users. In this sense, the objective of this work is to identify typical patterns of the curves of the indicator number of monthly attendance of the procedure "Reception with Risk Classification" in a sample of 50 EDs. For this, a quantitative and exploratory research was carried out that adopted the Machine Learning technique known as Cluster Analysis not supervised by the AGNES "AGlomerative NESting" method. 10 profiles (groups) of curves for the Reception with Risk Classification were identified in the selected EDs. Group 1 characterizes the standard profile of these EDs (76% of the total) and the other groups characterize the atypical patterns. The results were obtained using the free software R.

**Keywords:** Emergency departments · Indicator · Cluster analysis

## 1 Introduction

In Brazil, the National Humanization Policy (NHP) is a cross-cutting public policy that deals with the health work process as a whole, encompassing assistance and management, guaranteeing the protagonism of subjects and collectives, including the provision of services, care technologies and the construction of safe, harmonious environments that offer comfort and well-being to users. The overcrowding of emergency services represents a serious problem in the health system and the spontaneous demand of patients with simple diseases amplifies the number of attendance provided at these units. In view of these conditions, there is an urgent need to plan work routines and implement projects and proposals aimed at guaranteeing an efficient, dignified service, alleviating the pressure of this excessive demand. Thus, humanization is the only way to impact the effects that plague the service routine in the urgency and emergency network.

The fact of dealing with the National Humanization Policy, through a service that supports the quality of the services provided in the public emergency departments (ED), allows ordering the routine of medical care, improving the quality of the services provided to the population, mitigating the possibilities of users' dissatisfaction and dissatisfaction. In addition, it is possible to prioritize the care of patients in conditions of greater

© IFIP International Federation for Information Processing 2021
Published by Springer Nature Switzerland AG 2021
A. Dolgui et al. (Eds.): APMS 2021, IFIP AICT 631, pp. 342–350, 2021.
https://doi.org/10.1007/978-3-030-85902-2_37

severity, increasing the productivity and efficiency of the Unit's specialized teams, as well as reorganizing the flows and guaranteeing the care provided in a humanized way.

From this perspective, a Lean Thinking research is being developed in the emergency departments, in partnership between the Ministry of Health of Brazil and the Federal Fluminense University since the year 2020. This study aims to promote a new service culture that supports the improvement in the quality and efficiency of the services provided to users of the Unified Health System (SUS) attended at the public emergency departments (ED). The research project prioritizes the following points in the emergency and urgency network: care for patients in conditions of greater severity, reduction of the average length of stay of the patient in the ED, as well as the reorganization of the continuous flow of patients and the guarantee of the care provided in a humanized way.

Within the perspective of difficulty and subjectivity in assessing the quality of health systems, there is a concept in the literature that it is necessary to define appropriate metric criteria for evaluation, with reproducible and comparable indicators. Such criteria bring to the objective plan data that could be lost in the subjectivity of personal assessments, regarding the feeling of quality of the service received [6]. For this purpose, the indicators are used as a tool to quantify how the results of the emergency departments can be classified, with regard to the efficiency and quality of the services provided to its users.

In this sense, the objective of this work is to identify typical patterns of the curves of the indicator number of monthly attendance of the procedure "Reception with Risk Classification" in the EDs. For this, the Machine Learning technique known as Cluster Analysis was used unsupervised with the hierarchical group strategy (implemented by the method AGNES "AGlomerative NESting") [5].

## 2 Proposed Method

This section presents the classification of the research and the steps of the quantitative method for obtaining the results.

As for the approach and nature, this research is classified as applied quantitative research, as it aims to generate knowledge for practical application, aimed at solving specific problems, emphasizing objectivity in the collection and analysis of numerical data, as well as the use of statistics in this analysis [1].

As for the objectives, this is an exploratory research, as it aims to provide greater familiarity with the problem with a view to making it explicit, involving data analysis that stimulate understanding [1].

The steps of the proposed method are: Description of the problem; Characterization of the data; Application of Clusters Analysis and Analysis of Results.

## 3 Application of the Proposed Method

The steps of the proposed method are presented in this section. It is noteworthy that the free software R was used to obtain the results.

## 3.1 Description of the Problem

The choice of indicators is subject to the definition of the objective of the work to be carried out. Basically, the goal is the expected resolution of a problem that the manager faces in his daily routine. The indicators must lead the manager to act to achieve the expected result.

One of the problems faced by the manager in the urgency and emergency sectors, as in the case of EDs, is the issue of overcrowding of patients. Overcrowding is recognized as a concern, as it affects patients and healthcare professionals. The tension caused by the crowding of patients creates a deficit in the quality of emergency departments. Crowding has been associated with reduced access to emergency medical services, delays in attending to cardiac patients, increased patient mortality, prolonged patient transport time, inadequate pain management, violence from angry patients against staff, increased costs of patient care and decreased satisfaction with the doctor's work [2].

According to Wang et al. [7], Overcrowding degrades the quality of emergency departments due to the increase in ambulance movements, an increase in the rate of patients who leave the emergency room unattended, an extension of the patient's length of stay, a decrease in patient satisfaction.

As a result, one of the overcrowding indicators was analyzed, which is the monthly number of visits referring to the "Reception with Risk Classification" procedure in the EDs.

## 3.2 Data Characterization

This work is done in a sample of 50 emergency departments (ED) selected as defined in the Lean research project at the EDs.

The EDs are classified according to their size, according to Ordinance No. 10 OF JANUARY 3, 2017, Chapter V of the investment resource, Art. 13 [4]. The EDs enabled in investment until December 31, 2014, maintain the classification in sizes I, II, and III, for the specific purpose of concluding the financing of the approved investment, without prejudice to the granting of the cost, as provided for in Arts. 23 and 24 of this Ordinance, and in the following terms (Table 1):

**Table 1.** Definition of sizes applicable to EDs

| Size of EDs | Recommended population for the area covered by the EDs | Minimum number of observation beds | Minimum number of beds Emergency room |
| --- | --- | --- | --- |
| I | 50.000–100.000 habitants | 7 | 2 |
| II | 100.001–200.000 habitants | 11 | 3 |
| III | 200.001–300.000 habitants | 15 | 4 |

From the sample of 50 EDs, there are 3 EDs of Size I (6%), 19 EDs of Size II (38%) and 28 (56%) EDs of Size III. In addition, their types of procedures are:

Reception with Risk classification: Represents one of the interventions with decisive potential to reorganize the care of emergency services and implement networked health production.

Urgent care with observation up to 24 h in specialized care: It includes the initial examination and monitoring of patients in urgent situations. In this case, the service goes beyond the consultation, as the patient remains under observation for a maximum of 24 h.

Medical care in an emergency departments: It consists of the attendance in an emergency departments destined to assist patients affected by urgency and emergency situations, performing the initial care, stabilizing the patient and defining the responsible referral when necessary.

Orthopedic care with temporary immobilization: Comprises medical consultation with temporary immobilization, not including radiological examination.

The data referring to the number of visits for each type of procedure were monthly from January 2019 to April 2020, by selected ED.

It is noteworthy that the data analyzes were performed for the number of monthly attendance of the "Reception with Risk Classification", as this presents a more expressive quantity than the other procedures.

### 3.3 Application of Cluster Analysis

In order to identify curve patterns in the number of monthly attendance of the Reception procedure with Risk Classification in the EDs, the Machine Learning technique was used. The technique used was Cluster Analysis not supervised with the hierarchical clustering strategy (implemented by the AGNES "AGlomerative NESting" method). The results of this grouping are reported through a graphical representation known as a Dendrogram [3].

Figure 1 illustrates the dendrogram of the cluster analysis, where 10 groups (according to the cut in the dendrogram) of the number of monthly attendance curves for "Reception with Risk Classification" were adopted. It is noteworthy that the red rectangles in the dendrogram identify the EDs in the respective groups.

Table 2 shows the EDs and their respective size in each group. It is noteworthy that group 1 was not reported in this table because it has 38 EDs. In a next stage of the work, a Group Analysis will be carried out only for the EDs in group 1, in order to find more homogeneous groups.

### 3.4 Analysis of Results

This section was elaborated from the results obtained from the Cluster Analysis.

The groups and their respective monthly service number curves are shown in Fig. 2. Group 1 characterizes the pattern of the studied curves (76% of the curves) and the other groups characterize atypical behaviors of the monthly service number curves for the procedure "Reception with Risk Classification". It is noteworthy that to facilitate visual comparison, all graphics were placed on the same scale.

**Agglomerative Hierarchical Grouping**

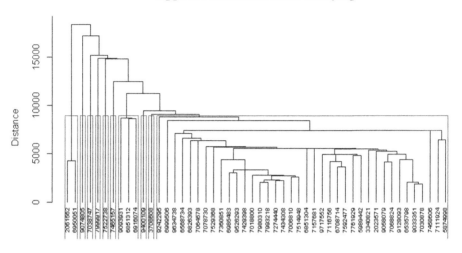

Identification code for each Emergency Department (Reception with Risk Classification)
hclust (*, "single")

**Fig. 1.** Dendrogram with 10 groups

**Table 2.** Identification (ID) Code of the Emergency Departments (ED) by group, federative unit, city and size

| Groups | Federative unit | City | ED ID Code | Size |
|---|---|---|---|---|
| 2 | Distrito Federal | C1 | 7465157 | III |
| 3 | Maranhão | C2 | 6851312 | III |
| 3 | São Paulo | C3 | 9093931 | II |
| 3 | São Paulo | C4 | 6916074 | III |
| 4 | São Paulo | C5 | 7522738 | II |
| 5 | São Paulo | C6 | 9400109 | II |
| 6 | São Paulo | C7 | 2061562 | II |
| 6 | São Paulo | C8 | 6950051 | II |
| 7 | São Paulo | C9 | 7038747 | III |
| 8 | São Paulo | C10 | 7999917 | II |
| 9 | São Paulo | C11 | 3708608 | I |
| 10 | São Paulo | C12 | 9074805 | III |

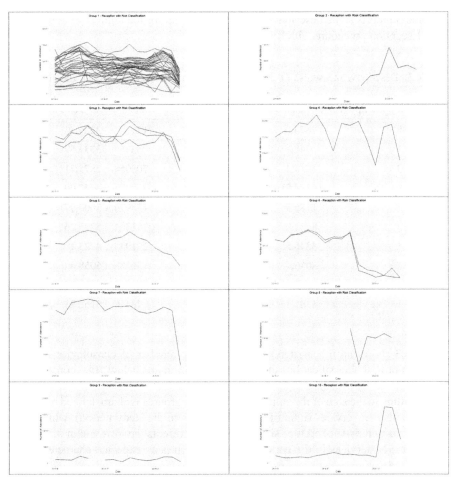

**Fig. 2.** Profiles of curves of the number of monthly attendance for the procedure "Reception with Risk Classification"

According to Fig. 2, it is noted that groups 2, 8 and 9 have a problem of lack of registration, as there are empty gaps in their respective graphs. However, group 9 presents an expected behavior of ED of size I (number of monthly care lower than the others).

Within the context of the analysis of indicators, the groups of 3, 4, 5, 6, 7 and 10 deserve to be highlighted, as they are considered atypical to the behavior of the other EDs when compared to group 1. For better management, to understand this behavior and making a decision is fundamental to the quality of the services provided. In this case, it is expected that the EDs of size III will have a greater amount of monthly assistance, as they were dimensioned for a greater number of services.

It is observed that all groups showed a drop in the number of attendance in April 2020 (last month reported), due to the Covid-19 effect.

Table 3 reports some descriptive measures by group for the number of attendances in the Reception procedure with Risk Classification.

**Table 3.** Descriptive measures of the number of reception assistance with risk classification

| Group | Number of EDs | Average | Minimum | Maximum | Sum | Standard deviation |
|---|---|---|---|---|---|---|
| 1 | 38 | 7674,0 | 0 | 15892 | 4159293 | 3229,13 |
| 2 | 1 | 6748,1 | 2181 | 14088 | 53985 | 3733,57 |
| 3 | 3 | 14751,1 | 4700 | 20233 | 708054 | 2875,21 |
| 4 | 1 | 16183,2 | 6209 | 21780 | 258931 | 4447,09 |
| 5 | 1 | 11185,9 | 3697 | 14765 | 178974 | 3000,97 |
| 6 | 2 | 8431,0 | 24 | 14829 | 261361 | 5795,26 |
| 7 | 1 | 18246,5 | 1379 | 21981 | 291944 | 4723,18 |
| 8 | 1 | 8055,4 | 2 | 12967 | 56388 | 5058,49 |
| 9 | 1 | 1167,6 | 413 | 2066 | 17514 | 510,75 |
| 10 | 1 | 4401,1 | 1600 | 17303 | 70418 | 5162,73 |

Table 3 shows an inconsistency in the monthly number of consultations for the Reception of Risk Classification, since it obtains minimums below 1000 consultations in a month, mainly in groups 1, 6, 8 and 9.

According to the sum of the entire period studied, group 3 is characterized by having a large number of services (total of 708054 services in the study period), which is an expected characteristic of EDs of size III. Group 9 presents the lowest sum of services (total of 17514 services in the period studied), which is an expected characteristic of EDs of size I.

In addition, the high variability of groups 6, 7 and 8 (standard deviation: 5795.26, 4723.18 and 5058.49, respectively) stands out, as they showed a sharp drop in the number of attendance at some point during the study period. Group 4 also enters the list with a heterogeneous service number, as it presented a curve behavior with three sudden drops in the number of mensal services.

Finally, from the point of view of the average number of services, groups 3, 4 and 7 have the highest averages. Highlight for groups 3 and 4 with EDs of size II that do not have the same physical and work team dimension as EDs of size III.

It is noteworthy that this work was developed from the state of the art that dealt with cluster analysis and indicators in emergency departments. However, among the researched articles, no article identified the typical pattern of the curves of the number of monthly attendance of the procedure "Reception with Risk Classification" in emergency departments.

# 4   Conclusion

In this work, the number of monthly assistance referring to the "Reception with Risk Classification" procedure of 50 emergency departments (ED) sampled was analyzed. The number of services is one of the overcrowding indicators that makes it possible to assess the quality of service in the urgency and emergency sectors.

For this, the Cluster Analysis technique was used, which made it possible to identify 10 standard curves of this indicator. From the results obtained, it stands out:

- Lack of registration of some information and inconsistency of some values. This situation may have been caused by the lack of a computerized system or by the absence of a protocol for data entry in the EDs;
- Presence of EDs of different sizes in the same group. This characteristic may indicate a problem with the flow of patients, since the size of the ED was dimensioned according to the recommended population for the respective coverage area;
- Suggestion to implement a system to monitor the length of stay (LoS) time in EDs.

Finally, it is recommended as future work: the construction of a computational platform for standardization in data recording; the use of other cluster analysis techniques to improve the accuracy of the method; and finally, a specific cluster analysis in group 1 EDs to create more homogeneous subgroups.

**Acknowledgment.**   The authors would like to thank the Brazilian Ministry of Health, Fluminense Federal University and Euclides da Cunha Foundation. This Research is part of a "Lean Project in UPAs 24 h" that has been funded by the Brazilian Ministry of Health (TED 125/2019, number: 25000191682201908).

# References

1. Garg, B.: Research Methodology Approaches and Techniques, 1st edn. Kobo Editions, Canada (2018)
2. Hoot, N.R., et al.: Measuring and forecasting emergency department crowding in real time. Ann. Emerg. Med. **49**(6), 747–755 (2007)
3. Johnson, R., Wichern, D.W.: Applied Multivariate Statistical Analysis, 6th edn. Prentice Hall, New Jersey (2007)
4. Brazilian Ministry of Health: PORTARIA N° 10 DE 3 DE JANEIRO DE (2017). http://bvsms. saude.gov.br/bvs/saudelegis/gm/2017/prt0010_03_01_2017.html. Accessed 14 Mar 2021
5. Hastings, S.N., et al.: Exploring patterns of health service use in older emergency department patients. Acad. Emerg. Med. **17**(10), 1086–1092 (2010)
6. Viola, D.C., Ara, M., et al.: Advanced units: quality measures in urgency and emergency care. Einstein (São Paulo, Brazil) **12**(4), 492–498 (2014)
7. Wang, H., et al.: The role of patient perception of crowding in the determination of real-time patient satisfaction at Emergency Department. Int. J. Qual. Health Care **29**(5), 722–727 (2017)
8. Moe, J., et al.: Identifying subgroups and risk among frequent emergency department users in British Columbia. J. Am. College Emerg. Phys. Open **2**(1), e12346 (2021)

9. Goodman, J.M., et al.: Emergency department frequent user subgroups: development of an empirical, theory grounded definition using population health data and machine learning. Families Syst. Health **39**(1), 55 (2021)
10. Fleury, M.J., et al.: Typology of patients who use emergency departments for mental and substance use disorders. BJPsych Open **6**(4) (2020)
11. Wong, A.H., et al.: Physical restraint use in adult patients presenting to a general emergency department. Ann. Emerg. Med. **73**(2), 183–192 (2019)
12. Shehada, E.R., et al.: Characterizing frequent flyers of an emergency department using cluster analysis. In: 17th World Congress on Medical and Health Informatics, MEDINFO 2019, pp. 158–162. IOS Press, Loyn (2019)

# Lean Healthcare in Reducing HAI an Integrative Literature Review

Laryssa Carvalho de Amaral$^{(\boxtimes)}$ (ID), Robisom Damasceno Calado (ID),
Luiza Werner Heringer Vieira (ID), and Sandra Maria do Amaral Chaves (ID)

Universidade Federal Fluminense, Rio das Ostras, Brazil
{laryssaamaral,robisomcalado,sandrachaves}@id.uff.br,
luizawheringer@gmail.com

**Abstract.** The aim of this study was to analyze the contributions of
the Lean Healthcare approach through the use of methods and tools
to reduce Healthcare-Associated Infections (HAI). The guiding question of the study was to investigate how can the Lean Healthcare approach contribute to the implementation of actions to mitigate the risk of
HealthCare-Associated Infections? Articles were collected from 2010 to
February 2021, in the Web of Science, Scopus, Emerald databases. An
analysis form was created. Of the 21 articles collected, 09 were selected,
according to the PRISMA®. As for Lean Healthcare methods and tools,
the use of standardized work, CTQ and SIPOC was identified as the
most used to mitigate the risk of HAI. The high LOS of patients in the
hospital increases the risk of HAI, with Lean Six Sigma results in the
reduction of hospital costs and lower rates of LOS, with the reduction
of the number of complications and infections. Standardized work has
reduced the rates of HAI, which reinforces the idea of the need to implement the Lean Healthcare approach, with a focus on respect for people
and consequently on the valorization of teamwork, which leads to change
in the organizational culture.

**Keywords:** Healthcare-Associated Infections · Lean · Lean Healthcare

## 1 Introduction

Healthcare-Associated Infections (HAI), as well as the increased resistance of
microorganisms to antimicrobials compromise the delivery of safe care to the
patient, being responsible for the greatest number of adverse events associated
with health care, with high morbidity and mortality. In view of this scenario,
the World Health Organization - WHO launched the first Global Challenge for
patient safety with the aim of awakening the world's view to the importance
of thinking about management models for reducing HAI as a vital element for
the delivering safe care to the patient. Therefore, one of the strategic goals
established by WHO is to promote hand hygiene [1,3,9,11].

HAI are infections that can be caused by exogenous or endogenous agents
[7]. In relation to exogenous causes, there is a need for a control of the hospital

© IFIP International Federation for Information Processing 2021
Published by Springer Nature Switzerland AG 2021
A. Dolgui et al. (Eds.): APMS 2021, IFIP AICT 631, pp. 351–361, 2021.
https://doi.org/10.1007/978-3-030-85902-2_38

environment, to ensure the safety of patients in preventing infections, which can be due to several causes, such as the lack of standardized procedures, the prolonged use of invasive devices, antibiotics and environmental hygiene, which can contribute to the spread of microorganisms [1,7].

When we think about the health care field, the Lean Healthcare approach, together with its tools, has been shown to be efficient to reduce the length of stay of patients in hospitals, seeking mechanisms for the rational use of resources, which can promote a better quality of life. and reduce costs [4].

HAI are the most common adverse events that affect millions of patients annually worldwide. The reduction in HAI is considered an indicator of the quality of health care provided [7]. Thus, the standardization of work consists of the execution of successive and defined activities for each professional, properly structured to be carried out efficiently. With the Lean Healthcare approach, standardized work is defined as one of its foundations [10].

There is a vast literature on the implementation of the Lean approach in the health area, with the presentation of several methods and techniques that can ensure the improvement of health care.

Just like SIPOC (i.e. Suppliers, Input, Process, Outputs and Customers) which enables the analysis of the client, the output and the data, to clarify the main characteristics of the process, providing visibility for the whole team in relation to the process, with the control and prevention of risks, contributing to the reduction of HAI [8].

Little is discussed about the context in which health professionals work, their conditions and workload, which can increase the chances of adverse events, which is one of the causes of HAI [12].

Therefore, the guiding question of this study is to investigate: How can the Lean Healthcare approach contribute to the implementation of actions to mitigate the risk of HealthCare-Associated Infections?

The objective of this article was to analyze the contributions of the Lean Healthcare approach through the use of methods and tools in reducing Healthcare-Associated Infections (HAI).

## 2    Method

The method used was the Integrative Literature Review. Three questions were elaborated, in order to reach the proposed objective: (1) What are the main problems for reducing HAI with the application of the Lean Healthcare? (2) How does the Lean Healthcare approach contribute to the identification of risks in cases of HAI? In the first stage, the keywords "Healthcare-Associated Infections", "Lean" and "Lean Healthcare" were defined, the filters for the selection of articles were applied, in the English language and in the period from 2010 to February 2021 and the databases were chosen: Web of Science, Scopus, Emerald and the strategy of identifying duplicates was used in the manager of Mendeley references from the collected records. In the second stage, after reading the titles and abstracts, records that do not answer the study questions and systematic reviews, letter to the editor, conference papers were excluded, and those that meet the scope of this study were included, and records collected from other sources through the snowball method. The gray literature records were excluded, and the articles were analyzed using a form indicating: references, actions to mitigate the risk of HAI, Lean Healthcare methods and tools and results obtained. The collected data were analyzed for presentation.

## 3   Results

Of the 21 articles collected, 9 were selected, according to three stages represented in the PRISMA® flowchart. It was verified through the List of Predatory Journals (https://predatoryjournals.com/journals/) that none of the selected articles appear in the gray literature (Fig. 1).

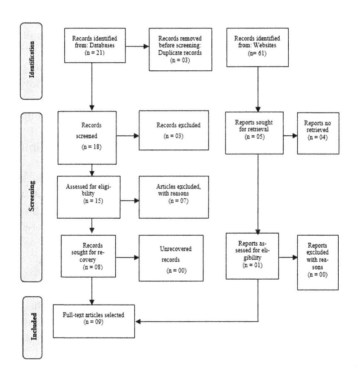

**Fig. 1.** Flowchart of the PRISMA 2020

The analysis of the articles through the form created for this systematic review enabled the construction of Tables 1 and 2. In the analyzed articles, it was found that in its objectives, the prevalence of HAI in hospitalized patients was the focus of the studies [2,8], with the use of the Lean Healthcare methodology in order to carry out the improvement procedures. Lean Six Sigma has been recognized as a method for improving processes and reducing HAI as noted in some articles [6–8,11,13].

The application of tool methods to mitigate risks, it was verified in the studies, through the Lean Six Sigma, it was possible to identify statistical data that evidence causes of HAI risks and even so, they had corrective actions reducing the prevalence of HAI and improvement in the processes of health [7,8].

With regard to actions to mitigate HAI, there was the application of Lean engineering techniques to develop a standard process that included physical steps and also standard communication elements from the provider for patients and family members and the creation of a physical environment to give support for changes [14]. In addition to creating programs to apply the tools of the Lean Healthcare approach, two articles [2,5], and the participation of a multidisciplinary team were described [1].

The length of stay of patients in the hospital increases the risk of HAI, therefore the reduction of the LOS indicator (ie, Length of Stay) with the Lean Six Sigma method, resulted in the lowest number of complications and infections, in addition to contributing to the reduction of costs in the hospitalization of patients [11].

Statistical tests were performed on the groups of patients and presented in graphs to provide a visualization of the decrease in the length of stay, from 13.14 days reduced to 9.21 days, that is, a reduction of 29.9%. The general population was also divided into subgroups according to six variables that potentially influence the length of stay [13].

Regarding the methods and tools of the Lean Healthcare approach, the use of standardized work, CTQ (Critical to Quality) and SIPOC was observed as the most used Lean tools to mitigate the risk of HAI. As seen in the studies: The rate of urinary tract infection was 25.4% to 6.5% with the use of standardized work [1]. The percentage of colonized patients was reduced from 0.36 to 0.19% [9] and reduced the number of hospitalizations by 20%, which generated the lowest number of HAI [8].

The Length of Stay with a reduction of 29.9% [13]. On average, the county's HAI had been reduced from 12.8% to 9.8 % during the year [5].

The participation of patients in critical practices throughout their journey allowed the adoption of innovative management, centered on the patient, for example, clinical protocols based on lines of care, with the application of standardized work tools, Lean and 3 P techniques (i.e., Production, Preparation and Process) [14].

The use of Strategic Management tools, such as Balanced Scorecard (BSC), associated with Lean and Six Sigma tools, such as PDSA and also TQM (i.e. Total Quality Management), contributed to the reduction of the hospital rate of HAI that was below the national average and consequently reduced the mortality rate [6].

**Table 1.** Objectives and actions to mitigate HAI risks related to Lean Healthcare methods (Source: Authors).

| Ref. | Study objectives | Actions to mitigate the risk of HAI |
|---|---|---|
| [2] | Determine the prevalence of HAI and associated risk factors, distribution of systemic AMU and AMR trend in Ferrara University Hospital (FUH), comparing prevalence data recorded in 2016 and 2018 | By the Lean Healthcare Management program that was carried out in FUH in 2018, establishing improvement actions in Rehabilitation and Surgical departments and the update of FUH standard operating procedure for prevention of catheter-associated urinary tract infection according to the latest available evidence-based guidelines. UTI significantly decreased in 2018 (from 25.4% to 6.5%), dropping behind gastrointestinal system infections and surgical site infections |
| [7] | Lean Six Sigma (LSS) has been recognized as an effective management tool for improving healthcare performance. Here, LSS was adopted to reduce the risk of healthcare-associated infections (HAI), a critical quality parameter in the healthcare sector | Lean Six Sigma proved to be a useful tool for identifying variables affecting the risk of HAI and implementing corrective actions to improve the performance of the care process. Finally, corrective measures to prevent HAI were implemented and monitored for 1 year after implementation |
| [8] | In this article, we report on the application of the Lean Six Sigma (LSS) methodology to reduce the number of patients affected by sentinel bacterial infections who are at risk of HAI | The LSS methodology was applied in the general surgery department by using a multidisciplinary team. The pre-intervention (January 2011 to December 2012) and postintervention (January 2013 to December 2014) phases were compared to analyze the effects of the methodology implemented. The methodology allowed the identification of variables that influenced the risk of HAI and the implementation of corrective actions to improve the care process, thereby reducing the percentage of infected patients |
| [14] | Our aim was to develop a standard process for room entry in the intensive care unit that improved compliance with hand hygiene and allowed for maximum efficiency | We recognized that hand hygiene is a single step in a substantially more complicated process of room entry. We applied Lean engineering techniques to develop a standard process that included both physical steps and also standard communication elements from provider to patients and families and created a physical environment to support this |

(*continued*)

Table 1. (*continued*)

| Ref. | Study objectives | Actions to mitigate the risk of HAI |
|---|---|---|
| [11] | Postoperative length of hospital stays (LOS) will be the main indicator to assess the improvements brought by the new antibiotic quantitatively | Antibiotic prophylactic protocol |
| [13] | The aim of this paper is to analyze the introduction of DTAP, employing Lean Thinking and Six Sigma methodology based on the DMAIC cycle | The adoption of a DTAP for patients with femur fracture is the resulting consequence of the improvement process regarding quality of care, introduced of this hospital with the goal of optimizing human resources and materials |
| [5] | The purpose, therefore, was to investigate processes and impact related to implementing two concurrent quality initiatives in a Swedish hospital | Programs called the Dynamic and Viable Organization (DVO) and a national initiative on stopping healthcare associated and hospital acquired infections (SHAI) |
| [6] | The purpose of this paper is to study critical practices for adopting improvement knowledge as a management innovation in a professional organization | Identifying the patient as the guiding principle and encouraging involvement and local change initiatives. Identifies critical practices for adopting a management innovation |
| [1] | To prevent environmental transmission of pathogens, hospital rooms housing patients on transmission based precautions are cleaned extensively and disinfected with ultraviolet (UV) light. We sought to improve room turnover efficiency to allow for UV disinfection | A multidisciplinary healthcare team participated in a 60 day before-and-after trial that followed the Toyota Production System Lean methodology. We used value-stream mapping and manual time studies to identify areas for improvement |

**Table 2.** Tools and results obtained distributed by the articles (Source: Authors).

| Ref. | Tools | Resulted |
|------|-------|----------|
| [2] | Standardized work | Results of this study contribute to reinforces the statement that HAI and AMR remain a high burden for healthcare systems, undermining patient safety in hospitals and causing high rates of morbidity, mortality and costs |
| [7] | DMAIC. SIPOC, CTQ, Ishikawa, cause-effect diagram, brainstorming sessions | The implementation of an LSS approach could significantly decrease the percentage of patients with HAI |
| [8] | DMAIC. SIPOC, CTQ, Ishikawa, cause-effect diagram, brainstorming sessions | The LSS is a helpful strategy that ensures a significant decrease in the number of HAI in patients undergoing surgical interventions. The implementation of this intervention in the general surgery departments resulted in a significant reduction in both the number of hospitalization days and the number of patients affected by HAI |
| [14] | Standardized work, Lean techniques and 3P (Production Preparation Process) | We observed meaningful improvement in the performance of the new standard as well as time savings for clinical providers with each room entry. We also observed an increase in room entries that included verbal communication and an explanation of what the clinician was entering the room to do |
| [11] | Lean Six Sigma: DMAIC. SIPOC, CTQ and box-plot | The results show that the LOS of patients treated with Ceftriaxone is lower than those who were treated with the association of Cefazolin plus Clindamycin, the difference is about 41%. Moreover, a lower number of complications and infections was found in patients who received Ceftriaxone |
| [13] | Lean Six Sigma: DMAIC, Critical to quality (CTQ), SIPOC, VSM, Ishikawa fishbone (root cause) diagram, Standardized Work and Box – plot | Statistical tests were performed on the groups and graphics were provided to visualize the decrease of LOS (29.9%). The overall population was also divided in subgroups according to six variables potentially influencing LOS |
| [5] | PDSA | Even though both initiatives shared the same improvement approach, there was no strong indication that they were strategically combined to benefit each other. The initiatives existed side by side with some coordination and some conflict |

(*continued*)

**Table 2.** (*continued*)

| Ref. | Tools | Resulted |
|------|-------|----------|
| [6] | Lean Six Sigma: PDSA; TQM | The hospital rate of Healthcare-Associated Infections (HAI) is below the national mean (7.1% mean at SkaS, 9.4% mean in Sweden). |
| [1] | Kaizen, 5S. VSM | We decreased room turnover time by half in 60 days by decreasing times between and during routine tasks. Utilizing Lean methodology and manual time study can help teams understand and improve hospital processes and systems |

Legend: DMAIC - Define, Measure, Analyze, Improve and Control; CTQ - Critical to Quality; SIPOC - Suppliers, Input, Process, Outputs and Customers; 3P - Production, Preparation and Process; VSM - Value Stream Mapping; PDSA - Plan, Do, Study and Act; TQM - Total Quality Management; A3 - A3 Report; 5S - Sort, Set in order, Shine, Standardize and Sustain; WHO - World Health Organization; DTAP - Diagnostic - Therapeutic Assistance Pathway

## 4   Discussion

Based on the results obtained, it was founded that the Lean approach contributes to reduce the risks of HAI, difficulties were pointed out in studied articles to obtain data for statistical analysis.

One of the main problems for reducing HAI with the application of Lean Healthcare is the difficulty of obtaining data to plan the actions necessary to mitigate the risk of HAI in hospitals. In the analyzed articles, the scarcity of data for statistical analysis and the use of Lean Six Sigma tools, such as the Pareto Graph and the use of tests to assess the correlations between HAI and specific parameters on the type of intervention, comorbidities, allergies and among other factors that can also increase the risk of HAI.

Another problem is related to the low adherence of health professionals to the use of clinical protocols, caused by work overload, as activities are carried out that do not add value to the patient, thus making adherence to the protocols difficult. Despite knowledge about the protocols and the risks of contamination, health professionals are not sensitive to the need to comply with the protocols, compromising patient safety and the risk of HAI. Standardized work reduces HAI, as found in the analysis of the articles with the results found.

Standardized work consists of a Lean tool that reduces HAI, as verified in the analysis of the articles, which point to the need for planning to implement actions that reduce the risk of HAI. Prevention actions were carried out based on the identification of risks in different sectors of hospitals, leading to a reorganization of patient-centered actions, which is corroborated by the Lean Healthcare approach.

Actions contribute to the development of a sustainable safety culture, responsibilities and actions, whose focus is on providing the best result and experience for patients, are based on the Lean Healthcare approach through the application of tools and methods that ensure the realization of its principles.

As verified in the articles, the applications of Lean methods and tools, such as the standardized work and SIPOC, corroborate the importance of the Lean approach in mitigating the risks of HAI and also the application of quality tools, such as the CTQ (Critical to Quality).

It was founded in the studies that the risk of HAI is increased due to the increase in the length of stay of patients in the hospital environment, as verified in the statistics presented in the studies through the LOS indicator and percentage of patients with HAI, thus showing the importance of the Lean approach that contributes to reducing the average length of stay of hospitalized patients.

Organizational culture and leadership are essential facilitators for patient involvement as a member of the healthcare team. Through culture it is possible to develop a favorable environment for the identification of improvements, inclusion of patients as members of the health team, determination of responsibilities, negotiation, teamwork and open and frequent communication. In this environment, on the other hand, visible leadership is needed, which is an example in the development of an environment that favors continuous learning, characterized by transparency in relationships, constant concern with failures promoting continuous improvement and reliability of care processes.

# 5   Conclusion

Through this systematic review of the literature, the authors of the present study concluded that the Lean Healthcare approach contributes to the reduction of HAI rates. The methods and tools of the Lean Healthcare approach are essential to redirect the actions that will mitigate the risk of HAI, particularly the standardized work, which despite the low adherence of health professionals, when implemented, resulted in a decrease in the length of stay of patients and therefore reduced HAI rates.

Considering that adherence to clinical protocols was one of the problems discussed in the analyzed articles, we saw that this reinforces the idea of the need to implement the Lean Healthcare approach that focuses on respect for people and, consequently, the valorization of teamwork and professionals of health, which leads to a change in the organizational culture.

It is noteworthy that there are still few studies in the literature on the Lean Healthcare approach aimed at the problem of HAI, which was evidenced by the scarcity in the literature, which reinforces the need to expand future studies, particularly related to the application of HFMEA, for being a tool that provides risk analysis, which we recommend for future research.

# 6   Limitations

The scarce literature on HAI with the application of the Lean approach became the main limitation in carrying out this study, particularly with a focus on the issue of patient safety in the face of HAI risks.

**Acknowledgements.** The authors would like to thank the Brazilian Ministry of Health, Fluminense Federal University and Euclides da Cunha Foundation. This Research is part of a "Lean Project in UPAs 24h" that has been funded by the Brazilian Ministry of Health (TED 125/2019, number: 25000191682201908).

# References

1. Ankrum, A.L., Neogi, S., Morckel, M.A., Wilhite, A.W., Li, Z., Schaffzin, J.K.: Reduced isolation room turnover time using lean methodology. Infect. Control Hosp. Epidemiol. **40**(10), 1151–1156 (2019)
2. Antonioli, P., Bolognesi, N., Valpiani, G., Morotti, C., Bernardini, D., Bravi, F., Di Ruscio, E., Stefanati, A., Gabutti, G.: A 2-year point-prevalence surveillance of healthcare-associated infections and antimicrobial use in Ferrara University Hospital, Italy. BMC Infect. Dis. **20**(1), 1–8 (2020)
3. Anvisa: Programa nacional de prevenção e controle de infecções relacionadas à assistência à saúde. https://www.gov.br/anvisa/pt-br/centraisdeconteudo/ publicacoes/servicosdesaude/publicacoes/pnpciras_2021_2025.pdf
4. Backman, C., et al..: Implementation of a multimodal patient safety improvement program "safetyleap" in intensive care units: a cross-case study analysis. Int. J. Healthc. Qual. Assur. **31**(2), 140–149 (2018)
5. Ernst, J., Schleiter, A.J.: Standardization for patient safety in a hospital department: killing butterflies with a musket? Qual. Res. Organ. Manag. Int. J. **13**, 368–383 (2018)
6. Hellström, A., Lifvergren, S., Gustavsson, S., Gremyr, I.: Adopting a management innovation in a professional organization: the case of improvement knowledge in healthcare. Bus. Process Manag. J. (2015)
7. Improta, G., Cesarelli, M., Montuori, P., Santillo, L.C., Triassi, M.: Reducing the risk of healthcare-associated infections through lean six sigma: the case of the medicine areas at the Federico II University Hospital in Paples (italy). J. Eval. Clin. Pract. **24**(2), 338–346 (2018)
8. Montella, E., Di Cicco, M.V., Ferraro, A., Centobelli, P., Raiola, E., Triassi, M., Improta, G.: The application of lean six sigma methodology to reduce the risk of healthcare-associated infections in surgery departments. J. Evalu. Clin. Practi. **23**(3), 530–539 (2017)
9. Nyström, M.E., Garvare, R., Westerlund, A., Weinehall, L.: Concurrent implementation of quality improvement programs: coordination or conflict? Int. J. Healthc. Qual. Assur. **27**(3), 190–208 (2014)
10. O'Reilly, K., et al.: Standard work for room entry: linking lean, hand hygiene, and patient-centeredness. Healthcare **4**, 45–51 (2016)
11. Organization, W.H., et al.: Report on the Burden of Endemic Health Care-Associated Infection Worldwide, WHO (2011)

12. Ricciardi, C., Fiorillo, A., Valente, A.S., Borrelli, A., Verdoliva, C., Triassi, M., Improta, G.: Lean six sigma approach to reduce LOS through a diagnostic-therapeutic-assistance path at A.O.R.N cardarelli. TQM J. **31**, 651–672 (2019)
13. Ricciardi, C., et al.: A health technology assessment between two pharmacological therapies through six sigma: the case study of bone cancer. TQM J. **32**, 1507–1524 (2020)
14. Team, P.E.A.: Engaging Patients in Patient Safety-a Canadian Guide. Canadian Patient Safety Institute (2017)

# Karakuri: A Proposal to Waste Reduce in the Health Service

Stephanie D'Amato Nascimento$^{(\boxtimes)}$ , Maria Helena Teixeira da Silva ,
Sérgio Crespo Pinto , Robisom Damasceno Calado ,
and Ricardo Rodrigo Alves

Instituto de Ciência e Tecnologia, Universidade Federal Fluminense,
Rio das Ostras, Brazil
{sd_amato,maria_helena,screspo,robisomcalado}@id.uff.br,
rralves@toyota.com.br

**Abstract.** The objective of this study is to analyze the adherence of
the Karakuri (low-cost automation) technology for eliminating Lean
wastes in healthcare units. To carry out this study reports of continuous improvement practices were analyzed from three Units located in
the State of São Paulo (Brazil). These practices of Lean Healthcare approach implementation were applied during the period from June to October 2020. The present literature on the term Karakuri was also analyzed
in four data research bases and it was made a study around correlate
cases. In conclusion, it is noted that Karakuri seems capable of eliminating the main wastes reported in the analyzed health units in question,
which are: defects, motion, waiting, and not using human talent.

**Keywords:** Karakuri · Lean healthcare · Lean wastes · Low-cost
automation

## 1 Introduction

### 1.1 Conceptualization

In the global pandemic scenario declared by the WHO (World Health Organization) in January 2020, worldwide attention has turned to the health area. The
increase in reports and publications related to the theme is considerable, not only
in the health area but also in the areas of management, technologies, and public
policies. Overcrowding in health services is a problem faced worldwide that has
gained a broader focus. In Brazil, SUS, the Unified Health System, stands out
as the only universal public health system that serves more than 100 million
people. According to IBGE (Brazilian Institute of Geography and Statistical), 7
out of 10 Brazilians depend on SUS for actions related to healthcare [8]. In the
public health there is a constant political and social pressure to increase service
levels and decrease costs [6].

In the context of the need to reduce waste, increase quality and productivity,
Lean Manufacturing (LM), or Lean Production, stands out with the "do more

© IFIP International Federation for Information Processing 2021
Published by Springer Nature Switzerland AG 2021
A. Dolgui et al. (Eds.): APMS 2021, IFIP AICT 631, pp. 362–372, 2021.
https://doi.org/10.1007/978-3-030-85902-2_39

with less" approach [5] LM is an approach from the Toyota Production System (TPS - Toyota Production System) whose focus is to reduce waste, shorten times and increase the quality of production processes [21]. The term Lean Manufacturing became popularly known in the mid-90s with the book "The Machine that Changed the World" [30] segments have sought Lean approach to improve their processes.

In the current context, automation can play an important role in creating value and increasing patient satisfaction [2]. According to Granlund and Wiktorsson [6], automation proves to be a successful way of reducing the time spent on attention to activities that do not add value to the health environment.

The term Karakuri (which means, in Japanese, "mechanism" or "trick") was used in the past to refer to automation dolls used as entertainment in Japan. Through the use of physic principles, the Karakuri devices execute intelligent movements. Today, the term is used to describe devices with similar characteristics that are being used to increase the productivity of the work environment in simple, easy-to-implement, and low-cost automation [24]. According to Bhanu and Kumar [3], Karakuri is an automation technique, part of the Lean strategy that provides low-cost automation solutions, with simple maintenance and a quick return on investment capable of bringing innovation at low cost.

Considering the above, this work aims to analyze the adherence of Karakuri automation technique for the elimination of waste associated with 3 Emergency Care Units (UPA) located in the state of São Paulo, Brazil. This proposal is formulated to achieve the following specific objectives:

– Analysis of waste associated with the 3 UPAs under study;
– Analysis of the wastes mentioned in the Karakuri literature;
– Presentation of correlated Karakuri application in literature;
– Comparison between the need and benefits from Karakuri application in healthcare units.

## 2    Theoretical Framework

### 2.1    The 8 Lean Wastes

According to Bahnu and Kumar [3], lean production works with a simple but vital concept of eliminating waste and providing good quality products and services. The Lean concept, (according to the authors) keeps production costs as low as possible while guaranteeing quality. The waste or Mura, in Japanese, can be taken as activities that do not add value to the customer. The authors add that waste elimination results in decreased inventories, a better quality of work, products, and improved services. Leading them to build a better relationship with customers and suppliers. According to Ohno [21], real efficiency improvement is achieved when no waste is produced, that is, when only the necessary quantity is produced, with the necessary manpower capacity.

The seven wastes of lean thinking, proposed by Ohno [21] are presented in Table 1 along with the eighth waste proposed later by Liker and Meier [13].

The waste addressed in Table 1 from Lean Manufacturing is brought to the Lean Healthcare through the examples of Costa and Godinho Filho [4].

**Table 1.** Lean Wastes definition and examples (Source: Adapted by the author from Ohno [21], Liker and Meier [13] and Costa and Godinho Filho [4])

| Waste | Definition by Ohno [21] and Liker and Meier [13] | Example in healthcare by Costa and Godinho Filho [4] |
|---|---|---|
| Overproduction | Produce earlier or in larger quantities than necessary | Perform unnecessary treatment |
| Waiting | Inoperative operations, processes, or workers waiting for the end of a process, lack of tools, supplies, parts, or information | Wait for a doctor's appointment |
| Transport | Parts or products moved unnecessarily | Transfer patients between rooms |
| Over processing | Perform unnecessary process that does not add value to the customer | Filling out unnecessary forms |
| Inventory | Raw material, finished products, or Work In Process beyond the quantity needed for a controlled system | Medicines out of stock or in excess |
| Motion | Extra movement by employees that does not add value to the final customers | Search for medicines |
| Defects | Need for rework or defective products | Inspect work already done looking for errors |
| Defects | Waste of time, ideas, improvements, skills, and creativity | (Not covered) |

## 2.2   Karakuri: A Lean Automation Technology

Karakuri dolls are the precursors of industrial robots [9]. The history of robotics in Japan began in the EDO period (1603 onwards) when several Karakuris were made with the simple intention of bringing fun to people. Despite its millenary history, Karakuri technology continues to be used today mainly in the industry. According to Murata et al. [20] the Karakuri's goal in the industry is to automate an objective operation, making it easier to perform and thus increasing productivity. Aiming this the Technology uses a simple mechanism based mainly on natural principles such as mechanical principles (levers, pulleys, gears, and cam connection mechanisms), hydromechanics, magnetism, sound, optics, and physical properties in general.

Karakuri technology consists of devices that produce work with low energy and without any motor or power unit, that is, without electricity. Weber [29] mentions, for example, that a Karakuri system of frames, can incorporate rollers, levers, pedals, and counterweights to move, by gravity, boxes of parts and components.

Currently, Karakuri is disseminated by multinational companies such as Toyota, which includes it as a global goal in their day-to-day. The company, cradle of Toyota Production System, finds in Karakuri several benefits such as energy savings, reduced emissions of carbon, ergonomic improvements, and use of human talents. The global initiative of training its employees in Karakuri Kaizen around the world can engage teams to propose solutions and implement them in their lines of work.

# 3   Methods and Data

## 3.1   Wastes Analysis in Health Units

For this study, the waste identified in reports from 3 UPAs located in the State of São Paulo (Brazil) was analyzed. We will call the UPAs as UPA 1, UPA 2, and UPA 3. In all, 88 reports of Lean improvement practices associated with the elimination of 8 wastes previously listed. Table 2 shows the number of times that each waste was associated with each unit analyzed.

**Table 2.** Wastes in analyzed ECU (UPAs).

| Waste category | UPA 1 | UPA 2 | UPA 3 | Total |
|---|---|---|---|---|
| Overproduction | 0 | 1 | 0 | 1 |
| Waiting | 7 | 7 | 9 | 23 |
| Transport | 0 | 0 | 0 | 0 |
| Over processing | 0 | 2 | 2 | 4 |
| Motion | 2 | 16 | 13 | 31 |
| Inventory | 2 | 1 | 6 | 9 |
| Defects | 11 | 10 | 19 | 40 |
| Unutilized talent | 12 | 4 | 2 | 18 |
| **Total** | **34** | **41** | **51** | **126** |

Through the Pareto chart in Fig. 1, it is possible to analyze that four categories are responsible for 88.9% of the waste of the analyzed UPAs, which are: defects, movement, waiting, and non-use of human talent. Therefore, by Pareto rule, we consider these to be the most important waste to be eliminated.

## 3.2   Karakuri – Theoretical Review of Eliminated Lean Wastes

In this research stage, a bibliographic search for the term "Karakuri" was carried out in the following Databases: Scopus, Compendex, Web of Science, Taylor and Francis. The first search identified a total of 86 conference papers and journal articles between the years 2001 and 2020. The documents were screened by the

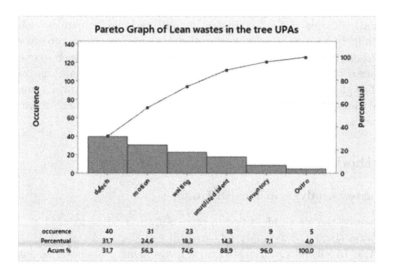

**Fig. 1.** Analysis of Wastes in Karakuri literature (Source: From author)

scheme presented in Fig. 2. As illustrated in the Figure, after the screening, 3 more documents were added from external fonts to enrich the analysis.

After the filtering and selection steps according to the image above, 19 articles were selected and analyzed for evidence of waste elimination. Table 3 lists the types of waste eliminated in each document analyzed. Articles 14 and 15 are purely mathematical/physical articles that do not address nor indicate the benefits of Karakuri in eliminating waste.

The Pareto chart on Fig. 3 lists the waste most often cited as disposed of after the Karakuri technique implementation in the analyzed articles. From the graph, it is possible to see that, according to the authors, the most eliminated waste was respectively: excessive movement, transportation, waiting, not using human talent, being responsible to 91,7% of total wastes related.

### 3.3   Karakuri Correlated Cases from Literature

From the literature review, the correlated cases was analyzed by 5 main questions as follows in Table 4.

### 3.4   Karakuri Health Examples

In the literature analyzed in previous steps, just one example was related to health use [11] however some examples of Karakuri are found in the area, despite not being popularly recognized as Karakuri. The Alcohol in Gel mechanical

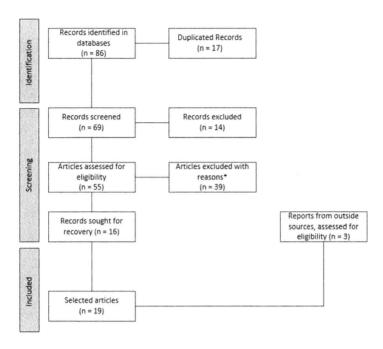

**Fig. 2.** Research screening process (Source: Adapted by the author from PRISMA [22].)

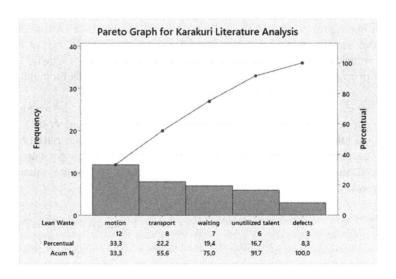

**Fig. 3.** Pareto graph of Karakuri literature analysis (Source: From author)

Table 3. Wastes in analyzed ED (UPAs).

| Source | Wastes associated |
|---|---|
| 1. Prasetyawan et al. [23] | Defects, transport, motion, and unutilized talent |
| 2. Anggrahini et al. [1] | Motion, waiting and transport |
| 3. Riegr and Mašín [26] | Motion and transport |
| 4. Heng et al. [7] | Motion and transport |
| 5. Madisa et al. [14] | Waiting and motion |
| 6. Yamamoto et al. [31] | Unutilized talent |
| 7. Kit et al. [12] | Motion, transport, and waiting |
| 8. Sawaguchi [27] | Unutilized talent |
| 9. Mašín and Riegr [15] | Unutilized talent and motion |
| 10. Rani et al. [24] | Waiting, motion, transport, and defects |
| 11. Katayama et al. [10] | Waiting, motion, and transport |
| 12. Weber [29] | Waiting and unutilized talent |
| 13. Murata and Katayama [19] | Unutilized talent |
| 14. Riegr e Mašín [25] | N/A |
| 15. Mašín e Riegr [16] | N/A |
| 16. Matsuidara [17] | Motion and transport |
| 17. Kijima et al. [11] | Waiting and unutilized talent |
| 18. Shamsudin et al. [28] | Motion |
| 19. Bhanu and Kumar [3] | Motion |

dispenser is a Karakuri device that allows the user to sanitize their hands with a pedal input, minimizing contamination risks through a simple, ergonomic, and low-cost solution, as they can be built with easily accessible materials. In this case, the process is facilitated, and the employee's exposure to biological risk is reduced.

Another example of this concept application in health is the device for adjusting the height of stretchers, where the principle of levers and cranks reduces the physical effort required to change the posture of bedridden patients. Thus, making the procedure more ergonomic, with less risk to the patient and the professional, in addition to allowing a smoother change in posture, ensuring the patient's well-being.

**Table 4.** Karakuri correlated cases

| Ref | Mechanism |
|---|---|
| [23] | The mechanism works through the gravitational potential energy (weight of the food mixture) added to the mechanical energy from a machine in the previous process, the pan with the preparation slides on the conveyor through rollers, ensuring the transfer of the mixture to the next station. The authors point out that the improvement had a low implementation cost, despite not delving into financial details |
| [1] | The energy used by the Karakuri comes from the gravitational potential energy (the weight of the pan). The pan is placed on a platform that "lowers" its weight and transfers it to a conveyor, guiding it to an automated oven. After the cooking time, the oven releases the pan on the other section of the platform, taking the pan to the next production step. The Karakuri was made of aluminum, PVC tubes, bearings, and others that guarantee the low cost of the technique |
| [26] | This Karakuri cart benefits from the weight transported, which releases a brake mechanism and allows movement until another brake mechanism stops the system on the opposite wheels. When the weight is removed, the cart returns to its initial position. The authors do not mention the costs for producing the solution but use lowcost materials for prototyping |
| [7] | The proposed solution is an inclined plane equipped with bearings that move and provide friction as needed to the plant trays. The structure also has a mechanism for allocating pumps for irrigation of the crop. The author does not address values; however, he justifies that the choice of method is due to the low-cost relative to other automation options |
| [14] | From the before/after image and the complementary text, it is possible to verify that the incorporated mechanism raised the workers' posture, acting mainly in the coil replacement system, where it is possible to verify the presence of a lever and bearings, indicating that the energy that moves the Karakuri system in question is mechanical energy. The authors point out the Karakuri as a low-cost technique, but authors do not detail them in the studied literature |
| [12] | Mechanical energy reaches the Karakuri through the activation of a pedal, which transforms it into potential energy, allowing the empty input box to be replaced by a full input box to supply the workstation. For the sliding of the boxes, bearings are used in an inclined plane. A lever system is used through the foot pedal. The author does not address specific costs for implementing the technique, however, he informs that its application in transport and supply systems is justified in view of the cost of other means of automation |
| [24] | The application in the manufacturing industry is justified to eliminate the need for electrical energy from the system (related to cost and supply failures). The new tipping system works through gravitational force, replacing the need for electricity. There was no substantial cost in the implementation of Karakuri, as equipment that already operated on electricity underwent adjustments to become a Karakuri solution |
| [17] | The Karakuri technique is applied through a tube, which uses gravitational energy and moves a box full of parts brought by a self-guided vehicle (AGV) to a specific rail. The author ponders the savings provided by the use of Karakuri but does not effectively address its costs |
| [11] | The device works behind the simulated coat of a horse. The complex Karakuri mechanism receives mechanical energy through a crank and transmits it through gears, shafts, and other elements, resulting in elliptical movements in each horse's legs and other additional movements such as swinging the hips, simulating the animal trot. Despite not addressing financial aspects, the author points out that the technique had a low cost |

# 4   Conclusion

The present study aimed to compare the wastes found in the Emergency Care Units (UPAs) analyzed and what the Karakuri technique can deliver as a benefit. By comparing the wastes from literature and wastes found in UPAs we conclude that the technique seems to contribute to eliminate the three main wastes found in the studied reports from UPAs (defect/rework, movement, waiting, and non-use of human talent). Thus, by the examples in the health area and by the correlated cases, it is indicated that Karakuri could bring benefits to the humanization, environment, ergonomic improvements, savings, and risk reduction when used as a low-cost automation technique in the Lean Healthcare scenario.

As a proposal for future work, it is suggested to study more deeply the interaction between Karakuri application in Health, its limitation, and steps, in order to explore the fact that there is only one article on the researched application of the method in the health area.

**Acknowledgements.** The authors would like to thank the Toyota Brazil Company for the contribution with this research.

The authors also would like to thank the Federal Fluminense University and the Fundação Euclides da Cunha, which made the research project funded by the Ministry of Health of Brazil viable; TED 125/2019, Number: 25000191682201908.

# References

1. Anggrahini, D., Prasetyawan, Y., Diartiwi, S.I.: Increasing production efficiency using Karakuri principle (a case study in small and medium enterprise). In: IOP Conference Series: Materials Science and Engineering, vol. 852, p. 012117. IOP Publishing (2020)
2. Benzidia, S., Ageron, B., Bentahar, O., Husson, J.: Investigating automation and AGV in healthcare logistics: a case study based approach. Int. J. Log. Res. Appl. **22**(3), 273–293 (2019)
3. Bhanu, M.V., Kumar, P.B.S.: Global study and implementation of Karakuri (2018)
4. Costa, L.B.M., Godinho Filho, M.: Lean healthcare: review, classification and analysis of literature. Prod. Plann. Control **27**(10), 823–836 (2016)
5. De Haan, J., Naus, F., Overboom, M.: Creative tension in a lean work environment: implications for logistics firms and workers. Int. J. Prod. Econ. **137**(1), 157–164 (2012)
6. Granlund, A., Wiktorsson, M.: Automation in healthcare internal logistics: a case study on practice and potential. Int. J. Innov. Technol. Manag. **10**(03), 1340012 (2013)
7. Heng, A.T.G., Mohamed, H.B., Rafaai, Z.F.B.M.: Implementation of lean manufacturing principles in a vertical farming system to reduce dependency on human labour. Int. J. Adv. Trends Comput. Sci. Eng. **9**(1), 512–520 (2020)
8. IBGE: Pesquisa Nacional de Saúde, Ministério do Planejamento, Orçamento e Gestão (2019). http://www.pns.icict.fiocruz.br/arquivos/Portaria.pdf
9. Iguchi, N., Uchiyama, J., Kimura, H., Hamashima, Y.: Development of a performance robot. Adv. Robot. **5**(1), 3–13 (1990)

10. Katayama, H., Sawa, K., Hwang, R., Ishiwatari, N., Hayashi, N.: Analysis and classification of Karakuri technologies for reinforcement of their visibility, improvement and transferability: an attempt for enhancing lean management. In: Proceedings of PICMET 2014 Conference: Portland International Center for Management of Engineering and Technology; Infrastructure and Service Integration, pp. 1895–1906. IEEE (2014)
11. Kijima, R., Hashimoto, K., Jiang, Y., Aoki, T., Ojika, T.: A development of VR KARAKURI horse riding system for exhaustive therapy and study on its validity. Trans. Virtual Reality Soc. Jpn. **6**(3), 157–164 (2001)
12. Kit, B.W., Olugu, E.U., Zulkoffli, Z.: Redesigning of lamp production assembly line (2018)
13. Liker, J.K., Meier, D.: Toyota Way Fieldbook. McGraw-Hill Education, New York (2006)
14. Madisa, I.M., Taib, M.F.M., Reza, N.A.: Implementation of Karakuri Kaizen to improve productivity and ergonomics in wire rope industry. In: Proceedings of the 9th International Conference on Industrial Engineering and Operations Management (2019)
15. Mašín, I., Riegr, T.: Dynamic characteristics of the Karakuri transport trolley (2016)
16. Mašín, I., Riegr, T.: Advanced modelling of the Karakuri mechanism. In: Proceedings of the 7th International Conference on Mechanics and Materials in Design, pp. 747–748 (2017)
17. Matsudaira, Y.: The continued practice of 'ethos': how Nissan enables organizational knowledge creation. Inf. Syst. Manag. **27**(3), 226–237 (2010)
18. Matsumura, R., Shiomi, M., Miyashita, T., Ishiguro, H., Hagita, N.: What kind of floor am i standing on? Floor surface identification by a small humanoid robot through full-body motions. Adv. Robot. **29**(7), 469–480 (2015)
19. Murata, K., Katayama, H.: Development of Kaizen case-base for effective technology transfer-a case of visual management technology. Int. J. Prod. Res. **48**(16), 4901–4917 (2010)
20. Murata, K., Wakabayashi, K., Watanabe, A., Katayama, H.: Analysis on integrals of lean module technologies-the cases of visual management, Poka-Yoke and Karakuri technologies. Res. Electron. Commer. Front. **1**(2), 21–29 (2013)
21. Ohno, T.: Toyota Production System: Beyond Large-Scale Production. CRC Press, Boca Raton (1988)
22. Page, M.J., et al.: The PRISMA 2020 statement: an updated guideline for reporting systematic reviews. BMJ **372**, 8 (2021)
23. Prasetyawan, Y., Agustin, A.A., Anggrahini, D.: Simple automation for pinneaple processing combining with Karakuri design. In: IOP Conference Series: Materials Science and Engineering, vol. 852, p. 012102. IOP Publishing (2020)
24. Rani, D., Saravanan, A., Agrewale, M.R., Ashok, B.: Implementation of Karakuri Kaizen in material handling unit. Technical report, SAE Technical Paper (2015)
25. Riegr, T., Mašín, I.: New approach to solving mathematical equation for damped oscillations by sliding (coulomb) friction at the Karakuri mechanism. In: IRF 2018: Proceedings of the 6th International Conference on Integrity-Reliability-Failure (2018)
26. Riegr, T., Mašín, I.: Solution of damped oscillations by coulomb friction at the Karakuri mechanism using MAPLE software. In: Medvecký, Š, Hrček, S., Kohár, R., Brumerčík, F., Konstantová, V. (eds.) Current Methods of Construction Design. LNME, pp. 383–390. Springer, Cham (2020). https://doi.org/10.1007/978-3-030-33146-7_44

27. Sawaguchi, M.: How does Japanese "Kaizen activities" collaborate with "Jugaad innovation"? In: 2016 Portland International Conference on Management of Engineering and Technology (PICMET), pp. 1074–1085. IEEE (2016)

28. Shamsudin, E., Darus, S.A.A.Z.M., bin Raja Zainal Abidin, R.M.H.: Implement Karakuri as a material handling in production sealer line (2019)

29. Weber, A.: Modular framing systems let engineers get creative (2014). https://www.assemblymag.com/articles/92483-modular-framing-systems-let-engineers-get-creative

30. Womack, J.P., Jones, D.T., Roos, D., Carpenter, D.: The Machine that Changed the World: [Based on the Massachusetts Institute of Technology 5-Million-Dollar 5-Year Study on the Future of the Automobile]. Rawson Associates (1991)

31. Yamamoto, Y., Sandström, K., Munoz, A.A.: Karakuri IoT. In: Advances in Manufacturing Technology XXXII: Proceedings of the 16th International Conference on Manufacturing Research, Incorporating the 33rd National Conference on Manufacturing Research, University of Skövde, Sweden, 11–13 September 2018, vol. 8, p. 311. IOS Press (2018)

# Home Healthcare Routing and Scheduling Problem During the COVID-19 Pandemic

Fatemeh Taghipour[1] , Reza Tavakkoli-Moghaddam[1(✉)] ,
and Maryam Eghbali-Zarch[2]

[1] School of Industrial Engineering, College of Engineering, University of Tehran, Tehran, Iran
{f.taghipour,tavakoli}@ut.ac.ir
[2] Department of Industrial Engineering, Faculty of Engineering,
Alzahra University, Tehran, Iran
m.eghbaliii@gmail.com

**Abstract.** Home healthcare routing and scheduling problems provide a better condition for elderly and disabled people. In the presence of the worldwide pandemic of coronavirus disease 2019 (COVID-19 pandemic), these problems have shown their efficiency and necessity to notice this new group of patients suffering from COVID-19. For this purpose, a new mathematical model is developed for a home healthcare routing and scheduling problem (HHCRSP) regarding the impact of the COVID-19 outbreak. The satisfaction and cost of assigning staff to patients with suspected or confirmed COVID-19 are considered. This problem considers travel time between patients and aims to minimize it. The proposed model is solved using GAMS optimization software. Computational experiments are considered for several test problems and a sensitivity analysis is conducted to validate the model performance.

**Keywords:** Home healthcare · Staff scheduling · Employee satisfaction · COVID-19

## 1 Introduction

One of the effective ways to use resources more efficiently and reduce costs is home healthcare services at home [1]. Furthermore, population aging and increasing life expectancy, government pressure on the health system to reduce costs lead to increasing demand for home healthcare services [2, 3]. On the other hand, in the presence of pandemics, SAR-COV-19 leads people to provide their needs at home as much as possible [4, 5]. Due to the increasing demand for home healthcare services, organizations seek to optimize activities and increase the quality of their services. In this context, one of the problems that have been considered by researchers in recent years is routing and scheduling for staff members [1]. These problems include the combination of nurses, patients, and nurses' routing and scheduling to patients, which complicates the problem and causes more attention of researchers.

A home healthcare problem is an extension of a vehicle routing problem with time windows [6] that aims to determine the optimal plan for staff to visit the patients considering many limitations, such as the availability of patients and nurses' time windows, staff workload, and skill requirement [7].

© IFIP International Federation for Information Processing 2021
Published by Springer Nature Switzerland AG 2021
A. Dolgui et al. (Eds.): APMS 2021, IFIP AICT 631, pp. 373–382, 2021.
https://doi.org/10.1007/978-3-030-85902-2_40

The primary purpose of this paper is to develop an optimization model considering the allocation of nurses to patients with suspected or confirmed COVID-19. Due to the change of lifestyle during the COVID-19 outbreak, it is obvious that nurses are not willing to service patients with suspected or confirmed COVID-19. Therefore, the satisfaction level of nurses has changed and affects the costs of the organization. The rest of this paper is organized as follows. Sections 2 and 3 present the literature review and the problem description. Section 4 solves an illustrative problem by the developed model. Finally, Sect. 5 concludes this research and presents opportunities for further research.

## 2 Literature Review

As mentioned in the previous section, the complexity, novelty, and increasing demand of a home healthcare routing and scheduling problem (HHCRSP) cause more attention to researchers. Most researchers have focused on the cost and travel time of staff, and a few numbers have focused on minimizing the number of staff. Braekers et al. [6] proposed a bi-objective model in a home healthcare scheduling problem (HHCSP) to minimize the routing and overtime costs and client inconvenience. They considered the hard time windows, preferences of staff and patients, and overtime of staff. Moussavi et al. [8] presented a multi-period HHCSP to minimize travel time or travel distance. Their model considered continuing care, time windows of staff and patients, determined staff numbers, and the route each staff should travel each period. The Gurobi MILP solver was used to solve and a meta-heuristic algorithm based on the decomposition of the formulation was used to simplify. Mankowska [9] presented a daily scheduling model for HHCSPs to minimize travel costs or maximize service quality. Their model involves synchronization and qualifications of staff members, temporal dependencies, and workload balance. To solve the model, a powerful heuristic method based on a sophisticated solution was proposed.

A few numbers of papers deal with the minimum number of staff, in a long-term period, it is more important than a single period [2]. In this regard, Allaoua et al. [10] developed a single-period integer linear programming model to minimize the number of staff assigned and considering skills and available time windows for staff. They developed a matheuristic method for large-sized problems. Hiermann et al. [11] proposed a single-period model for home healthcare problems to find the most convenient assignment for staff and patients at the same time. Then, they suggested meta-heuristics, such as memetic algorithm (MA), scatter search (SS), simulated annealing (SA) algorithm, and variable neighborhood search (VNS) to solve the model.

Other criteria that can be observed in the objective function of some papers are two types of satisfaction; staff and patients [12]. Rest and Hirsch [13] proposed an HHCSP to minimize travel time and waiting time of staff by considering patient hard time window, maximum working time, and rest time for staff members to satisfy staff. To solve this problem, they suggest tabu search (TS) algorithm. An example of patient satisfaction can find in Wirnitzer et al. [14] who proposed five mixed-integer programming formulations with different objective functions by considering the continuity of care, breaks, shift rotations, maximum working times, and patient or nurse preferences.

Lean management is one of the latest management systems in healthcare [15]. The results of lean implementation in management include reducing errors, prioritizing patients, and providing quality and more effective services to patients [16, 17]. One of the most important goals of lean is to save time and reduce waiting time in a queue [16, 18] and reduce costs, and increase productivity [16]. For example, Improta et al. [19] reduced the patient waiting time by using lean in the emergency room. Other goals of implementing this improvement in organizations have suggested maximizing the value and eliminating waste and possible response to the need to respond to rapid change [20, 21].

Young et al. [22] is one of the first lean studies to improve patient care. The authors introduce the three industrial approaches of pure thinking, the theory of limitations, and Six Sigma, how these three approaches relate to health care. Then they concluded that for these approaches to be effective, good leadership and participation of employees are needed. Also, by considering these approaches, they concluded that there are many standard features between them, and all of them need to recognize bottlenecks and use a suitable approach to eliminate them. Kim et al. [23] introduced a hospital that provides more helpful and better patient care using lean management. They also addressed the obstacles to using lean methods and ways to overcome them, But Poksinska [24] implemented and described the obstacles and problems.

Staff members are our valuable resources, especially in the presence of covid-19 so we should care for them most. Our main contribution is to reach an assignment that considers maximum cost and minimum satisfaction of staff who assign to covid-19 suspicious patients.

## 3   Problem Definition

In this section, a multi-period mathematical model is presented. A number of patients and staff, cost of assigned staff to the patients with suspected or confirmed COVID-19, and satisfaction of this staff are assumed. The objective of this problem is to minimize the travel time of each staff. The following notations are employed to develop a model for the HHCRSP.

**Sets**

| | |
|---|---|
| $i$ | Set of employees |
| $j$ and $k$ | Set of patients |
| $p$ | Set of periods per day |
| $d$ | Set of days |

**Parameters**

| | |
|---|---|
| $ni$ | Number of employees |
| $nj$ | Number of patients |
| $nd$ | Number of considered days |
| $npi$ | Maximum number of periods each staff can work |

$MS_{jpd}$    Matrix of the care services needed by the patients.
$CS_{jpd}$    Matrix of patients with suspected or confirmed COVID-19.
$DS_{jk}$    Travel time between patients $j$ and $k$.
$\alpha_i$    Disutility of giving care service of nurse $i$ in the vicinity of patients with suspected or confirmed COVID-19.
$Cost_i$    Extra cost for each employee if he/she is assigned to patients with suspected or confirmed COVID-19.
$Cost\ o$    Maximum cost of visiting patients with suspected or confirmed COVID-19.
$SS$    Minimum satisfaction level of the staff who assign to patients with suspected or confirmed COVID-19

## Variables

$X_{ijpd}$    1 if staff member $i$ is assigned to patient $j$ in period $p$ of day $d$; 0, otherwise.
$X'_{ijpd}$    1 if staff member $i$ is assigned to covid-19 suspicious patients $j$ in period $p$ of day $d$; 0, otherwise.
$DI_i$    Time of traveling staff member $i$ during all days.
$SA_{id}$    Satisfaction level of staff member $i$ during day $d$.
$SD_i$    Satisfaction level of staff member $i$ during all days.
$TI_{ijkd}$    1 if staff member $i$ has to visit patient $k$ after $j$ in day $d$; 0, otherwise.
$DID_{id}$    Time of traveling staff member $i$ in day $d$.
$DTI$    Total travel time of all staff members
$CTT_{id}$    Cost of staff member $i$ who assigned to patients with suspected or confirmed COVID-19 during day $d$.
$CO$    Cost of all staff's assignments to patients with suspected or confirmed COVID-19
$SO$    Satisfaction level of all staff members assigned to patients with suspected or confirmed COVID-19

The proposed mathematical programming model of the HHCRSP is presented as follows:

$$MinZ = DTI \tag{1}$$

s.t.

$$\sum_{i=1}^{ni} X_{ijpd} = MS_{jpd} \qquad \forall p \in P, j \in J, d \in D \tag{2}$$

$$\sum_{j=1}^{nj} X_{ijpd} \leq I \qquad \forall p \in P, j \in J, d \in D \tag{3}$$

$$\sum_{j=1}^{nj} \sum_{p=1}^{h} X_{ijpd} \leq npi \qquad \forall i \in I, d \in D \tag{4}$$

$$X_{ijpd} + \sum_{k \in J-j} X_{ik(p+1)d} \leq 1 \quad \forall i \in I, j \in J, p \in, d \in D \tag{5}$$

$$X_{ijpd} + \sum_{l \in I-i} X_{lj(p+1)d} \leq 1 \quad \forall i \in I, j \in J, p \in, d \in D \tag{6}$$

$$X'_{ijpd} \leq X_{ijpd} \times \left(2 - X_{ijpd}\right) \quad i \in I, j \in J, p \in, d \in D \tag{7}$$

$$\sum_{i=1}^{ni} X'_{ijpd} = CS_{jpd} \quad \forall j \in J, p \in, d \in D \tag{8}$$

$$\sum_{i=1}^{ni} \sum_{j=1}^{nj} TI_{ijkd} = 1 \quad \forall k \in J, d \in D \tag{9}$$

$$\sum_{j=1}^{nj} TI_{ikjd} = \sum_{j=1}^{n} TI_{ijkd} \quad \forall i \in I, k \in J, d \in D \tag{10}$$

$$TI_{ijkd} \leq X_{ijpd} \quad \forall i \in I, j \in J, k \in J, d \in D \tag{11}$$

$$DID_{id} = \sum_{j=1}^{nj} \sum_{k=1}^{nj} DIS_{jk} \times T_{ijkd} \quad \forall i \in I, d \in D \tag{12}$$

$$DI_i = \sum_{d=1}^{nd} DID_{id} \quad \forall i \in I \tag{13}$$

$$DTI = \sum_{d=1}^{nd} \sum_{i=1}^{m} DID_{id} \tag{14}$$

$$\sum_{j=1}^{nj} \sum_{p=1}^{npi} \alpha_i X'_{ijpd} + X_{ijpd} - X'_{ijpd} = SA_{id} \quad \forall i \in I, d \in D \tag{15}$$

$$SD_i = \sum_{d=1}^{t} SA_{id} \, \forall i \in I \tag{16}$$

$$\sum_{d=1}^{nd} \sum_{i=1}^{ni} SA_{id} \geq SS \tag{17}$$

$$\sum_{d=1}^{nd} \sum_{i=1}^{ni} SA_{id} = SO \tag{18}$$

$$\sum_{j=1}^{nj} \sum_{p=1}^{npi} cost_i \times X'_{ijpd} = CTT_{id} \qquad \forall i \in I, d \in D \tag{19}$$

$$\sum_{d=1}^{nd} \sum_{i=1}^{ni} CTT_{id} \leq Costo \tag{20}$$

$$\sum_{d=1}^{nd} \sum_{i=1}^{ni} CTT_{id} = CO \tag{21}$$

The objective function (1) minimizes the total travel time by all the staff members during a day. Eq. (2) ensures that only one staff member can assign to each patient who needs care services. Eq. (3) ensures that each staff member can only assign to one patient. Eq. (4) implies that each staff member cannot work more than the maximum number of periods per day. Eq. (5) prevents that staff members visit two different patients in two consecutive periods because the model considers travel time between visits. Eq. (6) implies continue of care if a patient needs more than one period. Eqs. (7) and (8) imply that employees are assigned to patients with suspected or confirmed COVID-19. Eqs. (9)–(11) indicate the sequence of the patients who are visited by an employee. Eq. (12) calculates the travel time of staff members in a day. Eqs. (13) and (14) calculate the travel time for staff member $i$ during all days and the total travel time of all staff. Eq (15) and (16) indicate the satisfaction level of staff member $i$ during a day and the satisfaction level of staff member $i$ during all days. Eq. (17) controls the minimum satisfaction level of the staff members assigned to the patients with suspected or confirmed COVID-19. Eq. (18) computes the satisfaction level of all staff members assigned to the patients with suspected or confirmed COVID-19. Eq. (19) calculates the cost of all staff's assignments to the patients with suspected or confirmed COVID-19 during a day. Eq. (20) indicates the budget constraint of visiting requirements for the patients with suspected or confirmed COVID-19. Eq. (21) calculates the cost for all staff members assigned to the patients with suspected or confirmed COVID-19.

## 4   Numerical Experiments

In this section, the proposed model is solved by GAMS software using the BARON solver for small-sized instances.

### 4.1   Case Study and Result

To validate the model introduced in this study, the model is applied on small-sized instances that each day includes 12 periods. Table 1 represents the value of parameters for each test problem. All the data are selected random numbers between 0 and 30.

## 4.2 Sensitivity Analysis

The model is solved by changing some parameters to determine sensitivity analysis. The number of patients with suspected or confirmed COVID-19 in the fourth instance of the last section is modified to investigate the changes in cost and satisfaction level of staff members servicing the patients with suspected or confirmed COVID-19. The computational results are shown in Figs. 1, 2, 3 and 4, which reveal that by increasing the number of patients with suspected or confirmed COVID-19, the cost of all staff's assignments to these people increases, and the satisfaction level of staff members will decrease.

**Table 1.** Description of each test problem

| No. of days | No. of patients | No. of staff | DTI | SO | CO | Cycle time |
|---|---|---|---|---|---|---|
| 1 | 4 | 2 | 63 | 1.11 | 3.5 | < 0 |
| 1 | 8 | 3 | 186 | 2.86 | 17.75 | 11s |
| 1 | 12 | 5 | 117 | 6.7 | 36.3 | 918s |
| 1 | 16 | 7 | 474 | 4.56 | 41.19 | 3240s |
| 2 | 4 | 2 | 148 | 2.44 | 4.5 | < 0 |
| 2 | 8 | 3 | 104 | 3.19 | 24.13 | 63s |
| 2 | 12 | 5 | 89 | 3.86 | 37.47 | 1920s |
| 2 | 16 | 7 | 397 | 4.32 | 46.25 | 8944s |
| 3 | 4 | 2 | 123 | 3.2 | 8.27 | < 0 |
| 3 | 8 | 3 | 179 | 2.5 | 54.46 | 71s |
| 3 | 12 | 5 | 517 | 1.95 | 98.72 | 3856s |
| 3 | 16 | 7 | 945 | 2.5 | 189.3 | 5470s |

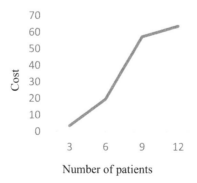

**Fig. 1.** Effects of changing the number of patients on the staff's cost ($D = 1$)

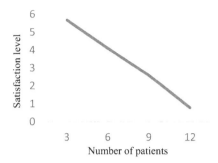

**Fig. 2.** Effects of changing the number of patients on the staff's satisfaction level ($D = 1$)

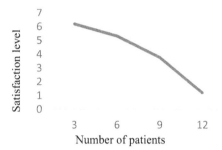

**Fig. 3.** Effects of changing the number of patients on the staff's cost ($D = 2$)

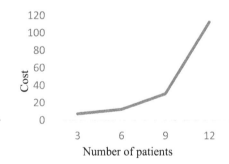

**Fig. 4.** Effects of changing the number of patients on the staff's satisfaction level ($D = 2$)

## 5   Conclusion

In this study, a mathematical model was proposed for a home healthcare routing and scheduling problem (HHCRSP). The impact of the Covid-19 pandemic is considered in the proposed model by controlling the cost and minimum satisfaction level of staff members, who are assigned to the patients with suspected or confirmed COVID-19. To validate the model performance, a set of small-sized instances were solved using GAMS software and a sensitivity analysis was conducted. For future studies, the model can be extended to a multi-objective model by considering more preferences of patients and staff. The model can be solved by heuristics and meta-heuristics for large-sized problems.

# References

1. Khodabandeh, P., Kayvanfar, V., Rafiee, M., Werner, F.: A bi-objective home health care routing and scheduling model with considering nurse downgrading costs. Int. J. Environ. Res. Public Health **18**(3), 900 (2021)
2. Fikar, C., Hirsch, P.: Home health care routing and scheduling: a review. Comput. Oper. Res. **77**, 86–95 (2017)
3. Shahnejat-Bushehri, S., Tavakkoli-Moghaddam, R., Momen, S., Ghasemkhani, A., Tavakkoli-Moghaddam, H.: Home health care routing and scheduling problem considering temporal dependencies and perishability with simultaneous pickup and delivery. IFAC-PapersOnLine **52**(13), 118–123 (2019)
4. Euchi, J.: Optimising the routing of home health caregivers: can a hybrid ant colony metaheuristic provide a solution? Br. J. Healthc. Manag. **26**(7), 192–196 (2020)
5. Euchi, J.: Do drones have a realistic place in a pandemic fight for delivering medical supplies in healthcare systems problems? Chin. J. Aeronaut. **34**(2), 182–190 (2021)
6. Braekers, K., Hartl, R.F., Parragh, S.N., Tricoire, F.: A bi-objective home care scheduling problem: analyzing the trade-off between costs and client inconvenience. Eur. J. Oper. Res. **248**(2), 428–443 (2016)
7. Goodarzian, F., Abraham, A., Fathollahi-Fard, A.M.: A biobjective home health care logistics considering the working time and route balancing: a self-adaptive social engineering optimizer. J. Comput. Des. Eng. **8**(1), 452–474 (2021)
8. Moussavi, S.E., Mahdjoub, M., Grunder, O.: A matheuristic approach to the integration of worker assignment and vehicle routing problems: application to home healthcare scheduling. Exp. Syst. Appl. **125**, 317–332 (2019)
9. Mankowska, D.S., Meisel, F., Bierwirth, C.: The home health care routing and scheduling problem with interdependent services. Healthc. Manag. Sci. **17**(1), 15–30 (2013). https://doi.org/10.1007/s10729-013-9243-1
10. Allaoua, H., Borne, S., Létocart, L., Calvo, R.W.: A matheuristic approach for solving a home health care problem. Electr. Notes Discrete Math. **41**, 471–478 (2013)
11. Hiermann, G., Prandtstetter, M., Rendl, A., Puchinger, J., Raidl, G.R.: Metaheuristics for solving a multimodal home-healthcare scheduling problem. CEJOR **23**(1), 89–113 (2013). https://doi.org/10.1007/s10100-013-0305-8
12. Cissé, M., Yalçındağ, S., Kergosien, Y., Şahin, E., Lenté, C., Matta, A.: OR problems related to Home health care: a review of relevant routing and scheduling problems. Oper. Res. Healthc. **13**, 1–22 (2017)
13. Rest, K.-D., Hirsch, P.: Daily scheduling of home health care services using time-dependent public transport. Flex. Serv. Manuf. J. **28**(3), 495–525 (2015). https://doi.org/10.1007/s10696-015-9227-1
14. Wirnitzer, J., Heckmann, I., Meyer, A., Nickel, S.: Patient-based nurse rostering in home care. Oper. Res. Healthc. **8**, 91–102 (2016)
15. Mazzocato, P., et al.: Complexity complicates lean: lessons from seven emergency services. J. Health Organ. Manag. **28**(2), 266–288 (2014)
16. Mazzocato, P., et al.: How does lean work in emergency care? A case study of a lean-inspired intervention at the Astrid Lindgren Children's hospital, Stockholm Sweden. BMC Health Serv. Res. **12**(1), 1–13 (2012)

17. Mazzocato, P., Savage, C., Brommels, M., Aronsson, H., Thor, J.: Lean thinking in healthcare: a realist review of the literature. BMJ Qual. Saf. **19**(5), 376–382 (2010)
18. Hydes, T., Hansi, N., Trebble, T.M.: Lean thinking transformation of the unsedated upper gastrointestinal endoscopy pathway improves efficiency and is associated with high levels of patient satisfaction. BMJ Qual. Saf. **21**(1), 63–69 (2012)
19. Improta, G., et al.: Lean thinking to improve emergency department throughput at AORN Cardarelli hospital. BMC Health Serv. Res. **18**(1), 1–9 (2018)
20. Womack, J.P., Byrne, A.P., Fiume, O.J., Kaplan, G.S., Toussaint, J.: Going Lean in Health Care. Institute for Healthcare Improvement, Cambridge (2005)
21. Jones, D., Mitchell, A.: Lean Thinking for the NHS, vol. 51, NHS Confederation, London (2006)
22. Young, T., Brailsford, S., Connell, C., Davies, R., Harper, P., Klein, J.H.: Using industrial processes to improve patient care. BMJ **328**(7432), 162–164 (2004)
23. Kim, C.S., Spahlinger, D.A., Kin, J.M., Billi, J.E.: Lean health care: what can hospitals learn from a world-class automaker? J. Hosp. Med. **1**(3), 191–199 (2006)
24. Poksinska, B.: The current state of Lean implementation in health care: literature review. Qual. Manag. Healthc. **19**(4), 319–329 (2010)

# Lean Transformation in Healthcare: A French Case Study

Anne Zouggar Amrani[(⊠)], Benjamin Garel, and Bruno Vallespir

Univ. Bordeaux, CNRS, IMS, UMR 5218, 33405 Talence, France
{anne.zouggar,bruno.Vallespir}@ims-bordeaux.fr,
Benjamin.GAREL@chu-martinique.fr

**Abstract.** Evidence from current publications mainly US, UK, Canada and recently Europe has highlighted the relevancy of Lean healthcare in improving patient pathway efficiency. The aim of this case study is to demonstrate how the French public hospital succeeded at using Lean thinking, and how, using the Lean tools with the team's involvement, it improved the quality of health services delivered to the patient. Emergency complaints dropped by 50%, length of stay reduced by 30%. This paper highlights how Lean was implemented to revolve around the patient, creating valuable results which improved the global performance, whilst maintaining employees' involvement and satisfaction. Two Lean experiences in hospital units are described showing the reduction of wastes and the possible monitoring of KPI to sustain Lean pillars Jidoka and JIT. These two first Lean experiences are not alone, more hospital services have been experiencing Lean approaches and more results will be communicated soon in wider publication.

**Keywords:** Lean implementation · Lean practices · Innovative organization · Lean drivers · Case study

## 1 Introduction

The first documented implementations of Lean thinking in healthcare were in the National Health Service NHS in the UK (2001) and in Virginia Mason Production System VMPS Hospital in the USA (2002). Although the Lean healthcare literature is rich in case studies such as from the Virginia Mason Medical Center USA hospital [1] ThedaCare; Brazilian hospitals [2] Swedish hospitals [3]; Denmark [4], Spain [5], India [6], and the UK [7, 8], successful French case studies are almost inexistent. [9] was the only French publication to deal with Lean thinking, and it approached the concept from an engineering view point. Furthermore, the study was a literature review, and not a case study. The aim of this article is to fill this void and to disseminate the impact of Lean implementation in French hospitals, which is gaining national interest. Grenoble hospital is among the first successful case studies in France, reporting significant results and findings. The objective, beyond the presentation and description of the case study, is to analyze how the applied Lean experiences utilized the Lean principles, how they contribute to waste elimination, and what the main drivers of success were, to help future healthcare interventions to successfully implement lean management.

© IFIP International Federation for Information Processing 2021
Published by Springer Nature Switzerland AG 2021
A. Dolgui et al. (Eds.): APMS 2021, IFIP AICT 631, pp. 383–390, 2021.
https://doi.org/10.1007/978-3-030-85902-2_41

## 2   Literature Review

When dealing with Lean in the organization of healthcare, patient satisfaction has been reported to be the main target to keep in mind – representing the client from the Lean point of view. The Lean transformation should be built around the patient and be made efficient for the patient [10]. Eliminating waste from patient pathways is a powerful initial analysis to improve the delivery of healthcare processes [11] and improve clinical processes [12, 13], reduce errors and inappropriate procedures [14], paying careful attention to the transformation of the mind-set and the organization. [15] evoke the importance of creating better service for patients through implementing "improvement culture", to create a mind-set focused on quality improvement. The study does not merely deal with cutting costs, but above all deals with the quality of processes, of care services, ensuring that quick response with time to care is qualified as the primary objective of healthcare services whilst maintaining security, safety for the patient along their pathways and reducing length of stay [7].

When trying to integrate Lean thinking, no matter the sector concerned (manufacturing, service, healthcare), the dominating idea in leading successful Lean conversion is "the cultural mindset" to put into place rather than the "tool box" [15, 16]. Moreover, leadership becomes a prevailing factor and consistent driving force that can shape an adequate problem-solving environment [8, 17, 18].

Authors in [8, 19, 20] have argued the need for new research and findings about Lean implementation. [9] found that 59% of articles published that deal with Lean in the healthcare sector do not describe, or only very little, the content of the implementation approach (qualified as having low methodological maturity), only results are reported without details from an engineering point of view. Articles do not provide any information to understand how it was applied. Among articles studying the Lean approach, a majority (70.7%) have struggled to provide a structured method and effective management guidance [9, 16, 21].

## 3   Problem Statement

The research methodology considered here is a qualitative single case study. It is an interesting and in-depth exploration of Lean healthcare to thoroughly address the following research questions: *How was lean introduced, implemented and sustained in public French hospital? Author in* [22] highlighted that case study design becomes relevant to focus studies on the questions "how" and "why". Using a unique case study, [23] allowed a review of the organization over an extended time period; it is not always possible to capture the changes using an alternative methodology. [24] analyzed more than 109 papers on Lean healthcare and found that the Lean implementation process and its sustainability remain under investigation.

The objective of this case study is to browse the different Lean techniques that hospitals have experienced in the Operating room and the Sterilization Service Department with the aim of revealing implementation with a possibility of transposition to other similar healthcare contexts.

# 4   Lean implementation in Grenoble Hospital

The Grenoble hospital is a middle-sized public French university hospital comprising 8000 medical and administrative staff, with almost 2000 beds and a budget of 600 million euros. It provides around 86 000 emergency admissions, 35 000 operations in operating rooms and 90 000 hospitals stays per year.

**Data Collection.** Data were essentially acquired from the former Quality/Performance Director of the hospital. Conference documents, implementation plans from internal elements were collected, reviewed and analyzed to track the necessary actions. Moreover, interviews with the quality director of the hospital reinforced the case description to allow a documented publication.

## 4.1   Sterilization Service Department (SSD): Jidoka, Crosstraining, Visual Management, and Pull Flow Implementation

The SSD is in charge of cleaning/sterilizing Medical Devices (MD) and surgery instruments required by the surgical teams in the Operating Rooms (OR). The SSD has to ensure the availability of MD to 120 care units and 39 ORs in the hospital. The materials are provided in different boxes with an average use rate of 2.5 boxes by surgical act. In fact, we reported 300 sterilized boxes/day (boxes of MD) and 800 sterilized bags/day (bags for specific instruments) provided for the different care units and ORs. The most recurring problem was the *unavailability of sterile surgical materials* in ORs. A long and variable Instrument Availability Delivery (IAD) was noticed. The IAD is the time between the moment when the used (dirty) instruments leave the OR and the moment when they come back again, clean and sterile. Besides the postponement of some operations by surgeons due to the unavailability of materials, there is also patient cancellation which causes re-planning. Usually several boxes (A type and B type) are opened at the same time making the remaining instruments non sterile, requiring again a sterilization cycle. When asked for the origin of this shortcoming and such high IADs, the SSD revealed that the boxes of instruments arrived at the SSD mixed up between the two box types (instruments of A type and B type) and it was time consuming to sort them before starting. The sterilization process begins with a full placement of MD to the correct place (type A or B), which is obviously an No Value Added (NVA) activity. The first improvement was to ensure the *pre-sorting of the MD* very early in the process, meaning "as soon as the operation is done" to ensure the Jidoka principle in Lean. It was assigned to the OR nurse who is able according to the standards and classification of tools to put them back in the right box before sending them to the SSD. The initial reaction was skeptical due to the added load of perceived work. The chief nurse plays an important role because after a VSM deployment the significant role of rearranging the instruments early on (by the end of surgery) was revealed to have a high potential for improving the system. Often less than five minutes were required by them to do that, moreover, the

OR nurse was able to recognize the instruments and their boxes quickly. If a nurse of lower level was in the operating block, a standardized document was provided to carry out these verifications. That first action reduces the NVA of sorting at SSD and allows the flow of sterilization to start easily and without mix ups. Another problem was the *de-synchronization of the cleaning process.* Some cleaning "operators" were launching different boxes of various MD in parallel creating a mixture of orders. An increased amount of Work In Progress (WIP) was noticed at the conditioning area because the last operator of the flow had too much work rearranging the MD per patient requirements. It created a queue and became bottlenecked, preventing the MD from arriving on time to the OR. It took them a long time to rearrange the MD per patient and constitute the batches to send to the OR. The generated Lean corrective action was to stop the cleaning process and deploy a multi-skills cleaning agent coming from upstream phases (cleaning unit) to help the rearrangement of the MD with the conditioning agent. Identified places with *visual paintings* on the ground based on Kanban system were also implemented to ensure the right positioning of the WIP coming to the conditioning area. When the cross-trained operators come from the cleaning area, they clean their hands, wear specific material and enter the conditioning area to rearrange the materials and the boxes. By increasing competencies with training, the multi-skills agent becomes able to increase the rate of instruments rearrangement per box at the conditioning area, smoothing the flow of outputs.

### 4.2  Biology Laboratory: Kanban and Poka Yoke Implementation

The Biology lab is in charge of more than 80% of analysis. A recurrent problem was a lack of reagents for vital medical analysis generating rushed orders and inducing stress to employees because of the criticality of some patient cases and the absolute necessity to obtain the reagents. One of the main pillars of the Toyota System is the "Pull System". How to create a continuous flow using reagents without ordering too much, and without forgetting to order at the right moment and undergo shortages? The Pull System taught us how to inform with visual signals the need to order. Kanban was suitable (Fig. 1): two bins were installed; one with the first reagents to use, and the second behind with reagents to use only when the first one is empty, the signal to launch the order. Another problem remained: as the reagents can be checked by many people in the hospital, often the parallel use of reagents in both bins makes the system fail. The Lean team idea was to put "tape with a Kanban card" on the second bin to avoid parallel use of the bins and ensure the intentional opening of the bin through removing the tape and putting the Kanban card in the collector so that the logistician can transmit the order for supply.

The cost dimension was rarely the objective (Table 1), even obvious financial gains became subsequent to restructured, organized and efficient processes in services. Keeping the patient focus leads the healthcare leaders to absorb and treat the patient flow. Being Lean means achieving wastes reduction as showed in Table 1. We sum up the different waste that was encountered.

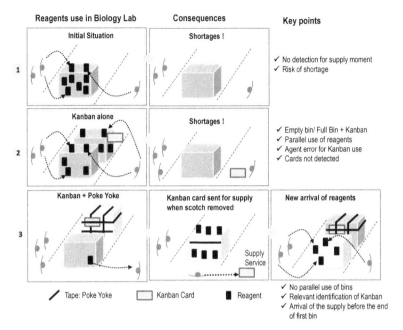

**Fig. 1.** Kanban and Poka Yoke in biology lab for continuous flow

**Table 1.** Wastes solved by lean transformation and KPI monitoring

| Service Unit | 08 Wastes ( W1..W8) | |
|---|---|---|
| Sterilization service | - Inventory of Mixed medical devices (W2)<br>- Waiting time at OR for medical devices (W4)<br>- Overproduction at cleaning unit (W5)<br>- Cleaning agent ability not fully used (W8) | W1: Transportation<br>W2: Inventory<br>W3: Motion<br>W4: Waiting |
| Biology Lab | - Stocks of reagents to avoid shortages (W2)<br>- Motions for getting reagents in emergency (W3)<br>- Risk of defective care if analysis missing (W7) | W5: Over Production<br>W6: Over processing<br>W7: Defect<br>W8:Non utilized Talents |

The different wastes were palliated thanks to the different Lean features of the project. The analysis as demonstrated in this paper did not start with waste analysis (like it is often done in the industry). In hospital, patient focus is the prevalent objective around which the analysis is gravitated. When different Lean actions are implemented through the described Lean experiences, we naturally witness a decrease of those different wastes. Moreover, the used KPI to monitor are really sustaining the two Lean pillars: Jidoka (doing the quality at the origin) and JIT (Pull system to not overload). The Table 2 is summarizing this idea.

**Table 2.** KPI monitoring based on the two pillars of Lean

| Lean Pillars | KPI sustaining Lean Pillars | |
| --- | --- | --- |
| | Sterilization Servic | Biology Lab |
| JIDOKA | • Medical devices reliability<br>• Increase rearranged medical devices | • Reagents availability<br>• No more shortages |
| JUST IN TIME | • Decrease Instruments Availability Delay (IAD)<br>• Speed up rearranged medical devices<br>• Decrease OR delay | • Relevant moment of reagents supply |

# 5   Conclusion

Before the Lean implementation, 70% of boxes were rearranged at the SSD and 30% underwent rearrangement in the OR. After, the SSD became an efficient service providing 98% of rearranged MD to the hospital. The WIP is now under control with visual management and paintings on the ground. U cells were suggested to respect the flow direction from the cleaning to sterilizing process. The variability of the IAD decreased significantly ($-50\%$), the quality of rearrangement increased (no missing instruments, to the right assignments), the non-conformity rate dropped by 30%, and the increase of SSD capacity rose about 10%. To conclude with the necessary changes in the SSD, it is important to remember that surgery cancellation due to missing MD is one of the main threats in the hospital.

In Biology Laboratory, the kanban solution was interesting required only a little training using one biologist and two agents to lead the team to understand the system and sustain the applicability of the Kanban in daily use. No more shortages were noted in the following years. Kanban and Poka Yoke (tape is error proof) installed in the Biology laboratory resulted in a much smoother ordering cycle, reducing mental stress for agents and no missing reagents for the whole hospital.

The strategic target in hospital is focusing on patient care. Around the patient are built the provided care processes and activities with an identification of different blocking points that slow down the patient flow either for a quality problem, security or human and technical resources availability. Interestingly, the Lean principles were relevantly adapted and transmitted through training to the staff. For hospital leaders and practitioners, this paper comes as a testimony and feedback containing foremost insights to sustain the Lean healthcare approach. From the Gemba to the Lean techniques and lean principles application, waste elimination, to finish with monitoring KPI, those steps can be used as a benchmark for other healthcare managers and practitioners in service field as Healthcare.

# References

1. Kenney, C.: Transforming Health Care: Virginia Mason's Pursuit of the Perfect Patient Experience. Productivity Press, New York (2011)
2. Toussaint, J., Gerard, R.: On the mend: revolutionizing healthcare to save lives and transform the industry. Lean enterprise institute Jun 15, 2010 - Business & Economics -180 (2010)
3. Holden, R.J., Hackbart, G.: From group work to teamwork: a case study of "lean" rapid process improvement in the ThedaCare Information Technology Department. IIE Trans. Healthc. Syst. Eng. **2**(3), 190–201 (2012)
4. Karstoft, J., Tarp, L.: Is Lean Management implementable in a department of radiology? Insights Imag. **2**(3), 267–273 (2011). https://doi.org/10.1007/s13244-010-0044-5
5. Boronat, F., Budia, A., Broseta, E., Ruiz-Cerdá, J.L., Vivas-Consuelo, D.: Application of lean healthcare methodology in a urology department of a tertiary hospital as a tool for improving efficiency. Actas Urológicas Españolas (English Edition) **42**(1), 42–48 (2018)
6. Prasanna, S.G.: Process improvement using value stream mapping – a lean thinking in Indian health care sector. J. Health Med. Nurs. **24**, 130–135 (2016)
7. Nicolay, C.R., et al.: Systematic review of the application of quality improvement methodologies from the manufacturing industry to surgical healthcare. Br. J. Surg. **99**(3), 324–335 (2012)
8. Leite, H., Radnor, N.B.Z.: Beyond the ostensible: an exploration of barriers to lean implementation and sustainability in healthcare. Prod. Plan. Control **31**(1), 1–18 (2020)
9. Curatolo, N., Lamouri, S., Huet, J.C., Rieutord, A.: A critical analysis of lean approach structuring in hospitals. Bus. Process. Manag. J. **20**(3), 433–454 (2014)
10. Dahlgaard, J.J., Pettersen, J., Dahlgaard-Park, S.M.: Quality and lean health care: a system for assessing and improving the health of healthcare organisations. Total Qual. Manag. Bus. Excell. **22**(6), 673–689 (2011)
11. Poksinska, B.: The current state of Lean implementation in health care: literature review. Qual. Manag. Healthc. **19**(4), 319–329 (2010)
12. Kaplan, G.S., Patterson, S.H., Ching, J.M., Blackmore, C.C.: Why lean doesn' work for everyone. BMJ Qual. Saf. **23**(12), 970–973 (2014)
13. Papadopoulos, T.: Continuous improvement and dynamic actor associations: a study of lean thinking implementation in the UK National Health Service. Leadersh. Health Serv. **24**(3), 207–227 (2011)
14. Young, T., Brailsford, S., Connell, C., Davies, R., Harper, P., Klein, J.H.: Using industrial processes to improve patient care. Br. Med. J. **328**(7432), 162 (2004)
15. Stentoft, J., Freytag, P.V.: Improvement culture in the public mental healthcare sector: evaluation of implementation efforts. Prod. Plan. Control **31**(7), 540–556 (2020)
16. Dobrzykowski, D.D., McFadden, K.L., Vonderembse, M.A.: Examining pathways to safety and financial performance in hospitals: a study of lean in professional service operations. J. Oper. Manag. **42**, 39–51 (2016)
17. Longoni, A., Pagell, M., Johnston, D., Veltri, A.: When does lean hurt? An exploration of lean practices and worker health and safety outcomes. Int. J. Prod. Res. **51**(11), 3300–3320 (2013)
18. Longenecker, C.O., Longenecker, P.D.: Why hospital improvement efforts fail: a view from the front line. J. Healthc. Manag. **59**(2), 147–157 (2014)
19. Costa, L.B.M., Filho. M.G.: Lean healthcare: review, classification and analysis of literature. Prod. Plan. Control **27**, 823–836 (2016)
20. Amrani, A.A., Ducq, Y.: Lean practices implementation in aerospace based on sector characteristics: methodology and case study. Prod. Plan. Control **31**, 1706197 (2020). https://doi.org/10.1080/09537287.2019.1706197

21. Peltokorpi, A.: How do strategic decisions and operative practices affect operating room productivity? Healthc. Manag. Sci. **14**(4), 370–382 (2011)
22. Yin, R.K.: Case Study Research: Design and Methods, vol. 5, Sage, Thousand Oaks (2009)
23. Zhang, S., Bamford, D., Moxham, C., Dehe, B.: Strategy deployment systems within the UK healthcare sector: a case study. Int. J. Product. Perform. Manag. **61**(8), 863–880 (2012)
24. D'Andreamatteo, A., Ianni, L., Lega, F., Sargiacomo, M.: Lean in healthcare: a comprehensive review. Health Policy **119**(9), 1197–1209 (2015)

# A Robust Home Health Care Scheduling and Routing Approach with Time Windows and Synchronization Constraints Under Travel Time and Service Time Uncertainty

Salma Makboul[1]($\boxtimes$), Said Kharraja[2], Abderrahman Abbassi[3], and Ahmed El Hilali Alaoui[4]

[1] Modelling and Mathematical Structures Laboratory, FST Fez, SMBA University, Fez, Morocco
salma.makboul@usmba.ac.ma
[2] University of Lyon, UJM-Saint-Etienne, LASPI, Roanne, France
[3] Faculty of Sciences, Marrakesh, Morocco
[4] Euromed University, 32 Rue Meknes, Fez, Morocco

**Abstract.** Home health care (HHC) services represent a set of medical services given to patients at their homes. The patients require a set of care that must be coordinated and treated by skilled caregivers corresponding to their needs. This study proposes an HHC routing and assignment approach based on a mixed-integer linear programming model that aims to minimize total route cost. The HHC approach takes into account a set of HHC specific constraints and criteria. Secondly, we propose a new robust counterpart HHC model under uncertainty based on the well-known budgeted uncertainty set. The robust counterpart HHC model deals with travel and service times uncertainty. The computational results compare the deterministic model with its robust counterpart model. The small and medium instances have been solved using TSP benchmarks with specific data concerning HHC problems. The models have been implemented using ILOG CPLEX Optimization Studio. The computational results of small and medium instances indicated the efficiency of the proposed approach. Robustness analysis of the obtained results was conducted using a Monte Carlo simulation and indicated the price of robustness. The increase of route cost in comparison with the risk of infeasibility shows the importance of the designed robust routes for HHC routing and scheduling problems.

**Keywords:** Home health care · Uncertainty · Mixed integer linear programming · Routing problem · Precedence constraints · Synchronization constraints

© IFIP International Federation for Information Processing 2021
Published by Springer Nature Switzerland AG 2021
A. Dolgui et al. (Eds.): APMS 2021, IFIP AICT 631, pp. 391–402, 2021.
https://doi.org/10.1007/978-3-030-85902-2_42

# 1   Introduction

Home health care (HHC) services are characterized as medical and paramedical services given to patients at home [7]. The patients may have various kinds of required care. Besides, the main benefit of HHC services is the reduction of the hospitalization rate [12]. Reports from the world health organization (WHO) have also suggested that elderly patients tend to receive medical treatment at home rather than traditional hospital care [9]. Due to the costly long-term stay in hospitals [4], it is preferable to let the patients stay as long as possible in their own homes. The routing and scheduling HHC problem is an extension of the vehicle routing problem (VRP) with specific side constraints that make the HHC problem more challenging to solve [5]. The problem consists of scheduling the patients according to the treatment and care needed and caregiver skills. The patient may have specific requests such as gender, language, and so on. [5]. The caregivers must visit the patients within their time window because they may not be available all day. Besides, some patients may need more than one care, and the set of care must be coordinated [14,17]. Sometimes, care required by some patients can involve the presence of more than one caregiver at the same time, which corresponds to the simultaneous synchronization of care [11]. The vast majority of the studies considered knowing the route travel time value in advance, but what happens is the opposite. The travel time can be affected by weather, traffic congestion, stop-and-go movement, and so on. Hence, we also must take into account the uncertainties to build efficient planning. The service time is related to the skill of caregivers and expertise or patient health. Therefore, the service time can also be affected by uncertainties.

The HHC problem has been the subject of several studies in the literature. For recent reviews of the HHC planning and scheduling problems, the reader is referred to [5,8,15]. The HHC process is divided into three decision levels: long-term strategic [2], tactical medium-term, and operational short-term. We cite some papers dealing with the operational level, which refers to the routing and assignment problems [5] and take into consideration the uncertainties. The assignment problem seeks to allocate nurses to patients while considering their skills and workload balancing. The problem of assignment and routing involves scheduling and assigning patient visits to caregivers. The main objective is to define the order and the time during which visits should be carried out [7]. [17] presented modeling for the constrained problem as a fixed partitioning issue and built a "branch and price" algorithm to solve it. Later, [4] presented a mixed-integer linear programming (MILP) model to balance caregivers' workloads and to decrease the waiting time between consecutive visits and considered the concept of pattern. Only a few studies take into account the uncertain factors relevant to the HHC [5]. [10] presented a stochastic setting where uncertainty occurs regarding where and when prospective patients require treatment. Later, [1] proposed a robust optimization (RO) approach for HHC in the chemotherapy context. [18] presented a model for the daily HHC routing and scheduling problem by taking into account travel and service time uncertainty from the RO perspective. In this study, we first propose an MILP model for the deterministic HHC problem to minimize route cost and take into consideration a set of HHC constraints: Time window, the precedence of care, the synchronization of care, the consistency between caregivers' skill and patients' requirement, caregivers lunch breaks, workload restrictions, and coordination between the patients assigned to each caregiver. Secondly, we propose a robust counterpart HHC model that aims to find the "robust" routes traveled by the caregivers. The robust HHC problem refers to the scheduling that represents the minimal cost that increases the chances of having a route feasible in practice when the travel and service times are subject to uncertainty. The routing problem considers travel and service times uncertainties and the most realistic scenario; not all the travel and service times will deviate from their nominal values. To the best of our knowledge, we are the first to study this case on an HHC decision system with a time window and synchronization constraints. We will evaluate how much it will cost to have the routes of the caregivers feasible in practice. The robust model is inspired by the novel robust counterpart vehicle routing problem with time window (VRPTW) presented in [16] based on the well-known budgeted uncertainty set introduced by [3]. The remainder of this paper is presented as follows. Section 2 presents the proposed deterministic model that aims to minimize route costs. Section 3 describes a brief review of the robust optimization (RO) and defines the robust counterpart under travel and service times uncertainties. The computational results are presented in Sect. 4, the deterministic and robust solutions are compared. Finally, Sect. 5 concludes the paper with conclusions and perspectives.

## 2    Problem Description and Mathematical Modeling

We consider a set of patients requiring a heterogeneous set of care. The HHC organization operates in a single area. The planning includes several constraints related to the HHC problems: Time window, the precedence of care, the synchronization of care, the consistency between caregivers' skill and patient requirement, caregivers lunch breaks, workload restrictions, and coordination between the patients assigned to each caregiver. The caregivers have three grades (advanced, medium, and usual skills). The assignment is done according to the required skill (i.e., a caregiver having "grade 2" can be assigned to a patient requiring medium skill and less). Each care has a service time. The care has an order that must be respected which refers to the precedence synchronization [11]. Without loss of generality, we consider the case where at most two caregivers can be required by a patient simultaneously. We first propose the following deterministic MILP, which aims to minimize route cost. A complete directed network $G = (I, E)$ is considered. $E$ defines the set of arcs, and $I$ corresponds to the cares of the patients. Each caregiver starts its tour from node 0 and ends it at node 0, referring to the HHC structure (depot). The lunch break is also referred to as dummy care (node 1) in the model. Therefore, each caregiver should visit care 1.

Parameters

| | | | |
|---|---|---|---|
| $K$ | Set of caregivers | $M$ | High value |
| $S$ | Set of cares that require two caregivers $(S \subset I)$ | $sp_i$ | Skill required by the care $i$ |
| $s_i$ | Service time at care $i$ | $sk_k$ | Skill of the caregiver $k$ |
| $t_{ij}$ | Traveling time from node $i$ to node $j$, $(i,j) \in E$ | $c_{ij}$ | Route cost between care $i$ and care $j$ |
| $[A_i, B_i]$ | Respectively the earliest and the latest service time for the care $i$ | $I_i$ | Is 1 if the care $i$ requires one caregiver and 2 if it requires two caregivers |
| $MAXWL$ | The maximum daily workload for the caregivers | $ord_i$ | Order of precedence of care $i$(if care $i$ has to be planned before care $j$ we have $(ord_i \geq ord_j)$) |

**Objective Function**

$$\min \sum_{(i,j)\in E} \sum_{k\in K} x_{ijk} c_{ij} \tag{1}$$

**Constraints**

$$\sum_{k\in K} y_{ik} = I_j \qquad \forall j \in I| \quad j \neq 0, j \neq 1 \tag{2}$$

$$\sum_{j\in I|j\neq 0} x_{0jk} \leq 1 \qquad \forall k \in K \tag{3}$$

$$\sum_{j\in I|j\neq 0} x_{j0k} \leq 1 \qquad \forall k \in K \tag{4}$$

$$\sum_{j\in I|j\neq 1} x_{1jk} \leq 1 \qquad \forall k \in K \tag{5}$$

$$\sum_{j\in I|j\neq 1} x_{j1k} \leq 1 \qquad \forall k \in K \tag{6}$$

$$y_{jk} = \sum_{i\in I|i\neq j} x_{ijk} \qquad \forall k \in K \quad \forall j \in I \tag{7}$$

$$y_{0k} = 1 \qquad \forall k \in K \tag{8}$$

$$y_{1k} = 1 \qquad \forall k \in K \tag{9}$$

$$\sum_{i\in I} x_{ijk} = \sum_{i\in I} x_{jik} \qquad \forall k \in K \quad \forall j \in I|j \neq 0, j \neq 1 \tag{10}$$

$$ord_i \geq ord_j x_{ijk} \qquad \forall k \in K \quad \forall(i,j) \in E|j \neq 0, i \neq 0, i \neq 1, j \neq 1 \tag{11}$$

$$y_{jk} = 0 \qquad \forall k \in K \quad \forall j \in I| \quad sp_j > sk_k, j \neq 0, j \neq 1 \tag{12}$$

$$AT_{jk} \geq A_j \qquad \forall k \in K \quad \forall j \in I|j \neq 0, j \neq 1 \tag{13}$$

$$AT_{jk} \leq B_j \qquad \forall k \in K \quad \forall j \in I|j \neq 0, j \neq 1 \tag{14}$$

$$AT_{jk} = AT_{jk'} \qquad \forall k, k' \in K \quad \forall j \in S \tag{15}$$

$$RT_{jk} \leq MAXWL \qquad \forall k \in K \quad \forall j \in I|j \neq 0, j \neq 1 \tag{16}$$

$$AT_{jk} \geq AT_{ik} + (s_i + t_{ij})x_{ijk} - (1 - x_{ijk})M \quad \forall k \in K \quad \forall (i,j) \in E|j \neq 0, j \neq 1 \quad (17)$$

$$RT_{jk} \geq RT_{ik} + (s_j + t_{ij})x_{ijk} - (1 - x_{ijk})M \quad \forall k \in K \quad \forall (i,j) \in E|j \neq 0, j \neq 1 \quad (18)$$

$$RT_{ik} \geq 0, \quad AT_{ik} \geq 0 \qquad \forall k \in K \quad \forall i \in I \quad (19)$$

$$x_{ijk} \in \{0,1\}, \quad y_{jk} \in \{0,1\} \qquad \forall k \in K \quad \forall j \in I \quad \forall i \in I \quad (20)$$

Objective (1) aims to minimize total route cost. Constraint (2) ensures the number of caregivers a care requires. Constraints (3) and (4) guarantee that at most one care can be done after/before the depot 0, respectively. Constraints (5) and (6) ensure that each caregiver has a lunch break (defined as a dummy care). Constraint (7) guarantees that if care is assigned to a caregiver, the care has a successor (linking between routing and assignment variables). Constraint (8) ensures that each caregiver is assigned to the depot (HHC structure). Constraint (9) assigns a lunch break to each caregiver (defined as dummy care). The classical flow conservation restrictions on the routing variables are the (10) constraint. Constraint (11) ensures the priority of cares (precedence synchronization). Constraint (12) guarantees that the care is assigned to the caregiver with the required skill. Constraints (13) and (14) guarantee that the start time of the care respects the time window of patients. Constraint (15) ensures the presence of two caregivers at the same moment for the care requiring simultaneous synchronization. Constraint (16) ensures that the maximum daily workload of each caregiver, represented as the number of service times and travel times, should be respected. Constraint (17) calculates the arrival time of the caregiver to care. Constraint (18) gives the elapsed time for the caregivers at each node. Constraints (19) and (20) enforce the binary variables of the model and the non-negativity restrictions.

## 3   Robust Formulation

In this section, we propose a modeling that deals with travel time and service time uncertainties. Robust optimization (RO) is an approach that aims to find robust solutions to optimization problems in which the data are uncertain without resorting to probabilistic distribution. RO has seen a resurgence of interest in recent decades with many contributions. Unlike stochastic approaches, RO models uncertain data using continuous or discrete sets of possible values, with no attached probability distribution. The proposed model provides robust solutions which are protected against uncertainty. We adapt the optimization approach developed by [16] inspired by [3] that relatively adds few variables and constraints compared to the duality formulation. We assume that the travel time and service time are uncertain values modeled as independent random variables $\tilde{t}_{ij}$ and $\tilde{s}_i$. The random variables fall within the symmetric and bounded ranges defined as follows: $\tilde{t}_{ij} \in [\bar{t}_{ij} - \hat{t}_{ij}, \bar{t}_{ij} + \hat{t}_{ij}]$ (where $\bar{t}_{ij}$ is the nominal travel time value and $\hat{t}_{ij}$ its deviation ($\hat{t}_{ij} \geq 0$)) and $\tilde{s}_i \in [\bar{s}_i - \hat{s}_i, \bar{s}_i + \hat{s}_i]$ (where $\bar{s}_i$ is the nominal service time value and $\hat{s}_i$ its deviation ($\hat{s}_i \geq 0$)). The decision variables are assumed to be nonnegative. The worst case will always be achieved at the right-hand side of the ranges $[\bar{t}_{ij} - \hat{t}_{ij}, \bar{t}_{ij} + \hat{t}_{ij}]$ and $[\bar{s}_i - \hat{s}_i, \bar{s}_i + \hat{s}_i]$. Hence, the ranges do not have to be symmetric. The normalized scale deviations $\epsilon_{ij}^t = \frac{\tilde{t}_{ij} - \bar{t}_{ij}}{\hat{t}_{ij}}$ and $\epsilon_i^s = \frac{\tilde{s}_i - \bar{s}_i}{\hat{s}_i}$ are random variables in $[0,1]$ (without loss of generality). The cumulative uncertainty of the random variable is bounded by the budget of robustness $\nabla^t$ and $\nabla^s$ which represents the number of travel time and service times affected by the uncertainties, respectively. The data uncertainty models are represented by the polyhedral uncertainty sets (21) and (22) as follows:

$$\nu^t = \{\tilde{t} \in \mathbb{R}_+^{|E|}| \quad \tilde{t}_{ij} = \bar{t}_{ij} + \hat{t}_{ij}\epsilon_{ij}^t, \quad \sum_{(i,j)\in E} \epsilon_{ij}^t \leq \nabla^t, \quad 0 \leq \epsilon_{ij}^t \leq 1, \quad \forall(i,j) \in E\}$$
(21)

$$\nu^s = \{\tilde{s} \in \mathbb{R}_+^{|I|}| \quad \tilde{s}_i = \bar{s}_i + \hat{s}_i\epsilon_i^s, \quad \sum_{i\in I} \epsilon_i^s \leq \nabla^s, \quad 0 \leq \epsilon_i^s \leq 1, \quad \forall i \in I\} \quad (22)$$

The budget of uncertainty refers to the number of parameters that are subject to uncertainty. The random variable $\epsilon_{ij}^t$ and $\epsilon_i^s$ are continuous from the interval $[0,1]$ and the budget of robustness bounds their sum (i.e., some travel and service times take their worst-case value, while others take their expected value). When $\nabla^t = 0$ and $\nabla^s = 0$, it refers to the deterministic case. The large budgets express more conservative solutions. Because of the structure of the uncertainty set, the robustness of a route can be defined explicitly using recursive equations [13].

**Time Window:** $AT_{j\tau\gamma}$ represents the earliest exact time from which the service can start at node $j$ when up to $\tau$ travel times and $\gamma$ service times reach their worst-case values. The recursion can compute it:

$$AT_{j\tau\gamma} = \begin{cases} A_0, & \text{if} \quad j = 0; \\ \max\{A_j, AT_{j-1\tau\gamma} + \bar{s}_{j-1} + \bar{t}_{j-1j}\}, & \text{if} \quad \tau = 0 \text{ and } \gamma = 0; \\ \max\{A_j, AT_{j-1\tau\gamma} + \bar{s}_{j-1} + \bar{t}_{j-1j}, \\ \quad AT_{j-1,\tau-1,\gamma-1} + \bar{s}_{j-1} + \hat{s}_{j-1} + \bar{t}_{j-1j} + \hat{t}_{j-1j}\}, & \text{otherwise.} \end{cases}$$

To be robust feasible, the route must satisfy:

$$AT_{j\tau\gamma} \leq B_j \quad \forall j \in I \quad \forall \tau \in \{0,1,...,\nabla^t\} \quad \forall \tau \in \{0,1,...,\nabla^s\} \quad (23)$$

$AT_{j\tau\gamma}$ is not necessarily given by the largest travel and service times deviations because the care only can start after when the time window of the patient opens.

**Caregiver Workload:** $RT_{j\tau\gamma}$ represents the largest elapsed time when the caregiver leaves node $j$ when up to $\tau$ travel times and $\gamma$ service times reach their worst-case values. It can be computed by the recursion:

$$RT_{j\tau\gamma} = \begin{cases} s_0, & \text{if} \quad j = 0; \\ RT_{j-1\tau\gamma} + \bar{s}_j + \bar{t}_{j-1j}, & \text{if} \quad \tau = 0 \text{ and } \gamma = 0; \\ \max\{RT_{j-1\tau\gamma} + \bar{s}_j + \bar{t}_{j-1j}, RT_{j-1,\tau-1,\gamma-1} + \bar{s}_j + \hat{s}_j + \bar{t}_{j-1j} + \hat{t}_{j-1j}\}, & \text{otherwise.} \end{cases}$$

To be robust feasible, the route must satisfy:

$$RT_{j\tau\gamma} \leq MAXWL \quad \forall j \in I \quad \forall \gamma \in \{0, 1, ..., \nabla^s\} \quad \forall \tau \in \{0, 1, ..., \nabla^t\} \qquad (24)$$

The robust model is giving by replacing constraints (13–19) by constraints (25–33)

$$RT_{jk\tau\gamma} \leq MAXWL \qquad \forall k \in K \quad \forall j \in I | j \neq 0, j \neq 1 \quad \forall \gamma \in \{0, 1, ..., \nabla^s\} \quad \forall \tau \in \{0, 1, ..., \nabla^t\} \quad (25)$$

$$AT_{jk\tau\gamma} \geq A_j \qquad \forall k \in K \quad \forall j \in I | j \neq 0, j \neq 1 \quad \forall \gamma \in \{0, 1, ..., \nabla^s\} \quad \forall \tau \in \{0, 1, ..., \nabla^t\} \qquad (26)$$

$$AT_{jk\tau\gamma} \leq B_j \qquad \forall k \in K \quad \forall j \in I | j \neq 0, j \neq 1 \quad \forall \gamma \in \{0, 1, ..., \nabla^s\} \quad \forall \tau \in \{0, 1, ..., \nabla^t\} \qquad (27)$$

$$AT_{jk\tau\gamma} = AT_{jk'\tau\gamma} \qquad \forall k, k' \in K \quad \forall j \in S \, \forall \gamma \in \{0, 1, ..., \nabla^s\} \quad \forall \tau \in \{0, 1, ..., \nabla^t\} \qquad (28)$$

$$AT_{jk\tau\gamma} \geq AT_{ik\tau\gamma} + (\overline{s}_i + \overline{t}_{ij})x_{ijk} - (1 - x_{ijk})M \quad \forall k \in K \quad \forall (i,j) \in E | j \neq 0, j \neq 1$$
$$\forall \gamma \in \{0, 1, ..., \nabla^s\} \quad \forall \tau \in \{0, 1, ..., \nabla^t\} \qquad (29)$$

$$AT_{jk\tau\gamma} \geq AT_{ik\tau-1\gamma-1} + (\overline{s}_i + \widehat{s}_i + \overline{t}_{ij} + \widehat{t}_{ij})x_{ijk} - (1 - x_{ijk})M \quad \forall k \in K \quad \forall (i,j) \in E | j \neq 0, j \neq 1$$
$$\forall \gamma \in \{1, ..., \nabla^s\} \quad \forall \tau \in \{1, ..., \nabla^t\} \qquad (30)$$

$$RT_{jk\tau\gamma} \geq RT_{ik\tau\gamma} + (\overline{s}_j + \overline{t}_{ij})x_{ijk} - (1 - x_{ijk})M \quad \forall k \in K \quad \forall (i,j) \in E | j \neq 0, j \neq 1$$
$$\forall \gamma \in \{0, 1, ..., \nabla^s\} \quad \forall \tau \in \{0, 1, ..., \nabla^t\} \qquad (31)$$

$$RT_{jk\tau\gamma} \geq RT_{ik\tau-1\gamma-1} + (\overline{s}_j + \widehat{s}_j + \overline{t}_{ij} + \widehat{t}_{ij})x_{ijk} - (1 - x_{ijk})M \quad \forall k \in K \quad \forall (i,j) \in E | j \neq 0, j \neq 1$$
$$\forall \gamma \in \{1, ..., \nabla^s\} \quad \forall \tau \in \{1, ..., \nabla^t\} \qquad (32)$$

$$RT_{jk\tau\gamma} \geq 0, \quad AT_{jk\tau\gamma} \geq 0 \quad \forall k \in K \quad \forall j \in I \quad \forall \gamma \in \{0, 1, ..., \nabla^s\} \quad \forall \tau \in \{0, 1, ..., \nabla^t\} \qquad (33)$$

## 4   Computational Results

The proposed models were implemented with ILOG CPLEX Optimization Studio and performed on Dell computer Intel Xeon CPU $E5 - 2667$ v4 3.20 GHz 64 GB RAM using [6] benchmarks, and some generated data specific to HHC problems. The maximum daily workload is fixed to 480 min. I.PXCYKZ denotes a given instance I with X patients, Y cares and Z caregivers. The number of required care characterizes each patient. Table 1 presents an example of required care by the patients. (S) refers to cares requiring simultaneous synchronization. (+) denotes if a patient requires care and (−) otherwise. The default values of travel and service times are considered nominal values. The deviations are computed as 0.2 of the default values.

**Table 1.** Data related to instance (1.P10C15K5)

| Patient | Care α | Care β | Care γ |
|---|---|---|---|
| 1 | (+) | (−) | (−) |
| 2 | (+) before β | (+) before γ | (+) |
| 3 | (+) before β (S) | (+) | (−) |
| 4 | (+) | (−) | (−) |
| 5 | (−) | (+) | (−) |
| 6 | (−) | (−) | (+) |
| 7 | (+) before β | (+) | (−) |
| 8 | (−) | (+) | (−) |
| 9 | (+) | (+) | (−) before α |
| 10 | (−) | (+) | (−) |

**Table 2.** Computational results

| Instance | $\nabla^t = 0$ $\nabla^s = 0$ | $\nabla^t = 1$ $\nabla^s = 0$ | $\nabla^t = 2$ $\nabla^s = 0$ | $\nabla^t = 4$ $\nabla^s = 0$ | $\nabla^t = 0$ $\nabla^s = 1$ |
|---|---|---|---|---|---|
| 1.P10C15K5 | 154.65 | 166.75 | 185.84 | 199.84 | 162.75 |
| 2.P10C15K5 | 160.52 | 172.64 | 185.82 | 196.80 | 170.36 |
| 3.P20C30K8 | 254.69 | 266.77 | 287.86 | 296.84 | 270.96 |
| 4.P20C30K8 | 231.59 | 245.80 | 262.82 | 299.83 | 265.80 |
| 5.P30C40K10 | 396.47 | 410.66 | 463.75 | 470.86 | 396.47 |
| 6.P30C40K10 | 386.51 | 402.69 | 430.70 | 486.89 | 412.74 |
| 7.P40C52K12 | 533.67 | 546.75 | − | − | 523.74 |
| 8.P40C52K12 | 598.66 | 617.71 | − | − | 590.74 |
| **Instances** | $\nabla^t = 0$ $\nabla^s = 2$ | $\nabla^t = 0$ $\nabla^s = 4$ | $\nabla^t = 1$ $\nabla^s = 1$ | $\nabla^t = 2$ $\nabla^s = 2$ | $\nabla^t = 4$ $\nabla^s = 4$ |
| 1.P10C15K5 | 179.98 | 179.54 | 188.26 | 220.98 | 220.36 |
| 2.P10C15K5 | 189.78 | 196.46 | 203.75 | 216.94 | 216.94 |
| 3.P20C30K8 | 301.58 | 307.96 | 280.63 | 311.94 | 315.78 |
| 4.P20C30K8 | 272.12 | 309.83 | 298.74 | 306.43 | 320.74 |
| 5.P30C40K10 | 410.66 | 433.75 | 390.75 | 454.74 | 454.74 |
| 6.P30C40K10 | 400.70 | 479.94 | 450.95 | 486.81 | 483.74 |
| 7.P40C52K12 | 540.78 | − | 550.10 | − | − |
| 8.P40C52K12 | − | − | 612.41 | − | − |

**Table 3.** Computational time (seconds)

| Instance | $\nabla^t = 0$ $\nabla^s = 0$ | | $\nabla^t = 1$ $\nabla^s = 0$ | | $\nabla^t = 2$ $\nabla^s = 0$ | | $\nabla^t = 4$ $\nabla^s = 0$ | | $\nabla^t = 0$ $\nabla^s = 1$ | |
|---|---|---|---|---|---|---|---|---|---|---|
| | CT | GAP | CT | GAP | CT | GAP | CT | GAP | CT | GAP |
| 1.P10C15K5 | 60.20 | + | 300.85 | 0.004 | 401.25 | 0.013 | 1684.69 | 0.122 | 432.25 | 0.002 |
| 2.P10C15K5 | 34.96 | + | 450.29 | 0.003 | 545.26 | 0.003 | 1858.36 | 0.070 | 514.95 | 0.045 |
| 3.P20C30K8 | 103.69 | + | 605.27 | 0.010 | 1505.63 | 0.075 | 1762.84 | 0.012 | 511.78 | 0.07 |
| 4.P20C30K8 | 150.78 | 0.001 | 640.98 | 0.010 | 1783.69 | 0.092 | 1907.45 | 0.052 | 597.65 | 0.193 |
| 5.P30C40K10 | 204.98 | 0.013 | 852.93 | 0.136 | 1902.75 | 0.312 | 1940.75 | 0.120 | 625.14 | 0.178 |
| 6.P30C40K10 | 340.78 | 0.156 | 1270.98 | 0.121 | 1974.95 | 0.152 | 2075.65 | 0.211 | 945.14 | 0.196 |
| 7.P40C52K12 | 500.96 | 0.163 | 1243.87 | 0.120 | − | − | − | − | 1432.85 | 0.142 |
| 8.P40C52K12 | 690.96 | 0.181 | 1783.87 | 0.320 | − | − | − | − | 1534.88 | 0.174 |
| **Instance** | $\nabla^t = 0$ $\nabla^s = 2$ | | $\nabla^t = 0$ $\nabla^s = 4$ | | $\nabla^t = 1$ $\nabla^s = 1$ | | $\nabla^t = 2$ $\nabla^s = 2$ | | $\nabla^t = 4$ $\nabla^s = 4$ | |
| | CT | GAP | CT | GAP | CT | GAP | CT | GAP | CT | GAP |
| 1.P10C15K5 | 620.10 | 0.09 | 1350.15 | 0.14 | 1763.08 | 0.013 | 1989.96 | 0.122 | 2989.33 | 0.32 |
| 2.P10C15K5 | 594.19 | 0.04 | 1474.08 | 0.19 | 1845.16 | 0.003 | 2201.66 | 0.15 | 2982.64 | 0.392 |
| 3.P20C30K8 | 1003.19 | 0.007 | 1635.72 | 0.23 | 1805.13 | 0.075 | 2485.36 | 0.19 | 3420.95 | 0.412 |
| 4.P20C30K8 | 1230.01 | 0.19 | 1647.87 | 0.20 | 1823.57 | 0.092 | 2547.25 | 0.23 | 3485.32 | 0.480 |
| 5.P30C40K10 | 1403.17 | 0.213 | 1687.37 | 0.30 | 1890.51 | 0.312 | 2340.23 | 0.29 | 3111.54 | 0.255 |
| 6.P30C40K10 | 1340.08 | 0.320 | 1770.78 | 0.39 | 2983.71 | 0.152 | 3005.65 | 0.32 | 3105.84 | 0.324 |
| 7.P40C52K12 | ∼ | 0.458 | − | − | 2304.32 | 0.241 | − | − | − | − |
| 8.P40C52K12 | − | − | − | − | 2140.79 | 0.413 | − | − | − | − |

**Objective Function:** Table 2 presents the obtained objective function value for each instance. The deterministic case (DC) value corresponds to ($\nabla^t = 0, \nabla^s = 0$). The objective function also increases as $\nabla^t$ and $\nabla^s$ increase. It can be explained by requiring its maximal value for all travel and service times. The objective function becomes insensitive to uncertainties for high values of $\nabla^t$ and $\nabla^s$ because it has already reached the maximum number of routes the caregiver will travel according to their workload.

**Computational Time:** Table 3 presents the computational time (CT) of the models in seconds. The uncertainties have a significant impact on CT. The CT of the robust models increases significantly compared to the deterministic counterpart. The robust model with a large budget is more challenging to solve compared with the deterministic counterpart. Most of the instances have not been solved to optimality due to the complexity of the problem. Especially the robust counterpart. The average gap concerning instance (5.P30C40K10) and instance (2.P10C15K5) are about 0.212% and 0.089%, respectively. We remark that the average gap is significantly small for the smallest instances compared to the largest ones. (+) refers to the problem that has been solved to optimality. (−) refers to when we get "out of memory". ($\sim$) is when the solver stops after more than 3600 seconds of running time. The CT increases when $\nabla$ is around 4; the number of routes that a caregiver can travel is reached, and all the chosen arcs and nodes require their maximum travel and service times, respectively.

### 4.1 Robustness Analysis

We evaluate the robust HHC solutions in terms of feasibility and robustness. We seek to find solutions that are immunized against real-life uncertainties. The extra cost incurred by such a robust solution should be compensated by the gain in terms of robustness or feasibility [16]. We consider two performance measures: The price of robustness (PoR) and the risk.

*Price of Robustness (PoR).* The PoR is defined as $\frac{z(\nabla)-z}{z}.100\%$ where $z(\nabla)$ is the optimal objective value for a given value of the budget of robustness $\nabla^s$ and $\nabla^t$, $z$ represents the optimal objective value according to the deterministic problem [16]. *Probability of constraint violation (Risk).* The main objective of robustness is to find solutions that are immunized against uncertainty. The routes provided by the robust solution are likely to be more feasible in real-life situations than deterministic solutions when route travel time or/and care time increase which is very common. To evaluate the risk, we use the Monte Carlo simulation and we generate 200 random uniform realization of service time in the range $[\bar{s}_i, \bar{s}_i + \hat{s}_i]$ and travel time in the range $[\bar{t}_{ij}, \bar{t}_{ij} + \hat{t}_{ij}]$. We aim to assess the number of times a given robust solution is infeasible out of the 200 realizations. The main steps of the simulation are given as follows:

1. **Input.** Solution $x(\nabla)$
2. For $\omega = 1$ to 200, do:
   - $\tilde{s}_i^\omega \sim U[\bar{s}_i, \bar{s}_i + \hat{s}_i], \forall i \in I$

$\tilde{t}_{ij}^{\omega} \sim U[\bar{t}_{ij}, \bar{t}_{ij} + \hat{t}_{ij}], \forall (i,j) \in E$

-Evaluate the feasibility of $x(\nabla)$ with $s_i^{\omega}$ and $t_{ij}^{\omega}$

3. **Output**. The empirical probability evaluating constraint violation of the solution

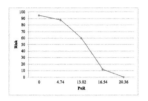 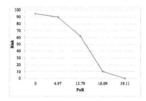 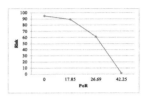

**Fig. 1.** Risk versus PoR ($\nabla^t = 0$ to $4$, $\nabla^s = 0$)    **Fig. 2.** Risk versus PoR ($\nabla^s = 0$ to $4$, $\nabla^t = 0$)    **Fig. 3.** Risk versus PoR ($\nabla^s = 0$ to $4$, $\nabla^t = 0$ to $4$)

For a small value of $\nabla^t$ and $\nabla^s$, the risk of having an infeasible route significantly increases (e.g., for the case $\nabla^s = 0$ and $\nabla^t = 0$, the risk of having infeasible routes increases by 95.98%, the risk is obtained from the Monte Carlo simulation). Hence, the deterministic solutions are not protected against uncertainties [16], and most routes will be infeasible if caregivers' travel and service times deviate from the nominal values. Besides, if we assume the default values and the caregiver travels within his worst-case travel time, the time window of his next care can be already closed, and the route will be infeasible. The robust solutions give an advantage in terms of cost and feasibility for most instances. Figure 1, Fig. 2 and Fig. 3 illustrate the risk versus the PoR for instance (P20C30K8). They show a trade-off between the risk of having constraints violation and not being too conservation. The cost is not dramatically increased to protect the solutions against constraints violation. Figure 2 shows that there is almost no risk ($risk = 0.05\%$) of constraint violation when the PoR is up to 19.11% ($\nabla^s = 0$ to $4$, $\nabla^t = 0$ and 20% deviation).

## 5    Conclusions and Perspectives

In this study, a new approach to solving the daily HHC routing and scheduling problem was proposed. The approach included a set of relevant HHC constraints. Also, a novel robust counterpart HHC model under travel and service time uncertainty was presented. Using the price of robustness gives the minimal cost that leads to an increase in the chances of having caregivers' routes feasible in practice. A robustness analysis using a Monte Carlo simulation was done to show the efficiency of the proposed approach. The proposed robust counterpart model was more challenging to solve than the deterministic model, especially for large budgets. The proposed approach gives good solutions compared to the approaches in the literature; the price of robustness was not dramatically increased even for large values of the budget of robustness. Comparing the increase of route cost and risk of solutions infeasibility using the simulation indicates the relevance of the obtained robust solutions and the importance of considering uncertainty to solving real-life planning problems. A column generation algorithm will be developed to solve the large instances of the problem with additional constraints and criteria, which will be future work.

# References

1. Ben Rabaa, F., Harbi, S., Amraoui, A.: Robust optimization for a home care scheduling problem. In: The 4th International Conference on Logistics Operations Management, GOL 2018, Lehavre France, 10–12 April, April 2018
2. Benzerti, Y.: Home health care operations management: applying the districting approach to Home Health Care. Ph.D. thesis, Ecole Centrale Paris (2012)
3. Bertsimas, D., Sim, M.: Robust discrete optimization and network flows. Math. Program. **98**, 49–71 (2003)
4. Cappanera, P., Scutella, M.: Joint assignment, scheduling, and routing models to home care optimization: a pattern based approach. Transp. Sci. **49**, 830–852 (2015)
5. Cisse, M., Yalcindag, S., Kergosien, Y., Sahin, E., Lente, C., Matta, A.: Or problems related to home health care: a review of relevant routing and scheduling problems. Oper. Res. Health Care **13**, 1–22 (2017)
6. Dumas, Y., Desrosiers, J., Gelinas, J., Solomon, M.M.: An optimal algorithm for the traveling salesman problem with time windows. Oper. Res. **43**, 367–371 (1995)
7. En-nahli, L., Allaoui, H., Nouaouri, I.: A multi-objective modelling to human resource assignment and routing problem for home health care services. IFAC-PapersOnLine **48**, 698–703 (2015)
8. Fikar, C., Hirsch, P.: Home health care routing and scheduling: a review. Comput. Oper. Res. **77**, 86–95 (2017)
9. Grenouilleau, F., Legrain, A., Lahrichi, N., Rousseau, L.M.: A set partitioning heuristic for the home health care routing and scheduling problem. Eur. J. Oper. Res. **275**, 295–303 (2019)
10. Groër, C., Golden, B., Wasil, E.: The consistent vehicle routing problem. Manuf. Serv. Oper. Manage. **11**, 630–643 (2009)
11. Haddadene, S.A., Labadie, N., Prodhon, C.: NSGAII enhanced with a local search for the vehicle routing problem with time windows and synchronization constraints. IFAC-PapersOnLine **49**, 1198–1203 (2016)
12. Lanzarone, E., Matta, A., Sahin, E.: Operations management applied to home care services: 2012 the problem of assigning human resources to patients. IEEE Trans. Syst. Man. Cybern. Part A Syst. Hum. **42**, 1346–1363 (2012)
13. Lee, C., Lee, K., Park, S.: Robust vehicle routing problem with deadlines and travel time/demand uncertainty. J. Oper. Res. Soc. **63**, 1294–1306 (2012)
14. Makboul, S., Kharraja, S., Abbassi, A., El Hilali Alaoui, A.: A three-stage approach for the multi-period green home health care problem with varying speed constraints. In: 13ème Conference Internationale De Modelisation, Optimisation et Simulation (MOSIM 2020), AGADIR, Maroc. AGADIR (virtual), Morocco, 12–14 November 2020, November 2020
15. Mascolo, M.D., Martinez, C., Espinouse, M.: Routing and scheduling in home health care: a literature survey and bibliometric analysis. Comput. Ind. Eng., 107255 (2021). https://doi.org/10.1016/j.cie.2021.107255

16. Munari, P., Moreno, A., Vega, J.D.L., Alem, D., Gondzio, J., Morabito, R.: The robust vehicle routing problem with time windows: compact formulation and branch-price-and-cut method. Transp. Sci. **53**, 1043–1066 (2019)
17. Rasmussen, M., Justesen, T., Dohn, A., Larsen, J.: The home care crew scheduling problem: preference-based visit clustering and temporal dependencies. Eur. J. Oper. Res. **219**, 598–610 (2012)
18. Shi, Y., Boudouhb, T., Grunder, O.: A robust optimization for a home health care routing and scheduling problem with consideration of uncertain travel and service times. Transp. Res. Part E Log. Transp. Rev. **128**, 52–95 (2019)

# Lean Healthcare Applied Systematically in the Basic Image Examination Process in a Medium-Sized Medical Clinic

Samuel Martins Drei[1]($\boxtimes$) ID, Paulo Sérgio de Arruda Ignácio[2] ID,
Antônio Carlos Pacagnella Júnior[2] ID, Li Li Min[3] ID,
and Thiago Augusto de Oliveira Silva[4] ID

[1] Fluminense Federal University, Niterói, Brazil
[2] School of Applied Sciences, Campinas State University, Limeira, Brazil
psai@unicamp.br, antonio.junior@fca.unicamp.br
[3] School of Medical Sciences, Campinas State University, Campinas, Brazil
limin@unicamp.br
[4] Institute of Exact and Applied Sciences, Federal University of Ouro Preto,
João Monlevade, Brazil
thiago@ufop.edu.br

**Abstract.** The objective of this paper is to propose a systematic application of Lean Healthcare in the activity of fetching the next patient in the X-ray examination process. The methodology used is based on two pillars: the first, called a survey, aims to map the process in which the focus activity is inserted and, thus, obtain the associated waste, identifying the root cause of what will be treated. Then, the proposed Lean pillar will propose a tool to remedy the waste in question, applying the necessary actions within a schedule and, finally, collecting the results and comparing them with the initial measurement. As a result, there is an improvement in the hospital processes from the Lean perspective, specifically in the activity of fetching the next patient, in the X-Ray examination, reducing the non-added value identified in this activity, that is, in the waiting time of the patient. As conclusions, it was possible to obtain concrete information about the implementation of this proposed systematic application, so that it can be replicated in other wastes, generating a systematic proposal, filling a gap in the literature, which covers the lack of fully structured studies with practical results of Lean Healthcare systematics in the medical clinic wing of hospitals.

**Keywords:** Lean Healthcare · Health services · Medical clinic · X-ray exam

## 1 Introduction

Given the development of the Toyota Production System (TPS), [1] stated that the production of small quantities with a variety of products was appropriate for the reality in Japan. As such ideals went against Mass Production, TPS focused on increasing efficiency with the methodical and complete elimination of waste. Thus, this system, later

© IFIP International Federation for Information Processing 2021
Published by Springer Nature Switzerland AG 2021
A. Dolgui et al. (Eds.): APMS 2021, IFIP AICT 631, pp. 403–412, 2021.
https://doi.org/10.1007/978-3-030-85902-2_43

named Lean Manufacturing or Lean Production, was inserted in the means of manufacturing based basically on waste reduction, through the elimination or mitigation of
non-added value [2]. Lean Production established a production method that, at the same
time, prioritized the elimination of waste without neglecting the quality of the resulting
product at the end of the process, ensuring cost reduction for the organization and the
approval of its consumer market.

In addition, over time, Lean Manufacturing has become part of other realities on the
market, such as the service sector, creating the concept known as Lean Management [3].
Thus, the Lean tool can be used in several areas of production, service provision or decision making, given that there is a plausible conduct of its application. Therefore, it was
reflected in the health environment, with the designation of Lean Healthcare. Managers
and politicians have struggled to limit rising costs, while still providing appropriated
health care [4]. Hospital's environment is very conducive to incorporating Lean, given
that such health organizations are complex places of great social importance, which must
provide quality services with a comprehensive restriction of resources [5]. Hospital environments in Brazil operate with a budget that is often less than necessary [6], so you have
to be even more competent in your processes to be able to serve your patients without
reducing the expected quality, after all this issue directly affects the health, well-being
and even the life of your patient. Many healthcare organizations have improved their
operations by adopting the lean healthcare approach [7–9].

With the increasing application of Lean Healthcare in recent years [10] the research
problem is configured to use, appropriately, the teachings and tools derived from Lean to
achieve a standard of application that can be replicated in hospitals in the future. So, the
objective of this paper is to propose a systematic application of Lean Healthcare in the
activity of fetching the next patient in the X-ray examination process, with the specific
objectives of: (i) surveying the current state of the X-ray examination in patients in the
medical clinic wing, (ii) use the tools and concepts available in the literature to support
the solution of the identified wastes and (iii) propose a viable solution aligned with Lean
Healthcare and collect its results.

The justification for this study is given by the benefits that Lean Healthcare is able
to bring to health services, focusing primarily on hospitals within the Unified Health
System in Brazil, especially in the medical clinic wing, as well as the importance of the
study hospital to its region.

It is important to highlight that, notoriously, other papers have already addressed
individual applications of Lean Manufacturing in health sectors, however this study is
part of a project branch, based on the non-identification of applications in the medical
clinic wing of hospitals [11], in order to formalize a systematic application of Lean
Healthcare in this wing, through the application proposed in different activities e.g. [12,
13].

Finally, some studies also bring systemic applications of Lean Healthcare [14], however they point out that it was not fully applied and the associated performance measures
were not elaborated. So, the proposed application is a novelty justified, given this gap,
through the complete application of a new proposed systematic, collecting the results of
the application, for future replications, especially in the medical clinic wing.

# 2 Theoretical Reference

## 2.1 Lean Healthcare

Lean manufacturing is primarily based on the elimination of waste, assisting the process of value creation for all stakeholders [15]. In hospitals, the main processes are those responsible for the diagnosis, treatment and convalescence of the patient. The other processes are supporters of the essential services [16, 17]. For [18] the application of the Lean philosophy in health is perfectly acceptable, and the first step is to include time and comfort as factors in the evaluation of the system. [3] states that one of the imperative points in the Lean is that the entire organization must be examined and included, in order to generate improvements, therefore, in this new paradigm, it is necessary to involve all people in the organization in what it is expected by the patient, and it is necessary to create a permanent flow of people, information and materials creating value, without incurring additional costs for the health organization in question.

In addition, in health care it is necessary to take into account the basis for defining the value for the patient, which is found in the six dimensions of care of the United States Institute of Medicine: (1) safe care, (2) efficient, (3) effective, (4) agile, (4) patient-centered and (5) fair. Everything that is not considered value to the patient is defined as activities that do not add value, or waste [19]. This requires the implementation of quality procedures in the scope of health as a mandatory requirement, which consists of a detailed analysis of the processes involved in carrying out a procedure, including all aspects [20]. The health system is constantly exposed to a seemingly inexhaustible flow of new advances aimed at improving the organizational and operational aspects of the health system, through changes in operations, organization or management principles, given the nature of this sector [4].

Therefore, for [21] the successful implementation of Lean Healthcare can help achieve some objectives, which include (i) Increase the distribution of power among people; (ii) Improved flow; (iii) Eliminate unnecessary expenses; (iv) Align resources and their demand; (v) Make it perfect the first time; (vi) Learn in practice; (vii) Identify problems more easily and (viii) Anticipation of tasks. By applying Lean, organizations can eliminate waste such as overproduction, waiting, transport processing, unnecessary inventory, unnecessary displacement and defects [15].

## 2.2 Lean Healthcare Applications

As well as Lean Manufacturing, different forms of application of Lean have also been developed within a healthcare environment, because, however much there is a basis for application in manufacturing, when talking about services, adaptation and even insertion of new points in the lean health methodology are necessary [22]. [23] evaluated whether the application of Lean techniques, in a rehabilitation service, reduced waste and added value to the client. Thus, the cost of stored material was reduced by 43% and consumption per treated patient by 19%, increasing the patient's dedication time by 7%.

In turn, [24] evaluated how the 5S management method creates changes in the workplace and in the process and results of health services, and how it can be applied in a low-resource setting, based on data from a pilot intervention of the 5S program. implemented in a hospital in Senegal. Thus, with regard to efficiency, it was possible to reduce the time to search for items, as well as decrease the movement of the team through the laboratories. Regarding patients, it was possible to reduce the waiting time and also improve the guidelines given to them. Finally, in relation to safety, there was an improvement in the sterilization process. [25] used a case study with a survey of the root causes using the 5 whys tool, questioning the professionals in the focus unit of the work, who had been in it for at least three months. Finally, [26] proposes to study the logistics problems in Brazilian hospitals through Lean methods, approaching different aspects of logistics and the procedures adopted in health institutions such as receiving, storage and transporting.

According to the current literature, emergency and surgical procedures have been the main targets when it comes to Lean implementation [14], so this paper focuses on the medical clinic, specifically on the hospitalization activity of the entrance process, once that there is no paper in medical clinic when it comes to Lean Healthcare [11]. Furthermore, this paper addresses this gap, proposing a systematic application of Lean Healthcare, already applied in a medium-sized Brazilian hospital, presenting positive results exposed in it.

## 3   Method

The methodology of this paper was divided into two pillars - called Lean Survey and Proposal, so that the application structure proposed in previous works [11, 13] and, thus, present a systemic application.

In the Survey, the following steps were applied, presented in the study of the process: (i) Mapping of the X-Ray Examination process, through observations made during the day-to-day of the hospital, establishing the times of each activity through the use of the Value Stream Mapping (VSM) establishing the aggregated and non-aggregated values of each activity; (ii) Identification of the waste that occurred in the activity of fetching the next patient, inserted in the X-Ray Examination process. Furthermore, on the waste and Lean proposal section, it was (iii) Established the causes of this waste, with the selection of the greatest number of occurrences to be remedied, using the Pareto Graph; and (iv) Identification of the root cause of the waste to be remedied, using the 5 whys tool.

Then, still in the waste and Lean proposal parte, having identified the root cause, the second pillar, Lean Proposal, began, which unfolded as follows: (i) Lean improvement proposal, related to the identified waste, using an appropriate lean manufacturing tool; (ii) Structure the action plan proposed by [22]; (iii) Apply the actions, following the topics of the schedule. Finally, is (iv) calculate the time savings that the solution has brought to the activity, in the results part.

The purpose of this proposal is that it be executed and made available to those responsible and applied in the future, not only in the activity of fetching the next patient, but in other waste identified in the medical clinic, thus developing a systematic application of Lean Healthcare, which can be used widely.

# 4 Development

## 4.1 Study of the Process

The medical clinic is a complex wing having more than one distinct flow. Thus, it was splinted into three types of flows to be better studied. They are: (i) entry in the medical clinic, (ii) general processes - which include procedures performed on all or most inpatients - and (iii) specific processes - which include specific processes for each patient, based on their needs.

This study covered the specific processes, represented by the flow of X-ray Examination, which exemplifies this process, from the request to the patient's final accommodation.

A current VSM was done about the X-ray Examination, as shows in Fig. 1, the VSM begins with the order of order of the X-ray arriving at the medical clinic, from that the responsible nurse identifies the patient who will go for the exam, the bed where he is and so, calls the stretch bearer for transport. When he arrives, he goes to the responsible nurse who delivers the order orders and sends him to the first patient's room to do the exam.

Thus, the activity of preparing the patient begins, which consists of taking him out of bed and placing him in a wheelchair. It is important to note that this flow was limited to patients who need this assistance, which represents a large part of the interns of the medical clinic. Then, stretch bearer directs the patient to the X-ray room, located on the first floor, passing through the hospital's access ramps.

Upon arriving at the room, the stretch bearer delivers the order to the X-ray employee, who begins to set up the machine and the patient for the exam. The stretch bearer waits in the waiting room and, when the exam is over, he returns with the patient to his room and begins the process with the next patient in line, considering that the exam orders are usually sent together to the medical clinic.

Based on this, the application focused on the specific activity of seeking the next patient, who has 40 min of non-added value and 28 min of added value, due to the long waiting time, that is, non-added value, surpassing the added value of the activity in 12 min and all other non-added values of the activities that comprise the process.

## 4.2 Waste and Lean Proposal

From the traced VSM and the daily observations made during the three-month period in the hospital, with a daily demand of 20 patients a day in this process, it was possible to raise the occurrences of non-added value in the fetching the next patient activity. Thus, the wait for the stretch bearer proved to be the most expressive, and almost unique occurrence, only coexisting with 16% of the times that the wait came from the nurse's specification, as shown in the Pareto Graph in Fig. 2.

So, the 5 whys tool was used in this first occurrence, to find its root cause, as shows in Table 1. Therefore, it was identified that the root cause of patients' waiting is due to the fact that the stretch bearer only fetches the next one when the exam is over. This happens because he delivers the patient to the X-ray employee and waits for the exam

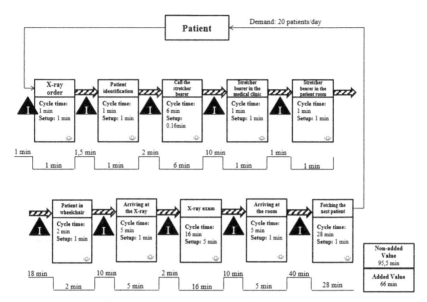

**Fig. 1.** VSM of the X-ray examination process.

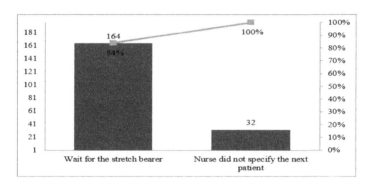

**Fig. 2.** Pareto diagram for the occurrences of waiting for fetching the next patient

to finish, however from the moment of delivery, the patient becomes the employee's responsibility, leaving the stretch bearer idle for the duration of the X-ray.

Thus, the stretch bearer is idle in the waiting room of the examination and the patient already identified by the nurse is waiting for the stretch bearer to pick him up, however he only arrives after the examination and travel back to the medical clinic.

Given its nature, the proposal is a new flow, specifically for the X-ray, in which the activity of fetching the next patient is placed in parallel with the activity of X-ray exam, so that, while the X-ray examination happens, the stretch bearer goes to the medical clinic and fetches the next patient, avoiding idleness and reducing their waiting time.

Figure 3 presents the necessary actions to fully achieve this proposal, following the parameters of the action plan of the *Lean Institute Brasil*.

**Table 1.** 5 whys for waiting for the stretch bearer

| N | Why | Because |
|---|-----|---------|
| 1 | Why is the patient waiting for the stretch bearer? | Because the stretch bearer is absent |
| 2 | Why is the stretch bearer absent? | Because he's in the x-ray room |
| 3 | Why is the stretch bearer in the x-ray room? | Because he still hasn't come back from the previous trip |
| 4 | Why didn't he come back from the previous trip? | Because he is waiting for the exam to end |
| 5 | Why is he waiting for the exam to end? | **Because he only fetches the next patient when the exam is over** |

| Action Plan | | Objective: Reduce the waiting time for the stretch bearer to fetches the next patient | | | | J | F | M | A | M | J | J | A | S | O | N | D | Department: Medical Clinic |
|---|---|---|---|---|---|---|---|---|---|---|---|---|---|---|---|---|---|---|
| Task n | Task | Metric | Responsable | Target Date | | | | | | | | | | | | | | Review |
| 1 | Meeting with the directors | Meeting held in the room with the directors of the hospital | Applicator 1 | Oct-19 | | | | | | | | | | | O/X | | | O |
| 2 | Change in the flow | Explain to the stretch bearer the new flow proposed | Applicator 1 | Oct-19 | | | | | | | | | | | O/X | | | O |
| 3 | Adaptation phase | Moment of adaptation for the new process | Applicator 1 | Nov-19 | | | | | | | | | | O/X | | | | O |
| 4 | X-ray flow | The new flow is incorporated | Applicator 1 | From December/2019 to February/2020 | | X | | | | | | | | | | O | | O |
| 5 | Results recording | Measurement made through observations to compare with initial values | Applicator 1 | Mar-20 | | | O/X | | | | | | | | | | | O |
| Prepared by: Applicator 1 and Applicator 2 | | | | | | | | | | | | Caption | | | | | |
| | | O Start date | | | | | | | | | | O On goal | | | | | |
| | | X Conclusion date | | | | | | | | | | V Below goal | | | | | |
| | | * Review | | | | | | | | | | X Problem | | | | | |

**Fig. 3.** Action plan and effects for fetching the next patient.

## 4.3  Results

These are the results related to the activity of fetching the next patient. Figure 4 shows the values obtained for this activity and the impact they had in relation to the time originally collected and the post-application, as well as the number of occurrences of the root cause.

| Fetching the next patient | | | | | | | | | |
|---|---|---|---|---|---|---|---|---|---|
| Initial Time | | | | | Time Reached | | | | |
| Total Time | Non-added value | Percentage of non-added value | Added value | Percentage of added value | Total Time | Non-Added value | Percentage of non-added value | Added value | Percentage of added value |
| 68 | 40 | 59% | 28 | 41% | 20 | 10 | 50% | 10 | 50% |
| Initial occurrences | | | | | Occurrences reached | | | | |
| 164 | | | | | 0 | | | | |

**Fig. 4.** Results obtained with the fetching the next patient

It is clear that the total time from 68 min to 20 min, this was mainly due to the decrease in the non-aggregated value from 40 min to 10 min, a decrease of 75%. Due to the activities in parallel, there was also a decrease in the total added value, which went from 28 to 10 min, causing an equal distribution between added and non-added value. In addition, the number of occurrences has decreased by 100%, completely eliminating the identified waste.

## 5 Conclusion

With this study, it is possible to recognize the positive results that a systemic application leads to waste in the process activities of a hospital's medical clinic, especially when it comes to Lean.

Regarding the objective of the study, it was possible to achieve it, since a systematic application of Lean Healthcare was proposed in the activity of looking for the next patient in the X-ray examination process.

In addition, the specific objectives were also fulfilled, considering that it was possible to survey the current status of the X-ray examination in patients in the medical clinic wing, through observations made in the day-to-day of the hospital and, mainly, with the use of VSM to establish the activity with the greatest amount of non-added value, in addition to pointing out other activities that have high waiting times, generating the possibility of applying the systematic again.

From the identification of the root cause and the establishment of what type of Lean waste was present in the activity, it was also possible to propose an adequate tool, improving the flow of patients in the process, reducing waiting time, without generating any cost for the hospital.

Furthermore, the proposed application obtained positive results from the Lean perspective, that is, it resulted in a reduction in wasted time, focused on the non-aggregated values of the studied activity. Since these results are promising, using a combination of adequate and systematized Lean Healthcare tools, it is possible to expand the application into more activities to establish the proposed systematic.

In addition, the present study contributes to the Lean Healthcare literature, as it focuses on a hospital ward that is not explored in other applications, that is, in the medical clinic. In addition, the application systematic is a novelty in the literature, as it is fully structured and has practical results, which can be replicated.

Finally, for future research, it is recommended to repeat the application, in order to systematize this proposal, in other medical clinic processes, replicating step by step what was built, to obtain positive results, from a Lean point of view, and thus, validate the fully systematic in the medical clinic.

## References

1. Ohno, T.O.: Sistema Toyota de Produção: além da produção em larga escala. Porto Alegre (1997)
2. Womack, J.P., Jones, D.T., Roos, D: The Machine That Changed the World: The Story of Lean Production, Free Press, New York (1990)

3. Liker, J.K.: O modelo Toyota: 14 princípios de gestão do maior fabricante do mundo. Bookman Editora (2016)
4. Colldén, C., Gremyr, I., Hellstrom, A.: A value-based taxonomy of improvement approaches in healthcare. J. Health Organ. Manag. 31(4), 445–458 (2017)
5. Raimundo, E., Dias, C., Guerra, M.: Logística de medicamentos e materiais em um hospital público do Distrito Federal. RAHIS-Revista de Administração Hospitalar e Inovação em Saúde, 12(2) (2015)
6. O'Dwyer, G., Oliveira, S., Seta, M.: Avaliação dos serviços hospitalares de emergência do programa QualiSUS. Cien. Saude Colet. 14(5), 1881–1890 (2009)
7. Kollberg, B., Dahlgaard, J., Brehmer, P.: Measuring lean initiatives in health care services: issues and findings. Int. J. Product. Perform. Manag. 56(1), 18 (2007)
8. Hicks, C., McGovern, T., Prior, G., Smith, I.: Applying lean principles to the design of healthcare facilities. Int. J. Prod. Econ. 170, 677–686 (2015)
9. Matthias, O., Brown, S.: Implementing operations strategy through lean processes within health care. Int. J. Oper. Prod. Manag 36(11), 1435–1457 (2016)
10. Brandao de Souza, L.: Trends and approaches in lean healthcare. Leader. Health Serv. 22(2), 121–139 (2009)
11. Drei, S.M., Ignácio, P.S.A.: Aplicação do Lean Healthcare na atividade de colocar a pulseira da clínica médica de um hospital de médio porte. Anais do XL Encontro Nacional de Engenharia de Produção (2020)
12. Drei, S.M., Ignácio, P.S.A.: Avaliação de um procedimento sistemático para o Lean Healthcare. Anais do XXXIX Encontro Nacional de Engenharia de Produção (2019)
13. Drei, S.M., Ignácio, P.S.A.: Lean healthcare applied in medicines' preparation in medical clinic at a medium-sized hospital. In: International Joint Conference on Industrial Engineering and Operations Management (2021)
14. Régis, T.K.O., Santos, L.C., Gohr, C.F.: A case-based methodology for lean implementation in hospital operations. J. Health Organ. Manag. 33(6), 656–676 (2019)
15. Krijnen, A. The Toyota way: 14 management principles from the world's greatest manufacturer. Act. Learn. Process. 4, 105–114 (2007)
16. Kollberg, B., Dahlgaard, J.J., Brehmer, P.O.: Measuring lean initiatives in health care services: issues and findings. Int. J. Prod. Perf. Manag. 56, 7–24 (2007)
17. de Vries, J., Huijsman, R., Aronsson, H., Abrahamsson, M., Spens, K.: Developing lean and agile health care supply chains. Supply Chain Manag. Int. J. 16(3), 8 (2011)
18. Womack, J.P., Jones, D.T.: A máquina que mudou o mundo. Gulf Professional Publishing, Houston (2004)
19. Min, L.L., Sarantopoulos, A., Spagnol, G.S., Calado, R.D.: O que é esse tal de Lean Healthcare. Pedro e João, São Carlos (SP) (2014)
20. de Oliveira, K.B., dos Santos, E.F., Junior, L.V.G.: Lean Healthcare as a Tool for Improvement: A Case Study in a Clinical Laboratory. In: Duffy, V.G., Lightner, N. (eds.) Advances in Human Factors and Ergonomics in Healthcare. AISC, vol. 482, pp. 129–140. Springer, Cham (2017). https://doi.org/10.1007/978-3-319-41652-6_13
21. Graban, M.: Lean Hospitals: Improving Quality, Patient Safety, and Employee Engagement. CRC Press, Boca Raton (2016)
22. Lean Institute Brasil. https://www.Lean.org.br/workshop/110/Lean-na-saude.aspx. Accessed 4 May 2021
23. Dávilla, S.P., González, J.T.: Mejora de la eficiencia de un servicio de rehabilitación mediante metodología Lean Healthcare. Rev. Calid. Asist. 30(4), 162–165 (2015)
24. Kanamori, S., Sow, S., Castro, M.C., Matsuno, R., Tsuru, A., Jimba, M.: Implementation of 5S management method for lean healthcare at a health center in Senegal: a qualitative study of staff perception. Glob. Health Action 8(1), 27256 (2015)

25. Siqueira, C.L., Siqueira, F.F., Lopes, G.C., Gonçalves, M.D.C., Sarantopoulos, A.: Enteral diet therapy: use of the lean healthcare philosophy in process improvement. Rev. Bras. Enferm. **72**, 235–242 (2019)
26. Gayer, B.D., Marcon, É., Bueno, W.P., Wachs, P., Saurin, T.A., Ghinato, P.: Analysis of hospital flow management: the 3 R's approach. Production (2020)

# Application of VSM for Improving the Medical Processes - Case Study

Katarzyna Antosz[1](✉) 📷, Aleksandra Augustyn[1],
and Małgorzata Jasiulewicz – Kaczmarek[2] 📷

[1] Faculty of Mechanical Engineering and Aeronautics, Rzeszow University of Technology,
Al. Powstancow Warszawy 12, 35-959 Rzeszów, Poland
katarzyna.antosz@prz.edu.pl

[2] Faculty of Management Engineering, Poznan University of Technology, Rychlewskiego 2,
60-965 Poznań, Poland

**Abstract.** Lean Manufacturing has been used in many types of organizations, including healthcare. In the healthcare area, lean healthcare is a management philosophy to develop a hospital culture characterized by increased patient, and other stakeholder, satisfaction through continuous improvements. The starting point for improving activities in the healthcare is identification of the problems and wastes in the processes. In this context VSM is a helpful tool. That why the purpose of this paper is to present the possibility of application the VSM for improving the selected process in the healthcare. Thanks to the application of this method, the necessary information on the sources of losses in the analyzed process was obtained.

**Keywords:** Lean Healthcare · Value stream mapping · Process improvement

## 1 Introduction

Global industry motivates companies to seek and implement more competitive production system. Many of them are implementing or planning to implement Lean Manufacturing (LM). This philosophy, developed in the 1990s, is based on the Toyota Production System (TPS) [1–3] and is mainly used in industry to increase the performance and efficiency of processes. Its assumptions are based on the paradigm of creating value from the customer's point of view and attempts to minimize and eliminate all kinds of waste in the processes. LM has been used in many types of organizations, including healthcare [4–7]. In the health sector, it is referred to as Lean Healthcare (LH), and in healthcare hospitals - Lean Hospitals (LHo). Lean Healthcare is based on five fundamental principles formulated by [8]. In the healthcare area, lean health care is a management philosophy to develop a hospital culture characterized by increased patient, and other stakeholder, satisfaction through continuous improvements, in which all employees (managers, doctors, nurses, laboratory people, technicians, office people etc.) actively participate in identifying and reducing non-value added activities (waste) [9].

© IFIP International Federation for Information Processing 2021
Published by Springer Nature Switzerland AG 2021
A. Dolgui et al. (Eds.): APMS 2021, IFIP AICT 631, pp. 413–421, 2021.
https://doi.org/10.1007/978-3-030-85902-2_44

The methods and tools of LH (or LHo) are, e.g.: 5Why, 5S, visual control and kanban, Poka Yoke, value stream mapping (VSM), work standardization [10, 11]. The result of their application in healthcare is the improvement of work organization, minimizing the time lost, for example, due to improperly identified medical documentation, medical materials and tools, and long waiting times [12]. Value stream mapping helps to minimize or eliminate unnecessary activities, which improving the quality and safety of patient service, as well as shortening the treatment time.

In this context VSM is healthcare is a potential, helpful tool to reduce patient time in the hospital and traditional management costs [13, 14]. The major benefits of VSM over other techniques are: (1) VSM helps visualize the process using iconic representation; (2) VSM helps identify the sources of waste in the process; (3) VSM shows the linkage between information flow and material flow; and (4) VSM helps form the basis of an implementation plan [15, 16]. That why the main goal of the paper is to present the possibility of use VSM to identify and analyze sources of waste in the process of patient rehabilitation in the healthcare. Given the purpose above, the paper is organized as follows: in Sect. 2 the work methodology is presented. Then, in Sect. 3 the results of VSM application for the selected processes are shown. Finally, the conclusions and direction of the future research are presented.

## 2  Research Methodology

The value stream mapping (VSM) method has been used to implement lean concepts in order to enhance services in rehabilitation department in the hospital. The research methodology included the following stages: (1) Selection of the area for analysis. (2) Collection of data from the process. (3) Development of the current state map. (4) Problem identification and analysis. (5) A proposal for improvement activities. (6) Development of the future state map. (7) Evaluation of the effectiveness of the proposed improvements. The goal of the Current State -VSM was to identify the causes of problems and wastage in order to propose solutions to decrease the lead-time of the particular service. Based on the aforementioned analysis and proposed improvements a Future State -VSM has been proposed. The activities of the selected process have been analyzed to identify: lead time (LT), value-added (VA) time, non-value-added ( NVA) time and necessary-non-value-added (NNVA) time. The formulas presented in [17], have been used to perform necessary calculations:

$$LT = \sum_{i=1}^{n} A_i \tag{1}$$

where:
$A_i$ – time of the $i^{th}$ activity, $i = 1, ..., n$;
$n$ – number of activities.

$$VA = \sum_{i=1}^{m} A_{VAi} \tag{2}$$

where:
$A_{VA\,i}$ – time of the $i^{th}$ value-added activity, $i = 1, ..., m$;
$m$ – number of value-added activities.

$$NVA = \sum_{i=1}^{m} A_{NVAi} \tag{3}$$

where:
$A_{VA\,i}$ – time of the $i^{th}$ value-added activity, $i = 1, ..., m$,
$m$ – number of value-added activities.

$$NNVA = \sum_{i=1}^{p} A_{NNVAi} \tag{4}$$

where:
$A_{NNVAi}$ – time of the $i^{th}$ necessary-non-value-added activity, $i = 1, ..., p$,
$i$ - number of necessary-non-value-added activities.

## 3  Results

### 3.1  Process Description

The subject of the analysis was the daytime rehabilitation process of patients in a selected hospital ward. The aim of the first stage was to identify "patient families". The medical procedures used in the unit (stages of treatment) and the illnesses experienced by the patient were used for identification. The Department of Therapeutic Rehabilitation admits patients after stroke (AS), after heart attack (AHA), with a shoulder injury (SHI), after breaking a leg (ABL) and arm (ABA) as well as with a spine injury (SPI). After analyzing the stages of treatment and the diseases of patients who go through the same stages of treatment, the families of patients can be identified (Table 1).

Table 1.  Identified "patient families"

| Treatment stage | Patient | | | | | |
|---|---|---|---|---|---|---|
| | AS | AHA | SHI | SPI | ABL | ABA |
| Self-assisted exercises | X | X | | X | | |
| Balance exercises | X | X | X | | X | X |
| Resistance exercises | | | | X | X | X |
| Upright standing | X | X | X | | X | |
| Restoring precision of movements | X | X | X | | X | X |
| Hospitalization | X | X | | | | |
| Isometric exercises | X | X | X | X | X | X |
| Mobilization | | | X | X | X | X |
| Active exercises | | | X | X | X | X |

Based on the similarities and treatment stages, two main groups of patients can be identified. The first one (green colour) is a group of patients after AS and AHA treatment steps are the same. The second family (red colour) consists of patients who regularly come to the Ward and try to achieve full fitness after breaking the lower and upper limbs. Value stream mapping was performed for the entire rehabilitation process of patients in group 1.

## 4   Current State Analysis - Current State Map (CS-VSM)

The data for the process analysis was collected by analyzing the day of the selected patient's stay at the Therapeutic Rehabilitation Department. The patient spends 24 h in the ward, but the day of his activity lasts about 10 h. Each day of the patient looks similar in terms of the exercises performed. The individual steps of the rehabilitation process are as follows [18]:

**Stage 1. Visit to the general practitioner.** The patient reports to his general practitioner. Describing his ailments, he receives a referral to a specialist doctor in order to perform the necessary tests.

**Stage 2. Waiting in line.** Waiting for the patient to see a specialist.

**Stage 3. Visit to a specialist doctor.** In conversation with the patient, the doctor conducts a medical interview in order to select the appropriate scope of tests.

**Stage 4. Waiting in line.** Waiting for the patient to perform the test.

**Stage 5. Performing the test.** The responsible person performs tests on the patient according to the doctor's instructions. After the test, he performs its description and orders to return to the specialist doctor.

**Stage 6. Waiting in line.** Waiting for the patient to see a specialist.

**Stage 7. Another visit to the specialist doctor.** On the basis of the performed examination, its description and an additional conversation with the patient, the doctor gives the final diagnosis to the patient and directs him to the hospital.

**Stage 8. Waiting.** The patient waits for a place to become available in the Rehabilitation Department.

**Stage 9. Admission to hospital.** The patient goes with the referral and full health documentation to the Admission Room. The patient is admitted to the hospital and referred to the ward.

**Stage 10. Admission to the ward.** At the Rehabilitation Department, the doctor on duty checks the patient's health records, establishing a plan of appropriate treatments and exercises.

**Stage 11. Waiting.** The patient is expected to wait for the next working day again. On the day of admission to the ward, he does not perform any procedures or exercises.

**Stage 12. Performing balance exercises.** The patient, together with the physiotherapist, goes to the rehabilitation rooms, where he performs a series of balance exercises.

**Stage 13. Performing resistance exercises.** Immediately after the balance exercises, the patient switches to resistance exercises.

**Stage 14. Waiting.** After prolonged physical exertion, the patient needs rest before carrying out any further exercises.

**Stage 15. Upright standing.** Standing upright is the most important exercise for this group of patients. They are designed to restore the smoothness of movements.

**Step 16. Waiting.** Bad work organization of employees means that the patient has to wait for the next stage of exercise.

**Stage 17. Performing exercises to restore precision of movements.** Performing exercises that restore the precision of movements by circling the shoulders, raising the hands, bending the elbows, rolling objects on the ground and grasping objects.

**Stage 18. Waiting.** Expectation of the patient, due to the lack of availability of exercise equipment at the moment.

**Stage 19. Performing isometric exercises.** The patient goes to the strengthening exercise rooms, where he performs a series of exercises that strengthen the spine, abdominal muscles, back muscles and limbs.

**Stage 20. Mobilization.** Immediately after the isometric exercises, the patient starts mobilization exercises.

**Stage 21. Waiting.** Waiting for a free exercise room.

**Stage 22. Active exercises.** At the end of the day, the patient performs exercises to improve muscular endurance and increase the efficiency of the circulatory and respiratory system.

**Stage 23. Decision to continue the patient's hospitalization.** Physiotherapists, together with the doctor on duty, assess the patient's health condition and its progress in treatment. Based on observation and discussion, they make a decision regarding his further hospitalization.

**Stage 24. Waiting.** The patient is waiting for the next day. On the next day, the patient repeats the early stages of treatment (12–23).

**Stage 25. Waiting in queue.** The patient is waiting in a queue for a check-up with a specialist doctor.

**Stage 26. Follow-up visit with a specialist doctor.** After hospitalization, the patient goes for a follow-up visit. The doctor re-assesses the patient's health, completes the medical documentation and issues the patient an discharge from the hospital.

**Stage 27. Release from hospital.** The doctor on duty completes the formalities related to the patient's release from the hospital and gives consent to leave the hospital.

Figure 1 shows a part of a current state value stream map (CS-VSM) of the selected process. In the process, waiting time appears as well as many problems.

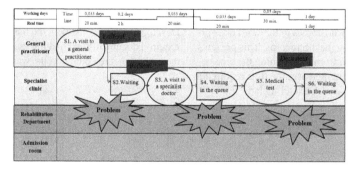

**Fig. 1.** The current state value stream map (CS-VSM) - a part

# 5  Problem Identification and Proposals of Improvements

The analysis of the current state map (CS-VSM) allowed for the identification of problems occurring in the process. For the identified problems, improvement actions were proposed (Table 2). The main goal of the proposed activities is to improve the analyzed rehabilitation process, which will allow to decrease the lead time (LT).

**Table 2.**  Problems and proposal for improvements - selected problems

| Name and description of the problem | Proposal of improvements |
|---|---|
| Problem 1: Incorrect patient registration system | |
| The patient arrives at the clinic several hours before the doctor arrives on duty to be sure that that it will be accepted. He registers and waits for the doctor to come to the office | Creation of a website or application, thanks to which each patient could register with a doctor at a suitable date and time |
| Problem 2. Long admission time to the hospital | |
| The patient waits for admission about 60 days. This is due to the limits in accordance with the agreement signed with the Polish National Health Fund (NHF) and the availability of places in the hospital | Increasing the number of places in the hospital and analyzing the possibility of increasing admission limits |
| Problem 3: One person dealing with carrying out and describing the test | |
| The person responsible for carrying out the tests is on duty alone, which extends the time of carrying out the test and describing it | Reorganization of employees' work. This action solves problem 3 and contributes to the deletion of steps 4, 6 and 7 of the process |
| Problem 4: Lack of a person responsible for coordinating the activities of the admission room staff | |
| There is no person responsible for supervising and coordinating the activities carried out by the staff of the admission room as part of the admission process. The consequence is poor organization of staff work and long process implementation time | Choosing a team coordinator. From among all the admission room employee, one person should be selected who would coordinate the work of employees and supervise the process of entering data into the database in order to improve this activity |
| Problem 5: Incomplete health records received from the patient | |
| The patient who comes to the ward does not have complete health records. This prevents the physician from determining the type of treatments and increases the time required to create a framework exercise plan | Developing a check list with the list of required documents and making it available to the patient before coming to the ward |

# 6   Development of the Future State Map (FS-VSM)

By solving the analyzed problems and eliminating mainly the waiting times, it will be possible to decrease the duration time of some activities from CS-VSM and finally the lead time (LT) (Table 3). Additionally, it will be possible to eliminate some NAV.

**Table 3.** The duration time of activities - CS-VSM and FS-VSM

| Stage | Value | CS-VSM | FS-VSM | Stage | Value | CS-VSM | FS-VSM |
|-------|-------|--------|--------|-------|-------|--------|--------|
| S1 | AV | 20 min | 20 min | S15 | AV | 1 h | 5 min |
| S2 | NAV | 12 min | 10 min | S16 | NAV | 30 min | 45 min |
| S3 | AV | 20 min | 20 min | S17 | AV | 45 min | 45 min |
| S4 | NAV | 20 min | 0 | S18 | NAV | 1 h | 0 |
| S5 | AV | 30 min | 30 min | S19 | AV | 45 min | 45 min |
| S6 | NAV | 1 day | 0 | S20 | AV | 30 min | 30 min |
| S7 | NAV | 10 min | 0 | S21 | NAV | 10 min | 0 min |
| S8 | NAV | 60 days | 40 days | S22 | AV | 45 min | 45 min |
| S9 | AV | 30 min | 20 min | S23 | NNVA | 30 min | 2h |
| S10 | AV | 45 min | 20 min | S24 | NAV | 2 h | 0 |
| S11 | NAV | 1 day | 1 day | S25 | NNVA | 2 h | 30 min |
| S12 | AV | 45 min | 45 min | S26 | AV | 30 min | 20 min |
| S13 | AV | 30 min | 30 min | S27 | NNVA | 20 min | 5 min |
| S14 | NAV | 1 h | 1 h | | | | |

Figure 2 shows a part of a future state value stream map (FS-VSM) of the selected process. In the process thanks to improvements the stages 4, 6 and 7 were removed, so are not included in the FS-VSM.

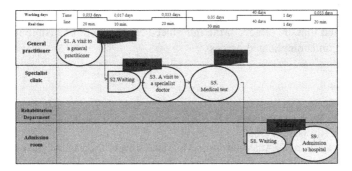

**Fig. 2.** The future state value stream map (FS-VSM) - a part

## 7  Assessment of the Effectiveness of the Implemented Activities

Using the indicators from Table 1 and the data collected from the process with the support of VSM, the rehabilitation process performance have been analyzed.

Table 4 presents the comparison of VA, NVA, NNVA and LT in the CS-VSM and FS-VSM. Is should be noted that the value of all indicators decreased. LT decreased about 35%.

**Table 4.** Comparison of VA, NVA, NNVA and LT - CS-VSM and FS-VSM

| Indicator | CS-VSM | FS-VSM |
|---|---|---|
| AV | 8 h 58 min | 8 h 40 min |
| NAV | 62 days 5 h 22 min | 41 days 2 h 30 min |
| NNAV | 2 h 50 min | 50 min |
| LT | ≈ 65 days | ≈ 42 days |

## 8  Conclusions

The aim of this article was to use VSM for analysis to identify and analyze sources of waste in the process of patient rehabilitation in the healthcare. Thanks to the application of the method, the necessary information on the sources of losses in the analyzed process was obtained. The use of VSM made it possible to identify areas of improvement not only for one process, but also for other services provided. Even partial of the proposed changes may contribute to increasing the efficiency and quality of the services provided. The proposed activities allow for the achievement of potential benefits in the form of shortening LT. However, one should bear in mind the potential difficulties and organizational and social limitations that may be related to the implementation of the proposed improvement activities. Before starting the implementation of changes, appropriate discussions and analyzes should be carried out in cooperation with the hospital management, the body authorized to supervise the activities of the institution. It should not be expected that the effects of the introduced changes will be visible immediately, as it is a long and complicated process.

## References

1. Jones, D.T., Roos, D., Womack, J.P.: Machine that Changed the World Simon and Schuster, Free Press, Riverside (1990)
2. Gornicka, D., Burduk, A., Jagodzinski, M.: Value stream mapping: an analysis of the versatility of applying the method in practice. In: Conference: 2nd International Conference on Economics and Management Innovations (ICEMI), 15–16 July 2017, Dhurakij Pundit University of Bangkok, Thailand International Conference On Economics And Management Innovations (ICEMI 2017), vol.1, No. (1), pp. 359–361 (2017)

3. Gola, A., Nieoczym, A.: Application of OEE coefficient for manufacturing lines reliability improvement. In: 4th International Conference on Management Science and Management Innovation (MSMI) 23–25 June 2017, Suzhou, Peoples R China Proceedings Of The 2017 International Conference On Management Science And Management Innovation (MSMI 2017). AEBMR-Advances in Economics Business and Management Research 31, pp. 189–194 (2017)
4. Henrique, D.B., Filho, M.G.: A systematic literature review of empirical research in Lean and Six Sigma in healthcare. Total Qual. Manag. Bus. Excell. 31(3–4), 429–449 (2020)
5. Mousavi Isfahani, H., Tourani, S., Seyedin, H.: Lean management approach in hospitals: a systematic review. Int. J. Lean Six Sigma 10(1), 161–188 (2019)
6. Fogliatto, F.S., Anzanello, M.J., Tonetto, L.M., Schneider, D.S., Muller Magalhães, A.M.: Lean-healthcare approach to reduce costs in a sterilization plant based on surgical tray rationalization. Prod. Plan. Control 31(6), 483–495 (2020)
7. Kahm, T., Ingelsson, P.: Stuck in the middle" first-line healthcare managers' responsibilities and needs of support when applying lean. Int. J. Qual. Serv. Sci. 12(2), 173–186 (2020)
8. Womack, J.P., Jones, D.T.: Lean Thinking – Lean Thinking. ProdPress, Poland (2008)
9. Dahlgaard, J.J., Pettersen, J., Dahlgaard-Park, S.M.: Quality and lean health care: a system for assessing and improving the health of healthcare organizations. Total Qual. Manag. Bus. Excell. 22(6), 673–689 (2011)
10. Smith, I., Hicks, C., McGovern, T.: Adapting Lean methods to facilitate stakeholder engagement and co-design in healthcare. BMJ 368, 35 (2020)
11. Sunder, M.V., Mahalingam, S., Krishna, M.S.N.: Improving patients' satisfaction in a mobile hospital using lean six sigma – a design-thinking intervention. Prod. Plan. Control 31(6), 512–526 (2020)
12. Wargin, J.J., Bishop, S.: Lean healthcare: rhetoric, ritual and resistance. Soc. Sci. Med. 71, 1332–1340 (2010)
13. Singh, S.K., Garg, S.K.: Value stream mapping: literature review and implications for Indian industry. Int. J. Adv. Manuf. Technol. 53(5–8), 799–809 (2011)
14. Jackson, T.L.: Mapping Clinical Value Streams, CRC Press, Boca Raton (2013)
15. Mazur, L.M., Chen, S.J.: Understanding and reducing the medication delivery waste via systems mapping and analysis. Health Care Manag. Sci. 11(1), 55–65 (2008)
16. Pilar, I.V.C., Julio, J.G.S., Juan, A.M.G., Jose, P.G.S.: Value stream mapping on healthcare. In: 2015 International Conference on Industrial Engineering and Systems Management (IESM), pp. 272–276. IEEE (2015)
17. Antosz, K., Stadnicka, D., Ratnayake, R.M.C.: Use of lean management philosophy in health sector: a VSM based case study. In: 2016 IEEE International Conference on Industrial Engineering and Engineering Management (IEEM), pp. 1523–1528. IEEE (2016)
18. Augustyn, A.: Analysis and improvement realization services in medical facility (in Polish). Unpublished work prepared under supervision of Katarzyna Antosz, Rzeszow (2020)

# DMAIC: A Proposed Method to Improve the Cleaning and Disinfection Process in Hospitals

Joana de Oliveira Pantoja Freire[1]([✉]) [iD], Robisom Damasceno Calado[2][iD], and Graciele Oroski Paes[1][iD]

[1] Universidade Federal do Rio de Janeiro, Rio de Janeiro, Brazil
[2] Universidade Federal Fluminense, Rio das Ostras, Brazil
robisomcalado@id.uff.br

**Abstract.** The hospital environment influences the chain of transmission of pathogenic microorganisms linked to the incidence of infections, making cleaning and disinfection (L&D) management measures necessary in order to contribute to patient safety. This study aimed to propose the viability of using the Lean Six Sigma approach in the management and improvement of the hospital terminal hygiene process. **Method:** Exploratory descriptive research, through a theoretical survey of tools used in the Lean Six Sigma approach. The DMAIC method, was used as a guide for this project hypothesis of management of the process of terminal hygienization of beds of a University Hospital in the city of Rio de Janeiro. **Results:** In the definition and organization phase of the project, the terminal sanitization process can be mapped through the VSM diagram and by an employee interview instrument during GEMBA. L&D quality assessment tools like fluorescent markers, ATP test and microbiological cultures can serve as pre and post indicators for possible improvement interventions. Tools such as: Current Reality Tree; GUT matrix; Prioritization matrix and 5W2H plan have proven to be good choices for the analysis and improvement phases. Finally, audits, visual and participative management, indicator reports for maintenance of actions can be done. **Conclusion:** The application of the DMAIC method of the Lean Six Sigma approach in hospital cleaning processes proved to be objective and feasible according to the proposed method, presenting itself as an alternative basis for future projects in the area.

**Keywords:** Disinfection · Cleaning hospital service · Action research · Health care associated infection

## 1 Introduction

Healthcare-Related Infections (HAIs) are one of the leading causes of death in the world, killing more people in the U.S. when compared to AIDS, breast

© IFIP International Federation for Information Processing 2021
Published by Springer Nature Switzerland AG 2021
A. Dolgui et al. (Eds.): APMS 2021, IFIP AICT 631, pp. 422–430, 2021.
https://doi.org/10.1007/978-3-030-85902-2_45

cancer and traffic accidents [2]. Patients with weakened immune systems, such as patients with hematological diseases undergoing chemotherapy, and who have undergone a greater number of invasive procedures, are the most susceptible to SAIs [3].

A study by the World Health Organization - WHO showed an increase in the prevalence of these infections in patients hospitalized in Intensive Care Units - ICU, surgical and orthopedic wards1. In developed countries, 5 to 10% of patients admitted to the ICU acquire an infection. In Brazil, of all recorded admissions, the infection rate is around 14% [2,19]. Another important data is that after adherence to bundled practice protocols, which is an evidence-based structured way of improving processes and patient care, a reduction of up to 70% of bloodstream infections was obtained [5].

There is evidence that the environment has a strong influence on the chain of transmission of microorganisms linked to the incidence of infections. Contaminated surfaces can serve as reservoirs for microorganisms, especially those resistant to antimicrobial agents, making the hospital environment, even indirectly, a substantial risk to the patient [1,10,14].

Hospital hygienization is achieved through cleaning and disinfection procedures. If in cleaning the focus is on the removal of organic matter, disinfection aims to remove or eliminate microorganisms in their vegetative form, not necessarily in sporulated forms, through chemical agents applied to inanimate surfaces, such as equipment and fixed surfaces, which were previously cleaned [1]. Surfaces close to the patient are touched more by the hands, and therefore, should be cleaned and disinfected more frequently. Evaluating and ensuring adequate sanitation of these surfaces has a positive impact on the treatment and recovery of patients' health [1,3].

On the other hand, terminal cleaning is defined as the process of cleaning and/or disinfecting all areas of the health care service to reduce dirt, microbial population, and the possibility of environmental contamination. However, it is carried out periodically, on a scheduled basis, and includes floors, walls, ceilings, and all furniture. It is carried out after the patient's discharge, transfer, or death [1].

Studies show that when a patient is admitted to a bed where there was previously a patient colonized by multi-drugs resistant microorganisms, there is a higher incidence of colonization by this same pathogen. Another reports that contaminated surfaces increased by more than 100% the risk of colonization of susceptible patients who occupied the same room as previously colonized patients [7].

It is common to find inadequate processes in terms of quality management of sanitation in the hospital environment mainly due to flaws in technique, routine, lack of monitoring, and qualified human resources, among others. Due to the importance of the issue, regulatory agencies recommend the use of monitoring methods of hospital cleaning/disinfection as an essential part of the institutional hospital infection control program [1].

The four main resources used for this purpose are: visual inspection, fluorescent markers, ATP testing, and microbiological cultures [6].

Visual inspection is the most commonly used method and sometimes the only one. Despite its low cost, it depends on a personal assessment and can be quite subjective. Studies have shown that surfaces were considered approved by visual inspection even before they were cleaned and disinfected, and it is not considered a good marker by itself [6].

The use of fluorescent products gives immediate feedback, is fast and easy to apply, but is laborious, because it must be marked without the staff noticing, and after cleaning, the places must be evaluated. Like visual inspection, it does not assess the microbial load [15].

The bioluminescence ATP test is a quick and objective method that provides a quantitative measurement. Its disadvantages include high cost, no identification of microorganisms, no cut-off point or standard, and the possibility of false-positive results [7].

The microbiological culture of the environment, considered the gold standard method, has high sensitivity and specificity and identifies the type of pathogen contaminating the area being evaluated. However, it is the least used and standardized method in institutions due to its high cost, necessary supplies, and delay in results, being used in cases of outbreaks and research [6].

Yearning for change and improvement, health services in general have sought new tools to manage and control their processes, identify problems and solve them effectively, such as Lean Six Sigma. The Six Sigma methodology is quantitative, structured, and focused on improving processes that already exist in the institution. It occurs with little or no investment and can be an excellent strategy for creating a culture of continuous improvement of results by increasing the quality of hospital hygiene processes [12].

The structured method often used in Six Sigma is the DMAIC (define-measure-analyze-improve-control) method, and statistical tools that assess opportunities for improvement [16, 17]. Lean Six Sigma has been contributing significantly to problem identification and resolution, process quality improvement, external (patients) and internal (healthcare professionals) customer satisfaction, and reduction of associated costs, operating expenses, and inventory [9].

Despite being highly recommended, using existing technologies and tools to understand and measure the sanitation process does not guarantee, by itself, a safe environment, free of pathogenic microorganisms. Considering that there are no indicators of quality in the cleaning of hospital surfaces in the study scenario of this research, the following question was defined: Is the application of Lean Six Sigma considered an effective strategy to improve the quality of processes related to hospital cleaning? Based on this premise, the study in question aimed to propose the feasibility of using the LEAN Six Sigma approach as a way of managing and improving the process hospital terminal sanitation.

## 2   Methods

This is an exploratory study, performed by a professional specialist in hospital infection control.

In order to facilitate the theory development through practical application, an Action Research methodology will be used. DMAIC method and action research study conducted in a hospital is a classic example of how LSS can bring bottom-line impact to an organization, alongside contributing to the process improvement mind-set in employees [8].

In a first moment, articles and case studies with a Lean Six Sigma approach were surveyed in literature, as well as frequently used quality tools. In parallel, the standard operating process for terminal hygiene of the beds was reviewed. A new literature search was performed, this time focusing on the existing cleaning and disinfection quality monitoring tools, to serve as indicators for this Lean Six Sigma project proposal. Finally, we tested the use of the DMAIC method as a guide to know the current state of the process, identify opportunities for improvement and effective interventions for the same.

The choice of this method was based on its structure, which allows the achievement of goals through accurate knowledge of the entire problem, avoiding hasty conclusions, and the spending of financial and human resources on ineffective actions [16]. In other words, this management method follows the effectiveness of process management through the diagnosis of an undesirable situation and the consequent search for solutions. Its main objective is to control and improve processes, services, and products in a continuous manner, with a beginning, middle, and end, bringing as an important aspect not only the solution of problems but also the issue of sustaining the proposed work [4].

Therefore, DMAIC depends on metrics (indicators) and reliable data to verify the quality of a process in statistical terms, evaluating what is being done, produced, if it is according to what is recommended or within the expected specifications [17].

## 3   Result

### 3.1   Definition Phase

To clearly define the scope of the project, it is necessary to assess the history of the problem and agree on the points of the project through a contract that should answer the following questions: What is the problem you want to solve? What is the goal to be achieved and what gains does it correspond to? Which process/stage is related to the problem in question?

It is worth remembering that this project covers the improvement of processes related to cleaning the hospital environment, which in turn covers the control of health care-related infections and patient safety, and is therefore consistent with the strategic planning of the study institution. This is an important aspect to start any Lean Six Sigma project. Other characteristics of a good project are that there must be a reliable database to measure the current state and the results after the intervention. In this case we will create these reliable and measurable indicators, and the solutions at the moment to solve the problem are not known to the institution.

In the definition phase the reasoning map compiles and documents project information in an objective and precise manner, creating a mental model and facilitating the conduction of the project. In this map it is already possible to foresee the possible tools to be used, to register the evolution of the actions in real time, and to promote the understanding of the project by people outside the team because it demonstrates how and why the data was collected. It allows contribution through ideas, shows which questions still need to be answered and which have already been completed [17].

The Project Agreement is the final deliverable of the definition phase containing a lot of important information, containing the LSS project planning, theme, the involved team, global goals from the problem, history, indicators, involved process.

### 3.2   Measurement Stage

In this stage, the problem defined in the previous stage must be analyzed in greater depth, with the help of statistics and process mapping. This moment is extremely important, because by knowing the problem in depth it will be possible to identify its possible causes.

Value Stream Mapping (VSM) is a structured diagram that will document all the processes involved from the beginning to the end of a hospital's terminal sanitation. The VSM allows you to know the process and record the duration of each step, as well as the conformities and non-conformities according to the Standard Operating Protocol - SOP guidance [20].

The data is collected where the process occurs (checklist) being essential in order to discover bottlenecks and potential improvements. This tool can capture the process "how it is" and not as "we suppose it to be" (Fig. 1).

The evaluation of the quality of room cleaning and disinfection can occur by using one or all of these three techniques: fluorescent markers, ATP test, and room culture. The choice of the suitable technique will vary according to either the human or the financial resources of each institution. It is worth mentioning that the researcher/evaluator does not interfere with the observations and tests in this initial phase.

Another way to identify problems and possible solutions is through observation and questions about the process to those who perform it, that is, to go to the "shop floor". This term, also known as "Gemba", is a Japanese term meaning "the place where things happen in manufacturing". In the context of LEAN it is the workplace where the process takes place [17]. Gemba can occur through an oral interview script, composed of open-ended questions.

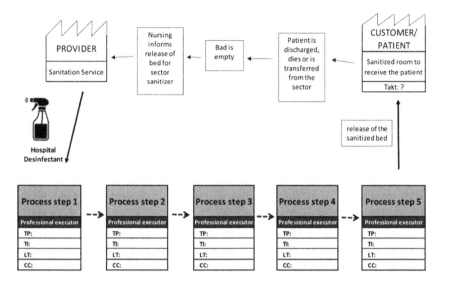

Customer supplier

Process flowchart: representes na operation of a process.

---▶ Push arrow: pulled system, demand generated by the cliente that push the process.

Takt – Takt Time: Process demand rate, pace.

TP – Total time of the process step.

TI – Time of interruptions during the process.

LT – Lead Time: Time between the start and end of a process without interruptions.

CC – Complete/Correct: representes the percentage of observations in that process that were complete and correct. (Source: author herself)

**Fig. 1.** VSM model for the terminal cleaning process according to institutional recommendation.

### 3.3   Analysis Stage

This stage aims to seek proof of the causes of the problem identified previously, through facts, such as records, graphs and data. The main job is to identify the root cause(s) of the problem, so that the next stage can intervene.

The Current Reality Tree (ARA) has its origin in the Theory of Constraints, which brings the concept that all the elements that constitute a system have an interdependent relationship among themselves, and that every system has a single cause for several effects that affect it. ARA will be used to identify a common causer that is generating unwanted effects, and this cause and effect relationship is created from the formation: "If... Then". That is, "If the hypothesis is true, then such an effect will occur" [13].

The GUT Matrix at this point will be used to prioritize the treatment of the problems found considering their severity, urgency and tendency. The severity (G) refers to the impact of a certain problem on people, things, processes, as well as the consequences of the permanence of this problem if it is not solved.

Urgency (U) refers to the time needed to solve the problem, and trend (T) is the ability of a problem to grow over time, whether it tends to increase or decrease for example [11].

In the DMAIC analysis phase, the relationship between the prioritized causes (variable Xs) in relation to the result of interest (variable Y) must be verified through statistical analysis.

If the goal is to correlate two variables (x and y) to find out if changing the X variable affects the behavior of the Y variable, for continuous data statistical tools can be used: Correlation; Scatter Diagram; and Simple Linear Regression. For discrete data I will use the Chi-square test.

If the focus is to demonstrate that the stratification of variable Y generates different results, i.e., that the stratification factors represent causation, then for continuous and discrete data you can use Hypothesis Tests, Test of Equality of Variance, ANOVA and Multi-Variance Plots [18].

### 3.4  Improvement Stage

In this phase the objective is to propose, test and execute solutions for the problems found. It is common to have many improvement ideas that can result in a voluminous action plan that is difficult to execute on time.

The Prioritization Matrix or Effort and Impact Matrix is a tool that uses scores to correlate the proposed solutions to the prioritization criteria. The most common criteria are: low cost, ease, speed, impact on the Root Cause, and low or high probability of causing undesired effects. A weight is defined for each criterion (e.g. 5 to 10 points) and a scoring legend is used for the relationship between each criterion and solution (e.g. 0 - no correlation; 1 - weak correlation; 3 - moderate correlation; 5 - strong correlation). Multiplying the value generated by the weight of the criterion and adding the values in the row gives the final result, that is, the improvement solutions that should be prioritized in the action plan [20].

To unfold the elected ideas into more tangible information, the 5W2H Action Plan will be used. This diagram-shaped model clearly and concisely lists what will be done (What), when it will be done (When), who will do it (Who), where it will be done (Where), why it will be done (Why), how it will be done (How), and the financial impact of the action (How Much) [11,20].

In the execution of the improvement plan the results are generated. The level of results depends on the quality of the actions and the level of execution of the proposed action plan. In this stage, the objectives are: to guarantee the execution of the actions outlined in the planning stage; to disclose the actions to everyone in order to obtain an alignment among the involved areas and to promote training if necessary to guarantee the actions.

### 3.5  Control Stage (Control)

After the implementation of the improvement actions it is necessary to evaluate and monitor the achieved results to ensure their sustainability. Some practices used in this stage are Audits; View Management Board; Reports sent to leaders with project indicators; and Participative Management with monthly strategy alignment meetings [4,16,17].

## 4   Conclusion

The application of the DMAIC method of the Lean Six Sigma approach in hospital sanitation processes proved to be objective and feasible according to the proposed step-by-step instructions, presenting itself as an excellent alternative basis for future improvement projects in the area. Once all the steps of the method were structured, it is possible to reproduce it for other health institutions with small adjustments according to the SOP guidance of each location and physical structure.

Finally, the paper has suggested that Action Research is a suitable research design for Lean Six Sigma/DMAIC projects allowing a collaborative approach to knowledge enhancement based on the researcher and research becoming equal in the pursuit of knowledge.

Actions to improve processes related to the cleaning of hospital surfaces are capable of promoting a safer environment for patients, companions, and healthcare professionals. Through the DMAIC method proposed in Lean Six Sigma projects it will be possible to identify possible losses such as: rework, unnecessary movement, inadequate spending of material, incorrect sanitizing technique, lack of standardization, among others.

It is also possible to obtain potential gains through the creation of performance indicators, standardization and optimization of the work process, creation of active listening and valorization of the hygiene professionals.

**Acknowledgements.** The authors would like to thank the Federal Fluminense University and the Fundação Euclides da Cunha, which made the research project funded by the Ministry of Health of Brazil viable; TED 125/2019, Number: 25000191682201908

## References

1. Agência Nacional de Vigilância Sanitária: Segurança do paciente em serviços de saúde: limpeza e desinfecção de superfícies (2010). https://www.gov.br/anvisa/pt-br/centraisdeconteudo/publicacoes/servicosdesaude/publicacoes/publicacoes/manual-de-limpeza-e-desinfeccao-de-superficies.pdf/view
2. Agência Nacional de Vigilância Sanitária: Programa nacional de prevenção e controle de infecçøes relacionadas à assistência à saúde (PNPCIRAS 2016–2020) (2016). https://www.gov.br/anvisa/pt-br/centraisdeconteudo/publicacoes/servicosdesaude/publicacoes/pnpciras-2016-2020.pdf
3. Barros, F.E., Soares, E., de Oliveira Teixeira, M.L., da Silva Castelo Branco, E.M.: Controle de infecções a pacientes em precaução de contato. J. Nurs. UFPE/Revista de Enfermagem UFPE **13**(4) (2019)
4. Buzzi, D., Plytiuk, C.F.: Pensamento enxuto e sistemas de saúde: um estudo da aplicabilidade de conceitos e ferramentas lean em contexto hospitalar. Revista Qualidade Emergente **2**(2) (2011)

5. Centers for Disease Control and Prevention: Healthcare-associated infections (HAI) progress report (2016). http://www.cdc.gov/hai/surveillance/progress-report/
6. Frota, O.P., Ferreira, A.M., Rigotti, M.A., Andrade, D.D., Borges, N.M.A., Ferreira, M.A.: Effectiveness of clinical surface cleaning and disinfection: evaluation methods. Revista brasileira de enfermagem **73** (2020)
7. Furlan, M.C.R., et al.: Correlação entre métodos de monitoramento de limpeza e desinfecção de superfícies ambulatoriais. Acta Paulista de Enfermagem **32**, 282–289 (2019)
8. Gibbons, P.M.: Improving overall equipment efficiency using a lean six sigma approach. Int. J. Six Sigma Competitive Adv. **2**(2), 207–232 (2006)
9. Honda, A.C., Bernardo, V.Z., Gerolamo, M.C., Davis, M.M.: How lean six sigma principles improve hospital performance. Qual. Manage. J. **25**(2), 70–82 (2018)
10. Khan, H.A., Ahmad, A., Mehboob, R.: Nosocomial infections and their control strategies. Asian Pac. J. Trop. Biomed. **5**(7), 509–514 (2015)
11. Kist, L.T., da Rosa, F.R., Moraes, J.A.R., Machado, E.L.: Diagnosis of hospital waste management in vale do rio pardo-rio grande do sul, Brazil. Revista de Gestão Ambiental e Sustentabilidade **7**(3), 554–569 (2018)
12. Maleyeff, J., Arnheiter, E.A., Venkateswaran, V.: The continuing evolution of lean six sigma. TQM J. (2012)
13. Nascimento, R.F.D.C., et al.: Gerir projectos com base na teoria das restrições: caso de estudo. Ph.D. thesis, Instituto Superior de Engenharia de Lisboa (2017)
14. Oliveira, A.C.D., Damasceno, Q.S.: Superfícies do ambiente hospitalar como possíveis reservatórios de bactérias resistentes: uma revisão. Revista da Escola de Enfermagem da USP **44**, 1118–1123 (2010)
15. Smith, P.W., Sayles, H., Hewlett, A., Cavalieri, R.J., Gibbs, S.G., Rupp, M.E.: A study of three methods for assessment of hospital environmental cleaning. Healthcare Infect. **18**(2), 80–85 (2013)
16. Spagnol, G.S., Calado, R.D., Sarantopoulos, A., Min, L.L.: Lean na Prática. GlobalSouth Press, New York (2018)
17. Werkema, C.: Criando a cultura lean seis sigma. Elsevier Brazil (2013)
18. Werkema, C.: Ferramentas Estatísticas Básicas do Lean Seis Sigma Integradas: PDCA e DMAIC. Elsevier (2016)
19. World Health Organization: Health care-associated infections fact sheet (2021). https://www.who.int/gpsc/country_work/gpsc_ccisc_fact_sheet_en.pdf.   Accessed 08 Jan 2021
20. Zattar, I.C., Silva, R.R.L.D., Boschetto, J.W.: Aplicações das ferramentas lean na área da saúde: revisão bibliográfica. J. Lean Syst. **2**(2), 68–86 (2017)

# New Trends and Challenges in Reconfigurable, Flexible or Agile Production System

# Proposal of a Methodology to Improve the Level of Automation of an Assembly Line

Hasnaa Ait Malek[1,2]([⊠]), Alain Etienne[1], Ali Siadat[1], and Thierry Allavena[2]

[1] LCFC, Arts et Métiers Paristech, Metz, France
[2] Stellantis, Vélizy Villacoublay, France
hasnaa.aitmalek@stellantis.com

**Abstract.** This paper's aim is to propose a new methodology for organizing and identifying the assembly operations that ought to be automated in an automotive assembly line. A state of the art in the matters of automation methods is presented to situate the research work and to analyze the different methods presented in the literature. As a result, there are some lacks in terms of methods that sought to help improve the level of automation. In the last part, the different requirements of the methodology are defined, which led to a proposal for a method that respects all the requirements and that allows not only the grouping of operations, but also the analysis of the automation and the line balancing. Finally, and to verify the proposal, three activities of the automation methodology have been applied on a Stellantis assembly line. The result of the study showed that it is possible to group several screwing operations.

**Keywords:** Automation · Assembly line · Methodology

## 1 Introduction

After the first Industrial Revolution, factories underwent changes in the way they were organized and worked [1].With the development of electronics and information technology, a great technological evolution took place; notably the automation of production lines.

Being in constant evolution, the automotive industry was the first to adopt this path [2]. Thus, and in order to remain competitive on the market, manufacturers have turned to automation to increase their production capacity and respond to product variability. Nevertheless, the automation of assembly lines remains a major challenge [3]. Although new technologies are highly developed, the use of highly automated processes is not necessarily the best solution [4]. Several criteria, such as profitability, cost, quality and ergonomics, are also involved.

The diversity of models and parts as well as the integration of new vehicles (hybrids, electric) are constraints to be taken into consideration to achieve an automation that meets these criteria. Thus, the objective of the research work undertaken is to define the principles and criteria, to identify and select the operations to be automated in order to develop a decision support method.

© IFIP International Federation for Information Processing 2021
Published by Springer Nature Switzerland AG 2021
A. Dolgui et al. (Eds.): APMS 2021, IFIP AICT 631, pp. 433–441, 2021.
https://doi.org/10.1007/978-3-030-85902-2_46

Today, since assembly operations are short in time, it is preferable to group them to optimize the profitability of automation. This leads to these following questions: What is the most appropriate approach to identify groupings of operations and automate them? What are the criteria for grouping operations in order to balance the operator and machines loads?

## 2   State of the Art on the Different Decision Support Methods

The literature in automation is technical and aims at developing new technologies rather than improving the level of automation of a production line. This section presents the different decision support methods found in the literature. It consists in seven methods that have been analyzed, while presenting their advantages and drawbacks. The aim is to study them in order to propose a method to improve the level of automation of an automotive assembly system. Thus, the methodology must be analytical (R1); it must allow a detailed study of all the operations, as well as the resources attached to them. The objective is to collect all the information related to the assembly process before proposing a solution. The methodology must also allow the study of several automation scenarios (R2), and through several criteria with different weighting, it must propose an automation scenario (R3). The method must give calculation rules (R4) for each of the criteria and must allow an analysis of grouping of operations (R5) and balancing of the assembly line (R6).

In the 2000s, Parasuraman, Sheridan and Wickens [5] proposed their method in the form of a flowchart containing several iterative steps. The first one is to identify the type of automation, then identify the different possible levels of automation. For each of them, one must examine its consequences on human performance: this is the first evaluation criterion. However, human performance is not the only important factor. Secondary evaluation criteria include the reliability of the automation and the costs of the proposed solutions. This method takes human performance as the evaluation criterion, and therefore does not give an opportunity to choose other preponderant criteria. On the other hand, for all criteria (primary and secondary), the method does not present calculation rules to quantify them, nor conditions to deduce the optimal automation scenario. And finally, the list of criteria is not complete, several parameters other than Man and costs must be taken into account.

In 2013, Konold and Reger [6] developed a method to identify the level of automation that should be adopted. It consists of 4 levels of assembly systems. The decision is made by analyzing several parameters such as production volume, cycle time, product life cycle etc. This method, as simple as it is, presents a single result that can be in the form of several levels of automation. However, it analyzes the entire process, and therefore does not study each workstation (or even operation by operation). As far as the input parameters are concerned, Konold and Reger do not give details on how they established the conditions for acceptance or rejection. It is also necessary to point out that these parameters are insufficient, and that the cost dimension is non-existent.

In 2005, Boothroyd [7] studied the assembly of an electrical outlet in order to identify the most suitable and cost-effective assembly process. The author calculated the cost of each component for six different assembly processes. For each possible assembly scenario, he calculated the unit cost of the product. He estimated the cycle time for each station, the cost of each machine/robot, the labor cost, the efficiency rate, the quality control cost. Combining all of this data, Boothroyd performed a comparison of the overall cost of production for each of the six possible assembly scenarios. This method has several advantages, including the fact that it allows to analyze the components of the product, but also to start from several hypotheses and evaluate them in a quantitative way. Nevertheless, Boothroyd has used standard costs for machines and robots, but these costs differ according to their technical characteristics (more precision, more reliability…) which could distort the results. In addition, the only criterion used is the direct cost. It goes without saying, this criterion is preponderant for the majority of companies, but it is not the only one: the choice can be made on the basis of other criteria (quality, ergonomics…).

In 2008, Almannai, Greenough and Kay [8] developed a decision support tool combining the Quality Function Deployment (QFD) method and the Failure Mode and Effect Analysis (FMEA) method. The methodology consists of 3 steps: the first one consists in identifying the criteria relating to the company's objectives using the QFD method. The second in, identifying the automation alternatives through a second multi-criteria analysis. Finally, the resulting solution is submitted to the FMEA which identifies the risks and potential failures associated with it. This method has several advantages such as relying on the company's objectives and decisions, as well as using evaluation criteria to confirm the best automation choice. However, the authors did not provide a list of criteria or a method for performing multi-criteria analysis. Nor did they provide any clarification on the definition of automation alternatives.

Kapp [9] has developed a method to evolve manufacturing systems to produce new products. This method is called USA (Understand, Simplify, Automate) [10] and consists of understanding the existing process, optimizing it (Simplify) then automating it. This method has several advantages, especially in terms of understanding and optimizing the process. Nevertheless, on the automation part, this method does not give a clear vision of the approach to follow in order to choose a profitable automation.

Lindström and Winroth [11] proposed methods called Dynamo and Dynamo ++, which was taken up and developed by Fasth and Stahre [12]. These methods consist of understanding the manufacturing process by studying the production flow, the functioning and constraints of each station and by identifying the operations and sub-operations carried out on the line. The analysis of all these data allows to assign the minimum and maximum possible level of automation to each operation. Once these degrees of automation have been identified, the authors proposed a list of criteria to evaluate each possibility; these criteria are chosen based on the company's strategy. These methods have several advantages, including the use of the definition of the level of automation that differentiates between physical and cognitive operations. The authors suggest that a good understanding of the functioning and constraints of a production process is beneficial before embarking on an automation approach. However, these approaches do not provide a decision aid for assigning possible levels of automation and choosing an optimal level. In addition, the evaluation criteria have not been addressed.

In 2008, Gorlach and Wessel [13] carried out a study of VolksWagen assembly lines at three different sites. By combining several criteria such as cost, productivity, quality and flexibility, the authors were able to determine whether or not automating these lines is cost-effective. This method proposes to study automation based on four criteria: cost, quality, productivity and flexibility. For each criterion, the authors proposed a series of parameters to be evaluated. These criteria are quantitative except the last one (flexibility); the authors did not explain how to quantify it. The authors did not propose a decision-support method neither on the choice of a technical solution, nor on the multi-criteria analysis to be carried out to choose the level that meets the criteria chosen beforehand.

In the Table 1, the methods found in the literature have been evaluated based on the requirements presented in Sect. 2. As a result, none of the methods analyzed in the literature fully address the problem. Especially since none of the methods deal with the grouping of operations (R5) and the balancing of the line (R6). Certainly, there are methods in the literature that deal with these two topics, but they don't focus on automation. In this paper, we have cited the best-known methodologies that deal with the subject of automation. It is therefore essential to present a methodology that meets the industrial requirements cited above. This section introduces the proposed method and explains its different steps.

**Table 1.** Evaluation of methods found in the literature based on the requirements presented in Sect. 2

| Methods | Requirements | | | | | |
|---|---|---|---|---|---|---|
| | R1 | R2 | R3 | R4 | R5 | R6 |
| Parasuraman & Sheridan's method [5] | | X | | | | |
| Konold's method [6] | X | X | X | | | |
| Boothroyd's method [7] | X | X | | X | | |
| Almannai's method [8] | | X | X | | | |
| Kapp's method [9] | X | | | | | |
| Dynamo & Dynamo ++ [11,12] | X | X | | | | |
| Gorlach's method [13] | X | X | X | | | |

With R2: the methodology allows the study of several automation scenarios, R3: the methodology proposes an automation scenario, R4: The methodology gives calculation rules, R5: the methodology allows an analysis of grouping of operations, and R6: the methodology allows an analysis of assembly line balancing.

## 3   Proposal of a New Decision Support Methodology

The proposed decision-support methodology is illustrated by a SADT in Fig. 1. This method consists of applying six main steps that aim at grouping operations, proposing and evaluating several automation scenarios while analyzing the line balancing. This method is mainly based on the experiences of the trade. This work will try to translate them into several constraints and rules to provide assistance to industrials.

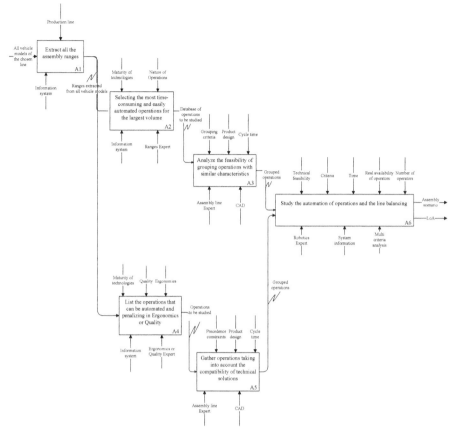

**Fig. 1.** Presentation of the proposed methodology to improve the level of automation

A1: This step consists of extracting all the assembly ranges from all the vehicle models on a production line. These ranges are extracted from the Stellantis information system.

A2: This step consists of creating a database presenting all the assembly operations with their duration and nature. The nature of the operations is the first selection criteria. For each type of operation, the level of maturity of the automation must be determined. The goal is to identify the longest and most easily automated nature of operations.

A3: This step consists of analyzing the possibility of grouping operations with the same nature and similar characteristics. To do so, six sub-steps illustrated in Fig. 2 are applied.

A3.1: The aim is to list all operations that have similar characteristics. This step requires analyzing the product and indicating, for each operation, the grouping criteria (qualitative and quantitative) established through discussions and meetings with Stellantis industrials. An example of these grouping criteria and codes is shown in Table 2. The available resources are the 3D model of the parts, experience and the knowledge rules.

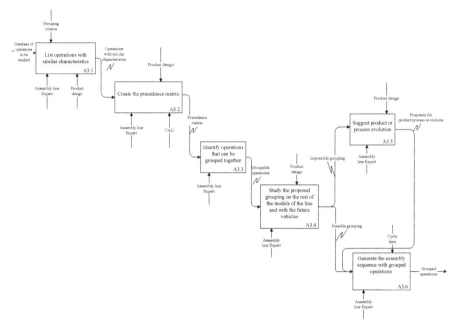

**Fig. 2.** Details of A3 - analysis of grouping of operations

**Table 2.** Sample codes and grouping criteria

| Code title | Grouping criteria |
| --- | --- |
| Grease/lubricate | Nature, quantity, characteristics of the lubricant, surface accessibility, area on the vehicle |
| Checking compliance | Nature of control, accessibility, area on the vehicle |
| Screw, tighten | Screw head, torque, socket, accessibility, area on the vehicle |

A3.2: The aim of this step is to study the planning of operations by drawing a precedence matrix.

A3.3: The aim of this step is to identify operations that can be performed in the same workstation and with similar characteristics.

A3.4: This step consists of analyzing the different products assembled in the line, but also the future ones, and identifying whether the grouping of operations is technically feasible for each of these products.

A3.5: If the grouping of operations proves to be impossible, evolutions of the product and/or production process can be proposed.

A3.6: The last step is to generate the assembly sequence with the grouped operations. The control point here is the cycle time.

A4-A5: These steps are an alternative of the previous process. The aim is to study and group together critical operations in ergonomics and/or quality (Same approach as A3).

A6: The sixth and final step of the methodology is to study the line balancing and evaluate the different automation scenarios through a multi-criteria analysis. The criteria chosen are cost, quality, ergonomics, flexibility and reliability of the means of production.

# 4   Example

The methodology to improve the level of automation of an assembly line was applied on a Stellantis vehicle assembly line. The application of the A1, A2 and A3 activity is detailed in the following sections.

## 4.1   A1-A2: Routing Extraction and Database Creation

The first step was to extract all the ranges. Then, we had to determine the scope and the nature of the operations to be studied.

The nature of the operations was classified according to the time and the level of maturity of the automation. Thus, it was decided to study the screwing operations. Then, a database was created by filling in, for each operation, the workstation number, the time and the grouping criteria. In the case of screwing, to be able to group the operations, it is necessary to have the same screw head, the same torque and the same socket. The screwing operations must be accessible, and must be done on the same vehicle area to make automation possible and profitable.

For confidentiality reasons, the extraction of the ranges, as well as the construction of the database will not be presented.

On the chosen production line, 272 screwing operations have been studied. The sockets are between 10 and 13 mm, the torques are between 2 and 70 Nm and the screw heads are Hexagonal or Torx.

## 4.2   A2-A3: Creation of the Database and Study of Grouping Criteria

Based on the grouping criteria for screwing operations (torque, screw head, socket, area on the vehicle, and accessibility), those to be studied first are the screwing operations that are accessible at the front and rear of the vehicle, have hexagonal head, 10 mm socket, subjected to a torque of 4 Nm.

Once the operations were identified, a matrix was drawn to study the different precedence constraints. The groupings of operations identified by applying the different stages of the A3 activity are the screw connections that concern: the lower bumper absorber and the headlamps at the front of the vehicle, as well as the central support for the rear bumper. In order to carry out the grouping of screw connections, it is essential to validate the pre-holding of the parts. This last criterion has not been previously taken into account in the methodology to simplify it and avoid setting up several constraints that will make the search for groupings difficult.

# 5   Discussion

One of the objectives of this research work is to propose a decision support method to increase the level of automation for an automotive assembly system with the aim of improving quality, ergonomics and cost.

To ensure good profitability, the solution is to group together operations that have the same characteristics while analyzing precedence constraints.

At this stage, a grouping of screwing operations has been identified manually. A model based on constraint programming is being developed to help the industrials process all the input data, and depending on the constraints imposed, the model will propose an optimal grouping of operations. The next step is to propose several automation scenarios by specifying the appropriate technology to perform the grouped operations automatically. The decision will be made through a multi-criteria analysis that will take into account the cost, ergonomics, quality, flexibility and reliability of the means of production.

The methodology allows to explain the exact steps to follow in order to process the different input data (factory architecture, product design, production lines, assembly operations characteristics...), and thus to propose groupings of operations to have a profitable automation. With the tools being developed (precedence matrix, constraint programming model, etc.), the industrials will save time in processing all this information. Indeed, the program will display proposals for grouping operations, and based on the current technologies, the profitability of the project and the various criteria mentioned above, the industrials will choose the grouping of operations that will respect all these criteria. In perspective, and to validate the methodology and illustrate the gains after its use, we will apply it on a grouping of operations already performed by Stellantis.

# 6   Conclusion

A review of the literature shows that there are few methods to help people make decisions about the level of automation. These approaches are based on few criteria and do not clarify the methods for calculating or estimating these criteria. Thus, an analysis of these different methods has been carried out while explaining the requirements that the method must meet to answer our problem. The methodology must study all the information related to the architecture of the plant, the sequence of operations or the resources used. The methodology must allow us to group operations together, but also to propose several automation scenarios. Based on several criteria, this method must study the line balancing and present the automation scenario that best satisfies all of these criteria.

To verify this methodology, it was manually applied on the Stellantis assembly line. The result is a grouping of 3 screwing operations. This paper presents the details of the application of A1, A2 and A3 activities. The next step will be to analyze the balancing of the line and to evaluate the choice of technologies and the automation scenario.

**Acknowledgements.** This research work was carried out as part of a CIFRE thesis between the Stellantis group and the Arts et Métiers de Metz.

This work takes place within the framework of the OpenLab Materials and Processes, which brings together the Arts et Métiers network, GeorgiaTech Lorraine and the Stellantis Group.

# References

1. Muhuri, P.K., Shukla, A.K., Abraham, A.: Industry 4.0: a bibliometric analysis and detailed overview. Eng. Appl. Artif. Intell. **78**, 218–235 (2019). https://doi.org/10.1016/j.engappai.2018.11.007
2. Jovane, F., Koren, Y., Boër, C.R.: Present and future of flexible automation: towards new paradigms. CIRP Ann. **52**, 543–560 (2003). https://doi.org/10.1016/S0007-8506(07)60203-0
3. Wiendahl, H.-P., et al.: Changeable manufacturing—classification. Des. Oper. CIRP Ann. **56**, 783–809 (2007). https://doi.org/10.1016/j.cirp.2007.10.003
4. Krüger, J., Nickolay, B., Heyer, P., Seliger, G.: Image based 3D surveillance for flexible man-robot-cooperation. CIRP Ann. **54**, 19–22 (2005). https://doi.org/10.1016/S0007-8506(07)60040-7
5. Parasuraman, R., Sheridan, T.B., Wickens, C.D.: A model for types and levels of human interaction with automation. IEEE Trans. Syst. Man Cybernet Part A: Syst. Hum. **30**, 286–297 (2000). https://doi.org/10.1109/3468.844354
6. Konold, P., Reger, H.: Praxis der Montagetechnik: Produktdesign, Planung. Springer-Verlag, Systemgestaltung (2013)
7. Boothroyd, G.: Assembly Automation and Product Design. CRC Press, Taylor & Francis, Boca Raton (2005)
8. Almannai, B., Greenough, R., Kay, J.: A decision support tool based on QFD and FMEA for the selection of manufacturing automation technologies. Robot. Comput.-Integr. Manuf. **24**, 501–507 (2008). https://doi.org/10.1016/j.rcim.2007.07.002
9. Kapp, K.: The USA Principle—The Key to ERP Implementation Success. APICS Perform. Advan. **7**, 62–67 (1997)
10. Singh, R.: Automatic Tool Changer. (2015).
11. Lindström, V., Winroth, M.: Aligning manufacturing strategy and levels of automation: a case study. J. Eng. Tech. Manage. **27**, 148–159 (2010). https://doi.org/10.1016/j.jengtecman.2010.06.002
12. Fasth, Å., Stahre, J.: Does Level of Automation need to be changed in an assembly system? - A case study. 9 (2008).
13. Gorlach, I., Wessel, O.: Optimal level of automation in the automotive industry. Eng. Lett. **16**, 141–149 (2008)

# Scalability and Convertibility Models and Approaches for Reconfigurable Manufacturing Environments

Abdelhak Dahmani$^{(\boxtimes)}$ and Lyes Benyoucef

Aix Marseille University, University of Toulon, CNRS, LIS, Marseille, France
Abdelhak.dahmani@lis-lab.fr

**Abstract.** The reconfigurable manufacturing system (RMS) is one of the newest manufacturing paradigms. In this paradigm, machine components, machine software, or handling units can be inserted, removed, modified, or interchanged as needed and, where appropriate, imposed by the need to adapt and adjust quickly and cost-effectively to changing requirements. RMS is considered to be a convenient processing paradigm for the manufacture of varieties as well as a scalable enabler for this variety. Considered as two of the six main RMS characteristics, in this paper, we review the most used models and solving approaches dedicated to scalability and convertibility in reconfigurable manufacturing environments. Moreover, we highlight the most critical research gaps.

**Keywords:** Scalability · Convertibility · Reconfigurable Manufacturing Systems

## 1 Introduction

In today's world, the responsiveness of manufacturing systems and cost-efficiency are the main factors affecting companies' competitiveness. Especially in recent years, companies face the problem that customers prefer a greater variety of unpredictable quantities. In this case, the best way will be to design a modern manufacturing system with new features that will allow them to improve their capabilities in the future. RMS is the cornerstone of this manufacturing model. The idea of (RMS) became well known at the end of the 20th century, introduced by [1] as a potential solution to the problems mentioned above, which can be applied to meet the needs of customers and the environment of demand for functions and production capacity. Thanks to its flexible structure and six key characteristics: modularity, integrability, customization, convertibility, scalability, and diagnosability [2].

In this research work, we focus on two major RMS characteristics, namely scalability and convertibility. We present the existing models and approaches. More precisely, Sect. 2 presents some research works dedicated to the six RMS characteristics. Section 3 analyses some models and approaches used to address scalability problems. Section 4 discusses some models of convertibility and their solving approaches. Section 5 highlights the most important research gaps in the field. Finally, Sect. 6 concludes the paper.

© IFIP International Federation for Information Processing 2021
Published by Springer Nature Switzerland AG 2021
A. Dolgui et al. (Eds.): APMS 2021, IFIP AICT 631, pp. 442–451, 2021.
https://doi.org/10.1007/978-3-030-85902-2_47

## 2 RMS Characteristics

Due to its six key characteristics, namely integrability, modularity, customization, convertibility, scalability, and diagnosability [2], RMS is considered one of the most appropriate paradigms to address today's industrial problems. Moreover, even if scalability has attacked researchers' intentions, convertibility is still an open question. In the following, we provide some recent definitions of the six characteristics [3].

- **Scalability (design for capacity changes):** The capability of modifying production capacity by adding or removing resources and/or changing system components.
- **Convertibility (design for functionality changes):** The capability of transforming the functionality of existing systems and machines to fit new production requirements.
- **Diagnosability (design for easy diagnostics):** The capability of real-time monitoring the product quality and rapidly diagnosing the root causes of product defects.
- **Customization (flexibility limited to part family):** System or machine flexibility around a part family, obtaining thereby customized flexibility within the part family
- **Modularity (modular components):** The compartmentalization of operational functions into units that can be manipulated between alternative production schemes
- **Integrability (interfaces for rapid integration):** The capability of integrating modules rapidly and precisely by hardware and software interfaces.

## 3 Scalability Models and Approaches

The scalable architecture enables the company to have a manufacturing infrastructure that meets current demand and improves production to respond quickly to customer demand. A review of the literature on the scalability of manufacturing systems shows that there are two main lines of research in this area [4]:

- Design of RMS focused on increasing their scalability level.
- Capacity planning using the scalability of RMS to adapt their production throughput to the existing demand.

   **Model 1**: [5] considered RMS scalability based on the transfer line (TL) type structure with batch production for mass production. They modeled conventional transfer line systems and the homogeneous paralleling flow line (PFL) system. Finally, they concluded that *the characteristic of cost-effective capacity changes in HPFL meets the scalability requirements of reconfigurable manufacturing systems.*

**1. Conventional Transfer Line System**
The system density function $P(\tau)$ is given by:

$$P(\tau) = \sum_{i=1}^{q} p_i(\tau) = \sum_{i=1}^{q} \frac{n_i}{n_s} \delta(\tau - \tau_i) \tag{1}$$

where $\tau_i (i \in 1, 2, \ldots, q)$ is the cycle time, $\delta$ is Dirac delta function, $q$ is the number of distinct cycle time, $n_i$ is the stage frequency for cycle time and $n_s = \sum_{i=1}^{q} n_i$, $p_i(\tau)$ is the stage density function.

## 2. Homogeneous Paralleling Flow Line (HPFL) System

The minimum required number of stations for stages with cycle time $\tau$ can be defined as:

$$\left[ g(\tau, v) = \frac{v}{T_a} \tau \right] \tag{2}$$

From Eqs. (1) and (2), the minimum number of required stations is given by:

$$n(\tau, v) = n_s P(\tau) g(\tau, v) = \sum_{i=1}^{q} n_i \delta(\tau - \tau_i) \left[ \frac{v}{T_a} \tau \right] \tag{3}$$

Here, the minimum number of required stations $N(v)$ for the entire system with demand "$v$" is given by:

$$N(v) = \int_0^\alpha n(\tau, v) d\tau = \int_0^\alpha \sum_{i=1}^{q} n_i \delta(\tau - \tau_i) \left[ \frac{v}{T_a} \tau \right] d\tau = \sum_{i=1}^{q} n_i \left[ \frac{v}{T_a} \tau_i \right] \tag{4}$$

where $T_a$: is the available working time of the system. If the cost of stations is identical and equal to $c$, then the total station cost of the system is given by:

$$C(v) = c \times N(v) \tag{5}$$

**Model 2**: In [6], the dynamic scalability process in RMS is modeled, where different capacity scalability policies are evaluated for different demand scenarios. The model represents capacity scalability as a scaling rate $SR(t)$ determined by the required capacity $RC(t)$ together with the scalability delay $SDT$ given by:

$$SR(t) = \frac{C(t) - RC(t)}{SDT} \tag{6}$$

where $C(t)$ is the capacity level at time $t$. Moreover, the inventory adjustment rate at time $t$, $AI(t)$ is controlled by the inventory gap between the desired inventory level $DI(t)$ and current inventory level $I(t)$ given by:

$$AI(t) = \frac{DI(t) - I(t)}{IAT} \tag{7}$$

where $IAT$ is the inventory adjustment time (i.e., the time required to react for inventory discrepancy between the current inventory level and the desired level).

Furthermore, the production rate $PR(t)$ is controlled by the capacity scalability level given by:

$$PR(t) = \frac{C(t) * U}{MUT} \tag{8}$$

where $U$ is the utilization level of the available capacity, and $MUT$ is the manufacturing unit time (i.e., used to switch from stock to rate to maintain dimensional balance). The customer orders fulfillment at time $t$, $OFR(t)$ is controlled by the shipment rate $ShR(t)$ and given by:

$$OFR(t) = \text{ShR}(t) = \min(\text{DSR}(t), \text{MSR}(t)) \tag{9}$$

where $MSR(t)$ is the maximum shipment rate at time $t$ (i.e., depends on the system's current inventory) and $DSR(t)$ is the desired shipment rate at time $t$.

**Model 3**: Due to the capacity scalability, [7] developed a dynamic model for the reconfigurable manufacturing systems whereby various system configurations, based on control theory and feedback analysis, can be implemented in response to changing demand. The proposed model contains two feedback loops. The first loop concerns the $WIP$ error, where the second loop addresses the production rate error. They established a transfer function (Eq. 10) for the system based on the schema bloc depicted in Fig. 1.

$$\frac{\text{PR}}{\text{Cap}^*} = \frac{G_W\left(T_D^{-1}+S\right)+G_C T_{LT}^{-1} T_D^{-1}}{S^2+S\left(T_D^{-1}+T_{LT}^{-1}+G_W\right)+(G_W T_{LT}+G_C+1)T_{LT}^{-1} T_D^{-1}} \tag{10}$$

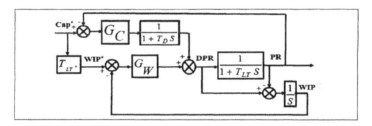

**Fig. 1.** Block Diagram for Dynamic RMS Model [7]

Finally, to illustrate the proposed dynamic approach, a simple numerical example of a reconfigurable machine shop with 40K goods per day is simulated, applying different controllers P, PD, PI, and PID. Where: $WIP^*$: Desired work in process level (parts), $WIP$: Actual $WIP$ level (parts), $DPR$: Desired production rate (parts/days), $PR$: Actual production rate (parts/days), $T_{LT*}$: Expected lead time (days), $T_{LT}$: Lead time (days), $G_W$: $WIP$ control gain (1/day), $Cap^*$: Desired capacity rate (parts/days), $G_C$: Capacity scalability control gain (parts/days) and $T_D$: Capacity installation delay time (days).

**Model 4**: To understand the relationship between management level and operational level in capacity scalability problem in RMS, [8] used the Shema bloc of Fig. 2 to establish transfer function for the system (Eq. 11), where the dynamic approach for capacity scalability and the classical static approach are compared.

The problem is solved using the multi-objective weighted sum method.

$$\frac{\text{PR}}{\text{Cap}^*} = \frac{G_C T_{LT}^{-1} T_D^{-1}}{S^2+S\left(T_{LT}^{-1}+T_D^{-1}\right)+(1+G_C)T_{LT}^{-1} T_D^{-1}} \tag{11}$$

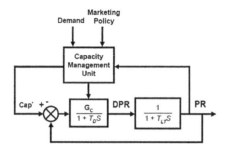

**Fig. 2.** Capacity Scalability System Block Diagram [8]

**Model 5**: [9] proposed a practical method to determine the most cost-effective system reconfiguration to meet new market demand. An optimization model is proposed for scalability planning, where a case study is illustrated to examine and validate the approach. They defined scalability $Sc$ as:

$Sc = 100$ – smallest incremental capacity in percentage.

As input for the model, they used: configuration information, stage characteristics, manufacturing machine reliability information, and demand, where scalability planning aims to minimize the number of machines needed (Eq. 12). A genetic algorithm-based approach is used to solve the problem.

$$\text{Min} \left( \sum_{i=1}^{L} (N_i + M[i]) \right) \tag{12}$$

**Model 6:** [10] used the mathematical model of [11] that encompasses the product complexity in the system scalability with a variable number of stages. A case study is presented to show the capabilities of the proposed model. Based on the scalability of the available system, the following issues have been discussed:

– Machines added per stage should be minimized (Eq. 13).
– Stages added (if required) for the new part be minimized (Eq. 14).
– Maximization of throughput (Eq. 15).

$$min \sum_{l=1}^{L} \Delta_l + \frac{TH_{new}}{TH(M,T,t)} \tag{13}$$

$$min \sum_{i=1}^{I} \sum_{l=1}^{L} S_{ij}(\Delta_l) \tag{14}$$

$$maxTH(M,T,t) \tag{15}$$

where $S$: existing machine stage, $M$: existing machines per stage, $C$: existing machine capabilities, and $TH$: throughput. They tested several scenarios of individual parts segregated into two main groups. The first group contributes towards the part manufacturing complexity, and the second group contributes mainly towards the product assembly complexity.

**Model 8:** [12] represented the dynamic capacity evolution by:

$$y_k = min(C_k, D_k) \tag{16}$$

$$C_{k+1} = C_k + X_{k-T} \tag{17}$$

where $C_k$ represents the firm's capacity level at time $k$, $X$ is the control input that defines the addition or removal of capacity, and $y_k$ represents the firm's sales. The delay time $T$ is limited to be a multiple of the time increment $K$. The capacity management cost is given by:

$$G_k(C_k) = E\{(\gamma_P - P)min(C_k, D_k) + \gamma_S max(0, D_k - C_k) + \gamma_H C_k\} \tag{18}$$

where $\gamma_p$: is the service costs per unit to produce, $P$: is the sold fixed price per unit, $\gamma_S$:is the penalty cost per unit, $\gamma_H$ is the holding or overhead cost, per unit of capacity at each period, $\gamma_N$ is the remaining capacity (i.e., can be sold for a salvage value per unit of terminal capacity), $\beta$ is the discount factor and $M_k(X_k)$ is the cost of expanding/subtracting capacity incurred at time $k$. Hence, to minimize the expected discounted cost, Eq. 19 is used.

$$\min_{X_0...X_{N-1}} \left\{ -\beta^N \gamma_N C_N + \sum_{k=0}^{N-1} \beta^k \left[ G_k(C_{k+1}) + M_k(X_k) \right] \right\} \tag{19}$$

In the end, they applied a feedback control system approach to the capacity management problem for sub-optimal solutions.

## 4  Convertibility Models and Approaches

In this section, we focus mainly on convertibility. It is defined in [3] as "*the capability to transform the functionality of existing systems and machines to fit new production requirements.*"

**Model 1:** The first mathematical model used to measure and quantify systems convertibility $C_S$ was proposed in [13]. The model combines three metrics, including configuration convertibility $C_C$, machine convertibility $C_M$, and material handling convertibility $C_H$, given by:

$$C_S = w_1 C_C + w_2 C_M + w_3 C_H \tag{20}$$

where $w_1$, $w_2$ and $w_3$ are the associated weights.
The metrics are evaluated using Eqs. 21, 22, 23, and 24.

$$C_C = \frac{R*X}{I} \tag{21}$$

where $R$: refers to the number of routing connections in each configuration, $X$: is the minimum number of replicated machines in the process plan at a particular stage, and $I$: is the minimum increment of conversion. Normalization of Eq. 21 is proposed:

$$C_{c\text{-normalized}} = 1 + \left[ \frac{\log\left(\frac{Cc'}{Cc'_{\text{serial}}}\right)}{\log\left(\frac{Cc'_{\text{parallel}}}{Cc'_{\text{serial}}}\right) \times \frac{1}{9}} \right] \tag{22}$$

$$C_M = \frac{\sum_{i=1}^{N} C'_M}{N} \qquad (23)$$

$$C_H = \frac{\sum_{i=1}^{N} C'_H}{N} \qquad (24)$$

Furthermore, [14] developed an index to measure RMS's reconfigurability using multi-attribute utility theory, where convertibility is considered using model 1 above. In [15], a NSGA II based-approach for machine selection is proposed based on three criteria, namely machine utilization, configuration convertibility, modeled using Eqs. 21 and 22, along with machines costs. Moreover, [16] measured the reconfigurability of manufacturing systems by quantifying RMS characteristics, where Eqs. 21 and 22 are employed to measure the convertibility (Cc).

According to [17], if more information is available about the number of processed variants (product family), model 1 needs an adaptation. To evaluate the configuration convertibility of a mixed model assembly line, a fourth parameter is introduced to cope with the system's variation and given by:

$$Y = \frac{Variants\ number}{Workstations\ number} \qquad (25)$$

The new configuration convertibility $C_{CY}$ is given by:

$$C_{CY} = C_{c,\ Normalized} * Y \qquad (26)$$

Besides, they proposed a model to quantify product convertibility $C_{PF}$ given by:

$$C_{PF} = a * b \qquad (27)$$

$$a = \frac{Total\ number\ of\ components\ with\ setting\ interfaces}{Total\ number\ of\ setting\ interfaces}$$
$$b = \frac{Total\ number\ of\ components\ with\ gripping\ interfaces}{Total\ number\ of\ gripping\ interfaces} \qquad (28)$$

Finally, the product convertibility $C_{PF}$ can be computed as:

$$C_{PF} = \frac{\sum_{k=1}^{l} C_{F_k}}{\sum_{k=1}^{l} I_{sF_k}} * \frac{\sum_{j=1}^{m} C_{F_j}}{\sum_{j=1}^{m} I_{gF_j}} \qquad (29)$$

where $F_i$ is the $i$ th component family, $l$ is the number of component families with setting interfaces, $m$ is the number of component families with gripping interfaces, $C_F$ is the number of components in $F_i$ and $I_{gF}$ is the number of gripping interfaces in $F_i$.

They used an example of an automobile industry plant to measure each product family's convertibility and justify selecting the plant that can easily manufacture variant x. Finally, the convertibility indicator equation is:

$$C_S = w_1 C_{CY} + w_2 C_{WS} + w_3 C_H + w_4 C_{PF} \qquad (30)$$

**Model 2:** [18] provided an analysis of automated assembly system design convertibility based on its equipment structures and layout, where system convertibility $C_S$ is defined as a "*sum of equipment convertibility $C_E$, and layout convertibility $C_L$*" and calculated using Eq. 31:

$$C_S = w_E C_E + w_L C_L \qquad (31)$$

– The equipment convertibility ($C_E$):

$$C_E = \frac{\sum_{k=1}^{N} C_{SS,k}}{N} \tag{32}$$

where $C_{SS,k}$ convertibility of sub-system $k$ (equipment level), $N$ is the number of sub-systems and $k$ sub-system.

– Layout convertibility ($C_L$):

$$C_L = \frac{L_A + L_C + L_R}{3} \tag{33}$$

where: $L_A$ is the autonomy index, $L_C$ is the connectivity index, and $L_R$ is the replication index. Finally, they presented a test case based on heuristics for battery module assembly to evaluate their approach.

## 5   Discussions

RMS is a highly active research field with several state-of-the-art contributions in various disciplines, including design, layout optimization, reconfigurable control, process planning/process generation, and production scheduling. From the above literature review, we can claim the following gaps:

1. Only linear models were considered for the dynamic cases to model scalability, ignoring non-linearity, which is more representative of reality. In contrast, no dynamic model has been shown in the literature for convertibility
2. Scalability has gained researchers' intention where several models are presented, while for convertibility, all the presented models are based on a single extended model.
3. Simulation-based optimization may be used to enable the application of alternative optimization methods. Furthermore, various simulation approaches such as discrete-event simulation can be utilized to mimic scalability, as well as other robust control mechanisms.
4. The system expansion and the addition of machines will increase waste and energy consumption, which were not considered in any model.
5. The number of employees and work shifts were not included, nor was the control system's adjustment, which raises the frequency of alerts. This lack of information and control will affect the product's quality and ultimately lead to yield loss.

## 6   Conclusion

Manufacturing firms must adjust their manufacturing methods to adapt to changing market demands. RMS is one method for accomplishing this goal by promoting its flexibility while optimizing efficiency and costs.

This research work tackled two main RMS characteristics that improve the system's overall performance: convertibility and scalability. Some models and approaches were presented and analyzed. Nevertheless, we found out that several models exist in the literature for scalability where only one model was proposed and extended to deal with convertibility. Moreover, there is a dearth of research works that integrate sustainability while modeling the characteristics.

# References

1. Koren, Y., et al.: Reconfigurable manufacturing systems. CIRP Ann. **48**(2), 527–540 (1999)
2. Koren, Y. (2006) General RMS characteristics. comparison with dedicated and flexible systems. In: Dashchenko, A.I. (eds.) Reconfigurable Manufacturing Systems and Transformable Factories, pp. 27–45. Springer, Berlin, Heidelberg (2006). https://doi.org/10.1007/3-540-293 97-3_3
3. Koren, Y., Gu, X., Badurdeen, F., Jawahir, I.S.: Sustainable living factories for next generation manufacturing. Procedia Manuf. **21**, 26–36 (2018)
4. Koren, Y.: The emergence of reconfigurable manufacturing systems (RMSs). In: Benyoucef, L. (ed.) Reconfigurable Manufacturing Systems: From Design to Implementation. SSAM, pp. 1–9. Springer, Cham (2020). https://doi.org/10.1007/978-3-030-28782-5_1
5. Son, S., Lennon Olsen, T., Yip-Hoi, D.: An approach to scalability and line balancing for reconfigurable manufacturing systems. Integr. Manuf. Syst. **12**(7), 500–511 (2001)
6. Deif, A.M., ElMaraghy, H.A.: Assessing capacity scalability policies in RMS using system dynamics. Int. J. Flex. Manuf. Syst. **19**(3), 128–150 (2007)
7. Deif, A.M., ElMaraghy, W.H.: A control approach to explore the dynamics of capacity scalability in reconfigurable manufacturing systems. J. Manuf. Syst. **25**(1), 12–24 (2006)
8. Deif, A.M., ElMaraghy, W.H.: Integrating static and dynamic analysis in studying capacity scalability in RMS. Int. J. Manuf. Res. **2**(4), 414–427 (2007)
9. Wang, W., Koren, Y.: Design principles of scalable reconfigurable manufacturing systems. IFAC Proceedings Volumes **46**(9), 1411–1416 (2013)
10. Hasan, S.M., Baqai, A.A., Butt, S.U., Ausaf, M.F., Zaman, U.: Incorporation of part complexity into system scalability for flexible/reconfigurable systems. Int. J. Adv. Manuf. Technol. **99**(9), 2959–2929 (2018). https://doi.org/10.1007/s00170-018-2654-x
11. Koren, Y., Wang, W., Gu, X.: Value creation through design for scalability of reconfigurable manufacturing systems. Int. J. Prod. Res. **55**(5), 1227–1242 (2017)
12. Asl, F.M., Galip Ulsoy, A..: Stochastic optimal capacity management in reconfigurable manufacturing systems. J. Manuf. Sci. Product. **6**(1–2), 83–88 (2004)
13. Maler-Speredelozzi, V., Koren, Y., Hu, S.J.: Convertibility measures for manufacturing systems. CIRP Ann. **52**(1), 367–370 (2003)
14. Gumasta, K., Gupta, S.K., Benyoucef, L., Tiwari, M.K.: Developing a reconfigurability index using multi-attribute utility theory. Int. J. Prod. Res. **49**(6), 1669–1683 (2011)
15. Goyal, K.K., Jain, P K., Jain, M.: Multiple objective optimization of reconfigurable manufacturing system. In :Advances in Intelligent and Soft Computing, pp. 453–460. Springer India (2012)
16. Mittal, K.K., Jain, P.K., Kumar, D.: Optimal configuration selection in reconfigurable manufacturing system. In: Deep, K., Jain, M., Salhi, S. (eds.) Decision Science in Action. AA, pp. 193–202. Springer, Singapore (2019). https://doi.org/10.1007/978-981-13-0860-4_14
17. Lafou, M., Mathieu, L., Pois, S., Alochet, M.: convertibility indicator for manual mixed-model assembly lines. Procedia CIRP **17**, 314–319 (2014)
18. Chinnathai, M.K., Alkan, B., Harrison, R.: Convertibility evaluation of automated assembly system designs for high variety production. Procedia CIRP **60**, 74–79 (2017)

19. Wang, W., Koren, Y.: Scalability planning for reconfigurable manufacturing systems. J. Manuf. Syst. **31**(2), 83–91 (2012)
20. Elmasry, S.S., Youssef, A.M.A., Shalaby, M.A.: Investigating best capacity scaling policies for different reconfigurable manufacturing system scenarios. Procedia CIRP **17**, 410–415 (2014)
21. Wang, W., Koren, Y.: design principles of scalable reconfigurable manufacturing systems. IFAC Proc. **46**(9), 1411–1416 (2013)
22. Elmasry, S.S., Youssef, A.M.A., Shalaby, M.A.: A cost-based model to select best capacity scaling policy for reconfigurable manufacturing systems. Int. J. Manuf. Res. **10**(2), 162–183 (2015)

# Changeable Manufacturing: A Comparative Study of Requirements and Potentials in Two Industrial Cases

Stefan Kjeldgaard$^{(\boxtimes)}$ ⓘ, Alessia Napoleone ⓘ, Ann-Louise Andersen ⓘ,
Thomas Ditlev Brunoe ⓘ, and Kjeld Nielsen ⓘ

Department of Materials and Production, Aalborg University, Aalborg, Denmark
stefank@mp.aau.dk

**Abstract.** Today's global manufacturing environment is characterized by intense competition in dynamic and uncertain markets. Consequently, manufacturers are required to accommodate a higher variety of products with frequent new introductions and shorter life-cycles in a rapid and cost-efficient way, to sustain competitiveness. In light of these requirements, changeable manufacturing systems appear promising. However, empirically founded research is limited in regard to how different requirements lead to different applications and resulting potentials in various industrial settings. Therefore, this paper presents a comparative study of requirements, enablers, and potentials of changeability in two industrial cases (i) a Danish manufacturer of capital goods for the energy sector, (ii) a Danish manufacturer of sporting goods for the maritime sector. The objective of the paper, is to generate insights which can support various industrial settings in the transition towards changeable manufacturing. Findings include: (i) in high-volume contexts, reconfigurability is suitable to accommodate a production mix with increasing dimensions of parts, with potential to improve equipment utilization to reduce capital expenses (ii) in global manufacturing contexts, reconfigurability is suitable to accommodate frequent changes of production location, with potential to improve demand proximity to gain a competitive advantage.

**Keywords:** Changeable manufacturing systems · Reconfigurable manufacturing · Changeability requirements · Changeability potentials · Changeability enablers

## 1 Introduction

More than 20 years ago, Reconfigurable Manufacturing Systems (RMSs) were introduced with the aim to combine the throughput of Dedicated Manufacturing Systems (DMSs) with the functionality range of Flexible Manufacturing Systems (FMSs) [1, 2]. DMSs are cost-effective when market requirements are stable, as they are designed to produce a single product or part at a high rate, usually through fixed automation [3, 4]. In contrast, FMSs are designed to produce a wide variety, usually at a lower rate [4].

© IFIP International Federation for Information Processing 2021
Published by Springer Nature Switzerland AG 2021
A. Dolgui et al. (Eds.): APMS 2021, IFIP AICT 631, pp. 452–461, 2021.
https://doi.org/10.1007/978-3-030-85902-2_48

To do this, FMSs embodies capital-intensive general-purpose flexibility which might not be needed as market requirements evolve [3]. While DMS and FMS are static systems, RMS can be adapted over time as its capacity and functionality can be changed to what is needed, when needed [3, 5]. Therefore, unlike DMS and FMS, RMS can dynamically meet evolving market requirements. Thus, enabling manufacturing companies to face the current context characterized by increasingly frequent and unpredictable market changes [6]. This capability of RMS is enabled by several characteristics which are presented with definitions in Table 1. The table also provides information on the relative importance of characteristic in accordance with Koren et al. [1], Koren [5] and Rösiö [7]. These authors either classify the characteristics as: Necessary (Ne), Core (Co), Basic (Ba), Critical (Cr), Supportive (Su) or Non-Categorized (NC).

**Table 1.** Enabling characteristics of reconfigurability.

| Characteristic | Definition | [1] | [5] | [7] |
|---|---|---|---|---|
| Convertibility | The ability to convert functionality to new products or parts | Ne | Co | Cr |
| Scalability | The ability to increase or decrease the rate of production | Ne | Co | Cr |
| Customization | The limitation of functionality to a product or part family | Ne | Co | Ba |
| Modularity | The grouping of functional elements to physical modules | | Co | Su |
| Integrability | The ability to integrate modules through standard interfaces | | Co | Su |
| Diagnosibility | The ability to detect and diagnose errors in reconfigurations | | Co | Su |
| Automatibility | The ability to increase or decrease the degree of automation | | | Nc |
| Mobility | The ability to move or relocate modules | | | Nc |

To ensure competitiveness in increasingly dynamic contexts, RMS deserves the interest of manufacturing companies. However, despite the general trend, it is not likely to think that all companies need RMS against DMS or FMS. Many researchers instead refer to changeability as a combination of capabilities associated to either RMS, DMS and/or FMS [8, 9]. To embed changeability into the design of a manufacturing system, it is essential that manufactures analyze their specific requirements. Then, hereafter, select and implement the appropriate type and extent of changeability by embodiment of the suitable classes and enablers in the appropriate manufacturing constituents [8].

However, previous research on this subject generally has a limited empirical focus on the industrial transition toward changeable manufacturing [4, 10]. One of the few industry-applicable tools allowing the identification of changeability requirements, suitable enablers, and resulting potentials is the "Participatory System Design Methodology for Changeable Manufacturing Systems" (PSDM) proposed by Andersen and ElMaraghy et al. [8]. The steps of the PSDM are: (i) identify relevant company data based on a questionnaire requiring the participation of stakeholders in the company, (ii) define patterns of change requirements, (iii) determine appropriate manufacturing paradigm, (iv) determine required change enablers, (v) determine existing change enablers, constraints, and manufacturing paradigm and (vi) recommended transition towards new manufacturing paradigm. Applying the PSDM allows manufacturers to identify their requirements and potentials in terms of combination of RMS, DMS and/or FMS capabilities. Although Andersen and ElMaraghy et al. [8] applied the PSDM in two industrial cases, to the best of the authors' knowledge, there is no further literature applying the PSDM in industry. Given the necessity to extend the empirical focus of research on changeability, this paper further applies the PSDM in two manufacturing cases in order to address the following research question: *"What are differences in requirements and potentials of changeable manufacturing in different manufacturing settings?"*.

The remainder of the paper is structured as follows: Sect. 2 outlines the case research method, Sect. 3 presents the case study findings, Sect. 4 presents the cross-case findings, and Sect. 5 presents conclusions and further research directions.

## 2  Methodology

The case study approach - following the directions of Eisenhardt [11] - has been adopted given the explorative nature of the research question and the aim of providing both empirical and theoretical insights on differences in requirements and potentials of changeability. The near-polar characteristics and contexts of the two cases included in the study are provided in Table 2. Eisenhardt [11] provides a framework to guide the process of building theory from case study research through a set of steps and activities. Details on how these activities are executed in the focal research, are presented in Table 3. Moreover, the table includes (i) a reference to where the results of executed activities are presented and (ii) an account of how and why an activity has been modified.

**Table 2.** Overview of company and context characteristics within each case.

| Characteristic | Case A | Case B |
| --- | --- | --- |
| Company type | Public-limited and large company | Privately held medium-sized company |
| Industry | Capital goods for the energy sector | Sporting goods for the maritime sector |
| Product | Large size, modular architecture | Medium-sized, integral architecture |
| Demand | Global and project-based demand | International and order-based demand |
| Competition | High degree of competition | Medium degree of competition |
| Position | Largest actor and market-leader | Second largest actor in niche market |
| Area of focus | Manufacturing of product module | Manufacturing of product family |
| Prod. location | Global manufacturing footprint | Local manufacturing facility |
| Prod. strategy | Make-to-stock and make-to-order | Make-to-order |

In terms of data collection, semi-structured interviews were conducted with various representatives from the companies in order to capture emergent themes and unique case features. In case A, this resulted in (i) three meetings of five, one and two hours with a product and manufacturing engineer at the headquarters (ii) one meeting of one hour with a supply chain planner during an online session (iii) one meeting of one hour with the lead of new product introductions at the mother-factory. In case B, two meetings of four and two hours were held with the production manager at the factory. In both cases, archival data were extracted from (i) ERP system, (ii) spread-sheets and (iii) presentations, to validate and enrich the qualitative statements related to the change requirements. Furthermore, direct observations from tours at the mother-factories and discussions with workers, aided the identification of existing enablers and constraints.

**Table 3.** Details on the execution of activities.

| Step | Description |
| --- | --- |
| Get started | The research question is defined with supportive motivation in Sect. 1 |
| Select cases | Two companies have been selected for the case-study. Due to theoretical reasons, the cases differ on their company and context characteristics in order to generate as many insights as possible from the cross-case comparison. Thereby, increasing the possibility to extend emergent theory and provide examples of polar types. This is needed in order to satisfy the research question sufficiently, despite the inclusion of a small sample size. Due to practical reasons, the companies should (i) be located in proximity to Denmark and (ii) have interest in the research topic of changeability |
| Craft protocol | The protocol provided by the PSDM have been applied with minor modifications for the joint collection and analysis of data in the focal research |
| Collect data | A combination of qualitative and quantitative data has been collected using multiple sources of evidence in order to strengthen the grounding of theory by triangulation of synergistic evidence. Details are provided in the text |
| Analyze data | Within-case analyses are provided in Sect. 3 where the PSDM has been applied. The cross case-comparison is provided in Sect. 4 |
| Shape hypothesis | The within-case analyses and cross-case comparison are used to shape the hypotheses of the focal research. These hypotheses constitute the emergent theory which is presented in Sect. 4 and summarized in Sect. 5 |
| Enfold literature | The emergent theory of the focal research is compared with the extant literature on the topic of changeability and reconfigurability e.g. the seminal works of Koren et. al. [1, 5]. Aforementioned, is provided in Sect. 4 where complementary and conflicting findings, in-between, are presented |
| Reach closure | Closure have been reached prematurely, where the inclusion of additional cases is expected to enhance the theoretical saturation in further research |

## 3 Findings

The results of applying the PSDM are presented in Fig. 1 for case A and in Fig. 2 for case B. The figures present the mapping of (i) the specified changeability requirements in the stakeholder domain, (ii) to appropriate manufacturing system paradigms in the functional domain, (iii) to existing constituents, enablers and constraints in the physical domain. These mapped connections between system design domains indicate gaps between (i) the appropriate and the existing paradigms, (ii) the changeability requirements and the changeability extent of existing constituents. In order to continuously and efficiently match the requirements, the identified gaps should be mitigated through embodiment of suitable enablers in the constituents where they are present. For both industrial cases, the identified suitable enablers of the appropriate paradigms to be embodied in existing constituents are presented in Table 4 along with the derived operational, tactical, and strategic potentials of the context-specific embodiment.

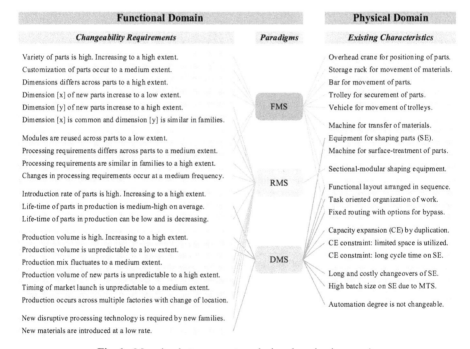

**Fig. 1.** Mapping between system design domains in case A.

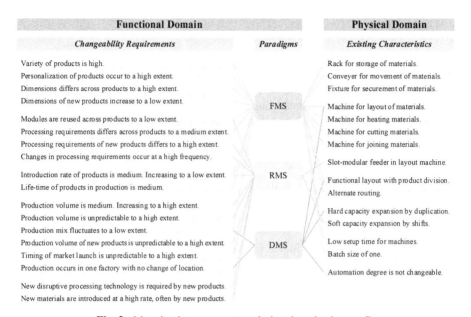

**Fig. 2.**  Mapping between system design domains in case B.

**Table 4.**  Potentials of recommended transition and embodiment in case A and case B.

|  | Case A | Case B |
|---|---|---|
| Suitable paradigm | The suitable paradigm is a mix of FMS and RMS. Flexibility is suitable to be retained for the handling equipment and production machines to cope with different materials and dimensions of parts. Reconfigurability is suitable to be embodied in the shaping equipment to cope with the increasing dimensions of parts that are similar within new families | The suitable paradigm is a mix of FMS and RMS. Flexibility is suitable to be retained for the handling equipment and majority of production machines to cope with different dimensions of products. Reconfigurability is suitable to be embodied in the layout machines to cope with the change of materials and processing technology |
| Suitable enablers | The shaping equipment is suitable for embodiment of sectional modularity, mobility, integrability and customization to enable convertibility between parts within families where dimension [x], materials and processing technology are common and dimension [y] is similar | The layout machine is suitable for embodiment of slot-modularity and integrability with respect to the material feeder, to enable convertibility between current and new generations of products where the materials and processing technology are different |
| Derived potential | Operational potential in terms of rapid and cost-efficient response to changes in production mix across multiple factories<br>Tactical/strategic potential in terms of: (i) increasing the proximity of production to demand to reduce transport cost and to achieve a competitive advantage (ii) increase the lifetime- and capacity utilization of shaping equipment to reduce capital costs (iii) negate the need for MTS to reduce inventory levels | Operational potential in terms of rapid and cost-efficient response to changes in production mix within the factory<br>Tactical/strategic potential in terms of: (i) reducing the time-to-market of new product generations requiring new materials and new processing technology (ii) increasing the lifetime- and capacity utilization of the layout machines to reduce capital expenses |

# 4 Discussion

A noteworthy cross-case finding is the relation between the industrial context and the change requirements related to the product, which results in a difference of suitable changeability classes to embody in the manufacturing system constituents. Despite geometrical dimensions and materials being primary determinants of product performance in both cases, the material mix is stable in case A as opposed to a high rate of change in case B. Moreover, the dimensions increase at a higher rate in case A whereas the range of dimensions is higher in case B. This result in flexibility being suitable to cope with the stable material mix in case A and differing dimensions in case B, whereas reconfigurability is suitable to cope with the increase of dimensions in case A and change of materials in case B. These differences in change requirements related to the product can be attributed to the following contextual differences:

- The highly competitive and industrialized B2B context of case A, which generates a higher clock-speed of dimensional increase (order winner) and need for the material mix minimizing the weight (order qualifier). The solution space of the latter is exhausted due to industrialization, where the solution is applied across competitors.
- The niche B2C context of case B where the (i) dimensions are dependent on the customers system, (ii) materials are selected in accordance with customers objective i.e. for casual or competitive purpose, generating a higher need for personalization.

Another noteworthy cross-case finding is the relation between the industrial context and the change requirements related to the geographical location of production, resulting in differing potentials of changeability. Despite global markets with fluctuating demand being supplied in both cases, there is a difference in terms of changes to the: number, location, functionality and capacity of factories. Production in case B is rooted in a single factory with no change of location. In contrast, production in case A is spread across multiple external and internal factories where each factory has a unique range of functionality and capacity that is changed several times per year. These changes are among others made in order to:

- Decrease the time and thus the cost of transporting the large-scale capital goods, by means of increasing the proximity of production to the location of the demand.
- Gain a competitive advantage to win project-based orders in competitive tendering schemes, by means of complying with requirements for localized manufacturing.

Although both case-companies remain competitive through the capability to *deliver the desired product, in the correct quantity, at the correct time, at the right place*, the findings emphasize the need for global manufacturers of large-scale capital goods to produce at the correct place as well. These findings extend the propositions of Koren [5] regarding the aforementioned capability required to remain competitive in the 21st century. Moreover, as reconfigurability is an enabler of the capability in case A, RMS is thus, not only aiding to provide *exactly the capacity and functionality needed, exactly when needed*, but also where needed. This enforces the propositions of Andersen and

ElMaraghy et al. [8] regarding that reconfigurability is a multi-dimensional and context-dependent capability where the enablers can be embodied in constituents in various ways depending on the context-specific drivers and potentials of reconfigurability.

Moreover, with regards to case-A, the capability to produce everything, everywhere, can be enabled to a higher extent through flexibility in the shaping equipment by means of additive manufacturing. However, as flexibility requires the functionality range to be pre-determined, an extensive range would need to be pre-specified to cope with the high rate of increase in dimensions without risking low capacity and lifetime utilization of the equipment. This is assessed by the company to be economically infeasible since a high production volume is required which necessitates a larger quantity of equipment if flexible as opposed to the reconfigurable counterpart. These findings thereby indicate that reconfigurability is suitable in high volume cases if the variety, extent of dimensional increase and rate of production location change, is high as well. This application of RMS in case A thereby goes beyond the intention of RMS being suitable in medium volume cases as originally proposed by Koren et al. [1].

With regards to the PSDM, it proved applicable in generating relevant input for further concept design in both cases, although the industrial context, unit of analysis, data foundation and degree of participation, differed. An insight gained throughout the process of applying the PSDM is that the degree to which the output was deemed relevant by the company stakeholders were higher when (i) the requirements and existing characteristics were respectively based on quantitate data and observation of the shop-floor, (ii) a high degree of participation, follow-up questions and free flow of thoughts were present during the collection and analysis of data. The latter stimulated the identification of the drivers for changes to the production location in case A, and the tactical/strategic potential of enabling these changes through reconfigurability.

Another insight gained, is that some patterns of requirements could not be identified as being suitable to be met by only one of the changeability classes. For example, the low reuse of part modules present in both cases constrains the possibility to achieve economies of scale with regards to part modules through DMS, thereby leading to FMS or RMS being suitable instead. Moreover, the existing manufacturing systems and their constituents could neither be classified as strictly being one of the changeability classes. For example, the layout machine in case B is dedicated to a material type while being flexible in its range of dimensions it can handle. This multi-dimensional aspect is not supported to be identified and mapped by solely applying the factual questions, provided by the PSDM. Therefore, the PSDM is proposed a modification where the questions lead to a degree and rate of change with respect to context-specific production parameters from which a gap with the related functionality and capacity range of existing manufacturing constituents could be mapped. The latter is expected to stimulate conceptual design to a higher extent by increasing the boundary of the solution space in terms of potential classes and enablers to embody in constituents to achieve the required type and extent of changeability and the resulting potentials.

# 5   Conclusion and Further Research

This paper contributes with insights from a comparative case-study on differences in requirements and potentials of changeability in two companies with different industrial settings of manufacturing. The primary insights gained are listed in the following:

- Reconfigurability is suitable to cope with increasing product parameters resulting from a high clock speed of industrialization in competitive B2B contexts, whereas flexibility is suitable to cope with a high extent of personalization in B2C contexts.
- Reconfigurability is suitable in high volume cases if: variety of parts, dimensional increase of new parts and rate of production location change, is high as well.
- Reconfigurability is suitable for global manufacturers of large-scale capital goods to gain a competitive advantage by enabling the capability to provide exactly the capacity and functionality needed, exactly when needed, exactly where needed.
- Changeability requirements can be enabled through the embodiment of various types of classes and enablers in existing constituents of manufacturing systems.
- Existing manufacturing systems and their constituents can embody multiple classes of changeability e.g. dedicated on one parameter and flexible on another parameter.

Future research should aim at applying the PSDM in additional cases to advance the theoretical saturation on differences in requirements and potentials of changeability. Moreover, future research should aim at providing a tool to map the degree and rate of change of context-specific production parameters from which gaps with the related functionality and capacity range of existing manufacturing constituents can be drawn. This tool is suggested, as the limited methods provided by research i.e. the PSDM constraints the boundary of the solution space of conceptual manufacturing system design in terms of limiting the possibility of multiple classes and enablers of changeability being able to meet the context-specific requirements for changeability. By accounting for the former, the suggested tool is expected to support the industrial transition towards changeable and reconfigurable manufacturing systems in brownfield contexts.

**Acknowledgements.** The research presented in this paper is supported by the REKON research project and the MADE research program. REKON is funded by the Danish Industry Foundation and MADE is funded by the Innovation Fund Denmark.

# References

1. Koren, Y., et al.: Reconfigurable Manufacturing Systems. CIRP Ann. Manuf. Technol. **48**(2), 527–540 (1999)
2. Koren, Y., Gu, X., Guo, W.: Reconfigurable manufacturing systems: principles, design, and future trends. Front. Mech. Eng. **13**(2), 121–136 (2018)
3. Koren, Y., Shpitalni, M.: Design of reconfigurable manufacturing systems. J. Manuf. Syst. **29**(4), 130–141 (2010)
4. Bortolini, M., Galizia, F.G., Mora, C.: Reconfigurable manufacturing systems: literature review and research trend. J. Manuf. Syst. **49**, 93–106 (2018)

5. Koren, Y.: The Global Manufacturing Revolution: Product-Process-Business Integration and Reconfigurable Systems. Wiley, Hoboken, New Jersey (2010)
6. Bi, Z.M., Lang, S., Shen, W., et al.: Reconfigurable manufacturing systems: the state of the art. Int. J. Prod. Res. **46**(4), 967–992 (2008)
7. Rösiö, C.: Supporting the Design of Reconfigurable Production Systems. (2012)
8. Andersen, A., ElMaraghy, H., ElMaraghy, W., et al.: A participatory systems design methodology for changeable manufacturing systems. Int. J. Prod. Res. **56**(8), 2769–2787 (2018)
9. Wiendahl, H., et al.: Changeable manufacturing-classification design and operation. CIRP Ann.Manuf. Technol. **56**(2), 783–809 (2007)
10. Andersen, A.-L., et al.: Tailored reconfigurability: a comparative study of eight industrial cases with reconfigurability as a key to manufacturing competitiveness. In: Benyoucef, L. (ed.) Reconfigurable Manufacturing Systems: From Design to Implementation, pp. 209–245. Springer, Cham (2020). https://doi.org/10.1007/978-3-030-28782-5_11
11. Eisenhardt, K.M.: Building theories from case study research. Acad. Manage. Rev. **14**(4), 532–550 (1989)

# A Systematic Approach to Development of Changeable and Reconfigurable Manufacturing Systems

Ann-Louise Andersen$^{(\boxtimes)}$ , Alessia Napoleone , Thomas Ditlev Brunoe ,
Bjørn Christensen , and Kjeld Nielsen

Department of Materials and Production, Aalborg University, Fibigerstraede 16,
9220 Aalborg East, Denmark
ala@mp.aau.dk

**Abstract.** The implementation of changeable and reconfigurable manufacturing systems and realization of benefits connected to rapid, efficient, and dynamic change of functionality and capacity is key to achieve manufacturing competitiveness. Therefore, this paper proposes a systematic methodology for the design and development of changeable and reconfigurable manufacturing systems, derived from design theory, reconfigurability theory, as well as practical experience. The methodology consists of a concrete course of actions that connects design phases and working steps based on the content of the design task. Furthermore, the paper addresses project-related and contextual aspects of reconfigurability development, which indicates how the proposed methodology should be adapted to the specific company and task at hand. Thus, the proposed methodology is intended for further validation in different types of manufacturing companies that are transitioning towards reconfigurability.

**Keywords:** Reconfigurable manufacturing · Reconfigurability · Changeable manufacturing · Engineering design · Manufacturing system

## 1 Introduction

Changeability and reconfigurability are widely recognized as principles that can be implemented and utilized in manufacturing systems for responding efficiently to product variety, rapid new product introductions, and demand fluctuations [1, 2]. Related benefits include reduced capital expenditures, shorter lead-time, higher capacity utilization, and prolonged life-time of systems, machines, and equipment [3]. Compared to traditional manufacturing systems with static, pre-planned, and built-in functionality and capacity, changeable and reconfigurable systems have dynamic abilities to change [4]. However, as these systems are not solely designed for their immediate and initial purpose, but rather in combination with properties having more long-term exposure, they represent complex engineering development tasks [5, 6].

© IFIP International Federation for Information Processing 2021
Published by Springer Nature Switzerland AG 2021
A. Dolgui et al. (Eds.): APMS 2021, IFIP AICT 631, pp. 462–470, 2021.
https://doi.org/10.1007/978-3-030-85902-2_49

In this regard, traditional approaches to manufacturing system design and development are largely insufficient for supporting a transition towards reconfigurability [7]. First of all, as changeability and reconfigurability can be categorized as non-functional system requirements that express the behavior of the system in the long-term, system design and evaluation need to be aligned with product evolutions that represent significant uncertainty [8, 9]. Secondly, translating uncertain and dynamic requirements into system designs that embed the reconfigurability enablers is a task that requires evaluation of multiple realization options, e.g. in terms of combination of flexibility and reconfigurability, level of implementation, and system constituents to implement these enablers [9, 10]. Furthermore, in practice, design of manufacturing systems is usually conducted largely iteratively, trial-and-error based, and with limited explicit consideration of reconfigurability characteristics [11]. Therefore, there is both an academic and industrial need for a systematic approach to engineering design and development of changeable and reconfigurable manufacturing systems. In this regard, a systematic engineering design methodology prescribes a concrete course of actions for the design of a technical system, usually in form of design phases, design activities, and their related sequence [12]. Thus, while design in general is a largely creative task, a design methodology can support this task by prescribing a general working procedure, while still being adaptable to the specific task at hand [12]. Moreover, while design and evaluation constitute the important parts of the system development process, the project-related and contextual aspects supporting the implementation of this process should also be addressed [11].

Based on these requirements, this paper proposes a systematic methodology for the design and development of changeable and reconfigurable manufacturing systems. The remainder of the paper is structured as follows: Sect. 2 outlines related research as background for the proposed methodology. Section 3 describes the phases, activities and supportive tools of the proposed design methodology, while Sect. 4 summarizes the contribution and outlines future developments of the methodology.

## 2   Related Research

Changeable manufacturing systems have been defined as systems that are able to accomplish early and foresighted adjustments on all process levels in an economically feasible way [15]. Thus, changeability can be regarded as an umbrella term for manufacturing that depending on the context is realized by the most economically feasible combination of a-priori flexibility and dynamic reconfigurability [15]. The concept of the Reconfigurable Manufacturing System (RMS) was originally introduced by Koren [13] as a manufacturing system paradigm combining the high efficiency of the Dedicated Manufacturing System (DMS) and the high flexibility of the Flexible Manufacturing System (FMS). The key feature of the RMS is the ability to be continuously changed in order to provide the exact functionality and capacity needed at the exact time needed [3]. In this regard, a successful reconfiguration is enabled by the three necessary RMS characteristics i.e. convertibility, scalability and customization, as well as the three supportive characteristics i.e. modularity, integrability, and diagnosability [3]. Thus, compared to the flexibility, which is pre-planned and built-in, reconfigurability is a dynamic capability to expand and change the flexibility boundaries of the system [14]. Depending on the company and manufacturing setting, the level and combination of flexibility and reconfigurability need to be matched with the specific changeability requirements. An overview of the fundamental aspects involved in this has been proposed in various forms [2, 4, 15], covering the following:

- Change drivers are external and internal factors that request or necessitate some response by the company and are represented by demand volatility and variety [4].
- Change focus/object is the impact of change drivers, which externally towards the market and customers covers mix, product, and volume, and internally is focused on the production processes, equipment, and organization [4, 15].
- Change strategy is the degree and importance of changeability to pursue [4, 15].
- Change extent/extension is the level, time, and effort involved in the selected change strategy, which have to be comparable with the expected benefits [15].
- Change enablers are the system properties that ensure changeability, e.g. reconfigurability characteristics such as modularity, integrability, etc. [15].
- Change utilization is the planning, training, and implementation of the changeability of the system [15].

When relating these fundamental aspects of changeability to the design of changeable manufacturing systems, key questions concern for instance the need and requirements for change, the optimal and feasible combination of flexibility and reconfigurability to select, as well as concrete and technical design of related enablers [4]. Thus, the aforementioned changeability aspects are both interrelated and related to different stages during manufacturing system design. In Fig. 1, a sequence of the changeability aspects is proposed, which outlines a relevant order of how these are involved during manufacturing system design [10, 15].

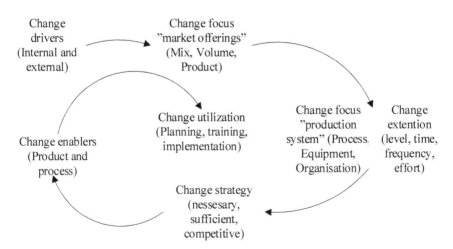

**Fig. 1.** Fundamentals of changeability and their sequence during system design.

Several methodologies and frameworks for the design of changeable and reconfigurable manufacturing systems have been proposed in previous research [16]. To mention a few, Schuh [17] focused on object-oriented design for changeable systems, while Benkamoun [6] proposed systemic design for changeability in manufacturing system architectures. Azab et al. [14] proposed a framework for the mechanics of change in a manufacturing system that is dynamically changeable and reconfigurable. Likewise, Francalanza et al. [18] proposed a methodology for designing changeable manufacturing systems based on a traditional design cycle approach. Tracht and Hogreve [19] proposed a framework for the design of reconfigurable systems, as well as the reconfigurations phase, while Deif and ElMaraghy [20] proposed a method for design of reconfigurable hardware, software, and human aspects of the manufacturing system. Abdi and Labib [21] focused particularly on designing for reconfigurability using AHP for justification analysis. Finally, Andersen et al. [10] focused on synthesizing all previous methods for design of reconfigurability into a number of design phases and activities. However, while the issue of designing changeable and reconfigurable manufacturing systems appears to attract increasing research attention [16, 22], there appears to be limited consideration of industry-applicability as a main principle in the existing methodologies [23]. In general, designing changeable and reconfigurable manufacturing systems involves decisions in various categories similar to any manufacturing system design activity irrespective of the level of changeability, i.e. decisions on process type, layout, capacity, machines, automation, organization and human resources, etc. [11]. However, in addition, design of changeability requires knowledge of the key enabling principles and competences in how to design and realize these [16]. In order to support this task, a structured system design methodology for changeable manufacturing should enable the following aspects, which is largely overlooked in previous research:

- Application in manufacturing companies to allow for a transition towards changeable and reconfigurable manufacturing, i.e. by allowing for a problem-directed approach and by being adaptable to different types of design processes in different types of companies [12].
- Application in already existing company approaches to manufacturing system design, e.g. by being applicable to traditional and widely used phases in engineering design, while also providing the necessary tools to apply in the phases [12, 16, 23].
- Guidance in the design and evaluation steps specifically involved in establishing requirements and solutions for changeable and reconfigurable manufacturing systems that are dynamically changeable [7, 16].
- Continuous evolution and utilization of changeability in the long-term, i.e. capture the continuous loop of designing, evaluating, and utilization changeability and reconfigurability in manufacturing [14].

## 3   Proposed Four-Step Development Approach

Based on the requirements elaborated in the previous section, this section describes the proposed systematic methodology for the design and development of reconfigurable manufacturing systems. In the methodology, design phases where the level of abstraction is gradually decreased is combined with knowledge of the fundamental aspects of changeability and their design sequence as outlined in Fig. 1. Furthermore, the methodology was tested and modified in an iterative way during five research projects on different aspects of reconfigurability development in five different case companies. Specifically, a first version of each phase was created based on theory, which was elaborated and changed during the interactions with the case companies. For instance, one research project in a specific case company focused largely on tools in phase 1 to clarify the company's potentials, while another research project in a second case company focused largely on developing and evaluating reconfigurable concepts. Thus, the different projects and case companies assisted in further developing and modifying each phase in order to meet the requirements stated in the previous section.

## 3.1  Development Process and Activities

The proposed systematic methodology consists of four phases, which are depicted in Fig. 2. The process starts with an initial clarification of the design tasks and changeability requirements, continues with conceptual design, detailed design, and finally implementation and utilization of changeability. Thus, these phases largely follow the traditional phases of engineering design in order to support easy application in manufacturing companies. However, one main difference is the addition of the last phase, where implementation and utilization of changeability takes place. This phase either results in a system in operation or a repetition of some design phases before once again resulting in a system in operation. Thus, the proposed methodology enables companies in supporting dynamic reconfigurations to various extents throughout the system's lifetime.

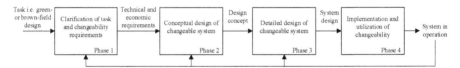

**Fig. 2.**  Four-phased design steps related to changeability and reconfigurability.

Each of the four phases contain both design and evaluation activities that support the realization of changeability and reconfigurability. Figures 3, 4, 5 and 6 depict the specific activities of each phase, while at the same time outlining the covered fundamental aspects of changeability as described in Fig. 1. Thus, phase 1 covers the clarification of the design task and changeability requirements through analysis of change drivers, external and internal change focuses, change extension, and finally the change strategy. Phase 2 and 3 focus on the necessary and supportive change enablers, while phase 4 covers the implementation and utilization of changeability. One of the main decisions in the design phase is to determine the optimal combination of flexibility and reconfigurability to reach the change requirements. This decision is primarily dealt with in phase 2, which evaluates different concepts build on the explicit change requirements from phase 1.

Furthermore, for each phase and activity, different design and evaluation tools should be applied. Generally, previous research suggests numerous different methods and tools that could potentially be applied during reconfigurability design [16, 23]. However, only tools ready for application after testing and further development in the case companies are included here. For brevity, tools and the availability of these are not elaborated further in this paper, however, some evident gaps exist that are elaborated further in Sect. 4.

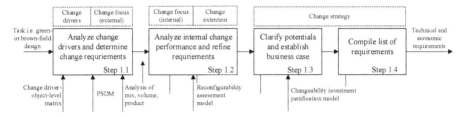

**Fig. 3.** Phase 1 containing clarification of task and changeability requirements.

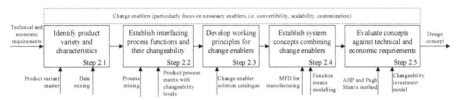

**Fig. 4.** Phase 2 containing conceptual design of changeable system.

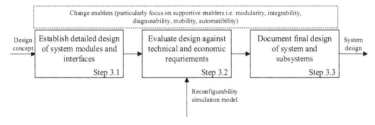

**Fig. 5.** Phase 3 containing detailed design of changeable system.

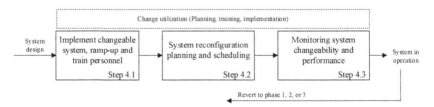

**Fig. 6.** Phase 4 containing implementation and utilization of changeability.

## 3.2  Project-Related and Contextual Aspects of Development

The proposed methodology incorporates the fundamental aspects of changeability and connects related design activities in a sequence going from initial clarification of change drivers to a changeable and reconfigurable system in operation. Thus, the methodology is intended as an addition to manufacturing development methods in companies that most often follow the same overall design phases, rather than as a stand-alone method that substitutes conventional design methodologies for manufacturing. Likewise, the methodology should be adapted to the specific company context and design task at hand. Specifically, the methodology can be adapted in the following ways related to how development projects are conducted and how the development process is implemented in a company:

- Design task: both a green-field and a brown-field design task can be accommodated by the methodology, i.e. designing a new manufacturing system for a new product or redesigning an existing manufacturing system for improving changeability performance. The alteration of the methodology here lies primarily in regards to whether existing equipment is considered in phase 2 or whether such constraints are not considered in the phases.
- Starting point: depending on the pre-existing knowledge within the company, emphasis can be on different development phases. Clearly, some companies need extensive focus on phase 1 e.g. in terms of clarifying needs and identifying the business case, while other companies can progress quicker to conceptual and detailed design.
- Development focus: development can take outset in a specific product and manufacturing system or alternatively be a pre-project that establishes solutions for changeability enablers that can be used in later specific development projects.
- Tools applied: depending on data availability and preferences within companies, some phases have different alternatives of which tools to apply. For instance, more advanced data mining techniques are possible to use in some phases, however, may not be feasible in all types of companies.

## 4   Conclusions and Future Research

In this paper, a systematic methodology for the development of changeable and reconfigurable manufacturing systems is proposed. Industrial application was a main objective in the creation of the methodology in order to support a wider industrial transition towards changeability and reconfigurability. The methodology consists of four phases roughly following traditional phases in engineering design; clarification of task and requirements, conceptual design, detailed design, as well as implementation and utilization of changeability. To summarize, the proposed methodology contributes with the following benefits:

- Allows for a problem-directed and company-specific approach, i.e. the starting point of the methodology can be both green-field and brown-field development, emphasis can be on different phases depending on company need, and the tools applied can also be adapted in various phases depending on data availability, etc.
- Enables direct application in a manufacturing company as a sequence of development activities and tools specifically targeting changeability and reconfigurability that can be "added" to the already existing engineering development methods.

- Provides a systematic guidance for the engineering designer, which ensures the appropriate match between changeability requirements and enablers.
- Ensures that manufacturing system development takes long-term and dynamic requirements into account, as well as captures the continuous loop of designing, evaluating, and utilizing changeability and reconfigurability.

Nevertheless, several further research directions are prevailing. First of all, the proposed methodology needs further testing and validation in manufacturing companies. This would also produce further insights on how to support companies in successful implementation of reconfigurability principles, as well as provide further insights on the contextual aspects of development. Secondly, well-proven and readily applicable supportive tools are still missing for some development activities. In particular future research should focus on maturing methods for establishing reconfigurable concepts, as well as focusing on supportive tools for phase 3 and 4.

**Acknowledgements.** The research presented in this paper is funded by the Danish Industry Foundation in connection to the project "Development of Reconfigurable Manufacturing" (REKON).

# References

1. Koren, Y.: The rapid responsiveness of RMS. Int. J. Prod. Res. **51**(23–24), 6817–6827 (2013)
2. Wiendahl, H., ElMaraghy, H.A., Nyhuis, P., et al.: Changeable manufacturing-classification, design and operation. CIRP Ann. **56**(2), 783–809 (2007)
3. Koren, Y.: General RMS characteristics. Comparison with dedicated and flexible systems. In: ElMaraghy, H.A. (ed.) Reconfigurable Manufacturing Systems and Transformable Factories, pp. 27–45. Springer, Heidelberg (2006). https://doi.org/10.1007/3-540-29397-3_3
4. Wiendahl, H.-P.: Systematics of changeability. In: Handbook Factory Planning and Design, pp. 91–118. Springer, Heidelberg (2015). https://doi.org/10.1007/978-3-662-46391-8_5
5. ElMaraghy, W., ElMaraghy, H., Tomiyama, T., et al.: Complexity in engineering design and manufacturing. CIRP Ann. Manuf. Technol. **61**(2), 793–814 (2012)
6. Benkamoun, N.: Systemic design methodology for changeable manufacturing systems Dissertation (2016)
7. Rösiö, C., Säfsten, K.: Reconfigurable production system design–theoretical and practical challenges. J. Manuf. Technol. Manage. **24**(7), 998–1018 (2013)
8. Farid, A.M.: An engineering systems introduction to axiomatic design. In: Farid, A.M., Suh, N.P. (eds.) Axiomatic Design in Large Systems, pp. 3–47. Springer, Cham (2016). https://doi.org/10.1007/978-3-319-32388-6_1
9. Francalanza, E., Borg, J., Constantinescu, C.: Development and evaluation of a knowledge-based decision-making approach for designing changeable manufacturing systems. CIRP J. Manuf. Sci. Technol. **16**, 81–101 (2017)
10. Andersen, A., ElMaraghy, H., ElMaraghy, W., et al.: A participatory systems design methodology for changeable manufacturing systems. Int. J. Prod. Res. **56**(8), 2769–2787 (2018)
11. Bellgran, M., Säfsten, K.: Production development: design and operation of production systems. Springer-Verlag, London (2009)

12. Pahl, G., Beitz, W.: Engineering design: a systematic approach. MRS Bull. **21**(8), 71 (2013). https://doi.org/10.1557/S0883769400035776
13. Koren, Y., Heisel, U., Jovane, F., et al.: Reconfigurable manufacturing systems. CIRP Ann. Manuf. Technol. **48**(2), 527–540 (1999)
14. Azab, A., ElMaraghy, H., Nyhuis, P., et al.: Mechanics of change: a framework to reconfigure manufacturing systems. CIRP J. Manuf. Sci. Technol. **6**(2), 110–119 (2013)
15. ElMaraghy, H.A., Wiendahl, H.P.: Changeability - an introduction. In: ElMaraghy, H.A. (ed.) Changeable and Reconfigurable Manufacturing Systems, pp. 3–24. Springer, London (2009). https://doi.org/10.1007/978-1-84882-067-8_1
16. Andersen, A., Brunoe, T.D., Nielsen, K., et al.: Towards a generic design method for reconfigurable manufacturing systems - analysis and synthesis of current design methods and evaluation of supportive tools. J. Manuf. Syst. **42**, 179–195 (2017)
17. Schuh, G., Lenders, M., Nussbaum, C., et al.: Design for changeability. In: ElMaraghy, H.A. (ed.) Changeable and Reconfigurable Manufacturing Systems, pp. 251–266. Springer, London (2009). https://doi.org/10.1007/978-1-84882-067-8_14
18. Francalanza, E., Borg, J., Constantinescu, C.: Deriving a systematic approach to changeable manufacturing system design. Procedia CIRP **17**, 166–171 (2014)
19. Tracht, K., Hogreve, S.: Decision making during design and reconfiguration of modular assembly lines. In: ElMaraghy, H. (ed.) Enabling Manufacturing Competitiveness and Economic Sustainability, pp. 105–110. Springer, Heidelberg (2012). https://doi.org/10.1007/978-3-642-23860-4_17
20. Deif, A.M., ElMaraghy, W.H.: A systematic design approach for reconfigurable manufacturing systems. In: ElMaraghy, H.A., ElMaraghy, W.H. (eds.) Advances in Design, pp. 219–228. Springer, London (2006). https://doi.org/10.1007/1-84628-210-1_18
21. Abdi, M.R., Labib, A.W.: A design strategy for Reconfigurable Manufacturing Systems (RMSs) using Analytical Hierarchical Process (AHP): a case study. Int. J. Prod. Res. **41**(10), 2273–2299 (2003)
22. Bortolini, M., Galizia, F.G., Mora, C.: Reconfigurable manufacturing systems: literature review and research trend. J. Manuf. Syst. **49**, 93–106 (2018)
23. Napoleone, A., et al.: Towards an industry-applicable design methodology for developing reconfigurable manufacturing. In: Lalic, B., Majstorovic, V., Marjanovic, U., von Cieminski, G., Romero, D. (eds.) APMS 2020. IAICT, vol. 591, pp. 449–456. Springer, Cham (2020). https://doi.org/10.1007/978-3-030-57993-7_51

# An Industry-Applicable Screening Tool for the Clarification of Changeability Requirements

Alessia Napoleone$^{(\boxtimes)}$ ⓘ, Ann-Louise Andersen ⓘ, Thomas Ditlev Brunoe ⓘ,
and Kjeld Nielsen ⓘ

Department of Materials and Production, Aalborg University, Aalborg, Denmark
alna@mp.aau.dk

**Abstract.** Manufacturing companies need changeability in order to adapt to change drivers, such as unpredictable market demand and increasingly relevant sustainability requirements. Specific change drivers determine different changeability requirements, thus leading to the need for different changeability enablers. Therefore, before starting the identification and design of changeability enablers, companies should effectively identify their changeability requirements. In this study, an industry-applicable screening tool for the clarification of changeability requirements is proposed. The tool allows companies to discern whether they need flexibility or reconfigurability enablers. The tool has been validated with industry experts and is ready to be disseminated in industry.

**Keywords:** Changeable manufacturing · Reconfigurability · Change drivers · Changeability requirements · Industry-applicable tool

## 1 Introduction

Manufacturing companies are exposed to increasingly frequent and unpredictable market changes, including the rapid introduction of new products, shorter product lifecycles, and constantly varying product demand [1, 2]. Moreover, the increasing need to be sustainable pressures companies to adapt products and manufacturing processes to new regulations and requirements [3, 4]. Changeability is a necessary ability for manufacturing companies to withstand such turbulent scenario in a quick and cost-effective way [5]. Focusing on individual factories - i.e. sets of manufacturing and logistics systems directly and indirectly responsible for the manufacturing of specific product groups - changeability is achieved recurring to reconfigurability and flexibility [5]. Both the concepts of flexibility and reconfigurability deal with modifications in manufacturing systems and their distinction is in the timing, cost, and number of steps necessary to implement modifications [6]. Specifically, flexibility allows fast adaptation within narrow corridors of change [7]. At some point during system lifecycle, the necessary flexibility may be already available or may be absent. If absent, it may be acquired and, to do so, the system

© IFIP International Federation for Information Processing 2021
Published by Springer Nature Switzerland AG 2021
A. Dolgui et al. (Eds.): APMS 2021, IFIP AICT 631, pp. 471–478, 2021.
https://doi.org/10.1007/978-3-030-85902-2_50

must be reconfigurable, i.e. already predisposed to afford such acquisition. Unlike flexibility, reconfigurability actions require higher, but adequate effort in terms of reasonable time and low costs in order to allow any change (thus, not within a predetermined range of change).

Both contextual and internal change drivers influence manufacturing companies. For example, a specific company might need to adapt processes to upcoming sustainability standards or adapt both products and processes to a new market need or output volume. However, how and to what extent change drivers affect the manufacturing company differs across companies [8] and determines distinctive and company-specific changeability requirements. Moreover, there is no universal way to achieve changeability, but many researchers have contributed to identify a variety of changeability enablers [9–12]. These enablers should be selected based on these distinctive changeability requirements of manufacturing companies [13]. The overall process of analysis of requirements and development of changeability enablers can thus be divided into three sequential sub-processes:

1  Identification of changeability requirements, based on change drivers: definition of companies' need for changeability and expected changes;
2  Assessment of the existing changeability enablers: analysis of the existing ability of the manufacturing system/s to meet changes and definition of changeability enablers to meet the requirements;
3  Development of the required changeability enablers: development of design concepts of changeability enablers, allowing the company to fill the gap between changeability requirements and existing changeability enablers.

Therefore, it is critical for manufacturing companies to start from the identification of changeability requirements (first-sub-process) in order to appropriately improve the existing changeability enablers or acquire the required ones. For this reason, this study addresses the following research question: *"What practical tool can be provided to manufacturing companies to allow the clarification of their distinctive changeability requirements?"*.

## 2  Literature Review

A structured literature review [14] has been conducted in the following four stages.

In stage one, the objective of the literature review has been defined: the identification and analysis of already existing tools for the clarification of changeability requirements.

In stage two, the literature search has been performed. The search database used to find literature is Scopus. The research domain has been defined using the following search string: ("changeability" OR "reconfigurability") AND ("manufacturing") AND ("need" OR "requirement" OR "change driver") AND ("assessment" OR "clarification" OR "identification" OR "specification"). Moreover, the following inclusion criteria have been applied: only (i) articles published within the last 10 years and (ii) written in English have been selected. In this way, 26 articles have been identified. After a preliminary analysis of their abstracts, 15 pertinent articles have been selected.

In stage three, the pertinent literature has been analysed in detail and described in an Excel database. Considering the three sub-processes of the analysis of requirements and development of changeability enablers introduced in Sect. 1, five articles focused on the identification of changeability requirements have been finally selected.

In stage four, the results of the analysis of the five articles focused on the identification of changeability requirements have been reported as these provide the theoretical ground of this study. Specifically, two of the five articles actually provided and described tools for the clarification of changeability requirements: these are Garbie and Parsaei [15] and Andersen et al. [13]. Among these two articles, Andersen et al. [13] not only provided a procedure for the specification of changeability requirements, but also demonstrated applicability in industry, in both existing and new systems. The remaining three articles still provided interesting insights for the development of the tool proposed in this study. Dit Eynaud et al. [11] directed to two already existing questionnaires (provided in Andersen et al. [12] and Maganha et al. [14]) that can be exploited to identify changeability requirements. Karl and Reinhart [16] listed and specified manufacturing resources-relevant influencing factors leading to changeability requirements and provided axes for mapping the requirements. Benkamoun et al. [17] provided an overall framework for designing changeability from the outset which clarifies concepts and terminology.

# 3 Methodology and Tool Development

The tool proposed and presented in this paper was developed in a research project that aims to develop and disseminate industry-applicable tools for the design of changeable and reconfigurable manufacturing. The project involves collaboration with various manufacturing companies transitioning towards changeable manufacturing.

The methodology adopted in this study consists in two steps: tool development and tool validation. Specifically, the two steps have been sequentially implemented in several iterations, following the Delphi Method and adjusting the tool based on the feedback of four of the companies participating in the project.

The development of the tool takes outset in the tool provided by Andersen et al. [13], which has been adapted based on the need to: (i) enhance the understandability in industry, and (ii) provide an exhaustive analysis of change requirements considering insights from other analysed literature.

Andersen et al.'s tool implies: (i) the collection of facts through a questionnaire and (ii) the interpretation of these facts by associating them to requirements in terms of flexible, dedicated or reconfigurable manufacturing systems. Thus, a company applying such tool gets an overview of how its change drivers lead to a combination of requirements belonging to flexible, dedicated and reconfigurable systems.

Andersen et al.'s tool has been modified as follows.

- With regard to the collection of facts, the questionnaire has been modified.

  - The questions of the questionnaire proposed in this study cover three areas as drivers of changeability requirements: (i) product, (ii) production, and (iii) technology and sustainability.

- In each of these areas, individual questions require respondents to indicate present or expected levels of change. Specifically, respondents need to select an option within five choices ranging within five levels: very low (1), low (2), medium (3), high (4), and very high (5).

• With regard to the interpretation of collected facts, the criteria for interpretation have also been modified.

  - The distinction in requirements associated to either flexible, dedicated or reconfigurable manufacturing systems has been overcome since, taking a changeability perspective, flexibility, reconfigurability and dedication are not opposite concepts but can overlap and coexist in changeability enablers. To prepare the ground for the following phase of identification of changeability enablers, the identification of changeability requirements should be as clear as possible.
  - The developed tool should allow a company to discern between short-term changeability requirements, associated for example to the need to conduct changeovers when switching across product variants, and long-term requirements, associated for example to the need to introduce new sustainable materials.
  - Thus, in the tool proposed in this study, change requirements have been classified according to the three categories identified by Tracht and Hogreve [18]. These categories are: (i) change requirements in product/part variants, (short-term changeability requirements); (ii) change requirements in production capacity, (mid-term changeability requirements); and (iii) change requirements in product features (long-term changeability requirements).
  - The collected facts can thus be mapped on three graphs, belonging to these three categories of change requirements.

As anticipated, the aforementioned changes to the tool have been validated following the Delphi Method. The involvement of the four companies through the Delphi Method has ensured that the developed tool can be both effective and easily used by companies. At each company, one main stakeholder – usually the production manager – has been involved. The Delphi Method has been conducted in five rounds: (i) one for each of the involved companies (where the feedback from individual stakeholders was exploited to improve both understandability and effectiveness of the tool), and a final common round (consisting in the finalization of the tool and the illustration of the results of its implementation at each of the companies).

## 4   Illustration of the Tool

The tool provided in this study consists of an Excel questionnaire which, once filled by relevant company's stakeholders, allows an automatic visualization of the results.

An extract of the questionnaire is provided in the following Fig. 1.

As shown in Fig. 1., stakeholders filling the questionnaire have the possibility to differentiate answers for selected product/part families, so to allow the comparison of the corresponding changeability requirements.

After collecting facts by answering to the questionnaire, companies get an automatic quantification of their changeability requirements and can visualize them on three graphs – corresponding to change requirements in product variants, product capacity, and product features -as exemplified in Fig. 2. When change requirements in product variants (short-term) prevail, companies need to consider flexibility enablers rather than reconfigurability enablers. Conversely, when change requirements in production capacity (mid-term) and/or product features (long-term) prevail, companies need reconfigurability enablers. In the example of Fig. 2, changeability requirements for Product A are very important, this is especially true for Variant and Product changeability; conversely, capacity changeability requirements look stable. With regard to Product B, changeability requirements do not appear important.

Thus, by simply answering to the questionnaire, manufacturing companies can assess their changeability requirements and understand not only whether they need variety, capacity or product changeability, but also to what extent the different categories are required. Indeed, as shown in Fig. 2, the possibility to see not only the average value of changeability requirements, but also the entire distribution of requirements on axes eventually allows deriving interesting observations regarding specific parameters particularly affecting the results of the analysis.

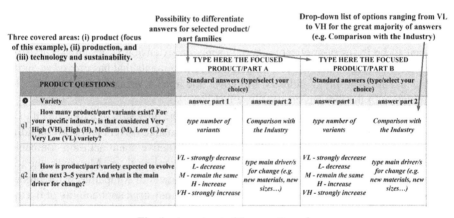

**Fig. 1.** An extract of the questionnaire

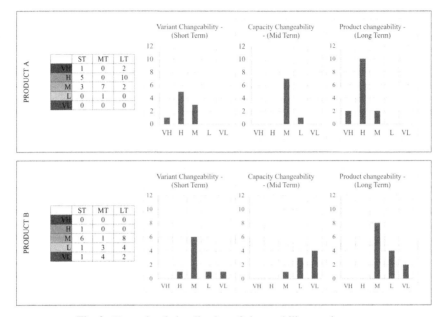

**Fig. 2.** Example of visualization of changeability requirements

## 5 Conclusions

In this study, a tool for the clarification of changeability requirements is provided and is meant to be industry-applicable. It represents an extension of an already existing procedure for the specification of changeability requirements that has proved scientific validity and applicability in industry. Moreover, the involvement of four companies for the validation of the tool has ensured its effectiveness.

The contribution for practitioners of this study is the provision of a tool aimed to be universally used by companies: it considers diverse change drivers and corresponding changeability requirements, so that disparate and distinctive manufacturing companies can practically use it to identify those specifically regarding them. Moreover, the tool can be autonomously used by companies. The involvement of companies in the process of building this tool has extensively contributed to the industrial applicability of the tool itself.

The theoretical contribution of this research lies in the attempt to provide a tool that allows to differentiate between short-term changeability requirements, which, can be addressed by companies by investing in flexibility enablers, and long-term changeability requirements, which – conversely – can be addressed by investing in reconfigurability enablers. To this regard, distinguishing between these two kinds of requirements is relevant because literature has addressed them in different ways, thus providing different solutions. As stressed in Sect. 1, unlike flexibility, reconfigurability solutions require higher, but adequate effort in terms of reasonable time and low costs in order to allow any change (thus, not within a predetermined range of change). Moreover, since the questionnaire also ensures the collection of facts related to expected product evolution due to the increasingly relevant sustainability requirements, (which lead to change requirements in product features, i.e. long-term changeability requirements), the tool also makes evident the impact that sustainability requirements have on the need of companies for reconfigurability enablers. To this end, further research should aim at applying the tool in a consistent number of manufacturing companies, so to quantitatively show the impact of sustainability requirements on the need for reconfigurability enablers.

As the tool allows companies to clarify the nature of their changeability requirements, future research should also aim at supporting companies in the following step, which is the identification of concrete instances of flexibility and reconfigurability enablers that would allow them accommodating the identified requirements.

**Acknowledgments.** This research was conducted within the REKON project funded by the Danish Industry Foundation.

# References

1. Echsler Minguillon, F., Schömer, J., Stricker, N., Lanza, G., Duffie, N.: Planning for changeability and flexibility using a frequency perspective. CIRP Ann. **68**(1), 427–430 (2019)
2. Koren, Y., Shpitalni, M.: Design of reconfigurable manufacturing systems. J. Manuf. Syst. **29**(4), 130–141 (2010)
3. Barwood, M., Li, J., Pringle, T., Rahimifard, S.: Utilisation of reconfigurable recycling systems for improved material recovery from e-waste. Procedia CIRP **29**(2015), 746–751 (2015)
4. Battaïa, O., Benyoucef, L., Delorme, X., Dolgui, A., Thevenin, S.: Sustainable and energy efficient reconfigurable manufacturing systems. In: Benyoucef, L. (ed.) Reconfigurable Manufacturing Systems: From Design to Implementation. SSAM, pp. 179–191. Springer, Cham (2020). https://doi.org/10.1007/978-3-030-28782-5_9
5. ElMaraghy, H.A., Wiendahl, H.-P.: Changeability – an introduction. In: ElMaraghy, H.A. (ed.) Changeable and Reconfigurable Manufacturing Systems, pp. 3–24. Springer, London (2009). https://doi.org/10.1007/978-1-84882-067-8_1
6. Terkaj, W., Tolio, T., Valente, A.: A review on manufacturing flexibility. Int. J. Prod. Res. **51**(19), 5946–5970 (2009)
7. Azab, A., ElMaraghy, H., Nyhuis, P., Pachow-Frauenhofer, J., Schmidt, M.: Mechanics of change: a framework to reconfigure manufacturing systems. CIRP J. Manuf. Sci. Technol. **6**(2), 110–119 (2013)
8. Rösiö, C.: Supporting the design of reconfigurable production systems. Ph.D. Dissertations, Mälardalen University Press (2012)
9. Napoleone, A., Pozzetti, A., Macchi, M.: Core characteristics of reconfigurability and their influencing elements. IFAC-PapersOnLine **51**(11), 116–121 (2018). Elsevier B.V.
10. Andersen, A.-L., Nielsen, K., Brunoe, T.D., Larsen, J.K., Ketelsen, C.: Understanding changeability enablers and their impact on performance in manufacturing companies. In: Moon, I., Lee, G.M., Park, J., Kiritsis, D., von Cieminski, G. (eds.) APMS 2018. IAICT, vol. 535, pp. 297–304. Springer, Cham (2018). https://doi.org/10.1007/978-3-319-99704-9_36
11. Dit Eynaud, A.B., Klement, N., Gibaru, O., Roucoules, L., Durville, L.: Identification of reconfigurability enablers and weighting of reconfigurability characteristics based on a case study. Procedia Manuf. **28**(2019), 96–101 (2019)

12. ElMaraghy, H.A.: Changing and evolving products and systems – models and enablers. In: ElMaraghy, H.A. (ed.) Changeable and Reconfigurable Manufacturing Systems, pp. 25–45. Springer, London (2009). https://doi.org/10.1007/978-1-84882-067-8_2

13. Andersen, A.-L., ElMaraghy, H., ElMaraghy, W., Brunoe, T.D., Nielsen, K.: A participatory systems design methodology for changeable manufacturing systems. Int. J. Prod. Res. **56**(8), 2769–2787 (2018)

14. Durach, C.F., Kembro, J., Wieland, A.: A new paradigm for systematic literature reviews in supply chain management. J. Supply Chain Manag. **53**(4), 67–85 (2017)

15. Garbie, I.H., Parsaei, H.R.: A role of reconfiguring manufacturing enterprise as a major requirement for sustainability. In: 62nd IIE Annual Conference and Expo, Orlando, Florida, pp. 73–82. Elsevier (2012)

16. Karl, F., Reinhart, G.: Reconfigurations on manufacturing resources: identification of needs and planning. Prod. Eng. **9**(3), 393–404 (2015). https://doi.org/10.1007/s11740-015-0607-x

17. Benkamoun, N., Kouiss, K., Huyet, A.-L.: An intelligent design environment for changeability management - application to manufacturing systems. In: 20th Proceedings of the International Conference on Engineering Design (ICED 15), vol. 1, pp. 1–10. Design Society, Milan (2015)

18. Tracht, K., Hogreve, S.: Decision making during design and reconfiguration of modular assembly lines. In: ElMaraghy, H.A. (ed.) Enabling Manufacturing Competitiveness and Economic Sustainability, pp. 105–110. Springer, Berlin, Heidelberg (2012). https://doi.org/10.1007/978-3-642-23860-4_17

# Impact of Different Financial Evaluation Parameters for Reconfigurable Manufacturing System Investments

Thomas Ditlev Brunoe$^{(\boxtimes)}$ 🆔, Alessia Napoleone 🆔, Ann-Louise Andersen 🆔, and Kjeld Nielsen 🆔

Department of Materials and Production, Aalborg University, Aalborg, Denmark
tdp@mp.aau.dk

**Abstract.** The need for frequently adapting manufacturing systems to dynamic market demand, short product lifecycles, technology evolution, and sustainability requirements is increasingly challenging manufacturing companies. To this end, Reconfigurable Manufacturing Systems (RMSs) are relevant, however, still far from their wide implementation in industry. In this regard, a main barrier to the introduction of reconfigurability is the justification of investments in developing and purchasing reconfigurable systems and equipment. Often such reconfigurable systems have high initial investments that provide returns in the mid/long-term, while at the same time being subject to and dependent on high uncertainty in product and market evolutions. Thus, uncertainty in different aspects of demand and the time-horizon of evaluations are main aspects in evaluating the financial benefits of reconfigurability. Therefore, this paper investigates how different choices of financial evaluation parameters affect the financial feasibility of reconfigurable manufacturing compared to more traditional manufacturing concepts. The findings of the paper provide valuable insights on how practitioners should proceed in adequately capturing the value of reconfigurability during investment decisions.

**Keywords:** Reconfigurable manufacturing · Changeable manufacturing · Reconfigurability · Financial evaluation · Financial simulation

## 1 Introduction

Designing manufacturing systems may have a huge impact on a company's competitiveness, since the design determines a wide range of performance measures ranging from cost to quality [1, 2] Reconfigurable manufacturing systems (RMSs) are widely recognized as key to competitiveness in highly volatile and uncertain manufacturing environments [3, 4]. Compared to both very dedicated manufacturing systems and highly flexible manufacturing systems where functionality and capacity boundaries are static, pre-planned, and a-priori, the RMS has the reconfigurability capability, i.e. it is dynamically changeable towards shifting functionally and capacity requirements [5]. Thus, RMSs have the exact capacity and functionally needed, exactly when needed, through

© IFIP International Federation for Information Processing 2021
Published by Springer Nature Switzerland AG 2021
A. Dolgui et al. (Eds.): APMS 2021, IFIP AICT 631, pp. 479–487, 2021.
https://doi.org/10.1007/978-3-030-85902-2_51

enablers of convertibility, scalability, modularity, integrability, customization, and diagnosability [6]. However, while benefits of the reconfigurability capability are obvious, reconfigurability principles are only scarcely implemented in the manufacturing industry [7, 8]. Furthermore, with only few available examples RMS introduction into companies, literature lacks best practices driving industrial companies in the transition toward this new industrial paradigm [9]. In this regard, one of the main barriers is to justify the investment in developing and purchasing reconfigurable systems and equipment [10–12]. Some of the main issues in justifying reconfigurability from a financial standpoint are:

- Reconfigurable system and equipment concepts are often more expensive in terms of initial development and investments costs, however, less expensive during the operating life-time if and when changes in system requirements occur [13].
- Reconfigurability allows for prolonging the life-time of manufacturing systems, going beyond the product or part that it was initially designed for [14]. However, this also means that surplus initial investments in reconfigurability should be cancelled out over longer time periods than traditionally dedicated manufacturing systems [15].
- Reconfigurability benefits such as easy conversion of functionality and scaling of capacity in small increments should match the specific change requirements in the manufacturing company, i.e. systems and equipment can be more or less reconfigurable depending on the specific need [16]. Thus, considering the specific uncertainty parameters to the company is needed in order to evaluate investments [13, 17].
- Reconfigurability of a manufacturing system allows for continuously meeting functionality and capacity demand, thereby mitigating risks of having either too much or too little capacity and functionality [6]. Thus, these longer-term benefits should be evaluated during investment decisions against the inherent market uncertainty [17].

Evidently, various trade-offs exist during evaluations and investment decisions related to reconfigurability. However, previous research provides only limited support for conducting explicit evaluations of these in a manufacturing company. Kuzgunkaya and ElMaraghy [11] proposed a fuzzy multi-objective mixed integer optimization model incorporating both financial and strategic perspectives on investing in both reconfigurable and flexible manufacturing systems. Furthermore, Niroomand et al. [18] investigated the impact of different reconfiguration characteristics for capacity investment strategies. Aiping et al. [19] proposed to use the "dynamic" real options analysis to deal with the disadvantages of traditional methods for the financial evaluation of RMS and to give the capacity to include in the evaluation both the uncertain and active decision. From a purely qualitative standpoint, Zhang et al. [20] performed an analytical comparison of dedicated, flexible, and reconfigurable manufacturing systems. Wiendahl and Heger [13] outlined the concept of a scenario approach to justifying the investments in changeable manufacturing, while Andersen et al. [17] proposed an investment evaluation model of changeable and reconfigurable manufacturing concepts utilizing a simulation approach to account for uncertainty in requirements.

While previous research presents some models for evaluating investments in reconfigurability, limited explicit consideration has targeted the exploration of uncertainty as a key premise for the justification of reconfigurability. Furthermore, previous research presents no explicitly investigations of which financial evaluation parameters that affects the feasibility of reconfigurability from a financial standpoint. However, uncertainty in terms of different aspects of demand volatility combined with the time-horizon of evaluations represent a main aspect in evaluating the financial benefits of reconfigurability compared to more conventional manufacturing concepts. Therefore, this paper addresses the following research question:

- How do different choices of financial evaluation parameters affect the evaluation of reconfigurable manufacturing concepts?

The remainder of the paper is structured as follows; Sect. 2 present the research methodology and outlines the simulation model used for addressing the research question. Section 3 presents the results, while Sect. 4 discusses implications of the results. Conclusively, the contribution is summarized and future work is outlined in Sect. 5.

## 2 Methodology

The research question is addressed by performing a number of simulations and comparing the financial performance of three different manufacturing system concepts for a particular manufacturing scenario. The manufacturing scenario is a case where a demand exists for 8,000 units/month, growing at a randomly distributed rate over a five-year period. When the demand grows, the manufacturing system will eventually need to scale to meet the demand. To address the research question, a base case simulation is performed, and following this, specific model parameters are changed to evaluate how this affects the indications of the economic evaluation.

The base case is a Net Present Value (NPV) calculation based on a Monte Carlo simulation, where the demand is characterized by a trend over a five-year period according to a triangular distribution, minimum 8.000 units/month, mode 10.000 units/month, maximum 16.000 units/month. Also, a normally distributed noise factor is applied to generate monthly demand fluctuations. The scenarios and the concepts applied in the simulation were based on data determined during a workshop as part of an industrial research project, related to reconfigurable manufacturing, for an actual development project. In the simulations, no turnover and thus profit figures have been included, since the turnover would not be different between the scenarios, as demand is always met. Furthermore, the net present value is the value of investments and cost, and thus a value of e.g. 1000 implies a negative cash flow of 1000 has occurred.rr.

The simulation period is five years, and a Monte Carlo simulation is performed, where 200 simulations runs are performed, each simulating the five years. For each simulation run, a demand scenario is generated, and within each month of the scenario, the actual period demand is calculated by applying the noise factor. A three-month running average is applied to the monthly demand, to simulate levelling capacity by stocking or doing back orders. Within each month, the simulation determines whether the current capacity is sufficient. If not, a scaling of capacity is performed. After the simulation, the net present value of investments and variable costs are calculated. Internal interest rate (IRR) was set to 10% annually. The simulation Model was implemented in a Python script, including the libraries Numpy, Math, and Pyplot.

Three concepts are synthesized, 1) one where manufacturing relies solely on manual work and does not require automation, and thus scales only by adding variable cost, and also has no capacity limit, 2) one dedicated concept, where a major investment is made to establish production, with lower variable cost, and scaling is done by replicating the production line, and 3) one reconfigurable concept, that requires a higher initial investment but can add capacity at a much lower cost. The characteristics of the manufacturing concepts are outlined in Table 1.

**Table 1.** Concept characteristics

| Concept | 1 – highly manual | 2 - dedicated | 3 – reconfigurable |
|---|---|---|---|
| Initial investment | No investment | 5.000.000 | 6.000.000 |
| Initial capacity | No upper limit | 10.000 units/month | 10.000 units/month |
| Scaling increments | Not relevant, pure manual work | 10.000 units/month | 2000 units/month |
| Scaling cost | Not relevant, pure manual work | 4.500.000 | 800.000 |

Once the base case simulation was performed, a number of simulations were performed changing model parameters and evaluating the impact on the simulation results. The following simulation parameters were changed: a) changing the simulation to a two-year timespan rather than five, b) removing monthly noise factor, c) removing the moving average, d) narrowing the triangular distribution, e) reducing the interest rate, f) increasing the interest rate.

## 3   Results

Simulating the base case and evaluating for the lowest NPV, showed that concept 2 – the dedicated concept performed best in 22 simulations, whereas concept 3, the reconfigurable concept, performed best in 178 cases, and concept 1 – the manual concept, never performed best. Figure 1 illustrates this in a scatter plot, plotting the net present value on the Y axis, and the final demand after five years on the X axis.

The results of the simulations when changing the model parameters are indicated in Table 2, showing the number of simulations, where each concept turned out with the lowest NPV for each model parameter change. The results are also illustrated in Fig. 2., where each simulation is illustrated as a scatter plot, as in Fig. 1.

The six simulations changing model parameters show significant impact on which concept appears more profitable. Changing the simulation time horizon to a two-year horizon clearly makes the highly manual concept and the dedicated concept more favorable, and the reconfigurable concept less favorable. Removing the monthly fluctuations shifts the dedicated concept to be slightly more favorable compared to the base case,

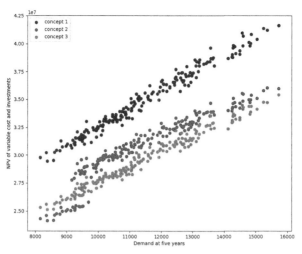

**Fig. 1.** Base case scatter plot of NPV of variable and investment cost vs. Demand after five years

**Table 2.** Number of simulations where each concept provided the lowest NPV

| Concept | 1 | 2 | 3 | Fig. ref |
|---|---|---|---|---|
| Base case | 0 | 22 | 178 | 1 |
| Two year simulation period | 140 | 60 | 0 | 2 (a) |
| No monthly fluctuations | 0 | 40 | 160 | 2 (b) |
| No moving average | 0 | 5 | 195 | 2 (c) |
| Narrow demand distribution | 0 | 101 | 99 | 2 (d) |
| Half IRR | 0 | 17 | 183 | 2 (e) |
| Double IRR | 0 | 24 | 176 | 2 (f) |

but still the reconfigurable concept performs significantly better. Removing the moving average on the monthly demand, which was introduced to provide the opportunity to level capacity rather than chasing the demand, made the reconfigurable concept even more favorable, turning out at the best performing concept in nearly all simulation runs. Narrowing the triangular distribution on the demand at five years to introduce less long-term uncertainty, made the dedicated concept perform better compared to the base case, and in this simulation the dedicated concept and the reconfigurable concept performed almost equally good. Changing the interest rate by doubling or halving it had little impact on the results, only making the dedicated concept perform slightly worse with a lower IRR, and slightly better with a double IRR.

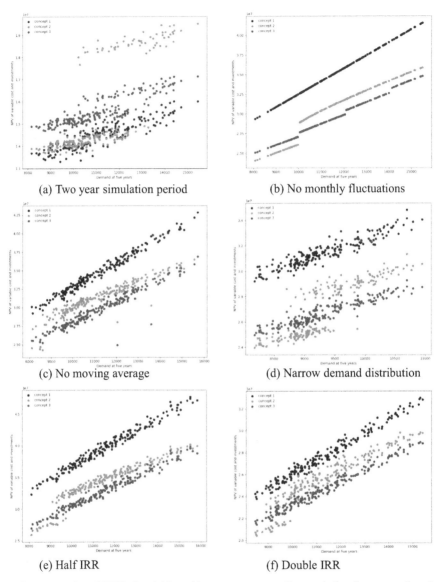

**Fig. 2.** Scatter plot of NPV of variable and investment cost vs. Demand after five years, changing model parameters

# 4   Discussion

Generally, the results indicate that making slight changes in some of the simulation parameters when performing can have a major impact on the evaluation results.

Previous research has suggested that short-term planning, and short-term evaluation of benefits from manufacturing concepts is a barrier towards implementing reconfigurable manufacturing [21]. This was confirmed by the results from the simulations, clearly showing that lower investment concepts performed better in short term simulations, which intuitively makes sense. However basing investment decisions on short term calculations obviously may prevent companies from achieving full benefits of longer-term investments, which may be more profitable over a longer time perspective. Likewise, company investment policies should support continuous investments in reconfigurability over a longer time-horizon, rather than targeting investment as a project by project matter covering only the initial and immediate usage of the system.

The results of the simulations further indicated that a higher degree of uncertainty in the demand made the reconfigurable concept perform better. This was observed when removing the monthly fluctuations, making the dedicated concept perform better. Furthermore, when the triangular distribution for the five-year demand was narrowed, the dedicated concept performed better. This emphasizes the importance of paying close attention to the uncertainty in demand when evaluating concepts, since it can make a significant difference in what concept performs better. It is the impression of the authors from previous research, and also found in literature [22] that few companies actually incorporate uncertainty when making predictions of demand in relation to planning new manufacturing systems, which then becomes a barrier towards implementing reconfigurable manufacturing that would otherwise have been an advantage to the companies. The simulation where the moving average was removed made reconfigurability even more favorable, which intuitively makes sense, since chasing the demand calls for more rapid changes, which is one of the main advantages of reconfigurable manufacturing. Hence, the ability to stock products or do back orders influence the need for short-term scalability, which should also be paid close attention to, since it will influence the actual need for scalability and thus reconfigurability. Finally, contrary to what might be expected, changing the interest rate had very little impact on which concept turned out as the most profitable.

## 5  Conclusion

In this paper, the impact of changing simulation parameters on the financial evaluation of manufacturing concepts, comparing reconfigurable manufacturing to traditional manufacturing concepts was evaluated. This was done by applying a Monte Carlo simulation on three manufacturing concepts with a stochastic demand, and calculating the net present value of the different concepts. By changing simulation parameters such as time span, uncertainty and interest rate, the impact of these on the attractiveness of reconfigurability was investigated. The results showed that changing these parameters has a major impact on which concept is predicted as the most profitable. This is particularly true for uncertainty and timespan, but to a lesser degree the interest rate that only a minor impact. The results presented in the paper can be used in future research, where financial evaluation of reconfigurable manufacturing is relevant, as well as in practice as a guidance on the importance of choosing the appropriate model assumptions. However, given that uncertainty by nature is difficult to estimate accurately, the results suggest that performing sensitivity analyses on the most impacting model parameters would provide valuable insight for practitioners in the transition towards reconfigurability. Future research on this topic will include also taking additional paradigms into consideration such as flexible manufacturing systems, matrix production, and swarm production. Furthermore, scaling manufacturing systems will usually happen over a period where also new products are introduced. This implies that systems are reconfigured to address new demand volumes, but also conversions due to new products requiring new functionality. Future research will aim at also incorporating this into the model, since these two aspects will clearly have influence on each other.

# References

1. Koren, Y., Gu, X., Guo, W.: Reconfigurable manufacturing systems: principles, design, and future trends. Front. Mech. Eng. **13**(2), 121–136 (2018)
2. Psarommatis, F., May, G., Dreyfus, P., et al.: Zero defect manufacturing: state-of-the-art review, shortcomings and future directions in research. Int. J. Prod. Res. **58**(1), 1–17 (2020)
3. Mehrabi, M.G., Ulsoy, A.G., Koren, Y.: Reconfigurable manufacturing systems and their enabling technologies. Int. J. Manuf. Technol. Manage. **1**(1), 114–131 (2000)
4. Mehrabi, M.G., Ulsoy, A.G., Koren, Y.: Reconfigurable manufacturing systems: key to future manufacturing. J. Intell. Manuf. **11**(4), 403–419 (2000)
5. ElMaraghy, H.A.: Flexible and reconfigurable manufacturing systems paradigms. int. J. Flex. Manuf. Syst. **17**(4), 261–276 (2005)
6. Koren, Y.: The rapid responsiveness of RMS. Int. J. Prod. Res. **51**(23–24), 6817–6827 (2013)
7. Maganha, I., Silva, C., Ferreira, L.M.D.: Understanding reconfigurability of manufacturing systems: an empirical analysis. J. Manuf. Syst. **48**, 120–130 (2018)
8. Maganha, I., Silva, C., Ferreira, L.M.D.F.: An analysis of reconfigurability in different business production strategies. In: Proceedings of the Conference on Manufacturing Modelling, Management and Control (2019). (in press)
9. Bortolini, M., Galizia, F.G., Mora, C.: Reconfigurable manufacturing systems: literature review and research trend. J. Manuf. Syst. **49**, 93–106 (2018)
10. Amico, M., Asl, F., Pasek, Z. et al.: Real options: an application to RMS investment evaluation. In: Dashcenko, A.I. (ed.) Reconfigurable Manufacturing Systems and Transformable Factories, pp. 675–693. Springer (2006). https://doi.org/10.1007/978-3-319-22756-6_34
11. Kuzgunkaya, O., ElMaraghy, H.A.: Economic and strategic perspectives on investing in RMS and FMS. Int. J. Flex. Manuf. Syst. **19**(3), 217–246 (2007)
12. Hollstein, P., Lasi, H., Kemper, H.: A survey on changeability of machine tools. In: ElMaraghy, H. (ed.) Enabling Manufacturing Competitiveness and Economic Sustainability, pp. 92–98. Springer (2012). https://doi.org/10.1007/978-3-642-23860-4_15
13. Wiendahl, H., Heger, C.L.: Justifying changeability. A methodical approach to achieving cost effectiveness. J. Manuf. Sci. Product. **6**(1–2), 33–40 (2004)
14. Koren, Y., Wang, W., Gu, X.: Value creation through design for scalability of reconfigurable manufacturing systems. Int. J. Prod. Res. **55**(5), 1227–1242 (2017)
15. Rösiö, C., Jackson, M.: enable changeability in manufacturing systems by adopting a life cycle perspective. In: Zäh, M.F. (ed.) Proceedings of the 3rd International Conference on Changeable, Agile, Reconfigurable and Virtual Production, pp. 612–621 (2009)
16. Andersen, A., ElMaraghy, H., ElMaraghy, W., et al.: A participatory systems design methodology for changeable manufacturing systems. Int. J. Prod. Res. **56**(8), 2769–2787 (2018)

17. Andersen, A., Brunoe, T.D., Nielsen, K., et al.: Evaluating the investment feasibility and indus-
    trial implementation of changeable and reconfigurable manufacturing concepts. J. Manuf.
    Technol. Manage. **29**, 449–477 (2018)
18. Niroomand, I., Kuzgunkaya, O., Bulgak, A.A.: Impact of Reconfiguration Characteristics for
    Capacity Investment Strategies in Manufacturing Systems. Int J Prod Econ **139**(1), 288–301
    (2012)
19. Li, A., Lv, C., Xu, L.: Analysis and Research of System Configuration and Economic
    Evaluation of Reconfigurable Manufacturing System, pp. 1727–1732. IEEE (2007)
20. Zhang, G., Liu, R., Gong, L. et al.: An analytical comparison on cost and performance
    among DMS, AMS, FMS and RMS. In: Dashcenko, A.I. (ed.) Reconfigurable Manufacturing
    Systems and Transformable Factories, pp. 659–673. Springer (2006). https://doi.org/10.1007/
    3-540-29397-3_33
21. Andersen, A., Nielsen, K., Brunoe, T.D.: Prerequisites and barriers for the development
    of reconfigurable manufacturing systems for high speed ramp-up. Procedia CIRP **51**, 7–12
    (2016)
22. Gielisch, C., Fritz, K., Noack, A., et al.: A product development approach in the field of
    micro-assembly with emphasis on conceptual design. Appl. Sci. **9**(9), 1920 (2019)

# A Hybrid Architecture for a Reconfigurable Cellular Remanufacturing System

Camilo Mejía-Moncayo⬤, Jean-Pierre Kenné⬤, and Lucas A. Hof$^{(\boxtimes)}$⬤

École de Technologie Supérieure, Montreal, Québec H3C 1K3, Canada
lucas.hof@etsmtl.ca

**Abstract.** Remanufacturing is a practice that postpones the product 'end-of-life,' returning the properties or features of a new product to a used product. This type of process represents an efficient circular economy strategy to extend product life, reducing its footprint. However, remanufacturing systems must overcome different challenges related to uncertainty and its impact on efficiency. In this sense, this study exposes a hybrid remanufacturing-manufacturing architecture with embedded features from cellular and reconfigurable manufacturing systems. The architecture was synthesized in a Mixed Integer Non-Linear optimization model, which defines the remanufacturing system configuring the cells and product families, balancing the workloads, establishing the scheduling sequence and quantifying the reconfigurability cost. The result is a self-adaptive system that maintains a continuous production rate by managing its capacity.

**Keywords:** Remanufacturing · Reconfigurability · Cellular manufacturing

## 1 Introduction

Circular Economy (CE) has become a feasible alternative to face environmental challenges and provide sustainable solutions. The application of CE strategies or 9 R's framework, which includes refuse, reduce, reuse, repair, refurbish, remanufacture, repurpose, recycle, and recover energy [5] promotes technical and natural loops [8]. In this sense, remanufacturing takes significant importance closing the loop [6] (see Fig. 1), in which products could recover its conditions as a new product and be returned to users [13], but with a reduced ecological footprint [6] and price [13]. Remanufacturing has other advantages as savings in labor, material, and energy costs, shorter production lead times, new market and product development opportunities, and even an improvement in the firm image by exposing a positive socially concerned image [4].

© IFIP International Federation for Information Processing 2021
Published by Springer Nature Switzerland AG 2021
A. Dolgui et al. (Eds.): APMS 2021, IFIP AICT 631, pp. 488–496, 2021.
https://doi.org/10.1007/978-3-030-85902-2_52

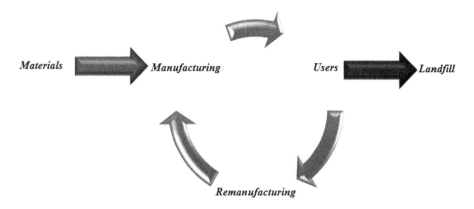

**Fig. 1.** Remanufacturing as a technical loop

Remanufacturing has significant advantages for the Original Equipment Manufacturers (OEMs), customers, and society. However there are barriers and challenges for OEMs to implement remanufacturing systems. In this sense, Matsumoto [9] identifies the following barriers: an effective collection of used products, development of an efficient remanufacturing process, customer acceptance of remanufactured products, and OEMs do not always have an incentive to remanufacture. In addition, uncertainty is a common condition of remanufacturing systems. The unknown state, quantity, and time to return used products produce a series of process inefficiencies.

The disassembly and assembly processes play an important role in remanufacturing [14]. Disassembling, cleaning, refurbishing, replacing parts (as necessary), reassembling the product, and quality control [13] are common operations to remanufacture a product. Remanufacturing itself is an enormous challenge for any company due to uncertainty in the products' conditions, quantity and the human intensive labor in disassembling [14]. This represents a significant effort of planning and operation by the inherent inefficiencies. Consequently, this challenge has to be faced beyond the traditional manufacturing paradigms to achieve a remanufacturing system design with self-adaptative capacity.

Cellular Manufacturing Systems (CMS) allow to decompose a complex factory into mini-factories or cells, which are manageable units in which groups of product families are manufactured [10,11]. This paradigm change greatly facilitates the administration and control of the production system, and at the same time, makes it more flexible [7,12]. Moreover, if we reinforce the system providing features like a reconfigurable manufacturing system, it could manage the demand changes by adding or reducing its capacity [2,3].

In this sense, the hypothesis defined for this study is: The mix of cellular and Reconfigurable manufacturing architectures in a remanufacturing system could help manage uncertainty in the quantity and product conditions. Hence if the capacity is increased or decreased, the system could keep a constant production rate, avoiding work in process stocks.

That is the core of this study, which describes a hybrid architecture for a cellular reconfigurable manufacturing/remanufacturing system that could adapt itself to different products in variable quantities. This study was synthesized as a Mixed Integer Non Linear Programing (MINLP) model that integrates cell formation, workload balancing, reconfigurability, and scheduling. A validation process was carried out through a case study modeled in GAMS [1].

The paper is organized as follows: the problem statement is outlined in Sect. 2. In Sect. 3 we introduce a case study. In Sect. 4 the modeling and programming results are analyzed and discussed. Finally, the conclusions of the proposed model and study will be discussed.

## 2   Problem Statement

The proposed hybrid architecture mixes cellular and reconfigurable manufacturing features for a remanufacturing system. Cellular manufacturing concepts simplify the system from a plant with complex and aggregated processes to a set of mini plants or cells with a reduced size and complexity. This is possible by defining product families, which share similar processes and product parts, which are assigned to one of the cells in which each product family will be remanufactured completely.

The fluctuations in used products return rates are common in remanufacturing, generating uncertainty. Thereby, it is difficult to forecast a return rate for the used products by the multiple factors involved in the use, status, and final disposal of them. In this sense, this study tries to manage this issue by balancing workloads; modifying cell capacity, adding or removing machines or workplaces to maintain a continuous production rate depending of the processed product. This feature allows the system to adapt to the return rate variations, making it reconfigurable and self-adaptable.

The architecture features were summarized in an optimization model which simultaneously integrates cells and product families formation, workloads balancing, system reconfigurability and production scheduling sequence as is described as follows:

- The strategy carried out for product families and cell definition considers the similarities between products and workplaces were modeled as Hamming distances among them. The minimization of this term allows grouping the workplaces into cells and products into families.
- The system's reconfigurability is achieved by balancing the workloads by adding or removing workplaces depending on cycle time and product operation times. This allows keeping the production flow constant in each cell.
- Production scheduling is modeled as the lateness, establishing it as the sum of the absolute differences between the product due date and the time to complete.

The model is subject to the following restrictions:

- For the definition of cells and product families, the model must guarantee that each workplace has to be assigned at least to one cell, allowing different cells to have the same workplaces to avoid movements between cells. The model restricts that each product must belong to a single product family, which at the same time must belong to a single cell.
- Workload balancing needs that the number of workplaces for each product must be greater or equal to "0" to avoid alternative solutions that exceed the budget for the purchase and installation of workplaces. The model restricts those costs to a maximum budget.
- To achieve feasible solutions in terms of production scheduling and reconfigurability. The solutions must avoid duplication in the production sequences precedence's, restrict precedence assignments only to products that belong to the same product family.

## 3  Development of the Case Study

The process carried out started from the definition of a case study with a unique solution, summarized at Table 1. The generation of the ideal case study is based on the following assumptions:

- The remanufacturing system is part of the Original Equipment Manufacturer OEM business strategy.
- The products to be remanufactured have been previously defined and the operations necessary to process them.
- The conditions of the returned products meet the minimum requirements to be remanufactured.
- The remanufacturing system is integrated into a reverse logistics system that allows the return of used products.
- There is a constant demand for the products that define the production sequence.

The analyzed case has 5 types of machines or workstations, 5 products, 5 periods, and 2 cells or product families. This case was generated following the steps described below:

1. We generated an ordered incidence matrix with a diagonal block in which each cell and product family could be identified.
2. We disordered randomly the incidence matrix.
3. The operation times were generated randomly, for each operation defined in incidence matrix.
4. Cycle times were generated, calculating the greatest common divisor from each product's columns of operation times.
5. The return rate of the used products was generated randomly.
6. We generated the due date following a specific sequence.

**Table 1.** Case study

| Incidence matrix | | Products | | | | |
|---|---|---|---|---|---|---|
| | | 1 | 2 | 3 | 4 | 5 |
| Workplaces | 1 | 1 | | 1 | | 1 |
| | 2 | | 1 | | 1 | |
| | 3 | 1 | | 1 | | 1 |
| | 4 | | 1 | | 1 | |
| | 5 | 1 | | 1 | | 1 |

| Operation times | | Products | | | | |
|---|---|---|---|---|---|---|
| | | 1 | 2 | 3 | 4 | 5 |
| | 1 | 450 | | 180 | | 360 |
| | 2 | | 210 | | 340 | |
| | 3 | 300 | | 90 | | 60 |
| | 4 | | 140 | | 170 | |
| | 5 | 150 | | 270 | | 120 |

| Return rate | | Products | | | | |
|---|---|---|---|---|---|---|
| | | 1 | 2 | 3 | 4 | 5 |
| Periods | 1 | 684 | 849 | 197 | 852 | 298 |
| | 2 | 197 | 467 | 140 | 757 | 564 |
| | 3 | 953 | 141 | 523 | 930 | 737 |
| | 4 | 600 | 652 | 577 | 629 | 848 |
| | 5 | 110 | 545 | 478 | 708 | 599 |

| Due date | | Products | | | | |
|---|---|---|---|---|---|---|
| | | 1 | 2 | 3 | 4 | 5 |
| | 1 | 102600 | 59430 | 120330 | 204270 | 138210 |
| | 2 | 167760 | 236960 | 180360 | 365650 | 214200 |
| | 3 | 357150 | 375520 | 404220 | 533620 | 448440 |
| | 4 | 538440 | 579260 | 590370 | 686190 | 641250 |
| | 5 | 657750 | 724340 | 700770 | 844700 | 736710 |

| | Products | | | | |
|---|---|---|---|---|---|
| Cycle time | 1 | 2 | 3 | 4 | 5 |
| | 150 | 70 | 90 | 170 | 60 |

The characteristics of the proposed hybrid manufacturing architecture were synthesized in a mathematical model, which was solved for the case study obtaining the schema shown in Fig. 2. In this schema, the system has two cells that have their own product family. Cell 1 is made up of 6 workplaces type 1, 2 workplaces type 3, and 3 workplaces type 5; that can be added or removed depending on whether it process product 1, 3, or 5, which belong to the family assigned to Cell1. Similarly, cell 2 processes products 2 and 4 and comprises 3 workplaces type 2 and 2 workplaces type 4.

The number of workplaces shown for each cell in Fig. 2 allows keeping the workloads balanced. It is possible to verify that its value is the result to divide the operation time by the cycle time as is showed in Table 2 and Fig. 3 for product 1. Finally, thanks to workload balancing and system reconfigurability it is possible to maintain a continuous production rate following the production sequence exposed in Fig. 4.

**Table 2.** Number of workplaces for product 1

| Workplace | Operation time | Cycle time | Number of workplaces |
|---|---|---|---|
| 1 | 450 | 150 | 3 |
| 3 | 300 | 150 | 2 |
| 5 | 150 | 150 | 1 |

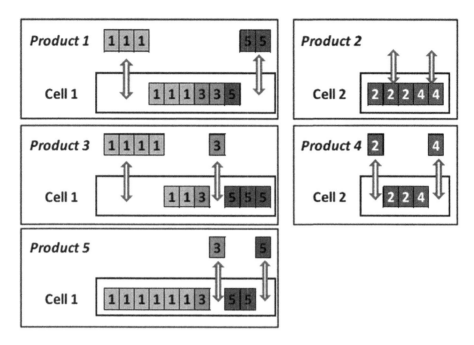

**Fig. 2.** Schema of proposed remanufacturing architecture for the case study

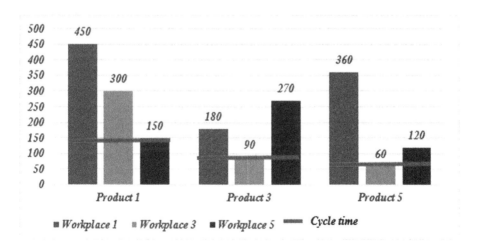

**Fig. 3.** Operation times for cell 1

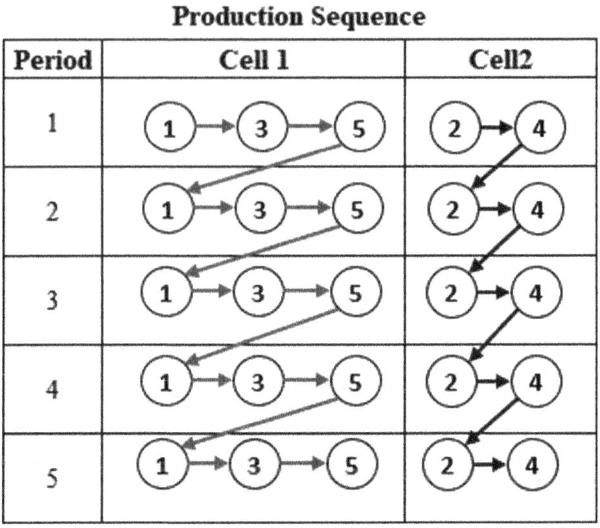

**Production Sequence**

Fig. 4. Production sequence per cell and period for the case study

# 4    Results Analysis and Discussion

In this study, hybrid architecture for remanufacturing is discussed, in which the mix of cellular and Reconfigurable manufacturing features are integrated into a remanufacturing system. The generation of manufacturing cells decomposes the system complexity, reducing the variation among the products processed in each cell. This allows each cell to process similar products (product family), sharing workplaces, tools, and equipment. The system reconfigurability is achieved by increasing or decreasing cell capacity depending on the product, which keeps a constant production rate by workload balancing. This feature allows it to manage the uncertainty by adapting the system capacity to the different product conditions and quantities.

The purpose of the hybrid architecture is achieved in its characteristics, supporting the hypothesis of this work as is described in the case study. First, the system's complexity is reduced by generating two cells that fully process the family of products assigned to them (see Fig. 2). This helps to control the system thanks to the disaggregation of the products return rate in each cell, reducing the impact of variability by quantity or condition of returned products in the throughout manufacturing system, only affecting the cell to which the variable product belong.

The workload balancing allows maintaining a constant flow through each cell, avoiding maintaining work in process; this contributes to the cost reduction derived from inventories, transport and material handling, and even represents benefits for stakeholders by reducing lead times.

Whatever manufacturing system must to be reconfigured when finish a product production and start another. In this architecture system's reconfigurability is a crutial feature which allows to make this task without spend a lot time, thanks to a flexible automated system which make it possible. However, reconfigurability could represent big challenges depending on each workplace's characteristics, by the dimensions, weight, the number of tools or equipment, or even the energy sources, or the management of polluted substances, which need to be analyzed depending on the processes needed, or the complexity of the processed product.

This proposal scope presents limitations by the reduced dimension of the case addressed. Subsequent works require the evaluation of different cases, including industrial data, to verify its applicability or the consideration of a multi-objective focus.

## 5    Conclusions

In this work, a hybrid manufacturing architecture for the design of reconfigurable cellular remanufacturing systems is exposed. This architecture includes the definition of manufacturing cells with their respective product family, thanks to the similarities of the products in the processes required to remanufacture them. The system's reconfigurability is achieved by a automated system which modify the system's capacity, adding or removing workplaces for keeping workload balanced in each cell. In this way, processing times are keeping less than or equal to each product's cycle time. This study is under development, and validation is ongoing with more complex cases and real industrial data.

## References

1. GAMS - Cutting Edge Modeling. https://www.gams.com/
2. Bortolini, M., Galizia, F.G., Mora, C.: Reconfigurable manufacturing systems: literature review and research trend. J. Manuf. Syst. **49**, 93–106 (2018)
3. Brunoe, T.D., Andersen, A.L., Nielsen, K.: Changeable manufacturing systems supporting circular supply chains. Procedia CIRP **81**, 1423–1428 (2019)
4. Bulmuş, S.C., Zhu, S.X., Teunter, R.: Capacity and production decisions under a remanufacturing strategy. Int. J. Prod. Econ. **145**(1), 359–370 (2013). https://doi.org/10.1016/j.ijpe.2013.04.052
5. van Buren, N., Demmers, M., van der Heijden, R., Witlox, F.: Towards a circular economy: the role of Dutch logistics industries and governments. Sustainability (Switzerland) **8**(7), 1–17 (2016). https://doi.org/10.3390/su8070647
6. Ellen Macarthur Foundation: Towards The Circular Economy. Technical report, Ellen Macarthur Foundation (2013). https://www.ellenmacarthurfoundation.org/assets/downloads/publications/Ellen-MacArthur-Foundation-Towards-the-Circular-Economy-vol.1.pdf
7. Irani, S.A. (ed.): Handbook of Cellular Manufacturing Systems. John Wiley & Sons Inc. (1999). https://doi.org/10.1002/9780470172476
8. Kirchherr, J., Reike, D., Hekkert, M.: Conceptualizing the circular economy: an analysis of 114 definitions (2017). https://doi.org/10.1016/j.resconrec.2017.09.005

9. Matsumoto, M., Yang, S., Martinsen, K., Kainuma, Y.: Trends and research challenges in remanufacturing. Int. J. Precis. Eng. Manuf. Green Technol. **3**(1), 129–142 (2016). https://doi.org/10.1007/s40684-016-0016-4

10. Mejía-Moncayo, C., Battaia, O.: A hybrid optimization algorithm with genetic and bacterial operators for the design of cellular manufacturing systems. IFAC-PapersOnLine **52**(13), 1409–1414 (2019)

11. Mejia-Moncayo, C., Rojas, A.E., Dorado, R.: Manufacturing cell formation with a novel discrete bacterial chemotaxis optimization algorithm. In: Figueroa-García, J.C., López-Santana, E.R., Villa-Ramírez, J.L., Ferro-Escobar, R. (eds.) WEA 2017. CCIS, vol. 742, pp. 579–588. Springer, Cham (2017). https://doi.org/10.1007/978-3-319-66963-2_51

12. Mejía-Moncayo, C., Rojas, A.E., Mura, I.: A discrete bacterial chemotaxis approach to the design of cellular manufacturing layouts. In: Gervasi, O., et al. (eds.) ICCSA 2018. LNCS, vol. 10960, pp. 423–437. Springer, Cham (2018). https://doi.org/10.1007/978-3-319-95162-1_29

13. Parkinson, H.J., Thompson, G.: Analysis and taxonomy of remanufacturing industry practice. Proc. Inst. Mech. Eng. Part E J. Process Mech. Eng. **217**(3), 243–256 (2003). https://doi.org/10.1243/095440803322328890

14. Tolio, T., et al.: Design, management and control of demanufacturing and remanufacturing systems. CIRP Ann. Manuf. Technol. **66**(2), 585–609 (2017). https://doi.org/10.1016/j.cirp.2017.05.001

# Assembly Line Balancing with Inexperienced and Trainer Workers

Niloofar Katiraee[1(✉)], Serena Finco[1], Olga Battaïa[2], and Daria Battini[1]

[1] Department of Management and Engineering, University of Padova, Stradella San Nicola 3, 36100 Vicenza, Italy
niloofar.katiraee@phd.unipd.it
[2] KEDGE Business School Campus Bordeaux, 33405 Talence, France

**Abstract.** In this paper, we present a simple assembly line balancing problem for two different sets of workers: trainer workers who are more experienced and likely older, and inexperienced workers who are usually younger and require more time to perform some tasks. Therefore, the main characteristic of this problem is that trainer workers are involved in helping and supporting inexperienced ones in executing some tasks which are more complicated to be carried out. Moreover, task times vary according to the stations where they can be performed due to different sets of equipment we can find in each of them. The problem is modelled as a linear program and solved optimally by applying it to a real-case application. The developed model can be successfully applied in order to help companies to manage a high level of turnover.

**Keywords:** Assembly line balancing · Inexperienced workers · Training activities · Employee turnover · Production system

## 1 Introduction

Fierce global competition forces manufacturing companies to strive for their operational efficiency. One of the key factors on the agenda of production managers, who constantly requires optimization, is the planning and use of resources. An enormous scientific effort is dedicated to developing intelligent and reliable algorithms that are based on the analysis of the data produced and recorded in the context of Industry 4.0 [1]. However, the implementation of such an approach in human-centred manufacturing environments, where human resources are highly involved in the manufacturing process, is not straightforward. For evident reasons, in practice, human factors cannot be ignored when it comes to planning manufacturing activities involving human operators. Since, in manufacturing, a high number of tasks are still human-centred and their performances largely depend more on workers than on machines [2]. This happens especially in big size highly customized product assembly systems [3]. The workforce may vary in terms of many factors, such as skill, age, gender and physical attributes and these differences can affect the overall performances of the production system in terms of time i.e. [4],

© IFIP International Federation for Information Processing 2021
Published by Springer Nature Switzerland AG 2021
A. Dolgui et al. (Eds.): APMS 2021, IFIP AICT 631, pp. 497–506, 2021.
https://doi.org/10.1007/978-3-030-85902-2_53

cost i.e. [5], and throughput i.e. [6]. Moreover, some companies face a high turnover employee problem. Therefore, the consideration of workers' differences and characteristics in production systems plays an important role, particularly in all human-centered fields like manual assembly systems [7]. In this study, we propose a linear programming model with aim of cost minimization, which allows to balance the assembly line by taking into account different workers in terms of experience. Workers can be experienced or inexperienced due to their years of working or age. This factor can impact the task processing time directly. Therefore, in this study, experienced workers can be assigned to stations, to assist or train inexperienced workers. Furthermore, the number of experienced workers has been considered limited in this paper and they mainly play a trainer's role to supervise and help inexperienced workers to perform tasks within a given cycle time due to this help. On the other hand, inexperienced workers may perform tasks (depending on the types of tasks) in higher time and for this reason, an increment time coefficient is taken into account. Moreover, the task time is not only depending on the workers, in this study but also it depends on the equipment of the station, meaning that each task may have a different duration according to the station where it is assigned.

Besides, as the experienced workers may perform tasks faster, they also may suffer from some functional incapacities simultaneously. For example, age can have a positive impact on workers' experience, while it can decline functional capacities. Therefore, ageing workers could use their skills to compensate for capacity reduction [8]. Hence, in this study, we aim to use and manage the current workers in an assembly line in a way to simply reduce the total cost but to consider workers' differences and the integration of experienced and inexperienced workers in the assembly line. In this way, consequently, inexperienced workers can be helped to promote themselves to be more trained and well-prepared for the future under the supervision of skilled workers. On the other hand, an appropriate condition can be guaranteed for experienced workers (e.g. ageing workers) to reduce their physical workload and prevent them from early retirement as also pointed as an issue by previous studies [9]. Therefore, this issue is timely due to the higher turnover in the workforce particularly considering Covid-19 emergency over the last year.

The remainder of this paper is organized as follows. Section 2 provides a literature review of studies related to assembly line balancing problem considering workers' skill and experience. Section 3 presents the problem description and model formulation. Section 4 describes the model application to a real case study and discusses the obtained results. Finally, Sect. 5 presents conclusions and future perspectives.

## 2  Literature Review

Workers can vary between each other in terms of, for example, skill level, age, gender and physical capability. Therefore, not every worker can perform every task at the same processing time since human characteristics may differ [10]. These differences can be influential widely in sections with a high percentage of workers' involvement, like an assembly line in production systems. Initially, the Simple Assembly Line Balancing Problem (SALBP) has been widely discussed in the literature by considering equal workers in terms of experience, age and wage i.e. [11]. Then, several studies and models have been proposed by including ergonomics, postures, physical fatigue during the assembly line design phase [12].

As the installation of an assembly line requires large investments, assembly line balancing problems can be distinguished based on objectives addressed through such as cost- or profit-oriented models. A particular stream of the cost-oriented models involves selecting and assigning workers with different skill levels to the tasks and stations. For example, [13] developed a model for balancing assembly line with the inclusion of both skilled workers (permanent) and unskilled workers (temporary) with a goal of minimization of temporary workers and consequently the total cost. In their studies, the unskilled workers are temporary workers who are applied to help skilled workers, when it is needed to increase production volume during some months due to the seasonal nature of the demand for its product. Moreover, [14] considered the balancing of an assembly line with consideration of both skilled and unskilled workers (in this case unskilled workers were not temporary) to minimize the total annual workstation costs and annual salaries of skilled and unskilled workers. They assumed that unskilled workers could be assigned to help skilled ones to reduce task processing time. Besides, concerning this issue, other studies have put attention on the mixed-model assembly line balancing with the inclusion of unskilled temporary workers to minimize the summation of workstation and workers' cost [15, 16]. It should be noted that studies that considered workers' differences in terms of disability also exist. Since the pioneering study in [17], the integration of workers with disabilities in assembly line balancing and worker assignment has received considerable attention among researchers.

In this study, our main aim is to balance a simple assembly line by taking into account both experienced and inexperienced workers in the assembly line. We consider the case where the number of experienced workers is limited, and inexperienced workers need to be assisted and trained by trainers and skilled workers. Furthermore, in this study all workers are permanent, and they are assigned to stations according to their skills. We assume that inexperienced workers can do some tasks in a standard task time, while they are not able to do other tasks in a standard time due to the task's skills demand. In this case, inexperienced workers can be helped and trained by experienced ones according to their availability.

## 3   Mathematical Formulation

Our model aims to balance a single assembly line by taking into consideration the inexperienced workers and trainer workers. In companies characterized by a high turnover level, workers involved in the assembly process might not have the appropriate experience to be able to execute all types of tasks. However, to face this problem training activities can be performed by some highly skilled workers that have been worked in the company for several years. In this case, they know all tasks as well as the product to assemble. In such a scenario, if trainers are involved, two benefits can be achieved: first, the cycle time can be reduced, then, inexperienced assemblers can improve their competencies. Therefore, here, a mathematical model is developed to integrate the two sets of workers with consideration of different corresponding costs and task times. The following assumptions are considered:

1) There are two sets of workers: assemblers and trainers.
2) Assemblers have a different level of experience; they can perform all tasks or just some of them.
3) Trainers are highly skilled, and they are involved in the assembly line to supervise or train inexperienced workers.
4) The number of trainers in the line is limited and they have higher salary cost.
5) In each workstation, either only one assembler is assigned to a workstation or a tandem of a trainer and one assembler.
6) The task time varies according to the station where it is performed and the workers performing it.
7) If a trainer is involved in a workstation the time required to perform some tasks can be reduced.
8) We have a maximum number of workstations and we can decide to open just some of them according to their operating cost.
9) The cycle time is known.
10) Precedence relations among tasks must be respected.

According to Table 1 and the assumptions previously made our model is defined as follows:

$$\text{O.F.} \quad \text{Minimize Cost} \tag{1}$$

Subject to

$$\sum_j \sum_w x_{ijw} = 1, \forall i = 1, .., I \tag{2}$$

$$x_{ijw} \leq z_{jw}, \forall i = 1, .., I; \forall j = 1, .., J, \forall w = 1, .., W \tag{3}$$

$$x'_{ijt} \leq z'_{jt}, \forall i = 1, .., I; \forall j = 1, .., J, \forall t = 1, .., T \tag{4}$$

$$\sum_w z_{jw} \leq v_j, \forall j = 1, .., J \tag{5}$$

$$\sum_j \sum_w j x_{ijw} \leq \sum_j \sum_w j x_{kjw}, \forall (i, k) \in A \tag{6}$$

$$\sum_i \sum_w t_{ijw} x_{ijw} - \sum_i \sum_t t'_{ijt} x'_{ijt} \leq Ctv_j \forall j = 1, .., J \tag{7}$$

$$\sum_j z_{jw} \leq 1 \forall w = 1, .., W \tag{8}$$

$$\sum_w z_{jw} \leq 1 \forall j = 1, .., J \tag{9}$$

$$\sum_j z'_{jt} \leq 1 \forall w = 1, .., W \tag{10}$$

$$\sum_t z'_{jt} \leq 1 \forall j = 1, .., J \tag{11}$$

**Table 1.** List of all indexes, parameters, variables and decision variables.

| *Indexes* | |
| --- | --- |
| $i,k$ | Index for tasks |
| $j$ | Index for workstation |
| $w$ | Index for workers |
| $t$ | Index for trainers |
| *Parameters* | |
| $I$ | Number of tasks |
| $J$ | Number of workstations |
| $W$ | Number of workers |
| $T$ | Number of trainers |
| CW | Cost of workers |
| CT | Cost of trainers |
| $C_j$ | Cost of workstation |
| Ct | Cycle time |
| A | Set of all precedence among tasks |
| $t_{ijw}$ | Processing time for task $i$ in workstation $j$ by worker $w$ (it can assume a value $\geq 1$ and $\infty$ if worker $w$ cannot perform the task |
| $t'_{ijt}$ | Reduction time for task $i$ if a trainer $t$ is involved in workstation $j$ (it assumes a value $< t_{ijw}$ and it is 0 if no reduction can be achieved) |
| *Variables* | |
| Cost | The total costs of the assembly system |
| $ST_j$ | The time of workstation $j$ |
| *Decision variables* | |
| $v_j$ | Boolean variable that assumes a value 1 if workstation $j$ is opened, 0 otherwise |
| $x_{ijw}$ | Boolean variable that assumes a value 1 if task $i$ is assigned to workstation $j$ to worker $w$, 0 otherwise |
| $x'_{ijt}$ | Boolean variable that assumes a value 1 if trainer $t$ is involved in workstation $j$ for task i, 0 otherwise |
| $z_{jw}$ | Boolean variable that assumes a value 1 if worker $w$ is assigned to workstation $j$, 0 otherwise |
| $z'_{jt}$ | Boolean variable that assumes a value 1 if trainer $t$ is assigned to workstation $j$, 0 otherwise |

$$x_{ijw}, x'_{ijt}, z_{jw}, z'_{jt}, v_j \in \{0; 1\} \tag{12}$$

$$Cost \in \mathbb{R} \tag{13}$$

Where:

$$Cost = \sum_j \left[ \sum_t CT\, z'_{jt} + \sum_w CWz_{jw} + C_j y_j \right]$$

The objective function defined in (1) minimizes the cost of the whole assembly line by including the trainer and assembler workers costs and the operating costs of opened workstations. Constraint (2) guarantees that each task is assigned to exactly one station and one assembler. Constraints (3) and (4) guarantee that a task i can be assigned to an assembler (3) or a trainer (4) at workstation j if this assembler or this trainer is assigned to the same workstation. Constraint (5) verifies that if station j is open there is at most one assembler assigned to it. Constraint (6) ensures the respect of precedence relations among tasks. Constraint (7) guarantees the respect of cycle time. Constraint (8) and (9) ensure that a worker can be assigned only to one workstation and, in each workstation, we can have at most an assembler worker. The same is defined with constraints (10) and (11) for trainer workers. Finally, constraints (12) and (13) are domain constraints for the decision variables.

## 4   Numerical Case and Discussion

We test our model by considering a real assembly system in which 71 tasks are performed to obtain the final product. The model is solved with Cplex V12.9.0 with default parameters. The computational experiments were conducted with Intel(R)Core™ i7-6700HQ 2.60 GHz and 16 GB RAM. The precedence constraints are shown in Fig. 1.

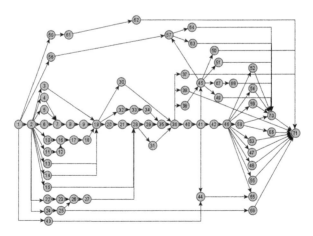

**Fig. 1.** Precedence constraints

Six problem instances have been solved, each one having a different cycle time. In all cases, we assume to have 2 trainer workers, 5 assemblers with different experience level: one worker having experienced just for some tasks, two unskilled with no experience, two having experience and requiring help just for some tasks.

As costs we consider the ratio between the salary of the trainer and the assembler and we assume it equal to 1.3. Then, we consider the ratio between the operating cost of a workstation and an assembler and we assume it equals to 3. We consider five values of cycle time: 35, 40, 45, 50, 55 and 60 min. The task times have as a mean value of 2.3 min with a standard deviation of 2.9 min. Moreover, tasks can be performed in a maximum of 5 workstations. Furthermore, the model is solved with a conventional approach for Assembly Line Balancing and Worker Assignment (ALBWP) without consideration

**Table 2.** The results of the applied model with and without consideration of trainer workers.

| CT [min] | Cost | Number of trainer workers | Number of assembler workers | Number of open stations | Station time |
|---|---|---|---|---|---|
| 35 | 18.6 | 2 | 4 | 4 | [34.904, 34.727, 34.959, 34.835] |
| | 23 | 0 | 5 | 5 | [34.918, 34.908, 34.377, 34.897, 34.939] |
| 40 | 17.3 | 1 | 4 | 4 | [39.872, 38.774, 39.989, 39.581] |
| | 20 | 0 | 5 | 5 | [33.603, 39.391, 39.669, 36.656, 37.956] |
| 45 | 14.6 | 2 | 3 | 3 | [44.62, 44.453, 43.863] |
| | 16 | 0 | 4 | 4 | [44.89, 44.418, 44.994, 44.968] |
| 50 | 13.3 | 1 | 3 | 3 | [49.858, 49.973, 49.963] |
| | 16 | 0 | 4 | 4 | [45.995, 43.145, 49.737, 44.922] |
| 55 | 13.3 | 1 | 3 | 3 | [54.378, 54.638, 51.904] |
| | 16 | 0 | 4 | 4 | [46.972, 41.824, 54.255, 46.23] |
| 60 | 12 | 0 | 3 | 3 | [59.732, 59.883, 59.154] |
| | 12 | 0 | 3 | 3 | [59.678, 59.942, 59.84] |

of trained workers and the reduction coefficient. In the conventional approach, it is assumed that we have just assembler workers with different skills (some of them are able to perform all tasks and others not).

Table 2 indicates the results of solving the model with and without consideration of trainers, respectively. The number of occupied stations, the number of trainers and assemblers, and finally the station times can be seen in Table 2.

The differences between the two models are significant in term of costs with the only exception of the instance with a cycle time of 60 min since results are equals for costs and workers assignment. For all other cases, a trainer or both are involved in helping some workers. Thanks to trainers the time among stations is more balanced despite the traditional ALBWP. Moreover, with our model the assembly process can be executed in the same cycle time in a lower number of workstations and consequently lower cost, and this represents a good achievement for companies.

Finally, we can see that in the case where only assemblers are assigned to stations to perform tasks (for example when the cycle time is 60), then cost for both traditional ALBWP and the model here proposed is similar. Moreover, station time differs just a bit. It is mainly due to the fact that for higher cycle time workers with some experience levels are preferable to inexperience ones since they might not require support from trainers. Moreover, even if they might require more time in executing some tasks the station time is enough to avoid the trainer involvement.

## 5   Conclusions

In this paper, we have proposed a linear programming model that assigns tasks to workstations and workers according to the time and their level of experience. With this model, we aim to minimize the total cost of an assembly line that includes the salary of the workers as well as the cost of each workstation. To improve the assembly line performance, we assume that highly skilled workers, called trainers, might be involved in helping inexperienced ones. In such a way, less-skilled workers can learn from trainers and station time might be reduced. The model here proposed is helpful for practitioners who face a high employees' turnover and need fast training for all new workers involved in the assembly phase to avoid efficiency reductions. We apply our model to a real case assembly line and we compare our results with those obtained with a traditional approach for assembly line balancing and worker assignment problem. Results are very promising since the use of trainers in the line help to not only reduce the cost in comparison to the traditional line but also to reduce the maximal workstation time creating an important potential for increasing the line throughput and efficiency. On the other hand, in this case, we can support ageing and experienced workers by involving them in training activities and preventing them from the excessive physical workload. For future research the effect of learning and accumulated experience can be added as expressed by previous studies [18, 19]. Furthermore, numerical experiments, the model will be tested on a large benchmark of problem instances to assess more precisely its behavior and performances. Since the problem is NP-hard by extension of simple assembly line balancing problem, large instances might not be solved in a reasonable amount of time and consequently, heuristic approaches will be developed to tackle such instances. Finally, a sensitivity analysis will be carried out to investigate better the impact of different parameters of the problem on the final result.

# References

1. Schuh, G., Reuter, C., Prote, J.P., Brambring, F., Ays, J.: Increasing data integrity for improving decision making in production planning and control. CIRP Ann. **66**(1), 425–428 (2017)
2. Calzavara, M., Battini, D., Bogataj, D., Sgarbossa, F., Zennaro, I.: Ageing workforce management in manufacturing systems: state of the art and future research agenda. Int. J. Prod. Res. **58**(3), 729–747 (2020)
3. Zennaro, I., Finco, S., Battini, D., Persona, A.: Big size highly customised product manufacturing systems: A literature review and future research agenda. Int. J. Prod. Res. **57**(15–16), 5362–5385 (2019)
4. Ramezanian, R., Ezzatpanah, A.: Modeling and solving multi-objective mixed-model assembly line balancing and worker assignment problem. Comput. Indust. Eng. **87**, 74–80 (2015)
5. Martignago, M., Battaïa, O., Battini, D.: Workforce management in manual assembly lines of large products: a case study. IFAC-PapersOnLine **50**(1), 6906–6911 (2017)
6. Buzacott, J.A.: The impact of worker differences on production system output. Int. J. Prod. Econ. **78**(1), 37–44 (2002)
7. Katiraee, N., Calzavara, M., Finco, S., Battini, D., Battaïa, O.: Consideration of workers' differences in production systems modelling and design: state of the art and directions for future research. Int. J. Prod. Res. **59**, 1–32 (2021).
8. Boenzi, F., Digiesi, S., Mossa, G., Mummolo, G., Romano, V.A.: Modelling workforce ageing in job rotation problems. IFAC-PapersOnLine **48**(3), 604–609 (2015)
9. Sgarbossa, F., Grosse, E.H., Patrick Neumann, W., Battini, D., Glock, C.H.: Human factors in production and logistics systems of the future. Ann. Rev. Control **49**, 295–305 (2020). https://doi.org/10.1016/j.arcontrol.2020.04.007
10. Battaïa, O., Dolgui, A.: A taxonomy of line balancing problems and their solution approaches. Int. J. Prod. Econ. **142**(2), 2259–277 (2013)
11. Boysen, N., Fliedner, M., Scholl, A.: A classification of assembly line balancing problems. Eur. J. Oper. Res. **183**(2), 674–693 (2007)
12. Otto, A., Battaïa, O.: Reducing physical ergonomic risks at assembly lines by line balancing and job rotation: a survey. Comput. Ind. Eng. **111**, 467–480 (2017)
13. Corominas, A., Pastor, R., Plans, J.: Balancing assembly line with skilled and unskilled workers. Omega **36**(6), 1126–1132 (2008)
14. Moon, I., Shin, S., Kim, D.: Integrated assembly line balancing with skilled and unskilled workers. In: Grabot, B., Vallespir, B., Gomes, S., Bouras, A., Kiritsis, D. (eds.) APMS 2014. IAICT, vol. 438, pp. 459–466. Springer, Heidelberg (2014). https://doi.org/10.1007/978-3-662-44739-0_56
15. Kim, D., Park, J., Moon, I.: Integrated mixed-model assembly line balancing with unskilled temporary workers. In: Umeda, S., Nakano, M., Mizuyama, H., Hibino, H., Kiritsis, D., von Cieminski, G. (eds.) APMS 2015. IAICT, vol. 460, pp. 324–331. Springer, Cham (2015). https://doi.org/10.1007/978-3-319-22759-7_38
16. Kim, D., Moon, D.H., Moon, I.: Balancing a mixed-model assembly line with unskilled temporary workers: algorithm and case study. Assembly Autom. **38**(4), 511–523 (2018). https://doi.org/10.1108/AA-06-2017-070
17. Miralles, C., Garcia-Sabater, J.P., Andres, C., Cardos, M.: Advantages of assembly lines in sheltered work centres for disabled. A case study. Int. J. Prod. Econ. **110**(1–2), 187–197 (2007)

18. Bukchin, Y., Cohen, Y.: Minimizing throughput loss in assembly lines due to absenteeism and turnover via work-sharing. Int. J. Prod. Res. **51**(20), 6140–6151 (2013)
19. Rabbani, M., Akbari, E., Dolatkhah, M.: Manpower allocation in a cellular manufacturing system considering the impact of learning, training and combination of learning and training in operator skills. Manage. Sci. Lett. **7**(1), 9–22 (2017)

# FMS Scheduling Integration for Mass Customization

Yumin He[1(✉)] and Milton Smith[2]

[1] Beihang University, Beijing 100191, People's Republic of China
heyumin@buaa.edu.cn
[2] Texas Tech University, Lubbock, TX 79409, USA

**Abstract.** In today's manufacturing and supply chain environments, many companies face challenge in responding to customers' requirements quickly and providing customized products quickly at low cost. Mass customization can help companies in providing customized products and services quickly and at a low price. Integrated decision-making has been found effective in many situations. This paper reviews the scheduling research of flexible manufacturing systems (FMSs). The FMS scheduling problem is part of the FMS production and operation management problem. Because the production management of FMSs is very difficult, the FMS scheduling problem is very complicated. Many researchers have investigated the FMS scheduling problem. The paper summarizes the FMS scheduling research with recent development. In addition, a framework of FMS scheduling integration for mass customization is developed based on the literature survey. A control flow of FMS part processing is designed as part of the framework. Further development of the FMS scheduling integration is suggested.

**Keywords:** Mass customization · Flexible manufacturing system · FMS scheduling · Scheduling integration

## 1 Introduction

Contemporary manufacturing and supply chain environments are dynamic and changing. It often occurs that customers' requirements need to be satisfied quickly and accurately at a low price [1]. Mass customization (MC) is aimed to provide customized products and services with low cost and high quality in changing environments [2]. Integrated decision-making has been found effective to adequately manage various conflicting objectives and to make coordination control [3–5].

Flexible manufacturing systems (FMSs) can be used to automate mass customization [2]. The production management of FMSs is more difficult than that of mass production lines and job shops [6]. Orders sent to a capacity constrained FMS might not be processed on time and excess-capacity parts have to be sent to a job shop [7]. He, Stecke, and Smith [8] investigated simultaneous robot and machine scheduling with part input sequencing in FMSs for mass customization. Interactions between FMS robot scheduling and machine scheduling with part input sequencing are found. He and Stecke [9] investigated the

© IFIP International Federation for Information Processing 2021
Published by Springer Nature Switzerland AG 2021
A. Dolgui et al. (Eds.): APMS 2021, IFIP AICT 631, pp. 507–515, 2021.
https://doi.org/10.1007/978-3-030-85902-2_54

problem of simultaneous FMS part input sequencing and robot scheduling and suggested the integration of simultaneous FMS part input sequencing and robot scheduling with operation scheduling.

This paper reviews the FMS scheduling research in the area of FMS part input sequencing, robot scheduling, and machine scheduling. The FMS scheduling research with recent development is presented. Based on the literature survey, a framework of FMS scheduling integration for mass customization is developed. A control flow of FMS part processing is designed as part of the framework.

## 2  FMS Scheduling

Many researchers have investigated the FMS scheduling problem. The FMS scheduling research is summarized in the following as FMS part input sequencing, sequencing and scheduling in FMSs with robot material handling, and machine scheduling with FMS material handling.

### 2.1  FMS Part Input Sequencing

FMSs can be classified as flexible flow systems (FFSs) and general flexible machining systems (GFMSs) [10]. FFSs include flexible assembly systems and flexible transfer lines. General flexible machining systems include both dedicated and nondedicated flexible machining systems. A flexible machining cell (FMC) is a single machine and its associated equipment [11].

FMS part input sequencing has been studied with the FMS production planning problems in earlier studies. For example, Stecke and Kim [12] developed a modified Johnson's algorithm for FFSs. Stecke [13] developed several approaches to solve the FMS part input sequencing problem. Research on FMS part input sequencing is summarized in Table 1.

### 2.2  Sequencing and Scheduling in FMS with Robot Material Handling

Researchers have studied sequencing and scheduling in FMS with robot material handling. For example, Sethi *et al.* [27] studied a real FMS with a robot, two or three machine tools, and a single part type to maximize throughput. Sriskandarajah *et al.* [28] scheduled a bufferless dual-gripper robot handling multiple part types to maximize throughput.

Dawande *et al.* [30] surveyed robot move sequencing and part scheduling in FMSs with robot material handling. Research on sequencing and scheduling in FMS with robot material handling is summarized in Table 2.

### 2.3  Machine Scheduling with FMS Material Handling

Researchers have studied machine scheduling with FMS material handling. In earlier studies, Blazewicz et al. [37] proposed a pseudo-polynomial dynamic programming approach to schedule machine and vehicle. Sabuncuoglu and Hommertzheim [38] proposed a dynamic dispatching algorithm to schedule machines and AGVs. Research on machine scheduling with FMS material handling is summarized in Table 3.

**Table 1.** Summary of research on FMS part input sequencing.

| No | Author | Subject | Objective | Year |
|---|---|---|---|---|
| [12] | Stecke, Kim | Part type selection, part input | Over & under load | 1991 |
| [13] | Stecke | Part input sequencing | Over & under load | 1992 |
| [14] | O'Keefe, Rao | Part input | Makespan, throughput | 1992 |
| [15] | Kim, Yano | Part type selection | Total tardiness | 1994 |
| [16] | Smith, Stecke | Part input sequencing | Balancing workload | 1996 |
| [17] | Leu | Order-input sequencing | Set up time | 1999 |
| [18] | Sawik | Sequential loading | Production time | 2000 |
| [19] | Sawik | Blocking scheduling | Completion time | 2001 |
| [20] | Kim, Lee, Yoon | Part input sequencing | Makespan | 2001 |
| [21] | Sawik | Simultaneous balancing and scheduling | Completion time | 2002 |
| [22] | Lacomme, Moukrim, Tchernev | Job-input sequencing | Makespan | 2005 |
| [23] | He, Smith | Part input sequencing | Production | 2007 |
| [24] | Gusikhin, Caprihan, Stecke | Input sequencing | In-sequence parts, buffer size | 2008 |
| [25] | Sawik | Batching scheduling, cyclic scheduling | Completion time | 2012 |
| [26] | He et al. | Part input sequencing | Production | 2015 |

**Table 2.** Summary of research on sequencing and scheduling in FMS with robot material handling.

| No | Author | Subject | Objective | Year |
|---|---|---|---|---|
| [27] | Sethi et al. | Sequencing parts and robot moves | Long run average throughput | 1992 |
| [28] | Sriskandarajah et al. | Robot move sequencing, part sequencing | Throughput rate | 2004 |
| [29] | Geismar et al. | Cyclic scheduling | Throughput | 2005 |
| [31] | Dawande, Pinedo, Sriskandarajah | Cyclic scheduling | Throughput | 2009 |
| [32] | Yildiz, Akturk, Karasan | Cyclic scheduling | Cycle time | 2011 |
| [33] | Zahrouni, Kamoun | Sequencing parts and robot activities | Cycle time | 2012 |

(*continued*)

**Table 2.** (*continued*)

| No | Author | Subject | Objective | Year |
|---|---|---|---|---|
| [34] | Che, Kats, Levner | Robotic flow shop scheduling | Cycle time, stability radius | 2017 |
| [35] | Gultekin, Coban, Akhlaghi | Cyclic scheduling | Throughput rate | 2018 |
| [36] | Foumani, Razeghi, Smith-Miles | Cyclic scheduling | Partial cycle time | 2020 |
| [9] | He, Stecke | Simultaneous part input sequencing and robot scheduling | Production | 2021 |

**Table 3.** Summary of research on machine scheduling with FMS material handling.

| No | Author | Subject | Objective | Year |
|---|---|---|---|---|
| [37] | Blazewicz *et al.* | Production scheduling, vehicle scheduling | Completion time | 1991 |
| [38] | Sabuncuoglu, Hommertzheim | Scheduling machines and AGVs | Mean flowtime, mean tardiness | 1992 |
| [39] | Ulusoy, Bilge | Simultaneous scheduling, of machines and AGVs | Makespan | 1993 |
| [40] | Bilge, Ulusoy | Simultaneous scheduling, of machines and AGVs | Makespan | 1995 |
| [41] | Abdelmaguid *et al.* | Simultaneous scheduling, of machines and AGVs | Makespan | 2004 |
| [42] | Deroussi, Gourgand, Tchernev | Simultaneous scheduling, of machines and AGVs | Makespan | 2008 |
| [43] | Babu *et al* | Simultaneous scheduling, of machines and AGVs | Makespan | 2010 |
| [44] | Lacomme, Larabi, Tchernev | Simultaneous scheduling, of machines and AGVs | Makespan | 2013 |
| [45] | Zheng, Xiao, Seo | Simultaneous scheduling, of machines and AGVs | Makespan | 2014 |
| [46] | Baruwa, Piera | Simultaneous scheduling, of machines and AGVs | Makespan | 2016 |
| [8] | He, Stecke, Smith | Part input sequencing, machine and robot scheduling | Total parts produced, robot utilization, mean flowtime | 2016 |
| [47] | Nouri, Driss, Ghedira | Simultaneous scheduling of machines and transport robot | Makespan | 2016 |

## 3  Framework of FMS Scheduling Integration

A framework of FMS scheduling integration for mass customization is developed based on the literature survey. The framework is illustrated in Fig. 1. The FMS is composed of CNC machines and a robot for material handling. The framework includes an information processing center that is composed of computers, servers, and tools to process information. The center can process data and information exchanged through internet, intranet, and extranet. It can also process data and information obtained from RFID. RFID technology is the significant advance in managing dynamic systems [48].

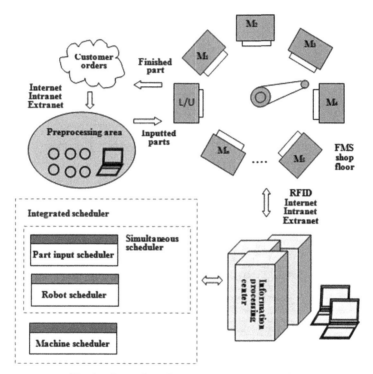

**Fig. 1.** Illustration of FMS scheduling integration.

The framework also includes a scheduler for FMS scheduling integration. The scheduler is composed of an FMS part input scheduler, an FMS robot scheduler, and an FMS machine scheduler. Scheduling algorithms found in the literature can be used as the schedulers. The algorithm for FMS part input sequencing developed in [23, 26] can be used as the FMS part input scheduler. The algorithm for FMS robot scheduling developed in [49] can be used as the FMS robot scheduler.

The combination of the FMS part input scheduler and the FMS robot scheduler can result in the simultaneous scheduler. The algorithm for simultaneous FMS part input sequencing and robot scheduling has been developed in [9]. The integrated scheduler for the FMS scheduling integration can be developed by integrating these individual schedulers. It can also be developed by combining the simultaneous scheduler with the FMS machine scheduler.

A control flow of FMS part processing is designed. The control flow is aimed to illustrate part flow and control decision making in the FMS. The control flow is explained in the following. Inputted parts are waiting for loading and unloading (L/U). If the L/U is available, a part is loaded to the FMS. Parts are waiting for the robot for moving. If the robot is available, a part is moved by the robot to a machine for next operation. Parts are waiting for machines for processing. If a machine is available, a part is processed by the machine. After an operation is finished, the part is checked. If the part does not finish all operations, the part is waiting for the robot for moving. Otherwise, the part is waiting for the L/U to be unloaded. The diagram of the control flow is illustrated in Fig. 2.

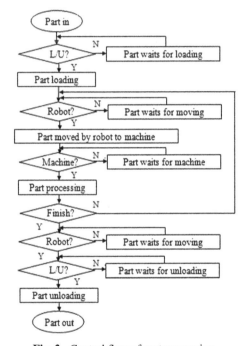

**Fig. 2.** Control flow of part processing.

# 4 Conclusion

In contemporary manufacturing and supply chain environments, companies often face challenge in satisfying customers' requirements quickly and accurately at a low price in dynamic and changing environments. Mass customization and integrated decision-making can provide help to companies to overcome the difficulty.

The FMS scheduling problem is part of the FMS production and operation management problem. Because the production management of FMSs is very difficult, the FMS scheduling problem is very complicated. Many researchers have investigated the FMS scheduling problem.

In this paper, the FMS scheduling research with recent development is reviewed. It is summarized as FMS part input sequencing, sequencing and scheduling in FMSs with robot material handling, and machine scheduling with FMS material handling. It is hoped that the review can provide researchers with a reference of the FMS scheduling.

This paper also develops a framework for FMS scheduling integration for mass customization. A control flow of FMS part processing is designed. The integrated scheduler for the FMS scheduling integration can be developed by integrating the FMS part input scheduler, the FMS robot scheduler, and the FMS machine scheduler. It can also be developed by combining the simultaneous scheduler with the FMS machine scheduler.

**Acknowledgment.** The authors would like to thank the session chair and the referees.

# References

1. Shapiro, B.P., Rangan, V.K., Sviokla, J.J.: Staple yourself to an order. Harv. Bus. Rev. **7–8**, 162–171 (2004)
2. Pine, B.J., II., Victor, B., Boynton, A.C.: Making mass customization work. Harv. Bus. Rev. **9–10**, 108–119 (1993)
3. Sawik, T.: Single vs. multiple objective supplier selection in a make to order environment. Omega: Int. J. Manage. Sci. **38**, 203–212 (2010)
4. Sawik, T.: Integrated supply, production and distribution scheduling under disruption risks. Omega: Int. J. Manage. Sci. **62**, 131–144 (2016)
5. Ivanov, D., Dolgui, A., Sokolov, B.: Robust dynamic schedule coordination control in the supply chain. Comput. Ind. Eng. **94**, 18–31 (2016)
6. Stecke, K.E.: Formulation and solution of nonlinear integer production planning problems for flexible manufacturing systems. Manage. Sci. **29**, 273–288 (1983)
7. Tetzlaff, U.A.W., Pesch, E.: Optimal workload allocation between a job shop and an FMS. IEEE Trans. Robot. Autom. **15**(1), 20–32 (1999)
8. He, Y., Stecke, K.E., Smith, M.L.: Robot and machine scheduling with state-dependent part input sequencing in flexible manufacturing systems. Int. J. Prod. Res. **54**, 6736–6746 (2016)
9. He, Y., Stecke, K.E.: Simultaneous part input sequencing and robot scheduling for mass customization. Int. J. Prod. Res. Online (2021). https://doi.org/10.1080/00207543.2021.189 4369
10. Rachamadugu, R., Stecke, K.E.: Classification and review of FMS scheduling procedures. Prod. Plan. Control **5**(1), 2–20 (1994)
11. Browne, J., Dubois, D., Rathmill, K., Sethi, S.P., Stecke, K.E.: Classification of flexible manufacturing systems. The FMS Magazine April, pp. 114–117 (1984)
12. Stecke, K.E., Kim, I.: A flexible approach to part type selection in flexible flow systems using part mix ratios. Int. J. Prod. Res. **29**, 53–75 (1991)
13. Stecke, K.E.: Procedures to determine part mix ratios for independent demands in flexible manufacturing systems. IEEE Trans. Eng. Manage. **39**(4), 359–369 (1992)
14. O'Keefe, R.M., Rao, R.: Part input into a flexible flow system: an evaluation of look-ahead simulation and a fuzzy rule base. Int. J. Flex. Manuf. Syst. **4**, 113–127 (1992)
15. Kim, Y.-D., Yano, C.-A.: A due date-based approach to part type selection in flexible manufacturing systems. Int. J. Prod. Res. **32**, 1027–1043 (1994)
16. Smith, T.M., Stecke, K.E.: On the robustness of using balanced part mix ratios to determine cyclic part input sequences into flexible flow systems. Int. J. Prod. Res. **34**, 2925–2941 (1996)

17. Leu, B.-Y.: Comparative analysis of order-input sequencing heuristics in a cellular flexible assembly system for large products. Int. J. Prod. Res. **37**, 2861–2873 (1999)
18. Sawik, T.: Simultaneous versus sequential loading and scheduling of flexible assembly systems. Int. J. Prod. Res. **38**(14), 3267–3282 (2000)
19. Sawik, T.: Mixed integer programming for scheduling surface mount technology lines. Int. J. Prod. Res. **39**(14), 3219–3235 (2001)
20. Kim, Y.-D., Lee, D.-H., Yoon, C.-M.: Two stage heuristic algorithms for part input sequencing in flexible manufacturing systems. Eur. J. Oper. Res. **133**, 625–634 (2001)
21. Sawik, T.: Balancing and scheduling of surface mount technology lines. Int. J. Prod. Res. **40**(9), 1973–1991 (2002)
22. Lacomme, P., Moukrim, A., Tchernev, N.: Simultaneous job input sequencing and vehicle dispatching in a single vehicle automated guided vehicle system: a heuristic branch-and-bound approach coupled with a discrete events simulation model. Int. J. Prod. Res. **43**, 1911–1942 (2005)
23. He, Y., Smith, M.L.: A dynamic heuristic-based algorithm to part input sequencing in flexible manufacturing systems for mass customization capability. Int. J. Flex. Manuf. Syst. **19**, 392–409 (2007)
24. Gusikhin, O., Caprihan, R., Stecke, K.E.: Least in-sequence probability heuristic for mixed-volume production lines. Int. J. Prod. Res. **46**, 647–673 (2008)
25. Sawik, T.: Batch versus cyclic scheduling of flexible flow shops by mixed integer programming. Int. J. Prod. Res. **50**(18), 5017–5034 (2012)
26. He, Y., Rachamadugu, R., Smith, M.L., Stecke, K.E.: Segment set-based part input sequencing in flexible manufacturing systems. Int. J. Prod. Res. **53**, 5106–5117 (2015)
27. Sethi, S.P., Sriskandarajah, C., Sorger, G., Blazewicz, J., Kubiak, W.: Sequencing of parts and robot moves in a robotic cell. Int. J. Flex. Manuf. Syst. **4**, 331–358 (1992)
28. Sriskandarajah, C., Drobouchevitch, I., Sethi, S.P., Chandrasekaran, R.: Scheduling multiple parts in a robotic cell served by a dual-gripper robot. Oper. Res. **52**(1), 65–82 (2004)
29. Geismar, H.N., Sethi, S.P., Sidney, J.B., Sriskandarajah, C.: A note on productivity gains in flexible robotic cells. Int. J. Flex. Manuf. Syst. **17**, 5–21 (2005)
30. Dawande, M., Geismar, H.N., Sethi, S.P., Sriskandarajah, C.: Sequencing and scheduling in robotics cells: recent developments. J. Sched. **8**, 387–426 (2005)
31. Dawande, M., Pinedo, M., Sriskandarajah, C.: Multiple part-type production in robotic cells: equivalence of two real world models. Manuf. Serv. Oper. Manag. **11**(2), 210–228 (2009)
32. Yildiz, S., Akturk, M.S., Karasan, O.E.: Bicriteria robotic cell scheduling with controllable processing times. Int. J. Prod. Res. **49**, 569–583 (2011)
33. Zahrouni, W., Kamoun, H.: Sequencing and scheduling in a three-machine robotic cell. Int. J. Prod. Res. **50**(10), 2823–2835 (2012)
34. Che, A., Kats, V., Levner. E.: An efficient bicriteria algorithm for stable robotic flow shop scheduling. Euro. J. Oper. Res. **260**, 964–971 (2017).
35. Gultekin, H., Coban, B., Akhlaghi, V.E.: Cyclic scheduling of parts and robot moves in m-machine robotic cells. Comput. Oper. Res. **90**, 161–172 (2018)
36. Foumani, M., Razeghi, A., Smith-Miles, K.: Stochastic optimization of two-machine flow shop robotic cells with controllable inspection times: From theory toward practice. Robot. Comput.-Integrat. Manuf. **61**, 1–20 (2020)
37. Blazewicz, J., Eiselt, H.A., Finke, G., Laporte, G., Weglarz, J.: Scheduling tasks and vehicles in a flexible manufacturing system. Int. J. Flex. Manuf. Syst. **4**, 5–16 (1991)
38. Sabuncuoglu, I., Hommertzheim, D.L.: Dynamic dispatching algorithm for scheduling machines and automated guided vehicles in a flexible manufacturing system. Int. J. Prod. Res. **30**(5), 1059–1079 (1992)
39. Ulusoy, G., Bilge, U.: Simultaneous scheduling of machines and automated guided vehicles. Int. J. Prod. Res. **31**, 2857–2873 (1993)

40. Bilge, U., Ulusoy, G.: A time window approach to simultaneous scheduling of machines and material handling system in an FMS. Oper. Res. **43**(6), 1058–1070 (1995)
41. Abdelmaguid, T.F., Nassef, A.O., Kamal, B.A., Hassan, M.F.: A hybrid GA/ heuristic approach to the simultaneous scheduling of machines and automated guided vehicles. Int. J. Prod. Res. **42**, 267–281 (2004)
42. Deroussi, L., Gourgand, M., Tchernev, N.: A simple metaheuristic approach to the simultaneous scheduling of machines and automated guided vehicles. Int. J. Prod. Res. **46**(8), 2143–2164 (2008)
43. Babu, A.G., Jerald, J., Haq, A.N., Luxmi, V.M., Vigneswaralu, T.P.: Scheduling of machines and automated guided vehicles in FMS using differential evolution. Int. J. Prod. Res. **48**(16), 4683–4699 (2010)
44. Lacomme, P., Larabi, M., Tchernev, N.: Job-shop based framework for simultaneous scheduling of machines and automated guided vehicles. Int. J. Prod. Econ. **143**, 24–34 (2013)
45. Zheng, Y., Xiao, Y., Seo, Y.: A tabu search algorithm for simultaneous machine/AGV scheduling problem. Int. J. Prod. Res. **52**(19), 5748–5763 (2014)
46. Baruwa, O.T., Piera, M.A.: A coloured petri net-based hybrid heuristic search approach to simultaneous scheduling of machines and automated guided vehicles. Int. J. Prod. Res. **54**, 4773–4792 (2016)
47. Nouri, H.E., Driss, O.B., Ghedira, K.: Simultaneous scheduling of machines and transport robots in flexible job shop environment using hybrid metaheuristics based on clustered holonic multiagent model. Comput. Ind. Eng. **102**, 488–501 (2016)
48. Dolgui, A., Proth, J.-M.: Special section on radio frequency identification. IEEE Trans. Industr. Inf. **8**, 688 (2012)
49. He, Y., Smith, M.L., Dudek, R.A.: Robotic material handler scheduling in flexible manufacturing systems for mass customization. Robot. Comput.-Integrat. Manuf. **26**, 671–676 (2010)

# Economic Design of Matrix-Structured Manufacturing Systems

Patrick Schumacher[✉], Christian Weckenborg, and Thomas S. Spengler

Technische Universität Braunschweig, Institute of Automotive Management and Industrial
Production, Chair of Production and Logistics, Mühlenpfordtstraße 23,
38106 Braunschweig, Germany
p.schumacher@tu-braunschweig.de

**Abstract.** Due to increasing product variety and uncertain demand for highly
individualized products, a rising need for flexibility of manufacturing systems
can be observed. In this context, the concept of matrix-structured manufacturing
systems (MMS) has attracted increasing consideration. MMS aim to achieve high
operational flexibility by implementing a flexible product flow between stations
with automated guided vehicles and by providing redundant resources for each
operation, thus eliminating constant cycle times and the serial arrangement of
stations. This paper investigates the design of MMS pursuing an economic objective. We formulate a mixed-integer program for the design of MMS. Introducing
a numerical example, we illustrate the effectiveness of our approach and derive
future research opportunities.

**Keywords:** Matrix-structured manufacturing system · Flexible manufacturing
system · Mass customization

## 1   Introduction

Manufacturing companies face different trends such as individualized customer demand,
an increased level of global competition, and rapid technological progress. These trends
result in a higher product variety and uncertain demand [1]. Thus, a change of the
production paradigm from mass production to personalized production can be observed
[2]. Conventional manufacturing systems (MS) such as mixed-model assembly lines or
job shops are either capable of handling high product volumes or a high product variety
but struggle to cope with both requirements simultaneously. Manufacturing companies
are therefore confronted with the challenge of designing MS which can reliably produce
a high volume of products with a high product variety in an economic way. Due to the
digitization in the manufacturing sector (frequently referred to as Industry 4.0), novel
technologies arise which may serve as drivers of increasing flexibility at reasonable costs
[3]. The use of automated guided vehicles (AGVs), for example, enables flexible routing
between stations of a MS. One concept of manufacturing systems that emerged from the
use of these new technologies is the concept of matrix-structured manufacturing systems
(MMS).

© IFIP International Federation for Information Processing 2021
Published by Springer Nature Switzerland AG 2021
A. Dolgui et al. (Eds.): APMS 2021, IFIP AICT 631, pp. 516–524, 2021.
https://doi.org/10.1007/978-3-030-85902-2_55

MMS describes the concept of a MS capable of producing high volumes of multiple products by using a flexible material flow through the manufacturing system while redundant resources allow for alternative paths through the MS. The basic elements of the MMS are stations, each of which is an autonomous subsystem and may operate at an individual pace. By allowing an individual pace for every station, a constant cycle time can be avoided. This is desirable as different processing times for different products lead to unbalanced utilization of stations and may result in starving or blocking of stations in conventional MS [4]. Different resources, i.e., human workers, autonomous machines, or collaborative robots operate in the stations of the MMS. Each resource is characterized by the capability to complete specific operations for certain products with a corresponding processing time. As the resources are used redundantly, not only a single station is capable to execute a specific operation for a product, but several stations are. Therefore, starving and blocking effects can be avoided [1]. To fully utilize the advantages of dynamic cycle times and redundant resources, a flexible material flow is required. AGVs transport the products through the MMS until their assembly is completed. Thus, products can skip stations that are not required for the assembly, which leads to routing flexibility as redundant resources in different stations can be used for the processing of specific operations [5]. By combining the advantages of job shops and assembly lines, MMS aim to efficiently produce a high volume of products with a high variety [1].

First pioneer implementations demonstrate the industries' interest in MMS. The mechanical engineering company KUKA AG advertises the concept of MMS [6] and the automotive company Audi AG already uses a MMS for the assembly of the Audi R8, resulting in an estimated efficiency gain of 20% in comparison with the former mixed-model assembly line [7]. However, the increased flexibility of MMS might result in higher production costs compared to more efficient means of production. Thus, an evaluation of the initial configuration decisions in the design phase of MMS becomes necessary. In this contribution, we, therefore, present a cost-oriented approach to this long-term planning problem, as these objectives have been commonly considered in the design of other MS [8]. Therefore, we formulate a mathematical optimization model to obtain cost-efficient initial designs for MMS. We illustrate the effectiveness of our approach by providing a numerical example.

The remainder of this contribution is structured as follows. In Sect. 2, a literature review considering related contributions is given. In Sect. 3, the decision-making situation of the problem is described in detail. Our developed mathematical model is presented in Sect. 4. A numerical example is presented in Sect. 5. Finally, we conclude our contribution in Sect. 6.

## 2   Literature Review

Although capacity-oriented approaches for MS design are more frequently found in the literature, cost-oriented approaches are increasingly used to evaluate the initial configuration of different types of MS [9]. As the initial design of a MS is a long-term planning problem, costs are usually subdivided by the capital cost and the operating cost. Capital costs are associated with the investment in the basic elements of the MS, whereas operating costs include the costs incurred during the operation of the MS [8, 10]. For MMS, the capital cost is associated with the number of stations that are opened and the resources that operate in the stations. The operating costs mainly consist of the costs occurring for the transportation of products between the stations and the resource usage costs. Additional costs as failure cost, idle-time cost, or reconfiguration cost are mainly investigated in tactical decisions regarding the design of MS and are therefore out of the scope of this contribution [9]. To the best of the authors' knowledge, a cost-oriented design approach for MMS has not yet been addressed in the literature. For comprehensive literature reviews on cost-oriented design approaches of MS, we refer to Hazır, Delorme, Dolgui [9] and Yelles-Chaouche, Gurevsky, Brahimi, Dolgui [8].

The first mathematical formalization of a problem that is related to the design of MMS is the Simple Assembly Line Balancing Problem (SALBP) [11]. In SALBP, operations are assigned to serially arranged stations considering capacity restrictions and precedence relations between production operations. As the assumptions of SALBP are very restrictive, many contributions generalize the problem to enable the consideration of practical aspects in their mathematical formulations. Those contributions are often referred to as General Assembly Line Balancing Problems (GALBP) [12]. Many of those generalizations are linked to the problem considered in this contribution. In Mixed Model Assembly Line Balancing, two or more products are produced in the same MS [12]. This is also a key characteristic of MMS. The multi-manned Assembly Line Balancing Problem additionally considers the parallel execution of operations by different resources positioned in the same station [13]. As several resources can be assigned to every station, the parallel execution of operations is also considered during the design of MMS. The Robotic Assembly Line Balancing Problem (RALBP) investigates the design of automated assembly lines by assigning automated robots to the stations while predominantly pursuing a minimization of used stations or minimization of capital cost for the assigned robots [14, 15]. Resource selection and assignment are also considered during the design of MMS. Thus, in GALBP several aspects of the optimal design of MMS are investigated. However, a serial connection between stations is always assumed. Therefore, GALBP approaches cannot be applied to the design of MMS as the problem of designing MMS requires flexible material transport between stations. For comprehensive literature reviews on GALBP, we refer to Baybars [12], Boysen, Fliedner, Scholl [16] and Boysen, Fliedner, Scholl [17].

Two MS concepts that include a flexible material flow often considered in the literature are Flexible Manufacturing Systems (FMS) and Reconfigurable Manufacturing Systems (RMS). A FMS can be described as a computer-controlled and integrated complex of machines, which operate in different stations connected by an autonomous material handling system. The operations of the different machines and the material flows are coordinated and controlled by the central computer [18]. In comparison to the concept of MMS, the stations in FMS are not operated independently and therefore a common cycle time cannot necessarily be avoided. Moreover, only machines are used as resources in FMS, making MMS more suitable for applications in industries with a high amount of manual operations. A RMS is composed of several reconfigurable machines that can be easily added, removed, or reconfigured to execute different operations during the lifecycle of the RMS due to the modular structure of the machines. The modular structure also allows scaling the RMS to cope with increasing or decreasing capacities. The machines are connected and are operated through control software [8]. In contrast to the concept of MMS, the resources in RMS operate without being grouped into stations. Thus, resources can only be used sequentially and not simultaneously. Moreover, as already observed for FMS, only machines are used as resources in RMS, making MMS more suitable for use in industries with a high amount of manual operations. Therefore, approaches for the design of FMS or RMS differ from our approach as they either exclude crucial assumptions of the concept of MMS or include assumptions that are too narrow. For comprehensive literature reviews on the design of FMS and RMS, we refer to Bortolini, Galizia, Mora [19] and Yelles-Chaouche, Gurevsky, Brahimi, Dolgui [8].

Only a few contributions are investigating on MMS explicitly. Greschke, Schönemann, Thiede, Herrmann [4] and Schönemann, Herrmann, Greschke, Thiede [1] elaborate on the concept of MMS and propose a simulation-based approach for the evaluation of MMS designs. However, those contributions do not investigate the optimal design of MMS. Hottenrott, Grunow [5] propose a mixed-integer linear program and a decomposition-based solution approach for the design of flexible segments for the assembly of products. Although the contribution investigates the design of MMS, the selection of different resources is not considered. Moreover, no cost-oriented objective is pursued. To the best of the authors' knowledge, a cost-oriented design approach for MMS, as presented in Sect. 3, has not yet been addressed in the literature.

## 3  Problem Description

We investigate the design of MMS, in which a set of $p \in P$ products with high product variety and a given demand $D_p$ for the respective products is to be produced in a certain time interval. During this time interval, the MMS may operate upmost $Q$ time units. The shop floor is represented by $l \in L$ locations, at which stations could be opened. The stations are of the same size and are arranged in a grid, i.e., they can be identified by height and length coordinates to determine the distance between the individual stations. The distance between two stations $l$ and $l'$ is denoted $d_{l,l'}$. For the production of all products, a set of $o \in O$ operations have to be executed, where every product requires a subset $O_p \in O$ operations to be executed while complying with the known precedence relations of each operation for each product. A set $r \in R$ of resources are available and can operate in the stations. Due to the standardized station design, every resource can be operated in every opened station, but only a maximum of $K$ resources can be assigned to each station. The capability of resources is limited. Whether a resource is capable of performing a specific operation is depicted in binary parameters $a_{r,o}$. The parameters $\beta_{r,o,p}$ indicate the processing time resource $r$ requires to process operation $o$ for product $p$. The products are transported through the shop floor by AGVs on sets $n \in N_p$ of routes, where each product may have its individual set. A route consists of several operations that are executed by the resources in a specific sequence. The binary parameters $e^p_{oqn}$ indicate whether operation $o$ precedes operation $q$ in route $n$. Every route consists of a sequence of all required operations for the corresponding product, complying with the known precedence relations of the operations. Common dummy start and end operations ($O^S$ and $O^E$) as well as common dummy start and end resources ($R^S$ and $R^E$) exist as start points and end points of every route. Moreover, cost rates induced by opening a station at a location ($ic$), induced by resources ($c_r$), and a cost rate for material transport ($ct$) are known.

We decide whether a station is opened at location $l$ (binary variables $x_l$) and whether a resource $r$ is assigned to location $l$ (binary variables $y_{l,r}$). The auxiliary variables $y_{l,r,l',s}$ are used for the connection of the variables $y_{l,r}$. Moreover, the volume of products $p$ assigned to the routes $n$ (variables $f_{pn}$) are determined. Finally, we determine the flows of units of product $p$ from operation $o$ performed by resource $r$ to operation $q$ performed by resource $s$ on route $n$ (variables $v_{orqspn}$). The objective is to minimize the sum of costs for stations and resources and costs for transportation of the products within the MMS.

We make six further assumptions: First, common entry and exit locations are required for the MMS to enable a stable connection to adjacent production stages. The exact locations of the entry and exit points are known. Second, we assume constant cost rates for stations, resources, and transportation, i.e., a linear depreciation of necessary investments and constant interest rates are assumed. Third, we assume that no failures occur while executing operations. Fourth, we assume the demand $D_p$ for each product is known. Fifth, we assume that operations executed in the same station can be executed simultaneously. Therefore, no scheduling of the operations inside of the stations is required. Finally, we assume that processing times and the demand for each product are deterministic.

## 4  Model Formulation

To provide a formalized description of the problem, a mathematical model formulation is developed in this section using the notation introduced in Sect. 3.

$$Minimize \sum_{l \in L} x_l \cdot ic + \sum_{l \in L} \sum_{r \in R} y_{l,r} \cdot c_r + \sum_{p \in P} \sum_{r \in R} \sum_{s \in R} \sum_{l \in L} \sum_{l' \in L} ct \cdot d_{l,l'} \cdot y_{l,r,l',s} \cdot \sum_{o \in O_p} \sum_{q \in O_p} \sum_{n \in N_p} v_{orqspn} \tag{1}$$

Subject to:

$$\sum_{r \in R} y_{l,r} \leq K \cdot x_l \qquad \forall l \in L \tag{2}$$

$$\sum_{l \in L} y_{l,r} \leq 1 \qquad \forall r \in R \tag{3}$$

$$y_{l,r,l',s} = y_{l,r} \cdot y_{l',s} \qquad \forall l \in L, \ l' \in L, r \in R, \ s \in R \tag{4}$$

$$\sum_{n \in N_p} \sum_{p \in P} \sum_{r \in R} \sum_{o \in O_p} \sum_{q \in O_p} v_{orqspn} \cdot \beta_{s,q,p} \leq Q \qquad \forall s \in R \tag{5}$$

$$\sum_{n \in N_p} \sum_{p \in P} \sum_{r \in R} \sum_{o \in O_p} v_{orqspn} \leq a_{s,q} \cdot \sum_{l \in L} y_{l,s} \cdot \sum_{p=1}^{P} D_p \qquad \forall s \in R, q \in O_p \tag{6}$$

$$\sum_{r \in R} \sum_{s \in R} v_{orqspn} = f_{pn} \cdot e_{oqn}^p \qquad \forall s \in R, q \in O_p, \tag{7}$$
$$p \in P, \mathrm{n} \in N_p$$

$$D_p = \sum_{n \in N_p} f_{pn} \qquad \forall p \in P \tag{8}$$

$$\sum_{r \in R} \sum_{o \in O_p} v_{orqspn} = \sum_{t \in R} \sum_{u \in O_p} v_{qsutpn} \qquad \forall q \in O_p, s \in R, \ p \in P, \mathrm{n} \in N_p \tag{9}$$

$$x_l \in \{0, 1\} \qquad \forall l \in L \tag{10}$$

$$y_{l,r} \in \{0, 1\} \qquad \forall l \in L, r \in R \tag{11}$$

$$y_{l,r,l',s} \in \{0, 1\} \qquad \forall l \in L, l' \in L, \tag{12}$$
$$r \in R, s \in R$$

$$v_{orqspn} \geq 0 \qquad \forall q \in O_p, o \in O_p, o \neq q, \tag{13}$$
$$p \in P, r \in R, s \in R, n \in N_p$$

$$f_{pn} \geq 0 \qquad \forall p \in P, n \in N_p \tag{14}$$

Objective (1) is to minimize the total costs of the initial configuration of a MMS. The total costs consist of the costs for opened stations, the costs for resources, and the transportation costs. Constraints (2) guarantee that resources can only be assigned to opened stations and that the maximum number of resources per station is complied with. Constraints (3) ensure that every resource is assigned to upmost one station. Constraints (4) link the decision variables $y_{l,r}$ to the auxiliary variables $y_{l,r,l',s}$. Constraints (5) ensure that the workload assigned to each resource is less than the maximum time for resources to execute operations. Constraints (6) assure that operations can only be assigned to resources that are deployed and capable of executing the specific operation. Constraints (7) link the variables $f_{pn}$ and $v_{orqspn}$. Constraints (8) guarantee demand fulfillment for every product and Constraints (9) ensure flow balance for every resource. Constraints (10)–(14) define the domains of the decision variables. To derive quantitative evidence on the described problem, we present a numerical example in the following section. Therefore, we implemented the model in Python 3.9 and solved it using the Python Gurobi API (version 9.0.0). The computations were run on a standard computer with AMD Ryzen 7 PRO 3700U CPU @ 2.3 GHz and 16 GB RAM.

## 5 Numerical Example

This study considers a numerical example consisting of $|O| = 12$ operations which need to be executed complying to given precedence relations for $|P| = 3$ different products. The demand $D_p$ for each product is assumed to be 10.000 units. Processing times to execute the operations vary depending on the product. Common dummy start and end operations as well as dummy resources exist for every product. Stations can be opened at $|L| = 9$ locations, forming a $3 \times 3$ grid. A maximum of $K = 2$ resources can be operated in each opened station. As the problem of designing a MMS is a long-term planning problem, we assume the life cycle of the MMS and all included elements to be five years. Therefore, the aggregated demand of five years is to be produced for each of the models. Assuming 230 workdays per year and one daily eight-hour shift, the maximum time for resources to execute operations $Q$ is 552,000 min. We restrict the number of available resources to $|R| = 12$, consisting of five assembly workers (resources $R_1$ - $R_5$), five automated assembly robots (resources $R_6$ - $R_{10}$), and a dummy resource for the dummy start and end operations each. While we assume human workers to be capable of executing 50% of the operations, the assembly robots are only capable of executing two out of the twelve required operations to accommodate their limited capabilities. We further assume that the processing times of the assembly robots are halved for each operation in comparison to the processing times of human workers. To derive the costs of workers, assembly robots, material transport, and opened stations we have to justify additional assumptions. Data to estimate the investment for opening a station are generally difficult to obtain as the data depend on the actual assembly processes. Therefore, we suppose costs of 35,000 EUR per opened station comprising for equipment and installation of the station itself as proposed in Weckenborg, Spengler [20]. Assembly robots normally have a basic price between 42,000 EUR and 67,000 EUR. [21] We assume costs of 70,000 EUR per automated assembly robot taking into account additional costs for installation. Further, we assume the investment in technologies to be fully depreciated during a five-year period. Costs per worker, therefore, result in 327,520 EUR in the same five-year period based on hourly labor costs reported in Eurostat [22]. The cost rate for transporting one product for one distance unit $ct$ is assumed as 0.10 EUR.

The optimal solution for this numerical example with an objective value of 1,519,560 EUR is shown in Fig. 1. By analyzing the depicted solution, the following beneficial design characteristics can be observed: In the found solution, a station is opened at five of the nine possible locations. The opened stations are located next to each other and no gaps between opened stations occur. The resulting design can be described as compact to reduce costs for material transport. Two resources are assigned to every opened station so that the maximum space for resource deployment of opened stations is always exploited. While all automated assembly robots (resources $R_6$ - $R_{10}$) are assigned to a station, only three assembly workers (resources $R_1$ - $R_5$) are deployed. The resources that are positioned in stations between the start resource $R^S$ and the end resource $R^E$ are used to execute operations for the entire maximum time $Q$ (resources $R_5$, $R_1$, $R_7$ and $R_{10}$). Accordingly, high utilization of resources with central positioning can be assumed. Finally, it can be observed that the operations for most of the products are executed along the same routes. Only three routes were used for the production of the three products.

The resulting design can be partially interpreted as serially arranged stations for every product as the model also allows a line to be the optimal design. However, the numerical example only investigates a rather small example consisting of 3 products and 9 stations, thus resulting in rather few possible designs. Nevertheless, the products

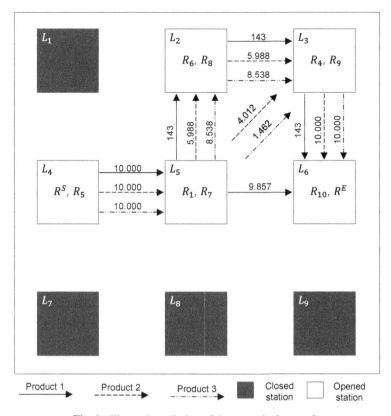

**Fig. 1.** Illustrative solution of the numerical example

are skipping not required stations, reducing short-term key figures like lead times and work in progress that would be increased if a strict serial arrangement would have been applied.

## 6  Conclusion

In the contribution at hand, we propose an optimization model for the economic design of MMS and illustrate our approach with a numerical example. Nevertheless, the design of MMS requires further research. In future work, the numerical example can be enhanced to examine the impact of increased product variety, available resources, the capabilities of resources, and changes in assumed cost rates on the economic design of MMS. Moreover, a comparison of the performance of derived configurations of MMS to more efficient means of production needs to be developed.

As we assumed the initial design of MMS to be a long-term planning problem, we decided to focus on a static and deterministic environment in the contribution at hand. However, as the stations of MMS are standardized and resources can be operated in every station, the configuration of MMS can easily be reconfigured. By considering dynamic demands and changing requirements for the MMS, the reconfigurability of MMS will also be the subject of our future research.

# References

1. Schönemann, M., Herrmann, C., Greschke, P., et al.: Simulation of matrix-structured manufacturing systems. J. Manuf. Syst. **37**, 104–112 (2015)
2. Koren, Y.: The Global Manufacturing Revolution: Product-Process-Business Integration and Reconfigurable Systems. Wiley, Hoboken N.J (2010)
3. Ivanov, D., Das, A., Choi, T.-M.: New flexibility drivers for manufacturing, supply chain and service operations. Int. J. Prod. Res. **56**(10), 3359–3368 (2018)
4. Greschke, P., Schönemann, M., Thiede, S., et al.: Matrix structures for high volumes and flexibility in production systems. Procedia CIRP **17**, 160–165 (2014)
5. Hottenrott, A., Grunow, M.: Flexible layouts for the mixed-model assembly of heterogeneous vehicles. OR Spectrum **41**(4), 943–979 (2019). https://doi.org/10.1007/s00291-019-00556-x
6. KUKA AG. https://bit.ly/3vsvl2n. Accessed on 13 March 2021
7. Handelsblatt. https://bit.ly/3rVGeHJ. Accessed on 13 March 2021
8. Yelles-Chaouche, A.R., Gurevsky, E., Brahimi, N., et al.: Reconfigurable manufacturing systems from an optimisation perspective: a focused review of literature. Int. J. Prod. Res. **58**, 1–19 (2020)
9. Hazır, Ö., Delorme, X., Dolgui, A.: A review of cost and profit oriented line design and balancing problems and solution approaches. Annu. Rev. Control. **40**, 14–24 (2015)
10. Touzout, F.A., Benyoucef, L.: Multi-objective sustainable process plan generation in a reconfigurable manufacturing environment: exact and adapted evolutionary approaches. Int. J. Prod. Res. **57**(8), 2531–2547 (2019)
11. Savelson, M.E.: The Assembly Line Balancing Problem. The Journal of Industrial Engineering **6**(3), 18–25 (1955)
12. Baybars, İ: A survey of exact algorithms for the simple assembly line balancing problem. Manage. Sci. **32**(8), 909–932 (1986)
13. Akagi, F., Osaki, H., Kikuchi, S.: A method for assembly line balancing with more than one worker in each station. Int. J. Prod. Res. **21**(5), 755–770 (1983)
14. Rubinovitz, J., Bukchin, J., Lenz, E.: RALB – a heuristic algorithm for design and balancing of robotic assembly lines. CIRP Ann. **42**(1), 497–500 (1993)
15. Michels, A.S., Lopes, T.C., Sikora, C.G.S., et al.: The Robotic Assembly Line Design (RALD) problem: model and case studies with practical extensions. Comput. Ind. Eng. **120**, 320–333 (2018)
16. Boysen, N., Fliedner, M., Scholl, A.: A classification of assembly line balancing problems. Eur. J. Oper. Res. **183**(2), 674–693 (2007)
17. Boysen, N., Fliedner, M., Scholl, A.: Production planning of mixed-model assembly lines: overview and extensions. Prod. Plan. Control **20**(5), 455–471 (2009)
18. Browne, J., Dubois, D., Rathmill, K., et al.: Classification of flexible manufacturing systems. FMS Magazine **2**, 14–27 (1984)
19. Bortolini, M., Galizia, F.G., Mora, C.: Reconfigurable manufacturing systems: literature review and research trend. J. Manuf. Syst. **49**, 93–106 (2018)
20. Weckenborg, C., Spengler, T.S.: Assembly line balancing with collaborative robots under consideration of ergonomics: a cost-oriented approach. IFAC-PapersOnLine **52**(13), 1860–1865 (2019)
21. RobotWorx. https://bit.ly/3d9GpKy. Accessed on 17 March 2021
22. Eurostat, https://bit.ly/2ODiYQm. Accessed on 17 March 2021

# Aggregate Planning for Multi-product Assembly Lines with Reconfigurable Cells

Mehmet Uzunosmanoglu[1,2]($\boxtimes$) , Birger Raa[1] , Veronique Limère[2,3] ,
Alexander De Cock[4] , Yogang Singh[1,2] , Angel J. Lopez[1,2] ,
Sidharta Gautama[1,2] , and Johannes Cottyn[1,2]

[1] Department of Industrial Systems Engineering and Product Design,
Ghent University, Ghent, Belgium
[2] ISyE CoreLab, Flanders Make, Lommel, Belgium
[3] Department of Business Informatics and Operations Management,
Ghent University, Ghent, Belgium
[4] CodesingS CoreLab, Flanders Make, Lommel, Belgium
https://www.FlandersMake.be

**Abstract.** This paper deals with aggregate planning of Reconfigurable Assembly Lines (RAL). The assembly line considered in this paper consists of hexagonal cells. These have multiple slots where processing modules can be inserted to perform certain operations. In addition, each cell has a single central slot where a central module can be inserted for inter-cellular and intra-cellular transportation of parts. Multiple products with different assembly sequences must be handled over multiple planning periods. An Integer Quadratic Programming (IQP) model is proposed to solve the following problems simultaneously: (i) assigning processing modules and a central module to the cells; (ii) installation of the cells and conveyors between the cells; and (iii) routing products, ensuring that availability of the resources is not exceeded. The assembly line should be reconfigured over time to adapt to possible product functionality and demand changes at minimum reconfiguration, operational and material handling costs while ensuring the demand is met within each period. The IQP model is implemented and solved for an illustrative problem and its extensions using Gurobi.

**Keywords:** Reconfigurable assembly lines · Aggregate planning ·
Integer quadratic programming · Optimisation models

## 1 Introduction

Market dynamics have evolved, with increased volatility of demand and variety of product functionalities, challenging the traditional manufacturing methods. Reconfigurable Manufacturing Systems (RMS) are proposed to fill the gap between the efficiency of Dedicated Manufacturing Systems (DMS) and the flexibility of Flexible Manufacturing Systems (FMS) to overcome the challenges with minimum cost [1].

© IFIP International Federation for Information Processing 2021
Published by Springer Nature Switzerland AG 2021
A. Dolgui et al. (Eds.): APMS 2021, IFIP AICT 631, pp. 525–534, 2021.
https://doi.org/10.1007/978-3-030-85902-2_56

Along with RMS systems, various optimization problems have appeared in the literature. E.g., for Reconfigurable Cellular Manufacturing Systems (RCMS), part grouping, cell formation, and cell loading problems are discussed [2,3]. For Reconfigurable Manufacturing Lines (RML), the number of workstations, the number of machines per workstation, and the operation assignments need to be decided [4]. Similarly, Reconfigurable Assembly Lines (RAL) face the same problem of determining the system configuration while ensuring the assembly sequence of parts [5]. The most common objective for these optimization problems is to minimize total costs, including e.g., investment, reconfiguration, operational, inventory, etc. Other aspects are often considered as well, such as reconfigurability, operational capability, and reliability [6]. An elaborate overview of the relevant literature is found in [7].

In this paper, aggregate planning for multi-product assembly lines with reconfigurable hexagonal cells is considered. Instead of workstations with parallel machines as in the traditional RML and RAL, cells are considered that have multiple slots for side modules. Every side module has specific assembly capabilities. Thus, different assembly operations can be performed within a single cell at the same time (if different side modules are assigned to the different slots in a cell), but also the same assembly operation can be performed in different cells at the same time (if there are multiple instances of the same side module and these are assigned to different cells). Furthermore, every cell needs a central module for moving parts and products between its side modules, but also between the cell and the conveyor connecting it to the other cells.

For the aggregate planning of these systems, 3 different problems are solved simultaneously: cell configuration, system configuration, and product routing. Cell configuration focuses on selecting side modules for the hexagonal cells to ensure that all assembly operations for all products can be performed, as well as selecting a central module for each cell. System configuration considers the decision of activating the cells and conveyors as well as positioning them, such that parts can visit the different cells along the assembly line, taking into account specific layout requirements and material handling costs. Product routing, finally, aims to find a routing for each part and product throughout the system. Note that it is possible for different instances of the same product type to follow a different routing, as long as the availability of the resources (side modules, central modules, conveyors) is not exceeded. For this integrated aggregate planning problem, a Quadratic Integer Programming model is developed and validated. The outcomes serve as an input for further detailed scheduling which is beyond the scope of this paper.

The rest of the paper is organized as follows. Section 2 presents the IQP model. An illustrative example is introduced, and multiple instances with different size are solved in Sect. 3. Finally, conclusions and plans for future work are given in Sect. 4.

# 2   Integer Quadratic Programming Model

This section introduces the integer quadratic programming model. Only some variables and constraints are shown for the side and central modules due to the space limitations in the paper. For the other resources, i.e., conveyors, transportation side modules, and the cells, the variables and constraints are similar to the ones shown in this section. Nevertheless, the missing variables and constraints are explained in words.

**Sets**

- $P$: Products to be assembled.
- $N_p$: Sequence of assembly operations for product $p \in P$.
- $T$: Periods in the planning horizon.
- $M$: Available side modules.
- $M(i)$: Side modules that can perform assembly operation $i \in N_p$.
- $C$: Available cell positions.
- $S$: Slots for side modules in a cell.

**Decision variables**

- $X_{k,c}^{p,i,t} \geq 0$: number of products $p$ having operation $i$ performed by side module $k$ at cell $c$ in period $t$.
- $Z_{s,k,c}^{t} \in \{0,1\}$: whether or not side module $k$ is assigned to slot $s$ of cell $c$ in period $t$.
- $ZCm_{k',c}^{t} \in \{0,1\}$: whether or not central module $k'$ is assigned to the central slot of cell $c$ in period $t$.
- $W_{s,k,c}^{t} \in \{0,1\}$: whether or not side module $k$ is in slot $s$ of cell $c$ for $t$ and $t+1$.
- $Y_{s,k,c}^{t} \in \{0,1\}$: whether or not side module $k$ is in slot $s$ of cell $c$ for $t$, but not $t+1$.
- $Q_{s,k,c}^{t} \in \{0,1\}$: whether or not side module $k$ is in slot $s$ of cell $c$ for $t+1$, but not $t$.

**Parameters**

- $d^{p,t}$: Demand for product $p$ in period $t$.
- $a_k$: Available processing time of side module $k$ per period.
- $b_k^{p,i}$: Processing time for task $i$ of product $p$ using side module $k$.
- $o_k^{p,i}$: Processing cost for task $i$ of product $p$ using side module $k$.
- $f_k, g_k$: Insertion/removal cost of side module $k$.
- $o_{k'}^{p,i}$: Cost for moving product $p$ at task $i$ using central module $k'$.

**IQP Model**

$$\text{minimize} \quad \sum_{t \in T} \sum_{p \in P} \sum_{i \in N_p} \sum_{k \in M(i)} \sum_{c \in C} o_k^{p,i} \cdot X_{k,c}^{p,i,t} \quad \textbf{(0.1)}$$

$$+ \sum_{t=1}^{|T|-1} \sum_{s \in S} \sum_{c \in C} \sum_{k \in M} \left( f_k \cdot (W_{s,k,c}^t + Y_{s,k,c}^t + Q_{s,k,c}^t) + g_k \cdot Y_{s,k,c}^t \right) \quad \textbf{(0.2)}$$

$$+ \sum_{t \in T} \sum_{k' \in CM} \sum_{c \in C} \left( \sum_{p \in P} \sum_{i \in N_p} \sum_{k \in M(i)} 3 \cdot o_{k'}^{p,i} \cdot X_{k,c}^{p,i,t} \right) \cdot ZCm_{k',c}^t \quad \textbf{(0.3)}$$

subject to

$$\sum_{k \in M} Z_{s,k,c}^t \leq 1 \quad (\forall c \in C, \forall s \in S, \forall t \in T) \tag{1}$$

$$\sum_{k \in M} Z_{3,k,c}^t + \sum_{k \in M} Z_{1,k,c+1}^t \leq 1 \quad (\forall c = 1, \ldots, |C| - 1, \forall t \in T) \tag{2}$$

$$\sum_{c \in C} \sum_{s \in S} Z_{s,k,c}^t \leq 1 \quad (\forall k \in M, \forall t \in T) \tag{3}$$

$$2 \cdot Z_{s,k,c}^t + Z_{s,k,c}^{t+1} = 3 \cdot W_{s,k,c}^t + 2 \cdot Y_{s,k,c}^t + Q_{s,k,c}^t$$
$$(\forall s \in S, \forall k \in M, \forall c \in C, \forall t \in T \backslash \{t_{|T|}\}) \tag{4}$$

$$W_{s,k,c}^t + Y_{s,k,c}^t + Q_{s,k,c}^t \leq 1 \quad (\forall s \in S, \forall k \in M, \forall c \in C, \forall t \in T \backslash \{t_{|T|}\}) \tag{5}$$

$$\sum_{p \in P} \sum_{\{i \in N_p | k \in M(i)\}} X_{k,c}^{p,i,t} \cdot b_k^{p,i} \leq a_k \cdot \sum_{s \in S} Z_{s,k,c}^t \quad (\forall k \in M, \forall c \in C, \forall t \in T) \tag{6}$$

$$\sum_{k \in M(i)} \sum_{c \in C} X_{k,c}^{p,i,t} = d_p^t \quad (\forall p \in P, \forall i \in N_p, \forall t \in T) \tag{7}$$

$$\sum_{c=1}^{c_1} \sum_{k \in M(i)} X_{k,c}^{p,i,t} \geq \sum_{c=1}^{c_1} \sum_{k \in M(i+1)} X_{k,c}^{p,i+1,t}$$
$$(\forall p \in P, \forall i < |N_p|, \forall c_1 \in C, \forall t \in T) \tag{8}$$

The objective function includes 4 different cost components. (0.1) is the operational cost of the side modules, whereas (0.2) consists of the configuration and reconfiguration costs of side modules. Note that using the same notation, configuration and reconfiguration costs of central modules, conveyors, and transportation side modules are easily calculated. As the third component, the operational cost of central modules is considered in (0.3), taking into account that every assembly operation results in three movements. Note that this term is quadratic. The fourth cost component, not shown in the model, is the total cost of using the conveyors.

In constraints (1), it is ensured that a slot can be utilized by at most one side module. Similar constraints can be written for the central modules by using appropriate indices. Also, an additional constraint ensures that the 5th slot of a cell is allocated to either a side module or a transportation side module depending on the decision made for the insertion of additional conveyors. Note that it is already assumed to have an operational conveyor between two adjacent cell positions for all the cell position pairs in the problem, as the part source and part exit need to be connected regardless of the number of open cells. But, their capacity might be insufficient to provide inter-cellular transportation. Then, a decision is made for each adjacent cell pair to insert an additional conveyor.

Constraints (2) are the problem-specific overlapping constraints. Some side module slots in adjacent cells are overlapping meaning that side modules cannot be inserted into these slots at the same time. E.g., slots I7 and I8 are overlapping (see Fig. 1).

Constraints (3) ensure each side module instance can be inserted into at most one slot during a period. Again, similar constraints can be written for the central modules just by changing the variable.

Constraints (4) identify which of the three reconfiguration scenarios involving insertion and removal of side modules applies. These constraints can also be written for all the other resources by just changing the variables appropriately. Related to these constraints, constraints (5) select one of these scenarios for each resource. These constraints are again duplicated for all the resources. These constraints help define the resource reconfiguration costs included in the objective function (0.2).

Constraints (6) ensures that the availability of the side modules is not exceeded if the side modules are inserted into the slots.

Constraints (7) impose that the demand for any product must be met in any period by performing all of its assembly operations. Since it is allowed to perform the same assembly operation in different cells and by using different side modules, the sum of the number of times an assembly operation is performed by any of these side modules must be equal to the period's demand. Along with this, constraints (8) ensure that parts can be assembled in different cells but only as long as they follow a one-directional flow, i.e., if assembly operation $i$ of product $p$ is performed at a certain cell, then operation $i + 1$ of that same product $p$ can only be performed within the same cell or successor cells.

In addition, the following constraints are included in the model, but not shown here due to space limitations: constraints to ensure that an assembly operation of a product can only be performed in a cell that has a central module that can handle the parts or sub-assemblies involved in that assembly operation; constraints to ensure that the total number of movements a central module performs does not exceed its availability; constraints to ensure that the additional conveyor between cells 2 and 3 can only be inserted if the additional conveyor between cells 1 and 2 is already inserted; and constraints to make sure that if an additional conveyor between two cells is inserted, an additional transportation side module must be inserted to those cells as well. Finally, two sets of constraints are added for ensuring that the capacity of conveyors is not exceeded and if necessary, additional conveyors are inserted.

## 3   Illustrative Problem and Results

In this section, an illustrative problem is introduced and the results are discussed.

**Problem Setup.** The example consists of a layout with 3 possible cell locations in each of which the side and central module slots are numbered as seen in Fig. 1. Also, the possible conveyor positions between the cells as well as 4 conveyor instances are shown. In addition, 5 types of side modules including a single type of transportation side module T4, and 2 types of central module instances are shown. The side modules perform different assembly operations, while the transportation side module T4 is used along with the central modules for moving the parts between the cells and conveyors. Note that there are some overlapping side module slots on adjacent cells which means that the overlapping side modules cannot be used at the same time, e.g., I7 on cell 1 and I8 on cell 2.

In Table 1, the resource specifications are given. ID is the name of the resource type where the letter shows the assembly operation name and the number is the variant ID, e.g., A1 and A2 are two different resource types that perform the same assembly operation A. Instances is the number of available resources of that type. Note that in the example, all of the assembly operations are performed by assembling two parts. Skill time is calculated as minutes, and skill cost is incurred each time the resource is used. Insertion and removal costs are assumed to be symmetrical, and they are used to calculate the reconfiguration cost.

As can be seen in Table 2, there are 3 different assembly operations A, B, and C that are necessary to assemble 4 different products that each have a different sequence of assembly operations to be performed. Furthermore, the demand for the products in 2 periods is given.

Note that at the beginning of the first period, it is assumed that there is no prior configuration for both the cells and the system (referred to as 'greenfield'). At the beginning of the second period, however, it is possible that a reconfiguration occurs depending on the configuration used during the first period (referred to as 'brownfield').

**Results.** The IQP model is implemented and solved with Gurobi for Python version 3.7. The program ran on a PC with an Intel(R) Core i7-10610U CPU @ 1.80GHz and 16 GB RAM capacity.

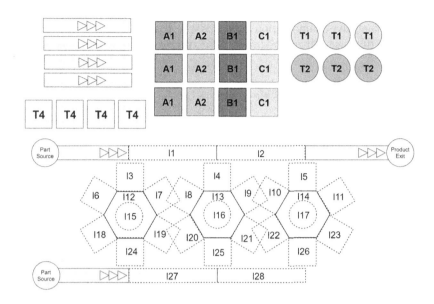

**Fig. 1.** Assembly system layout and available resources.

**Table 1.** Resource specifications.

| ID | Instances | Skill time | Skill cost | Migration cost | |
|----|-----------|------------|------------|--------|--------|
| | | | | Insert | Remove |
| Cell | 3 | – | – | 400 | 400 |
| A1 | 3 | 02:00 | 10 | 100 | 100 |
| A2 | 3 | 04:00 | 20 | 50 | 50 |
| B1 | 3 | 03:00 | 10 | 100 | 100 |
| C1 | 3 | * | * | 100 | 100 |
| T1 | 2 | 00:24 | 10 | 150 | 150 |
| T2 | 2 | 01:00 | 30 | 150 | 150 |
| T4 | 6 | – | – | 100 | 100 |
| T5 | 4 | 02:00 | 10 | 100 | 100 |

**Table 2.** Products, parts, assembly sequence, and demand.

| ID | Parts | $N_p$ | Demand |
|----|-------|-------|--------|
| P1 | 5,5,4 | $\{C, A\}$ | $\{2, 8\}$ |
| P2 | 1,2,3,4 | $\{A, B, A\}$ | $\{2, 3\}$ |
| P3 | 1,2,3,5,5,4 | $\{A, B, C, C, A\}$ | $\{6, 2\}$ |
| P4 | 2,6 | $\{C\}$ | $\{6, 4\}$ |

For the instance with 4 products, 3 possible cell locations, and 2 time periods, it is found that the configuration shown in the Fig. 2 is the only configuration that is used during both periods. Therefore, no reconfiguration is necessary.

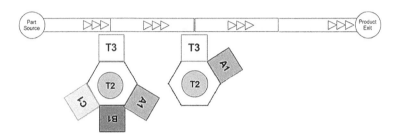

**Fig. 2.** Resulting configuration

To test the model with different instances, the illustrative problem is extended by increasing the number of periods and products. First, the number of periods is increased by generating new demand for each product randomly. However, this is performed in a way that the resource availability is sufficient to meet the product demands as the resource pool is kept the same during the experiments. For this reason, the random demands are drawn from $U(0, 10)$ where U is a discrete uniform distribution.

Secondly, the number of products is increased by generating random demands as described previously, and by generating random-length assembly operation sequences. To generate these sequences, first, the length is randomly defined by drawing a number from $U(1, 5)$ which ensures consistency with the products considered in the illustrative problem. Then, the sequences are filled randomly with the three assembly operations A, B, and C.

All the instances shown in Table 3 are generated and solved while the random seed is fixed. The first nine instances are generated by increasing the number of periods, and keeping the number of products fixed. Next eight instances are generated by increasing the number of products, and keeping the number of periods fixed.

As it is seen in the first nine rows in Table 3, the number of periods affects the solution time dramatically. Increasing it from 2 to 4 increases the solution time by a factor of 13, whereas the increase from 4 to 8 increases the solution time by a factor of 22. In addition, depending on the definition of the periods, e.g., hours, days, months, 10 periods can be considered as a short time horizon for the planning problem considered in this paper. Therefore, it is easy to see that for the problem instances where the periods represent short time intervals, and where it is necessary to generate a plan for a long time horizon, the problem can not be solved in an acceptable time with the current solution approach.

Increasing the number of products affects the solution time dramatically as well. Along with the fact that the model size increases with the number of products, it is observed that making such an aggregate plan including reconfiguration

**Table 3.** Results

| ID | $|P|$ | $|T|$ | CPU time (s) |
|----|-------|-------|--------------|
| 1  | 4  | 2  | 0.94   |
| 2  | 4  | 3  | 3.53   |
| 3  | 4  | 4  | 12.32  |
| 4  | 4  | 5  | 16.29  |
| 5  | 4  | 6  | 81.35  |
| 6  | 4  | 7  | 183.27 |
| 7  | 4  | 8  | 270.56 |
| 8  | 4  | 9  | 454.16 |
| 9  | 4  | 10 | 556.60 |
| 10 | 5  | 2  | 1.71   |
| 11 | 6  | 2  | 4.08   |
| 12 | 7  | 2  | 7.18   |
| 13 | 8  | 2  | 45.23  |
| 14 | 9  | 2  | 76.31  |
| 15 | 10 | 2  | 124.14 |
| 16 | 11 | 2  | 258.36 |
| 17 | 12 | 2  | –      |

is harder with resources fixed while the total demand is increasing. Therefore, this is reflected in the solution time as well. Likewise, the problem becomes infeasible when there are more products than the resources can process, as can be seen in row 17.

## 4  Conclusions and Future Work

In this paper, aggregate planning of a RAL with hexagonal reconfigurable cells is considered. The aggregate planning is performed by solving three problems simultaneously: cell configuration, system configuration, and product routing. The demand for multiple products across multiple periods is met with minimum operational and reconfiguration costs while considering the limited availability of all the resources in the RAL. An IQP model is proposed and solved with Gurobi for an illustrative example. Small problem instances are solved to optimality in limited time using Gurobi. However, for larger instances, this is no longer the case. Therefore, (meta)heuristic approaches will be developed as the next step.

Along with the development of other solution methods, the model will be generalized further. This involves considering automatic guided vehicles (AGVs) or other material handling systems for inter-cellular transportation, and extending the number of cell positions to more than three. In addition, as opposed to a fixed assembly operation sequence for each product type considered in this paper, a flexible sequence of operations (with precedence constraints) is often possible in practice. Therefore, it is planned for the future to adopt this flexibility by allowing alternative sequences in the model. Finally, more extensive computational experiments will be conducted.

**Acknowledgement.** This research was funded by the Flanders Make SBO project AssemblyRecon.

# References

1. Koren, Y., et al.: Reconfigurable manufacturing systems. In: Dashchenko, A.I. (eds.) Manufacturing Technologies for Machines of the Future, pp. 627–665. Springer, Heidelberg (1999). https://doi.org/10.1007/978-3-642-55776-7_19
2. Yu, J.-M., Doh, H.-H., Kim, H.-W., Kim, J.-S., Lee, D.-H., Nam, S.-H.: Iterative algorithms for part grouping and loading in cellular reconfigurable manufacturing systems. J. Oper. Res. Soc. **63**(12), 1635–1644 (2012)
3. Eguia, I., Molina, J.C., Lozano, S., Racero, J.: Cell design and multi-period machine loading in cellular reconfigurable manufacturing systems with alternative routing. Int. J. Prod. Res. **55**(10), 2775–2790 (2016)
4. Dou, J., Dai, X., Meng, Z.: Optimisation for multi-part flow-line configuration of reconfigurable manufacturing system using GA. Int. J. Prod. Res. **48**(14), 4071–4100 (2009)
5. Bryan, A., Jack Hu, S., Koren, Y.: Assembly system reconfiguration planning. J. Manuf. Sci. Eng. **135**(4), 13 p., 041005 (2013). https://doi.org/10.1115/1.4024288
6. Ashraf, M., Hasan, F.: Configuration selection for a reconfigurable manufacturing flow line involving part production with operation constraints. Int. J. Adv. Manuf. Technol. **98**(5–8), 2137–2156 (2018)
7. Yelles-Chaouche, A., Gurevsky, E., Brahimi, N., Dolgui, A.: Reconfigurable manufacturing systems from an optimisation perspective: a focused review of literature. Int. J. Prod. Res. 1–19 (2020). https://doi.org/10.1080/00207543.2020.1813913

# Integrated Workforce Allocation and Scheduling in a Reconfigurable Manufacturing System Considering Cloud Manufacturing

Behdin Vahedi-Nouri[1]([✉]) [iD], Reza Tavakkoli-Moghaddam[1] [iD], Zdenek Hanzalek[2] [iD], and Alexandre Dolgui[3] [iD]

[1] School of Industrial Engineering, College of Engineering, University of Tehran, Tehran, Iran
b.vahedi@ut.ac.ir
[2] IID-CIIRC, Czech Technical University in Prague, Prague, Czech Republic
[3] IMT Atlantique, LS2N – CNRS, La Chantrerie, Nantes, France

**Abstract.** The reconfigurable manufacturing system (RMS) has been acknowledged as an effective manufacturing paradigm to tackle high volatility in demand types and amounts. However, the reconfiguration needs an amount of time and leads to some level of resource wastage. Accordingly, a high frequency in the system's reconfiguration may have a negative impact on its performance. In this regard, this paper investigates the advantage of using cloud manufacturing (CMfg) resources in enhancing the performance of an RMS system. A novel mathematical model is developed for the integrated workforce allocation and production scheduling problem utilizing the CMfg under a non-permutation flow shop setting. This model simultaneously makes decisions on the utilization of the CMfg capacity for performing some jobs, and for the remaining jobs, determination of machines' configurations for each job, scheduling of the jobs on the machines, and allocation of operators to machines as well. This model aims to minimize the sum of job processing costs, overtime costs, and the cost of utilizing the CMfg resources. Finally, a computational experiment is conducted, which shows a promising improvement in the total cost of the production system by utilizing the CMfg capacity.

**Keywords:** Scheduling · Reconfigurable manufacturing system · Workforce · Cloud manufacturing · Mathematical modeling

## 1 Introduction

Over the recent decades, manufacturers have been tackling some emerging challenges, including high volatile market, mass customization, shortening of the product life cycle, globalization, and digital commerce. Accordingly, numerous attempts have been made by academia and manufacturers to survive and flourish in such situations. The reconfigurable manufacturing system (RMS) introduced in the 1990s is among the most effective and practical solutions to these challenges. Benefiting from flexible and changeable nature,

© IFIP International Federation for Information Processing 2021
Published by Springer Nature Switzerland AG 2021
A. Dolgui et al. (Eds.): APMS 2021, IFIP AICT 631, pp. 535–543, 2021.
https://doi.org/10.1007/978-3-030-85902-2_57

in RMSs, the functionality and capacity of a manufacturing system can be adjusted with diversity and variability of market demands in a cost-effective and timely manner [1, 2].

RMSs are formed based on some key modules, which can be modified, rearranged, added, or removed to fulfill the market requirements. However, reconfigurations lead to some level of interruption in the production flow, which may result in time and resource wastage. In some cases, the manufacturing systems can even experience material wastage when test samples are needed after a reconfiguration, as well as a temporary quality decline. As a result, frequent reconfigurations in a short period diminish the advantages of RMSs in high dynamic environments [3]. For this reason, devising some strategies to cover this shortcoming of RMSs and enhance their capability is of vital importance.

Recently, thanks to new technological advancements, including cloud computing, cyber-physical system (CPS), and internet of things (IoT), a novel manufacturing paradigm, so-called cloud manufacturing (CMfg), has been introduced in the Industry 4.0 era. In CMfg systems, diverse distributed manufacturing resources can be integrated and pooled to provide services to geographically distributed customers [4, 5]. Several successful implementations of this paradigm exist in the market, like Plethora, Fictive, and Xometry, to name a few. Considering the massive potential of CMfg systems in resource sharing, integrating the RMS with the CMfg concept can bring about a more productive, sustainable, and resilient system that can effectively alleviate its shortages. The nature of using the CMfg resources is similar to subcontracting or outsourcing. However, due to the advantages of the CMfg, it is now possible to find the right resources and have a price quotation within a short time [6]. As a result, these resources can be effectively incorporated into short-term planning.

Scheduling as a prominent pillar of production management has an influential role in the optimized utilization of resources. It determines the right time, resources, and workforces for the production of products respecting the performance metrics of manufacturing companies, e.g., the total costs, makespan, and total tardiness. Despite the importance of the workforce in the performance, profitability, and endurance of manufacturing companies, this factor has not been incorporated in the scheduling of RMSs so far. For this reason, this research focuses on a scheduling problem in an RMS benefiting the CMfg capacity and considering workforce requirements.

Although there is a rich literature on different areas of RMSs, scheduling problems in an RMS are in their early stages. In this regard, Ren et al. [7] studied cyclic scheduling in a flow shop problem with reconfigurable machines and developed a mixed-integer linear programming (MILP) model for it. Abbasi and Houshmand [8] sought to find the optimum sequence and size of batches and the system configuration for an RMS, in which the arrival of orders follows a Poisson distribution. Bensmaine et al. [9] provided a heuristic algorithm for joint process planning and scheduling of an RMS consisting of reconfigurable machine tools in an open shop setting. A multi-objective particle swarm optimization (MOPSO) algorithm was devised by Dou et al. [10] for an integrated configuration design and scheduling in a multi-part flow line. Ghanei and AlGeddawy [11] focused on the sustainability aspect in the joint layout and scheduling in an RMS. Mahmoodjanloo et al. [12] proposed an efficient differential evolution algorithm for a flexible job shop scheduling problem with reconfigurable machine tools.

In this paper, a scheduling problem in a reconfigurable flow shop environment consisting of a reconfigurable machine in each production stage is explored. The contributions of this paper can be summarized as follows: 1) Integrating operator allocation decisions to the production scheduling in the RMS; 2) Employing the CMfg capacity in the problem to improve the performance of the RMS; 3) Developing a novel mathematical model for the problem.

In the next section, the problem and its assumptions are first described, and then, the corresponding mathematical model is presented. Afterward, in Sect. 3, a computational experiment is conducted to illustrate the impact of utilizing the CMfg resources in the RMS. Finally, Sect. 4 concludes the paper.

## 2  Problem Definition

In this paper, a non-permutation flow shop environment with reconfigurable machines is studied. In this case, the requested jobs visit the same order of production stages; however, their sequence on each stage may be different. Each stage consists of one reconfigurable machine with a few available configurations, which can perform a subset of jobs. In other words, there is at least one configuration eligible for processing a specific job in each stage. A machine reconfiguration may be needed to perform a job, that its required time depends on the current and new configuration. Moreover, there is a set of heterogeneous operators with different capabilities and skill levels to work on stages. As a result, the processing time of a job in a stage depends on the assigned operator and the configuration with which the job is performed.

The planning horizon is one day with a specified normal and overtime working duration. Tackling the negative impact of excessive and non-productive reconfigurations, the possibility of performing some jobs by the CMfg resources is regarded. Using the instant quoting engine of the CMfg system, the cost of performing each job is known quickly before the planning. This problem aims to determine the jobs performed by the CMfg, assign operators to stages, determine a configuration for processing each job (not performed by the CMfg) on all stages, and specify the sequence of these jobs on each stage simultaneously. The respected objective function is to minimize the total cost, which comprises job processing cost, overtime cost, and cost of utilizing the CMfg capacity. The developed MILP model and its notation are elaborated below.

*Sets and Indices:*

$S$     Set of stages
$s$     Index of stage
$I$     Set of jobs
$i, i'$   Indices of job
$O$     Set of operators
$o$     Index of operator
$K_s$    Set of configurations for stage $s$
$k, k'$   Indices of configuration, $k = b$ indicates the configuration at the beginning

## *Parameters:*

$\Omega$    Normal working duration
$\Theta$    Allowed overtime
$\Gamma_i$    Cost of job $i$ in the CMfg system
$\Pi_{sk}$    Processing cost in stage $s$ with configuration $k$ (per unit of time)
$\Phi_{so}$    Overtime cost rate for stage $s$ with operator $o$ (per unit of time)
$\Psi_{so}$    Capability of operator $o$ to work on stage $s$ (a binary parameter)
$SC_{os}$    Skill coefficient of operator $o$ in stage $s$
$P_{isk}$    Normal processing time of job $i$ in stage $s$ with configuration $k$
$T_{skk'}$    Reconfiguration time on stage $s$ from configuration $k$ to $k'$
$E_{isk}$    Eligibility of configuration $k$ on stage $s$ to perform job $i$ (a binary parameter)
$L$    A very large positive number

## *Variables:*

$v_{os}$    1 if operator $o$ works on stage $s$; 0, otherwise
$x_{isk}$    1 If job $i$ in stage $s$ is performed with configuration $k$; 0, otherwise
$z_i$    1 If job $i$ is performed within the CMfg system; 0, otherwise
$y_{ii's}$    1 If job $i'$ is performed after job $i$ in stage $s$; 0, otherwise
$oc_s$    Overtime cost of stage $s$
$c_{is}$    Completion time of job $i$ in stage $s$
$pc_{is}$    Processing cost of job $i$ in stage $s$

## *MILP model:*

$$\sum_{i\in I}\sum_{s\in S} pc_{is} + \sum_{s\in S} oc_s + \sum_{i\in I}\Gamma_i z_i \tag{1}$$

s.t

$$\sum_{o\in O} v_{os} = 1, \forall s \in S \tag{2}$$

$$\sum_{s\in S} v_{os} = 1, \forall o \in O \tag{3}$$

$$\sum_{k\in K_s|E_{isk}=1} x_{isk} = 1 - z_i, \quad \forall i \in I, \ \forall s \in S \tag{4}$$

$$y_{ii's} + y_{i'is} \leq 1 + z_i + z_{i'}, \quad \forall i, i' \in I \ \& \ i \neq i', \ \forall s \in S \tag{5}$$

$$y_{ii's} + y_{i'is} \geq 1 - z_i - z_{i'}, \quad \forall i, i' \in I \ \& \ i \neq i', \ \forall s \in S \tag{6}$$

$$c_{is} \geq SC_{os}P_{isk} + T_{sbk} - L(2 - x_{isk} - v_{os}),$$
$$\forall i \in I, \ \forall s \in S, \ \forall o \in O, \ \forall k \in K_s|E_{isk} = \Psi_{so} = 1 \tag{7}$$

$$c_{is} \geq SC_{os}P_{isk} + T_{sbk} - L(2 - x_{isk} - v_{os}),$$
$$\forall i \in I, \ \forall s \in S, \ \forall o \in O, \ \forall k \in K_s | E_{isk} = \Psi_{so} = 1 \& s \neq 1 \tag{8}$$

$$c_{i's} \geq c_{is} + SC_{os}P_{i'sk'} + T_{skk'} - L(4 - x_{isk} - x_{i'sk'} - v_{os} - y_{ii's}),$$
$$\forall i, i' \in I, \ \forall s \in S, \ \forall o \in O, \ \forall k, k' \in K_s | E_{isk} = E_{i'sk'} = \Psi_{so} = 1 \& i \neq i' \tag{9}$$

$$c_{is} \leq \Omega + \Theta, \quad \forall i \in I, \ \forall s \in S \tag{10}$$

$$oc_s \geq \Phi_{so}(c_{is} - \Omega) - L(1 - v_{os}), \quad \forall i \in I, \ \forall s \in S, \ \forall o \in O | \Psi_{so} = 1 \tag{11}$$

$$pc_{is} \geq SC_{os}P_{isk}\Pi_{sk} - L(2 - x_{isk} - v_{os}),$$
$$\forall i \in I, \ \forall s \in S, \ \forall o \in O, \ \forall k \in K_s | E_{isk} = \Psi_{so} = 1 \tag{12}$$

$$v_{os}, x_{isk}, z_i, y_{ii's} \in \{0, 1\}, \quad oc_s, c_{is}, pc_{is} \in \mathbb{R}^+ \tag{13}$$

The objective function (1) is the sum of the cost of processing jobs inside the factory, overtime cost, and the cost of allocating jobs to the CMfg resources. Constraints (2) and (3) ensures that one operator is allocated to a stage, and an operator only works on one stage, respectively. Constraint (4) determines the configuration of a job in a stage if the job is processed in the factory. Constraints (5) and (6) specify the sequence of jobs in stages if processed in the factory. Constraints (7)–(9) calculate the completion time of the jobs inside the factory for each stage based on the respected configurations, assigned workers, and sequence of jobs. Constraint (10) limits the working duration. Overtime cost in each stage is specified in Constraint (11). The processing cost of each job in each station is calculated by Constraint (12). Finally, Constraint (13) defines the domains of the variables.

## 3   Computational Experiments

In this section, an instance with eight jobs, four operators, four stages, and three configurations for each machine in all stages is considered. The operators capable of working on each stage, the possible configuration for performing the jobs in the different stages, and the initial configuration of each stage are presented in Table 1. The other parameters, described in Table 2, are randomly generated. It should be noted that to set overtime cost rate ($\Phi_{so}$), 20% of the average processing cost rate (i.e., 70$) as the additional overhead cost during the overtime, plus 50% of the average hourly wage of a worker (i.e., 20$) concerning his/her skill level, as the additional wage, are accounted. Moreover, since the prices in the CMfg are very competitive, the minimum and maximum possible cost of a job, processed inside the system, plus 10% transportation cost, is regraded for the cost range of the job in the CMfg. The proposed model is solved by the CPLEX solver in GAMS 24.1 on a PC with Processor Intel (R) Core (TM) i5-7300HQ 2.5 GHz and 12 GB of RAM.

The instance is solved under two strategies: 1) without considering the CMfg capacity and 2) by considering the CMfg capacity, and the corresponding Gantt charts are depicted in Figs. 1 and 2, respectively. In these charts, the stages and their allocated operators

**Table 1.** Capable operators for stages and the required configuration for each job per station

| Stage | Operators | Jobs {Configurations} | | | | | | | | Configuration at the beginning |
|-------|-----------|------|------|------|------|------|------|------|------|---|
| | | i1 | i2 | i3 | i4 | i5 | i6 | i7 | i8 | |
| s1 | {o1/o2} | {k1} | {k2} | {k1} | {k2} | {k2} | {k1} | {k3} | {k2} | k2 |
| s2 | {o2/o3} | {k4/k5} | {k5} | {k5} | {k5} | {k5} | {k6} | {k4} | {k5} | k5 |
| s3 | {o1/o3} | {k9} | {k8} | {k8} | {k7} | {k9} | {k8} | {k7} | {k8} | k8 |
| s4 | {o1/o4} | {k12} | {k11/k12} | {k11} | {k11} | {k10} | {k11} | {k11} | {k10} | {k12} | k12 |

**Table 2.** Other parameters of the instance

| Parameters | Unit |
|------------|------|
| $P_{isk} = U[0.5, 1]$ | Hour |
| $T_{skk'} = U[0.75, 1.5]$ | Hour |
| $\Omega = 8$ | Hour |
| $\Theta = 4$ | Hour |
| $\Pi_{sk} = U[40, 100]$ | $ |
| $SC_{os} = U[0.9, 1.1]$ | — |
| $\Phi_{so} = 0.2 * 70 + \dfrac{0.5*20}{average_{s' \in S|\Psi_{os'}=1}(SC_{os'})}$ | $ |
| $\Gamma_i = 1.1 * U\left[\sum_{s \in S} \min_{k \in K_s, o \in O|E_{isk}=\Psi_{os}=1} SC_{os}P_{isk}\Pi_{sk}, \sum_{s \in S} \max_{k \in K_s, o \in O|E_{isk}=\Psi_{os}=1} SC_{os}P_{isk}\Pi_{sk}\right]$ | $ |

are shown on the vertical axis. For example, s1-o2 indicates operator 2 was allocated to stage 1. Also, each job is represented by a unique color and number. Reconfigurations are also characterized by k*−k* format. For instance, in Fig. 1, the first reconfiguration on stage 1 is from configuration k2 to configuration k1 (k2−k1). Accordingly, all jobs before this reconfiguration are performed by configuration k2, and the following jobs before the subsequent reconfiguration are processed by configuration k1. As it is clear, utilizing the CMfg capacity considerably improves both the total cost (8.9%) and the makespan (24.6%).

To assess the impact of jobs' costs in the CMfg on the justification of utilizing the CMfg, a sensitivity analysis is conducted and shown in Fig. 3. In this figure, the percentage of jobs processed by the CMfg, and the percentage of changes in the total cost by changing the jobs' costs in the CMfg (concerning the values of $\Gamma_i$ provided in Table 2) are depicted in a combined graph. For instance, when the costs of the jobs in the CMfg are increased 15%, 25% of jobs are performed by the CMfg, and the total cost is increased by 4.4%. Based on the result, even by increasing this cost up to 40%, one job is performed by the CMfg, and even in this case, using this strategy is effective and viable (i.e., the total cost is still lower than the case without using the CMfg). Finally, by increasing this cost to 45%, no job is sent to the CMfg.

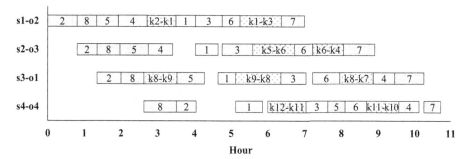

**Fig. 1.** Gantt chart of the instance without considering the CMfg capacity, objective = 1342.49$

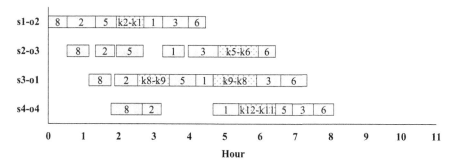

**Fig. 2.** Gantt chart of the instance considering the CMfg resources, objective = 1232.27$

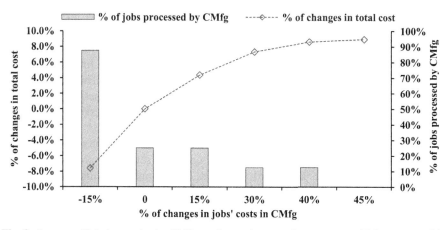

**Fig. 3.** Impact of jobs' costs in the CMfg on the total cost and percentage of jobs processed by CMfg

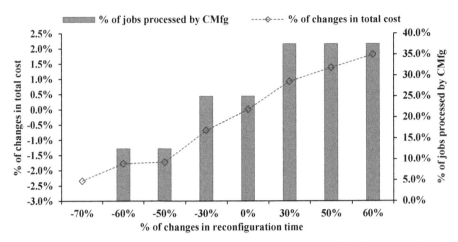

**Fig. 4.** Impact of reconfiguration time on the total cost and percentage of jobs processed by the CMfg

Moreover, the impact of the reconfiguration time on the total cost and the percentage of jobs processed by the CMfg is depicted in Fig. 4. Accordingly, despite a 60% decline in the reconfiguration times (concerning the values of $T_{skk'}$ reported in Table 2), the utilization of the CMfg resources is still viable. Furthermore, due to the possibility of using the CMfg, a 60% increase in the reconfiguration times only leads to a roughly 2% rise in the total cost. It means that incorporating the CMfg resources in RMSs can considerably enhance its performance.

## 4   Conclusion

This paper dealt with an integrated workforce allocation and production scheduling problem in a reconfigurable flow shop environment. To improve the proposedof the average processing system regarding excessive and non-productive reconfiguration, utilizing the cloud manufacturing (CMfg) capacity was proposed. A novel mixed-integer linear programming (MILP) model was developed for the problem to simultaneously make decisions on using the CMfg capacity, operator allocation, machine reconfiguration, and job scheduling. Afterward, an instance was considered to assess the viability of the proposed strategy. Based on the result, utilizing the CMfg resources can considerably reduce the total cost and makespan of the system.

To extend this research, we will investigate a multi-day horizon with night disruptions and transport times to the CMfg, and provide an efficient solution approach, especially for solving medium- and large-sized problems. Incorporating the uncertainty of the parameters (e.g., processing times) and exploring the dynamism of the production system (e.g., dynamic arrival of jobs) can also be considered for future research in this topic.

**Acknowledgment.** This work was supported by the European Regional Development Fund under the project AI&Reasoning (reg. no. CZ.02.1.01/0.0/0.0/15_003/0000466).

# References

1. Koren, Y., Gu, X., Guo, W.: Reconfigurable manufacturing systems: principles, design, and future trends. Front. Mech. Eng. **13**(2), 121–136 (2017). https://doi.org/10.1007/s11465-018-0483-0
2. Bortolini, M., Galizia, F.G., Mora, C.: Reconfigurable manufacturing systems: literature review and research trend. J. Manuf. Sys. **49**, 93–106 (2018)
3. Huang, S., Wang, G., Yan, Y.: Delayed reconfigurable manufacturing system. Int. J. Prod. Res. **57**, 2372–2391 (2019)
4. Liu, Y., Wang, L., Wang, X.V., Xu, X., Zhang, L.: Scheduling in cloud manufacturing: state-of-the-art and research challenges. Int. J. Prod. Res. **57**, 4854–4879 (2019)
5. Vahedi-Nouri, B., Tavakkoli-Moghaddam, R., Hanzálek, Z., Arbabi, H., Rohaninejad, M.: Incorporating order acceptance, pricing and equity considerations in the scheduling of cloud manufacturing systems: Matheuristic methods. Int. J. Prod. Res. **59**, 2009–2027 (2021)
6. Hasan, M., Starly, B.: Decentralized cloud manufacturing-as-a-service (CMaaS) platform architecture with configurable digital assets. J. Manuf. Sys. **56**, 157–174 (2020)
7. Ren, S., Xu, D., Wang, F., Tan, M.: Timed event graph-based cyclic reconfigurable flow shop modelling and optimization. Int. J. Prod. Res. **45**, 143–156 (2007)
8. Abbasi, M., Houshmand, M.: Production planning and performance optimization of reconfigurable manufacturing systems using genetic algorithm. Int. J. Adv. Manuf. Tech. **54**, 373–392 (2011)
9. Bensmaine, A., Dahane, M., Benyoucef, L.: A new heuristic for integrated process planning and scheduling in reconfigurable manufacturing systems. Int. J. Prod. Res. **52**, 3583–3594 (2014)
10. Dou, J., Li, J., Xia, D., Zhao, X.: A multi-objective particle swarm optimisation for integrated configuration design and scheduling in reconfigurable manufacturing system. Int. J. Prod. Res. Article in Press, 1–21 (2020)
11. Ghanei, S., AlGeddawy, T.: An integrated multi-period layout planning and scheduling model for sustainable reconfigurable manufacturing systems. J. Adv. Manuf. Sys. **19**, 31–64 (2020)
12. Mahmoodjanloo, M., Tavakkoli-Moghaddam, R., Baboli, A., Bozorgi-Amiri, A.: Flexible job shop scheduling problem with reconfigurable machine tools: An improved differential evolution algorithm. Appl. Soft Comput. **94**, Article No. 106416 (2020)

# A Bi-objective Based Measure for the Scalability of Reconfigurable Manufacturing Systems

Audrey Cerqueus$^{(\boxtimes)}$ ⓘ and Xavier Delorme ⓘ

Mines Saint-Etienne, Univ Clermont Auvergne, CNRS, UMR 6158 LIMOS,
Institut Henri Fayol, 42023 Saint-Etienne, France
{audrey.cerqueus,delorme}@emse.fr

**Abstract.** The reconfigurable manufacturing systems aim to efficiently respond to demand changes. One of the key characteristics of these systems is the scalability, i.e. the ability to modify the volume of the throughput in order to fit to the demand variability. The design of the RMS has a high impact on its scalability. In the literature, there are only few indicators to evaluate the scalability of a system and most of them are a posteriori measures. In this article, we propose a new measure to assess the scalability since the design phase of the RMS. We present experimental results on state-of-the-art instances to validate our approach. They show that the proposed measure evaluates accurately the scalability.

**Keywords:** Scalability · Reconfigurable manufacturing systems · Multi-objective indicator

## 1 Introduction

In a context of high volatility of market conditions and increased customization leading to smaller batches, manufacturing companies have to react quickly and efficiently to changes in order to remain competitive. Reconfigurable Manufacturing Systems (RMS) have been introduced in [10] to answer to this need of adaptability. Basically the main purpose of RMS is to manage shorter product lifecycles while keeping longer production system lifecycles.

RMS are production systems composed of serial stages with identical parallel resources, which can be automated (for example Computer Numerical Control (CNC) or Reconfigurable Manufacturing Tools (RMT)), or other resources such as workers or cobots. A gantry and a conveyor are generally used to move the products in this grid of stations [9]. A RMS layout is schematized in Fig. 1.

The efficiency of RMS relies on six key features: modularity (ability to reuse machines and tools), integrability (ability to rapidly and efficiently connect new modules), diagnosability (ability to identify automatically a problem in the production system), convertibility (ability to change the system for new products),

© IFIP International Federation for Information Processing 2021
Published by Springer Nature Switzerland AG 2021
A. Dolgui et al. (Eds.): APMS 2021, IFIP AICT 631, pp. 544–552, 2021.
https://doi.org/10.1007/978-3-030-85902-2_58

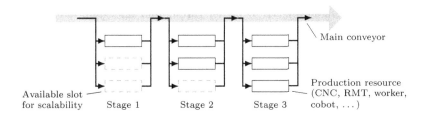

**Fig. 1.** RMS layout as seen by [9]

customization (ability to produce different parts in a family) and scalability (ability to adapt the volume of production). The four first ones are mainly related to technological issues while the two last ones are related to organizational issues. [11] states that scalability might be the most important feature to deal with the uncertain demand or ramp-up phase. This adaptation can be achieved by two levers, either (a) adding or removing parallel resources on the stations, or (b) processing a full reconfiguration of the system by changing the tools used in RMT or CNC machines [15]. A scalable system of good quality must be able to adapt quickly, incrementally (in small steps) and cost-effectively in order to provide at any time the exact capacity needed. In this paper, we focus on the first lever, which is the one allowing for the shorter and less expensive reconfiguration times, and we propose a new indicator to assess the scalability of an RMS, integrating these three aspects, at a strategic level and independantly of the initial state of the system.

The remaining of the paper is organized as follows: Sect. 2 presents the related literature, the proposed scalability measure is explained in Sect. 3 and experimental results are analyzed in Sect. 4. Section 5 concludes this paper.

## 2 Related Literature

The articles dealing with scalability of RMS can be separated in two categories, according to the decision level: operational or strategical.

The first category contains works dealing with the planning of the configurations. They focus on how to use the scalability of RMS to adjust production throughput over time to face the evolution of demand. [3,4] present a dynamic method to assess different reconfiguration policies on various scenarios of demand evolution. This method deals with each reconfiguration independently. [19] presents a heuristic method to minimize the number of machines required for successive reconfigurations. An extension of this method has been presented in [11], integrating buffers in RMS. The scalability level of different configurations of RMS are evaluated a posteriori with the production throughput gain when adding a given number of machines and rebalancing the line. Since the method is a heuristic, it is difficult to evaluate if the throughput comes from the scalability level of configurations or from the performance the optimization method. Based on the experimental results obtained, the authors stated that a lower number of stages leads to higher throughput and gain, mainly because it leads to a more reliable system. [5] has also used simulation to study the advantages of using RMS when dealing with unreliable production systems. [6] proposed a new approach for production planning to realize capacity scalability and functionality changes in planning processes. A prototypical application is developed to prove the applicability of their method. [8] presents a simulation-based method to optimize the production planning of RMS taking into account the variations of demand. [7] uses Petri nets to model an auto-adaptive RMS based on multi-agents to adjust production capacity.

The second category of studies considers the scalability of the RMS at the design phase. [10] and [18] studied the impact of different system configurations on throughput and scalability. [17] questioned the link between the balancing of the production systems and its productivity and scalability. They highlighted that unbalanced RMS can generate smaller steps of capacity changes. More recently, [13] proposed a classification of the main root causes leading to con-vertibility and scalability. Also, whereas RMS have been initially introduced for discrete manufacturing, [1] proposed to extend the definition of scalability to integrate process manufacturing and to calculate the average wasted capacity for a given curve of demand. In [19], a metric based on the smallest possible incremental capacity change is presented to evaluate a priori scalability on the first reconfiguration. However this metric is dependent on the current state of the system and does not allow to take into account the subsequent reconfigurations. Indeed, [15] states there is a need for new performance measures of scalability.

Finally, some works try to integrate the two decision level. In [12], the prob-lem of design multi-product and scalable RMS for multiple production periods is addressed, minimizing design and reconfiguration costs while fulfilling demand. The authors presented two approaches: a up- and downgrading method based on approximate demand in each period; and RMT selections and reconfigurations based on long-term demand estimations. A similar problem of design and recon-figuration planning was considered in [2] where the authors use the scalability of a mono-product RMS to minimize the energy cost.

## 3   Hypervolume Based Indicator for Scalability

In this study, we consider reconfigurations based on the same assignment of tasks to the stages. A configuration is thus defined by these two pieces of information: the balancing and the number of resources in each stage, and a reconfiguration consists in varying the number of resources on the stages, e.g. by switching on/off some resources. The cycle time of a stage is the workload (i.e., the sum of the processing time of tasks assigned to it) divided by the number of resources assigned to the stage. The takt time of a configuration is defined by the stage with the highest cycle time. Obviously, the only interesting configurations are those with the highest productivity for the same number of resources. A set of configurations can be derived for each feasible balancing, i.e. such that each tasks is assigned to one and only one stage, respecting the precedence constraints. For a given balancing, these configurations can be obtained by an iterative method: starting with one resource on each stage and gradually adding a resource in the bottleneck stages (i.e. the stages with the highest cycle time).

Knowing the whole set of available configurations, we can sort them by increasing number of resources and calculate the smallest capacity increment starting from each configuration by looking at the resources gap with the next configuration. We can thus obtain the average value of this stability measure among all the configurations, however it would not take into account some important features such as the available range of productivity or the efficiency of configurations. Actually, a scalable balancing should ideally provide a set of highly efficient configurations (i.e., with few idle times) covering a large range of possible market demands with a small increment between them. These characteristics are very similar to the ones sought from a set of trade-offs in the fields of multiobjective optimization and we can thus use the same metrics. Here, we will focus on one of the most used multi-objective metric which is the hypervolume [20]. For an optimization problem with multiple objectives to minimize, the hypervolume is the area above the set of points to evaluate and below a reference point (point such that it is not possible to have highest value along any objective). It is denote $\mathcal{H}$ in the following.

Two objectives are considered in our scalability measure: the takt time (to assess the productivity of the configurations) and the number of resources (to evaluate their cost). These two criteria are to be minimized. A reference point for the hypervolume computation can easily be determined since a configuration cannot have a takt time strictly greater than the sum of the processing times and the number of resources in bounded (otherwise there would be an infinity of configurations derived from a balancing).

To evaluate the quality of a set of configurations, we compare the objective values obtained for the specific set associated with a balancing (denoted by $F_a$), with the best possible values, i.e. with the whole set of feasible balancings and all derived configurations ($F_r$, it is a Pareto front [14]):

$$HV = 1 - \frac{\mathcal{H}(F_a)}{\mathcal{H}(F_r)}.$$

$HV$ is in [0,1] and a low value indicates that the configurations derived from the considered balancing are close from the best possible values ($F_r$) and thus indicates that this set of configurations is of good quality.

## 4   Experimental Results

We conducted experiments to evaluate the proposed indicator and to compare it with, on one hand, some classical line balancing indicators (takt time, number of stages, idle time and smoothness index computed on the configuration with one resource per station) and on the other hand the average scalability value from [19] on all configurations. This scalability indicator for a configuration is either the smallest incremental capacity in percentage (i.e. the number of bottleneck stages over the total number of resources in the system) if it is possible to add resources in all bottleneck stations, or the number of stages otherwise (corresponding to the creation of a whole new line).

For an instance, this comparison is done for all feasible balancings, obtained by a total enumeration under the following assumptions:

- There cannot be more than three resources per stage.
- The total number of resources for a configuration cannot exceed 50% of the number of tasks.
- At most 50% of the tasks can be assigned to the same stage.

The 10 smallest instances from [16] are used as a benchmark. We limited the time for the enumeration to 5 min, on a computer with an Intel Core i7, with a 2.60 GHz processor with 16 GB of RAM. Under these conditions, only the 5 smallest instances completed the process under the time limit. For the other, we gradually reduced the number of operations, following a random order, until the resulting instance finished the process under the time limit. Table 1 summarizes the initial number of operations of the considered instances and the size of the instances solved by our process.

**Table 1.** Description of the benchmark instances size (number of tasks). The table on the left are the instances for which the execution finished within the time limit and the one on the right contains the instances that needed to be reduced for the experiments.

| Instance name | Size | Instance name | Original size | Reduced size |
|---|---|---|---|---|
| Mertens | 7 | Mitchell | 21 | 14 |
| Bowman8 | 8 | Roszieg | 25 | 13 |
| Jaeschke | 9 | Heskia | 28 | 11 |
| Jackson | 11 | Buxey | 29 | 12 |
| Mansoor | 11 | Sawyer30 | 30 | 11 |

Table 2 shows the correlation between the classical line balancing indicators and both the average scalability indicator from [19] and the proposed hypervolume metric $HV$. Since all indicators are to be minimized, the high score indicates a strong correlation. The table shows that neither the average scalability nor the $HV$ indicator are correlated with the classical line balancing measures, except with the number of stages. The strong correlation with the number of stages can be explained by the fact that a low number of stages most likely allows to derive more configurations within the limit of number of resources per stage or in the system. The $HV$ indicator is highly impacted by the number of configurations in the set. When computing the average scalability indicator, there is always a configuration for which the indicator corresponds to the creation of a new line (thus with a particularly high value) which has less impact if the set of configurations is large. However, the number stages alone does not give any guaranty on the productivity of the configurations. It is also interesting to note that the average scalability is almost unrelated to the smoothness index which is the most usual indicator used to assess if a solution is well-balanced. Thus this table shows that the classical line balancing indicators cannot be used to evaluate the scalability of a RMS.

**Table 2.** Correlation of the classical SALBP indicators with the scalability indicators

|  | takt time | nb of stages | Idle time | Smoothness |
|---|---|---|---|---|
| avg Scalability | −0.372 | 0.951 | 0.237 | 0.093 |
| $HV(F^a)$ | −0.220 | 0.981 | 0.432 | 0.292 |

**Table 3.** Correlation of the hypervolume with the average scalability from [19]

| Instance | avg Scalability | Instance | avg Scalability |
|---|---|---|---|
| Mertens | 0.920 | Mitchell | 0.895 |
| Bowman8 | 0.959 | Roszieg | 0.922 |
| Jaeschke | 0.953 | Heskia | 0.946 |
| Jackson | 0.927 | Buxey | 0.922 |
| Mansoor | 0.954 | Sawyer30 | 0.942 |

Table 3 shows a strong correlation between the average scalability indicator from [19] and the proposed hypervolume metric $HV$, for each instance of the benchmark, with low variability. This means that on most of the balancings, the two measures agree of the assessment of the scalability, but not on all of them. Indeed, for a balancing with a high number of configurations derived, the average scalability indicator will be high, but the $HV$ indicator can be quite low if the takt time of these configurations is high (when the stages are unbalanced, the productivity is then low).

**Table 4.** Minimum and average value of the hypervolume metric among all balancings

| Instance | min HV | avgHV | Instance | min HV | avgHV |
|---|---|---|---|---|---|
| Mertens | 0.0% | 0.3% | Mitchell | 4.4% | 72.3% |
| Bowman8 | 3.1% | 59.9% | Roszieg | 4.8% | 72.4% |
| Jaeschke | 3.6% | 48.3% | Heskia | 5.8% | 70.6% |
| Jackson | 6.6% | 61.7% | Buxey | 8.4% | 72.0% |
| Mansoor | 7.4% | 61.9% | Sawyer30 | 5.8% | 71.1% |
|  |  |  | Average on all instances | 4.99% | 59.05% |

Finally, Table 4 shows the minimum and average values of the hypervolume indicator $HV$ among all the balancings for each instance. Since the indicator $HV$ is based on the ratio of the hypervolume on the set of configurations derived from the same balancing and the set of best possible values, the minimum $HV$ evaluates the quality of the optimal balancing (according to the $HV$ metric) with respect to a case where every reconfiguration would be possible. Here we have an average gap of 5% which corresponds to the cost of the assumption to only consider sets of configurations based on the same balancing. This cost seems really low by comparison with the additional costs of reconfiguration times associated with a change of tools. The average $HV$ evaluates the quality of a random balancing which is significantly higher (nearly 60%). This highlights the potential gains of optimizing this indicator during the design phase.

# 5   Conclusion

Scalability is a one of the main characteristics of RMS but its evaluation at the design step has not received a lot of attention. Only two works have considered this issue: (a) [19] has defined a measure based on the smallest increment required to increase the capacity starting from for a given configuration, and (b) [13] has tried to identify the core characteristics leading to the scalability. In this paper, we propose a new measure based on a classical multi-objective metric to assess the scalability level of a balancing by taking into account all the configurations which can be achieved.

The preliminary results show that this approach can be viewed as an extension of the smallest increment measure which is not dependent on a specific initial state. It also reveals that the usual line balancing criteria are mainly unrelated with the scalability, enhancing the need for dedicated metrics which could be used to optimize the scalability at the design step. This conclusion fits with the statement of [17] on the possibility to have a good scalability level in some unbalanced RMS. It is also interesting to note that, similarly to the conclusions reported in [11], in our setting a lower number of stages is strongly and positively correlated with the scalability even without taking the reliability into account.

A first perspective of this work would be to use the proposed measure to characterize what makes a system scalable. In addition, this work focuses on only one of the RMS characteristics. Even if a priori the different characteristics seem to be independant, it would be interesting to study their interaction. Finally, the development of methods to design RMSs optimizing their scalability together with classical performance measures would be needed.

# References

1. Accorsi, R., Bortolini, M., Galizia, F.G., Gualano, F., Oliani, M.: Scalability analysis in industry 4.0 manufacturing. In: Scholz, S., Howlett, R., Setchi, R. (eds.) Sustainable Design and Manufacturing 2020. Smart Innovation, Systems and Technologies, vol. 200, pp. 161–171. Springer, Singapore (2021). https://doi.org/10.1007/978-981-15-8131-1_15
2. Cerqueus, A., Gianessi, P., Lamy, D., Delorme, X.: Balancing and configuration planning of rms to minimize energy cost. In: Lalic, B., Majstorovic, V., Marjanovic, U., von Cieminski, G., Romero, D. (eds.) APMS 2020. IAICT, vol. 592, pp. 518–526. Springer, Cham (2020). https://doi.org/10.1007/978-3-030-57997-5_60
3. Deif, A.M., EIMaraghy, W.H.: A control approach to explore the dynamics of capacity scalability in reconfigurable manufacturing systems. J. Manuf. Syst. **25**(1), 12–24 (2006)
4. Deif, A.M., ElMaraghy, H.A.: Assessing capacity scalability policies in rms using system dynamics. International Journal of Flexible Manufacturing Systems **19**, 128–150 (2007)
5. Gola, A., Pastuszak, Z., Relich, M., Sobaszek, L., Szwarc, E.: Scalability analysis of selected structures of a reconfigurable manufacturing system taking into account a reduction in machine tools reliability. Eksploatacja i Niezawodnosc - Maintenance and Reliability **23**(2), 242–252 (2021)
6. Hees, A., Schutte, C.S.L., Reinhart, G.: A production planning system to continuously integrate the characteristics of reconfigurable manufacturing systems. Prod. Eng. **11**, 511–521 (2017). https://doi.org/10.1007/s11740-017-0744-5
7. Hsieh, F.S.: Design of scalable agent-based reconfigurable manufacturing systems with petri nets. Int. J. Comput. Integr. Manuf. **31**(8), 748–759 (2018)
8. Hu, Y., Guan, Y., Han, J., Wen, J.: Joint optimization of production planning and capacity adjustment for assembly system. Procedia CIRP **62**, 193–198 (2017)
9. Koren, Y., Gu, X., Guo, W.: Reconfigurable manufacturing systems: principles, design, and future trends. Front. Mech. Eng. **13**(2), 121–136 (2017). https://doi.org/10.1007/s11465-018-0483-0
10. Koren, Y., Hu, S.J., Weber, T.W.: Impact of manufacturing system configuration on performance. Ann. CIRP **47**(1), 369–372 (1998)
11. Koren, Y., Wang, W., Gu, X.: Value creation through design for scalability of reconfigurable manufacturing systems. Int. J. Produ. Res. **55**(5), 1227–1242 (2017)
12. Moghaddam, S.K., Houshmand, M., Saitou, K., Fatahi Valilai, O.: Configuration design of scalable reconfigurable manufacturing systems for part family. Int. J. Prod. Res. **58**(10), 2974–2996 (2020)
13. Napoleone, A., Andersen, A.-L., Pozzetti, A., Macchi, M.: Reconfigurable manufacturing: a classification of elements enabling convertibility and scalability. In: Ameri, F., Stecke, K.E., von Cieminski, G., Kiritsis, D. (eds.) APMS 2019. IAICT, vol. 566, pp. 349–356. Springer, Cham (2019). https://doi.org/10.1007/978-3-030-30000-5_44
14. Pareto, V.: Manuel d'économie politique. F. Rouge, Lausanne (1896)
15. Putnik, G., Sluga, A., ElMaraghy, H., Teti, R., Koren, Y., Tolio, T., Hon, B.: Scalability in manufacturing systems design and operation: state-of-the-art and future developments roadmap. CIRP Ann. - Manuf. Technol. **62**, 751–774 (2013)
16. Scholl, A.: Balancing and Sequencing of Assembly Lines. Physica-Verlag HD, Contributions to Management Science (1999)

17. Son, S.Y., Olsen, T.L., Yip-Hoi, D.: An approach to scalability and line balancing for reconfigurable manufacturing systems. Integr. Manuf. Syst. **12**(7), 500–511 (2001)
18. Spicer, P., Koren, Y., Shpitalni, M., Yip-Hoi, D.: Design principles for machining system configurations. CIRP Ann. **51**(1), 275–280 (2002)
19. Wang, W., Koren, Y.: Scalability planning for reconfigurable manufacturing systems. J. Manuf. Syst. **31**(2), 83–91 (2012)
20. Zitzler, E., Thiele, L., Laumanns, M., Fonseca, C.M., da Fonseca, V.G.: Performance assessment of multiobjective optimizers: an analysis and review. IEEE Trans. Evol. Comput. **7**(2), 117–132 (2003)

# Digital Twin Framework for Reconfigurable Manufacturing Systems: Challenges and Requirements

Emna Hajjem[1], Hichem Haddou Benderbal[2(✉)] (iD), Nadia Hamani[1] (iD), and Alexandre Dolgui[2] (iD)

[1] University of Picardie Jules Verne, 48 Rue d'Ostende, 02100 Saint-Quentin, France
{emna.hajjem,nadia.hamani}@u-picardie.fr
[2] IMT Atlantique, LS2N-CNRS, Nantes, France
{hichem.haddou-ben-derbal,alexandre.dolgui}@imt-atlantique.fr

**Abstract.** Due to the rapid development of new generation information technologies (such as IoT, Big Data analytics, Cyber-Physical Systems, cloud computing and artificial intelligence), Digital twins have become intensively used in smart manufacturing. Despite the fact that their use in industry has attracted the attention of many practitioners and researchers, there is still a need for an integrated and detailed Digital Twin framework for Reconfigurable Manufacturing Systems. To investigate related works, this manuscript reviews the existing Reconfigurable Manufacturing Systems Digital Twin frameworks. It also presents a classification of several studies based on the Digital Twin framework features and properties, the used decision-making tools and techniques as well as on the manufacturing system characteristics. The paper ends with a discussion and future challenges to put forward a structured and an integrated Reconfigurable Manufacturing Systems - Digital Twin framework.

**Keywords:** Digital Twin (DT) · Modular framework · Reconfigurable Manufacturing System (RMS) · State of the art · Industry 4.0

## 1 Introduction

Recent manufacturing companies are facing, nowadays, an increasing individualization demand in products and high-frequency market changes driven by global competition. To remain competitive, these enterprises must be flexible enough to allow multiple variations of production sequences by adapting changes in the manufacturing system for the production and provision of new customized/individualized products. In this new market environment, traditional manufacturing systems fail to realize customized production with large-scale manufacturing efficiency. Thus, new solutions were proposed in order to overcome the afore-mentioned challenges. For instance, we can mention Reconfigurable Manufacturing System (RMS) introduced by Koren et al. [10]. In RMS, hardware and software components can be added, removed, modified or interchanged

© IFIP International Federation for Information Processing 2021
Published by Springer Nature Switzerland AG 2021
A. Dolgui et al. (Eds.): APMS 2021, IFIP AICT 631, pp. 553–562, 2021.
https://doi.org/10.1007/978-3-030-85902-2_59

to quickly improve production capacity and functionality of manufacturing systems within a part family in response to the sudden changes in demand. This enhancement is possible thanks to the use of RMS comprising reconfigurable machines, reconfigurable controllers as well as methodologies for their systematic design and rapid ramp-up. Such systems combine the advantages of Dedicated Manufacturing Lines (DML) and Flexible Manufacturing Systems (FMS) to cope with fluctuant market demands. On the other hand, the concept of Digital Twin (DT) together with advanced technologies continues to evolve as it expands to achieve intelligent manufacturing. In a reconfigurable environment, characterized by unpredictable changes in demand and flexibility, the DT can be adequately applied to solve problems facing RMS. In fact, DTs are characterized by the seamless integration between the cyber and physical spaces [15], which can help to achieve a flexible decision-making process providing appropriate RMS configuration [2]. Despite the increasing popularity of the DT research and the various frameworks introduced to integrate it in manufacturing environments, few research works considered its incorporation into RMS [20]. Furthermore, the application of a DT framework in manufacturing requires the consideration of different narratives and components, leading to a lack of clearness of the proposed application methods and application frameworks [20]. Therefore, in this paper, we review the existing works on DT, by highlighting the importance of smart manufacturing in identifying the main common narrative about DT frameworks and develop a new DT modular framework for RMS in further study. The remainder of this paper is organized as follows: Sect. 2 presents the history and definitions of the digital twin. Section 3 describes the used search protocol, while Sect. 4 discusses the recently-developed approaches and frameworks for DT related to manufacturing. Section 5 classifies these approaches. Then the obtained findings are discussed in Sect. 6. Finally, Sect. 7 concludes the paper and presents some research issues for future works.

## 2    Background and Definitions

The DT concept can be related to NASA's Apollo project (performed in the late 1960s) in which the vehicle on the ground mirrored the vehicle in space. DT was used for training purposes as well as to simulate solutions for critical situations [14]. Subsequently and in early 2000's, Michael Grieves [7] introduced the concept of DT for product lifecycle management composed of three main parts: physical product in real space, virtual product in virtual space and the two-way connections of data and information, which ties the two spaces together. Later on, Glaessgen and Stargel [5] defined a DT as "an integrated multiphysics, multiscale, probabilistic simulation of an as-built vehicle or system that uses the best available physical models, sensor updates, fleet history, etc., to mirror the life of its corresponding flying twin." In 2014, Grieves and Vickers [6] defined DT as: a set of virtual information constructs that fully describes a potential or actual physical manufactured product from the micro-atomic level to the macro-geometrical level. At its optimum, any information that could be obtained from inspecting a physical manufactured product can be picked up from its DT. Hence, the DT concept was expanded from product to the wide manufacturing system [1, 16, 21]. To summarize, the concept of DT is a high-fidelity virtual representation of a physical system with a seamless and real-time data flow between the two spaces (physical space and virtual space). It reflects a new data-driven vision that combines real time data analytics, optimization and simulation.

# 3   Methodology

A literature review was conducted to answer our research question:

**RQ:** What are the future challenges and the main requirements to put forward a structured and integrated RMS-DT modular framework?

To source the related relevant papers, we used the Scopus database with the following keywords: "Digital Twin, framework", "Digital Twin" and "Reconfigurable Manufacturing System". We only focused on these keywords to narrow the results to papers that proposed frameworks related to manufacturing (RMS) and to eliminate those related to products. In fact, we did not put any restriction on the type of the manuscript or the year of its publication, but we only considered English papers. We collected 50 papers that were carefully read and analyzed. Besides, we cited some works that have proposed a DT framework for smart manufacturing systems or for RMS. Due to space limitation, only 12 papers were kept for the classification described in Sect. 5. The selection was based on the complexity of the developed framework and the impact factor of the selected publications.

# 4   Existing Approaches

The literature presented a number of different narratives about DT [15]. These narratives diverge in terms of features and components that they cover from 3D to 8D frameworks to specified simulation [6, 12, 13, 22]. However, they are often included in each other. In this context, Abid khan et al. [8] built a common narrative about DT by proposing a six-dimensional DT-framework. The DT framework, named spiral DT, presented two new dimensions: spirally and dynamicity. The former refers to a regular enhancement of the physical product and/or the manufacturing system, whereas the latter ensures a consistent and a continuous synchronization of the virtual product with the corresponding physical product. Furthermore, the authors presented a secure and reliable data management blockchain-based solution called twin chain.

One of the key challenges for DTs is data. To provide an efficient and complete data support for the application of workshop DTs, Kong et al. [9] developed a data construction method relying on the functional requirements of manufacturing system analyzed according to the characteristics of the manufacturing data. Authors argued that most existing techniques neglected data preparation. Thus, they included three modules into their framework, namely data representation module, data organization module and data management module.

Moreover, DTs have large applications areas not only to the single robot, product or asset, but also to the full-length manufacturing cell. In this context, Zhang et al. [18] considered an autonomous manufacturing cell as implementation scenario. They proposed a data-driven and knowledge-driven framework for DT Manufacturing Cell (DTMC). This latter, can support autonomous manufacturing by applying an intelligent perceiving, simulating, understanding, predicting, optimizing and controlling strategy thanks to three key enabling technologies used to support this strategy namely: DT model, dynamic knowledge bases and knowledge-based intelligent skills.

Other studies focused on digital assembly as key technology in the design and manufacturing of complex products. For instance, Yi et al. [17] suggested a DT reference model for intelligent planning of assembly process and provided an application framework for DT based assembly. The goal of their work was to improve the assembly quality and efficiency of complex products. The framework comprises three layers: (i) physical space layer for total-elements components of product assembly station, (ii) interaction layer for communication connection and data processing, and (iii) virtual space layer for DT-based assembly application service platform. Other researchers, like Strak et al. [13], used a test environment of smart manufacturing cells to define an 8-dimension DT model. The authors' goal was to investigate the methodological, technological, operative, and business aspects of developing and operating Digital Twins for the ramp-up activities of the smart factory cell. In the same perspective, Francis et al. [4] studied the needs and opportunities to enable data-driven digital twins in the domain of smart manufacturing. They developed a generic data-driven framework incorporating machine learning and process mining techniques as well as continuous model improvement and validation of automated construction of digital twins for smart factories. This work is based on the idea that building digital twins in real-time, based on data collected from IoT devices in smart factories, will allow achieving some purposes such as up-to-date robust models, supporting flexible and reconfigurable factory layouts, integrated ongoing model validation, etc.

On the other hand, some important attributes for nowadays manufacturing systems are flexibility, responsiveness, cost efficiency, reliability, scalability and ability to be easily reconfigurable in order to counter the increasingly mass customization or mass individualization paradigm. In this context, Zhang et al. [19] developed a DT-enabled reconfigurable modelling approach to permit the automatic reconfiguration of the DT-based manufacturing system. Authors proposed a five-dimensional model-driven reconfigurable DT system framework. Besides, by mapping the physical and virtual entities, the latter can derive some of the capabilities and dependencies of the DT. Furthermore, researchers used expandable model structure and optimization algorithms to suggest a reconfigurable strategy. The aim of their study was to fulfill different granularities and targets requirements in terms of reconfigurability. In the same context, Haddou-Benderbal et al. [2] provided a conceptual RMS-DT modular framework as a way to integrate DT into RMS. In fact, the DT framework aims at providing a holistic system visibility and a flexible decision-making process to achieve the needed RMS responsiveness and to improve its performances during its operating phase by continuously collecting real time data from the RMS components. Subsequently, these data can be stored, processed and analyzed by using information analytics as well as simulation and optimization module blocs. Based on the afore-mentioned ideas, DT can quickly provide critical decisions, such as the appropriate RMS configuration, to efficiently cope with sudden changes occurring in market driven by global competition.

In terms of reconfiguration, Leng et al. [11] suggested a novel digital twin-driven approach for rapid reconfiguration of automated manufacturing systems and fast optimizing process. The DT comprises two parts: the semi-physical simulation that maps data of the system and provides input data to the second part, which is optimization. The results of the optimization part are fed back to the semi-physical simulation for verification. The proposed approach could realize rapid alteration of manufacturing system capacity and fast integration of multiple processes into the existing systems, allowing the manufacturers to launch new product orders rapidly. From the perspective of 5-dimension DT model, Cheng et al. [3] introduced a DT enhanced Industrial Internet (DT-II) reference framework towards smart manufacturing. The interplay and relationship between DT and Industrial Internet were first discussed. The sensing/transmission network capability, which is one of the main characteristics of Industrial Internet, urges the implementation of DT, to provide DT with a means of data acquisition and transmission. Conversely, with the capability of high-fidelity virtual modelling and simulation computing/analysis, DT can greatly enhance the simulation computing and analysis of Industrial Internet. The related research and application of DT for product life cycle management were pointed out in Zheng, Yang, & Cheng [20] papers. Zheng et al. [20] presented both a broad sense and a narrow sense definition of the DT concept and characteristics. Based on the mentioned definitions, authors introduced a DT framework for product lifecycle management. This framework includes an information-processing layer with three main function modules, namely data storage, data processing and data mapping.

As we can notice from the literature review and to the best of our knowledge, there is a lack of research works around holistic and unified DT Frameworks that integrate the characteristics of RMS as well as software reconfiguration. In this context, the goal of this paper is to attempt and find the common components that should be used in DT framework, in a later research stage, to establish our future RMS-DT Framework.

## 5   Classification Analysis

In this section, we classify the existing approaches, as mentioned hereafter in Table 1, based on the **DT-framework features and properties**, the used **decision-making** tools and techniques as well as the **manufacturing system** characteristics. As the literature contains multiple understanding, interpretations and applications of DTs frameworks, we present, in this study, a classification analysis to select the researchers' common points of view about the DT-framework. The main three parts of the DT are the physical part, the digital part and data space. Hence, we classify the existing approaches mentioned earlier relying on the properties of these three parts **as follows:**

a) **Features:**

- Dimensions (P1): represent the numbers of layers that constitute the DT-framework.
- Human Machine Interface (HMI) (P2): concerns the capability of DT to be set and its ability to transfer data after being processed.
- Virtual Reality/Augmented Reality/Mixed Reality (VR/AR/MR) (P3): to describe the manufacturing system in the virtual and collaborative environment.

b) **Physical part and Digital part:**

- Physical part: the usage of process model (P4): it shows whether a process model of the physical part was used to implement the simulation model of the virtual part or not.

- Digital part: Model character (P5): the DT model is divided into two categories: static model (S) or dynamic model (D). Model scope (P6) represents either a single entity or a whole system.
- Physical part and Digital part (P7): describe the relationship between the digital part and the physical part that may be directly bound (B) or independent (I).

c) **Data space:**

- Data input (P8): differentiates between raw data (R) (sensors, RFID (Radio Frequency Identification) or other collection devices) and processed data(P) (by analytic software).
- Processing and transferring data (P9): transfer data from one point (i.e., physical space) to another one (i.e., virtual space).
- Data repository (P10): a data set is isolated to be mined for data reporting and analysis.
- Data driven decision-making (P11): consists in taking organizational decisions based on actual data.
- Data exchange (P12): from information management system (ERP, MES, …)
- Big Data treatment (P13): using cloud, fogs, blockchain (Twinchain), etc.

Soon after, we determined the used:

a) **Decision-making tools and techniques:** Simulation (P14), Optimization (P15) and Artificial Intelligence (AI) - Machine Learning (ML) (P16).

At last, we focused on the:

b) **Manufacturing System characteristics:** cited the applied approaches: Reconfigurability/changeability (P17), Sustainability (P18), Modularity (P19) and Scalability (P20).

## 6 Discussion

Table 1 demonstrates the differences between the DT frameworks presented in literature and illustrate the main features, proprieties and technologies required to propose modular RMS-DT framework. The goal of reference frameworks, is to help at different levels, identifying the development requirements and constraints. This identification is possible by matching the user's requirements with the application of enabling technologies evaluation methodologies. We can conclude, from the literature that:

- At least, **5-dimension frameworks** should be adopted to cover all the required properties and integrate all the advancing techniques and enabling technologies that can facilitate the functioning of the manufacturing system.
- The majority of the DT frameworks contain a **bi-directional data link** between the digital part and the physical one. Furthermore, all papers described a DT **employed**

**Table 1.** Classification analysis

| References | Features | | | Physical Part/Digital Part | | | | Data space | | | | | | Decision making | | | Manufacturing system | | | |
|---|---|---|---|---|---|---|---|---|---|---|---|---|---|---|---|---|---|---|---|---|
| | P1 | P2 | P3 | P4 | P5 | P6 | P7 | P8 | P9 | P10 | P11 | P12 | P13 | P14 | P15 | P16 | P17 | P18 | P19 | P20 |
| Li et al. [12] | 5 | | | * | D | S | B | R | * | * | * | | * | * | | | | | | |
| Khan et al. [8] | 6 | | | * | D | E | I | R | * | * | | | * | * | | | | * | | |
| [Zhang et al. 18] | 5 | * | | * | D | S | I | R/P | * | * | | * | * | * | | | | | | |
| [Yi et al. 17] | 5 | * | | * | D | S | I | R/P | * | * | * | * | * | * | | * | | | | |
| [Kong et al. 9] | 3 | | * | * | D | S | B | R | * | * | * | | * | * | | | | | | |
| Stark et al. [13] | 8 | * | | * | D | S | I | R/P | * | * | * | | * | * | | * | * | | | |
| Jiewu et al. [11] | | | | * | D | S | I | R | * | * | | | * | * | | | * | | * | |
| Zhang et al. [18] | 4 | * | * | * | D | S | I | R/P | * | * | * | * | * | * | | * | * | | | |
| [Francis et al. 4] | 3 | * | | * | D | S | I | R | * | * | * | | * | * | * | | * | | | |
| Benderbal et al. [2] | 3 | | | * | D | S | I | R/P | * | * | * | | * | * | * | | * | | | |
| Cheng et al. [3] | 5 | | | * | D | S | I | R/P | * | * | * | * | * | * | | | * | | | |
| Zheng et al. [20] | 3 | | | * | D | S | I | R/P | * | * | * | | * | * | * | | | | | |

to process and transfer data, while only few of publications dealt with the role of DT in storing them. Therefore, a data repository can be incorporated, allowing the manufacturing companies to create their own knowledge base. It is clear that the basis of DT integration is Data and information incorporation, which represents an important challenge for manufacturers [23]. It is used/adapted by DT users according to their requirements even if was initially it developed for such requirement. Consequently, to face this challenge, DTs must be designed from the outset for such eventuality. The RMS-DT framework must deal with several aspects of the different production lines that usually represent complex and heterogeneous areas. Hence, a modular aspect should be considered for such additional purposes. Furthermore, the conceptual elements of most DT should be physically bound to their counterparts to ensure a bidirectional communication between all the DT parts.

– **Information management system**, such as Manufacturing Execution System (MES) and Enterprise Resource Planning (ERP), can help a corporation become more self-aware by linking business areas, a single source of information, and accurate, real-time data reporting.

– **Human-machine interface** should be used to ensure the best interaction between the physical part and virtual part of the DT. This interface is usually used is to attain an intelligent and data-driven decision-making within the smart manufacturing environment.

– **Sustainability** and sustainable manufacturing are becoming very crucial issues in new manufacturing environment. However, few works considered sustainable development and performance evaluation as parts of their established DT frameworks. Hence, Digital Twin-driven sustainable manufacturing should be critically analyzed.

– The ultimate goal of RMS and changeable manufacturing system is to ensure responsiveness with a customized flexibility. Hence, there is a need to consider the **coupling** of the manufacturing systems **flexibility** with **scalability** and **modularity**. This coupling must be also taken into account in RMS-DT frameworks.

– **The reconfiguration of the software part of the RMS** in "plug & play" manner is still relatively an emerging idea. However, it represents the path to be followed in order to produce high-quality product and to meet the market requirements.

Each paper focused on developing a different DT. Thus, the research challenge is to form a structured and an integrated RMS-DT framework that assesses the sustainability performance of the whole manufacturing system and ensures the reconfiguration of both the hardware part and software part to provide a flexible decision-making process.

## 7   Conclusion and Further Works

With the increasing individualization demands in product and mass customization, Reconfigurable Manufacturing System has become more and more important. RMS with DT has the functions of intelligent sensing, simulation, which makes the production more efficient, and more intelligent. In this paper, a review for the existing DT frameworks defined in literature was presented. After that, a classification analysis was performed to extract the most common narrative between them. For future research perspectives, a DT modular framework for RMS will be introduced. The concept of sustainability of our RMS-DT framework was integrated to account for the crucial and urgent need for more sustainable manufacturing. Additionally, the integration of AI in RMS-DT framework can improve Digital Twins performance by applying machine learning (ML) for prescriptive analytics in production, which will facilitates flexible and robust decision making in the design stage as well as in the real time operation stages. Finally, an interesting direction of work is to study how other RMS characteristics can be integrated using this RMS-DT framework describe this model general use to meet the needs of the Industry 4.0.

# References

1. Barbieri, C., West, S., Rapaccini, M., et al.: Are practitioners and literature aligned about digital twin. In: 26th EurOMA Conference Operations Adding Value to Society (2019)
2. Benderbal, H.H., Yelles-Chaouche, A.R., Dolgui, A.: A digital twin modular framework for reconfigurable manufacturing systems. In: Lalic, B., Majstorovic, V., Marjanovic, U., von Cieminski, G., Romero, D. (eds.) APMS 2020. IAICT, vol. 592, pp. 493–500. Springer, Cham (2020). https://doi.org/10.1007/978-3-030-57997-5_57
3. Cheng, J., Zhang, H., Tao, F., et al.: DT-II: digital twin enhanced Industrial Internet reference framework towards smart manufacturing. Robot. Comput. Integrat. Manuf. **62**, 101881 (2020)
4. Francis, D.P., Lazarova-Molnar, S., Mohamed, N.: Towards data-driven digital twins for smart manufacturing. In: Selvaraj, H., Chmaj, G., Zydek, D. (eds.) ICSEng 2020. LNNS, vol. 182, pp. 445–454. Springer, Cham (2021). https://doi.org/10.1007/978-3-030-65796-3_43
5. Glaessgen, E., Stargel, D.: The digital twin paradigm for future NASA and US Air Force vehicles. In: 53rd AIAA/ASME/ASCE/AHS/ASC Structures, Structural Dynamics and Materials Conference 20th AIAA/ASME/AHS Adaptive Structures Conference 14th AIAA. p. 1818 (2012)
6. Grieves, M.: Digital twin: manufacturing excellence through virtual factory replication. White Paper **1**, 1–7 (2014)
7. Grieves, M., Vickers, J.: Digital twin: mitigating unpredictable, undesirable emergent behavior in complex systems. In: Kahlen, F.-J., Flumerfelt, S., Alves, A. (eds.) Transdisciplinary perspectives on complex systems, pp. 85–113. Springer, Cham (2017). https://doi.org/10.1007/978-3-319-38756-7_4
8. Khan, A., Shahid, F., Maple, C., Ahmad, A., Jeon, G.: Towards Smart manufacturing using spiral digital twin framework and twinchain. IEEE Trans. Indust. Inf. , 1–1 (2020). https://doi.org/10.1109/TII.2020.3047840
9. Kong, T., Tianliang, H., Zhou, T., Ye, Y.: Data construction method for the applications of workshop digital twin system. J. Manuf. Syst. **58**, 323–328 (2021). https://doi.org/10.1016/j.jmsy.2020.02.003
10. Koren, Y., Xi, G., Guo, W.: reconfigurable manufacturing systems: principles, design, and future trends. Front. Mech. Eng. **13**(2), 121–136 (2017). https://doi.org/10.1007/s11465-018-0483-0
11. Leng, J., Liu, Q., Ye, S., et al.: Digital twin-driven rapid reconfiguration of the automated manufacturing system via an open architecture model. Robot. Comput. Integrat. Manuf. **63**, 101895 (2020)
12. Li, L., Mao, C., Sun, H., Yuan, Y., Lei, B.: Digital twin driven green performance evaluation methodology of intelligent manufacturing: hybrid model based on fuzzy rough-sets AHP, multistage weight synthesis, and PROMETHEE II. Complexity **2020**, 1–24 (2020). https://doi.org/10.1155/2020/3853925

13. Stark, R., Fresemann, C., Lindow, K.: Development and operation of Digital Twins for technical systems and services. CIRP Annals **68**(1), 129–132 (2019). https://doi.org/10.1016/j.cirp.2019.04.024

14. Rosen, R., von Wichert, G., Lo, G., Bettenhausen, K.D.: About the importance of autonomy and digital twins for the future of manufacturing. IFAC-PapersOnLine **48**(3), 567–572 (2015). https://doi.org/10.1016/j.ifacol.2015.06.141

15. Tao, F., Zhang, H., Liu, A., et al.: Digital twin in industry: state-of-the-art. IEEE Trans. Indust. Inf. **15**(4), 2405–2415 (2018)

16. Tao, F., Cheng, J., Qi, Q., Zhang, M., Zhang, H., Sui, F.: Digital twin-driven product design, manufacturing and service with big data. Int. J. Adv. Manuf. Technol. **94**(9–12), 3563–3576 (2017). https://doi.org/10.1007/s00170-017-0233-1

17. Yi, Y., Yan, Y., Liu, X., Ni, Z., Feng, J., Liu, J.: Digital twin-based smart assembly process design and application framework for complex products and its case study. J. Manuf. Syst. **58**, 94–107 (2021). https://doi.org/10.1016/j.jmsy.2020.04.013

18. Zhang, C., Zhou, G., He, J., Li, Z., Cheng, W.: A data- and knowledge-driven framework for digital twin manufacturing cell. Procedia CIRP **83**, 345–350 (2019). https://doi.org/10.1016/j.procir.2019.04.084

19. Zhang, C., Wenjun, X., Liu, J., Liu, Z., Zhou, Z., Pham, D.T.: A reconfigurable modeling approach for digital twin-based manufacturing system. Procedia CIRP **83**, 118–125 (2019). https://doi.org/10.1016/j.procir.2019.03.141

20. Zheng, Y., Yang, S., Cheng, H.: An application framework of digital twin and its case study. J. Ambient Intell. Hum. Comput. **10**(3), 1141–1153 (2018). https://doi.org/10.1007/s12652-018-0911-3

21. Ding, K., Chan, F.T.S., Zhang, X., Zhou, G., Zhang, F.: Defining a digital twin-based cyber-physical production system for autonomous manufacturing in smart shop floors. Int. J. Prod. Res. **57**(20), 6315–6334 (2019). https://doi.org/10.1080/00207543.2019.1566661

22. Gabor, T., Lenz, B., Marie, K., Michael, T.B., Alexander, N.: A simulation-based architecture for smart cyber-physical systems. In: 2016 IEEE International Conference on Autonomic Computing (ICAC), pp. 374–379. IEEE (2016)

23. Qi, Q., et al.: Enabling technologies and tools for digital twin. J. Manuf. Syst. **58**(B), 3–21 (2019)

# Mathematical Model for Processing Multiple Parts on Multi-positional Reconfigurable Machines with Turrets

Olga Battaïa[1], Alexandre Dolgui[2(✉)], Nicolai Guschinky[3], and Fatme Makssoud[4]

[1] Kedge Business School, Talence, France
[2] IMT Atlantique, LS2N, Nantes, France
alexandre.dolgui@imt-atlantique.fr
[3] Operatoinal Research Laboratory, United Institute of Informatics Problems,
National Academy of Sciences, Minsk, Belarus
[4] Lebanese University, Beirut, Lebanon

**Abstract.** In this paper, we propose a new mathematical model for the combinatorial optimization problem of batch machining at multi-positional machines with turrets where the parts are sequentially machined on $m$ working positions. Sequential activation is realized by the use of turrets. Constraints related to the design of machining of turrets and working positions, as well as precedence constraints related to operations are given. The objective of the optimization is to minimize the total cost. The paper provides the problem definition, all aspects of the mathematical modelling and the model has been validated by presenting the case of an industrial example.

**Keywords:** Batch machining · Reconfigurable rotary machine · Mixed Integer Programming · Optimization

## 1 Introduction

The problem of managing product variety is one of important issues in manufacturing. This industry is facing new challenges like shorter product lifecycles and increasing demand turbulence. The actual market is considered as volatile since customer demand and product design as well as its expected functionalities evolve rapidly. Manufacturing companies are required to adapt to this evolution in the short term and if possible designed to be usable for a large variety of parts. In our previous study, we considered this issue of processing multiple parts in different modes in machining systems: batch machining [1] and mixed-model execution [2]. In this paper, we consider the design of multi-positional reconfigurable machines with turrets.

The concept of Reconfigurable Manufacturing Systems (RMS) has been introduced in [3] with the objective to provide efficient solutions for managing volatile market demand and rapid changes in product design. According to a recent state-of-the-art study [4], the assessment of reconfigurability level is realized on the basis of composite

© IFIP International Federation for Information Processing 2021
Published by Springer Nature Switzerland AG 2021
A. Dolgui et al. (Eds.): APMS 2021, IFIP AICT 631, pp. 563–573, 2021.
https://doi.org/10.1007/978-3-030-85902-2_60

metrics for the main RMS attributes [5, 6] or global reconfigurability indices [7]. The main attributes include modularity [8], integrability [6], diagnosability [5], convertibility and customization [6], scalability [9]. In terms of principal performances of RMS, the researchers distinguish responsiveness [10], system complexity [10], reliability [9] and quality [11].

Due to its impact on all decision levels, the reconfiguration has been addressed in system design problems [12], layout problems [13], process planning [8], setup planning [14], scheduling [15], etc. The existing studies in the literature concern both the level of individual reconfigurable machine tools [12] and reconfigurable flow lines [16, 32]. In terms of the design options for reconfiguration, some formulations are limited to the choice from a set of available elements [16], other generalized formulations include the possibility to introduce new elements in the reconfigurable system [1].

In this study, we focus on combinatorial aspects of the design process and present a detailed model with the objective of its reproduction by other scholars. The novelty of this contribution is an original mathematical model developed for a manufacturing system that has not been studied in the literature yet. Here below we present the description of the manufacturing system considered.

Multi-position reconfigurable machines are equipped with several working positions. In this study, we denote by $m$ the number of working positions. In each position, several processing modules (spindle heads or turrets) can be installed to process the operations assigned to that position. They are activated sequentially or simultaneously. Sequential activation is carried out using turrets. A turret regroups several machining modules that are activated by rotating the active one. Simultaneous machining is possible if the machining modules can be applied to different sides of the part and work in parallel. The number of processing modules and the order of their activation at each work position are configurable. Horizontal and vertical spindle heads and turrets are available to access different sides of workpieces in working position. Finally, the machining module can handle multiple machining operations. The tools to be installed are selected depending on the machining operations assigned to the module. Several cutting tools can be installed in one module, for example, Fig. 1 shows a horizontal turret with 5 machining modules, where the module has two cutting tools.

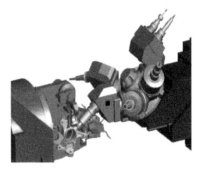

**Fig. 1.** A horizontal turret with 5 machining modules, one of them has 2 tools.

In order to help designers to take optimal decisions concerning the machining on multi-positional machine with turrets, we develop a new mathematical model for the case where multiple parts are machined on such machines. The rest of the paper is organized as follows. In Sect. 2, we detail the problem description and present a new mathematical problem for the defined problem. In Sect. 3, we run numerical experiments on industrial problem instances in order to validate the proposed mathematical model. Conclusions are given in Sect. 4.

## 2 Problem Definition and MIP Formulation

There are $d_0$ types of parts to be machined, each type is noted as $d = 1, 2, ..., d_0$. The demand for each part $d$ is defined by $O^d$. Parts are located at the loading position in a given sequence and they are processed simultaneously one per working position in the order of their loading. The rotary transfer machine is reconfigured after the end of processing of $O^d$ parts of type $d$, i.e. the fixtures of parts are changed and some spindles are mounted or dismounted if necessary.

Let $\mathbf{N}^d$ be the set of machining operations needed for machining of elements of the $d$-th part $d = 1, 2, ..., d_0$. Each machining operation is located on one side of the part, and in total we note by $n_d$ sides the number of sides required machining for part $d$. We denote as $N_s^d, s = 1, 2, ..., n_d$, the set of operations to be performed on the $s$-th side of part $d$. The part $d$ can be located at machine in different orientations $\mathbf{H}(d)$. The orientation of the part defines which sides are accessible for horizontal and vertical machining modules. The types "vertical" and "horizontal" are denoted in this study by index $j = 1,2, j = 1$ for "vertical" machining modules and $j = 2$ for "horizontal" machining modules. Matrix $\mathbf{H}(d)$ can be represented by where $h_{rs}(d)$ is equal $j, j = 1,2$ if the elements of the $s$-th side of the part $d$ can be machined by spindle head or turret of type $j$.

The complete set of operations $\mathbf{N}$ to be realized in the manufacturing system can be obtained by merging all operations required for all parts, i.e. $\mathbf{N} = \bigcup_{d=1}^{d_0}$. All operations $p \in \mathbf{N}$ are characterized by the following parameters:

- the length $\lambda(p)$ of the working stroke for operation $p \in \mathbf{N}$, i.e. the distance to be run by the tool in order to complete operation $p$;
- range $[\gamma_1(p), \gamma_2(p)]$ of feasible values of feed rate which characterizes the machining speed;
- set $H(p)$ of feasible orientations of the part (indexes $r \in \{1, 2, ..., r_d\}$ of rows of matrix $\mathbf{H}(d)$) for execution of operation $p \in N_s^d$ by spindle head or turret of type $j$ (vertical if $h_{rs}(d) = 1$ and horizontal if $h_{rs}(d) = 2$).

Let subset $N_k, k = 1,...,m$ contain the operations from set $\mathbf{N}$ assigned to the $k$-th working position. Let sets $N_{k1}$ and $N_{k2}$ be the sets of operations assigned to working position $k$ that are concerned by vertical and horizontal machining, respectively. Finally, let $b_{kj}$ be the number of machining modules (not more than $b_0$) of type $j$ installed at the $k$-th working position and respectively subsets $N_{kjl}, l = 1,...,b_{kj}$ contain the operations from set $N_{kj}$ assigned to the same machining module. This assignment has to respect the technological constraints that emanate from the machining process required. They can be grouped in three following families.

Each feasible design solution has to satisfy the following technical and technological constraints. The *precedence constraints* can be specified by a directed graph $G^{OR} = (\mathbf{N}, D^{OR})$ where an arc $(p, p') \in D^{OR}$ if and only if the operation $p$ has to be executed before the operation $p'$. It should be noted that if such operations $p$ and $p'$ belong to different sides of the part then they cannot be executed at the same position without violating the precedence constraint. The *inclusion constraints* are given by undirected graphs $G^{SP} = (\mathbf{N}, E^{SP})$, $G^{SM} = (\mathbf{N}, E^{SM})$, $G^{ST} = (\mathbf{N}, E^{ST})$, and $G^{SS} = (\mathbf{N}, E^{SS})$ where the edge $(p, p') \in E^{SP}$ $((p, p') \in E^{ST}, (p, p') \in E^{SM}, (p, p') \in E^{SS})$ if and only if the operations $p$ and $q$ must be executed at the same position, in the same machining module, by the same turret or the same spindle. The *exclusion constraints* are defined by undirected graphs $G^{DP} = (\mathbf{N}, E^{DP})$, $G^{DM} = (\mathbf{N}, E^{DM})$, $G^{DT} = (\mathbf{N}, E^{DT})$, and where the edge $(p, p') \in E^{DP}$ $(p, p') \in E^{DM}$, $(p, p') \in E^{DT})$, if and only if the operations $p$ and $p'$ cannot be executed on the same position, same machining module or the same turret. It is assumed that infeasible combinations of part orientations are given by a set $E^{DH}$, each element of which $e = \{(d_1, r_1), (d_2, r_2), \ldots, (d_k, r_k)\}$ represents a collection of pairs (part number $d$ and row number of $\mathbf{H}(d)$) that prohibit simultaneously orientation $r_1$ for part $d_1$, orientation $r_2$ for part $d_2$, and orientation $r_k$ for part $d_k$. Obviously, the set $E^{DH}$ includes $\{(r', d',), (r'', d'')\}$ if there exist $p \in N_{s'}^{d'}$, $s' \in \{1, \ldots, n_d'\}$, $q$, $s'' \in \{1, \ldots, n_d''\}$ such that $(p, q) \in E^{SS} \cup E^{SM} \cup E^{ST}$ and $h_{r's'}(d') \neq h_{r''s''}(d'')$.

We can built set $\mathbf{N}'$ based on graph $G^{SSM} = (\mathbf{N}, E^{SSM} = E^{SS} \cup E^{SM})$. Let $G_i^{SSM} = (N_i^{SSM}, E_i^{SSM})$, $i = 1, \ldots, n^{SSM}$, be connectivity components of $G^{SSM}$ including isolated vertices. Only one vertex (operation) $\wp_i$ is chosen from each $N_i^{SSM}$, let $X(p) = \wp_i$ for all $p \in N_i^{SSM}$ and included into $\mathbf{N}'$.

Let us introduce the following notation:

$X_{pq}$    decision variable which is equal to 1 if the operation $p$ from $\mathbf{N}'$ is assigned to the block $q = 2(k-1)b_0 + (j-1)b_0 + l$, i.e. $l$-th machining module of spindle head or turret type $j$ at the $k$-th position;

$Y_{kj}^{ds}$    auxiliary variable which is equal to 1 if at least one operation from $N_s^d$ is assigned to spindle head or turret of type $j$ at the $k$-th position;

$Y_{kjl}^d$    auxiliary variable which is equal to 1 if at least one operation for machining elements of the $d$-th part is executed in the $l$-th machining module of spindle head or turret type $j$ at the $k$-th position;

$Y_{kjl}$    auxiliary variable which is equal to 1 if the $l$-th machining module of spindle head or turret type $j$ is installed at the $k$-th position;

$Y_{1min}$    auxiliary variable which is equal to $k$ if $k$ is the minimal position covered by vertical spindle head or turret;

$Y_{1max}$    auxiliary variable which is equal to $k$ if $k$ is the maximal position covered by vertical spindle head or turret;

$Y_1$    auxiliary variable which is equal to 1 if the vertical spindle head or turret is installed;

$Z_k$    auxiliary variable which is equal to 1 if at least one operation is assigned to the $k$-th position;

$h_r^d$    auxiliary variable which is equal to 1 if elements of the $d$-th part are machined with the $r$-th orientation;

$F_{kjl}^d$    an auxiliary variable for determining the time of execution of operations from $\mathbf{N}^d$ in the $l$-th machining module of spindle head or turret type $j$ at the $k$-th position;

$F_k^d$    an auxiliary variable for determining the time of execution of operations from $\mathbf{N}^d$ at the $k$-th position;

$F_d$    an auxiliary variable for determining the time of execution of all the operations from $\mathbf{N}^d$;

$T_k^d$    an auxiliary variable which is equal to $F^d$ if the $k$-th position exists and 0 otherwise;

$\tau_a$    is an additional time for advance and disengagement of tools.

We calculate in advance parameters $t_{pp'} = \max((\lambda p), \lambda(p'))/\min(\gamma_2(p), \gamma_2(p')) + \tau^a$. They represent the minimal time necessary for execution of operations $p$ and $p'$ in the same machining module. It is assumed that $(p, p') \in E^{DM}$ if $\min(\gamma_2(p),\gamma_2(p')) < \max(\gamma_1(p),\gamma_1(p'))$.

For each operation $p \in \mathbf{N}$, we calculate a set $B(p)$ of block indices from $\{1,2,\ldots,2m_0b_0\}$ and a set $K(p)$ of position indices from $\{1,2,\ldots,m_0\}$ where operation $p \in \mathbf{N}$ can be potentially assigned.

Let $I(k) = [2(k-1)b_0 + 1, 2kb_0]$, $I(k,j) = [2(k-1)b_0 + (j-1)b_0 + 1, 2(k-1)b_0 + jb_0]$, and $I(k,j,l) = [2(k-1)b_0 + (j-1)b_0 + l, 2(k-1)b_0 + (j-1)b_0 + l]$, respectively.

## 2.1 Objective Function

Let $C_1$, $C_2$, $C_3$, and $C_4$ be the relative costs for one position, one turret, one machining module of a turret, and one spindle head respectively. Since the vertical spindle head (if it presents) is common for several positions its size (and therefore the cost) depends on the number of positions to be covered, $C_5$ is the relative cost for covering one additional position by vertical spindle head. The objective function aims in minimizing the total cost that includes the cost of all positions, all turrets, all machining modules, all spindle heads and all positions covered by the vertical spindle head. The total cost is calculated as the multiplication of the cost coefficients by the number of corresponding equipment used in the line. The objective of the design optimisation problem considered in this paper is to minimize this cost.

$$\text{Min } C_1 \sum_{k=1}^{m_0} Z_k + C_4 \sum_{k=1}^{m_0} Y_{k21} + (C_2 + 2C_3 - C_4) \sum_{k=1}^{m_0} \sum_{j=1}^{2} Y_{kj2}$$
$$+ C_3 \sum_{k=1}^{m_0} \sum_{j=1}^{2} \sum_{l=3}^{b_o} Y_{kjl} + C_4 Y_1 + C_5(Y_{1\,\max} - Y_{1\,\min}) \tag{1}$$

## 2.2 Assignment Constraints

Equations (2) provide assignment of each operation from $\mathbf{N}'$ exactly to one machining module.

$$\sum_{q \in B(p)} X_{pq} = 1; \, p \in \mathbf{N}' \tag{2}$$

Expressions (3) are used to model precedence constraints.

$$\sum_{\substack{q \in \bigcup_{k-1}^{k-1} \bigcup_{j=1}^{2} I(k',j') \cap B(p)}} q X_{\chi(p')q} + \sum_{q \in I(k',j') \cap B(p)} \leq \sum_{q \in I(k',j') \cap B(p')} (q-1) X_{\chi(p')q'} (p,p') \in D^{OR};$$

$$p, p' \in \mathbf{N}; k \in K(p'); j = 1, 2 \tag{3}$$

Expressions (4) are used to model inclusion constraints for working positions.

$$\sum_{q \in I(k) \cap B(p)} X_{\chi(p)q} = \sum_{q \in I(k) \cap B(p')} X_{\chi(p')q'}; \ (p,p') \in E^{SP};$$

$$p, p' \in \mathbf{N}; k \in K(p) \cap K(p') \tag{4}$$

Expressions (5) are used to model inclusion constraints for turrets.

$$\sum_{q \in I(k,j) \cap B(p)} X_{\chi(p)q} = \sum_{q' \in I(k,j) \cap B(p')} X_{\chi(p')q'};$$

$$p, p' \in \mathbf{N}; k \in K(p) \cap K(p'); j = 1, 2 \tag{5}$$

Expressions (6)–(8) are used to model exclusion constraints for working positions, turrets, and machining modules, respectively

$$\sum_{q' \in I(k) \cap B(p)} X_{\chi(p)q} + \sum_{q' \in I(k) \cap B(p')} X_{\chi(p')q'} \leq 1, \ (p,p') \in E^{DP};$$

$$p, p' \in \mathbf{N}; k \in K(p) \cap K(p') \tag{6}$$

Expressions (7) are used to model exclusion constraints for turrets.

$$\sum_{q \in I(k,j) \cap B(p)} X_{\chi(p)q} + \sum_{q \in I(k,j) \cap B(p')} X_{\chi(p')q'} + Y_{kj2} \leq 2; \ (P,P') \in E^{DT};$$

$$k \in K(p) \cap K(p'); j = 1, 2 \tag{7}$$

Expressions (8) are used to model exclusion constraints machining modules.

$$X_{\chi(p)q} + X_{\chi(p')q} \leq 1; (p,p') \in E^{DM}; p, p' \in \mathbf{N}; q \in B(p) \cap B(p') \tag{8}$$

Equations (9) prohibit assignment of operations from $N_s^d$ to machining modules of type $j$ if there is no feasible orientation of part $d$ for such an execution.

$$X_{\lambda(p)q} = 0; p \in; d = 1, .., d_0; s = 1, \ldots, n_d;$$

$$k \in K(p); \{h_{rs}(d) = j | r = 1, \ldots, r_d\} = \emptyset; q \in I(k,j) \cap B(p) \tag{9}$$

Equations (10) guarantee assignment of operations from $N_s^d$ to the same type of spindle head or turret.

$$\sum_{\substack{q \in B(p) \cap \bigcup_{k \in K(p)} I(k,j)}} X_{\chi(p)q} = \sum_{\substack{q' \in B(p') \cap \bigcup_{k \in K(p')} I(k,j)}} X_{\chi(p')q'}; p, p' \in N_s^d;$$

$$j = 1, 2; d = 1, \ldots, d_0; s = 1, \ldots, n \tag{10}$$

Constraints(11)–(15) define the existence of machining module $l$ of type $j$ at position $k$.

Constraints (11) initialize variable $Y_{kjl}^d$ when one operation for machining elements of th $d$-th part is executed in the $l$-th machining module of spindle head or turret type $j$ at the $k$-th position.

$$Y_{kjl}^d \leq \sum_{p \in N^d, q \in I(k,j,l) \cap B(p)} X_{\chi(p)q}; d = 1, \ldots, d_0;$$

$$k = 1, \ldots, m_0; j = 1, 2; l = 1, \ldots, b_0 \tag{11}$$

Constraints (12) verifies the number of operations assigned to $Y_{kjl}^d$

$$\sum_{p \in N^d, q \in I(k,j,l) \cap B(p)} X_{\chi(p)q} \leq \left| N^d \right| Y_{kjl}^d; d = 1, \ldots, d_0;$$

$$k = 1, \ldots, m_0; j = 1, 2; l = 1, \ldots, b_0 \tag{12}$$

Constraints (13) initialize variable $Y_{kjl}^d$ when if the $l$-th machining module of spindle head or turret type $j$ is installed at the $k$-th position.

$$Y_{kjl} \leq \sum_{d=1}^{d_o} Y_{kjl}^d; \ k = 1, \ldots, m_0; j = 1, 2; l = 1, \ldots, b_0 \tag{13}$$

Constraints (14) limits the number of machining modules installed at the $k$-th position.

$$\sum_{d=1}^{d_0} Y_{kjl}^d \leq d_0 Y_{kjl}; k = 1, \ldots, m_0; j = 1, 2; l = 1, \ldots, b_0 \tag{14}$$

Constraints (15) verify that variables $Y_{kjl}$ are initialized sequentially.

$$Y_{kjl-1} \geq Y_{kjl}; \ k = 1, \ldots, m_0; j = 1, 2; l = 2, \ldots, b_0 \tag{15}$$

Expressions (16)–(24) are used to calculate $Z_k, k = 1, \ldots m_0, Y_1, Y_{1\min}$ and $Y_{1\max}$.

$$Y_{k12} + Y_{k21} \leq 1; k = 1, \ldots, m_0 \tag{16}$$

$$Y_1 \leq \sum_{m=1}^{m_0} Y_{k11} \tag{17}$$

$$\sum_{m=1}^{m_0} Y_{k11} \leq m_0 Y_1 \tag{18}$$

$$Z_k \leq Y_{k11} + Y_{k21}; k = 1, \ldots, m_o \tag{19}$$

$$Y_{k11} + Y_{k21} \leq 2Z_k; k = 1, \ldots, m_o \tag{20}$$

$$(m_0 - k + 1) Y_{k11} + Y_{1\min} \leq m_0 + 1; k = 1, \ldots, m_0 \tag{21}$$

$$Y_{1\,\text{max}} \geq kY_{k11}; k = 1, ...., m_0 \tag{22}$$

$$Y_{1\text{max}} \leq m_0 Y_1 \tag{23}$$

$$Y_{1\text{min}} \leq m_0 Y_1 \tag{24}$$

Constraints(25)–(30) provide the choice of feasible orientation of each part $d$.

$$Y_{kj}^{ds} \leq \sum_{p \in N_s^d, q \in I(k,j) \cap B(p)} X_{\chi(p)q}; d = 1, \ldots, d_0; s = 1, \ldots, n_d;$$
$$k = 1, \ldots, m_0; j = 1, 2 \tag{25}$$

$$\sum_{p \in N_s^d, q \in I(k,j) \cap B(p)} X_{\chi(p)q} \leq |N_s^d| Y_{kj}^{ds}; d = 1, \ldots, d_0;$$
$$s = 1, \ldots, n_d; k = 1, \ldots, m_0; j = 1, 2 \tag{26}$$

$$\sum_{s=1}^{n_d} Y_{k1}^{ds} \leq 1; d = 1, \ldots, d_0; k = 1, \ldots, m_0 \tag{27}$$

$$h_r^d \geq 1 - \sum_{r=1}^{r_d} \sum_{j=1,\, j \neq rs}^{2} Y_{kj}^{ds}; d = 1, \ldots, d_0; r = 1, \ldots, n_d \tag{28}$$

$$\sum_{r=1}^{r_d} h_r^d = 1; d = 1, \ldots, d_0 \tag{29}$$

$$\sum_{(r,d) \in e} h_r^d \leq |e| - 1, e \in E^{DH}, k = 1, \ldots, m_0 \tag{30}$$

## 2.3 Time Calculation

Expressions (31)–(34) are used for estimation of execution time of operations from $\mathbf{N}^d$ by the $l$-th machining module, vertical spindle head and at the $k$-th position respectively.

$$F_{kjl}^d \geq t_{pp} X_{\chi(p)q}; p \in \mathbf{N}^d; j = 1, 2; d = 1, \ldots, d_0; k = 1, \ldots, m_0;$$
$$l = 1, \ldots, b_0; q \in I(k, j, l) \cap B(p) \tag{31}$$

$$F_{kjl}^d \geq t_{pp'}(X_{\chi(p)q} + X_{\chi(p')q-1}); p, p' \in \mathbf{N}^d; j = 1, 2; d = 1, ..., d_0;$$
$$k = 1, ..., m_0; 1 = 1, ..., b_0; q \in I(k, j, l) \cap B(p) \cap B(p') \tag{32}$$

$$F_{k11}^d \geq (\lambda(p')/\gamma_2(p') + \tau^a)(X_{\chi(p)q} + X_{\chi(p')q} - 1); p' \in \mathbf{N}^d; p \in \mathbf{N};$$
$$d = 1, \ldots, d_0; k, k' = 1, \ldots, m_0; k \neq k' \text{ or } p' \notin \mathbf{N};$$
$$q \in I(k, 1, 1) \cap B(p); q' \in I(k', 1, 1) \cap B(p') \tag{33}$$

$$F_k^d \geq \sum_{l=1}^{b_o} F_{kjl}^d + 2\tau^8 Y_{kj2} + \tau^8 \sum_{l=3}^{b_o} Y_{kjl} + b_0 \tau^8 \left( Y_{kj}^d - 1 \right);$$
$$d = 1, \ldots, d_0; k = 1, \ldots, m_0; j = 1, 2 \tag{34}$$

Expressions (35)–(37) provide the required productivity for the problem. Bound constraints for decision variables are straightforward a they are not presented here because of the limited article size.

$$F_k \geq F_k^d + \tau^r; \quad d = 1, \ldots, d_0; k = 1, \ldots, m_0 \tag{35}$$

$$T_k^d \geq F_d - T_0(1 - Z_k); d = 1, \ldots, d_0; k = 1, \ldots, m \tag{36}$$

$$\sum_{d=1}^{d_o} (F^d O^d + \sum_{k=1}^{m_o} T_k^d - F^d) \leq T_0 \tag{37}$$

## 3  Numerical Experiment

The purpose of this study is to evaluate the effectiveness of the mixed integer linear programming proposed model. It was tested on 25 industrial problem instances presented in Table 1 taken from mechanical parts for automotive industry. In this table |N| is the number of operations, OSP is the order strength of precedence constraints, DM, DT, DP, SS, and SM are the densities of graphs $G^{DM}$, $G^{DT}$, $G^{DP}$, $G^{SS}$, and $G^{SM}$ respectively. Experiments were carried out on ASUS notebook (1.86 Ghz, 4Gb RAM) with academic version of CPLEX 12.2. Columns Cost and time are respective the optimal cost of the solution and the solution time in seconds. As it can be seen all industrial problems have been rapidly solved by the proposed model. The solution time takes several seconds, the longest solution time to obtain the optimal solution is less than 5min. This provides a substantial help in decision making for designers.

**Table 1.** Parameters of industrial problems and results

| Test | N | OSP | DB | DG | DP | SSD | SB | Cost | Time |
|------|-----|-------|-------|-------|-------|-------|-------|------|-------|
| 1 | 92 | 0.011 | 0.234 | 0.339 | 0.125 | 0.012 | 0.021 | 67 | 4.4 |
| 2 | 52 | 0.02 | 0.434 | 0.697 | 0.299 | 0.027 | 0.02 | 56 | 0.3 |
| 3 | 82 | 0.013 | 0.237 | 0.21 | 0 | 0.014 | 0.008 | 49 | 1.4 |
| 4 | 88 | 0.034 | 0.297 | 0.238 | 0 | 0.012 | 0.026 | 49 | 1.1 |
| 5 | 90 | 0.039 | 0.309 | 0.246 | 0 | 0.011 | 0.034 | 49 | 1 |
| 6 | 116 | 0.01 | 0.173 | 0.277 | 0.046 | 0.006 | 0.008 | 74 | 2.5 |
| 7 | 70 | 0.012 | 0.185 | 0.164 | 0.004 | 0.008 | 0.011 | 62 | 1.6 |
| 8 | 74 | 0.024 | 0.22 | 0.182 | 0.001 | 0.008 | 0.01 | 77 | 266.1 |
| 9 | 40 | 0.026 | 0.515 | 0.636 | 0.164 | 0.021 | 0.026 | 67 | 0.8 |
| 10 | 48 | 0.014 | 0.363 | 0.369 | 0.078 | 0.018 | 0.018 | 63 | 2.4 |
| 11 | 44 | 0.023 | 0.013 | 0.091 | 0.101 | 0 | 0.025 | 71 | 2.6 |
| 12 | 92 | 0.011 | 0.234 | 0.339 | 0.125 | 0.012 | 0.021 | 67 | 15.8 |

(*continued*)

**Table 1.** (*continued*)

| Test | N | OSP | DB | DG | DP | SSD | SB | Cost | Time |
|------|-----|-------|-------|-------|-------|-------|-------|------|------|
| 13 | 52 | 0.02 | 0.434 | 0.697 | 0.299 | 0.027 | 0.02 | 56 | 1.6 |
| 14 | 116 | 0.01 | 0.174 | 0.275 | 0.043 | 0.006 | 0.011 | 89 | 58.2 |
| 15 | 70 | 0.014 | 0.185 | 0.164 | 0.004 | 0.008 | 0.011 | 93 | 3.8 |
| 16 | 40 | 0.026 | 0.515 | 0.636 | 0.164 | 0.021 | 0.026 | 63 | 5.3 |
| 17 | 74 | 0.024 | 0.22 | 0.182 | 0.001 | 0.008 | 0.01 | 74 | 69 |
| 18 | 40 | 0.026 | 0.515 | 0.636 | 0.164 | 0.021 | 0.026 | 67 | 1.8 |
| 19 | 92 | 0.011 | 0.234 | 0.339 | 0.125 | 0.012 | 0.021 | 67 | 1 |
| 20 | 78 | 0.013 | 0.24 | 0.176 | 0.039 | 0.007 | 0.008 | 68 | 3.1 |
| 21 | 80 | 0.013 | 0.234 | 0.175 | 0.078 | 0.007 | 0.009 | 63 | 9.9 |
| 22 | 116 | 0.01 | 0.174 | 0.275 | 0.043 | 0.006 | 0.011 | 89 | 11.8 |
| 23 | 70 | 0.014 | 0.185 | 0.164 | 0.004 | 0.008 | 0.011 | 65 | 2.4 |
| 24 | 48 | 0.014 | 0.363 | 0.369 | 0.078 | 0.018 | 0.018 | 63 | 1.5 |
| 25 | 74 | 0.024 | 0.22 | 0.182 | 0.001 | 0.008 | 0.01 | 87 | 35.3 |

## 4   Conclusion

We proposed a new mathematical model for the combinatorial optimization problem of processing multiple parts at multi-positional machines with turrets. A comprehensive mixed integer linear programming model has been developed for this optimization problem and it includes all technical and technological constraints, as well as productivity constraints and some preferences of the designers. The objective of the optimization is to minimize the total cost of the machining system. The numerical tests realized on 25 industrial problems showed that the proposed model is capable to find the optimal cost and the design solution in acceptable short time. The future research will be devoted to the extension of this study to the case of a flow line equipped with several multi-positional machines.

**Acknowledgements.** This work is supported by the region Pays de la Loire, France.

## References

1. Battaïa, O., Dolgui, A., Guschinsky, N.: Optimal cost design of flow lines with reconfigurable machines for batch production. Int. J. Prod. Res. **58**(10), 2937–2952 (2020)
2. Battaïa, O., Dolgui, A., Guschinsky, N.: Integrated process planning and system configuration for mixed-model machining on rotary transfer machine. Int. J. Comput. Integr. Manuf. **30**(9), 910–925 (2017)
3. Koren, Y., et al.: Reconfigurable manufacturing systems. CIRP Ann. Manuf. Technol. **48**(2), 527–540 (1999)

4. Bortolini, M., Galizia, F.G., Mora, C.: Reconfigurable manufacturing systems: literature review and research trend. J. Manuf. Syst. **49**, 93–106 (2018)
5. Gumasta, K., Kumar Gupta, S., Benyoucef, L., Tiwari, M.K.: Developing a reconfigurability index using multi-attribute utility theory. Int. J. Prod. Res. **49**(6), 1669–1683 (2011)
6. Farid, A.M.: Measures of reconfigurability and its key characteristics in intelligent manufacturing systems. J. Intell. Manuf. **28**(2), 353–369 (2014)
7. Goyal, K.K., Jain, P.K., Jain, M.: Optimal configuration selection for reconfigurable manufacturing system using NSGA II and TOPSIS. Int. J. Prod. Res. **50**(15), 4175–4191 (2012)
8. Chaube, A., Benyoucef, L., Tiwari, M.K.: An adapted NSGA-2 algorithm based dynamic process plan generation for a reconfigurable manufacturing system. J. Intell. Manuf. **23**(4), 1141–1155 (2012)
9. Bruccoleri, M., Pasek, Z.J., Koren, Y.: Operation management in reconfigurable manufacturing systems: reconfiguration for error handling. Int. J. Prod. Econ. **100**(1), 87–100 (2006)
10. Youssef, A., ElMaraghy, H.: Availability consideration in the optimal selection of multiple-aspect RMS configurations. Int. J. Prod. Res. **46**, 5849–5882 (2008)
11. Singh, R.K., Khilwani, N., Tiwari, M.K.: Justification for the selection of a reconfigurable manufacturing system: a fuzzy analytical hierarchy based approach. Int. J. Prod. Res. **45**(14), 3165–3190 (2007)
12. Youssef, A.M., ElMaraghy, H.A.: Modelling and optimization of multiple-aspect RMS configurations. Int. J. Prod. Res. **46**(22), 4929–4958 (2006)
13. Haddou Benderbal, H., Benyoucef, L.: Machine layout design problem under product family evolution in reconfigurable manufacturing environment: a two phase-based AMOSA approach. Int. J. Adv. Manuf. Technol. **104**(1–4), 375–389 (2019)
14. Borgia, S., Pellegrinelli, S., Petro, S., Tolio, T.: Network part program approach based on the STEP-NC data structure for the machining of multiple fixture pallets. Int. J. Comput. Integr. Manuf. **27**, 281–300 (2014)
15. Dou, J., Li, J., Su, C.: Bi-objective optimization of integrating configuration generation and scheduling for reconfigurable flow lines using NSGA-II. Int. J. Adv. Manuf. Technol. **86**(5–8), 1945–1962 (2016)
16. Ashraf, M., Hasan, F.: Configuration selection for a reconfigurable manufacturing flow line involving part production with operation constraints. Int. J. Adv. Manuf. Technol. **98**(5–8), 2137–2156 (2018)

# Production Management in Food Supply Chains

# Food Exports from Brazil to the United Kingdom: An Exploratory Analysis of COVID-19 Impact on Trade

João Gilberto Mendes dos Reis[1]([⊠]) [iD], Sivanilza Teixeira Machado[2] [iD], and Emel Aktas[3] [iD]

[1] RESUP – Supply Chain Research Group, Postgraduate Studies in Production Engineering/Postgraduate Studies in Business Administration, Universidade Paulista - UNIP, Dr. Bacelar. 1212, São Paulo, SP 04026002, Brazil
joao.reis@docente.unip.br
[2] Instituto Federal de Educação Tecnológica – IFSP, Mogi das Cruzes. 1501, Suzano, SP 08673010, Brazil
sivanilzamachado@ifsp.edu.br
[3] Centre for Logistics and Supply Chain Management, Cranfield University, Cranfield MK430AL, UK
emel.aktas@cranfield.ac.uk

**Abstract.** Brazil and the UK have been strategic partners throughout the years. While the former has traded minerals and food products – mainly agricultural commodities – the latter has been critical to infrastructure development in the South American country. However, the Crisis of COVID-19 Pandemic altered the scenario of international food production and distribution. This article aims to analyse Brazilian food exports to the UK in 2019 and 2020 to identify the impact of COVID-19 on the trade flows. To do so, we collected data from the Brazilian Ministry of Economy regarding the exports between the two nations and performed an exploratory investigation using graphical and quantitative analysis. The results suggest that the Pandemic crisis rose Brazilian exports of cereal and grains to the UK by around 50%, and the shortage of these items in the internal market has increased consumer prices by more than 60% during 2020.

**Keywords:** Food supply chains · International trade · Market fluctuations · Network analysis

## 1 Introduction

Brazil and the United Kingdom have been partners for centuries, first through Portugal that always had strong relations with the UK, where it commercialised agricultural items obtained from the colony in trade for textile products, and second directly with the country after the opening of Brazilian ports [1].

In the XIX century, during the Victorian years, the UK was the main actor responsible for the Brazilian development, building railways and infrastructure. The British reign

© IFIP International Federation for Information Processing 2021
Published by Springer Nature Switzerland AG 2021
A. Dolgui et al. (Eds.): APMS 2021, IFIP AICT 631, pp. 577–584, 2021.
https://doi.org/10.1007/978-3-030-85902-2_61

sent materials and technicians and provided loans for the construction of Brazil. On the other hand, the country provided mineral – mainly gold – and agricultural products to attend to the necessities of the UK and its colonies [2].

Nowadays, the trade relations between these two countries are completely different. First, the bilateral relations were established by the European Union for some decades, and second, due to the Brexit process that is redefining agreements between nations. In a nutshell, Brazil is the primary source of food products coming from crops while the UK is responsible for trading more complex items. Brazil also plays a vital source of the UK's organic food imports [3]. This does not mean that in this trade the UK never sells food products to Brazil, but they are so specific and in small quantities.

In 2019, the world was stricken by the worst pandemic crisis since the Spanish flu one hundred years ago. The virus spread fast across the globe, destroying economies, collapsing healthcare systems, and killing approximately 2.9 million people [4]. Even though vaccines are developed and deployed, the world remains menaced, and differences among countries are exposed [5]. The fact is that right now, the crisis is still far from finishing since the virus keeps changing and spreading through the human hosts [6].

In this research, the question we seek to answer is whether Brazilian exports to the UK have been affected by the crisis brought by the COVID-19 Pandemic. Therefore, the purpose of this article is to investigate Brazilian exports to the UK in 2019 and 2020 (before and during the Pandemic), considering the volume of trade in kilograms and monetary value in USD to verify behaviour changes in this relation.

To do so, we collected data provided by the Brazilian Ministry of Economy about trade relations between countries and analysed these data using the well-known spreadsheet editing software Microsoft Excel® v.16, the social network analysis software UCINET® 6 for Windows, and Graphviz. Based on a quantitative and graphical analysis, the results indicate an increase in the volume of exports of cereal and grains as a consequence of the Pandemic crisis.

## 2 Methodology

To achieve our objective, we collected the volume of exports (quantity in kilograms and monetary value in the US dollars - USD) from Brazil to the United Kingdom using the international trade system of the Ministry of Economy of Brazil. The system is called ComexStat [7] and provided us with all items exported during 2019 and 2020 to the UK.

The data were organised using a spreadsheet in Microsoft Excel® v.2019. The food products were gathered in groups according to their characteristics: (1) Beverages; (2) Cereal and Grains; (3) Chocolate; (4) Coffee and Tea; (5) Cookies; (6) Dairy Products; (7) Edible Oils; (8) Fish; (9) Fruits; (10) Ice Cream; (11) Juice; (12) Meat (broiler, cow, and pig); (13) Sauce; (14) Soup; (15) Sugar; and (16) Vegetables. Note that the juice category is not included in beverages because we wanted to observe its influence in trade since it is a great Brazilian commodity, mainly characterised by orange juice [8].

We imported the data to a social network tool, the software UCINET 6 for Windows version 6.698®, establishing the degree centrality [9] of the network for both in and out of the nodes (indegree and outdegree) in Tables 1 and 2.

**Table 1.** Outdegree values

| ID | Outdeg (Kg) | | nOutdeg | | Outdeg (USD) | | nOutdeg | |
|---|---|---|---|---|---|---|---|---|
| | 2019 | 2020 | 2019 | 2020 | 2019 | 2020 | 2019 | 2020 |
| Beverages | 2433576 | 47309092 | 0.003 | 0.036 | 2151285 | 28470864 | 0.003 | 0.039 |
| Cereal and Grains | 540866816 | 791614592 | 0.599 | 0.595 | 168871792 | 253548288 | 0.274 | 0.348 |
| Chocolate | 13301 | 32356 | 0.000 | 0.000 | 63602 | 106339 | 0.000 | 0.000 |
| Coffee and Tea | 54745952 | 45442128 | 0.061 | 0.034 | 135768736 | 114640424 | 0.221 | 0.157 |
| Cookies | 53585 | 76947 | 0.000 | 0.000 | 168389 | 242960 | 0.000 | 0.000 |
| Dairy Products | 21038 | 27773 | 0.000 | 0.000 | 50251 | 53301 | 0.000 | 0.000 |
| Edible Oil | 127814 | 134087 | 0.000 | 0.000 | 546241 | 405334 | 0.001 | 0.001 |
| Fish | 64297 | 12761 | 0.000 | 0.000 | 391897 | 187584 | 0.001 | 0.000 |
| Fruits | 163770464 | 166545024 | 0.181 | 0.125 | 148558368 | 146575232 | 0.241 | 0.201 |
| Ice_Cream | 41881 | 55427 | 0.000 | 0.000 | 143664 | 187832 | 0.000 | 0.000 |
| Juice | 13911968 | 16220615 | 0.015 | 0.012 | 20076384 | 21610106 | 0.033 | 0.030 |
| Meat (Broiler/Cow/Pig) | 42851400 | 46794640 | 0.047 | 0.035 | 111590784 | 97820944 | 0.181 | 0.134 |
| Sauce | 930700 | 1396641 | 0.001 | 0.001 | 1237667 | 2406870 | 0.002 | 0.003 |
| Soup | 843 | 2509 | 0.000 | 0.000 | 5135 | 16349 | 0.000 | 0.000 |
| Sugar | 79215288 | 207667744 | 0.088 | 0.156 | 22934314 | 59026808 | 0.037 | 0.081 |
| Vegetables | 3506831 | 6818260 | 0.004 | 0.005 | 2765565 | 4104541 | 0.004 | 0.006 |

**Table 2.** Indegree values

| ID | Indeg (Kg) | | nIndeg | | Indeg (USD) | | nIndeg | |
|---|---|---|---|---|---|---|---|---|
| | 2019 | 2020 | 2019 | 2020 | 2019 | 2020 | 2019 | 2020 |
| Brazil/UK | 902555712 | 1330150656 | 1.000 | 1.000 | 615324160 | 729403648 | 1.000 | 1.000 |

Afterwards, using the software open source Graphviz [10], we plotted four network graphics based on the trade data: (1/2) Exports in monetary value for 2019 and 2020; and (3/4) Exports in kilograms for 2019 and 2020. The groups were considered network nodes, and they were connected to a node of Brazil/UK exports based on the above-mentioned relations.

A graphical network approach was adopted to allow us to represent the differences among flows in a more comprehensive view. At the same time, it provided us with the opportunity to have access to some interesting representations of graph theory [11, 12]. The graphics are shown in the next section together findings discussion.

## 3 Results and Discussion

The networks comparing exports in 2019 and 2020 from Brazil to the UK in Free on Board (FOB) in USD currency is presented in Figs. 1 and 2. The strength of lines indicates the degree of the trade.

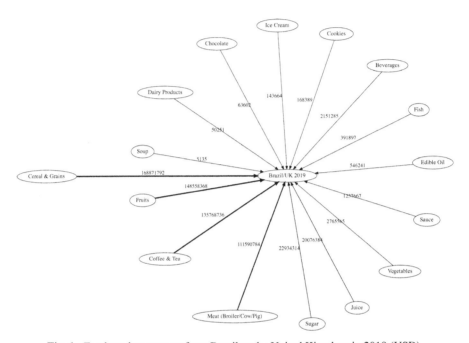

**Fig. 1.** Food product exports from Brazil to the United Kingdom in 2019 (USD).

The results did not show any difference among exports value position of food item groups between years for the five places; it was the same order: 1. Cereal and Grains, 2. Fruits, 3. Coffee and Tea, 4. Meat, and 5. Sugar. However, we identify that the exports in USD grew 15% from 2019 to 2020. The highlights were Beverages (1,223%), Soup (218%), Sugar (157%), Sauce (94%), Chocolate (67%) and Grains (50%). On the other hand, a reduction occurred in Fish (−52%), Edible oil (−26%), Coffee and Tea (−16%) and Meat (−12%).

After that, we performed the same analysis but now considering the weight of exports in kilograms. The results can be seen in Figs. 3 and 4.

Considering the weight instead of monetary value, the order of importance for the five places was the same: 1. Cereal and Grains, 2. Fruits, 3. Sugar, 4. Coffee and Tea, and 5. Meat. The volume of beverages converted to kilograms increased by 1,884%, Soup 198%, Sugar 162%, Chocolate 143%, Vegetables 94%, Sauce 50%, and Cereal and Grains 46%. On the flipside, Fish decreased 80%, Coffee and Tea decreased 17%.

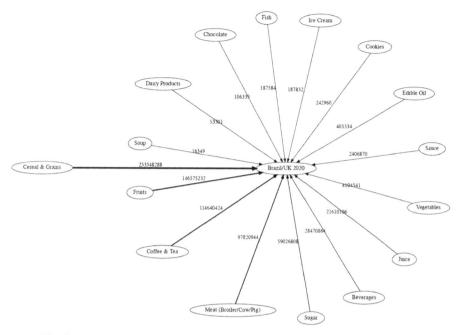

**Fig. 2.** Food product exports from Brazil to the United Kingdom in 2020 (USD).

Analysing the results, we note that beverages (by large composed of alcoholic drinks) performed a higher increase in quantity than in monetary value, indicating a better price per unit due to the volume. Considering the COVID-19 impact, the increase can represent a rise in alcoholic beverage consumption motivated perhaps by the anxiety and stress created by the Pandemic in the UK population. Garnett et al. [13] surveyed 30,375 adults in the UK from 21st March to 4th April 2020, demonstrating that 48.1% of respondents reported drinking about the same, and 26.2% reported drinking more than usual over the past week. Indeed, we cannot be conclusive about that on account of the shortness of the research, but it suggests a relation between the Pandemic and the rise in alcohol consumption. The same phenomenon has been observed in Brazil, China, and Germany [14].

Moving for the most relevant items on trade between Brazil and the United Kingdom (beverages correspond to just 2% of trade by USD), cereal and grains correspond to 50% of the quantity and 29% of the monetary value. We observe an increase of 46% in cereal and grains trade to the UK. Developed countries have demonstrated remarkable robustness and resilience over their food supply chains during the COVID-19 crisis, maybe as a learning of the costly mistakes of the 2007–8 food price crisis [15]. The big issue is that the actions of rich nations to support their population are raising the internal prices in developing countries [16] where farms see this as an opportunity to attend international markets as one way to reduce their losses over the years [17]. There is a risk of food security in developing countries not from disruptions to supply chains but rather from the devastating effects of COVID-19 on jobs and livelihoods [15].

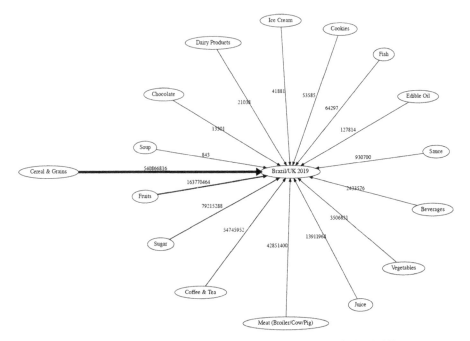

**Fig. 3.** Food product exports from Brazil to the United Kingdom in 2019 (kilograms).

In Brazil, for instance, the economy collapsed due to the reduction of external investments, the aggravation of sanitary and the political crisis that resulted in a loss of 29% of the value of the Brazilian Real (BRL) against the US dollar in 2020 [18]. Therefore, the rise of international demand for food items aligned with competitive Brazilian prices caused by the devaluation of the currency created a shortage in the internal market. Moreover, the necessity of foreign agricultural inputs, such as fertilizer, and rebuying agricultural grains to the animal protein industry, make food prices in Brazil soar by 15% in 2020, and around 60% for cereal and grains [19].

Regarding fruits that are a valuable Brazilian commodity, they did not show significant differences between one year to another and remained a valuable Brazilian export item to the UK with 20% of the monetary value received in 2019 and 2020. However, the UK ranked as one of the most important partners and tends to increase its demand in the following years [20].

An unexpected result was the highest reduction of the coffee and tea market in Brazil. This market corresponds to 17% of the monetary value in the period but reduced by 16%, indicating that this result could be better. The same occurred with the meat market that declined 14% in value and 9% in volume. Indeed, these results may be influenced by the other partners trying to acquire the same commodities and not necessarily a reduction in the UK market. However, we can exclude the enormous impact of COVID-19 in these commodity markets.

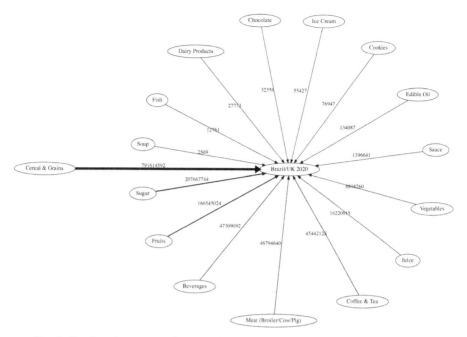

**Fig. 4.** Food product exports from Brazil to the United Kingdom in 2020 (kilograms).

## 4    Conclusions

The present article analysed the exports of food items between Brazil and the United Kingdom to identify possible effects of the COVID-19 Pandemic in the movement of these goods between the two countries. There is a long history between the two trading partners as a result of the influence of the UK in Portugal and afterwards in Brazil. The UK was critical in the process of Brazilian independence and development, mainly in the XIX century and the first half of the XX century. This process established a rich cultural exchange and economic trade.

Our results indicate an increase in exports of agricultural commodities (cereal and grains) and more complex food items such as beverages, soup, vegetables. The big issue for Brazil is that these items contribute little to the total trade. Besides that, the results allow us to suppose some impacts of the Pandemic in the behaviour of the UK consumer. Another significant result refers to an increase in food prices in Brazilian internal markets. It occurred as a consequence of the rise in international demand, as demonstrated by the Brazil/UK figures, and due to the devaluation of the Brazilian currency against the US dollar in 2020.

The limitations of the work include the sample size, the focus on two nations only, and the difficulty to capture in detail the COVID-19 impact. Moreover, two years is a small period to identify whether the behaviour will remain, or it is just a punctual situation. However, the study allows to create an exploratory analysis and establish a path to continue the research based on the assumptions here established.

# References

1. Fausto, B., Fausto, S.: A Concise History of Brazil. Cambridge University Press, New York (2014)
2. Graham, R.: Britain and the Onset of Modernization in Brazil 1850–1914. Cambridge University Press, New York (1968)
3. Barrett, H.R., Browne, A.W., Harris, P.J.C., Cadoret, K.: Organic certification and the UK market: organic imports from developing countries. Food Policy **27**, 301–318 (2002). https://doi.org/10.1016/S0306-9192(02)00036-2
4. Worldmeter. https://www.worldometers.info/coronavirus/
5. BBC News: Covid vaccine tracker: how's my country and the rest of the world doing? https://www.bbc.com/news/world-56025355
6. WHO: COVID-19 vaccines. https://www.who.int/emergencies/diseases/novel-coronavirus-2019/covid-19-vaccines
7. Ministry of Economy: Comexstat. http://comexstat.mdic.gov.br/pt/home
8. Odilla, F.: O que o caminho do suco de laranja brasileiro até as prateleiras britânicas revela sobre os desafios do Brexit. https://www.bbc.com/portuguese/brasil-48146698
9. Opsahl, T., Agneessens, F., Skvoretz, J.: Node centrality in weighted networks: generalizing degree and shortest paths. Soc. Netw. **32**, 245–251 (2010). https://doi.org/10.1016/j.socnet.2010.03.006
10. Graphviz: Graphviz. https://graphviz.org/
11. Borgatti, S.P., Everett, M.G., Johnson, J.C.: Analyzing Social Networks. SAGE, Los Angeles (2013)
12. Alejandro, V.A., Norman, A.G.: Manual Introdutório à Análise de Redes Sociais. Universidad Autonoma Del Estado de Mexico, Tlalpan (2005)
13. Garnett, C., Jackson, S., Oldham, M., Brown, J., Steptoe, A., Fancourt, D.: Factors associated with drinking behaviour during COVID-19 social distancing and lockdown among adults in the UK. Drug Alcohol Depend. **219**, 108461 (2021). https://doi.org/10.1016/j.drugalcdep.2020.108461
14. Garcia, L.P., Sanchez, Z.M., Garcia, L.P., Sanchez, Z.M.: Alcohol consumption during the COVID-19 pandemic: a necessary reflection for confronting the situation. Cad. Saúde Pública. **36** (2020). https://doi.org/10.1590/0102-311x00124520.
15. OCDE: Food Supply Chains and COVID-19: Impacts and Policy Lessons. http://www.oecd.org/coronavirus/policy-responses/food-supply-chains-and-covid-19-impacts-and-policy-lessons-71b57aea/
16. Malpass, D.: COVID crisis is fueling food price rises for world's poorest. https://blogs.worldbank.org/voices/covid-crisis-fueling-food-price-rises-worlds-poorest
17. Adami, A.: Exportações do agronegócio brasileiro em meio à pandemia do coronavírus. https://www.cepea.esalq.usp.br/br/opiniao-cepea/exportacoes-do-agronegocio-brasileiro-em-meio-a-pandemia-do-coronavirus.aspx
18. Reuters: How will the Brazilian real behave against the dollar in 2021? https://labsnews.com/en/articles/economy/exchange-rate-imbalance-between-the-brazilian-real-and-the-dollar-will-persist-in-2021-say-analysts/
19. IBGE: Tabela 7060: IPCA - Variação mensal, acumulada no ano, acumulada em 12 meses e peso mensal, para o índice geral, grupos, subgrupos, itens e subitens de produtos e serviços (a partir de janeiro/2020). https://sidra.ibge.gov.br/tabela/7060
20. Macshane, G.: Brazil eyes opportunities to strengthen fresh-fruit ties with UK post-Brexit. https://www.producebusinessuk.com/brazil-eyes-opportunities-to-strengthen-fresh-fruit-ties-with-uk-post-brexit/

# Application of Hybrid Metaheuristic Optimization Algorithm (SAGAC) in Beef Cattle Logistics

Marco Antonio Campos Benvenga⬤ and Irenilza de Alencar Nääs[(✉)] ⬤

Paulista University, São Paulo, Brazil

**Abstract.** The study objective was to evaluate the performance of SAGAC in optimizing a linear mathematical model in whole variables to determine the most cost-effective solution in transporting cattle for slaughter. The model determines the choice of refrigerator truck, road (route), and an open-truck in a scripting process. The tests performed with the SAGAC algorithm for optimizing the proposed model were compared with the results obtained, under similar conditions, by the branch-and-bound method for solving entire problems and solving a problem optimally. After the first twenty-two experimental trials, for comparison between the two methods, nine more experimental trials were carried out, with an increase in the degree of complexity, only with the SAGAC algorithm. The results obtained in the first twenty-two experimental trials demonstrate an equivalent performance between the two methods, showing that the SAGAC algorithm, even though it is not a technique that guarantees optimal results, in this case, was also able to find them. The nine final experiments performed only by SAGAC showed satisfactory results, with an evolutionary curve of exponential behavior.

**Keywords:** Meat production · Optimization · Algorithm · Logistics · Transportation

## 1 Introduction

Brazil has a substantial production capacity in agricultural activity sectors due to the available agricultural area. A large part of Brazilian agribusiness production represents an important share in the country's GDP and the balance of our exports, where agriculture has fundamental importance.

Brazil is the world's largest exporter of beef. Health control, knowledge and technology, and the country's natural aspects are pointed out as the keys to this product's success in the market [9]. The development of the food sector and market causes numerous organizational changes and structural in the chain, acting on the agents of production, transformation, trade, and distribution.

As a significant exporter of agribusiness products, Brazil's position has ensured the intense professionalization of the main objective of meeting markets' requirements with high safety standards, such as Europe and the United States. About 150 countries import

© IFIP International Federation for Information Processing 2021
Published by Springer Nature Switzerland AG 2021

A. Dolgui et al. (Eds.): APMS 2021, IFIP AICT 631, pp. 585–593, 2021.
https://doi.org/10.1007/978-3-030-85902-2_62

Brazilian products of animal origin. Of a total of around US\$16 $10^9$ of Brazilian exports in 2018, nearly 45% is beef, 42% chicken meat, 10% pork, and 3% other types of meat such as turkey, goose, sheep, and duck [8]. Brazilian meat production from 1994 to 2016 presents an increase in cattle, pork, and chicken production of 85% (nearly 3% per year), 162% (4% per year), and 285% (6% per year), respectively [1].

A factor considered in beef cattle transportation is minimizing the animals' stress during the movement between the farm and the slaughterhouse. Transport might cause a decrease in the meat's quality due to possible injuries caused by the vehicles' displacement and the stress of travel time [12].

Up to 3% of the live animal's weight loss occurs in the loading and unloading procedure, one of the main losses in the first hours and kilometers traveled in the transport [2]. The size of the animals, the type of truck used for transportation, the distances between the points of origin and destination (routes), the state of conservation of the road pavement, and the use of trucks with greater load capacity are the main determining factors of hematoma causes [6]. Other factors causing damage to animals' carcasses during transport for slaughter, such as transport cost, carrier density, loading, and accidents with vehicles, have also been studied [7].

The objective of this study is to compare the performance of the SAGAC hybrid metaheuristic algorithm with that of the Branch-and-bound [10] algorithm in a logistical process of routing cattle load from farms to slaughterhouses using trucks, determining the best possible combination of slaughterhouse factors, route, and truck that promotes the best value paid to producers.

## 2 Methods

### 2.1 Process Model

The model presented in the sequence was proposed by Ribeiro et al. (2018). The data to detail the mathematical model and develop the computational tool capable of assisting the rural producer in the decision-making process are as follow:

N = Total number of cattle to be transported for slaughter; G = estimated weight of cattle (in arrobas); F = Number of slaughterhouses available for slaughter; R = number of routes that can be used for the transport of livestock; C = number of trucks available for the transportation of cattle; T = Number of trucks to be effectively used in cattle transport; CFi = Slaughter capacity of the slaughterhouse i (cattle heads); Ai = Price paid by the slaughterhouse i for each amount of cattle (Reals); Dj = Distance between the farm and the refrigerator by route j (km); REj = estimation of weight reducers (per km) for route j (%); CCk = capacity of truck k (cattle heads); RCk = estimation of weight reducers (per km) for truck k (%); Hk = Freight price of truck k (Reals).

The estimated price (Reals) is calculated (Eq. 1) for the payment of the cattle transported from the farm to the processing plant $i$ using route $j$ on the truck $k$:

$$P_{ijk} = GA_i - GA_i D_j (RE_j + RC_k) = GA_i (1 - D_j (RE_j + RC_k)) \tag{1}$$

Be $y_{ijk}$ the binary decision variable defined by:

$y_{ijk}$ = {1 if the cattle are transported to the refrigerator $i$ by route $j$ using truck $k$ and, 0 otherwise}.

The entire decision variable for the problem is given by $x_{ijk}$ = number of cattle heads sent to the processing plant $i$ by route $j$ employing truck $k.$ The mathematical model of whole linear programming for the problem is described in Eq. 2 through 9, namely:

$$\text{Maximize}: f = \Sigma(Pijkxijk) - \Sigma(Hkyijk) \tag{2}$$

Subject to:

$$\Sigma xijk = N \tag{3}$$

$$\Sigma xijk \leq CFi(i = 1..) \tag{4}$$

$$xijk \leq CCkyijk(i = 1.., j = 1..R, k = 1..C) \tag{5}$$

$$\Sigma yijk \leq 1(k = 1..) \tag{6}$$

$$\Sigma yijk = T \tag{7}$$

$$xijk \geq 0 \text{ and } integer \tag{8}$$

$$yijk = 0/1 \tag{9}$$

## 2.2  The Hybrid Metaheuristic Algorithm (SAGAC)

The model is formed using two algorithms, the Simulated Annealing (SA) and the Genetic Algorithm (AG), with the inclusion of a mechanism (function) that promotes acceleration in the convergence (AGAC) of the obtained results. The SA algorithm acts on the generation of individuals who make up the modified genetic algorithm's initial population (AGAC). With the use of the SA algorithm, it is possible to have a good quality initial population composition, that is, pre-optimized individuals.

The routine behavior of the AGAC algorithm promotes Convergence Acceleration in which, after crossing, there is an assessment of the individuals (Sons) generated and a check for quality improvement concerning the individuals of the elite group of the population. If such development does not occur, the individual (s) of the offspring (ren) is (are) discarded, the individual (parent) of the worst quality is exchanged for another individual in the elite group who is closest and is better than the individual (Father) who was changed. After the individual's change (Father), a new crossing occurs for the missing child's generation (s). This sequence of steps will be repeated until both offsprings meet the criteria for improvement or the stipulated number of attempts is reached. Figure 1 shows the flowchart of the SAGAC hybrid algorithm.

With each cycle of processing of the Simulated Annealing (SA) algorithm, the best result (individual) is stored to compose the initial population used by the modified Genetic Algorithm with convergence acceleration mechanism (AGAC).

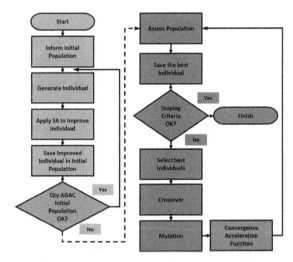

**Fig. 1.** Flowchart of the hybrid SAGAC algorithm

## 2.3 Setup Parameters of SAGAC

In the case of SAGAC, the variables that influence the algorithm's behavior are its processing parameters [3, 8, 11].

Parameters of SA

1. Initial Temperature = 100; it is the number of cycles that will be processed in an algorithm repetition loop;
2. TDS = 1; it is related to the Temperature Decay Scheme - defines how the temperature is decreased and the number of iterations performed for each temperature.
- the Temperature Decay Function is represented in Eq. 10.

$$[[\text{Temp}]]\_(i + 1) = [[\text{Temp}]]\_i - 1 \tag{10}$$

- Number of Iterations at each Temperature = 1.

Parameters of AGAC.

$1^{st}$ Population size = 100; $2^{nd}$. Generations Qty. = 1000; $3^{rd}$. Elitism = 10%; $4^{th}$. Mutation = 7%; and, for the case of the Convergence Acceleration Genetic Algorithm (AGAC); $5^{th}$. Qty of attempts to generate children in the elite = 1.

The optimization experiments (maximization of the cattle transport payment) were carried out based on the results developed by [10]. Was performed computational tests with the mathematical model proposed for 22 problems of shipping cattle from slaughterhouses. In all tests, the number of refrigerators, the number of routes (or roads), and the number of trucks available were respectively 3, 4, and 5. The other data used related to the sign's price, the routes, the trucks, price reducers, and freight charges are shown in Table 1, in [10], 22 computational experiments.

**Table 1.** Data used in [10], 22 experiments.

| Data | | | | | |
|------|---|---|---|---|---|
| N | 60 | | | | |
| G | 15 | | | | |
| F | 3 | | | | |
| R | 4 | | | | |
| C | 5 | | | | |
| T | 3 | | | | |
| Cfi | Cf1 = 40 | Cf2 = 40 | Cf3 = 40 | | |
| Ai | A1 = 120.00 | A2 = 110.00 | A3 = 130.00 | | |
| Rej | Re1 = 0.1 | Re2 = 0.2 | Re3 = 0.1 | Re4 = 0.1 | |
| CCk | CC1 = 12 | CC2 = 12 | CC3 = 12 | CC4 = 24 | CC5 = 28 |
| RCk | RC1 = 0.03 | RC2 = 0.03 | RC3 = 0.03 | RC4 = 0.02 | RC5 = 0.01 |
| Hk | H1 = 500.00 | H2 = 500.00 | H3 = 500.00 | H4 = 500.00 | H5 = 500.00 |

## 3  Results and Discussion

Analysis and comparison of the results obtained by SAGAC and those of other techniques used the optimize the same process.

Figure 2 shows the average convergence of the results obtained by SAGAC in the experimental trials. It represents the percentage of the evolution of the values obtained by the SAGAC method from the first Generation of the Genetic Algorithm to the last generation processed.

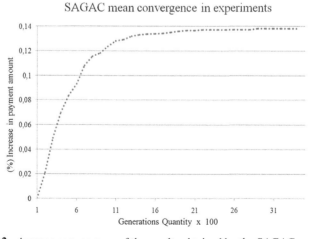

**Fig. 2.** Average convergence of the results obtained by the SAGAC method

Table 2 presents the results obtained by the two methods: branch-and-bound [10] and the SAGAC method. The first twenty-two results are sequenced descending because they are more easily located in their source [10]. The average of the results obtained between comparative tests (1 and 22) shows that the SAGAC algorithm presented a better performance (81.117,75) than that found by [10] (81.078,49). From experimental assay 23, the results data were obtained only by the SAGAC method.

**Table 2.** Comparative of the results obtained (Payment, $)

| Assay | N | T | [10] | SAGAC | Comparison |
|---|---|---|---|---|---|
| 22 | 12 | 1 | 18724.00 | 18724.00 | 0 |
| 21 | 19 | 1 | 29938.00 | 29938.00 | 0 |
| 20 | 20 | 1 | 31540.00 | 31540.00 | 0 |
| 19 | 26 | 1 | 41152.00 | 41152.00 | 0 |
| 18 | 33 | 2 | 51776.00 | 51686.00 | −90 |
| 17 | 35 | 2 | 54944.00 | 54818.00 | −126 |
| 16 | 38 | 2 | 56969.00 | 59516.00 | 2547 |
| 15 | 40 | 2 | 62864.00 | 62648.00 | −216 |
| 14 | 47 | 2 | 72014.00 | 72522.00 | 508 |
| 13 | 52 | 2 | 79604.00 | 79932.00 | 328 |
| 12 | 54 | 3 | 82896.00 | 82896.00 | 0 |
| 11 | 58 | 3 | 88824.00 | 88824.00 | 0 |
| 10 | 60 | 3 | 91812.00 | 91812.00 | 0 |
| 9 | 65 | 4 | 99097.00 | 98839.00 | −258 |
| 8 | 67 | 4 | 102139.00 | 101725.00 | −414 |
| 7 | 70 | 4 | 106054.00 | 106054.00 | 0 |
| 6 | 72 | 4 | 108940.00 | 108940.00 | 0 |
| 5 | 75 | 4 | 113269.00 | 113269.00 | 0 |
| 4 | 78 | 5 | 117590.00 | 116954.30 | −635.7 |
| 3 | 80 | 5 | 120476.00 | 119696.60 | −779.4 |
| 2 | 84 | 5 | 125181.20 | 125181.20 | 0 |
| 1 | 86 | 5 | 127923.50 | 127923.50 | 0 |
| 23 | 90 | 6 | | 135345.10 | |
| 24 | 100 | 6 | | 149333.00 | |
| 25 | 110 | 6 | | 162557.40 | |
| 26 | 120 | 6 | | 174860.20 | |

*(continued)*

**Table 2.** (*continued*)

| Assay | N | T | [10] | SAGAC | Comparison |
|---|---|---|---|---|---|
| 27 | 130 | 7 | | 188159.20 | |
| 28 | 140 | 7 | | 199558.60 | |
| 29 | 152 | 8 | | 213975.20 | |
| 30 | 176 | 9 | | 241093.20 | |
| 31 | 204 | 10 | | 273953.20 | |
| Average between 1 and 22 tests | | | 81.078.49 | 81.117.75 | |

Figure 3 shows the behavior of the results obtained by the SAGAC method in the thirty-one tests performed. The behavior of the SAGAC algorithm results indicates a correlation between the number of animals transported and the amount paid to producers according to the routes indicated by the SAGAC algorithm. Results obtained between experimental tests 1 and 22 by the SAGAC algorithm, which were compared with the results obtained by [10], have a degree of correlation of 99.97%. This correlation can be explained because the results presented by tests 1 to 22 were the optimum results or very close to the optimum. It can be inferred that the results obtained by the SAGAC algorithm between tests 23 and 31, which presented a degree of correlation of 99.94%, must be excellent or are very close to the optimum, thus confirming the excellent performance of the SAGAC algorithm.

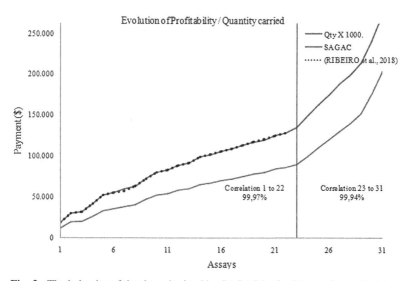

**Fig. 3.** The behavior of the data obtained by SAGAC in the 31 experimental trials.

## 4   Final Remarks

A hybrid metaheuristic optimization algorithm (SAGAC) was presented in this work as an alternative to solve logistics problems, more specifically, the selection of the set of factors: refrigerator, route, truck, and quantity of cattle transported, in order to maximize the value of the payment to the producer. This algorithm can optimize combinatorial analysis problems in which the factual solution spaces are too large (NP-Hard), making the application of deterministic algorithms that, in turn, ensure an unfeasible optimal solution.

From the results performed by the SAGAC algorithm, in the twenty-two comparative trials with the branch-and-bound deterministic algorithm, SAGAC obtained a higher mean difference of 0.14%. Nine experimental trials (from 23 to 31) with a higher degree of complexity were conducted. The resultant data showed exponential progression with an $R^2 = 0.927$, suggesting that profitability increases when the number of cattle transported increases.

**Acknowledgment.** The first author wishes to thank the Coordination of Superior Studies (Capes) for the scholarship.

## References

1. Embrapa, Cias - Central de Inteligência de Aves e Suínos. https://www.embrapa.br/qualid ade-da-carne/carne-em-numeros-2. Accessed 7 July 2020
2. Barnes, K., Smith, S. e Lalman, D. Managing shrink and weighing conditions in beef cattle. 2007. Oklahoma CooperativeExtension Service, ANSI-3257, Oklahoma StateUniversity. Disponível em. http://pods.dasnr.okstate.edu/docushare/dsweb/Get/Rendition-7449/ANSI-3257web.pdf. Acessed 21 Apr 2017
3. Kirkpatrick, S., Gelatti, C.D.; Vecchi, M.P.: Optimization by simulted annealing. Sci. New Ser. **220**(4598), 671–680 (May 1983)
4. Linden, R.: Algoritmos Genéticos – Uma importante ferramenta de inteligência computacional, 2a.edn. Brasport (2008)
5. MAPA, Ministério da Agricultura, Pecuária e Abastecimento: AGROSTAT - Estatisticas de Comércio Exterior do Agronegócio Brasileiro, Acesso em 07 de julho de 2020, Disponível em (2020). http://indicadores.agricultura.gov.br/agrostat/index.htm
6. Mendonça, F.S., et al.: Pre-slaughtering factors related to bruises on cattle carcasses. Anim. Prod. Sci. **58**(2), 385–392 (2016)
7. Miranda-de la Lama, G.C., Villarroel, M., e María, G.A.: Livestock transport from the perspective of the pre-slaughter logistic chain: a review. Meat Sci. **98**(1), 9–20 (2014)
8. Mitchell, T.M.: Machine Learning. McGraw-Hill Science, New York (1997)
9. Nääs, I.A.I., Mollo Neto, M.I., Canuto, S.A.I.,Waker, R.I., Oliveira, D.R.M.S.I.I., Vendram-etto, O.I.: Brazilian chicken meatproduction chain: a 10-year overview, Braz. J. Poul. Sci. (Revista Brasileira de Ciência Avícola) **17**(1) (2015)
10. Ribeiro, J.F.F., Oliveira, M.M.B., Filho, M.A.C.: Um modelo para a logística do abate do gado de corte, Pesquisa Operacional para o Desenvolvimento (2018)., ISSN:1984-3534
11. Santana, J.C.C., Mesquita, R.A.., Tamborgi, E.B., Librantz, A.F.H., Benvenga M.A.C.: Obtenção da Condição Ótima do Processo de Hidrólise do Amido de Mandioca por Amilases de Aspergillusniger, XVIII SINAFERM – Simpósio Nacional de Bioprocessos (2011)

12. Schwartzkopf-Genswein, K.S., Faucitano, L., Dadgar, S., Shand, P., González, L.A. e Crowe, T.: Road transport of cattle, swine and poultry in North America and its impact on animal welfare, carcass and meat quality: a review. Meat Sci. **92**(3), 227–243 (2012)
13. USDA - United States Department of Agriculture: Agricultural Projections to 2026, Report Interagency Agricultural Projections Committee USDA Long-term Projections, 100 p (2017)

# Integrated Workforce Scheduling and Flexible Flow Shop Problem in the Meat Industry

Beatrice Bolsi[1]($\boxtimes$) (ID), Vinícius Loti de Lima[2] (ID), Thiago Alves de Queiroz[3] (ID), and Manuel Iori[1] (ID)

[1] Department of Sciences and Methods for Engineering, University of Modena and Reggio Emilia, 42122 Reggio Emilia, Italy
{beatrice.bolsi,manuel.iori}@unimore.it
[2] Institute of Computing, University of Campinas, Campinas, SP 13083-852, Brazil
v.loti@ic.unicamp.br
[3] Institute of Mathematics and Technology, Federal University of Catalão, Catalão, GO 75704-020, Brazil
taq@ufcat.edu.br

**Abstract.** We address a problem from a meat company, in which orders are produced in two stages, consisting of preparing meats on benches and allocating them to conveyors to be packed in disposable trays. In an environment where machines are unrelated, the company has to take daily decisions on the number and start time of working periods, the number of workers and their allocation to machines, and the scheduling of activities to satisfy the required orders. The objective of the problem is to minimize, in a lexicographic way, the number of unscheduled activities, the weighted tardiness, and the total production cost. To solve the problem, we propose a multi-start random constructive heuristic, which tests different combinations of number of workers in the machines and for each combination produces many different schedules of the orders. The results of our computational experiments over realistic instances show that the heuristic is effective and can support the company on its daily decisions.

**Keywords:** Food industry · Two-stage flexible flow shop problem · Workforce schedule · Multi-start random constructive heuristic

## 1 Introduction

Production scheduling is one of the most common classes of problems faced by companies seeking to optimize their manufacturing system. In those problems, a set of jobs has to be scheduled in a set of machines while satisfying a set of practical constraints and optimizing a given objective. In this paper, we study a scheduling problem faced daily by a meat producing company. The company receives daily a set of orders to be produced in a single day. Each order is

© IFIP International Federation for Information Processing 2021
Published by Springer Nature Switzerland AG 2021
A. Dolgui et al. (Eds.): APMS 2021, IFIP AICT 631, pp. 594–602, 2021.
https://doi.org/10.1007/978-3-030-85902-2_63

associated with a due date and is produced by following up to two stages: in the first stage, the meat is processed (cut) on a given bench, and in the second stage it is sent to a conveyor to be packed into disposable trays. Benches and conveyors are seen as heterogeneous parallel machines, and their productivity depends on the number of workers operating each. Hence, the company needs to derive a new production plan every day, comprising the number of workers to operate each machine and the scheduling of operations composing the final orders. The scheduling problem faced by the company (which has a number of operational constraints that are formally discussed in Sect. 2) is a generalization of the well-known two-stage flexible (or hybrid) flow shop problem (see, e.g., [1]). The overall problem considers, in lexicographic way, the minimization of the number of unscheduled orders, the weighted tardiness, and the production costs.

The literature on flexible flow shop problems is broad. For surveys on general flexible flow shop variants, we refer the reader to [2,3], and [4]. Concerning two-stage variants, in [5], the authors investigate the influence of repetitive scheduling in an environment with a single machine in the first stage and multiple lines in the second stage. In [6], we find an application related to a sterilization plant, aiming at reducing the number of tardy jobs and the makespan, while respecting sequence-independent setup times and jobs processed in parallel batches. The work of [7] handles a system with unrelated parallel machines on each stage and task tail group constraints, aiming at minimizing the total tardiness. The authors propose a new scheduling rule able to outperform twelve dispatching rules from the literature. In [8], the authors study environments composed of a single machine either in the first or in the second stage and propose several solution procedures for such problems. The work of [9] deals with a problem in the glass-ceramic industry, whose objective is to minimize the makespan and energy consumption. Besides an integer linear programming model, they propose constructive, tabu search, and ant colony optimization algorithms.

In this paper, we propose a multi-start random heuristic to solve the integrated workforce scheduling and two-stage flexible flow shop problem faced by the company. The heuristic iteratively tests different combinations for the number of workers, and, for each combination, it generates the production schedule by following a constructive heuristic. At each iteration, it assigns orders to benches and conveyors by following a list of priorities. By means of extensive computational experiments based on realistic instances, we can show that the heuristic is effective in finding good-quality solutions and can provide a quick support to the company on its daily decisions.

The remaining of the paper is organized as follows. Section 2 provides a formal description of the problem. Section 3 presents the proposed solution method. Section 4 provides the results of the computational experiments. Finally, Sect. 5 presents concluding remarks and some directions for future research.

## 2   Problem Definition

In this section, we present a detailed description of the problem. The company receives a set $\mathcal{O}$ of orders to be scheduled in a workday, by following several operational constraints. Each order is expected to be produced before its due date, which corresponds to the time in which it has to be shipped to its final destination. Then, the problem is to determine the workforce and production schedule of a workday.

**A Workday.** A workday is defined over a time horizon of 24 hours, and may have up to two disjoint working periods. The two periods are subject to time-related constraints, such as minimum and maximum start time, end time, and duration. Moreover, there is a minimum and a maximum number of workers that can be hired for each period. Thus, a first set of decisions involves the number of working periods, as well as their start time, end time, and hired workers.

**A Period in a Workday.** After deciding the duration and number of workers for each period, production-related decisions must be made. The production in each period is composed by two stages. Each stage has a set of heterogeneous parallel machines whose speed is variable. We assume workers have the same efficiency and the speed of each machine is proportional to the number of workers operating it. In addition, workers cannot be reallocated to different machines during a period. Thus, another decision is to determine the number of workers operating each machine during the period.

In our case study, first-stage machines are benches on which the meat is prepared, whereas second-stage machines are conveyors where the products are distributed to be packed into disposable trays that finally receive a stamp with the product information.

**Details on the Production in a Period.** The last set of decisions considers the scheduling of the orders in each period. Each order in $\mathcal{O}$ is associated with: a final product identification code; a due date; a type of disposable tray (referred to as the order family); a type of stamp; a net weight; a raw product type, which has a productivity measure on each machine of the first and second stage where it can be processed. In addition, some orders have to be processed in a single stage, whereas others have to be processed (sequentially) on both stages. For orders that have to be processed in both stages, the product quality is preserved by imposing a maximum waiting time (e.g., 60 min or so) for the buffer between the two stages. We assume the buffer has unlimited size. All machines have setup times: first-stage and second-stage machines must undergo setups whenever the raw product is changed and whenever different tray or stamp types are required, respectively. Another machine-related constraint imposes a fixed transportation time between the first and second stage.

Summarizing, the decisions of the overall problem involve the distribution of the working periods, the number of workers on each period, the number of workers operating each machine in each period, and the production scheduling. The quality of a solution is measured by three objectives to be minimized according to a lexicographic order: (1) the total number of unscheduled orders; (2) the total weighted tardiness; and, (3) the total production cost. Although machine setups are not formally included in the objective, they are highly disliked by the company and it is always preferable to keep their number as small as possible.

# 3   Proposed Heuristic

To solve the problem under consideration, we propose a multi-start random heuristic. First, let us describe function $\mathtt{schedule}(\mathcal{O}, L)$, which schedules orders in $\mathcal{O}$ to a period that is already set (i.e., its time window is already fixed and workers are already allocated to machines). The list $L$ has a priority coefficient $l_o$ for each order $o \in \mathcal{O}$. The orders are iteratively scheduled by choosing the best order to be scheduled at each iteration based on the priority list in use, which considers, for instance, the creation of setups, the due dates, the coefficients, the size of the gaps created between productions. Whenever necessary, the sequence in $\mathcal{O}$ is used to break ties. Orders with early due dates that are processed in a single stage only are always chosen to be scheduled first. Whenever we are about to schedule orders that are processed in both stages, we first choose the second-stage machine with earliest available time, then we choose the best order to be scheduled in that machine following one of the priority lists, and, finally, we choose the best first-stage machine to schedule the chosen order, also based on a priority list.

The function $\mathtt{schedule}$ is used within a constructive heuristic for the integrated problem. The overall structure of this heuristic is presented in Algorithm 1. The algorithm creates several combinations of first- and second-period schedules by testing different combinations of number of workers and by iteratively updating the list of priority coefficients. The algorithm begins by testing different combinations of number of workers for the first period, and for each combination, the workers are allocated to machines by means of a function that consider a balance of the workload. After creating the schedule for the first working period, the heuristic follows by testing different possible schedules of the remaining orders in the second period. Again, for the second period all possible combinations of number of workers are tested and the workload based function is used again to allocate the workers to the machines. Then, different schedules for the second working period is created for each combination of workers. During the process of creating a schedule for a period, after fixing the number of workers for any of the two periods, we produce up to $\Pi_{\mathrm{list}}$ different schedules, which are iteratively obtained by updating the list $L$ and calling the function $\mathtt{schedule}$. Recall that $L$ is used in the internal of $\mathtt{schedule}$ as an order-priority quantifier. In this way, $L$ is updated in order to avoid tardiness, by increasing the priority coefficients of tardy orders in the current schedule.

Finally, our multi-start random heuristic is obtained by running Algorithm 1 for $\Pi_{\mathrm{shuffle}}$ iterations. In each of these iterations, we perform a reshuffle of the order sequence in $\mathcal{O}$. Recall that the sequence in $\mathcal{O}$ is used in a tie-breaking process in the internal of the $\mathtt{schedule}$ function, and, thus, it is expected to produce an impact in the final solution.

---
**Algorithm 1.** Heuristic for the integrated problem
---

1: $L \leftarrow (0, 0, ..., 0) \in \mathbb{Z}^{|\mathcal{O}|}$. // list of priority coefficients associated with $\mathcal{O}$
2: **for all** combinations of number of workers of stage one and two of the first working period **do**
3:     Assign workers to machines of the first working period by considering a workload balance
4:     **for** $i \leftarrow 1, \ldots, \Pi_{list}$ **do**
5:         $t_1 \leftarrow$ **schedule**$(\mathcal{O}, L)$
6:         $\mathcal{O}' \leftarrow \mathcal{O}$ minus the orders scheduled in the working period $t_1$
7:         $L' \leftarrow L$ updated for the orders in $\mathcal{O}'$
8:         **for all** combinations of number of workers of stage one and two of the second working period **do**
9:             Assign workers to machines of the second working period by considering a workload balance
10:            **for** $j \leftarrow 1, \ldots, \Pi_{list}$ **do**
11:                $t_2 \leftarrow$ **schedule**$(\mathcal{O}', L')$
12:                Update the best solution if its (lexicographic) objective function is worse than that of the current solution $(t_1, t_2)$
13:                **if** there is tardiness in the working period $t_2$ **then**
14:                    Increase the coefficient of priority in $L'$ of the orders with tardiness
15:                **else**
16:                    Break the loop
17:                **end if**
18:            **end for**
19:            **if** there is tardiness in the working period $t_1$ **then**
20:                Increase the coefficient of priority in $L$ of the orders with tardiness
21:            **else**
22:                Break the loop
23:            **end if**
24:        **end for**
25:    **end for**
26: **end for**

---

# 4   Computational Experiments

We implemented the multi-start random heuristic algorithm presented in Sect. 3 in C++. The algorithm was tested on 12 realistic instances, each representing one day of production at the company. Some randomization was applied to all instances in order to meet the company's privacy requirements. The tests were executed in a computer with an Intel Core i7 1.2 GHz processor and 8 GB of RAM. The goal of the experiments is to study the impact of different parameters in the performance of the heuristic. The parameters under consideration are:

(i) the range of workers available for each working period;
(ii) the number $\Pi_{shuffle}$ of random shuffles of the sequence $\mathcal{O}$;
(iii) the number $\Pi_{list}$ of iterations related to the update of the priority coefficients of the orders to try to prevent tardiness.

For each parameter configuration, we evaluate the three objectives of the problem plus the number of setups in the second stage.

We present in Table 1 the results related to parameter (i), which is the number of workers hired in each working period. We consider three configurations: $L$, meaning the Lowest fixed number of workers we may have; $H$, meaning the Highest fixed number of workers; and, $R$, meaning a Range/variable number of workers from $L$ to $H$. These results were obtained with $\Pi_{\text{shuffle}} = 25$ and $\Pi_{\text{list}} = 5$, that are values defined after preliminary trial and error tests. By observing the results, we can note that the higher is the number of workers, the better is the solution. However, the best results are obtained when the heuristic has a range of workers (situation $R$) to decide on. For all instances, configurations $R$ and $H$ have provided the same number of unscheduled orders and weighted tardiness (except for $I11$, where $R$ is better), but with $R$ the production cost is always equal or smaller. The number of setups in stage two is small and comparable among all configurations. The average computational time in seconds for $L$, $H$ and $R$ is 9.2, 8.7 and 1666.3, respectively, so $R$ requires a consistently larger amount of time to achieve the improved solutions.

As better solutions are obtained with configuration $R$, the following results consider only this configuration. In Table 2, we present the results we obtained when evaluating parameter (ii), the number of $\Pi_{\text{shuffle}}$ iterations. We attempted three values, namely 1, 25, and 100. When $\Pi_{\text{shuffle}}$ is set to 1, the heuristic looses its multi start characteristic. Hence, we expect to have better solutions as $\Pi_{\text{shuffle}}$ increases, because we may change the sequence in which orders are scheduled. In these experiments, we have set $\Pi_{\text{list}} = 5$.

**Table 1.** Results for different (i) values of works in each working period.

| Inst. | Unsch. Jobs | | | Weight. Tardiness | | | Prod. Cost | | | N. Setups | | |
|-------|---|---|---|---------|---------|---------|-----|-----|-----|---|---|---|
|       | L | H | R | L | H | R | L | H | R | L | H | R |
| I01 | 2 | 0 | 0 | 68795 | 0 | 0 | 192 | 223 | 206 | 2 | 2 | 2 |
| I02 | 2 | 0 | 0 | 1182 | 0 | 0 | 280 | 314 | 301 | 2 | 2 | 2 |
| I03 | 0 | 0 | 0 | 7811 | 0 | 0 | 317 | 326 | 319 | 2 | 2 | 2 |
| I04 | 0 | 0 | 0 | 209663 | 31738 | 31653 | 210 | 244 | 240 | 2 | 1 | 2 |
| I05 | 3 | 0 | 0 | 25212 | 2414 | 2414 | 229 | 270 | 266 | 2 | 2 | 2 |
| I06 | 13 | 1 | 1 | 156181 | 37174 | 37174 | 241 | 272 | 272 | 1 | 1 | 1 |
| I07 | 3 | 0 | 0 | 26965 | 0 | 0 | 259 | 285 | 285 | 1 | 2 | 2 |
| I08 | 0 | 0 | 0 | 0 | 0 | 0 | 91 | 95 | 92 | 1 | 2 | 2 |
| I09 | 0 | 0 | 0 | 124397 | 46 | 46 | 291 | 344 | 344 | 2 | 2 | 2 |
| I10 | 3 | 0 | 0 | 48977 | 0 | 0 | 318 | 332 | 328 | 1 | 2 | 2 |
| I11 | 12 | 2 | 2 | 252213 | 255366 | 76081 | 329 | 380 | 373 | 1 | 1 | 1 |
| I12 | 15 | 2 | 2 | 119295 | 197832 | 197832 | 340 | 395 | 395 | 1 | 1 | 1 |

Observing Table 2, the heuristic returns better solutions when more iterations are available. This means that restarting with possible different orders has a positive impact on the final solution even if the order sequence is the last criterion in the list of priorities. With 25 and 100 iterations, the heuristic is able to obtain the same number of unscheduled orders for all instances, although better weighted tardiness and production costs are achieved with 100 iterations for 3 instances (see $I06$, $I09$, and $I11$). The number of setups is relatively small and within the target value indicated the company (which is 3). The average computational time in seconds for $\Pi_{shuffle}=1$, 25, and 100 is 0.4, 7.6 and 30.8, respectively, so all three configurations are fast.

**Table 2.** Results for different (ii) numbers of $\Pi_{shuffle}$ iterations.

| Inst. | Unsch. Jobs | | | Weight. Tardiness | | | Prod. Cost | | | N. Setups | | |
|-------|---|----|-----|--------|--------|--------|-----|-----|-----|---|----|-----|
|       | 1 | 25 | 100 | 1      | 25     | 100    | 1   | 25  | 100 | 1 | 25 | 100 |
| I01   | 0 | 0  | 0   | 525    | 0      | 0      | 213 | 206 | 206 | 1 | 2  | 2   |
| I02   | 1 | 0  | 0   | 1118   | 0      | 0      | 304 | 301 | 301 | 1 | 2  | 2   |
| I03   | 0 | 0  | 0   | 91     | 0      | 0      | 334 | 319 | 319 | 1 | 2  | 2   |
| I04   | 0 | 0  | 0   | 51765  | 31653  | 31653  | 238 | 240 | 240 | 2 | 2  | 2   |
| I05   | 0 | 0  | 0   | 18895  | 2414   | 2414   | 268 | 266 | 266 | 2 | 2  | 2   |
| I06   | 4 | 1  | 1   | 117886 | 37174  | 16019  | 283 | 272 | 277 | 1 | 1  | 2   |
| I07   | 0 | 0  | 0   | 244    | 0      | 0      | 295 | 285 | 285 | 2 | 2  | 2   |
| I08   | 0 | 0  | 0   | 0      | 0      | 0      | 92  | 92  | 92  | 2 | 2  | 2   |
| I09   | 0 | 0  | 0   | 1348   | 46     | 36     | 332 | 344 | 316 | 2 | 2  | 2   |
| I10   | 0 | 0  | 0   | 0      | 0      | 0      | 347 | 328 | 328 | 1 | 2  | 2   |
| I11   | 2 | 2  | 2   | 76081  | 76081  | 40502  | 373 | 373 | 371 | 1 | 1  | 1   |
| I12   | 3 | 2  | 2   | 99445  | 197832 | 197832 | 399 | 395 | 395 | 1 | 1  | 1   |

Finally, we present in Table 3 the results for parameter (iii), the $\Pi_{list}$ number of iterations to change the order coefficient of priority. For these results, we selected configuration $R$ and $\Pi_{shuffle} = 100$, and tested $\Pi_{list}$ equal to 1, 5, and 20. Overall, we have conclusions similar to the previous ones, meaning that higher values of $\Pi_{list}$ can provide better solutions, especially if comparing the weighted tardiness and the production cost. The average computational time in seconds for $\Pi_{list}=1$, 5, 20 is 4.7, 30.8 and 177.1, respectively, which is again fast and compatible with a real-world use of the algorithm by a decision maker.

**Table 3.** Results for different (iii) numbers of $\Pi_{\text{list}}$ iterations.

| Inst. | Unsch. Jobs | | | Weight. Tardiness | | | Prod. Cost | | | N. Setups | | |
|-------|---|---|----|--------|--------|--------|-----|-----|-----|---|---|----|
| | 1 | 5 | 20 | 1 | 5 | 20 | 1 | 5 | 20 | 1 | 5 | 20 |
| I01 | 0 | 0 | 0 | 22069 | 0 | 0 | 211 | 206 | 206 | 1 | 2 | 2 |
| I02 | 0 | 0 | 0 | 308 | 0 | 0 | 313 | 301 | 301 | 1 | 2 | 2 |
| I03 | 0 | 0 | 0 | 100883 | 0 | 0 | 327 | 319 | 318 | 1 | 2 | 2 |
| I04 | 0 | 0 | 0 | 117752 | 31653 | 31650 | 222 | 240 | 241 | 2 | 2 | 2 |
| I05 | 0 | 0 | 0 | 23168 | 2414 | 2411 | 265 | 266 | 264 | 1 | 2 | 1 |
| I06 | 2 | 1 | 1 | 83453 | 16019 | 9137 | 285 | 277 | 275 | 1 | 2 | 2 |
| I07 | 0 | 0 | 0 | 162 | 0 | 0 | 287 | 286 | 286 | 2 | 2 | 2 |
| I08 | 0 | 0 | 0 | 0 | 0 | 0 | 92 | 92 | 92 | 2 | 2 | 2 |
| I09 | 0 | 0 | 0 | 99604 | 36 | 0 | 302 | 316 | 292 | 2 | 2 | 1 |
| I10 | 0 | 0 | 0 | 13829 | 0 | 0 | 321 | 328 | 328 | 0 | 2 | 2 |
| I11 | 2 | 2 | 2 | 90988 | 40502 | 11029 | 365 | 371 | 366 | 1 | 1 | 1 |
| I12 | 3 | 2 | 2 | 99445 | 197832 | 197832 | 399 | 395 | 395 | 1 | 1 | 1 |

# 5   Concluding Remarks

We have proposed a multi-start random heuristic for a complex integrated work-force scheduling and two-stage flexible flow shop problem from a meat production system. The heuristic schedules orders by following a list of priorities and considering different workers on machines in order to reach solutions with all orders scheduled, no tardiness, and the smallest possible production cost. The computational experiments have indicated that better solutions can be achieved when allowing more shuffles in the vector of orders ($\Pi_{\text{shuffle}}$), more iterations to handle the orders with tardiness ($\Pi_{\text{list}}$), and allowing the heuristic to decide on the number of works to set in each working period. In addition, even these more expensive settings have allowed practical computational time.

In future research, we are interested in studying how dynamic changes in assignments of workers to machines may affect productivity. Another direction is related to the extension of the proposed heuristic to include a local search step and possibly a tabu list in order to escape from local optima solutions. Finally, the proposed algorithm has been recently deployed to the company and, as attested by the company, it has been producing satisfactory results. We are currently working with the company to provide, as future research, an extensive comparison regarding the impact of the algorithm on their production system.

**Acknowledgements.** Vinícius Loti de Lima would like to thank the support of the São Paulo Research Foundation (process number 2017/11831-1). Thiago Alves de Queiroz acknowledges support by the National Council for Scientific and Technological Development (process number 311185/2020-7).

# References

1. Emmons, H., Vairaktarakis, G.: The Hybrid Flow Shop. ISOR, vol. 182, pp. 161–187. Springer, Boston (2013). https://doi.org/10.1007/978-1-4614-5152-5_5
2. Linn, R., Zhang, W.: Hybrid flow shop scheduling: a survey. Comput. Ind. Eng. **37**(1), 57–61 (1999)
3. Ruiz, R., Vázquez-Rodríguez, J.A.: The hybrid flow shop scheduling problem. Eur. J. Oper. Res. **205**(1), 1–18 (2010)
4. Tian-Soon Lee, Y.T.L.: A review of scheduling problem and resolution methods in flexible flow shop. Int. J. Ind. Eng. Comput. **10**(1), 67–88 (2019)
5. Tsubone, H., Suzuki, M., Uetake, T., Ohba, M.: A comparison between basic cyclic scheduling and variable cyclic scheduling in a two-stage hybrid flow shop. Decis. Sci. **31**(1), 197–222 (2000)
6. Rossi, A., Puppato, A., Lanzetta, M.: Heuristics for scheduling a two-stage hybrid flow shop with parallel batching machines: application at a hospital sterilisation plant. Int. J. Prod. Res. **51**(8), 2363–2376 (2013)
7. Li, Z., Chen, Q., Mao, N., Wang, X., Liu, J.: Scheduling rules for two-stage flexible flow shop scheduling problem subject to tail group constraint. Int. J. Prod. Econ. **146**(2), 667–678 (2013)
8. Hwang, F., Lin, B.: Survey and extensions of manufacturing models in two-stage flexible flow shops with dedicated machines. Comput. Oper. Res. **98**, 103–112 (2018)
9. Wang, S., Wang, X., Chu, F., Yu, J.: An energy-efficient two-stage hybrid flow shop scheduling problem in a glass production. Int. J. Prod. Res. **58**(8), 2283–2314 (2020)

# Optimization Strategies for In-Store Order Picking in Omnichannel Retailing

Xiaochen Chou[1,2], Nicola Ognibene Pietri[2], Dominic Loske[3,5(✉)],
Matthias Klumpp[4,5,6], and Roberto Montemanni[2]

[1] Dalle Molle Institute for Artificial Intelligence, Lugano, Switzerland
[2] University of Modena and Reggio Emilia, Modena, Italy
[3] University of Murcia, Murcia, Spain
[4] Georg-August-University of Göttingen, Göttingen, Germany
`dominic.loske@fom-net.de`
[5] FOM University of Applied Sciences, Essen, Germany
[6] Fraunhofer IML Dortmund, Dortmund, Germany

**Abstract.** The COVID-19 pandemic is changing consumer behavior and accelerating the interest for online grocery purchases. Hence, traditional brick-and-mortar retailers are developing omnichannel solutions enabling online purchases in parallel to normal activities. Buy-Online-Pick-up-in-Store concepts are flourishing in this context, and they are the topic of this work.

In this paper we propose a novel application of the sequential ordering problem to model products picking throughout the store shelves. The result is an optimized picking sequence that however takes also into account the characteristics of the goods (fragility, weight, etc.). The aim is to preserve goods integrity while allowing the pickers to optimize their route through the shop. The approach is exemplified on historical online orders of a real German shop.

**Keywords:** In-store order picking · Omnichannel Grocery Retailing · Sequential ordering problem

## 1 Introduction

Traditional factors such as a growing range of products, volatile demand behavior, or scarce logistics and sales space have been affecting grocery retailing for years. Thus, research has predominantly focused on demand and supply chain planning [1], the in-store backroom sizing problem [2], efficient in-store processes [3], as well as last-mile distribution [4]. Along with the COVID-19 pandemic, existing parameters were ex-tended through changing consumer behavior, switching from offline to online purchases [5]. As a consequence, the design of efficient operations to fulfill this online demand is a recent challenge for traditional brick-and-mortar (B&M) retailers that focused on offline sales during the last decades [6]. For the design of these omnichannel operations, [7] differ

© IFIP International Federation for Information Processing 2021
Published by Springer Nature Switzerland AG 2021
A. Dolgui et al. (Eds.): APMS 2021, IFIP AICT 631, pp. 603–611, 2021.
https://doi.org/10.1007/978-3-030-85902-2_64

between three typologies: (1) An integrated distribution center for online and offline orders that enable bulk and single unit picking and delivery, (2) distribution centers exclusively utilized to fulfill online orders, and (3) grocery retailers using their B&M structures for online order fulfillment, especially on the store-level to fulfill online and offline demands. These stores are named Buy-Online-Pick-up-in-Store concepts (BOPS) and are the argument of the remainder of this paper. We position our research at the interface of BOPS together with the optimization of operational efficiency in in-store order picking.

Observing the existing literature, BOPS and order picking are well-examined fields of research in marketing and operations management. Form a marketing perspective, focusing on, e.g., coupon promotions [8], customer behavior [9], or pricing strategies [10]. Regarding order picking, an extensive body of research investigates the human factor [11], batch assignment [12], and storage assignment [13] with the aim of optimizing these laborious and costly operations. However, approaches optimizing in-store order picking operations are scarce. In the existing literature, the grocery store order picking problem is treated as an open Traveling Salesman Problem (TSP) [14] on an underlying graph representing the location of the items within the shop. Therefore, the obtained solutions are the shortest tours from the store entrance to the cashiers going through all the nodes associated with goods of a given shopping list [15]. In a BOPS context, this approach makes sense since in-store logistics is performed for the most part by human operators and is considered a major cost-driver. It can account for up to 40% of all the activities executed during working hours [16]. Hence, a solution minimizing the time required for order picking can reduce the impact of BOPS-related movements on the overall logistics. However, when designing the layout of a store, the factors considered are normally revenue maximization [7,17] and customer satisfaction [18]. Thus, the product attributes analyzed are visibility, position to maximize impulsive purchases, variety, and availability. Characteristics like fragility or dimensions are simply left out of the equation when articles are located within a shop. As a consequence, operating according to shortest tours among the locations of the items without considering their characteristics might lead to product damaging (e.g., storing bottles ending on top of fresh fruit) or to a situation where the products can be placed into the final bags only after the whole tour is finished, causing a rearranging over-head.

The objective of this study is to explore how adding precedence constraints among the products to avoid the aforementioned issues alters the performance of the TSP solver. The resulting optimization model to represent the order picking is then a Sequential Ordering Problem (SOP) [19,20]. We will investigate the time overhead associated with the (more realistic) SOP solutions with respect to TSP solutions.

## 2    Problem Description

The idea behind the operational problem treated in this paper can be outlined as follows: in order to implement efficient in-store picking operations for online orders, the shortest path analysis is required for a given set of articles to be picked, given the store layout settings (topology of the shop and positioning of the products in the shelves). This operation has been proven effective in improving the pickers' efficiency in the BOPS model. However, a rearrangement of the articles is often required at the cashier to reposition fragile articles and/or to save final packaging space. Therefore, we propose to score the articles by characteristics and divide them into several sub-groups with different priorities. By adding these precedence constraints, the in-store picking problem can be solved as an SOP. In the following section, we discuss the scoring model in Sect. 3.1 in detail and explain how to solve the in-store picking problem as an SOP in Sect. 3.2.

# 3   Methodology

## 3.1   The Scoring Model

An order usually contains the following information for each of its articles: name of and producer, quantity, volume/weight, and the type of packaging. In order to have a complete set of data with normalized units of measurement, we select size, heaviness, and resistance (internal and external) as the main characteristics to be considered during an in-store picking tour. Size and heaviness are represented by the *volume* and *mass* of a product, that are the characteristics usually considered in determining the picking order for items. Another important characteristic is fragility, which is influenced by the internal and external resistance to pressure. Therefore, *density* and *packaging type* are considered as further attributes. All the attributes contribute to a weighted sum, and to have a reliable result, it is important to normalize the different attributes in the scoring system.

The scores for the packaging type are assigned based on estimated sturdiness. The scores for mass, volume, and density are defined by ad-hoc piecewise functions to take into account that the impact of these attributes varies based on their value. For example, when the volume of an article is small, it will not affect the picking order significantly. Hence, the associated coefficient is expected to be extremely value. On the contrary, when the volume of an article exceeds a certain threshold, people tend to collect it first, as it is a large item. In this case, the coefficient of volume should amplify its magnitude. The scoring function of the attributes we have derived by simulating real-world scenarios in common sense, are defined as follows:

$$SizeSC = \begin{cases} 1, & \text{if } Volume \leq 0.15L \\ 1 + [(Volume - 0.05)/0.10], & \text{if } 0.15L < Volume \leq 0.85L \\ 3 + [(Volume - 0.15)/0.10], & \text{if } 0.85L < Volume \leq 1.50L \\ 11 + 2 \cdot [Volume/0.50], & \text{if } Volume > 1.50L \end{cases}$$

$$HeavSC = \begin{cases} 1 + [Mass/0.10], & \text{if } Mass \leq 1Kg \\ 3 + 2 \cdot [Mass/0.25], & \text{if } 1Kg < Mass \leq 2Kg \\ 5 \cdot [Mass/0.25] - 20, & \text{if } Mass > 2Kg \end{cases}$$

$DensSC = 2 + 2 \cdot [Density/0.30]$
$PackSC =$ bag 2; tube 4; pack 6; carton 7; glass 10; can 14; bottle 18

With the scores defined in the same standard unit, the attributes are then combined in a weighted sum model. The priority of an article increases as the result of the weighted sum increases (e.g., the article with the highest output value has to be collected first). To achieve the proper model, the coefficients (weights) of the attributes have manually adjusted based on real-world simulations until satisfying picking lists were obtained. The resulting coefficients are incorporated in the following formula, which implement our scoring system:

$$Score = SizeSC \cdot 0.15 + HeavSC \cdot 0.15 + DensSC \cdot 0.30 + PackSC \cdot 0.40$$

An example of a random order and the attributes associated with its articles is presented in Table 1. The attributes of each article are shown together with the score received by the article according to the scoring system.

**Table 1.** An example of an order with the scores attributed to the articles

| Product | Volume | Mass | Density | Packaging | Score |
|---|---|---|---|---|---|
| 1L Bottle | 1.00 | 1.02 | 1.02 | Bottle | 13.24 |
| 2L Bottle | 2.00 | 2.05 | 1.03 | Bottle | 15.86 |
| Energy Drink | 0.50 | 0.52 | 1.04 | Can | 10.02 |
| Chips | 1.00 | 0.13 | 0.13 | Bag | 3.73 |
| Eggs | 0.90 | 0.38 | 0.42 | Carton | 6.54 |
| Milk | 1.00 | 1.05 | 1.05 | Carton | 8.94 |
| Instant Noodles | 0.40 | 0.12 | 0.30 | Package | 4.61 |
| Pasta | 1.40 | 0.50 | 0.36 | Package | 6.95 |
| Jam | 0.20 | 0.37 | 1.85 | Glass | 9.38 |
| Wine Bottle | 0.75 | 1.20 | 1.60 | Bottle | 14.09 |

## 3.2   The Route Planner

With the scoring model described in Sect. 3.1, articles in the shopping list have different priorities to be picked. The in-store order picking problem formalized in this work can therefore be modelled as a Sequential Ordering Problem. Given a weighted graph, the SOP is a TSP with precedence constraints between pairs of vertices. The objective is to find a Hamiltonian Path with minimum cost that satisfies the precedence constraints.

An instance of the SOP is represented by two graphs: a cost graph and a precedence graph. The cost graph is a weighted directed graph $G = (V, E)$ where $V$ is the set of vertices (corresponding to articles in our case), $E = \{(i, j)|i, j \in V\}$ is the set of edges, and each edge $(i, j) \in E$ has a cost $t_{ij}$. The precedence graph is a directed graph $H = (V, P)$ where $V$ is the same set of vertices. When an edge $(i, j) \in P$, vertex $v_i$ must precede vertex $v_j$ in any feasible solution. A Hamiltonian Path is a permutation of the full set of vertices that each vertex appears exactly once in the given graph. For a solution $S$, the cost is defined as $\sum_{(i,j) \in S} t_{ij}$. Given a start and a final vertex, the objective of the SOP is to find the Hamiltonian path with minimum cost that satisfies all the precedence constraints. It corresponds to the picking order in our application. However, if we strictly follow the precedence constraints, a lot of back-and-forth movements are expected, and a long time might be required to complete the task. This inefficiency is not the original aim of using the SOP model, but rather a side effect to the benefits of a SOP solution. Therefore, we propose also a relaxed SOP model to have a trade-off between priorities and total picking time. We define a limited number of precedence classes based on the scores. Precedences among articles in a same class are relaxed, assuming they are basically equivalent in terms of score. The number of classes can be adjusted according to the actual situation. In this work, we divide the articles into four precedence classes, in such a way that class 1 contains articles with a score up to 5.00; class 2 the articles with a score between 5.00 and 8.50; class 3 the articles with a score between 8.50 and 12.00; class 4 the articles with a score higher than 12.00.

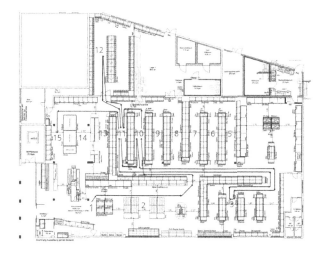

**Fig. 1.** Layout of a real store with numbers corresponding to zones containing products of different categories and a path indicating the optimal relaxed SOP solution of a given shopping list.

## 4    Experimental Simulation

In this section, we present methods and results for an experimental simulation of a real store in German. The layout of the store is presented in Fig. 1, with the numbers corresponding to areas/zones containing products of different categories. A path representing the optimal relaxed SOP tour for a given shopping list, is also depicted. Zone 1 includes the entrance and Zone 15 includes the cashier desks, they are the fixed starting and ending points in our model. Each article is associated with the zone containing it, and zones are used to estimate travel times. The travel time $t_{ij}$ from zone $i$ to zone $j$ is calculated assuming a walking speed of 0.85 m/s and considering the shortest path. The travel time between two articles in the same zone is set to zero. The time to pick up articles is not considered, since it has no influence on the total time spent moving in the shop. In the simulation, we compare the time required to pick all the articles on the shopping lists when the given lists are ordered in different ways. Ten representative online orders, consisting of 36.6 items on average and with a standard deviation of 6.12, are considered. The original shopping lists received are in random order (no optimization solution). The second picking sequence is calculated as the optimal TSP solution [14] (this solution ignores fragility and dimensions issues). The third list considered is ordered according to the optimal SOP solution, strictly following precedence constraint. For example, in the shopping list presented in Table 1, we follow the descending order of the Score value to pick the corresponding item, with 2L bottle being the first and chips being the last to be picked. This solution only optimizes the articles' sequence but not the picking path. The fourth list is ordered according to the relaxed SOP solution (Sect. 3.2, see also Fig. 1), which optimizes a trade-off between the characteristics of the articles and the total travel time. The SOP optimal solutions can be obtained by the algorithm described in [19]. To summarize, four scenarios are analyzed: (1) no optimization; (2) articles ordered according to the TSP solution; (3) articles ordered according to the SOP solution; (4) articles ordered according to the relaxed SOP solution.

The simulation results are reported in Fig. 2. We can observe that TSP solutions performs the best in terms of travel time, as expected. Knowing the disadvantages of TSP solutions in terms of sequencing of the articles, our main interest is to estimate the extra time required by the SOP solutions, which takes into account mainly the correct sequencing of the articles. The results from Fig. 2 shows that with the optimal SOP solutions increase the time by 150% on average compared to TSP solutions, while relaxed SOP solutions increase the time by only 71%, while retaining an operationally acceptable sequencing of the article. This indicates that relaxed SOP solutions are considerable choice for this problem.

Let us consider order 3 as an example to understand the characteristics of the different solutions. The order contains 46 different articles from four different zones. The articles are mainly located in Zones 10, 11, and 12, but one is in Zone 3 (Fig. 1). The list is ordered as the customer added the articles to the shopping cart. Therefore, when no optimization is considered, the picker first heads to Zone 11, then travels back and forth among Zones 11, 12, and 10 and finally goes to Zone 3 to pick the last item before going to the cashier at Zone 15. It takes 349 s travel time to finish the task, and yet more time is required to sort

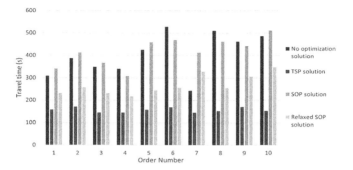

**Fig. 2.** Simulation of travel times for 10 representative online orders in four different scenarios.

out the messy shopping cart at the cashier (there is also a risk of damaging the goods, if the picker is not careful enough). When the TSP solution is considered, the picker visits Zones 3, 10, 11, 12 in this order. It takes only 145 s travel time to finish the task, but some more time is required to reorder the items for the subsequent packing process, and extra care had to be taken by the picker not to damage any product. When the list is ordered according to the SOP solution, heavy and non-fragile articles are picked first, leaving light, small and fragile articles for the end. The picker travels back and forth among Zones 11, 10 and 12, and goes to zone 3 at a certain point in the middle of the process. It takes 367 s travel time to complete the task (even more than the unsorted list), but in this case no extra time is needed to rearrange the articles, since packing is already sorted. When the list is ordered according to the relaxed SOP solution, the has considerably fewer moves back and forth among the zones. It takes only 231 s travel time, and there is no need to rearrange the articles for a final packing.

In conclusion, the relaxed SOP solution shows a good compromise between being very fast with unsorted (and potentially damaged) goods and producing a ready-to-go package already while picking articles. Furthermore, it is worth to mention that "scan as you pick" devices are becoming popular in grocery stores. The receipts are ready at the end of the tour and the articles can be directly packed in bags during picking. In this case, the benefit of SOP-based model is even more straightforward.

## 5   Conclusions

The research on the optimization strategies for in-store order picking presented in this paper aspires to contribute to the improvement of omnichannel in-store logistics operations for brick-and-mortar grocery retails. We proposed an optimization approach for the in-store order picking problem in the context of a Buy-Online-Pick-up-in-Store system relying on human workforce. We discussed the benefit of reordering shopping lists according to an ad-hoc precedence model instead, which prevents potential dam-ages to the goods during the in-store picking phase and also makes the packaging task trivial, with benefit also for customers, that have articles sorted in a correct order inside their bags. Experimental simulations on a real-world store have been per-formed, demonstrating the practicality and applicability of the approach, which has been shown to provide a good trade-off between efficiency and customer satisfaction. With the increasing popularity of co-called "scan as you pick" devices, even further benefits from this model can be foreseen.

# References

1. Hübner, A., Kuhn, H., Sternbeck, M.G.: Demand and supply chain planning in grocery retail: an operations planning framework. Int. J. Retail Distrib. Manage. **41**(7), 512–530 (2013)
2. Pires, M., Camanho, A., Amorim, P.: Solving the grocery backroom sizing problem. Int. J. Prod. Res. **58**(18), 5707–5720 (2020)
3. Reiner, G., Teller, C., Kotzab, H.: Analyzing the efficient execution of in-store logistics processes in grocery retailing. The case of dairy products. Prod. Oper. Manage. **22**, 924–939 (2013)
4. Hübner, A., Kuhn, H., Wollenburg, J.: Last mile fulfilment and distribution in omni-channel grocery retailing. Int. J. Retail Distrib. Manage. **44**(3), 228–247 (2016)
5. Wang, Y., Xu, R., Schwartz, M., Chen, X.: COVID-19 and retail grocery management: insights from a broad-based consumer survey. IEEE Eng. Manage. Rev. **48**(3), 202–211 (2020)
6. Boysen, N., Koster, R. de, Füßler. D.: The forgotten sons. Warehousing systems for brick-and-mortar retail chains. Eur. J. Oper. Res. **288**(2), 361–381 (2021)
7. Wollenburg, J., Hübner, A., Kuhn, H., Trautrims, A.: From bricks-and-mortar to bricks-and-clicks. Int. J. Phys. Distrib. Logist. Manage. **48**(4), 415–438 (2018)
8. Li, Z., Yang, W., Jin, H.S., Wang, D.: Omnichannel retailing operations with coupon promotions. J. Retail. Consum. Serv. **58**, 102324 (2021)
9. Kim, K., Han, S.-L., Jang, Y.-Y., Shin, Y.-C.: The effects of the antecedents of "Buy-Online-Pick-up-In-Store" service on consumer's BOPIS choice behaviour. Sustainability **12**(23), 9989 (2020)
10. Kong, R., Luo, L., Chen, L., Keblis, M.F.: The effects of BOPS implementation under different pricing strategies in omnichannel retailing. Transp. Res. Part E Logist. Transp. Rev. **141**, 102014 (2020)
11. Chou, X., Loske, D., Klumpp, M., Gambardella, L.M., Montemanni, R.: In-store picking strategies for online orders in grocery retail logistics (2021, Submitted)
12. Matusiak, M., de Koster, R., Saarinen, J.: Utilizing individual picker skills to improve order batching in a warehouse. Eur. J. Oper. Res. **263**(3), 888–899 (2017)
13. Ogasawara, A., Ishigaki, A., Yasui, S.: Adaptive storage reassignment in order picking systems to picker learning and change of demand. Procedia Manuf. **39**, 1623–1632 (2019)
14. Grosse, E.H., Calzavara, M., Glock, C.H., Sgarbosssa, F.: Incorporating human factors into decision support models for production and logistics. Curr. State Res. IFAC-PapersOnLine **50**(1), 6900–6905 (2017)
15. Pardines, I., Lopez, V.: Shop&Go: TSP heuristics for an optimal shopping with smartphones. Sci. China Inf. Sci. **56**(11), 1–12 (2013). https://doi.org/10.1007/s11432-013-5013-4

16. Liebmann, H.P., Zentes, J.: Handelsmanagement. Vahlen, Munich (2001)
17. Ozgormusa, E., Smith, A.E.: A data-driven approach to grocery store block layout. Comput. Ind. Eng. **139**, 105562 (2020)
18. Filipe, S., Henriques Marques, S., Fátima Salgueiro. M. de: Customers' relationship with their grocery store: direct and moderating effects from store format and loyalty programs. J. Retail. Consum. Serv. **37**, 78–88 (2017)
19. Mojana, M., Montemanni, R., Di Caro, G., Gambardella, L.M.: A branch and bound approach for the sequential ordering problem. Lect. Notes Manage. Sci. **4**(1), 266–27 (2012)
20. Montemanni, R., Smith, D.H., Rizzoli, A.E., Gambardella, L.M.: Sequential ordering problems for crane scheduling in port terminals. Int. J. Simul. Process Model. **5**(4), 348–361 (2009)

# Digital Twin Application for the Temperature and Steam Flow Monitoring of a Food Pasteurization Pilot Plant

Giovanni Tancredi⬤, Eleonora Bottani⬤, and Giuseppe Vignali⁽⊠⁾⬤

Department of Engineering and Architecture, University of Parma, Parco Area delle Scienze
181/A, 43124 Parma, Italy
giuseppe.vignali@unipr.it

**Abstract.** In this paper, the development of a Digital Twin of a beverages pasteurization system for temperature monitoring, using NI Lab-VIEW control system toolkit, is described. A cyber-physical production system, composed of a real-time simulation tool and a controller, has been set up and tested on a pilot plant set up in a university laboratory. The paper shows how the software platform, together with the hardware, has been implemented in a traditional system, not (yet) ready for Industry 4.0 technologies, and therefore underlines the main issues occurred during its development. The Digital Twin includes a CompactRIO controller and a set of probes, connected with the controller. The aim of the Digital Twin is to monitor the machine status, with a particular attention to the temperature reached by the service water and the required steam flow. To demonstrate the effectiveness of this system, a set of experimental tests has also been carried out.

**Keywords:** Digital twin · Industry4.0 · Temperature monitoring · Pasteurization system · Food industry

## 1 Introduction

The Industry 4.0 paradigm introduced many key-enable technologies, able to transform a traditional manufacturing process in an advanced cyber-physical production system. Among these, the Digital Twin (DT) appeared as one the most innovative and disruptive tools [1]. The main advantages of DT models encompass the possibility to prevent, anticipate and solve problems relating to the real system in a timely manner, to reduce the time and costs associated with simulations and analyses, to test a process before its implementation and to make effective decisions using data provided in real time [3]. DT is also a key technology of Industry 4.0 for solving safety issues [4]. On the basis of these consideration, a research project has been funded by the Italian national institute for insurance against industrial injuries (INAIL), for developing a DT approach for reducing injuries and hazards in industrial plants. The rationale behind this project stems from an attempt to deepen the current technological evolution of the DT approach, characterized by the use of real time data for modelling an industrial process, with a particular focus

© IFIP International Federation for Information Processing 2021
Published by Springer Nature Switzerland AG 2021
A. Dolgui et al. (Eds.): APMS 2021, IFIP AICT 631, pp. 612–619, 2021.
https://doi.org/10.1007/978-3-030-85902-2_65

of its application in manufacturing processes [2]. This paper describes the first activities of the project, whose aim is to delineate the architecture of a DT model that will be next used for reproducing the plant functioning in unusual operating conditions, thus allowing for testing the implications for safety and implementing appropriate countermeasures.

The scene of this study is a pilot plant which consists in a pasteurization system equipped with a counter-flow tube-in-tube heat exchanger used for pre-warming the fluid foods (process fluids), through the transfer of heat by water (service fluid), preheated with the steam provided by a heat steam generator. For this kind of plant, traditional control systems are typically used [5], while DT models, which are more user-friendly, are rarely available. This equipment has been already studied in a previous research about the implementation of safety control on fluid pressures [6]. The design phase aims to create a dynamic model of the pilot plant able to simulate the behavior of the machine, monitor its control parameters (in particular, the temperatures), and display the machine status on a user interface.

The scope of this work is therefore to design and validate a DT model built using Labview, for the purpose of monitoring the process water temperature and the steam inlet and solve possible safety issue generated during the pilot plant tests.

## 2  Digital Twin Architecture

A DT dynamic model capable of continuously simulating a beverage pasteurization process and retrieving the parameters of interest has been developed. The pilot plant is equipped with pressure and temperature sensors that provide real data for monitoring the machine status. The developed model consists in a hardware part and a software one. The hardware consists of a set of probes installed on the pilot plant that send the signals via 4–20 mA current loop to a data acquisition system. This latter has been implemented in Labview environment in both simulation and real time acquisition module, using a cDAQ 9133 chassis in which an analogue input module (NI 9208) has been installed. The data acquisition (DAQ) module captures the signal from three temperature sensors (PT100) that measure the inlet temperature of the service fluid and the outlet temperature of the process fluid. These probes have a measurement range of 0°–100°C and transfer the data to the DAQ by means of an analog signal in the range 4–20 mA. The software part was designed via G-code in Labview.

### 2.1  Software and Model Description

The software part consists of a user interface, called front panel, and a block diagram. The front panel shows the interface by which the user can monitor the system and vary the input parameters. For representing the pilot plant, three additional tabs (panels) were developed, called respectively Model/Controller tab, Machine status tab, and Data Acquisition tab. The block diagram (Fig. 1) shows the G-code in which the system equations are written.

The methodology adopted for the development of the DT consists in three distinct steps: (1) development of the simulation model of the pilot plant; (2) acquisition of the real data from the pilot plant; and (3) model validation.

For the development of the simulation model, the following features have been set: (i) the geometry of the tube in tube heat exchanger consists of six pipes of length equal to 4 m, with an annular section (diameter 0.076 m) in which the process fluid (hot water) flows, and an internal section (diameter 0.038 m) in which the fluid food flows. The thickness of both pipes is 3 mm (see [2]); (ii) as far as the output temperature of the fluid food, it is assumed that the product enters the system at ambient temperature ($\approx$25 °C) and exits at about 60 °C after recirculation; (iii) the flow rate of the service fluid is 12 m$^3$/h; (iv) the outlet pressure of the fluid food from the exchanger has been set approximately at zero (0.005 bar).

Additional assumptions made include the following ones: (i) the outer surface of the exchanger is perfectly insulated. Thus, the thermal power absorbed by the fluid food equals that furnished by the service fluid; (ii) the density and heat exchange values of the water have been set at those at 25 °C temperature.

Based on these considerations, a thermodynamic model has been created using Labview as reported in the grey central block of Fig. 1.

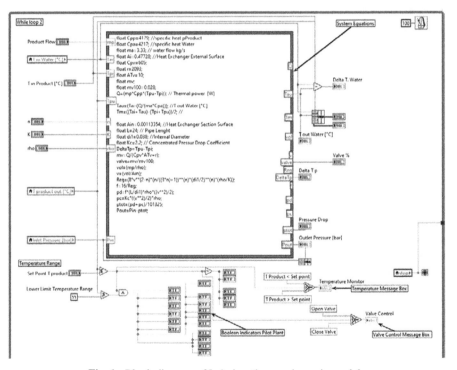

**Fig. 1.** Block diagram of Labview thermo-dynamic model.

The conceptual design of the DT aims to replicate the behavior of the physical plant taking into account the status of the service and process fluids flowing inside it. After having created the physical model, the first activity is the process simulation by setting some parameters in Labview code and reproducing the signal acquisition by the cDAQ

9133 and the data acquisition 9208 modules. By simulating the signal acquisition from real sensors, the model behavior can be easily evaluated.

The process consists in various steps in which the temperature is simulated as perceived by the PT100 through the cDAQ module, as summarized below:

- Setting of the rheology and heat capacity parameters of the service water and the food fluid;
- Simulating the sensor signal [mA] for to the service water and fluid flow entering the heat exchanger;
- Converting the signal from [mA] to [°C];
- Simulating the sensor signal in [mA] for the pressure at the outlet of the heat exchanger;
- Converting the signal from [mA] to [bar];
- Calculating of the water and food fluid outlet temperature;
- Calculating the pressure of the inlet of the food fluid product.

### 2.1.1  The Model/Controller Tab

The Model/controller tab (bottom part of Fig. 1) shows the model that reproduces the status of the fluids, together with two message boxes: the first one, named Temperature Monitor, shows if the product temperatures meet the set point or not by displaying a message. The second one refers to the modulating intake steam valve, called control valve, and shows the percentage of the requested opening of the valve section for providing the needed steam to reach the desired outlet temperature of the process fluid.

The pilot plant diagram follows a Boolean logic that help visualize the machine status by colors, that change from blue to red; "blue" denotes a normal operating condition of the machine, while "red" indicates an alert status.

### 2.1.2  The Machine Status Tab

The Machine status tab contains the controls and indicators and the graphic representation of the temperature and pressure signals. Based on the hypotheses made for the calculations of the output variable, the system plots the output temperature on a linear graph. An indicator shows the $\Delta T$ of the water, which varies according with the variation of the parameters related to the fluid food, and in particular, to the product flow and the inlet/outlet temperatures of the product. The $\Delta T$ of the fluid food refers to the range of T of the product and therefore reflects the thermal power exchanged; a variation in this value implies a change in the temperature of the outlet service fluid.

### 2.1.3  The Data Acquisition Tab

The Data acquisition tab concerns the signal acquisition through the DAQ assistant, or the simulation of the input signals to evaluate the variation of the controlled parameters in a simulated environment, without connecting the system to the physical plant. The tab contains the Start Simulation button that allows switching from the simulation of the data in the software environment, to the real data acquisition thought the cDAQ hardware. The simulation tool also allows creating an analogue signal in current and then reproducing the sensors behavior characterized by a current loop 4–20 mA. The correct temperature and pressure values can be obtained by means of specific calibrations. Sinusoids made with the DAQ assistant are displayed within the waveform chart, while the sampling frequency and the samples per cycle represent input data that the operator can change according with the level of accuracy and number of data required when simulating different cases. The data about the inlet temperature of the service fluid and about the inlet and outlet temperature and pressure of the process fluid can be displayed by simulating the process.

## 3   Simulation Tests

According to [7], simulation is an effective tool to be used for investigating the behavior of a system anytime it cannot be immediately observed or reproduced in practice. For the plant under examination, this is exactly the case. Indeed, the plant can be observed in its normal operating conditions (which will be detailed in the experimental tests), but obviously, it cannot work outside its normal operating conditions without jeopardizing its safety or the safety of people working on the plant. For reproducing these anomalous situations, simulation was used.

### 3.1   Initial Setting

The first aspect investigated is the variation in the water temperature during the heat exchange phase. To evaluate the temperature reached by the water, the thermal power exchanged within the process fluid and the service one was assumed to vary according with the heat capacity, the flow and the $\Delta T$ between the inlet and the outlet of the heat exchanger. The system acquires the input parameters and calculate the outlet temperature of the water at each variation of the input. The evaluation of the needed thermal power can also be useful to calculate the required flow of steam at the inlet. For the evaluation of the intake valve's opening percentage, it was assumed that the maximum steam value provided is 100 kg/h, according to the real capacity of the installed electric steam generator; this corresponds to about 0.0277 kg/s, which represents the input parameter defined as mv100 in the G-code developed. The ratio between the steam flow evaluated using the thermal heat equation and the mv100 value represents the intake valve's opening percentage. During the testing phase in the simulation environment, the input parameter was simulated as a continuous signal that fluctuates between the real operating conditions, as shown in Table 1. For the remaining input parameters, steady conditions were instead assumed; in particular: the inlet product temperature was set at 50 °C; the product flow was set at 0.4 kg/s; the set point of the product temperature was set at 60 °C.

**Table 1.** Initial setting of the simulation test.

|  | T max[°C] | T min [°C] | Sample Rate [Hz] |
|---|---|---|---|
| Inlet water temperature | 100 | 90 | 0.33 |
| Outlet product temperature | 60 | 55 | 0.33 |

## 3.2  Results with Simulated Data

The system provides three different outcomes based on the assumptions made. A first result can be seen in the message box, which shows whether the product temperature is higher (>) or lower (<) than the set point. As second outcome, the machine status monitor displays the tube-in-tube part of the pilot plant in blue color if the product temperature is in a range between the lower limit (55 °C) and the set point, or in the red one if the set point is overcome. For the valve control, a message box displays the intake valve status (valve opened vs. valve closed), while a numeric indicator shows the percentage of opening required for the valve to match the steam flow needed to reach the desired output temperature of the water. An example of the results obtained is shown in Fig. 2, referring to the case in which the set point of the product temperature is overcome.

As shown in that figure, the system requires the valve to be opened by about 40% to meet the set point; the temperature monitor displays the correct message, and the Boolean indicators turn red as expected. In the opposite situation, i.e., if a small temperature gap is observed between the inlet and outlet of the product, the percentage of valve's opening (and therefore the steam flow required) would instead be low. To evaluate the system's response in out-of-range cases, a test has been run by varying the product temperature, with a simulated signal temperature higher than the set point and one lower than the minimum threshold.

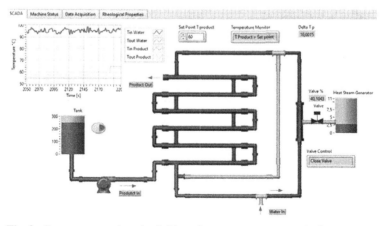

**Fig. 2.** System status when the fluid product temperature set point is overcome.

## 4    Experimental Tests

To assess the software robustness and validate the system model, a testing phase was carried out in laboratory settings. The signals of the three input parameters were acquired on the physical plant through the cDAQ hardware. The sample rate adopted was 0.33 Hz and the number of samples for each run was set at 1. The parameters assumed in the steady state conditions are the inlet product temperature (52 °C), the product flow (0.4 kg/s) and the set point of the product temperature (60 °C). The main outcomes are related to the behavior of the system in real environments. As shown in Fig. 3, the input signals acquired show a "T_in Water" of 91.25 °C and a "T_out product" of 57.10 °C. With the data acquired and the initial setting, the Temperature Monitor (Fig. 4) shows the correct message, and the machine status is displayed in blue. The ΔTp indicator led to a valve opening of about 38.87% while the valve control state displays the Open Valve status, required to reach the set point.

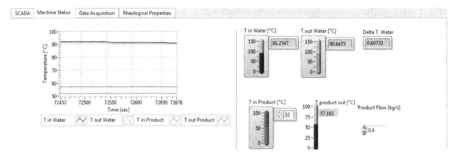

**Fig. 3.**  Real data acquired by the system (T_in water, T_out product).

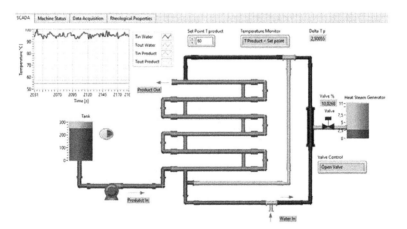

**Fig. 4.**  DT model with the acquisition of real signals from the field.

# 5   Conclusions and Future Research

The present paper aimed at developing and testing a DT application for monitoring the temperature of the service fluid flowing in a real plant and the amount of steam needed for its heating. The model has been implemented in a Labview environment; tests carried out both in simulation and laboratory environments, showed that the model reproduces the system functioning with good precision. Compared to traditional control systems, the proposed DT model is more user-friendly and can be easily learnt in its functioning by any employee working on the plant, which is expected to encourage its practical usage. Moreover, its future link with safety issues, mentioned earlier in the manuscript, will represent an innovative approach for reducing injuries and hazards in industrial plants. Starting from this study, future research activities will introduce additional features of the control system, by acting on the PLC when some parameters exceed the safety condition (e.g., excessive water temperature or pressure). From a more practical point of view, it is planned to complement this study by installing an internet connection in the laboratory environment, for real time monitoring of the plant by the employees.

**Acknowledgements.** This research was funded by the Italian National Institute for Insurance against Accidents at Work (INAIL), under grant program BRIC 2018, project ID12.

# References

1. Tao, F., Cheng, J., Qi, Q., Zhang, M., Zhang, H., Sui, F.: Digital twindriven product design, manufacturing and service with big data. Int. J. Adv. Manuf. Technol. **94**(9–12), 3563–3576 (2018)
2. Uhlemann, T.H.-J., Lehmann, C., Steinhilper, R.: The Digital twin: realizing the cyber-physical production system for Industry 4.0. Procedia CIRP **61**, 335–340 (2017)
3. Kritzinger, W., Karner, M., Traar, G., Henjes, J., Sihn, W.: Digital twin in manufacturing: a categorical literature review and classification. IFACPapersOnLine **51**(11), 1016–1022 (2018)
4. Bevilacqua, M., et al.: Digital twin reference model development to prevent operators' risk in process plants. Sustainability **12**(3), 1088 (2020)
5. Shaikh, N.I., Prabhu, V.: Model predictive controller for cryogenic tunnel freezers. J. Food Eng. **80**(2), 711–718 (2007)
6. Bottani, E., Vignali, G., Tancredi, G.: A digital twin model of a pasteurization system for food beverages: tools and architecture. In: Proceedings of the IEEE International Conference on Engineering, Technology and Innovation (ICE/ITMC), vol. 1, pp. 1–8. Institute of Electrical and Electronics Engineers Inc., Cardiff (2020)
7. Harrison, J.R., Lin, Z., Carroll, G.R., Carley, K.M.: Simulation modeling in organizational and management research. Acad. Manag. Rev. **32**(4), 1229–1245 (2007)

# Investigating the Role of Institutional Frameworks in Food Waste Reduction at the Retailer Interface in the European Union

Yvonne Rachael Owasi[1]([✉]) and Marco Formentini[2] [iD]

[1] Audencia Business School, Nantes, France
yvonnerachael.owasi@audencia.com
[2] Department of Information Engineering and Computer Science (DISI),
University of Trento, Trento, Italy
marco.formentini@unitn.it

**Abstract.** The role of institutional frameworks in curbing and preventing food wastage cannot be overstated. Recently, policy makers have increased their interest in helping supply chain actors to reduce food waste by initiating policies at national and local level, but also opened questions on how such initiatives are affecting supply chain actors' practices towards sustainable food supply chains. Based on this concern, this study set out to investigate the role institutional frameworks have to play in tackling the problem of food wastage at the retailer-supplier interface within the European Union (EU). To this end, the study mainly focuses on the issue of food wastage in France by taking a look at the impact of the French food waste law on the retail-supplier activities and relationship.

This qualitative study underlines that the French food waste law has led to initiatives that are promoting circular economy and closed loop food supply chain, though the policies are being limited by behavioural aspects of actors along the food supply chains. Therefore, this research calls for a better understanding of behaviour of actors and collaboration along the supply chain through the use of information technology to have a full positive impact of the policy interventions towards sustainable production and distribution of food.

**Keywords:** Food waste · Institutional framework · Closed loop supply chains · Food · Supply chain sustainability · Circular economy · Behavioural aspects

## 1 Introduction and Theoretical Background

The extent of food waste in the world is a call of concern both at European and global level, that's why it has received high consideration across different disciplines, policymakers and institutions. In the past decades, food supply chains have become globalised and thus longer. Actors along the supply chain design their activities to satisfying consumers with quality products while minimising overall cost and waste. Despite efforts to optimise food supply chains, current research shows that food waste remains unacceptably high along the food supply chains.

© IFIP International Federation for Information Processing 2021
Published by Springer Nature Switzerland AG 2021
A. Dolgui et al. (Eds.): APMS 2021, IFIP AICT 631, pp. 620–629, 2021.
https://doi.org/10.1007/978-3-030-85902-2_66

The research in food supply chain production and distribution has been conscientiously investigating the quantification, causes and recovery of food waste; for instance, Parfitt et al. [1] focus on the quantification and causes of food waste across all supply chains worldwide. Other authors focus on food waste on a particular region, country or sector, such as Teller et al. [2] who focus on western Europe food retail. However, the studies do not focus on retailers influence on food waste across the food supply chains and how the macro environment can play a role in helping actors achieve closed loop supply chain. Since retailers play a pivotal role to link the upstream and downstream sides in a supply chain, they have a significant influence on food waste generation across the supply chain [3].

Therefore, focusing on the retail and their link to the upstream side of the supply chain and the role regulations play to prevent and reduce retail interface food waste justified the study. The study took a general outlook at the state of the art of various intervention in the EU and then narrowed to how the French food-waste law is impacting food supply chain actors in France. Therefore, the study aims to answer the research question: *How is the French food waste law impacting the retailer-supplier activities and relationship towards sustainable food production and distribution?*

When we take a look at factors of food waste, they can be divided into two, those that occur along the supply chain and factors from the surroundings of the food supply chain [4]. Elements that occur along the supply chain are mostly due to decisions and actions taken by managers and staffs. Factors from the surrounding of the food supply chain can include natural constraints and megatrends in the industry [5], including government regulations.

In 2012, 460 million tons of food went to waste at the retail level in the EU [6]. Several studies suggest that the food wasted at the retail level is still fit for human consumption yet 795 million people are undernourished mostly in developing countries as per FAO data of the period 2014 to 2016 [7]. Besides the food security problem contribution of food waste, it also contributes to environmental impact. 170 million tons of $CO_2$ is estimated to be produced along food supply chains in the EU [4]. Food waste also hurts the performance of food supply chains as it leads to financial loses and reputation issues thus sustainability issues.

Retail stores play a vital pivot role as they are gatekeepers who act as a link to other actors in the chain. In Europe, the food retail sector holds the largest market share and therefore dominant position that has triggered a change in food market structure where power has shifted from food suppliers to food retailers [8]. In the EU, direct retail sales to consumers on agricultural products are estimated at 54% of total sales while farmers account for only 2% [9]. Example, according to Statista [10], in France, E.Leclerc holds the largest grocery market share of 21.6% making it the most extensive and dominant supermarket in France, followed by Carrefour with 19.8% share. The waste generated at the retail level seems to be lower than other stages of the supply chain. However, retailers' measures to reduce food waste are at the expense of other actors on the supply chain who end up bearing the economic and moral burden of waste as a consequence of retailers' market power through activities such as take back agreements [11].

However, the European Union (EU) is against the practices that lead to dispropor-tionate retail power. Suppliers are expected to report any misuse of power. However, such cases are rarely reported and are also difficult to prove. 87% of the EU suppliers fear to raise their concerns to retailers, and they do not report either, 65% of them stated they fear retaliation [12].

EU parliament and member state governments have come up with policies and pro-cedures to help food supply chains to be more efficient. Each policy initiative based on a specific approach. According to [13], policy approaches are classified as suasive or regulatory and targeted either towards the market or public, as shown in Fig. 1. A suasive approach is a policy approach that aims towards behavioural changes through the provision of information, while a regulatory approach uses penalties where public authorities mandate performance to be achieved [13].

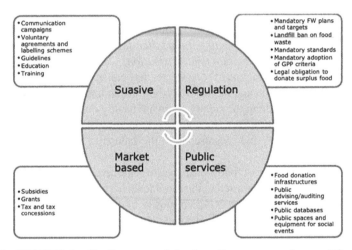

**Fig. 1.** FUSIONS classification approach for the policy measures. Source: [13, p. 22].

The EU initiatives in regards to food waste are the EU circular economy that targets to reduce food waste by 30% in 2025 and 50% by 2050 [14]. Eu waste directive encourages its members to create programmes that prevent waste by concentrating on significant environmental impacts and keeping in mind the product life cycle [15].

The government of France in 2016 initiated a law [16] (i.e., LOI n° 2016-138) that required supermarkets that have a floor space exceeding 400 m$^2$ to collaborate with charities in-order to donate surplus food within 48 h before expiry [17]. A law was also passed in 2012, forcing the private sector to recycle organic waste. In 2020 the government extended the 2016 food waste policy to wholesalers with an annual turnover of 50 million EUR. In Italy, there is the Good Samaritan Law (i.e., Law n. 155/2003) that is in charge of ensuring charities that collect the food are responsible for proper storage and expiry date [18]. Further, Italy introduced in 2016 the PINPAS law (i.e., Legge 19 Agosto 2016, n. 166), which gives food supply chain actors flexible ways to donate the food to charities and food bank with a tax incentive; however, there is not a penalty [18].

Despite many different approaches, there is still limited knowledge of the impact of these regulations on food supply chain actors. Few researchers have conducted a comparison of the various law among countries such as [14] studies the role of the food waste hierarchy in addressing policy and research by comparing the French and Italian food waste policies. [19] discusses the implication of the EU policies and legislation on food waste. The studies, however, do not cover the impact of this legislation on the activities and relationship of the supply chain actors. This study, thus, aimed to investigate this gap.

## 2    Methodology

To answer the research question, qualitative research method was used in this study because it makes possible the explanation of the relationships, variations, contextual factors, and norms prevalent at the retailer-supplier interface of the food supply chain.

Since this study focused on understanding relationship phenomena and how the law impacts them, semi-structured interviews were used because open-ended questions and probing are best suited to help the researcher get an in-depth knowledge of aspects. Another importance of qualitative surveys is the richness of the data collected, the possibility of coding in a logical way, and the possibility of interpretation of data in a reliable way [20]. The semi-structured interviews deployed in the course of this study were limited to retailers and suppliers in France. Limiting the focus to France is because of the narrow investigation area for the study settled on France.

The study took an exploratory approach. Multiple embedded case studies are used to help understand the impact of French food waste law on the retailer supplier activities and relationship since its initiation in 2016. Following the advice of [21], the study accounts for construct validity, internal validity, external validity, and reliability throughout the research process. Key informants (i.e., managers) from the supply chain and logistic departments were interviewed face to face and through skype because they have the in-depth data of the processes in the company. Based on their responses, further questions were asked to them until sufficient data was collected. Total interviews conducted were eight from five companies.

A systematic and rigorous analysis was conducted through interpretative phenomenological analysis and text interpretation. The main themes coded on both the supplier and retailer interview guide. Coding was done manually as the number of the interviews allowed it. The process helped in identifying broad issues and patterns. A visualised table of the grouped data from suppliers and retailer was created for in-depth analysis to answer the research question.

## 3    Findings

The study highlights a novel of contribution on how the law enforced on food supply chain is impacting sustainability and points of improvements on the laws targeting supply chain actors' activities thus filling the research gap on impacts of regulations. The study found out that the French food waste law has led to the development of new business models that help in the redistribution of close to Date Limit Consumption (DLCs) food to consumers, and thus, closed loop food supply chains. The new businesses are helping in the redistribution of surplus food by ensuring the food circulate within the economy and reach large masses of individuals thus minimal wastage.

These new businesses have also helped reduce the power retailers had in the market; however, retailers still have power. Because of the reduced power, suppliers can negotiate better deals than before. The law has also increased the awareness of food waste across the supply chain, and now actors can discuss this subject, and some have initiated green programs aimed towards circular supply chain such as conversion of unsold products into animal feed.

The study further found out that retailers have now increased the pressure to suppliers to be able to produce with the highest DLCs in order to improve the potential of them having less unsold food, this is because the law reduced 48 h in which they could still sale these products and the retailers have no logistical cost support. With the increased pressure at the supply side, this has increased the amount of waste on their side since they have to meet the quality standards of retailers in the shortest time as possible. Therefore, despite the increased awareness along the supply chain, the law fails to address the problem of overproduction. Further, the supplier indicated that France has a lot of product references and shorter DLCs, due to this contrast, food is waste.

Though the French Food waste law may not be an applicable model to all countries, it is clear that the law plays a vital role in helping the supply chain actors reduce food waste and thus foster sustainability through circular economy. Each country has to impose policy based on the sector that has a significant influence on the supply chain and taking into consideration the behaviour and interest of the supply chain actors. Tackling food waste needs a system-wide approach that enables behavioural change and collaboration among the actors along the supply chain.

Table 1 summarizes the benefits and existing limitations of the law.

**Table 1.** Benefits and existing limitations of the law.

| Benefits | Limitations |
|---|---|
| Increased awarness among actors along the supply chain. Actors can now discuss the subject of food waste | The law does not take into consideration the influence of retailers in the market and how the retailers' reaction to the law might impact the other actors along supply chain |
| The law has led to the development of New business models that are providing a new channel of redistribution for the suppliers or retailers | The law doesn't take into consideration the product life cycle. The high number of food references in France contradicts the aim of the law to reduce food waste. High number of references leads to overproduction and thus food waste |
| Increased quantity and quality of donations to the food banks and charities and thus less edible food going to waste | No stringent quality checks of the food donated as some of the food donated in the food banks are passed expiry date |
| Due to the development of the new business models, the power of retailers has reduced but the retailers still have the power in the food supply chain | Lack of logistic cost support to retailers |

## 4 Discussion

To understand the complexity of food waste occurring at different stages of the supply chain, both analysis of food waste along supply chains and the policy intervention suggestions should go beyond individualistic accounts of one actor in the supply chain.

The discussion is presented by answering the research question by outlining various themes that emerged. These themes are discussed by the use of theoretical lenses of agency theory and social exchange theory to help understand the findings. Eisenhardt [22] defines Agency theory as organization relationships that involve a principal and an agent who engage in cooperative behaviour, though with differing goals and attitudes toward risk. Social Exchange Theory (SET) is argued by [23] as a theory that provides insights on social phenomena related to social interactions in a business environment.

In this study, the principal is the retailer, the environment factor is the government that creates the legislation, and the agent is the supplier. The unit of analysis is the French food waste law impact at the retailer-supplier interface. The French food waste law is regulatory and market-based as it provides a penalty if not followed. The government expects the retailers to donate the food nearing expiry within 48 h before the date limit consumption (DLC), and this leads to a difference in agreement in priorities and methods. Retailers aims are to increase profits and reduce costs; however, redistributing to donation centres increases their logistic costs. The differing goals and objects among the actors have pushed retailers to increase pressure on their suppliers to provide products with the highest DLC according to the study findings.

Eisenhardt [22] indicates that organizations have an uncertain future and when environmental effects such as governmental regulation occur, they affect the outcome as they are viewed as risks. Therefore, outcome uncertainty and unwillingness to accept risk affect the contracts between the agent and the principal [22]. The study finding affirms this assumption of agency theory; we see that retailers have increased pressure on suppliers to be able to produce and distribute to them (retailers) with the highest DLC to reduce their risk of penalties due to waste. This behaviour further, affirms agency theory contribution of risk implications, where the risk-averse principal will push the risk to the agent.

Despite the increased pressure by retailers, suppliers are working to ensure they meet the retailers demand to be able to produce and distribute at the highest DLCs as they fear retailers' retaliation. A study conducted by Devin & Richards [11] also found out that suppliers did not report the misuse of power by retailers even when systems to help the suppliers were in place. The reciprocity behaviour by suppliers affirms SET premise that seeking rewards and avoiding punishment is the primary motivation for interaction among parties in social exchange [24].

The increased pressure by retailers to suppliers is due to coercive power, where suppliers have to give in the demands of the retailers' or they will risk renegotiation of contracts, or another supplier who can meet this demand is selected. The power imbalance between retailers and suppliers in the food supply chain, especially in Europe is because the point of sale is mostly at the retail stores [12]. However, the study found out that retail power has been reduced by the new business models that cropped up as the result of the law.

The new businesses have also resolved the challenge of take-back agreements that lead to increased food waste at the supplier side, this new business has offered a channel for suppliers to sale their products returned from retailers. Moreover, to some extent, solved the challenge of costly shelf space at retail stores that limited the amounts of products suppliers could display. Furthermore, these new business models have provided the traditional supply chain actors with a channel through which they can redistribute their surplus food to the consumers.

The agency costs incurred by the government in order to reduce food waste has born fruits of increased awareness of food waste among actors. Agency cost is the cost that arises from agents misusing their position and the cost of trying to discipline and monitoring them to ensure they do what is expected [22]. Example of evidence of improved awareness among actors in the supply chain is the Carrefour and other supermarkets initiative in France. Carrefour extends the use before dates to avoid the confusion that it creates to customers, who think it is an expiry date, example yoghurt best before date was extended from 7 days to 10 days with no changes to quality or ingredient [25].

Supply chain actors along the food supply chains tend to focus on internal waste reduction this result to sub-optimal results on the entire supply chain [5]. Though the new business structures are helping in the redistribution of the surplus food along the supply chain, supply chain actors such as retailer and suppliers should work to enforce strong collaborations. For the French food waste to have full effect on food reduction, collaboration among retailers and suppliers is called for, as a collaboration among actors will lead to a compromise of self-interest, improve product imbalance and thus optimal result along the supply chain. The collaboration will also help solve the underlying problem of overproduction that leads to food waste and technological solutions such as Artificial Intelligence can be put in place to connect both the supplier and the retailer data centers to be able to collect real time data that helps in accurate demand forecasting and thus solve the problem of overproduction.

Figure 2 shows a model framework of the impact of the French Food waste law on the relationship between the suppliers and retailers based on Social Exchange Theory (SET) and Agency Theory (AT). The figure is a pictorial representation of what has been discussed in Sect. 4.

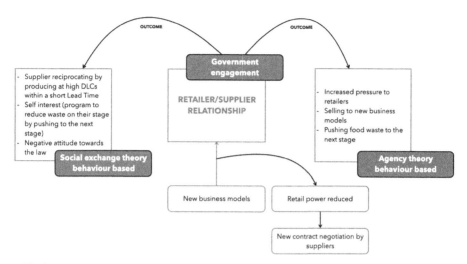

**Fig.2.** Model of French Food waste impact on supplier relationship based on SET and AT.

# 5    Relevance and Contribution

## 5.1   Theoretical Implications

The study adds to the several investigations carried out on food waste along the supply chain by filling the research gap of the impact of the regulations on supply chain sustainability. This study provided insights on the effectiveness of the regulations introduced in the supply chain to help actors reduce food waste.

It also contributes to supporting Giordano et al. [14] study, where the author reviewed the laws and compared the French and Italian law. This study provides a primary data collection from some of the actors in France and provides insights on the impact of the French Food waste law from a primary data point of view. Further, the study helped to support the agency theory and the social exchange theory premises and how they are critical when it comes to decision making that involves the interaction of actors along the supply chain.

This study further contributes to enhancing the understanding of close loop and sustainable supply chain, where food is reused and redistributed through new business structures and not for profit organization. This also contributes to supplier retailer relationship and how supply chain power can be reduced in the food supply chains in Europe if more players with different structure enter the market, thus fostering collaboration along the supply chains.

## 5.2   Managerial Implications

The findings suggest that despite organizations having a transactional relationship with their partners along the supply chain, it is essential that managers learn how to mitigate behavioral uncertainty along the supply chain. Managers should bear in mind how supply chain actors respond to transactional costs dilemmas which leads to abnormal behaviour. Therefore, managers must analyze their partners' behaviour and be prepared with counterbalancing remedies which might help in creating a business environment with a trusting atmosphere and reduce the negative impacts of behavioral interests.

Managers should take into account the agency theory and social exchange theory to be able to factor social, economic, and behavioral aspects when entering into contract decision making. Managers should also bear in mind that regulations that come in place in the market to help in the fight against food waste along the supply chain might not work for all supply chain actors. Therefore, managers need to leverage supply chain collaboration with their partners to be able to be effective and sustainable.

## References

1. Parfitt, J., Barthel, M., Macnaughton, S.: Food waste within food supply chains: quantification and potential for change to 2050. Philos. Trans. Royal Soc. London Ser. B: Biol. Sci. **365**(1554), 3065–3081 (2010)
2. Teller, C., Holweg, C., Reiner, G., Kotzab, H.: Retail store operations and food waste. J. Clean. Prod. **185**, 981–997 (2018)
3. Cicatiello, C., Franco, S., Pancino, B., Blasi, E., Falasconi, L.: The dark side of retail food waste: evidences from in-store data. Resour. Conserv. Recycl. **125**, 273–281 (2017)
4. Ocicka, B., Raźniewska, M.: Food waste reduction as a challenge in supply chains management. LogForum **14**(4), 549–561 (2018)
5. Mena, C., Adenso-Diaz, B., Yurt, O.: The causes of food waste in the supplier–retailer interface: evidences from the UK and Spain. Resour. Conserv. Recycl. **55**(6), 648–658 (2011)
6. Cicatiello, C., Franco, S.: Disclosure, and assessment of unrecorded food waste at retail stores. J. Retail. Consumer Serv. **52**, 101932 (2020)
7. Food and Agricultural Organization (ed.). Building Resilience for Food and Food Security. FAO (2017)
8. Fofana, A., Jaffry, S.: Measuring oligopsony power of UK salmon retailers. Mar. Resour. Econ. **23**(4), 485–506 (2008)
9. EU Agricultural Economic Briefs: https://ec.europa.eu/info/sites/info/files/food-farming-fisheries/trade/documents/agri-market-brief-04_en.pdf. Last accessed March 2020
10. Statista Research Department: http://www.statista.com/statistics/535415/grocery-market-share-france/. Last accessed March 2020
11. Devin, B., Richards, C.: Food waste, power, and corporate social responsibility in the Australian food supply chain. J. Bus. Ethics **150**(1), 199–210 (2016). https://doi.org/10.1007/s10551-016-3181-z
12. Ghosh, R., Eriksson, M.: Food waste due to retail power in supply chains: evidence from Sweden. Glob. Food Sec. **20**, 1–8 (2019)
13. FUSIONS: Recommendations and guidelines for a common European food waste policy framework, https://doi.org/10.18174/392296. Last accessed March 2020
14. Giordano, C., Falasconi, L., Cicatiello, C., Pancino, B.: The role of food waste hierarchy in addressing policy and research: a comparative analysis. J. Cleaner Prod. **252**, 119617 (2020)
15. Directive 2008/98/EC of the European Parliament and of the Council of November 19, 2008, on waste and repealing certain Directives (Text with EEA relevance). http://data.europa.eu/eli/dir/2008/98/oj/eng. Last accessed March 2020
16. LOI n° 2016-138 du 11 février 2016 relative à la lutte contre le gaspillage alimentaire, 2016-138. https://www.legifrance.gouv.fr/jorf/id/JORFTEXT000032036289. Last accessed March 2020

17. Chrisafis, A.: French law forbids food waste by supermarkets. The Guardian. https://www.theguardian.com/world/2016/feb/04/french-law-forbids-food-waste-by-supermarkets. Last accessed March 2020
18. Strefowa: Situation on food waste in Italy. Reduce Food Waste in Europe. http://www.reducefoodwaste.eu/situation-on-food-waste-in-italy.html (2016, 2019). Last accessed March 2020
19. FUSIONS: Review of EU legislation and policies with implications on food waste.pdf. ISBN: 978-94-6257-525-7 (2015)
20. Howard, L., Bruce, L.B.: Qualitative Research Methods for the Social Sciences, Global Edition. Pearson (2017)
21. Yin, K.: Case Study Research: Design and Methods. Sage, Thousand Oaks (2014)
22. Eisenhardt, K.M.: Agency theory: an assessment and review. Acad. Manag. Rev. **14**(1), 57–74 (1989)
23. Cropanzano, R., Anthony, E.L., Daniels, S.R., Hall, A.V.: Social exchange theory: a critical review with theoretical remedies. Acad. Manag. Ann. **11**(1), 479–516 (2017)
24. Griffith, D.A., Harvey, M.G., Lusch, R.F.: Social exchange in supply chain relationships: The resulting benefits of procedural and distributive justice. J. Oper. Manag. **24**(2), 85–98 (2006)
25. Carrefour Group Homepage: Action 11 combating waste. https://www.carrefour.com/en/group/food-transition/food-waste (2017). Last accessed March 2020

# Predicting Exports Using Time Series and Regression Trend Lines: Brazil and Germany Competition in Green and Roasted Coffee Industry

Paula Ferreira da Cruz Correia[1]([✉])(iD), João Gilberto Mendes dos Reis[1](iD), Emerson Rodolfo Abraham[1](iD), and Jaqueline Severino da Costa[2](iD)

[1] RESUP - Research Group in Supply Chain Management - Postgraduate Program in Production Engineering, Universidade Paulista, São Paulo, Brazil
paulafecruz@gmail.com
[2] Agroindustrial Management Department, Universidade Federal de Lavras - UFLA, Lavras, Brazil

**Abstract.** Trade is essential for countries development. In Brazil, coffee has been one of the most important export items and by large is commercialized as a green or roasted bean. The aim of this article is to establish a prediction mode for coffee exports using time series and trend lines. To do so, we collected the exportation volume from the two main export countries in each segment: Brazil and Germany. A five-year forecasting was produced using regression curves provided by Microsoft Excel. Our results indicated that polynomial fits best and this function is consistent with agricultural production that is conditioned to edaphoclimatic factors.

**Keywords:** Coffee · International trade · Time series forescasting

## 1 Introduction

Trade is an essential part in daily basis of humanity and entails knowledge, negotiation and regulation [1,2]. In 2019, 18.9 USD billion were exported worldwide being agricultural products responsible for 1.8 USD billion [3].

Agribusiness guarantee food security and is an important income source for developing countries [4]. Among many kind of products, coffee has been one of the most consumed items globally. The bean usually is commercialized as green beans or dehydrated form, but also can be sold benefited or roasted [5,6].

The two major world players in the sector are Brazil and Germany. In 2019, Brazil green coffee accounted for 4.5 million USD and 0.9 million USD in Germany. Regarding benefited form, the scenario is reversed. While Germany receive 1.4 million USD, Brazil obtained 9.765 thousand USD [7].

Previous literature have been studying different aspects from coffee supply chains. Lee and Bateman [8] analyzed the demand for fair trade and organic

© IFIP International Federation for Information Processing 2021
Published by Springer Nature Switzerland AG 2021
A. Dolgui et al. (Eds.): APMS 2021, IFIP AICT 631, pp. 630–636, 2021.
https://doi.org/10.1007/978-3-030-85902-2_67

coffee, its price range, and its demand by category. Naegele [9] studied the role of fair trade in the marketing of coffee beans in industrialised countries. Vogt [10] dealt with the definitions of the quality of the coffee bean and how these can be evaluated in different ways according to the type of market. Conceição [11] investigated the added value to the agro-industrial chain of Brazilian coffee [11]. However, we could find articles that investigate specifically prediction models in coffee industry.

To fill this gap, the article aims to estimate a prediction model for coffee exports. In this regard, we use time series applied to Brazil and Germany exports and forecast the volume for five years based on historical series. Our hypothesis is that the production of green and roasted coffee bean can be predicted using time series and regression trend line. These analysis are essencial to better understand the coffee supply chain.

The article is divided into sections. First, the introduction brings a scenario of the coffee bean trade and how two important players stand out in this area. Next, Sect. 2 presents materials and methods where the regression method is described to track trends and make predictions, and the steps followed in the construction of the work are presented. Then, the third section shows the results and discussions. Finally in Sect. 4 we remark the final considerations and perspectives.

## 2   Materials and Methods

### 2.1   Materials and Data

As previously mentioned the focus of the paper is to estimate a time series prediction for the export trade of green and roasted coffee beans considering Brazil and Germany historical series. To this regard, the study was conducted according to the following steps:

1. First, the export values of green and roasted bean coffee of Brazil and Germany were collected. Data were obtained from the FAOSTAT system of the United Nations Food and Agriculture Organization of the United States [7], considering the period between 1979 and 2019;
2. Second, the data were processed using Microsoft Excel©. Regressions models were performed to track trends and make predictions for five years. All trend lines were estimated. The prediction was made using the value closest to 1 of the coefficient of determination - a number that demonstrates the relationship between the variables analyzed [12];
3. Third, regression analysis was used to determine the relationship of each year analyzed (41 years) and the corresponding USD volume;
4. Finally, the predictions were analysed and discussed in the results section.

### 2.2   Method

Time series investigates the behavior over the years of historical series using different methods such as linear function, exponential, logarithmic, polynomial and power. It checks trends, seasonality, and randomness in a dataset [12, 13]. The most appropriate model for a specific set of data is one which presents the coefficient of determination (R) (Eq. 1) that aims to explain the relationship between variables, being considered appropriate the closer to one. Any measure above 0.7 is satisfactory and calculated based on the ratio between the explained and the total variance; the mean absolute error (MAE) (Eq. 2); and the mean squared error (MSE) (Eq. 3) are calculated according to the equations [13–15].

$$R = \sum (\widehat{y} - y)^2 / \sum (y - \overline{y})^2 \tag{1}$$

$$MAE = \sum |y - \widehat{y}| / n \tag{2}$$

$$MSE = \sum (y - \widehat{y})^2 / n \tag{3}$$

## 3    Results and Discussion

### 3.1    Brazil and Germany Exports

An overview of coffee exports - green coffee beans and roasted coffee - of Brazil and Germany in USD value between 1979 and 2019 is presented in Fig. 1.

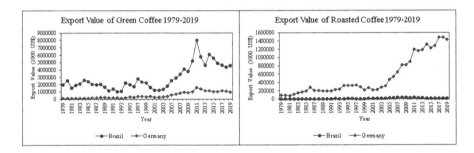

**Fig. 1.** Green and roasted coffee exported between 1979–2019 in USD by Brazil and Germany

Brazil leads the green coffee market while Germany is the main player in roasted coffee market. Note that the peak in roast coffee segment is almost the double of the peak in green coffee segment. In other terms it means that Germany domains premium market while Brazilian sell is based on volume in commodity market [16, 17].

Germany is renowned for supplying quality coffee beans, mainly to Europe. Its privileged location and the existing trade agreements reinforce its sovereignty, making the country a reference in coffee industry. Brazil, on the other hand, appears timidly roasted coffee sales. The country is recognised for exports of commodities and to advance for a most relevant sector of the markets [18], although timid, is a breakthrough for the country.

Despite these figures, Germany buys 20% of the volume of Brazilian green coffee. Considering that the country is not a producer of coffee beans due to climate, there is a clear indication that Brazil is one of the main sources of raw material to Germany coffee industry [19].

Another important aspect is that Brazil coffee industry has not yet achieved commercial advantages for the export of products with higher added value. It is evident that the world's largest producer of green coffee beans has difficulties in adding value to the product and chooses to export commodity which reflects the need of investments in the productive sector [11].

## 3.2 Time Series Regression Model

As mentioned previously, our article investigates a prediction model for green and rosted coffee based on the figures of Brazil and Germany. Table 1 presents the results of Brazil and Table 2 of Germany, as well. Functions excluded means that model is not fit to the research.

**Table 1.** Trend lines for Brazil's green coffee and roasted coffee

| Function | Equation | $R^2$ | MAE | MSE |
|----------|----------|-------|-----|-----|
| Green coffee | | | | |
| Linear | y = 98588x + 791196 | 0.507 | 885083 | 1.32217E+12 |
| Exponential | y = 1E+06e0.0309x | 0.461 | 2861506 | 1.08711E+13 |
| Logarithmic | y = 973966ln(x) + 152625 | 0.264 | 1601363 | 4.54925E+12 |
| Polynomial | y = 5006.3x2 − 111678x + 2E+06 | 0.653 | 723891 | 1.01876E+12 |
| Power | y = 1E+06x0.2917 | 0.220 | 1098970 | 2.26624E+12 |
| Roasted coffee | | | | |
| Linear | y = 584.51x − 3976.8 | 0.517 | 4963 | 44715747 |
| Logarithmic | y = 6976.3ln(x) − 11105 | 0.393 | 10976 | 188292819 |
| Polynomial | y = −5.4821x2 + 814.76x − 5626.9 | 0.522 | 8316 | 116393863 |

To find the appropriate trend line, the value which best approximates to 1 of the coefficient of determination ($R^2$) was considered. For the scenario of green coffee beans in Brazil the trend line is polynomial with a coefficient of determination of 65.34%. For Germany, the appropriate trend line would be the exponential with a coefficient of determination of 89.33%. However, analyzing the numbers, it was found that the situation is not appropriate and we chose to fit the country with the polynomial trend line with a coefficient of determination 81.10%, which is more similar to the situation of the country's export market.

Note that in the scenario of roasted coffee beans Brazil has a polynomial trend line with a coefficient of determination of 52.19% and Germany also has a polynomial trend line with a 93.93% coefficient of determination. In the case

**Table 2.** Trend lines for Germany's green coffee and roasted coffee

| Function | Equation | $R^2$ | MAE | MSE |
|---|---|---|---|---|
| Green coffee | | | | |
| Linear | y = 31011x − 145219 | 0.763 | 156232 | 4.1783E+10 |
| Exponential | y = 78519e0.071x | 0.893 | 7.86154E+12 | 9 |
| Polynomial | y = 734.22x2 + 173.47x + 75780 | 0.811 | 120837 | 3.3346E+10 |
| Power | y = 29935x0.8826 | 0.737 | 182038 | 6.9956E+10 |
| Roasted coffee | | | | |
| Linear | y = 34265x − 193105 | 0.787 | 179287 | 4.4629E+10 |
| Exponential | y = 83859e0.0696x | 0.892 | 5.01286E+12 | 3 |
| Logarithmic | y = 368502ln(x) − 498471 | 0.486 | 579805 | 4.7617E+11 |
| Polynomial | y = 1428.6x2 − 25738x + 236915 | 0.939 | 93241 | 1.2684E+10 |
| Power | y = 33035x0.8605 | 0.728 | 208763 | 8.758E+10 |

of Germany, the relationship between the studied variables is much higher than in the other scenarios.

Polynomial trend lines are more complex and have specific characteristics [12]. They are connected with coffee beans supply chains in which the market is subject to edaphoclimatic changes [20–22]. As remarked in 2011, Fig. 1, for green coffee beans, we realize that arose an increase in the production of Brazilian coffee beans reflected in the market as a whole.

### 3.3 Five-Year Prediction

The forecasts of trend lines based exports for the five-years were made for green and roasted coffee bean for both Germany and Brazil. Table 3 shows the USD values and Fig. 2 compares both countries in both scenarios.

**Table 3.** Five-year coffee prediction in USD

| Coffee type | Country | USD/year | | | | | Growth |
|---|---|---|---|---|---|---|---|
| | | 2020 | 2021 | 2022 | 2023 | 2024 | |
| Green | Brazil | 6,140,637 | 6,454,495 | 6,778,365 | 7,112,248 | 7,456,143 | 18% |
| | Germany | 1,378,230 | 1,440,812 | 1,504,863 | 1,570,382 | 1,637,369 | 16% |
| Roasted | Brazil | 38,263 | 39,544 | 40,836 | 42,138 | 43,452 | 12% |
| | Germany | 1,675,969 | 1,771,662 | 1,870,213 | 1,971,620 | 2,075,885 | 19% |

The five-years predictions allows us to exemplify our model application based on polynomial trend line. Note that according of model Brazil (18%) and Germany (19%) should maintain the lead in their correspondent markets. However,

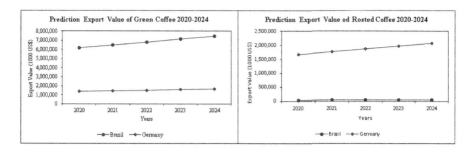

**Fig. 2.** Green and roasted coffee five-year prediction in USD for Brazil and Germany

the growth of Germany (16%) in green coffee market is higher than Brazil in roasted coffee segment. A logical conclusion is an increase of participation of Germany in both analyzed segments.

Finally, the results suggest that Brazilian producers and country government must focus on premium market to obtain a real profitability or will remain a raw material supplier subject in the fluctuations of commodity market. There is a growth forecast as coffee beverage consumption is increasing worldwide. Several studies have been pointing an increase in the consumption of specialty, gourmet and organic coffees [8–11,23] confirming the necessity of a change for players willing to meet these new demands.

## 4    Final Remarks and Outlook

This article aimed to estimate the time series for the export trade of green and roasted coffee beans considering Brazil and Germany scenario making a prediction for five years using an historical series of 41 years.

The results indicated that for both scenarios: green bean exports and roasted bean exports, the trend lines are polynomials which is consistent with agricultural production. The agricultural production of green coffee bean is conditioned by climate and relief factors which directly influence the quantity and quality of the production. Note that may be influenced by market demand that by opting for different types of coffee beans leverages production to meet demand.

In relation to the forecasts made for exports of green and roasted coffee beans for five years, it can be seen that the values are coherent with the growth in coffee drink consumption and its specificities that have been increasing over time. As a study limitation, the analysis of only 2 countries may not show the true world scenario, therefore the addition of other countries would bring better overview.

**Acknowledgments.** This study was financed in part by the Coordenação de Aperfeioamento de Pessoal de Nível Superior Brasil (CAPES). Finance Code 001.

# References

1. Ortigoza, S.A.G.: Paisagens do Consumo: São Paulo, Lisboa, Dubai e Seul. Editora UNESP, São Paulo (2010)
2. Secretaria Municipal de Educação e Esporte de Goiânia. https://sme.goiania.go.gov.br/conexaoescola/ensino_fundamental/historia-do-comercio-escambo/
3. World Trade Organization: WTO Data. https://data.wto.org/
4. Maranhão, R.L.A., Vieira Filho, J.E.R.: Inserção Internacional do Agronegócio Brasileiro. IPEA, Brasília (2017)
5. Food and Agriculture Organization of the United Nations. http://www.fao.org/home/en/
6. International Coffee Organization. http://www.ico.org/
7. Food and Agriculture Organization of the United Nations (2021). http://www.fao.org/faostat/en/#data/TP
8. Lee, Y., Bateman, A.: The competitiveness of fair trade and organic versus conventional coffee based on consumer panel data. Ecol. Econo. **184**, 106986 (2021)
9. Naegele, H.: Where does the Fair Trade money go? How much consumers pay extra for Fair Trade coffee and how this value is split along the value chain. World Dev. **133**, 105006 (2020)
10. Vogt, M.A.B.: Developing stronger association between market value of coffee and functional biodiversity. J. Environ. Manag. **269**, 110777 (2020)
11. Conceição, J.C.P.d., Ellery Junior, R.G.d., Conceição, P.H.Z.: Cadeia agroindustrial do café no brasil: Agregação de valor e exportação. Boletim de Economia e Política Internacional (24), 37–47 (2019)
12. Mc Fredries, P.: Fórmulas e Funções no Microsoft Excel. Ciência Moderna, Rio de Janeiro (2005)
13. Abraham, E.R., et al.: Time series prediction with artificial neural networks: an analysis using Brazilian soybean production. Agriculture **10**(10), 475 (2020)
14. Escolano, N.R., Espin, J.J.L.: Econometría: Series temporales y modelos de ecuaciones simultáneas. Editorial UMH, Alicante (2016)
15. Pecar, B., Davis, G.: Time series based predictive analytics modelling: using MS excel, Gloucestershire (2016)
16. Angeloni, G., et al.: What kind of coffee do you drink? An investigation on effects of eight different extraction methods. Food Res. Int. **116**, 1327–1335 (2019)
17. Conselho dos Exportadores de Café do Brasil. https://www.cecafe.com.br/publicacoes/
18. Mendonça, V.U.V.M.: Caracterização da Atividade de Exportação de Commodities Agrícolas no Brasil, São Paulo (2015)
19. Kleemann, J., et al.: Quantifying interregional flows of multiple ecosystem services - a case study for Germany. Glob. Environ. Change **61**, 102051 (2020)
20. Badmos, S., Fu, M., Granato, D., Kuhnert, N.: Classification of Brazilian roasted coffees from different geographical origins and farming practices based on chlorogenic acid profiles. Food Res. Int. **134**, 109218 (2020)
21. Läderach, P., et al.: Systematic agronomic farm management for improved coffee quality. Field Crop Res **120**(3), 321–329 (2011)
22. World Intellectual Property Organization: World intellectual property report 2017: intangible capital in global value chains. World Intellectual Property Organization, Geneva (2017)
23. Hindsley, P., McEvoy, D.M., Morgan, O.A.: Consumer demand for ethical products and the role of cultural worldviews: the case of direct-trade coffee. Ecol. Econ. **177**, 106776 (2020)

# Scheduling of Parallel Print Machines with Sequence-Dependent Setup Costs: A Real-World Case Study

Manuel Iori[1] , Alberto Locatelli[1]([⊠]) , and Marco Locatelli[2]

[1] Department of Sciences and Methods for Engineering, University of Modena and
Reggio Emilia, Reggio Emilia, Italy
{manuel.iori,alberto.locatelli}@unimore.it
[2] Department of Engineering and Architecture, University of Parma, Parma, Italy
marco.locatelli@unipr.it

**Abstract.** In the present work, we consider a real-world scheduling problem arising in the color printing industry. The problem consists in assigning print jobs to a heterogeneous set of flexographic printer machines, as well as in finding a processing sequence for the sets of jobs assigned to each printer. The aim is to minimize a weighted sum of total weighted tardiness and total setup times. The machines are characterized by a limited sequence of color groups and can equip additional components (e.g., embossing rollers and perforating rolls) to process jobs that require specific treatments. The process to equip a machine with an additional component or to clean a color group takes a long time, with the effect of significantly raising the setup costs. Furthermore, the time required to clean a color group between two different jobs depends directly on the involved colors. To tackle the problem, we propose a constructive heuristic followed by some local search procedures that are used one after the other in an iterative way. Extensive tests on real-world instances prove that the proposed algorithm can obtain very good-quality solutions within a limited computing time.

**Keywords:** Parallel machine scheduling · Flexographic printing process · Sequence-dependent setup costs · Food packaging industry

## 1 Introduction

Over the past years, the food packaging has undergone a remarkable evolution. The choice of materials and graphic design for the food packages has become more and more important to capture and establish a strong and lasting bond

We acknowledge financial support from Istituto Stampa s.r.l. (Italy). The work originates from the daily activity of Istituto Stampa s.r.l., a company whose headquarters is located in Reggio Emilia (Italy). The company operates in the field of packaging industry by producing and printing packaging materials for food products since 1933.

© IFIP International Federation for Information Processing 2021
Published by Springer Nature Switzerland AG 2021
A. Dolgui et al. (Eds.): APMS 2021, IFIP AICT 631, pp. 637–645, 2021.
https://doi.org/10.1007/978-3-030-85902-2_68

with the customers (see, e.g., Paine and Paine [5]). As a result, the color printing industry has had to adapt to the increasingly demanding requirements from the food industry and its costumers. Flexography and rotogravure printing are the most used technologies for printing flexible food packages. Rotogravure printing used to guarantee a better quality, but now flexography has reached a comparable quality. In addition, flexography is less expensive, faster, and allows for printing on almost any material such as paper, plastic, aluminum foil, and cellophane.

A flexographic printer machine is characterized by a limited sequence of color groups, by a set of additional components (e.g., embossing rollers and perforating rolls) that can be mounted to process jobs that require specific treatments, and by its washing system that can be manual or automatic.

As reported by Schuurman and Van Vuuren [7], the printing process of a printing job consists of two phases: the *machine setup phase* and the *printing phase*. The first phase consists in preparing the machine for printing a new job and may require to remove material from the machine, to equip the machine with an additional component, and to wash and refill color groups. If two consecutively scheduled printing jobs require significantly different color overlays, then substantial down times are incurred. The process to wash a color group and to refill it with another ink takes a long time and depends directly on the specific colors involved. The time required for the *machine setup phase* is job dependent, machine dependent, and job sequence-dependent. Indeed, it depends on the colors and additional components required by the new printing job, on the colors and additional components available in the printing machine and required by the previous printing jobs, and by the washing system type of the machine. On the other hand, the printing phase is just job dependent and machine dependent. More specifically, it depends on print volumes and complexity of the job, on the type of printed material, and on the printing speed of the machine.

Motivated by a real-world application, we consider a scheduling problem consisting in assigning printing jobs to a heterogeneous set of parallel flexographic printer machines, as well as in finding a processing sequence for the sets of jobs assigned to each printer. In accordance with the company, the objective is to minimize the weighted sum of total weighted tardiness and total setup times. We refer to this scheduling problem as the *Parallel Print Machine with Setup Costs Problem* (PMSCP). The PMSCP is a variant of the unrelated parallel machines with sequence-dependent setup times (see Pinedo [6]), in which the optimization criterion is the minimization of a weighted sum of total weighted tardiness and total setup times. Using the three-field notation for scheduling problems by Graham et al. [2], the PMSCP can be denoted as the $R|ST_{sd}|\alpha WT + \beta TST$ (see Allahverdi [1]), where $WT$ denotes the total weighted tardiness, $TST$ the total setup time, and $\alpha$ and $\beta$ are two input parameters. The PMSCP is strongly $\mathcal{NP}$-hard since it is a generalization of the weighted tardiness scheduling problem (see Lenstra et al. [3]). In this work, we solve the problem by means of a constructive heuristic followed by some local search procedures. The resulting algorithm is capable of finding good-quality solutions within a limited computing time on several instances derived from the real-world case study.

The remainder of the paper is organized as follows. In Sect. 2, we formally describe the PMSCP. In Sect. 3, we present the heuristic algorithm. In Sect. 4, we provide the outcome of our computational experiments. Finally, in Sect. 5, we draw some concluding remarks.

## 2  Problem Description

In this section, the PMSCP is formally described. The following sets are defined:

- $M = \{1, \ldots, m\}$: set of printing machines;
- $J = \{1, \ldots, n\}$: set of printing jobs;
- $C$: set of colors available;
- $A$: set of additional components (i.e., embossing rollers and perforating rolls) available to the printing company.

A printing job $j \in J$ is characterized by the following elements:

- $M_j \subseteq M$: subset of machines capable of printing $j$;
- $n_j$: number of colors of $j$;
- $C_j = \{c_1, \ldots, c_{n_j}\} \subseteq C$: subset of colors required by $j$;
- $A_j \subseteq A$: subset of additional components required by $j$;
- $v_j$: length of $j$, measured in linear meters;
- $d_j$: due date of $j$;
- $w_j$: tardiness penalty assigned to $j$ for each day of tardiness.

A flexographic printing machine $i \in M$ is characterized by:

- $J_i \subseteq J$: subset of jobs that can be printed by machine $i$;
- $\theta_i$: average printing speed of machine $i$;
- $p_{ij}$: processing time of job $j$ on machine $i$;
- $g_i$: number of color groups of machine $i$;
- $\ell_i$: washing system type of machine $i$, automatic ($\ell_i = 1$) or manual ($\ell_i = 0$).

The state of machine $i \in M$, after having finished job $j \in J$, is characterized by:

- $A_j^i \subseteq A$: subset of additional components installed on $i$;
- $d^i$: availability date of machine $i$, i.e., the moment in which the machine will finish processing $j$.

The state of a machine $i$ depends not only on the last processed job $j$, but also on all the jobs processed by $i$ before $j$. For instance, if the job processed before $j$ requires an additional component $a \in A$, whereas $j$ does not require it, then the component might be simply left on the machine. If, instead, $j$ requires a different component $a'$, then $a$ must be switched with $a'$. For this reason, the state of machine $i$, after having finished job $j$, has a great impact on $s_{jk}^i$, which is the setup time on machine $i \in M$, when processing job $k \in J$ after having finished job $j \in J$, and which might also depend on the sequence of jobs previously assigned to the machine. This complicates the computation of a heuristic solution and makes the PMSCP even more challenging to solve in practice.

Each machine $i \in M$ is available from Monday to Friday for 8 working hours per day. It can be simply paused at the end of each work shift and resumed the day after without any penalty cost (even during the *printing phase*).

A sequence of jobs $S^i = (j_1^i, \ldots, j_{n_i}^i)$ is feasible for machine $i \in M$ if $i \in M_j$ $\forall j \in S^i$. A feasible sequence $S^i$ of jobs defines a schedule for machine $i \in M$ and can be used to compute start and completion times of all jobs. Let the $T_j$ denote the number of days of tardiness of a job $j$ in $S^i$. Because of the sequence-dependent setups based on the colors and on the additional components, even computing $T_j$ is NP-hard (see, Meunier and Neveu [4], for a similar problem). In our work, we decided to adopt a simple but quick and good-enough evaluation of the setup costs (and hence of the tardiness), as described below in the next section.

A set of feasible sequences $S = \{S^i : i \in M\}$ is a feasible schedule for $J$ if it forms a partition of $J$. The PMSCP consists in finding a feasible schedule that minimizes $f = \alpha WT + \beta TST$, where $WT = \sum_{i \in M} \sum_{j \in S^i} w_j T_j$ and $TST = \sum_{i \in M} \sum_{h=1}^{n_i-1} s_{j_h^i j_{h+1}^i}^i$. With a slight abuse of notation, given a feasible sequence of jobs $S^i = (j_1^i, \ldots, j_{n_i}^i)$ for machine $i \in M$, we let $f(S^i)$ denote the value $\alpha \cdot \sum_{j \in S^i} w_j T_j + \beta \cdot \sum_{h=1}^{n_i-1} s_{j_h^i j_{h+1}^i}^i$.

## 3   Heuristic Approach

Given a machine $i \in M$ and two jobs $j, k \in J_i$, the setup cost $s_{jk}^i$ is evaluated as follows. A color washing time is added to $s_{jk}^i$ for each color $c \in C_k \backslash C_j$ (in the case $\ell_i = 0$) or if $|C_k \backslash C_j| \geq 1$ (in the case $\ell_i = 1$). Indeed, if a printer machine equips an automatic washing systems, the color groups can be washed in parallel, otherwise they have to be washed one by one. Moreover, if there is a $c \in C_k \backslash C_j$ corresponding to white or a special varnish, an additional penalty time is considered. Indeed, the process to wash a color group and to refill it with white ink or a special varnish requires complete cleaning of the corresponding color group with the effect of a significantly raising of the setup costs. Finally, for each additional component $a \in A_k \backslash A_j^i$ (i.e., set of additional components required by $k$ that are not mounted in $i$ after the finish of job $j$), a component changing time is added to $s_{jk}^i$. On the other hand, the time required by machine $i$ to print job $j \in J_i$ is calculated by multiplying $v_j$ by $\theta_i$. Given a feasible sequence of jobs $S^i$, the time required for the *machine setup phase* and the *printing phase* are calculated, for each job $j \in S^i$ as described above. Thus, since the working time of the machines is known and the jobs in $S^i$ are performed one after another, start and completion times of all jobs in $S^i$ can be directly calculated.

A constructive greedy heuristic algorithm (CGHA) to solve the PMSCP is presented in Algorithm 1. The proposed method assigns, at each iteration, a print job to a first available machine, generating, for each machine $i \in M$, a feasible sequence of jobs $S^i$. From Step 5 to Step 11, a set $L_i$ of jobs is created by selecting from $J_i$ the jobs behind schedule, i.e., $\{j \in J_i : d_j < d^i\}$. If there is no job behind schedule, $L_i$ is created selecting from $J_i$ the feasible jobs that can only be printed by machine $i$, i.e., $\{j \in J_i : |M_j| = 1\}$. If $L_i$ is still empty, then we set $L_i = J_i$. At Step 13, the greedy phase of the algorithm adds to $S^i$ the most convenient job $j \in L_i$. At step 16, the state of machine $i$ is updated. At Step 18, the obtained output $S = \{S^i : i \in M\}$ is a feasible schedule for $J$ by construction.

```
1  input: M = {1,...,m}, J = {1,...,n};
2  S^i = ∅ (i ∈ M);
3  while J ≠ ∅ do
4  │   select a machine i ∈ M such that d^i = min_{i'∈M}{d^{i'} : J_{i'} ≠ ∅} and J_i ≠ ∅;
5  │   L_i := {j ∈ J_i : d_j < d^i};
6  │   if (L_i = ∅) then
7  │   │   L_i = {j ∈ J_i : |M_j| = 1};
8  │   end
9  │   if (L_i = ∅) then
10 │   │   L_i = J_i;
11 │   end
12 │   let S^i_j be the sequence obtained adding job j at the end of sequence S^i;
13 │   select a job j ∈ L_i such that f(S^i_j) = min_{j'∈L_i} f(S^i_{j'});
14 │   add j at the end of sequence S^i;
15 │   remove job j from J;
16 │   update d^i, A^i_j;
17 end
18 return S = {S^i : i ∈ M}.
```
**Algorithm 1:** Constructive greedy heuristic algorithm (CGHA)

In order to search for improved solutions, three local search procedures, namely LS1, LS2, and LS3, are applied to the initial solution $S$ obtained by the CGHA.

LS1 is based on an intra-machine swap neighborhood, which selects a machine $i \in M$, two job positions $h$ and $g$ ($1 \leq h < g \leq n_i$) in the sequence of jobs $S^i$, and then generates a new sequence $S^i_{(h,g)} = (j^i_1, \ldots, j^i_g, \ldots, j^i_h, \ldots, j^i_{n_i})$ by switching job $j^i_h$ with job $j^i_g$. The generated solution is better then the current one if $f(S^i_{(h,g)}) < f(S^i)$.

LS2 is based on an inter-machine insertion neighborhood, which selects two machines $i, i' \in M$ and two job positions $h$ ($1 \leq h \leq n_i$) and $g$ ($1 \leq g \leq n_{i'}$) in the sequences of jobs $S^i$ and $S^{i'}$, respectively, and, if $i' \in M_{j^i_h}$, removes job $j^i_h$ from $S^i$ (generating $S^i_{(j^i_h,h-)} = (j^i_1, \ldots, j^i_{h-1}, j^i_{h+1}, \ldots, j^i_{n_i})$) and inserts $j^i_h$ before the $g$-th position of $S^{i'}$ (generating $S^{i'}_{(j^i_h,g+)} = (j^{i'}_1, \ldots, j^{i'}_{g-1}, j^i_h, j^{i'}_g, \ldots, j^{i'}_{n_{i'}})$). The generated solution is better then the current one if $f(S^i_{(j^i_h,h-)}) + f(S^{i'}_{(j^i_h,g+)}) < f(S^i) + f(S^{i'})$.

Given a feasible sequence of jobs $S^i = (j^i_1, \ldots, j^i_{n_i})$ and two job positions $h$ and $g$ ($1 \leq h < g \leq n_i$) of $S^i$, a sub-sequence of consecutive jobs $S^i_{(h-g)} = (j^i_h, \ldots, j^i_g)$ is $t$-maximal if $s^i_{j_{h-1}j_h} > t$ (or $h = 1$), $s^i_{j_g j_{g+1}} > t$ (or $g = n_i$), and $s^i_{j,j+1} \leq t$ (for each $j = h, \ldots, g - 1$). Stated in another way, a $t$-maximal sub-sequence is a maximal sequence of consecutive jobs to be executed on machine

$i$ whose setup times do not exceed $t$. By definition, a $t$-maximal sub-sequence $S^i_{(h-g)} = (j^i_h, \ldots, j^i_g)$ of $S^i$ is composed by at least of two jobs.

In LS3, firstly, for each $i \in M$, a list $L_i$ of all $t$-maximal sub-sequences of $S^i$ is created. LS3 is based on an intra-machine sub-sequence insertion neighborhood, which selects a machine $i \in M$, a $t$-maximal sub-sequence $S^i_{(h-g)}$ in $L_i$, a job position $k$ ($1 \leq k < h$ or $g < k \leq n_i$), and then generates a new sequence $S^i_{k,(h-g)} = (j^i_1, \ldots, j^i_{k-1}, j^i_h, \ldots, j^i_g, j^i_k \ldots, j^i_{n_i})$ by inserting the sub-sequence $S^i_{(h-g)}$ before the $k$-th position of $S^i$. The generated solution is better then the current one if $f(S^i_{k,(h-g)}) < f(S^i)$.

All three local search procedures operate in a first-improvement manner. For each $i \in M$, all the possible moves are considered, evaluating all the solutions in a neighborhood. As soon as an improving solution is found, the current solution $S$ is updated and the local search is re-executed on $S$. If, instead, no improving solution is found, then the local search returns the current local optimal solution.

All the described methods are combined in an improved single heuristic algorithm (ISHA). ISHA finds a first feasible solution $S$ by means of CGHA. Then, after the constructive phase is completed, $S$ is brought to a local minimum by sequentially invoking LS1, LS2, and LS3, in this order.

## 4   Computational Results

In this section, computational experiments are performed on various instances encountered in industry practice. In particular, the solutions obtained by means of the heuristic methods described in Sect. 3 are compared among them. All

**Table 1.** Computational results of CGHA and ISHA

| Inst. | $|J|$ | $|M|$ | CGHA | | | | ISHA | | | |
|---|---|---|---|---|---|---|---|---|---|---|
| | | | $WT$ | $TST$ | $f.o.$ | $run\ t.$ | $WT$ | $TST$ | $f.o.$ | $run\ t.$ |
| I01 | 61 | 10 | 4812 | 8460 | 7001 | 0.05 | 2587 | 7635 | 5616 | 4.61 |
| I02 | 90 | 10 | 9322 | 12895 | 11466 | 0.07 | 5203 | 11725 | 9116 | 19.19 |
| I03 | 87 | 10 | 19127 | 12620 | 15223 | 0.07 | 13632 | 10890 | 11987 | 13.42 |
| I04 | 106 | 10 | 15987 | 15520 | 15707 | 0.11 | 9241 | 13680 | 11904 | 20.92 |
| I05 | 96 | 11 | 18110 | 13530 | 15362 | 0.08 | 15375 | 12125 | 13425 | 12.66 |
| I06 | 89 | 10 | 8112 | 12730 | 10883 | 0.08 | 3598 | 10555 | 7772 | 14.45 |
| I07 | 89 | 10 | 18792 | 14125 | 15992 | 0.09 | 15187 | 11850 | 13185 | 13.04 |
| I08 | 78 | 10 | 4312 | 10315 | 7914 | 0.09 | 2159 | 9625 | 6639 | 8.77 |
| I09 | 94 | 10 | 15796 | 13510 | 14425 | 0.08 | 12436 | 12170 | 12277 | 17.54 |
| I10 | 104 | 10 | 7260 | 14335 | 11505 | 0.10 | 5027 | 12380 | 9439 | 19.19 |
| I11 | 95 | 10 | 26136 | 13940 | 18819 | 0.08 | 21327 | 12300 | 15911 | 14.16 |
| I12 | 103 | 10 | 18898 | 13400 | 15599 | 0.09 | 14456 | 12190 | 13096 | 16.86 |
| AVG | | | 13889 | 12948 | 13324 | 0.08 | 10019 | 11427 | 10864 | 14.57 |
| GAP | | | | | | | $-33\%$ | $-12\%$ | $-19\%$ | |

the algorithms are coded in Python version 3.9.2 and run on a computer with Intel(R) Core (TM) i7-10510U with CPU 1.80 GHz and RAM 16 GB, using Windows 10 Pro 64-bit.

We consider 12 real-world instances, each corresponding approximately to a week of production. Some randomization was applied to all instances in order to meet the company privacy requirements. The number of jobs spans from 61 to 106, while the number of machines is 10 or 11. According to the company interests, the parameters $\alpha$ and $\beta$ of the objective function are set to 0.4 and 0.6, respectively. The parameter $t$ in the intra-machine sub-sequence insertion moves is set to 125, since it has been seen that, with this value, LS3 produces good-quality solutions. In Table 1, the solutions obtained by means of CGHA and ISHA are compared. Columns $|J|$ and $|M|$ refer to the size of the instances in terms of number of jobs and number of machines, respectively. Entries in columns $WT$, $TST$, and $f.o.$ exhibit the sum of total weighted tardiness, the total setup times, and the value of the objective function, respectively. Finally, entries in columns $run\ t.$ exhibit the run-time required to solve the instances (expressed in seconds). The mean of the percentage gaps from the solution values obtained by means of CGHA are reported in the row GAP, while entries in row AGV exhibit the average values of each column. Table 1 shows that CGHA is able to find a solution in a very short time (on average, less than a tenth of a second) and, in terms of quality, it is similar to the solution produced by the company. ISHA is always able to significantly improve the quality of the initial solution found by means of CGHA. More precisely, the value of the objective function improves by an average of 19%, with an average percentage improvement of the weighted tardiness and of total setup times equal to 33% and 12%, respectively.

**Table 2.** Comparative evaluation of the different locals search procedures

| Inst. | $|J|$ | $|M|$ | CGHA | | CGHA+LS1 | | CGHA+LS2 | | CGHA+LS3 | | ISHA | |
|---|---|---|---|---|---|---|---|---|---|---|---|---|
| | | | $f.o.$ | $run\ t.$ | $f.o.$ | $run\ t.$ | $f.o.$ | $run\ t.$ | $f.o.$ | $run\ t.$ | $f.o.$ | $run\ t.$ |
| I01 | 61 | 10 | 7001 | 0.05 | 5809 | 1.46 | 6565 | 3.86 | 6784 | 0.41 | 5616 | 4.61 |
| I02 | 90 | 10 | 11466 | 0.07 | 9585 | 7.81 | 10791 | 8.58 | 10838 | 1.06 | 9116 | 19.19 |
| I03 | 87 | 10 | 15223 | 0.07 | 12656 | 8.47 | 14083 | 4.79 | 13741 | 3.11 | 11987 | 13.42 |
| I04 | 106 | 10 | 15707 | 0.11 | 12438 | 9.84 | 14567 | 13.52 | 14058 | 1.26 | 11904 | 20.92 |
| I05 | 96 | 11 | 15362 | 0.08 | 14073 | 2.42 | 14379 | 12.73 | 14556 | 0.67 | 13425 | 12.66 |
| I06 | 89 | 10 | 10883 | 0.08 | 8304 | 5.64 | 9163 | 12.36 | 9916 | 0.51 | 7772 | 14.45 |
| I07 | 89 | 10 | 15992 | 0.09 | 14229 | 3.64 | 14643 | 10.44 | 14656 | 0.78 | 13185 | 13.04 |
| I08 | 78 | 10 | 7914 | 0.09 | 6998 | 3.12 | 7490 | 12.13 | 7751 | 0.28 | 6639 | 8.77 |
| I09 | 94 | 10 | 14425 | 0.08 | 12695 | 12.04 | 13619 | 14.41 | 13583 | 0.50 | 12277 | 17.54 |
| I10 | 104 | 10 | 11505 | 0.10 | 10217 | 4.76 | 10443 | 19.41 | 10686 | 0.45 | 9439 | 19.19 |
| I11 | 95 | 10 | 18819 | 0.08 | 16397 | 4.03 | 13046 | 14.05 | 18029 | 0.34 | 15911 | 14.16 |
| I12 | 103 | 10 | 15599 | 0.09 | 13708 | 5.96 | 14273 | 18.62 | 14491 | 0.51 | 13096 | 16.86 |
| AVG | | | 13324 | 0.08 | 11426 | 5.76 | 11922 | 12.08 | 12424 | 0.82 | 10864 | 14.57 |
| GAP | | | | | −14% | | −10% | | −6% | | −19% | |

Table 2 highlights the results of the second round of experiments, which aims at analyzing the impact of the different local searches (described in Sect. 3) with respect to the initial solution computed by CGHA. In CGHA+LS1, CGHA+LS2, and CGHA+LS3, after the constructive phase is completed, the initial solution is brought to a local minimum by means of LS1, LS2, and LS3, respectively. The results demonstrate that LS1 is able to produce on average the largest improvement in therms of quality solution. Indeed, the value of the objective function improves by 14% with LS1, by 10% with LS2, and by 6% with LS3. This result reveals the strongly sequence-dependent nature of the problem. Indeed, thanks to the intra-machine swap moves of LS1, each sequence of jobs can be re-sorted, obtaining a large improvement of the initial solution. The slight improvements brought by LS3 are obtained in a very short time (on average, less than one second), thus LS3 is able to provide a good trade-off between the neighborhood size (hence time needed to explore it) and its effectiveness. On the other hand, LS2 is the most time-consuming local search, reflecting the fact that the LS2 neighborhood size is the largest one.

## 5    Conclusions

In this paper, we presented a local search approach for the PMSCP. Firstly, we proposed a constructive greedy procedure (CGHA) for the generation of a good initial solution and then a local search-based approach (ISHA) which combines CGHA with three different local search procedures. Experiments over real-world instances confirmed that the proposed methods are able to find good-quality solutions and can provide a quick support to the company on its weekly decisions. More precisely, the computational results show that CGHA is able to find solutions in a very short time, which, in therms of quality, are similar to the solutions produced by the company. Furthermore, ISHA is always capable to significantly improve the quality of the initial solution (by an average of about 20%) within small computing times (less than 20 s). These good results were confirmed in practice. Indeed, ISHA is currently used during the weekly production scheduling by the company, which attested that ISHA is capable of producing good quality results.

It is worthwhile to remark that the computing times of all the proposed approaches, including the full ISHA approach, are currently not a major issue. Indeed, taking into account that the planned activities cover about one week, larger computing times are feasible and further refinements of the proposed approaches are possible. In particular, as a possible topic for future research, we are interested in introducing some perturbation mechanism inside ISHA to help the search to escape from local optima solutions and to bring diversification into the search.

# References

1. Allahverdi, A.: The third comprehensive survey on scheduling problems with setup times/costs. Eur. J. Oper. Res. **246**(2), 345–378 (2015)
2. Graham, R., Lawler, E., Lenstra, J., Kan, A.: Optimization and approximation in deterministic sequencing and scheduling: a survey. Ann. Discrete Math. **5**(C), 287–326 (1979)
3. Lenstra, J., Rinnooy Kan, A., Brucker, P.: Complexity of machine scheduling problems. Ann. Discrete Math. **1**(C), 343–362 (1977)
4. Meunier, F., Neveu, B.: Computing solutions of the paintshop-necklace problem. Comput. Oper. Res. **39**(11), 2666–2678 (2012)
5. Paine, F.A., Paine, H.Y.: A Handbook of Food Packaging. Springer, Boston (2012)
6. Pinedo, M.: Scheduling. Springer, New York (2012). https://doi.org/10.1007/978-1-4614-2361-4
7. Schuurman, J., Van Vuuren, J.H.: Scheduling sequence-dependent colour printing jobs. S. Afr. J. Ind. Eng. **27**(2), 43–59 (2016)

# Use of Paraconsistent Logic Evidential Annotated Eτ in Logistic Systems

Liliam Sayuri Sakamoto$^{(\boxtimes)}$ , Jair Minoro Abe, Luiz Antonio de Lima ,
Jonatas Santos de Souza , Nilson Amado de Souza ,
and Angel Antonio Gonzalez Martinez

Paulista University, 1212, Dr. Bacelar Street, São Paulo, SP, Brazil

**Abstract.** Traditional Logistic Systems related to the modules of routes and vehicle tracking were observed, which use classical logic. In this study, whose main objective is a proposal for digital transformation with the implementation of the use of Paraconsistent Logic Evidential Annotated Eτ for optimization and reduction exposure to the risk of the cargo transport process in these types of systems. The methodology used para-analyzer algorithm, with selection of specialists in decision-making issues to improve the safety of the routes and minimize the risks exposed in the cargo's path from shipment to the destination. An exploratory research was used surveys through applied logical questionnaires. The main results were compared to the Case Study of the Volkswagen Modular Consortium - Trucks and buses, from a logistical point of view, in which the three main factors resulting from this study can prove an alignment regarding the implementation of this innovation. In addition, data on incidents occurred at the Agrobusiness Company regarding the Routes and Tracking Systems were presented, proving the risk of the loads and the need for digital transformation in these types of systems.

**Keywords:** Non-classic logic · Annotated evidential paraconsistent logic · Tracking systems · Route systems

## 1 Introduction

### 1.1 General Context

The objective of this study is to analyze the traditional Logistics Systems in the use of vehicle tracking modules that use processes based on classical logics, with a proposal to implement the use of the Paraconsistent Logic Evidential Annotated Eτ for the optimization and reduction of exposure to the risk of the cargo transport process.

In many areas, many companies have not yet realized how the digital transformation paradigm provides a profound change in the layers of the business, in which companies that wish to obtain a competitive advantage will need to adapt themselves in the implementation of innovations. There is no doubt that a "digital gap" has been growing among those that have become stagnant and those that are adapting [1]. And in the field of Logistics this could not be different, as most of these companies use Logistics

© IFIP International Federation for Information Processing 2021
Published by Springer Nature Switzerland AG 2021
A. Dolgui et al. (Eds.): APMS 2021, IFIP AICT 631, pp. 646–654, 2021.
https://doi.org/10.1007/978-3-030-85902-2_69

Systems with tracking models, some with the support of mobile applications to define routes for the transportation of various loads. These routes can be designed manually or automatically, but these route models are quite predictable [2], mainly because they use models based on classical logic. The support or monitoring tools generally consider only the classic approach, that is, they are based on the duality of whether it is a risky situation or not, however there is no study that involves the contradictions and the doubts that in most times are despised.

This dependence also causes the need for the digital transformation of this resource, which can be of its own IT structure or for a service provision by a specialized company. As [3] approaches, information technology and innovation help supply within a Value Chain.

The use of the Paraconsistent Logic Evidential Annotated Eτ with the practical application of the Para-Analyzer algorithm for decision making, that is, with the implementation of artificial intelligence can be one of the important points for innovation and digital transformation of logistics.

The study is structured starting with a bibliographic review of the Paraconsistent Logic Evidential Annotated Eτ and the Logistic Supply Chain, after presenting the methodology of the Para-analyzer algorithm, and in the discussions, comparisons were made with the Volkswagen Modular Case Study - Trucks and buses from the point from a logistical point of view, presented by [3] and with real situations of Tracking Systems and Route System of an agribusiness cargo transport company.

## 2 Background

### 2.1 Paraconsistent Logic Evidential Annotated Eτ in Logistic

The Paraconsistent Logic Evidential Annotated Eτ belongs to the class of non-classical evidential logics that arose specifically in logical programming, according to [4].

Paraconsistent Logic Evidential Annotated Eτ is a family of non-classical logics that emerged in the late 90s of the last century in logical programming [5].

Annotated logics constitute a class of paraconsistent logic. Such logics are related to certain complete lattices, which play an important role. A knowledge specialist on the subject to be addressed issues a quantitative opinion ranging from 0.0 to 1.0. These values are respectively the favorable evidence that is expressed by the symbol $\mu$ and the opposite evidence by $\lambda$.

Programs can be built using paraconsistent logic, making it possible to treat inconsistencies in a direct way. With this resource, one must apply in big data, expert systems, object-oriented database, representation of contradictory knowledge, with all the implications in artificial intelligence [5].

With the uncertainty and certainty degrees, they can get the following 12 output states (Table 1): extreme states, and non-extreme states [7].

**Table 1.**  Extreme and non-extreme states

| Extreme states | Symbol |
|---|---|
| True | V |
| False | F |
| Inconsistent | T |
| Paracomplete | ⊥ |
| Non-extreme states | Symbol |
| Quasi-true tending to inconsistent | QV → T |
| Quasi-true tending to paracomplete | QV → ⊥ |
| Quasi-false tending to inconsistent | QF → T |
| Quasi-false tending to paracomplete | QF → ⊥ |
| Quasi-inconsistent tending to true | QT → V |
| Quasi-inconsistent tending to false | QT → F |
| Quasi-paracomplete tending to true | Q⊥ → V |
| Quasi-paracomplete tending to false | Q⊥ → F |

Some additional control values are:

- Vscct = maximum value of uncertainty control = Ftun
- Vscc = maximum value of certainty control = Ftce
- Vicct = minimum value of uncertainty control = −Ftun
- Vicc = minimum value of certainty control = −Ftce

All states are represented in the next Figure (Fig. 1).

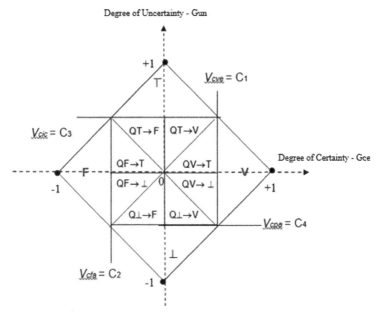

**Fig. 1.** Extreme and Non-extreme states that represent Table 1. Source: [7]

## 2.2 Logistic Supply Chain

Despite the operational environment of a logistical supply chain, it has the economic limiting factor and not only technological, according to [6], which affirms the need for speed of availability and consistency of delivery.

When assessing Supply Chain management, it is necessary to understand the impacts of decision making on processes, organization, and the social context [3].

In the Transport process, it is necessary to detail the functionalities and characteristics of the participants effectively. In which the transport structure can be rail, road, waterway, pipeline, or air [6]. In this study the focus will be on the type of road transport that can be monitored through routes determined automatically or manually.

There are also specialized transport services for: parcel services, intermodal transport, and non-operational intermediaries [6].

Generally, they can be specialized in precious cargo (jewelry, watches, electronic equipment) or valuable (paper money, pharmaceuticals), and these loads are often criminally intercepted and subtracted. Therefore, the need for a digital transformation and optimization of the routing process, which can come from projects focused on big data, IoT - Internet of Things and everything stored in a cloud [2].

In this transport sector, the financial factor is evaluated through the distance of the route, the weight and density of the product, totaling its total cost [6].

The current situation leads to the improvement of cost minimization so that there is a greater demand for product delivery. With the prioritization of a shorter, safer route, with less fuel consumption and less labor time, and with a reduced chance of subtraction or theft, generating implicit value in product delivery, and may even focus on applications furniture too.

The control must be carried out by the Transport Administration area, both in the operational scope and in the management of complaints [6]. Burglary incidents prior to final delivery can lead to financial loss, internal fraud, or theft. With the digital transformation in conjunction with the use of logic, it is possible to optimize results and increase the revenue of these logistics companies, this is a suggested solution that leads to the digital transformation expressly said within a logistics chain.

## 3   Methodology

In the first phase, a bibliographic review was carried out on the aspects of Logistics and the need for digital transformation, complemented using the Paraconsistent Logic Evidential Annotated Eτ be implemented to aid decision making [7].

The Paraconsistent Logic Evidential Annotated Eτ considers a proposition being represented by annotation values or states (Table 1). According to this concept, an algorithm called para-analyzer was created [5].

The research problem was: How to manage unstable and risky systems for cargo transportation?

The defined proposition was: Do logistics systems need to undergo digital transformation?

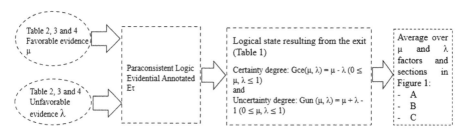

**Fig. 2.**  Para-analyzer algorithm function. Source: Adapted from [7]

Knowledge Engineers - KE, Farms users (foreman, veterinarian), IT analysts and Logistics analysts were selected, each of whom received a form to answer: Degree of favorable evidence μ and Degree of unfavorable evidence λ for each of the factors, within the transport process specifically for the delimitation of routes [8].

This Fig. 2 present, an algorithm from para-analyzer for digital transformation proposition [5].

**Table 2.** Chosen factors.

| Order | Factors |
|---|---|
| 1 | Are the routes standardized within the logistics systems? |
| 2 | Are routes frequently changed within logistics systems? |
| 3 | Can logistics systems detect blind spots on routes? |

**Table 3.** Average over $\mu$ and $\lambda$ factors – Part I

| KE[*] | Factor | Weight | Favorable degree of evidence $\mu$ | Unfavorable degree of evidence $\lambda$ |
|---|---|---|---|---|
| Farm user 1 | 1 | 1 | 1.0 | 0.1 |
| Farm user 2 | 1 | 1 | 0.9 | 0.2 |
| TI analyst 3 | 1 | 1 | 0.9 | 0.3 |
| TI analyst 4 | 1 | 1 | 0.8 | 0.2 |
| Logistic analyst 5 | 1 | 1 | 0.8 | 0.1 |
| Logistic analyst 6 | 1 | 1 | 0.9 | 0.2 |

[*] Knowledge engineers

**Table 4.** Average over $\mu$ and $\lambda$ factors – Part II.

| KE[*] | Factor | Weight | Favorable degree of evidence $\mu$ | Unfavorable degree of evidence $\lambda$ |
|---|---|---|---|---|
| Farm user 1 | 2 | 2 | 0.7 | 0.3 |
| Farm user 2 | 2 | 2 | 0.9 | 0.4 |
| TI analyst 3 | 2 | 2 | 0.9 | 0.1 |
| TI analyst 4 | 2 | 2 | 0.8 | 0.2 |
| Logistic analyst 5 | 2 | 2 | 0.7 | 0.2 |
| Logistic anayist 6 | 2 | 2 | 0.8 | 0.3 |

[*] Knowledge engineers

## 4   Analysis and Discussion

The algorithm is composed of a set of information collected through a research form for analysis of decision making.

It is observed the practical application of the para-analyzer algorithm and the proposal of this study, compared Case Study of the Volkswagen Modular Consortium - Trucks and buses from the logistical point of view presented by [3] which were addressed:

Figure 3 - A: represents the analysis carried out by the specialists regarding the appropriate scenario factor. For in this section, prove the proposition. For the average of

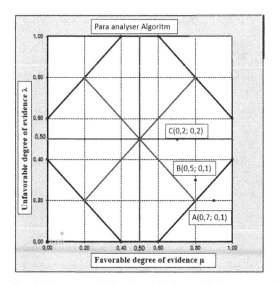

**Fig. 3.** Graphical representation of Tables 2, 3 and 4. Source: Authors.

favorable evidence is 0.7 and favorable evidence 0.1, in this case the result is real since it already exists in the current context.

Comparing with the Case Study of the Volkswagen Modular Consortium - Trucks and buses from a logistical point of view, presented by [3], where the strategy was to combine all aspects developed in the automobile industry between the years 1995 and 1996 and the Supply Chain mainly with respect to relative Computing Systems.

Figure 3 - B: represents the analysis carried out by the specialists regarding the appropriate scenario factor. For in this section, check the proposition as well. For the average of favorable evidence is 0.5 and favorable evidence 0.1, in this case the result is viable, since it already exists in the current context.

In the Volkswagen Case Study [3], the use of routes in three categories was approached according to the value of the parts for the specific storage areas, implying that there is a change in routes in the logistics system, thus remaining as well according to Factor 2.

In Fig. 3 - C: it represents the analysis carried out by the specialists regarding the appropriate scenario factor, in which case the chosen one was the need for digital transformation. For in this section, where there is no definition of when and how to prove that this functionality can be implemented. For the average of favorable evidence is 0.7 and favorable evidence 0.4, in this case the result is inconsistent since this context should have been more detailed.

In the Volkswagen Case Study [3], it was commented on the need for greater investments for innovation and technology, due to the increase in cargo theft and costs related to these occurrences. This indicates that Factor 3 agrees and must be followed up for a longer time for a result to be effectively assertive, since it was inconsistent, especially if it occurs many times (Table 5).

**Table 5.** Average over μ and λ factors – Part III

| KE[*] | Factor | Weight | Favorable degree of evidence μ | Unfavorable degree of evidence λ |
|---|---|---|---|---|
| Farm manager 1 | 3 | 3 | 0.6 | 0.5 |
| Farm manager 2 | 3 | 3 | 0.7 | 0.4 |
| TI manager 3 | 3 | 3 | 0.5 | 0.5 |
| TI manager 4 | 3 | 3 | 0.7 | 0.4 |
| Logistic manager 5 | 3 | 3 | 0.7 | 0.5 |
| Logistic manager 6 | 3 | 3 | 0.5 | 0.2 |

[*] Knowledge engineers

## 4.1 Comparisons with Real Data

We used information from a agrobusiness company, here as "ABC", which has headquarters in São Paulo and four farms, two in Minas Gerais and two others in Mato Grosso. This company uses Tracking system and Router System. The Tracking system in cargo trucker are monitoring by an outsourcing NOC – Network Operation Center, otherwise Router System is management by Service Desk in headquarters.

There are, on average, 300 employees, its capital is of family origin, and it was requested that its name and period of analyzed data remained secret.

A survey about critical incidents in the analyzed period of 7 months was conducted, starting in January/201x, and finishing in July/201x. According to a chart that summarizes the total analyzed critical incidents (shutdown).

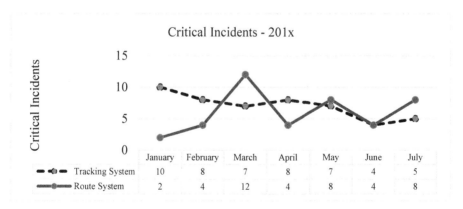

**Fig. 4.** Amount of occurrence of Critical Incident in the period of 7 months from Agrobusiness "ABC". Source: Authors.

The Routing System is unable to monitor 100% of the loads during its routes because of the instability of the interface with the ERP Logistics platform developed internally.

While the Tracking System shows a drop in coverage at times when it passes through shady places, a situation reported by the NOC to the Agrobusiness company.

Analyzing Fig. 4, we can see that both the Tracking System and the Routing System presented several failures during the analyzed period, which indicates the need for an adaptation so that there is no risk in cargo transportation.

## 5  Final Considerations

To make an analysis of several factors and make a combination between its varied possibilities for decision making, according to in several areas, as well as in logistics with support of digital transformation and Paraconsistent Logic Annotated Evidential Eτ for optimization of safe routes.

This study presents the three results of the analyzes of the specialists in comparison with the issues addressed in the Volkswagen Case Study and it was observed that there is an alignment between them.

On the other hand, the Agrobusiness company's Tracking Systems and Routing Systems present us with weaknesses that expose cargo transportation to risk due to the need for an update or digital transformation, perhaps with the implementation of Paraconsistent Logic Annotated Evidential Eτ.

All these scenarios bring a greater or lesser degree of risk of evidence for a digital transformation. Where it can be seen, that the greater the degree of granularity presented, the more appropriate the result to be mapped. The Evidential Annotated Paraconsistent Logic Eτ assists logistics with a direction, in which risk contexts can be identified and prevented.

## References

1. Sampaio, R: Vantagem digital: Um guia prático para a transformação digital, pp. 47–49. Alta Books Editora (2018)
2. Veras, M.: Gestão da tecnologia da informação: sustentação e inovação para a transfomação digital. Brasport, Rio de Janeiro (2019)
3. Bertaglia, P.R.: Logística e gerenciamento da cadeia de abastecimento/Paulo Roberto Bertaglia – 3. ed., Saraiva, São Paulo, p. 528 (2016)
4. Da Costa, N.C.A, Abe, J.M., Subrahmanian, V.S.: Remarks on annotated logic, Zeitschrift f. Math. Logik und Grundlagen d. Math. **37**, 561–570 (1991)
5. Abe, J.M., Silva Filho, J.I.d.: Celestino, Uanderson. Araújo, Hélio Corrêa de. Lógica Paraconsistente Anotada Evidencial Eτ. Comunicar.
6. Bowersox, D.J., et al.: Gestão logística da cadeia de suprimentos. AMGH Editora (2013)
7. Abe, J.M. (ed.): Paraconsistent Intelligent-Based Systems. ISRL, vol. 94. Springer, Cham (2015). https://doi.org/10.1007/978-3-319-19722-7
8. Akama, S. (ed.): Towards Paraconsistent Engineering. ISRL, vol. 110. Springer, Cham (2016). https://doi.org/10.1007/978-3-319-40418-9
9. Calado, A.M.F., et al.: Alguns dos erros mais comuns na tomada de decisão. Instituto Superior de engenharia de Coimbra, Coimbra (2007)
10. Banzato, E.: Tecnologia da informação aplicada à logística. INSTITUTO IMAM (2016)
11. Rocco, A., et al.: Estimação de estados em Sistemas Elétricos de Potência com técnicas baseadas em Lógia Fuzzy e Paraconsistente. Revista Seleção Documental n°27. Ed. Paralogike, Santos (2012)

# Selecting the Sustainable Fresh Food Surface Transport Array Using Analytic Hierarchy Process

Irenilza de Alencar Nääs[1]($\boxtimes$) iD, Nilsa Duarte da Silva Lima[1] iD,
Manoel Eulálio Neto[1,2] iD, and Gilson Tristão Duarte[1,2] iD

[1] Production Engineering, Paulista University, São Paulo, SP 04026-002, Brazil
{irenilza.naas,gilson.duarte}@docente.unip.br,
nilsa.lima@stricto.unip.br
[2] Centro Universitário Santo Agostinho, Teresina, PI 64019-625, Brazil

**Abstract.** The present study investigates the various arrays of the sectors involved in the logistic of fresh food transportation and distribution, considering the sustainability of the overall process, applying the AHP technique. The focus was on Brazil's fresh food distribution centers and how it is distributed in large cities. The criteria applied in two levels were selected from the literature, and the judgment was made by three experts using an online AHP platform. The final computation was considered the best array with a very high (90.4%) degree of agreement between the participants. The choice of Local food represented 72.1% in the concept of high sustainability. Choosing local foods must not be feasible in large countries or in countries that depend on food imports. However, for fresh food production, the local food production benefits go beyond economic costs, as it helps reduce greenhouse gas emissions, improve the carbon footprint of consumers, encourage sustainable agriculture, and have the shortest traceability.

**Keywords:** Food logistics · Food distribution · Food supply chain

## 1 Introduction

There are various initiatives to reduce the number of intermediaries in the food supply chain and geographically relocate production and consumption [1, 2]. These initiatives share the idea of meeting the many criteria of a sustainable food system. The expansion of the local food chains concept allows the renewal of the relations between the city and country [3]. Within this idea, the environmental impact is taken into account for potential tradeoffs in overall sustainability aspects [4].

Although climate change can be related to other GHG (Nx, NOx, HC, CO) emissions to the environmental impact in food supply chains [5], carbon dioxide emission has been intensely used to estimate the impact in several developed countries [5, 6]. Other ways of evaluating environmental impact are the use of the Life Cycle Assessment (LCA) approach [7], the water footprint [8], the food miles concept [9], and the energy efficiency [10].

© IFIP International Federation for Information Processing 2021
Published by Springer Nature Switzerland AG 2021
A. Dolgui et al. (Eds.): APMS 2021, IFIP AICT 631, pp. 655–660, 2021.
https://doi.org/10.1007/978-3-030-85902-2_70

The transport of fresh food in Brazil occurs from the producing region to regional distribution centers (Ceasa). Those distribution centers are usually located in large urban areas in each state. The fresh food supply chain is essential for management efficiency and effectiveness, given the complexity of dealing with this product. There is a substantial increase in distribution obstacles due to difficulties in ensuring product quality [11].

The analytical hierarchy process (AHP) is a multi-criteria decision-making methodology that includes qualitative information with the available quantitative data. AHP is a method for decision-making in complex environments in which various criteria are arbitrated in prioritizing and choosing alternatives. It transforms a complex problem into a multi-level hierarchical structure of objectives, criteria, sub-criteria, and alternatives [12]. AHP has been applied to numerous fields of knowledge, including project selection, transportation, and manufacturing [13–15]. AHP is a valuable tool for backing decision-making, mainly when professionals deal with complex and collaborative systems [16].

The present study aimed to evaluate the various arrays of the sectors involving the logistic of fresh food transportation and distribution considering the sustainability of the process, applying the AHP technique.

## 2   Material and Methods

The present study was carried out in two stages. First, we collected information regarding how the fresh food supply chain actors are related and the options available to have separate arrays. Second, a multi-criteria method (AHP) was applied to select the most sustainable array arrays of fresh food transportation and distribution.

The overall approach had the goal to check some alternatives to decrease the environmental impact on the logistic of the fresh food distribution in Brasil (Fig. 1).

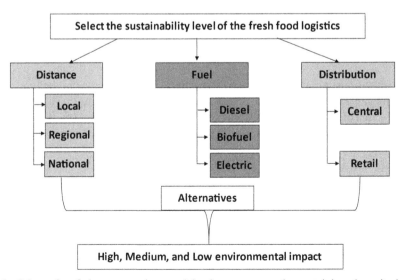

**Fig. 1.** Schematic of data processing used in the present study containing the criteria and alternatives.

The three first-level criteria were the Distance (from the production area to the distribution center), the Fuel (used in the transportation both in the production area to the city), and the Distribution (related to the product transportation distribution center to the retail). The second level in the distance criterion, we considered the sub-criterion of how far the food production was from the distribution center as Local (when close to the distribution center), Regional (when the farms are relatively within the same region of the distribution center), and National (when the production area is distant from the distribution center).

A set of pairwise comparison matrices (A; Eq. 1) was built. Each element on an upper level was used to compare the elements in the level below [ ] Saaty. The decision of relative weights (wi) of all pairs of the n elements are included as a number ($a_{ij}$) in a square matrix A (the comparison matrix):

$$A = (a_{ij}) \quad (i, j = 1, 2, 3, 4 \ldots n) \tag{1}$$

where $a_{ij} = w_i/w_j$ and $a_{ij} = 1/a_{ji}$.

The parameters for the pairwise comparison followed a 1–9 scale [12], where: 1 = not a priority; 2 = no to moderate priority; 3 = moderate priority; 4 = moderate to high priority; 5 = high priority; 6 = high to very high priority; 7 = very high priority; 8 = very high to greatest priority; 9 = highest priority. The weight was given to the (i, j)th position of the pairwise comparison matrix selected to support comparisons within a restricted range with sufficient sensitivity. The reciprocal of the appointed number was given to the (j, i)$^{th}$ position. The highest eigenvalue ($\lambda$max) was used to determine the consistency index (CI), shown in Eq. (2).

$$CI = (\lambda_{max} - n)/(n - 1) \tag{2}$$

where CI = consistency index; $\lambda_{max}$ = highest eigenvalue; and n = dimension of the matrix.

The matrix is entirely consistent when CI = 0, but results are acceptable when CI $\leq$ 0.1. The weights of element i were compared to element j and assigned to the (i, j)$^{th}$ position of the pairwise comparison matrix within a limited range with sufficient sensitivity.

Three experts were selected to judge the criteria and were assigned the task using the online software Business Performance Management Singapore (BPMSG) [17]. It is a web-based tool to support rational decision-making based on the AHP. It allows defining a hierarchy of criteria for a decision problem, calculating priorities, and evaluating a set of decision alternatives against those criteria. Most information on the selected criteria was learned from cited references described in the current literature [2–5, 18].

## 3   Results and Discussion

The AHP group consensus was very high (90.4%). The scores shown in Fig. 2 are the matrix results when computing the degree of importance of each selection criterion.

The most sustainable fresh food array was Local food (75.4%), followed by Retail (66.7%), Electric fuel source (40.9%), and Biofuel (34.2%). The remaining alternatives

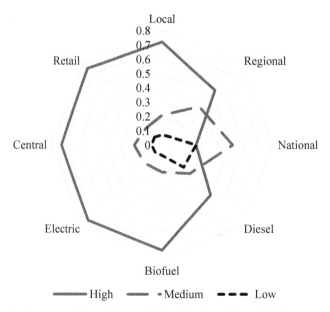

**Decision Hierarchy**

| Level 0 | Level 1 | Level 2 | Glb Prio. | High | Medium | Low |
|---|---|---|---|---|---|---|
| Select fresh food logistic alternative | Distance 0.607 | Local 0.754 | 45.8% | 0.721 | 0.208 | 0.071 |
| | | Regional 0.146 | 8.9% | 0.540 | 0.379 | 0.081 |
| | | National 0.100 | 6.1% | 0.245 | 0.510 | 0.245 |
| | Fuel 0.265 | Diesel 0.250 | 6.6% | 0.495 | 0.284 | 0.221 |
| | | Biofuel 0.342 | 9.1% | 0.736 | 0.190 | 0.073 |
| | | Electric 0.409 | 10.8% | 0.749 | 0.180 | 0.072 |
| | Distribution 0.128 | Central 0.333 | 4.3% | 0.725 | 0.200 | 0.075 |
| | | Retail 0.667 | 8.5% | 0.757 | 0.165 | 0.078 |
| | | | 1.0 | 66.9% | 23.8% | 9.3% |

**Fig. 2.** Scores were found on the different criteria in the different levels used to select the food logistic alternative for the best sustainable solution.

were all below this percentage. The choice of Local food represented 72.1% in the concept of high sustainability. The Electric truck result represented 74.9% in the concept of high sustainability, and the Distribution Retail was 75.7%, while Central was 72.5%. Apparently, the most significant difference in accounting for sustainability concept is the Distance that appears to be the most important criterion.

Figure 3 present the pairwise comparisons between the tested alternatives. They were associated with the criteria used to select the most sustainable array of choices during fresh food transportation.

**Fig. 3.** The pairwise comparisons between the tested alternatives with the answer of the group judgment.

Less polluting fuels and locally produced food, which supply retail and food distribution centers, are better choices to ensure food security (access and availability and rural development), less environmental impact of transport, and socio-economic impacts. In this study, the results showed that the decision making regarding the choices are justified by the higher level of sustainability for fresh food logistics, with higher percentages for local food, taking into consideration the distance that is one of the main factors in choosing the type of fuel in food logistics.

The relationship between fuel type and distribution depends on the supply of fuels and the distance and volume transported to distribution centers and retail [18]. A limitation concerning electric fuel sources is the low supply, mainly in regions with fewer inhabitants or cities that are further away from large centers and less developed. This scenario may change with the development of technologies that contribute to energy security by reducing dependence on fossil fuels implemented for rural and urban economic development. The choices made for food distribution with more sustainable criteria contribute to this vision of rural and urban development less dependent on fossil fuels and local food supply [7, 10]. The justification for choosing local foods is the benefits that go beyond economic costs, help to reduce greenhouse gas emissions, improve the carbon footprint of consumers, encourage sustainable agriculture, and have the shortest traceability.

## 4 Conclusions

Future distribution of food supply systems should consider the overall arrangement of demand and supply on sustainability grounds. New technologies should support the application of the concepts to legitimate the UN 2030 agenda.

## References

1. Edwards-Jones, G., Milà i Canals, L., Hounsome, N., et al.: Testing the assertion that 'local food is best': the challenges of an evidence-based approach. Trends Food Sci. Technol. **19**(5), 265–274 (2008). https://doi.org/10.1016/j.tifs.2008.01.008
2. Mariola, M.: The local industrial complex? Questioning the link between local foods and energy use. Agric. Hum. Values **25**, 193–196 (2008)
3. DuPuis, M., Goodman, D.: Should we go 'home' to eat? Towards a reflexive politics in localism. J. Rural Stud. **21**, 359–371 (2005)
4. Rothwell, A., Ridoutt, B., Page, G., Bellotti, W.: Environmental performance of local food: tradeoffs and implications for climate resilience in a developed city. J. Clean Prod. **114**, 420–430 (2016)
5. Ligterink, N.: Real-world vehicle emissions. The International Transport Forum. Discussion Paper No. 2017-06 (2017)
6. Dente, S.M.R., Tavasszy, L.: Policy-oriented emission factors for road freight transport. Transp. Res. Part A **61**, 33–41 (2018)
7. Hall, G., et al.: Potential environmental and population health impacts of local urban food systems under climate change: a life cycle analysis case study of lettuce and chicken. Agric. Food Secur. **3**, 6 (2014)
8. Page, G., Ridoutt, B., Bellotti, W.: Carbon and water footprint tradeoffs in fresh tomato production. J. Clean Prod. **32**, 219–226 (2012)

9. Watkiss, P.: The validity of food miles as an indicator of sustainable development. Final Report produced for DEFRA. ED50254 Issue 7 (2005). http://library.uniteddiversity.coop/Food/DEFRA_Food_Miles_Report.pdf
10. Mundler, P., Rumpus, L.: The energy efficiency of local food systems: a comparison between different modes of distribution. Food Policy **37**, 609–615 (2012)
11. Kumar, S., Himes, K.J., Kritzer, C.P.: Risk assessment and operational approaches to managing risk in global supply chains. J. Manuf. Technol. Manag. **25**, 873–890 (2014)
12. Saaty, T.L.: Decision making with the analytic hierarchy process. Int. J. Serv. Sci. **1**(1), 83 (2008). https://doi.org/10.1504/IJSSCI.2008.017590
13. Poh, K., Ang, B.: Transportation fuels and policy for Singapore: an AHP planning approach. Comput. Ind. Eng. **37**, 507–525 (1999)
14. Büyüközkan, G., Feyzioğlu, O., Nebol, E.: Selection of the strategic alliance partner in logistics value chain. Int. J. Prod. Econ. **113**, 148–158 (2008)
15. Nguyen, A.T., Nguyen, L.D., Le-Hoai, L., Dang, C.N.: Quantifying the complexity of transportation projects using the fuzzy analytic hierarchy process. Int. J. Proj. Manag. **33**, 1364–1376 (2015)
16. Rezaei, J., Ortt, R.: Multi-criteria supplier segmentation using a fuzzy preference relations based AHP. Eur. J. Oper. Res. **225**, 75–84 (2013)
17. BPMSG: Business Performance Management Singapore. Online software. https://bpmsg.com/
18. Duarte, G.T., Nääs, I.A., Innocencio, C.M., Cordeiro, A.F.S., Silva, R.B.T.R.: Environmental impact of the on-road transportation distance and product volume from farm to a fresh food distribution center: a case study in Brazil. Environ. Sci. Pollut. Res. **26**, 33694–33701 (2019)

# Sustainability in Production Planning
and Lot-Sizing

# A Partial Nested Decomposition Approach for Remanufacturing Planning Under Uncertainty

Franco Quezada[1,2(✉)], Céline Gicquel[3], and Safia Kedad-Sidhoum[4]

[1] Sorbonne Université, LIP6, 75005 Paris, France
franco.quezada@lip6.fr
[2] Universidad de Santiago de Chile (USACH), LDSPS, Industrial Engineering
Department, Santiago, Chile
franco.quezada@usach.cl
[3] Université Paris Saclay, LRI, 91190 Gif-sur-Yvette, France
celine.gicquel@lri.fr
[4] CNAM, CEDRIC, 75003 Paris, France
safia.kedad_sidhoum@cnam.fr

**Abstract.** We seek to optimize the production planning of a three-echelon remanufacturing system under uncertain input data. We consider a multi-stage stochastic integer programming approach and use scenario trees to represent the uncertain information structure. We introduce a new dynamic programming formulation that relies on a partial nested decomposition of the scenario tree. We then propose a new extension of the recently published stochastic dual dynamic integer programming algorithm based on this partial decomposition. Our numerical results show that the proposed solution approach is able to provide near-optimal solutions for large-size instances with a reasonable computational effort.

**Keywords:** Stochastic lot-sizing with remanufacturing · Multistage stochastic integer programming · Stochastic dual dynamic programming

## 1 Introduction

Remanufacturing is defined as a set of processes transforming used products into like-new finished products, mainly by rehabilitating damaged components. By reusing the materials and components embedded in used products, remanufacturing both contributes in reducing pollution emissions and natural resource consumption, making production processes more environment-friendly. However, remanufacturing systems involve several complicating characteristics, among which a high level of uncertainty in the input data needed to make planning decisions. This uncertainty mainly comes from a lack of control on the return flows of used products, both in terms of quantity and quality, and from the difficulty of forecasting the demand for remanufactured products.

© IFIP International Federation for Information Processing 2021
Published by Springer Nature Switzerland AG 2021
A. Dolgui et al. (Eds.): APMS 2021, IFIP AICT 631, pp. 663–672, 2021.
https://doi.org/10.1007/978-3-030-85902-2_71

The present work investigates production planning for a remanufacturing system involving three production echelons: disassembly of used products into parts, refurbishing of used parts and reassembly into like-new products. We consider uncertainties related to the quantity and quality of used products returned by customers, the demand for remanufactured products, and the production costs. We propose to handle this problem through a multi-stage stochastic programming approach in which production decisions are not made once and for all but rather adjusted over time according to the actual realizations of the uncertain parameters. We assume that the underlying stochastic input process has a finite probability space and we represent the information on the evolution of the uncertain parameters by a discrete scenario tree.

This problem was previously investigated in [2] and [3]. Quezada et al. [3] formulated the problem as a large-size mixed-integer linear program and proposed a customized branch-and-cut algorithm based on new valid inequalities to solve it. Quezada et al. [2] later investigated the use of the Stochastic Dual Dynamic integer Programming (SDDiP) algorithm recently presented in [4] to solve the problem. This algorithm relies on a full decomposition of the problem into a large number of small deterministic sub-problems. Although the above mentioned solution approaches were successful at providing near optimal solutions for small to medium size instances, some numerical difficulties were encountered to solve instances involving large-size scenario trees.

The present work thus discusses a new solution approach for this problem which uses a partial nested decomposition. The main idea consists in partially decomposing the problem into a set of medium-size stochastic sub-problems, each one defined on a sub-tree of the initial scenario tree. A new extension of the SDDiP algorithm exploiting this partial decomposition is then proposed.

The remaining part of this paper is organized as follows. Section 2 describes the problem under study and introduces a mixed-integer linear programming formulation. Section 3 briefly presents the proposed partial nested decomposition approach. Computational results are reported in Sect. 4. Finally, conclusions and directions for further works are discussed in Sect. 5.

## 2   Problem Description and Modeling

**Production System.** We consider a remanufacturing system comprising three main production echelons: disassembly, refurbishing and reassembly. We seek to plan the production activities in this system over a horizon comprising a discrete set $T = \{1, .., T\}$ of periods. The system involves a set $\mathcal{I}$ of items. Among these ones, item $i = 0$ represents the used products returned by customers in limited quantity at each period. A used product is composed of $I$ parts. Let $\alpha_i$ be the number of parts $i$ embedded in a used product. The returned products are first disassembled to obtain a set $\mathcal{I}_r = \{1, ..., I\}$ of recoverable parts. Due to the usage state of the used products, some of the parts obtained during disassembly have to be discarded. In order to reflect the variations in the quality of the used products, the yield of the disassembly process, i.e. the proportion of parts which will be recoverable, is assumed to be part-dependent and time-dependent. The recoverable parts are then refurbished on dedicated refurbishing processes. The set of $\mathcal{I}_s = \{I+1, ..., 2I\}$ of serviceable parts obtained after refurbishing are reassembled into remanufactured products which have the same bill-of-material as the used products. These remanufactured products, indexed by $i = 2I+1$, are used to satisfy the dynamic demand of customers.

The system comprises a set $\mathcal{P} = \{0, ..., P+1\}$ of production processes. Here, $p = 0$ corresponds to the disassembly process, $p \in \{1, ..., P\}$ corresponds to the process refurbishing the recoverable part indexed by $i = p$ into the serviceable part indexed by $i + I$ and $p = P + 1$ corresponds to the reassembly process. Note that the system comprises one individual refurbishing process for each item embedded in the used product and, hence, we have $P = I$. All these processes are assumed to be uncapacitated. However, the system might not be able to satisfy the customer demand on time due to part shortages if there are not enough used products returned by customers or if their quality is low. In this situation, the corresponding demand is lost incurring a high penalty cost to account for the loss of customer goodwill. Moreover, some used products and recoverable parts are allowed to be discarded. This option might be useful in case more used products are returned that what is needed to satisfy the demand for remanufactured products and in case there is a strong unbalance between the part-dependent disassembly yields leading an unnecessary accumulation in inventory of the easy-to-recover parts (Fig. 1).

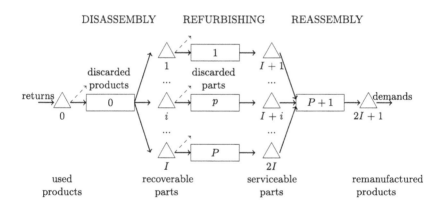

**Fig. 1.** Illustration of studied remanufacturing system

**Uncertainty.** We focus on the situation in which the input data needed to optimize the production plan for this system are subject to uncertainty and propose to handle this stochastic problem using a multi-stage stochastic integer programming approach. We assume that the evolution of the uncertain parameters can be represented by a discrete scenario tree $\mathcal{V}$ comprising a set $\mathcal{S} = \{1, \ldots, \Sigma\}$ of decision stages. A decision stage $\sigma$ may correspond to one or several planning periods: let $\mathcal{T}^\sigma$ be the set of time periods belonging to stage $\sigma$. Each node $n \in \mathcal{V}$ corresponds to a single period $t^n$ and a single stage $\sigma^n$. Let $\mathcal{V}^t$ be the set of nodes belonging to period $t$. Each node $n$ has a unique predecessor node denoted $a^n$ and represents the state of the system that can be distinguished by the information unfolded up to stage $\sigma^n$. At any non-terminal node of the tree, there are one or several branches to indicate future possible outcomes of the random variables from the current node. Let $\mathcal{C}(n)$ be the set of children of node $n$. The probability associated with the state represented by the node $n$ is denoted by $\rho^n$. A scenario is defined as a path in the tree from the root node to a leaf node and represents a possible outcome of the stochastic input parameters over the whole planning horizon.

Each node $n \in \mathcal{V}$ corresponds to a realization of the stochastic input parameters. Let $r^n$ be the quantity of collected used products, $d^n$ be the customers demand and $\pi_i^n$ be the proportion of recoverable parts $i \in \mathcal{I}_r$ obtained by disassembling one unit of returned product at node $n \in \mathcal{V}$. As for the costs, we have the setup cost $f_p^n$ for process $p \in \mathcal{P}$, the unit inventory cost $h_i^n$ for part $i \in \mathcal{I}$, the unit lost-sales penalty cost $l^n$, the unit cost $q_i^n$ for discarding item $i \in \mathcal{I}_r \cup \{0\}$ and the unit cost $g^n$ for discarding the unrecoverable parts obtained while disassembling one unit of returned product at node $n \in \mathcal{V}$.

**Mixed-Integer Linear Programming Formulation.** In order to build a mathematical model for the problem, we introduce the following decision variables at each node $n \in \mathcal{V}$: $x_p^n$ the quantity of parts processed on process $p \in \mathcal{P}$, $y_p^n \in \{0,1\}$ the setup variable for process $p \in \mathcal{P}$, $s_i^n$ the inventory level of part $i \in \mathcal{I}$, $w_i^n$ the quantity of part $i \in \mathcal{I}_r \cup \{0\}$ discarded and $\ell^n$ the lost sales of remanufactured products. This leads to the following MILP model.

$$\min \sum_{n \in \mathcal{V}} \rho^n F^n(x^n, y^n, s^n, w^n, l^n) \tag{1}$$

$$x_p^n \leq M_p^n y_p^n \qquad\qquad \forall p \in \mathcal{P}, \forall n \in \mathcal{V} \tag{2}$$

$$s_0^n = s_0^{a^n} + r^n - x_0^n - w_0^n \qquad\qquad \forall n \in \mathcal{V} \tag{3}$$

$$s_i^n = s_i^{a^n} + \pi_i^n \alpha_i x_0^n - x_i^n - w_i^n \qquad\qquad \forall i \in \mathcal{I}_r, \forall n \in \mathcal{V} \tag{4}$$

$$s_i^n = s_i^{a^n} + x_{i-P}^n - \alpha_i x_{P+1}^n \qquad\qquad \forall i \in \mathcal{I}_s, \forall n \in \mathcal{V} \tag{5}$$

$$s_{2I+1}^n = s_{2I+1}^{a^n} + x_{P+1}^n - d^n + \ell^n \qquad\qquad \forall n \in \mathcal{V} \tag{6}$$

$$s_i^{a^1} = 0 \qquad\qquad \forall i \in \mathcal{I}, \forall n \in \mathcal{V} \tag{7}$$

$$\ell^n, s_i^n \geq 0 \qquad\qquad \forall i \in \mathcal{I}, \forall n \in \mathcal{V} \tag{8}$$

$$x_p^n \geq 0, y_p^n \in \{0,1\} \qquad\qquad \forall p \in \mathcal{P}, \forall n \in \mathcal{V} \tag{9}$$

The objective function (1) aims at minimizing the expected total cost, over all nodes of the scenario tree. The total cost at node $n$, $F^n(x^n, y^n, s^n, w^n, l^n) = \sum_{p \in \mathcal{P}} f_p^n y_p^n + \sum_{i \in \mathcal{I}} h_i^n s_i^n + l^n \ell^n + \sum_{i \in \mathcal{I}_r \cup \{0\}} q_i^n w_i^n + g^n x_0^n$, is the sum of the setup, inventory holding, lost sales and disposal costs. Constraints (2) link the production quantity variables to the setup variables. Constraints (3)–(6) are the inventory balance constraints. Without loss of generality, we assume that the initial inventories are all set to 0, i.e., $s_i^{a^1} = 0$ for each $i \in \mathcal{I}$. Finally, Constraints (8)–(9) provide the domain of the decision variables.

In what follows, we denote by $X^n$ the subset of constraints (2)–(9) related to node $n$.

# 3    Partial Nested Decomposition Approach

In order to solve large-size instances of Problem (1)–(9), we propose a new solution approach based on a partial nested decomposition of the original stochastic problem into a series of smaller stochastic sub-problems.

**Partial Decomposition.** The proposed approach relies on a partial decomposition of the scenario tree $\mathcal{V}$ into a series of smaller sub-trees. This decomposition is obtained by first partitioning the set of decision stages $\mathcal{S} = \{1, \ldots, \Sigma\}$ into a series of macro-stages $\mathcal{G} = \{1, \ldots, \Gamma\}$, where each macro-stage $\gamma \in \mathcal{G}$ contains a number of consecutive stages denoted by $\mathcal{S}(\gamma)$. We let $t(\gamma)$ (resp. $t'(\gamma)$) represent the first (resp. the last) time period belonging to macro-stage $\gamma$. For a given macro-stage $\gamma$, each node $\eta$ belonging to the first time period in $\gamma$, i.e. each node $\eta \in \mathcal{V}^{t(\gamma)}$, is the root node of a sub-tree defined by the set of nodes $\mathcal{W}^\eta = \cup_{t=t(\gamma),\ldots,t'(\gamma)} \mathcal{V}^t \cap \mathcal{V}(\eta)$. Here, $\mathcal{V}(\eta)$ denotes the sub-tree of $\mathcal{V}$ rooted in $\eta$, $\mathcal{W}^\eta$ is thus the restriction of $\mathcal{V}(\eta)$ to the nodes belonging to macro-stage $\gamma$. Let $\mathfrak{L}(\eta) = \mathcal{W}^\eta \cap \mathcal{V}^{t'(\gamma)}$ be the set of leaf nodes of sub-tree $\mathcal{W}^\eta$. Finally, we denote by $\mathfrak{U} = \cup_{\gamma \in \mathcal{G}} \mathcal{V}^{t(\gamma)}$ the set of sub-tree root nodes induced by $\mathcal{G}$.

We then define the following sub-problem, denoted $\mathscr{P}^\eta(s^{a^\eta})$, for each sub-tree $\mathcal{W}^\eta, \eta \in \mathfrak{U}$. $\mathscr{P}^\eta(s^{a^\eta})$ focuses on optimizing the production plan for the nodes belonging to sub-tree $\mathcal{W}^\eta$ given the entering stock level of each part $i \in \mathcal{I}$, $s_i^{a^\eta}$, imposed by the parent node $a^\eta$.

$$Q^\eta(s^{a^\eta}) = \min \sum_{n \in \mathcal{W}^\eta} \rho^n F^n(x^n, y^n, s^n, w^n, l^n) + \sum_{\ell \in \mathfrak{L}(\eta)} \sum_{m \in \mathcal{C}(\ell)} Q^m(s^\ell) \quad (10)$$

$$(x^n, y^n, s^n, w^n, l^n) \in X^n \qquad\qquad \forall n \in \mathcal{W}^\eta \quad (11)$$

The objective value $Q^\eta(s^{a^\eta})$ denotes the optimal value of $\mathscr{P}^\eta(s^{a^\eta})$ as a function of the entering stock level $s^{a^\eta}$. It comprises two terms. The first term is related to the total expected production cost over all nodes $n \in \mathcal{W}^\eta$. The second term called the expected cost-to-go function represents the expected future costs, i.e. the costs which will have to be paid at the forthcoming decision stages, incurred by the production decisions made within sub-tree $\mathcal{W}^\eta$.

The expected cost-to-go function at leaf node $\ell \in \mathfrak{L}(\eta)$ is defined as the expected value of $Q^m(\cdot)$ over all the children $m$ of leaf node $\ell$ in the initial scenario tree $\mathcal{V}$, i.e. over all $m \in \mathcal{C}(\ell)$. This gives $\mathcal{Q}^\ell(\cdot) = \sum_{m \in \mathcal{C}(\ell)} Q^m(\cdot)$. The expected future costs of the decisions made in $\mathcal{W}^\eta$ are thus computed as the sum, over all nodes $\ell \in \mathfrak{L}(\eta)$, of $\mathcal{Q}^\ell(s^\ell)$. Note that for all nodes belonging to the last period of the planning horizon, i.e. for all $\ell \in \mathcal{V}^T$, $\mathcal{Q}^\ell(\cdot) \equiv 0$.

**Extended Stochastic Dual Integer Programming Algorithm.** The reformulation of Problem (1)–(9) using the dynamic programming recursion described by Equations (10)–(11) enables to develop a solution approach based on the Stochastic Dual integer Programming (SDDiP) algorithm recently presented by [4]. Basically, this algorithm will solve Problem (1)–(9) by solving a sequence of sub-problems (10)–(11) in which each expected cost-to-go function $\mathcal{Q}^\ell(\cdot)$ is iteratively approximated by a piece-wise linear function.

Note that a key assumption for developing such algorithm is that the scenario tree satisfies the stage-wise independence property. When there are several time periods per decision stage, this property can be defined as follows. For any two nodes $m$ and $m'$ belonging to stage $\sigma - 1$ and such that $t^m = t^{m'} = \max\{t, t \in \mathcal{T}^{\sigma-1}\}$, the set of nodes $\cup_{t \in \mathcal{T}^\sigma} \mathcal{V}^t \cap \mathcal{V}(m)$ and $\cup_{t \in \mathcal{T}^\sigma} \mathcal{V}^t \cap \mathcal{V}(m')$ are defined by identical data and conditional probabilities. This property enables us to significantly reduce the number of expected cost-to-go functions for which a piece-wise linear approximation must be build. Namely, in this case, the stochastic process can be represented at macro-stage $\gamma$ by a set $\mathcal{R}^\gamma = \{1, \ldots, R^\gamma\}$ of independent realizations. Each realization $\mathcal{X}^{\gamma,\zeta}$ corresponds to a subtree describing one of the possible evolutions of the uncertain parameters over periods $t(\gamma), \ldots, t'(\gamma)$. Let $\mathfrak{L}(\gamma, \zeta)$ denote the set of its leaf nodes. The expected cost-to-go functions thus depend on the macro-stage rather than on the node, i.e. we have $\mathcal{Q}^m(\cdot) \equiv \mathcal{Q}^\gamma(\cdot)$, for all $m \in \mathcal{V}^{t'(\gamma)}$, so that only one expected cost-to-go function has to be approximated per macro-stage. Moreover, we can define a single sub-problem $\mathscr{P}^\gamma$ per macro-stage and each sub-problem $\mathscr{P}^\eta, \eta \in \mathfrak{U}$, will be described as $\mathscr{P}^{\gamma^\eta}(s^{a^\eta}, \mathcal{X}^{\gamma^\eta,\zeta})$ where $\mathcal{X}^{\gamma^\eta,\zeta}$ is the realization corresponding to $\mathcal{W}^\eta$.

Each iteration $\upsilon$ of the extended SDDiP algorithm comprises a sampling step, a forward step and a backward step. In the sampling step, a subset of $W$ scenarios is sampled from the scenario tree. Let $\Omega_\upsilon = \{\omega_\upsilon^1, \ldots, \omega_\upsilon^w, \ldots, \omega_\upsilon^W\}$ be the set of sampled scenarios, $\omega_\upsilon^w$ be the set of nodes belonging to scenario $w$ at iteration $\upsilon$.

In the forward step, the algorithm proceeds stage-wise from macro-stage $\gamma = 1$ to $\Gamma$ by solving, for each sampled scenario $\omega^w$ and each macro-stage $\gamma$, the problem $\hat{\mathscr{P}}^\gamma(s^m, \mathcal{X}^{\gamma,\zeta^{w,\gamma}})$, which uses an approximate expected cost-to-go function, where $m = \omega^w \cap \mathcal{V}^{t'(\gamma-1)}$ is the node in the sampled scenario $\omega^w$ belonging to the last period of $\gamma$. At the end of this step, a statistical upper-bound of the problem is computed as the weighted average over all sampled scenarios.

In the backward step, the algorithm proceeds stage-wise from macro-stage $\gamma = G$ to macro-stage 1. Thus, for each scenario $w = 1, \ldots, W$, each node $m \in \omega_\upsilon^w \cap \mathcal{V}^{t'(\gamma)}$ and each realization $\zeta \in \mathcal{R}^{\gamma+1}$, it solves a suitable relaxation of Problem $\hat{\mathscr{P}}^{\gamma+1}(s^m, \mathcal{X}^{\gamma+1,\zeta})$. This relaxation is then used to improve the representation of the approximate cost-to-go function $\mathcal{Q}^\gamma(\cdot)$ through the generation of a new cut. Finally, the sub-problem solved at macro-stage $\gamma = 1$ provides a

---

**Algorithm 1:** Extended SDDiP algorithm

---

1  Initialize $LB \leftarrow -\infty, UB \leftarrow +\infty, v \leftarrow 1$

2  **while** *no stopping criterion is satisfied* **do**

3     **Sampling step**

4     Randomly select $W$ scenarios $\Omega_v = \{\omega_v^1, ..., \omega_v^W\}$

5     **Forward step**

6     **for** $w = 1, ..., W$ **do**

7        **for** $\gamma = 1, ..., \Gamma$ **do**

8           Solve $\hat{\mathscr{P}}^{\gamma}(s^m, \mathcal{X}^{\gamma,\varsigma})$ for $m = \omega_v^w \cap \mathcal{V}^{t'(\gamma-1)}$

9           Record $S_v^\ell$ for $\ell = \omega_v^w \cap \mathfrak{L}(\gamma, \zeta_v^{w,\gamma})$

10       **end**

11       $v^w \leftarrow \sum_{n \in \omega_v^w} F^n(x^n, y^n, s^n, w^n, l^n)$

12    **end**

13    $\hat{\mu} \leftarrow \sum_{w=1}^W v^w$ and $\hat{\sigma}^2 \leftarrow \frac{1}{W-1} \sum_{w=1}^W (v^w - \hat{\mu})^2$

14    $UB \leftarrow \hat{\mu} + z_{\alpha/2} \frac{\hat{\sigma}}{\sqrt{W}}$

15    **Backward step**

16    **for** $\gamma = \Gamma - 1, ..., 1$ **do**

17       **for** $w = 1, ..., W$ **do**

18          Let $m = \omega_v^w \cap \mathcal{V}^{t'(\gamma)}$

19          **for** $\zeta \in \mathcal{R}^{\gamma+1}$ **do**

20             Solve the linear relaxation of $\hat{\mathscr{P}}^{\gamma}(s^m, \mathcal{X}^{\gamma,\varsigma})$ and collect the coefficients of the strengthened Benders' cut

21             Solve the Lagrangian relaxation of $\hat{\mathscr{P}}^{\gamma}(s^m, \mathcal{X}^{\gamma,\varsigma})$ and collect the constant value of the strengthened Benders' cut

22          **end**

23       **end**

24       Add the generated cut to the current approximation of $\mathcal{Q}^{\gamma+1}$

25    **end**

26    $LB \leftarrow \hat{Q}_{v+1}^1(0)$

27    $v \leftarrow v + 1$

28 **end**

---

lower bound for the overall problem. The algorithm stops when the upper and lower bounds are close enough, according to a convergence criteria.

As a synthesis, the main steps of the proposed extended SDDiP algorithm are summarized in Algorithm 1.

**Original vs Extended SDDiP Algorithm.** The reader is referred to [4] for a detailed description of the original version of SDDiP algorithm. The proposed extended version differs from the original one with respect to two key features.

First, the original version of the SDDiP algorithm uses a full decomposition of the stochastic problem into small deterministic sub-problems, i.e. a decomposition in which each macro-stage $\gamma$ corresponds to a single decision stage $\sigma$, whereas we propose to use a partial decomposition. Using a partial nested decomposition instead of a full one may positively impact the computational efficiency of the SDDiP algorithm. Namely, it reduces the number of expected cost-to-go functions for which a piece-wise linear approximation must be built. Furthermore, each solved sub-problem covers a larger portion of the planning horizon so that the solution obtained at a given iteration of the algorithm will tend to be less myopic and thus of better quality. All this may contribute in accelerating the global convergence of the algorithm. Yet, the sub-problems to be solved at each iteration will be MILPs of larger size. The computational effort needed to solve them will thus also be larger. It is however possible to decrease it to some extent by using polyhedral approaches such as the one presented in [3].

Second, Zou et al. [4] showed that the finite convergence of the SDDiP algorithm is guaranteed when the state variables, i.e. the variables linking the decision stages to one another, are restricted to be binary. In Problem (1)–(9), the state variables are the continuous inventory level variables $s^n$. In this case, Zou et al. [4] suggest to use a binary approximation of the continuous state variables, which requires the introduction of a large number of additional binary variables in the problem formulation. In the proposed extended version of this algorithm, as done e.g. by Hjelmeland et al. [1] and Quezada et al. [2], we keep continuous state variables. In this case, the finite convergence of the algorithm is not theoretically guaranteed but, as this approximation leads to a significant reduction of the computational effort required at each iteration of the algorithm, it may positively impact the solution quality in practice.

## 4   Computational Results

In this section, we focus on assessing the performance of the proposed extended SDDiP algorithm by comparing it with the one of a stand-alone mathematical programming solver using formulation (1)–(9) and the one of the original SDDiP algorithm proposed by Zou et al. [4].

We randomly generated instances following the same procedure as the one used in [3]. We used various scenario tree structures and various values for the return over demand ratio $r/d$. Regarding the scenario tree structure, we used only balanced trees with $\Sigma \in \{4, 6, 8, 12\}$ stages, a constant number $b \in \{1, 2, 3, 5\}$ of time periods per stage and a constant number $R \in \{3, 5, 10, 20\}$ of equi-probable realizations per stage. We considered 8 possible combinations for these parameters, leading to instances involving between 1000 and 3.2 millions scenarios.

The partial decomposition of the problem was obtained by using a partition of the set of decision stages $\mathcal{S}$ in which each macro-stage corresponds to a constant number $G \in \{1, 2\}$ of stages. As for the stopping criteria, the algorithm stops when the lower bound does not improve after 30 iterations, or when 1000 iterations have been carried out. All the algorithms were implemented in C++ using the Concert Technology environment. The MILP and LP sub-problems embedded into the SDDiP algorithm were solved using CPLEX 12.8. All computations have been carried out on the computing infrastructure of the Laboratoire d'Informatique de Paris VI (LIP6), which consists of a cluster of Intel Xeon Processors X5690. We set the cluster to use two 3.46 GHz cores and 24 GB RAM to solve each instance. We imposed a time limit of 1800 s.

Table 1 displays the numerical results. Each line corresponds to a given combination of $\Sigma, R, b$, and $r/d$ and provides the average results for the related 20 instances. For each combination, Table 1 displays the gap between the lower bound and the upper bound ($|UB - LB|/UB$) found before some stopping criterion is reached and the average computation time. The label "$*$" represents the case when CPLEX in default mode could not report any gap within the imposed time limit.

Results from Table 1 show that the proposed extended SDDiP algorithm provides solutions of a significantly improved quality within the allotted computation time. Namely, the average gap over all tested instances is decreased from 54.50% (resp. 37.21%) when directly solving Problem (1)–(9) with CPLEX solver (resp. when solving it with the original SDDiP algorithm) to 4.98% when using the proposed extended version with $G = 2$ stages per macro-stage.

We note that a large part of this improvement can be explained by the use of continuous (rather than binary) state variables. This can be seen by comparing the average gap obtained with the original SDDiP algorithm, 37.21%, with the one obtained when using the proposed extension with $G = 1$, i.e. with a full decomposition of the problem, 6.85%. The use of a partial decomposition based on $G = 2$ stages per macro-stage then further improves the solution quality by decreasing the average gap from 6.85% to 4.98%.

**Table 1.** Numerical results

| Instance | | | | CPLEX | | SDDiP | | Extended SDDiP $G = 1$ | | Extended SDDiP $G = 2$ | |
|---|---|---|---|---|---|---|---|---|---|---|---|
| $r/d$ | $\Sigma$ | $b$ | $R$ | Gap | Time | Gap | Time | Gap | Time | Gap | Time |
| 1 | 4 | 1 | 10 | 0.25 | 1,800.56 | 11.63 | 1,418.64 | 3.45 | 754.98 | 0.97 | 599.16 |
| | | | 20 | 6.57 | 1,801.32 | 11.75 | 1,367.78 | 4.16 | 1,041.48 | 3.33 | 1,009.13 |
| | 6 | 1 | 10 | 76.89 | 1,818.76 | 13.97 | 1,803.22 | 3.76 | 1,246.89 | 3.02 | 1,498.62 |
| | | | 20 | * | * | 16.26 | 1,804.41 | 4.22 | 1,300.51 | 3.78 | 1,826.17 |
| | 8 | 2 | 5 | 92.02 | 1,837.44 | 24.70 | 1,814.66 | 2.77 | 1,547.97 | 2.23 | 1,791.41 |
| | | 5 | 5 | * | * | 17.58 | 1,840.89 | 2.37 | 1,683.59 | 2.70 | 1,832.97 |
| | 12 | 1 | 3 | 86.29 | 1,865.56 | 27.24 | 1,809.93 | 3.09 | 1,546.61 | 2.46 | 1,723.32 |
| | | 3 | 3 | * | * | 34.49 | 1,829.69 | 2.49 | 1,785.61 | 1.87 | 1,726.96 |
| | Average | | | 52.40 | 1,824.73 | 19.70 | 1,711.15 | 3.29 | 1,363.46 | 2.54 | 1,500.97 |
| 3 | 4 | 1 | 10 | 0.31 | 1,000.40 | 33.81 | 1,274.02 | 7.83 | 762.06 | 1.54 | 826.66 |
| | | | 20 | 7.96 | 1,801.40 | 28.91 | 1,276.01 | 6.64 | 1,039.12 | 3.84 | 942.09 |
| | 6 | 1 | 10 | 81.22 | 1818.23 | 44.30 | 1,804.47 | 8.87 | 1,202.04 | 5.94 | 1,135.28 |
| | | | 20 | * | * | 53.67 | 1,796.29 | 14.27 | 1,206.11 | 11.84 | 1,269.60 |
| | 8 | 2 | 5 | 96.13 | 1,837.22 | 68.13 | 1,818.03 | 10.17 | 1,525.91 | 6.71 | 1,682.07 |
| | | 5 | 5 | * | * | 65.15 | 1,830.02 | 11.79 | 1,800.69 | 12.60 | 1,824.09 |
| | 12 | 1 | 3 | 95.09 | 1,869.61 | 70.48 | 1,805.16 | 11.96 | 1,650.80 | 8.19 | 1,683.99 |
| | | 3 | 3 | * | * | 73.23 | 1,848.63 | 13.32 | 1,801.59 | 8.68 | 1805.90 |
| | Average | | | 56.63 | 1,825.60 | 54.94 | 1,683.37 | 10.61 | 1,373.54 | 7.42 | 1,393.46 |

# 5   Conclusion and Perspectives

We studied production planning for a remanufacturing system under uncertain input data and investigated a multi-stage stochastic integer programming approach. We proposed to use a new extension of the SDDiP algorithm recently introduced by Zou et al. [4] to solve the problem. Computational experiments carried out on large-size randomly generated instances suggested that the proposed extended algorithm significantly outperforms both the original SDDiP algorithm and the mathematical programming solver CPLEX using an extensive MILP formulation.

Note that, in the proposed extended SDDiP algorithm, we generate strengthened Benders' cuts to under approximate the expected cost-to-go functions. These cuts are linear inequalities for which part of the coefficients are obtained by solving the linear relaxation of the corresponding sub-problems and by recording the dual values of the constraints linking the sub-problems to one another. Nonetheless, by noticing that these dual values vary according to the linear relaxation formulation used for each sub-problem, it is possible to generate different strengthened Benders' cuts by using different formulations. Thus, an interesting research direction will be to exploit the current knowledge about the polyhedral structure of the sub-problem to iteratively strengthen the linear relaxation formulation of these sub-problems. For example, valid inequalities introduced in [3] can be used to strengthen the linear relaxation of each sub-problem in order to generate additional strengthened Benders' cuts, which might positively impact the performance of the algorithm.

Finally, note that we assumed in our problem modeling uncapacitated production processes. Extending the present work in order to account for production resources with limited capacity could thus be an interesting direction for further research.

**Acknowledgments.** This work was partially funded by the National Agency for Research and Development (ANID) / Scholarship Program / DOCTORADO BECAS CHILE/2018 - 72190160

# References

1. Hjelmeland, M.N., Zou, J., Helseth, A., Ahmed, S.: Nonconvex medium-term hydropower scheduling by stochastic dual dynamic integer programming. IEEE Trans. Sustain. Energy **10**(1), 481–490 (2018)
2. Quezada, F., Gicquel, C., Kedad-Sidhoum, S.: Stochastic dual dynamic integer programming for a multi-echelon lot-sizing problem with remanufacturing and lost sales. In: 2019 6th International Conference on Control, Decision and Information Technologies (CoDIT), pp. 1254–1259. IEEE (2019)
3. Quezada, F., Gicquel, C., Kedad-Sidhoum, S., Vu, D.Q.: A multi-stage stochastic integer programming approach for a multi-echelon lot-sizing problem with returns and lost sales. Comput. Oper. Res. **116**, 104865 (2020)
4. Zou, J., Ahmed, S., Sun, X.A.: Stochastic dual dynamic integer programming. Math. Program. **175**, 461–502 (2018). https://doi.org/10.1007/s10107-018-1249-5

# A Lot-Sizing Model for Maintenance Planning in a Circular Economy Context

Ernest Foussard[1,2]($\boxtimes$), Marie-Laure Espinouse[1], Grégory Mounié[2], and Margaux Nattaf[1]

[1] Univ. Grenoble Alpes, CNRS, Grenoble INP (Institute of Engineering Univ. Grenoble Alpes), G-SCOP, 38000 Grenoble, France
{ernest.foussard,marie-laure.espinouse,margaux.nattaf}@grenoble-inp.fr
[2] Univ. Grenoble Alpes, Inria, CNRS, Grenoble INP (Institute of Engineering Univ. Grenoble Alpes), LIG, 38000 Grenoble, France
gregory.mounie@imag.fr

**Abstract.** The transition towards Circular Economy is a crucial issue of European environmental policies. It requires a complete overhaul of production systems, in order to improve product lifecycles and to reduce the ecological footprint. In this context, maintenance is key to extend the products durability. This study addresses maintenance planning optimization within the Circular Economy framework.

An original lot-sizing model for tactical maintenance planning on a single-machine with multiple components is presented. The main features of this model are the consideration of the component health index and the global budget on environmental impact. The computational limits of the model and the impact of the budget constraint are assessed through experimentations.

**Keywords:** Circular economy · Lot-sizing · Maintenance planning · Mixed-integer linear programming

## 1 Introduction

Recently, the sustainability of production systems became a key concern. Indeed, there is an increasing pressure to implement environmental policies in the industry. The Circular Economy (CE) stands as one of the most promising paradigms to reform non-sustainable production systems. It consist of keeping products in a closed-loop by promoting the reusing and the remanufacturing of products and/or components. The goal is to reduce resource entries and waste generation of the supply chain. CE also encourages to extend the lifetime of products by repairing them. Lately, many industrials, researchers and policy-makers have demonstrated a growing interest in CE. One of the latest examples is the new

This work has been partially supported by the LabEx PERSYVAL-Lab (ANR-11-LABX-0025-01) funded by the French program Investissement d'avenir.

© IFIP International Federation for Information Processing 2021
Published by Springer Nature Switzerland AG 2021
A. Dolgui et al. (Eds.): APMS 2021, IFIP AICT 631, pp. 673–682, 2021.
https://doi.org/10.1007/978-3-030-85902-2_72

European Union action plan for CE, which aims at shifting the current production system towards a more sustainable and less resource-dependent one [3]. A key concept in CE is Performance Economy (PE), also known as Functional Economy. It consists in providing services rather than selling goods: the producer keeps ownership of his machines and sells the service. He thus remains responsible for the service level throughout the lifecycles of the machines. Laundromats or bike sharing services are examples of such business model. Preventive maintenance, i.e. any operation performed on a product with the aim to increase its durability, is central in this framework as it allows to keep the product in a satisfying service level for a longer time period.

In the field of optimization, the subject of CE is unequally covered. There are extensive studies about reverse logistics [2], sustainable manufacturing [6] and the implementation of CE in production planning [10]. However, to the best of our knowledge, the topic of maintenance planning in a CE context is yet to be studied. The purpose of this work is to address and open up new perspectives for maintenance planning optimization in the context of CE.

According to [11], preventive maintenance policies can essentially be divided into two categories: time-based maintenance and condition-based maintenance. Since end of life repurposing and recycling are key operations in CE, a special attention is paid in this work to the state of a machine component by component throughout its lifecycle. A conditional approach based on a precise measure of the component condition over time is therefore more suitable. Many studies use the virtual age of the machine or its components for that purpose: in both [5] and [8], components have a stochastic failure rate depending on their age. Breakdowns require to perform costly corrective maintenance operations and preventive maintenance operations restore the age of a component to zero. Fewer research works deal with the case of imperfect maintenance. In [11], the case of imperfect maintenance operations with percentage age reduction is investigated. In [7], on the other hand, the Equipment Health Index (EHI) is used to keep track of the condition of the machine under a time-based preventive maintenance policy. This indicator is used to determine whether a machine can safely process a job or not. The problem studied in this work contains an adaptation of the health index to machines with multiple components. A component with a low health index is more likely to break or cause overconsumption than a component in perfect state. The preventive maintenance framework proposed is very general, both perfect and imperfect maintenance, corresponding respectively to full or partial regeneration of one or several component health, are considered in the studied problem.

A more general version of the problem is investigated in [4]. The problem considers large planning horizon. Thus, different level of decision are mixed into the model (e.g. short term maintenance and long-term environmental strategy). A multi-objective mixed integer linear program (MILP) is designed. This model involves complex objectives depending on multiple parameters, and therefore computation times can be a major issue with realistic instances and low time granularity. The study of the multicriteria aspect of the problem is also challenging, as it involves four objectives functions that cannot be directly compared. This issue is overcome by separating tactical and operational decision-making into two sub problems and considering a single economic objective with a budget constraint on the environmental impact. The tactical problem consists in defining a rough mid-term maintenance policy. At the operational level, precise short-term resource allocation and maintenance schedules are defined. A model for the tactical problem is presented in this work.

This separation is inspired by tactical level and operational level production planning problems. Thus, the problem presented in this article shares many structural similarities with tactical production planning problems. As a consequence, the MILP is close to classical multi-item capacitated lot-sizing models with key differences in the way demand and setup are handled.

In this article, a modelization of the environmental aspects within a budget constraint is investigated. Such constraints have been extensively covered in recent lot-sizing studies. Four types of carbon emission constraints for lot-sizing models were first proposed in [1], namely global, periodic, cumulative and rolling constraints. These four categories were later reused as a mean of classifying such models in a literature review [10]. This idea can be extended to emissions in the broadest sense, to encapsulate any cost or output according to [9]. A similar approach is chosen for the model presented below, with a global constraint on environmental impact. The main advantage of such constraint is that it behaves as an epsilon constraint, and is therefore very suitable for a multicriteria approach. The impact of the environmental budget on the shape of the solutions is presented in this work. The main original feature in this constraint is the dependancy on the health index, to capture phenomenons such as overconsumption or increased waste emissions due to a worned component of the machine.

In Sect. 2, the problem and its parameters are precisely defined. In Sect. 3, a multi-item lot-sizing model variant is presented. In Sect. 4, some experimental results on generated instances are presented, computational limits are assessed and the impact of the environmental budget constraint is evaluated. Section 5 concludes this work and new research opportunities are suggested.

## 2  Problem Statement

Let us consider a single machine with multiple components $g \in \mathcal{G} = \{1, ..., G\}$. Since we deal with tactical-level decision making, the order of magnitude of the time horizon is typically one year, sampled in time periods $t \in \mathcal{T} = \{1, ..., T\}$ of medium granularity, e.g. one week.

Due to the context of PE, based on service exchanges, a different model for demand satisfaction, taking into account the availability $A_t$ of the machine is proposed. The demand $d_t$ is supposed to be homogeneous over each time period, and the fraction of the demand ultimately satisfied is proportional to the availability rate of the machine.

Each component is subject to a deterministic degradation per unit of demand satisfied $r_g$ during the exploitation of the machine, and the state of a component at the beginning of a period $t$ is represented by its health index $H_{gt}$. The health index ranges from 100 (perfect condition) to 0 (unusable). The initial state of the components is denoted $H_g^0$.

A set of maintenance operations $m \in \mathcal{M} = \{1, \ldots, M\}$ is available and can be used to restore the health of components by $reg_{gm}$. Two types of maintenance operations are considered in this work: refurbishment restoring the health of one component to 100 and partial maintenance actions restoring the health of one or several components by a fraction of its total health index. Each maintenance operation has a duration $p_m$ representing a fraction of the time period where the machine is unavailable. The economic cost can be decomposed into the cost of each scheduled maintenance operation $cp_m$, maintenance setup costs $cf$ and the opportunity cost corresponding to the unmet demand, which depends on the demand $d_t$, the benefit from demand satisfaction $pud$, and the availability rate of the machine $A_t$. Thus, the objective function is the sum of all maintenance costs, setup costs and costs of unsatisfied demand over the time horizon.

Furthermore, the environmental impact of the planning is taken into account and bounded within a global constraint. For the purpose of this work, the environmental impact of maintenance operations are not considered, but may be taken into account in further developments. The environmental cost of the planning captures over-pollution due to the exploitation of a worn component, namely over-consumption, increased emissions... This cost is modeled as a function of the health index of the components $H_{gt}$ and of the availability $A_t$.

$$c_{env} = f\left((H_{gt})_{g,t}, (A_t)_t\right) \leq budget$$

For the sake of simplicity, and with the aim of solving the problem by the means of mixed integer linear programming, a simple formulation based on the product of the wear level (i.e. the opposite of the health index $100 - H_{gt}$) with the availability rate is proposed.

$$c_{env} = \sum_{t \in \mathcal{T}} \sum_{g \in \mathcal{G}} c_g \left(100 - H_{gt}\right) \cdot A_t \leq budget$$

where $c_g$ represents an environmental penalty coefficient per component. A linearization of this constraint is proposed in Sect. 3.

To this day, the complexity of problem remains an open question. However, due to the structural similarities with multi-item capacitated lot-sizing problems, it is expected to be NP-hard.

In order to clarify some of the key aspects of the problem statement, a small example is proposed on a single-component machine over a time horizon $T = 5$. A single maintenance operation is available, with a duration $p = 0.5$ and a regeneration $reg = 50$. The demand $(d_t)$ is the following: $[0.5, 0.5, 1, 0.5, 0.5]$ and the degradation per unit of satisfied demand is $r = 40$. Finally, the economic costs are $cp = 1$, $cf = 10$ and $pud = 40$ and the environmental coefficient is $c = 1$. The evolution of the health index for two solution are presented in Fig. 1. Maintenance occurs at $t = 3$ for Solution 1, represented by a solid line, and at time $t = 4$ for Solution 2. Both solutions are identical until $t = 3$, thus Solution 2 is represented by a dashed line from $t = 3$.

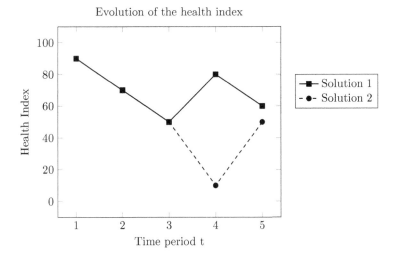

**Fig. 1.** Two solutions of the example instance

At $t = 1$, no maintenance operation are scheduled. Therefore, the machine is fully available $A_1 = 1$ and no maintenance costs or opportunity costs are charged. The value of the degradation is thus $d_1 \cdot r \cdot A_1 = 20$, and thus the new health index is 70.

The impact of the machine availability is visible in period $t = 3$: in the case of Solution 1, the maintenance operation of duration 0.5 is scheduled, therefore the machine availability $A_3$ is equal to 0.5. Under the hypotheses of the problem, half of the demand is not satisfied, thus the opportunity cost is then equal to $oc = pud \cdot d_3 \cdot (1 - A_3) = 20$. The total economic cost is then equal to $cf + cp + oc = 31$. The health index is also impacted by the maintenance, the variation is equal to the sum of the maintenance regeneration and of the degradation while the machine is available which results in an increase of health index by $reg - d_3 \cdot r \cdot A_3 = 30$. The availability also impacts the environmental cost, equal to $c \cdot (100 - 50) \cdot A_3 = 25$, which is higher in the case of Solution 2 since more demand is satisfied due to the absence of maintenance: 50.

The remaining part of the calculations is not detailed. Finally, Solution 1 yields an economic cost of 31 and an environmental cost of 125, while Solution 2 has an economic cost of 21 and an environmental cost of 185. These differences result from the fact that in Solution 1, the maintenance operation is scheduled early during a high demand period, while in Solution 2, the maintenance operation is scheduled late during a low demand period. With a budget of 150, Solution 1 would be feasible, but not Solution 2.

## 3   Model

In our framework, the health index of a component can be seen as a potential volume of demand which can be satisfied until the component is out of use.

Thus, maintenance operations allow to refill the component health. This characteristic is the foundation of the comparison with lot-sizing models presented below.

The model is indexed by time, components and maintenance operations. There are four sets of decision variables: $H_{gt} \in \mathbb{R}$ represents the health index, $X_{gt} \in \mathbb{R}$ represents the regeneration value. Binary variables are: $Y_{mt} = 1$ if maintenance operation $m$ is scheduled and 0 otherwise; $Z_t = 1$ if any maintenance operation is scheduled at time $t$ and 0 otherwise. The availability of the machine $A_t$ is expressed as a function of $Y_{mt}$: $A_t = 1 - \sum_{m=1}^{M} p_m Y_{mt}$. The budget constraint is linearized by introducing a new set of variables $B_{gmt}$ which represent the product $H_{gt} Y_{mt}$. Compared to classical lot-sizing models, $H_{gt}$ is analog to the inventory level, $X_{gt}$ to the production level, $Y_{mt}$ and $Z_t$ to the production setup variables.

$$\min. \sum_{t=1}^{T} cf Z_t + \sum_{m=1}^{M} \sum_{t=1}^{T} (cp_m + pud \cdot d_t \cdot p_m) Y_{mt} \tag{1}$$

$$X_{gt} + H_{gt} = H_{g,t+1} + d_t r_g \left(1 - \sum_{m=1}^{M} p_m Y_{mt}\right) \qquad \forall t \in T \backslash T, \ \forall g \in \mathcal{G} \tag{2}$$

$$X_{gt} \leq \sum_{m=1}^{M} reg_{gm} Y_{mt} \qquad \forall t \in T, \ \forall g \in \mathcal{G} \tag{3}$$

$$1 - \sum_{m=1}^{M} p_m Y_{mt} \geq 0 \qquad \forall t \in T \tag{4}$$

$$Y_{mt} \leq Z_t \qquad \forall t \in T, \ \forall m \in \mathcal{M} \tag{5}$$

$$H_{g1} = H_g^0 \qquad \forall g \in \mathcal{G} \tag{6}$$

$$H_{gt} + X_{gt} \leq 100 \qquad \forall t \in T, \ \forall g \in \mathcal{G} \tag{7}$$

$$\sum_{t=1}^{T} \sum_{g=1}^{G} c_g \left(100 - H_{gt} + \sum_{m=1}^{M} p_m (B_{gmt} - 100 \cdot Y_{mt})\right) \leq budget \tag{8}$$

$$0 \leq B_{gmt} \leq 100 \cdot Y_{mt} \qquad \forall t \in T, \ \forall g \in \mathcal{G}, \ \forall m \in \mathcal{M} \tag{9}$$

$$0 \leq H_{gt} - B_{gmt} \leq 100 \cdot (1 - Y_{mt}) \qquad \forall t \in T, \ \forall g \in \mathcal{G}, \ \forall m \in \mathcal{M} \tag{10}$$

$$B_{gmt} \geq 0 \qquad \forall t \in T, \ \forall g \in \mathcal{G}, \ \forall m \in \mathcal{M} \tag{11}$$

$$X_{gt}, H_{gt} \geq 0, \qquad \forall t \in T, \ \forall g \in \mathcal{G} \tag{12}$$

$$Y_{mt}, Z_t \in \{0, 1\} \qquad \forall t \in T, \ \forall m \in \mathcal{M} \tag{13}$$

The objective (1) is defined by the sum of the setup costs at each period, costs of maintenance operations and the opportunity costs due to unavailability of the machine. Equation (2) represents the conservation of the components health index over time and states the relation between the health index, the wear rate and the repair volume of each component. This constraint is equivalent to the flow balance constraint in multi-item lot-sizing problems, with the notable exception of the demand term which depends on the decision variable $Y_{mt}$. Equation (3) and Eq. (5) show the relation between maintenance setup variables $Y_{mt}$ and $Z_t$, and repair volume $X_{gt}$. These constraints are equivalent to the setup activation constraints in lot-sizing models. The health index of the components is initialized in (6). The Eq. (7) sets the maximum value for health index and regeneration quantity to 100.

Equation (4) ensures the non-negativity of the availability of the machine and thus limits the number of maintenance operations planned during the same period depending on their duration.

The environmental budget constraint is enforced through the remaining Eqs. (8), (9) and (10). In this linearized formulation, it involves the product variable $B_{gmt}$, expressed in Eqs. (9) and (10). Finally, the Eqs. (11), (12) and (13) state the domain of each variable.

# 4    Results and Performance Evaluation

## 4.1    Experimental Framework

The data set generation was mainly inspired by the case of laundromat washing machines, as this is frequently studied in CE and done previously in [4]. The behavior and the performances of the model are assessed on randomly generated instances of various size, according to the following protocol. The model is designed for tactical decision-making. Therefore, the horizon $T = 52$ is chosen to represent one year with a one-week granularity.

Three families of maintenance based on different characteristics are studied:

- F1 (resp F2): Single-component targeted maintenance operations. One maintenance per component is considered. The component is fully restored (resp. half restored).
- F3: Partial maintenance operations, divided into two groups of size $M = G$: operations of the first (resp. second) group regenerate 80% (resp. 20%) of the components by 20 (resp 80).

Ten instances are then generated per family with $G = 8$, and ten sub-instances are extracted from them with $G = 4$ and $G = 6$ components. These number of components are usually enough to cover the most critical parts of a laundromat washing machine. The costs and duration of maintenance operations are generated as follows. In average these values are proportional to the amount of regeneration of the maintenance. Let $S_m = \sum_{g \in \mathcal{G}} reg_{gm}$, then $cp_m \sim U([0.5 \cdot S_m, 1.5 \cdot S_m])$. For the purpose of these tests, the bounds for the generation of $p_m$ are computed such that one maintenance operation usually immobilizes the machine for 30% up to 60% of a period. Thus:

$p_m \sim U([0.004 \cdot S_m, 0.006 \cdot S_m])$.

The fix cost $cf$ is then defined as 10% of the average maintenance cost. To illustrate the variations of the demand over time, $d_t$ is set to 0.75 during school vacations (based on the French calendar) and is set to 1 otherwise. Indeed, students can represent an important fraction of the users of PE services such as laundromats. Thus, demand might be lower during school vacations. Since the machine is not necessarily new at the beginning, $H_g^0$ is generated using integer uniform distributions on the respective interval $[50, 100]$. In real situations, this model might be run multiple times during the lifecycle of the machine. To illustrate the fact that some components deteriorate faster than others, the values for

$r_g$ are generated on a large interval $[2, 10]$. Finally, the value of $pud$ is arbitrarily set to 1, as we mainly focus on parameters related to maintenance operations. This makes maintenance costs and opportunity costs equally as significant.

In a first set of experiments, the impact of the instance size is assessed. The coefficients $c_g$ are drawn arbitrarily in the interval $[0.5, 1.5]$ such that some components have more impact than others. The environmental budget is set to 1 000 000 so the global budget constraint is virtually inactive at first.

The impact of the budget constraint is assessed in a second set of experiments restrained to the case of $F1$ with $G = 4$: the results obtained in first instance are used as baselines for comparison. The baseline environmental cost $env\_cost$ is extracted from them. New instances are then built with budgets equal to 50% and 75% of the baseline cost.

The model has been implemented in OPL using IBM ILOG CPLEX Optimization Studio 12.10. All the experiments were done on a laptop running on Ubuntu 20.04.2 LTS, with 16 GB of RAM and one Intel Core i7-10610U CPU @ 1.80 GHz × 8. The time limit is set to 10 min per experiment.

### 4.2   Impact of Instances Size and Maintenance Operation Types

The purpose of the first set of experiments is to evaluate the computational limits of the approach based on the number of components and nature of the maintenance operations. The results per maintenance family are summed up in Table 1. The first and second columns show the characteristics of the instances. The third column contains the arithmetic mean of the computation time for instances which are solved to optimality in less than ten minutes. The average relative gap, computed as the ratio of the sum of the absolute gaps over the sum of the objectives for non-optimal (resp. all) solutions is provided in column four (resp. five). On the sixth and seventh columns, the minimal and maximal gaps are provided. Finally, the eighth column indicates the number of instances solved to optimality per maintenance family and number of components.

**Table 1.** Impact of instances size and maintenance operation types.

|    | G | Avg time (opt) | Avg gap (non-opt) | Avg gap | Gap min | Gap max | # opt |
|----|---|----------------|-------------------|---------|---------|---------|-------|
| F1 | 4 | 64.6 s | 2.4% | 0.4% | – | 2.4% | 9/10 |
|    | 6 | 347.9 s | 9.2% | 7.9% | – | 14.5% | 2/10 |
|    | 8 | – | 15.1% | 15.1% | 7.7% | 20.1% | 0/10 |
| F2 | 4 | 91.8 s | 3.7% | 0.9% | – | 7.1% | 8/10 |
|    | 6 | – | 5.6% | 5.6% | 2.3% | 11.0% | 0/10 |
|    | 8 | – | 8.0% | 8.0% | 5.5% | 12.2% | 0/10 |
| F3 | 4 | 104.7 s | 1.1% | 0.1% | – | 1.1% | 9/10 |
|    | 6 | 99.1 s | 2.7% | 0.3% | – | 2.7% | 9/10 |
|    | 8 | 60.4 s | 4.4% | 1.3% | – | 7.7% | 7/10 |

Looking at the average computation time and average gap, it appears that the nature of maintenance operations has a major impact on the performance of the model. The optimum is rarely found for the largest instances of $F1$ and the gap can be significant. For the case of 8 components, the gap is 7.7% in the best case and 20.1% in the worst case. On the other hand, it appears that in the case $F3$, most instances can be solved quite accurately or to optimality, including the largest ones.

It appears that in the case of $F1$, and $F2$ to a lesser extent, the computation time increases with the number of components and the quality of solutions decreases. In the case of $F3$, the increase is not as significant. It is likely that another unidentified parameter comes into play in this specific case.

The study of the interactions between maintenance of different families could be the object of further developments.

## 4.3    Impact of the Environmental Budget

The second set of experiments allows to evaluate the impact of the global environmental constraint on the computation time and the nature of the solutions. The results are summarized in Table 2. The first column indicates the bound on the environmental cost. The second, third and sixth columns are the same as in Table 1. In the fourth column (resp. fifth column) the ratio of the sum of the increases of the objective values (resp. number of maintenance operations scheduled) over the sum of the objective values of the baseline instance solutions is presented.

**Table 2.** Impact of the environmental budget constraints.

| Budget | Avg time (opt) | Avg gap (non-opt) | Avg obj. incr. | Avg maint. incr. | # opt |
|---|---|---|---|---|---|
| 50% | – | 12.8% | +48.5% | +57.6% | 0/10 |
| 75% | 76.6 s | 5.2% | +7.1% | +12.0% | 8/10 |
| 100% | 82.8 s | 4.6% | +0.0% | +0.0% | 9/10 |

The results presented in columns 2, 3 and 6 show that the computation time and quality of solution worsen when the budget is tighter. Most likely, the main reason for this problem results from the expression of the environmental cost as its linearization involves a large number of variables and constraints.

In this model, economic and environmental performance appear to be antagonistic. For each of the solutions obtained on these instances, the budget constraint is tight. This antagonism is clearly visible on column 4, with a 48.5% increase of the economic objective when the budget is equal to 50% of the baseline environmental cost. This increase in the economic cost is correlated with the increase of the number of maintenance operations as shown in column 5.

These results are consistent with respect to the expectations. Reducing the environmental costs requires to keep the components in a better condition. Thus more maintenance operations are scheduled and the economic cost increases.

# 5  Conclusion

In this article, the problem of maintenance planning in a PE context is investigated. A lot-sizing model for tactical decision-making, involving component health level and a global budget constraint on the environmental impact is proposed. Experiments on realistic sized randomly generated instances confirmed that this model can provide suitable solutions in a short amount of a time. The impact of an environmental constraint on economic profit was also illustrated on these instances. However, the model is yet to be assessed on industrial instances.

In this work the environmental impact is treated as a constraint. The integration of the environmental impact as a second objective and computation of a Pareto front could be explored and compared in future works. Finally, a natural direction for further investigation is the study of an operational counterpart of this problem, and the interaction between the solutions obtained in both situations.

# References

1. Absi, N., Dauzère-Pérès, S., Kedad-Sidhoum, S., Penz, B., Rapine, C.: Lot sizing with carbon emission constraints. EJOR **227**(1), 55–61 (2015)
2. Agrawal, S., Singh, R.K., Murtaza, Q.: A literature review and perspectives in reverse logistics. Resour. Conserv. Recycl. **97**, 76–92 (2015)
3. European Commission: New Circular Economy Action Plan For a cleaner and more competitive Europe. COM(2020) 98 EU - Communication (2020)
4. Foussard, E.: Maintenance planning for circular economy: laundromat washing machines case. G-SCOP - Laboratoire des sciences pour la conception, l'optimisation et la production, Technical report (2021). https://hal.archives-ouvertes.fr/hal-03151214/
5. Géhan, M., Castanier, B., Lemoine, D.: Integration of maintenance in the tactical production planning process under feasibility constraint. In: Grabot, B., Vallespir, B., Gomes, S., Bouras, A., Kiritsis, D. (eds.) APMS 2014. IAICT, vol. 438, pp. 467–474. Springer, Heidelberg (2014). https://doi.org/10.1007/978-3-662-44739-0_57
6. Giret, A., Trentesaux, D., Prabhu, V.: Sustainability in manufacturing operations scheduling: a state of the art review. J. Manuf. Syst. **37**, 126–140 (2015)
7. Kao, Y.-T., Dauzère-Pérès, S., Blue, J., Chang, S.-C.: Impact of integrating equipment health in production scheduling for semiconductor fabrication. Comput. Ind. Eng. **120**, 450–459 (2018)
8. Moghaddam, K.S., Usher, J.S.: Preventive maintenance and replacement scheduling for repairable and maintainable systems using dynamic programming. Comput. Ind. Eng. **60**(4), 654–665 (2011)
9. Retel Helmrich, M.J., Jans, R., van den Heuvel, W., Wagelmans, A.P.M.: The economic lot-sizing problem with an emission capacity constraint. EJOR **241**(1), 50–62 (2015)
10. Suzanne, E., Absi, N., Borodin, V.: Towards circular economy in production planning: challenges and opportunities. EJOR **287**(1), 168–190 (2020)
11. Yang, L., Ye, Z., Lee, C.-G., Yang, S., Peng, R.: A two-phase preventive maintenance policy considering imperfect repair and postponed replacement. EJOR **274**(3), 966–977 (2019)

# An Integrated Single-Item Lot-Sizing Problem in a Two-Stage Industrial Symbiosis Supply Chain with Stochastic Demands

Cheshmeh Chamani[1,2]([✉]) [iD], El-Houssaine Aghezzaf[1,2] [iD],
Abdelhakim Khatab[3] [iD], Birger Raa[1] [iD], Yogang Singh[1,2] [iD],
and Johannes Cottyn[1,2] [iD]

[1] Department of Industrial Systems Engineering and Product Design,
Ghent University, Ghent, Belgium
{Cheshmeh.Chamani,ElHoussaine.Aghezzaf,Birger.Raa,
Yogang.Singh,Johannes.Cottyn}@ugent.be
[2] Industrial Systems Engineering (ISyE), Flanders Make, Ghent, Belgium
[3] Laboratory of Industrial Engineering, Production and Maintenance,
Lorraine University, Metz, France
abdelhakim.khatab@univ-lorraine.fr
http://www.FlandersMake.be

**Abstract.** We consider a two-stage supply chain in which two production plants are collaborating in an industrial symbiosis to satisfy their respective stochastic demands. We formulate the production planning problems of these two plants as an integrated capacitated lot-sizing problem, in which the second production plant uses as an alternative raw material a by-product obtained as a residue from the production of the first plant. The goal is to minimize the overall total cost in the supply chain, including production and inventory of the final product and by-product transfer costs, while meeting the stochastic demands. First, a natural formulation of the problem is proposed, and is solved using the Sample Average Approximation (SAA) method. The analysis of the gaps exhibits however quite large optimality gaps. To improve these optimality gaps, a plant location like reformulation for this integrated lot-sizing problem is developed. The analysis has been carried out again to evaluate both formulations' performances in terms of the optimality gaps and computational times, both when items demands follow Gamma and Normal distributions. The analysis indicates that despite having a computational time of on average 1.7 times higher than the main formulation, the plant location reformulation provides better optimality gaps on average 22% improved and better ranges for upper and lower bounds under stochastic demands.

**Keywords:** Lot-sizing · By-product · Industrial symbiosis · Sample average approximation

© IFIP International Federation for Information Processing 2021
Published by Springer Nature Switzerland AG 2021
A. Dolgui et al. (Eds.): APMS 2021, IFIP AICT 631, pp. 683–693, 2021.
https://doi.org/10.1007/978-3-030-85902-2_73

# 1   Introduction and a Brief Literature Review

As the world's population increases, so does the demand for products and other consumption goods. Consequently, the demand for raw materials keeps growing. However, the supply of essential raw materials is now reaching its limit. This awareness forces today's societies and supply chains to consider more creative approaches to use and reuse available resources in a more sustainable manner. Circular economy is an emerging concept that facilitates moving away from the traditional linear economic model based on the take-make-consume-throwout pattern. Circular economy is a production and consumption model, which involves sharing, leasing, reusing, repairing, refurbishing, and recycling existing materials and products as long as this is possible to extend products life-cycles. Moving towards a more circular economy will bring about significant economic, environmental, and social benefits. The transition from a linear to a circular economy requires, however, a fundamental change in production and consumption systems beyond waste recycling and resources use efficiency.

Production processes have a high impact on product life, supply, resource use, and waste generation. When manufacturing a product usually a collateral flow that is generally considered as waste is generated. A practice that can potentially benefit the transition to a circular economy is to generate and capture value by converting these waste streams (through further processing) into a useful by-products. This practice of turning produced residues, considered as waste, from an industrial process into by-products for another process is generally known as *by-product synergy* [3]. Optimizing jointly the production in both processes while explicitly considering the operational synergy between the original product and the by-product generates value for the involved parties in the supply chain. This paper focuses on a basic building block of the underlying production planning problem to provide evidence for this fact.

Various versions of this planning problem have been investigated in the literature. Suzanne et al. [4] carried out a comprehensive review on mid-term production planning in the context of circular economy and reverse logistics. The review presents an overview of the mathematical formulations, and their related solution methods, for production planning problems that arise under disassembly for recycling, in product to raw material recycling, and in by-products and co-production settings. The single-item lot-sizing problem arising in a single production unit, generating a by-product during the production process of the main product to satisfy a deterministic demand, has been investigated by [5]. The authors proved that this problem is NP-Hard when the by-product inventory capacity is time dependent, and proposed a pseudo-polynomial time dynamic programming algorithm to solve it. They also showed that when the inventory capacity is time independent the problem can be solved in polynomial time also via dynamic programming. A capacitated lot-sizing problem involving two collaborating production plants under different collaboration settings in an industrial symbiosis, and under deterministic demand is discussed in [2]. The objective of the model is to minimize total costs associated with supply, disposal, production, inventory, and symbiosis. Our current research investigates the integrated capacitated lot-sizing problem for a two-plants supply chain, collaborating in an industrial symbiosis, in which the second plant uses as raw material a by-product obtained as a residue generated by the production in the first plant to satisfy their respective stochastic demands. In the next section, we present a mathematical model of this planning problem, which we extended further to deal with demand uncertainty.

## 2  Problem Statement and Model Formulation

We consider a centralized supply chain containing two production plants cooperating in an industrial symbiosis to minimize the supply chain's total costs. We investigate the underlying capacitated lot-sizing problem, called IS-SCLSP for short in the rest of the paper, where a primary plant generates a by-product that is used by the secondary plant to satisfy their respective stochastic demands. Over a planning horizon of T periods, the IS-SCLSP determines when and how much each plant should produce while satisfying the demand $d$. The amount of by-product generated in a rate $\alpha$ is moved forward to the second plant from its inventory level $J_t$ hold in the first plant and its demand from the second plant. The second plant consumes the by-product at a rate $\beta$ as raw material. The production system involves fixed setup cost $f_t^n$ and unitary production cost either from raw material $p_t^n$ or by-product $\hat{p}_t$. The finished product's surplus quantity can be stored in inventory at a unitary holding cost $h_t^n$ from period $t$ to $t+1$. The transportation of the by-product is performed at unitary cost $q_t$. The three main decisions posed by IS-SCLSP are: (1) when to produce the finished product $y_t^n$, (2) how much to produce of the finished product $x_t^n$, and (3) when and how much by-product to transfer $w_t$. Accordingly, all other related decisions, namely inventory levels of the main products $I_t^n$ and the by-product $J_t$, are implied.

Before proceeding to the problem modelling we list hereafter the assumptions that have been made: (1) Both production plants produce a single type of product; (2) Initial inventory of the main products as well as of the by-product is null at the beginning of the planning horizon; (3) Production in the first plant is made entirely of purchased raw material, and the resulting by-product can be consumed as raw material to satisfy part of the demand in the second plant; (4) There is a limit on the production capacity in each plant at each planning period; (5)There is no limit on by-product inventory level; (6) No backorder is allowed, and all the demands should be satisfied within their desired due date; (7) Processing the finished product in the second plant using the by-product is more cost effective than when only raw material is used.

**Summary of the Parameters:**

- $T$: Number of periods in the planning horizon.
- $N$: Number of production plants in the supply chain.
- $t$: Index for discrete period of the planning horizon.
- $n$: Index for production plants in the supply chain.
- $p_t^n$: Unit production cost of finished product using only raw material in plant $n$ in period $t$.

– $\hat{p}_t$: Unit production cost of finished product in plant 2, using by-product as raw material.
– $f_t^n$: Fixed set-up cost for plant $n$ in period $t$.
– $d_t^n$: Demand of product for plant $n$ in period $t$.
– $h_t^n$: Unit inventory cost of the finished product in plant $n$ in period $t$.
– $q_t$: Unit transportation cost of by-product.
– $\alpha$: Generation rate of the by-product from the production in the plant 1.
– $\beta$: Production rate of the finished product for plant 2 per unit of by-product used.
– $cap_t^n$: Production capacity level of plant $n$ in planning period $t$.

**Summary of the Variables - Natural Formulation:**

– $x_t^n$: Production amount in plant $n$ in planning period $t$.
– $y_t^n$: Set up variable in plant $n$, assuming value 1 if production takes place in planning period $t$.
– $I_t^n$: Inventory level of the finished product in plant $n$ at the end of planning period $t$.
– $J_t$: Inventory level of generated by-product in plant 1 at the end of planning period $t$.
– $w_t$: Amount of by-product that is transferred from plant 1 to plant 2 during the planning period $t$.

**Summary of the Variables - Plant Location Reformulation:**

– $x_{t't}^n$: Fraction of the demand of plant $n$ in period $t$ that is satisfied from the production in plant $n$ during period $t'$ not involving the by product.
– $w_{t't}$: Fraction of the demand of plant 2 in period $t$ that is satisfied form the production in plant 2 during period $t'$ involving the by-product.

### 2.1   The Natural Formulation of IS-SCLSP

The integrated capacitated lot-sizing problem for a two-plants supply chain exchanging a by-product can naturally be formulated as a (deterministic) mixed-integer linear program given below:

$$\text{Minimize } Z = \sum_{n=1}^{N}\sum_{t=1}^{T}(f_t^n \cdot y_t^n + p_t^n \cdot x_t^n + h_t^n \cdot I_t^n) + \sum_{t=1}^{T}(q_t \cdot w_t) + \sum_{t=1}^{T}(\hat{p}_t \cdot \beta \cdot w_t) \quad (1)$$

Subject to

$$I_{t-1}^1 + x_t^1 = d_t^1 + I_t^1, \quad \forall t \in T, n = 1 \quad (2)$$

$$I_{t-1}^2 + x_t^2 + \beta \cdot w_t = d_t^2 + I_t^2, \quad \forall t \in T, n = 2 \quad (3)$$

$$J_{t-1} + \alpha \cdot x_t^1 = w_t + J_t, \quad \forall t \in T, n = 1 \quad (4)$$

$$x_t^1 \leq cap_t^1 \cdot y_t^1, \quad \forall t \in T, n = 1 \quad (5)$$

$$(x_t^2 + \beta \cdot w_t) \leq cap_t^2 \cdot y_t^2, \quad \forall t \in T, n = 2 \quad (6)$$

$$I_t^n, x_t^n, J_t, w_t \geq 0 \quad y_t^n \in \{0,1\}, \quad \forall t \in T, n \in N \quad (7)$$

The objective function (1) minimizes the sum of the fixed and variable production costs, inventory holding costs of the finished products, transportation cost of the by-product, and processing of finished product in plant 2 involving the by-product as the raw material. Constraints (2) and (3) represent the inventory flow of the finished product in the two production plants respectively. Constraint (4) expresses the flow conservation of the by-product in the first plant. Inequalities (5) and (6) ensure that the fixed setup costs in the two plants are paid and that production capacities are not exceeded. Non-negativity and binary requirements on the variables are expressed through (7).

## 2.2 Reformulation of IS-SCLSP

As the result section below will show, the optimality gaps achieved through implementing SAA on the main problem for different sample sizes are relatively high. Adopting the reformulation idea from [1], we provide a reformulation based on a plant location problem for the primary model to provide better LP-relaxation gaps compared to the formulation in the original variables.

$$\text{Minimize } Z = \sum_{n=1}^{N}\sum_{t=1}^{T} f_t^n \cdot y_t^n + \sum_{n=1}^{N}\sum_{t=1}^{T}\sum_{t'=1}^{t}\left(p_{t'} + \sum_{r=t'}^{t-1} h_r\right)x_{t't}^n \cdot d_t^n$$

$$+ \sum_{t=1}^{T}\sum_{t'=1}^{t}\left(\hat{p}_{t'} + \sum_{r=t'}^{t-1} h_r\right)w_{t't} \cdot d_t^2 + \sum_{t'=1}^{t}\sum_{t=1}^{T} q_{t'} \cdot \frac{1}{\beta} \cdot w_{t't} \cdot d_t^2 \qquad (8)$$

Subject to

$$\sum_{t'=1}^{t} x_{t't}^1 = 1, \quad \forall t \in T, n = 1 \tag{9}$$

$$\sum_{t'=1}^{t} (x_{t't}^2 + w_{t't}) = 1, \quad \forall t, t' \in T, t' \le t \tag{10}$$

$$\sum_{t=t'}^{T} d_t^1 \cdot x_{t't}^1 \le cap_{t'}^1 \cdot y_{t'}^1, \quad \forall t' \in T, n = 1 \tag{11}$$

$$\sum_{t=t'}^{T} d_t^2 \cdot (x_{t't}^2 + w_{t't}) \le cap_{t'}^2 \cdot y_{t'}^2, \quad \forall t' \in T, n = 2 \tag{12}$$

$$\sum_{t'=1}^{t}\sum_{r=t'}^{T} \alpha \cdot x_{t'r}^1 \cdot d_r^1 - \sum_{t'=1}^{t}\sum_{r=t'}^{T} \frac{1}{\beta} \cdot w_{t'r} \cdot d_r^2 \ge 0, \quad \forall t \in T \tag{13}$$

$$0 \le x_{t't}^n, w_{t't} \le 1, \quad y_t^n \in \{0,1\}, \quad \forall t, n \tag{14}$$

In this reformulation, the variables $x_{t't}^n$ and $w_{t't}$ represent the combined fraction of the demand $d_t^2$ of the second plant in period $t$ that is satisfied by production in period $t'$ in the second plant not involving the by-product, ($x_{t't}^n$), and that involving the by-product, $w_{t't}$, respectively. Constraints (9) and (10) ensure that demands in both plants should be fully satisfied. Constraints (11) and (12) assure that fixed set-up costs are paid and the capacity limit is respected in case of production. Constraint (13) indicates that the amount of demand satisfied through consuming by-products in the second plant must not exceed the available by-product generated in plant one. Non negativity and binary requirements are expressed by (14).

# 3    The Sample Average Approximation (SAA) Procedure

The SAA procedure is based on solving $M$ samples of the stochastic problem, under a limited number of $S$ scenarios taken from the original distributions. Assuming each replication is optimally solved, the $M$ problem's average objective value provides a statistical lower bound to the original problem. The $M$ problems' optimal solutions are then reevaluated under a more extensive set of scenarios ($S' \geq S$) to estimate their actual objective function values. The solution which achieves the lowest estimated cost is assumed to be the best upper bound to the original problem. The methodology presented in the following are derived from [7].

*Summary of SAA Methodology*

1. Generate $M$ samples each of size $S$, $(\xi_m^i)$ $i = 1, ..., S$, $m = 1, ..., M$.
2. For each sample m solve the problem $Z_S^m = min_{x \in X}\{c^T x + \frac{1}{S} \sum_{i=1}^{S} Q(x, \xi_m^i)\}$.
3. set $\hat{x}^{*m}$ as the candidate solution.
4. Equation $\bar{z}_S = \frac{1}{M} \sum_{m=1}^{M} z_S^m$ provides a statistical estimate for the lower bound.
5. For $S' \geq S$, $\hat{z}_{S'}(\hat{x}^m) = min\{c^T \hat{x}^m + \frac{1}{S'} \sum_{i=1}^{S'} Q(\hat{x}^m, \xi^i)\}$.
6. Select $\hat{x}^{*m} = \{\hat{x}^m : \hat{z}_{S'}(\hat{x}^m) = min_{1 \leq i \leq m} \hat{z}_{S'}(\hat{x}^i)\}$.
7. The optimal gap would be $\frac{\hat{z}_{S'} - \bar{z}_S}{\hat{z}_{S'}}$.

Our problem is a two-stage stochastic capacitated lot sizing-by-product in which the first stage variables are the setup variable before demand for both plants are revealed. After the demand is revealed, the second stage variables determine the amount of production, the inventory level of the main product, and the amount of by-product that will be transferred. Below, we provide details on our implementation of the SAA procedure for our problem. The objective function aims to minimize the first stage costs and the mean of second stage variables costs under a defined set of scenarios. The stochastic formulation considering a limited set of $M$ samples each of $S$ scenarios (IS-SCLSP-SAA) is defined as follows, subject to Constraints (2)–(7):

$$\text{IS-SCLSP-SAA:} \quad min \sum_{n=1}^{N} \sum_{t=1}^{T} f_t^n \cdot y_t^n + \frac{1}{S} \cdot [\sum_{i=1}^{S} \varphi(y, \xi_m^i)| (2) - (7)] \quad (15)$$

$$\varphi(y, \xi_m^i) = min \sum_{n=1}^{N} \sum_{t=1}^{T} (p_t^n \cdot x_t^n + h_t^n \cdot I_t^n) + \sum_{t=1}^{T} (q_t \cdot w_t + \hat{p}_t \cdot \beta \cdot w_t) \quad (16)$$

# 4    Computational Experiments

We conduct our computational experiments on the data set described in [6] for CLSP with deterministic demand. As some of our required information was not defined in this data set, we randomly generated the data associated with production and transportation cost to solve our defined problem. The research aims to optimize total costs associated with setup, production, inventory, and by-product transportation cost for two collaborating production plants during a six-week planning horizon. It has been assumed that the generation rate of by-product from production in plant 1 ($\alpha$) is equal to 0.3, and the production rate of the finished product in plant 2 through consumption per unit of by-product ($\beta$) is equal to 0.5. The available production capacity at each planning period is calculated as follows $cap_t^n = \sum_t^T d_t^n * \gamma$, in which a predefined percentage ($\gamma = 0.85$) of cumulative demand will be satisfied from the same period to the end of the planning horizon. We set the scale of gamma distribution for demand to $\lambda = 1$. Each model described in this paper was implemented using the Julia 1.4.2 programming language and solved using Gurobi 9.0.2. We performed all experiments on an Intel(R) Core(TM) i7-7700HQ CPU @ 2.80 GHz, with 32 GB of RAM.

## 4.1   The Sample Average Approximation Procedure

We implemented the SAA approach on our primary formulation. For one replication, we run the experiment for different sample sizes varying from 10 to 100, assuming that demand follows gamma distribution with scale 1. Table 1 shows the results. The first and second columns represent the lower and upper bounds of the problem, respectively. The gap in the third column refers to the relative optimality gap computed as $(\frac{UB-LB}{UB}) * 100$. The fourth column represents the computational time. As it can be seen, the optimality gap for different sample sizes is relatively high. To obtain a better optimality gap, we conduct the same experiments on a reformulated version of the main problem discussed in Sect. 2.2. We carried out an analysis of the effect of reformulation on improving the optimality gap. As we can observe from Table 1, there is a reduction in the gap but a slight increase in the computational time for the reformulated version. As shown in the reformulated section in Table 1 scenario size 100, has the slightest optimality gap among other sample sizes. Therefore, 100 will be the selected number of scenarios for testing the SAA procedure under different sample sizes.

Table 2 and 3 present the results of the SAA procedure for 100 scenarios under different sample sizes. The sample size $s'$ to obtain the upper bound is considered to be 500. Due to statistical computation of the bounds, it is probable that the computed lower bound exceeds the upper bound. In this regard, to still have a good indication of how good the solution is, we consider the lower bound to be the minimum of all the lower bounds and the upper bound to be the average of all the upper bounds achieved under different sample sizes in the fourth and fifth columns, respectively, and this for both formulations. For ten replications, we obtain the best optimality gap. Therefore we set the number of scenarios and sample size to 100 and 10 respectively to compare over two probability distribution outcomes on obtained optimality gap and variance of lower bound. Figure 1 illustrates the difference in achieved optimality gaps for both reformulations applied for sample sizes 10, 20, 30, 40, 50 and 60, respectively, each of 100 scenarios. As we can see, sample size 10 in plant location formulation has the best performance in reducing the optimality gap. On average, the plant location formulation provides almost 22 % reduction in optimality gap compared to the natural formulation. The comparison in computational time for both formulations with the mentioned sample sizes is depicted in Fig. 2. On average, the plant location formulation takes 1.7 times longer time. Table 4 explains the comparison of the SAA performance in terms of the optimality gap and the variance in the obtained objective function lower bounds. The experiments were conducted with demand data following Gamma and Normal distributions with scale and standard deviation set to 0.5, 1, 1.5 and 2. As it can be seen, there is a significant difference in the optimality gap as well as in the obtained lower bound variance, which indicates the importance of better estimating the probability distribution of the demand parameters.

**Table 1.** Results of the SAA for $M = 1$, with a gamma distribution of scale $= 1$

| Size | Main formulation | | | | Plant-location formulation | | | |
|------|--------|--------|---------|-------|--------|--------|---------|-------|
|      | LB | UB | Gap (%) | CT(s) | LB | UB | Gap (%) | CT(s) |
| 10 | 15 510 | 15 565 | 0.35 | 6.10 | 15 368 | 15 431 | 0.40 | 9.40 |
| 20 | 15 540 | 15 585 | 0.28 | 5.70 | 15 337 | 15 412 | 0.48 | 9.50 |
| 30 | 15 479 | 15 583 | 0.66 | 6.0 | 15 439 | 15 457 | 0.11 | 9.60 |
| 40 | 15 597 | 15 630 | 0.21 | 5.30 | 15 375 | 15 388 | 0.08 | 9.20 |
| 50 | 15 553 | 15 589 | 0.23 | 6.10 | 15 353 | 15 423 | 0.45 | 10.0 |
| 60 | 15 434 | 15 560 | 0.80 | 4.90 | 15 387 | 15 396 | 0.05 | 9.50 |
| 70 | 15 487 | 15 574 | 0.55 | 5.10 | 15 333 | 15 349 | 0.10 | 10.0 |
| 80 | 15 564 | 15 597 | 0.21 | 5.01 | 15 390 | 15 392 | 0.01 | 9.60 |
| 90 | 15 576 | 15 623 | 0.30 | 5.40 | 15 428 | 15 453 | 0.16 | 9.50 |
| **100** | **15 582** | **15 606** | **0.10** | 4.90 | **15 385** | **15 385** | $\approx$**0.0** | 9.60 |

**Table 2.** Results for SAA implemented for main formulation with 100 scenarios and a gamma distribution of scale $= 1$

| Rep | LB | UB | $Min_{LB}$ | $\overline{UB}$ | Gap (%) | $\sigma_{LB}^2$ | CT (s) |
|-----|----|----|-----------|-----------------|---------|-----------------|--------|
| **10** | **15579** | **15566** | **15520** | **15585** | **0.41** | **101.0** | **34.8** |
| 20 | 15590 | 15546 | 15487 | 15584 | 0.62 | 95.8 | 68.5 |
| 30 | 15589 | 15556 | 15503 | 15589 | 0.55 | 60.5 | 102.5 |
| 40 | 15593 | 15552 | 15495 | 15588 | 0.59 | 71.2 | 161.0 |
| 50 | 15576 | 15541 | 15495 | 15588 | 0.70 | 44.2 | 170.0 |
| 60 | 15575 | 15543 | 15492 | 15584 | 0.59 | 24.0 | 201.0 |

**Table 3.** Results for SAA implemented for plant location formulation with 100 scenarios with gamma distribution of scale = 1

| Rep | LB | UB | $Min_{LB}$ | $\overline{UB}$ | Gap (%) | $\sigma^2_{LB}$ | CT (s) |
|---|---|---|---|---|---|---|---|
| **10** | **15411** | **15391** | **15381** | **15412** | **0.20** | **54.3** | **92** |
| 20 | 15413 | 15393 | 15336 | 15409 | 0.47 | 77.5 | 188 |
| 30 | 15409 | 15336 | 15333 | 15408 | 0.48 | 50.5 | 270 |
| 40 | 15413 | 15356 | 15324 | 15404 | 0.51 | 49.3 | 377 |
| 50 | 15409 | 15356 | 15317 | 15407 | 0.58 | 34.9 | 434 |
| 60 | 15407 | 15359 | 15329 | 15407 | 0.50 | 41.9 | 529 |

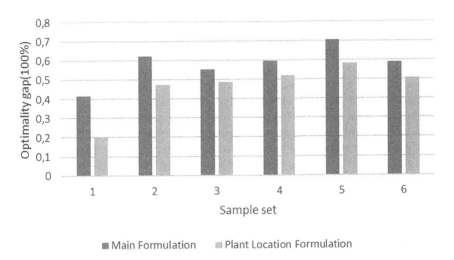

**Fig. 1.** Comparison between optimality gaps provided by each of the formulations

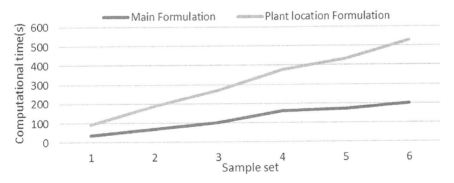

**Fig. 2.** Comparison between computational times for each of the formulations

**Table 4.** Results of the SAA procedure for Gamma and Normal distributions with 10 samples and 100 scenarios

| Gamma distribution | | | | | Normal distribution | | | | |
|---|---|---|---|---|---|---|---|---|---|
| Scale | $Min_{LB}$ | $\overline{UB}$ | Gap (%) | $\sigma^2_{LB}$ | $\sigma$ | $Min_{LB}$ | $\overline{UB}$ | Gap (%) | $\sigma^2_{LB}$ |
| 0.5 | 15368 | 15401 | 0.21 | 104 | 0.5 | 15404 | 15406 | 0.01 | 0.61 |
| 1 | 15520 | 15585 | 0.41 | 101 | 1 | 15399 | 15404 | 0.03 | 1.27 |
| 1.5 | 15324 | 15406 | 0.41 | 278 | 1.5 | 15396 | 15405 | 0.05 | 3.75 |
| 2 | 15289 | 15402 | 0.41 | 627 | 2 | 15397 | 15407 | 0.06 | 4.84 |

## 5   Conclusion and Perspectives

This paper investigates a new version of the capacitated lot-sizing problem, under demand uncertainty, arising in a supply chain transitioning to circular economy business model. The problem is formulated as an integrated capacitated lot-sizing problem for a supply chain involving two collaborating plants in form of an industrial symbiosis. In this model, the second plant uses as raw material a by-product obtained as a residue generated by the production in the first plant. In this analysis, it is assumed that the product demand of two cooperating plants follow gamma and normal probability distributions. To solve the resulting problem the Sample Average Approximation (SAA) method was implemented. The SAA approach's performance in terms of optimality gaps has been investigated in the original model and its plant-location reformulation version. The results indicate that the reformulated version has better performance in terms of optimality gap and upper and lower bounds but leads to higher computational time in comparison with the original formulation. Some extensions of the supply chain as well as the solution methodology are currently investigated.

## References

1. Barany, I., Van Roy, T., Wolsey, L.A.: Uncapacitated lot-sizing: the convex hull of solutions. In: Korte, B., Ritter, K. (eds.) Mathematical Programming at Oberwolfach II, vol. 22, pp. 32–43. Springer, Heidelberg (1984). https://doi.org/10.1007/BFb0121003
2. Daquin, C., Allaoui, H., Goncalves, G., Hsu, T.: Collaborative lot-sizing problem for an industrial symbiosis. IFAC-PapersOnLine **52**(13), 1325–1330 (2019)
3. Lee, D.: Turning waste into by-product. Manuf. Serv. Oper. Manage. **14**(1), 115–127 (2012)
4. Suzanne, E., Absi, N., Borodin, V.: Towards circular economy in production planning: challenges and opportunities. Eur. J. Oper. Res. **287**(1), 168–190 (2020)
5. Suzanne, E., Absi, N., Borodin, V., van den Heuvel, W.: A single-item lot-sizing problem with a by-product and inventory capacities. Eur. J. Oper. Res. **287**(3), 844–855 (2020)

6. Trigeiro, W.W., Thomas, L.J., McClain, Jo.: Capacitated lot sizing with setup times. Manage. Sci. **35**(3), 353–366 (1989)
7. Verweij, B., Ahmed, S., Kleywegt, A.J., Nemhauser, G., Shapiro, A.: The sample average approximation method applied to stochastic routing problems: a computational study. Comput. Opt. Appl. **24**(2), 289–333 (2003)

# Three-Phase Method for the Capacitated Lot-Sizing Problem with Sequence Dependent Setups

François Larroche[1,2(✉)], Odile Bellenguez[1], and Guillaume Massonnet[1]

[1] IMT Atlantique, LS2N, La Chantrerie, 4 rue Alfred Kastler, 44307 Nantes, France
`francois.larroche@imt-atlantique.fr`
[2] VIF, 10 Rue de Bretagne, 44240 La Chapelle-sur-Erdre, France

**Abstract.** This paper focuses on an industrial lot-sizing and scheduling problem that arises in the food industry and includes lost sales, over-times and sequence-dependent setups on parallel machines. We propose a preliminary version of a three-phase iterative approach to optimize separately the affectation, the sequencing and the production of items. Our first numerical results suggest that with some additional improvements, this approach could be use in real-life by planners to reduce their costs.

**Keywords:** Lot-sizing · Iterative method · Scheduling

## 1 Introduction

This paper presents a new heuristic to solve a lot-sizing and scheduling problem that arises in practical cases from the food industry. The problem we address is large and complex, which real-life applications often involve multiple production machines and specific constraints, which makes this problem complex to solve and motivates the planners to use optimization tools in order to reduce the costs incurred during their production process. However, modeling such complex systems leads to large mathematical formulations that are (in general) intractable with commercial solvers for Mixed Integer Programs (MIP). As a consequence, we aim to develop new approaches that enable us to obtain solutions to this problem in a quick and efficient fashion. Our work focuses on a multi-item capacitated lot-sizing problem including lost sales, safety stock, overtimes and sequence dependent setups on parallel machines. The discrete lot-sizing is a classical problem in the Operations Research literature since the work of Wagner and Whitin [13] and it has since been enriched with various additional considerations in order to model more accurately practical situations. Limited production capacity is among the most popular extensions of the original problem and has been the topic of multiple surveys, see e.g. [9] and [12]. The model we consider further generalizes this setting to include lost sales and shortage costs, as in [2] where the authors introduce new classes of valid inequalities. Parallel resources production is more and

© IFIP International Federation for Information Processing 2021
Published by Springer Nature Switzerland AG 2021
A. Dolgui et al. (Eds.): APMS 2021, IFIP AICT 631, pp. 694–702, 2021.
https://doi.org/10.1007/978-3-030-85902-2_74

more common in the literature, with a distinction between identical [3] and non-identical [8] machines. We focus on the latter in this paper.

The integration of lot-sizing and scheduling constraints greatly increases the complexity of the problems, which often leads researchers to propose heuristic solutions. Notable ones include [7], who develop a MIP-based neighborhood search heuristics to address a lot-sizing and scheduling problem on parallel machines. [4] present an industrial problem in which setup times depend on the sequence of production and propose a solution procedure based on sub-tour elimination and patching. [6] reviews the main modeling techniques for this class of problems and compare the efficiency of several solution methods. Iterative heuristics has been studied in the literature for the capacitated lot-sizing and scheduling problem in [11]. In [5], the authors used a two-level method to deal with the lot-sizing and the scheduling phases separately. In the related field of Production Routing problem, [1] use a two-phase method to dissociate the production and the distribution part.

In this paper, we propose a three-phase iterative method for the production problem described above, that has been introduced in [10] and is referred to as CLSSD-PM for *multi-item capacitated lot-sizing problem with lost sales, safety stock, overtimes, and sequence-dependent setups on parallel machines*. Section 2 describes the notations and assumptions. In Sect. 3, we present an original three-phase iterative method, with a quick description of each phase. In Sect. 4, we succinctly present the preliminary results obtained on a set of instances generated from industrial data. Finally, we propose some ideas to improve the procedure as well as perspectives for future research in the Sect. 5.

## 2   Notations and Assumptions

The problem studied in this paper is directly related to the problem originally presented in [10]. For an extensive presentation, the reader can refer to this paper. In this problem, we consider $N$ different items that can be produced over a discrete finite horizon of $T$ periods and $M$ non-identical parallel machines. We denote $\mathcal{N}$, $\mathcal{M}$ and $\mathcal{T}$ the set of items, machines and periods, respectively. For all items $i \in \mathcal{N}$, we consider a deterministic demand $d_t^i$ in each period $t \in \mathcal{T}$, which is either satisfied from on-hand inventory or lost. We denote $\tau_m^i$ the production time of one unit of item $i$ on machine $m$. Sequence-dependent setup times are denoted $\lambda_m^{ij}$ for a given machine $m$ and for each pair $(i, j)$ of items. Note that we allow asymmetric setup times but assume they satisfy the triangle inequality. For each machine $m \in \mathcal{M}$, we consider a time-dependent capacity $C_{mt}$ corresponding to the total time available in period $t$. However, in some circumstances this capacity can be exceeded up to a larger value $\overline{C}_{mt}$ which is the "true" available time capacity in period $t$. In that case, $\overline{C}_{mt} - C_{mt}$ corresponds to the maximum overtime allowed. Finally, we also impose that every production of item $i$ have to be greater than a minimum production quantity denoted $q_{min}^i$. The objective is to minimize the costs incurred in the production problem. We denote $p_{mt}^i$ the cost of producing one unit of item $i$ in period $t$ on machine $m$. In addition, we denote $c_{mt}$ the per-unit of time usage cost of machine $m$ during period $t$. For each unit of overtime, an additional cost $\bar{c}_{mt}$ is applied. We define the safety stock as an "optimal" inventory level to reach in each period. Inventory-related cost are then defined with respect to its value, namely a deficit cost $h_t^{i-}$ and an overstock cost $h_t^{i+}$ for each unit below or above the safety stock level. Finally, for each unmet unit of demand for item $i$ in period $t$, a shortage cost $l_t^i$ is incurred.

Throughout the remainder of this paper, we use the following decision variables:

$x_{mt}^i \in \mathbb{R}_+$  Quantity of item $i$ produced in period $t$ on machine $m$

$y_{mt}^i \in \{0,1\}$  binary variables equal to 1 if item $i$ is affected to machine $m$ in period $t$

$U_{mt} \in \mathbb{R}_+$  Time usage of machine $m$ in period $t$

$O_{mt} \in \mathbb{R}_+$  Time in overtime of machine $m$ in period $t$

$L_t^i \in \mathbb{R}_+$  Quantity of lost sales for item $i$ in period $t$

$I_t^i \in \mathbb{R}_+$  Inventory of item $i$ on hand at the end of period $t$

$I_t^{i+} \in \mathbb{R}_+$  Overstock (based on safety stock value) of item $i$ at the end of period $t$

$I_t^{i-} \in \mathbb{R}_+$  Safety stock deficit of item $i$ at the end of period $t$

$z_t^i \in \{0,1\}$  Binary variable equals to 1 if the stock of item $i$ is null at the end of period $t$

## 3    Three-Phase Iterative Approach

In this section we present a three-phase iterative method to solve the problem under study. The central idea is to decompose the original problem into three (smaller) subproblems we can solve sequentially, where the output of a phase serves as an input for the following one. Below is an overview of one iteration of the heuristic:

- The first phase minimizes the costs induced by production and inventory management, considering virtual *sequence-independent* setup times per item. Capacity restrictions are also relaxed based on the information provided by the other phases.
- The second phase proposes production sequences for the items affected to each machine and period.
- The last phase decides the quantity to produce and pushes its output information into the first phase of the next iteration as an additional constraint.

The procedure loops through these different phases until a stopping criteria (either the time limit or a given number of iterations without improvement) is met.

**First Phase: Assignment.** We describe the first phase with a mathematical model based on an aggregate formulation. In this phase, we do not consider the sequence dependent setup times, which are approximated. For that purpose, we introduce $ST_{mt}^i$ the setup time for item $i$ for machine $m$ at period $t$ (set to 0 at the first iteration). The aggregate formulation is expressed with the following MIP:

$$\min \sum_{t\in\mathcal{T}}\sum_{m\in\mathcal{M}} (c_{mt}U_{mt} + \bar{c}_{mt}O_{mt})$$

$$+ \sum_{t\in\mathcal{T}}\sum_{i\in\mathcal{N}} \left( h_t^{i+}I_t^{i+} + h_t^{i-}I_t^{i-} + l_t^i L_t^i + \sum_{m\in\mathcal{M}} p_{mt}^i x_{mt}^i \right) \qquad (1)$$

$$\text{s.t.} \quad U_{mt} = \sum_{i\in\mathcal{N}} (\tau_m^i x_{mt}^i + ST_{mt}^i y_{mt}^i) \qquad \forall m \in \mathcal{M}, \forall t \in \mathcal{T} \qquad (2)$$

$$U_{mt} \leq C_{mt} + O_{mt} \qquad \forall m \in \mathcal{M}, \forall t \in \mathcal{T} \qquad (3)$$

$$I_t^i = S_t^i + I_t^{i+} - I_t^{i-} \qquad \forall i \in \mathcal{N}, \forall t \in \mathcal{T} \qquad (4)$$

$$I_t^i = I_{t-1}^i + \sum_{m\in\mathcal{M}} x_{mt}^i + L_t^i - d_t^i \quad \forall i \in \mathcal{N}, \forall t \in \mathcal{T} \qquad (5)$$

$$I_t^i \leq D_{tT}^i(1 - z_t^i) \qquad \forall i \in \mathcal{N}, \forall t \in \mathcal{T} \qquad (6)$$

$$L_t^i \leq z_t^i d_t^i \qquad \forall i \in \mathcal{N}, \forall t \in \mathcal{T} \qquad (7)$$

$$x_{mt}^i \leq D_{t-1,T}^i y_{mt}^i \qquad \forall i \in \mathcal{N}, \forall m \in \mathcal{M}, \forall t \in \mathcal{T} \qquad (8)$$

$$x_{mt}^i \geq q_{\min}^i y_{mt}^i \qquad \forall i \in \mathcal{N}, \forall m \in \mathcal{M}, \forall t \in \mathcal{T} \qquad (9)$$

$$L_t^i, I_t^i, I_t^{i+}, I_t^{i-} \geq 0 \qquad \forall i \in \mathcal{N}, \forall t \in \mathcal{T} \qquad (10)$$

$$U_{mt}, O_{mt} \geq 0 \qquad \forall m \in \mathcal{M}, \forall t \in \mathcal{T} \qquad (11)$$

$$y_{mt}^i \in \{0,1\} \qquad \forall i \in \mathcal{N}, \forall m \in \mathcal{M}, \forall t \in \mathcal{T} \qquad (12)$$

$$z_t^i \in \{0,1\} \qquad \forall i \in \mathcal{N}, \forall t \in \mathcal{T} \qquad (13)$$

$$x_{mt}^i \geq 0 \qquad \forall i \in \mathcal{N}, \forall m \in \mathcal{M}, \forall t \in \mathcal{T} \qquad (14)$$

The objective (1) is to minimize the cost of production including line usage costs, lost sales and inventory stocks. Constraints (2) define the total working time of each machine which is equal to the production time and the approximated setup times. Constraints (3) define the time usage and overtime variables. Constraints (4) define the inventory stock from the safety stock, the overstock and the deficit. Constraints (5) are the inventory flow conservation equations through the planning time horizon. Constraints (6) and (7) ensure that lost sales for item $i$ occur only when the corresponding stock is null. Constraints (8) use the cumulative demand over the remainder of the horizon as an upper bound on the quantity of each item produced on each machine in a given period. Constraints (9) force the production to be higher than its minimum value. Finally constraints (10) and (14) define the domain of specific variables.

An additional constraint is necessary to link the information from the following phases after the first iteration. We denote $\tilde{y}_{mt}^i$ and $\tilde{x}_{mt}^i$ as the value of the corresponding variables $y_{mt}^i$ and $x_{mt}^i$ decided in the previous iteration. In addition, we also define for each machine $m$ and period $t$ the residual capacity $Cap_{mt}^{res}$ as the remaining available machine time after the previous iteration of the algorithm. Before the first iteration, we initialize these values as follows:

$\widetilde{y}_{mt}^i = 0$, $\widetilde{x}_{mt}^i = 0$ and $Cap_{mt}^{res} = \overline{C}_{mt}$. The following constraints (15) then indicate that for each item, the production time is bounded by the last solution but can be increased or decreased depending on items that are added to/removed from the assignment derived in the previous iteration and the residual capacity. Parameters $\sigma_{mt}^i$ then correspond to the proportion of the production time available that is allocated to item $i$.

$$\tau_m^i x_{mt}^i + (1 - \widetilde{y}_{mt}^i)ST_{mt}^i y_{mt}^i \leq \tau_m^i \widetilde{x}_{mt}^i + \left( \sum_{\substack{j \neq i \\ \widetilde{x}_{mt}^j \neq 0}} (1 - y_{mt}^j)(\tau_m^j \widetilde{x}_{mt}^j + ST_{mt}^j) \right.$$

$$\left. + Cap_{mt}^{res} - \sum_{\substack{j \neq i \\ \widetilde{x}_{mt}^j = 0}} (\tau_m^j q_{\min}^j + ST_{mt}^j)y_{mt}^j \right)\sigma_{mt}^i \quad (15)$$

$$\forall i \in \mathcal{N}, \forall m \in \mathcal{M}, \forall t \in \mathcal{T}$$

**Second Phase: Sequencing.** From the first phase, we obtain the assignment of items to each pair machine-period $(m, t)$. The goal of this phase is then to solve $M \times T$ sequencing problems with a straightforward mathematical formulation adapted from the TSP problem. For conciseness we do not detail it in this paper.

**Third Phase: Production.** The last phase solves the production problem. We denote $\Lambda_{mt}$ the sum of setup times on machine $m$ in period $t$ obtained from the previous phase. If $\Lambda_{mt} + \sum_{i \in \mathcal{N}} q_{\min}^i \widetilde{y}_{mt}^i > \overline{C}_{mt}$, i.e. the capacity is not sufficient to carry out the setup times and the minimum production for the current affectation, we set $\Lambda_{mt} = 0$ and $\widetilde{y}_{mt}^i = 0$ for all $i \in \mathcal{N}$. Otherwise, the values $\Lambda_{mt}$ and $\widetilde{y}_{mt}^i$ remain the same. Note that since the assignment decisions $\widetilde{y}_{mt}^i$ are fixed in the first phase, they are considered as parameter in this step and thus the only binary variables considered are $z_t^i$ to ensure a First-Come First-Serve discipline, which make the problem easy to solve.

$$\min \sum_{t \in \mathcal{T}} \sum_{m \in \mathcal{M}} (c_{mt}U_{mt} + \bar{c}_{mt}O_{mt})$$

$$+ \sum_{t \in \mathcal{T}} \sum_{i \in \mathcal{N}} \left( h_t^{i+} I_t^{i+} + h_t^{i-} I_t^{i-} + l_t^i L_t^i + \sum_{m \in \mathcal{M}} p_{mt}^i x_{mt}^i \right) \quad (16)$$

$$\text{s.t. } U_{mt} = \sum_{i \in \mathcal{N}} \tau_m^i x_{mt}^i + \Lambda_{mt} \qquad \forall m \in \mathcal{M}, \forall t \in \mathcal{T} \quad (17)$$

$$U_{mt} \leq C_{mt} + O_{mt} \qquad \forall m \in \mathcal{M}, \forall t \in \mathcal{T} \quad (18)$$

$$O_{mt} \leq \overline{C}_{mt} - C_{mt} \qquad \forall m \in \mathcal{M}, \forall t \in \mathcal{T} \quad (19)$$

$$\text{Constraints } (4)-(7)$$

$$x_{mt}^i \leq D_{t-1,T}^i \widetilde{y}_{mt}^i \qquad \forall i \in \mathcal{N}, \forall m \in \mathcal{M}, \forall t \in \mathcal{T} \quad (20)$$

$$x_{mt}^i \geq q_{\min}^i \widetilde{y}_{mt}^i \qquad \forall i \in \mathcal{N}, \forall m \in \mathcal{M}, \forall t \in \mathcal{T} \quad (21)$$

$$\text{Constraints } (10), (11), (13), (14)$$

**Update the Parameters of the First Phase.** At the end of the third phase, there is a crucial update step on the input parameters of the first phase. The algorithm updates the individual setup times based on their marginal contribution to the production sequence computed in the second step. That is, for each item $i$ in the sequence of production on machine $m$ in period $t$, the setup time $ST_{mt}^i$ corresponds to the additional time needed to insert this item is the sequence based on its direct predecessor and successor. On the other hand, we use the cheapest insertion time for all items that are not already included in the current sequence. This idea is described more in details in [1]. Current affectation decisions are stored in the variables $\tilde{y}_{mt}^i$, while *a priori* production quantities $\tilde{x}_{mt}^i$ are directly extracted from the production decisions $x_{mt}^i$ in the third phase. The three-phase algorithm is presented in flowchart 1.

## 4   Experimental Results

In this section, we present the experimental results obtained with our three-phase method. Our instances are designed from actual data from industrial cases. We derived 48 different instances combining the following parameters: $N \in \{20, 30, 40\}$, $T \in \{15, 30\}$, $M \in \{1, 2\}$. Note that these preliminary tests are not representative of the industrial reality, where the number of different items can be significantly larger than this benchmark. We use IBM Ilog CPLEX v12.10.0 to solve each phase and set a time limit of 900 s as a stopping criterion. If $UB$ corresponds to the best feasible solution found by the method tested, the gap is defined as follows:

$$\text{Gap} = 100 \cdot \frac{UB - LB}{LB}, \tag{22}$$

where $LB$ is the best lower bound obtained after 4 h with CPLEX. Table 1 presents a Gap comparison between the 3-phase method and a straightforward MIP resolution in 900 s by CPLEX. We set $\sigma_{mt}^i = 1/N$ in constraints (15). Although the results suggest that our method needs refinement before we consider applying it on larger instances, it shows some promises as a decision support tools for practitioners. In particular for larger instances, the 3-phase method seems to be more robust to obtain adequate solution. We observe that the computational time is mainly decided by the first and the second phase while the third one is solved quickly. For instances with $N \geq 30$, the second phase take a significant amount of the total computational time (Fig. 1).

## 5   Perspectives

We study a problem encountered in the food industry. Since the problem is too complex to obtain good solution using a straightforward MIP resolution, we base our approach on an iterative resolution from a decomposition into three smaller subproblems. Our first preliminary results suggest that some enhancement must

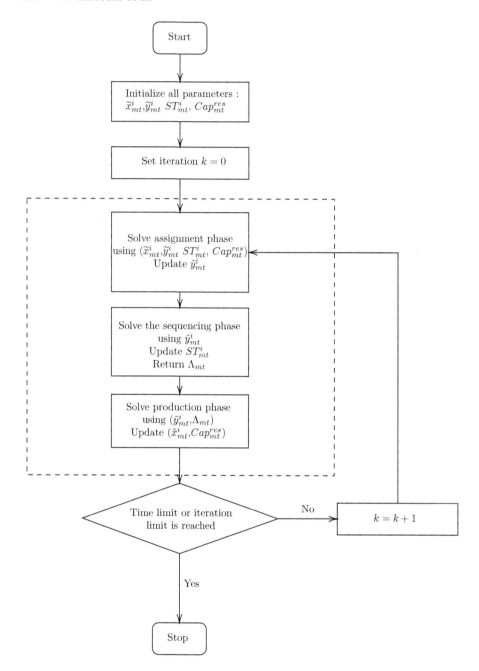

**Fig. 1.** Flowchart of the 3-phase approach

**Table 1.** Average gaps obtained with the 3-phase method

| Problem size | Avg. gap (%) | |
|---|---|---|
| N-M-T | 3-phase | CPLEX |
| 20-1-15 | 77.57 | 0.57 |
| 20-1-30 | 47.51 | 3.15 |
| 20-2-15 | 37.4 | 1.28 |
| 20-2-30 | 73.9 | 2.01 |
| 30-1-15 | 47.72 | 3.1 |
| 30-1-30 | 97.23 | 17.68 |
| 30-2-15 | 60.64 | 6.95 |
| 30-2-30 | 118.42 | 27.29 |
| 40-1-15 | 289.74 | 585.16 |
| 40-1-30 | 372.61 | 2051.43 |
| 40-2-15 | 112.14 | 16.76 |
| 40-2-30 | 154.04 | 1053.87 |

be implemented to improve the method. To that end, we could consider the addition of diversification mechanisms in the first phase to speed up the convergence. Another direction under investigation rely on the clustering approach introduced in [10] to simplify the second phase and compute efficiently good production sequences. Another research direction is to use heuristics or MIP-heuristics in the first and second phases to speed up the resolution.

# References

1. Absi, N., Archetti, C., Dauzère-Pérès, S., Feillet, D.: A two-phase iterative heuristic approach for the production routing problem. Transp. Sci. **49**(4), 784–795 (2015)
2. Absi, N., Kedad-Sidhoum, S.: The multi-item capacitated lot-sizing problem with setup times and shortage costs. Eur. J. Oper. Res. **185**(3), 1351–1374 (2008)
3. Beraldi, P., Ghiani, G., Grieco, A., Guerriero, E.: Rolling-horizon and fix-and-relax heuristics for the parallel machine lot-sizing and scheduling problem with sequence-dependent set-up costs. Comput. Oper. Res. **35**(11), 3644–3656 (2008)
4. Clark, A.R., Morabito, R., Toso, E.A.V.: Production setup-sequencing and lot-sizing at an animal nutrition plant through ATSP subtour elimination and patching. J. Sched. **13**(2), 111–121 (2010). https://doi.org/10.1007/s10951-009-0135-7
5. Dauzère-Péres, S., Lasserre, J.B.: Integration of lotsizing and scheduling decisions in a job-shop. Eur. J. Oper. Res. **75**(2), 413–426 (1994)
6. Guimarães, L., Klabjan, D., Almada-Lobo, B.: Modeling lotsizing and scheduling problems with sequence dependent setups. Eur. J. Oper. Res. **239**(3), 644–662 (2014)
7. James, R.J.W., Almada-Lobo, B.: Single and parallel machine capacitated lotsizing and scheduling: new iterative MIP-based neighborhood search heuristics. Comput. Oper. Res. **38**, 1816–1825 (2011)

8. Kaczmarczyk, W.: Modelling set-up times overlapping two periods in the proportional lot-sizing problem with identical parallel machines. DMMS **7**(1–2), 43 (2013)
9. Karimi, B., Fatemi Ghomi, S., Wilson, J.: The capacitated lot sizing problem: a review of models and algorithms. Omega **31**(5), 365–378 (2003)
10. Larroche, F., Bellenguez, O., Massonnet, G.: Clustering-based solution approach for a capacitated lot-sizing problem on parallel machines with sequence-dependent setups. Int. J. Prod. Res. (2021, Submitted)
11. Mateus, G.R., Ravetti, M.G., de Souza, M.C., Valeriano, T.M.: Capacitated lot sizing and sequence dependent setup scheduling: an iterative approach for integration. J. Sched. **13**(3), 245–259 (2010). https://doi.org/10.1007/s10951-009-0156-2
12. Quadt, D., Kuhn, H.: Capacitated lot-sizing with extensions: a review. 4OR **6**(1), 61–83 (2008). https://doi.org/10.1007/s10288-007-0057-1
13. Wagner, H.M., Whitin, T.M.: Dynamic version of the economic lot size model. Manag. Sci. **5**(1), 89–96 (1958)

# Optimization for Lot-Sizing Problems Under Uncertainty: A Data-Driven Perspective

Paula Metzker[1,2]([✉]), Simon Thevenin[1], Yossiri Adulyasak[2],
and Alexandre Dolgui[1]

[1] IMT Atlantique, LS2N - UMR CNRS 6004, 44307 Nantes, France
{paula.metzker-soares,simon.thevenin,alexandre.dolgui}@imt-atlantique.fr
[2] HEC Montréal, GERAD, Montreal, Canada
yossiri.adulyasak@hec.ca

**Abstract.** In a manufacturing context, the lot-sizing problems (LSP) determine the quantity to produce over a planning horizon. Often, the parameters used in the LSP models are unknown when the decisions are made, and this uncertainty has a critical impact on the quality of the decisions. However, the large amount of data that can nowadays be collected from the shop floor allows inferring information on the LSP parameters and their variability. Therefore, a recent research trend is to properly account for the uncertainty in the LSP optimization models. This work presents a survey on data-driven optimization approaches for the LSPs. We also provide a comparison of some promising optimization methodologies in the context of data-driven modeling of LSPs.

**Keywords:** Data-driven optimization · Lot-sizing problem · Optimization under uncertainties

## 1 Introduction

The lot-sizing problem (LSP) [16] determines the production lots over a planning horizon that minimize overall costs and maintains a satisfactory level of service. Due to its practical importance, the LSPs attracted a wide range of research from the manufacturing and mathematical optimization communities. In fact, production and distribution systems are settled in chaotic environment where production, quality, sales, purchasing, logistics, corporate, technical, accounting and marketing department are constantly affected by unexpected events. Thus, LSPs become inadequate to meet the needs of the industry if they are not simple enough to be adapted to changes in the environment [8].

The authors of this paper wish to thank the Region Pays de la Loire (www.paysdelaloire.fr) in France and the Canada Research Chair in Supply Chain Analytics (www.chaireanalytique.hec.ca) for financial support of this research.

© IFIP International Federation for Information Processing 2021
Published by Springer Nature Switzerland AG 2021
A. Dolgui et al. (Eds.): APMS 2021, IFIP AICT 631, pp. 703–709, 2021.
https://doi.org/10.1007/978-3-030-85902-2_75

Production and distribution systems face various sources of uncertainties (demand, lead time, production yield, among others) that affect the costs and service level associated with the lot-sizes. Traditionally, these systems dampen these uncertainties by changing parameters of the planning systems, such as the safety stock, safety lead-time, and re-planning frequency. Advances in computing technologies and the massive availability of data led to the design of data-driven optimizations to directly incorporate the uncertainties within the LSP, such as stochastic programming (SP) [5], robust optimization (RO) [1], and distributionally robust optimization (DRO) [17].

While SP models often seeks to minimize the expected costs over the distribution of the uncertain LSP parameters, RO models minimize the overall costs with regard to the worst-case realization of the LSP uncertainties. Finally, DRO extends stochastic optimization by taking into account the uncertain probability distributions of the unknown parameter. Even if these methodologies aim to optimize the LSPs by mitigating uncertainties, continual modification on LSP parameters leads to constant update of the production plans [7]. To overcome this issue, a data-based perspective of optimization methodologies emerges as a rather new and promising approach to compute production plans that are flexible to changes, and whose impacts due to unexpected events is more controllable.

The data-driven models often rely on a statistical analysis of the available data [3]. Bertsimas et al. [4] propose a data-driven approach based on sample approximation algorithms to choose the decision rules that perform best from the perspective of the worst-case within a stochastic process. Jiang and Guan [10] propose a data-driven methodology to obtain robust solutions from a chance-constrained problem with inaccurate probability distributions of the uncertain parameters. Then, they proposed a data-driven approach to solve the LSP under demand uncertainty based on sample average approximation algorithm [11]. Ning et al. [12] propose some artificial intelligence techniques have been investigated for labeling the available uncertain data and to compute near-optimal solutions through a data-driven approach. In addition, they present a data-driven via the DRO methodology. A state-of-the-art in data analytics and machine learning methods for process manufacturing in the light of big data approaches is presented in [13]. Zhao et al. [19] proposes a data-driven approach based on the kernel density estimation to represent the uncertain parameters into the optimization problems based on information from the historical data.

Although data-driven optimization emerges as a rather novel methodology to deal with non-deterministic optimization problems, a data-driven perspective of the LSPs is still missing in the literature. This has inspired us to develop a survey on data-driven optimization approaches applied to the LSPs. We are not interest in an exhaustive literature review, but in a survey of the existing literature on the data-driven optimization for LSP models. The remainder of this paper is organized as follows: Sect. 2 presents the application of the different optimization approaches starting from the data up to implementable decisions. Section 3 provides the advantages and issues of these methods in terms of computation time, tractability, flexibility in handling unforeseen events, and robustness of the solution. Section 4 gives the main research areas of the data-driven optimization to handle the LSPs. Finally, Sect. 5 summarizes the main findings of this work.

## 2    Data-Driven Optimizations for the LSPs

The main steps of the data-driven optimization (DDO) are: **i.** the definition of the uncertain parameters distribution characteristics; **ii.** the analysis and processing of available data (eventually coming from various sources) to learn how to represent the uncertainties; **iii.** the formulation and modeling of the problem within the perspective of a chosen optimization method. Figure 1 gives a schematic view on this methodology, and its steps are described in the rest of this section.

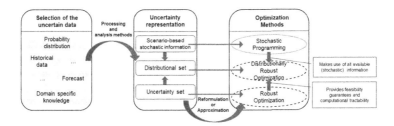

**Fig. 1.** DDO framework

**i. Selection of uncertain data:** On the one hand, ignoring uncertainties in LSPs leads to sub-optimal decisions. On the other hand, the inclusion of uncertainties increases the model complexity, and the solution might require large computation time and memory consumption. Consequently, the decision-maker must carefully analyze the historical data, forecast, probability distributions of data, experts insights, domain-specific knowledge, and any other available information to select the type of uncertainties to include in the optimization model. The decision-maker may consider the parameters whose value cannot be estimated accurately, and whose variance affects the decisions.

**ii. Uncertainty representation:** aims to incorporate partial information obtained from the uncertainties into the optimization methodology. For this, some data processing and analysis methods are used to manipulate uncertainties and extract as much useful and accurate information as possible. For the SP models, this step estimates the probability distributions with some statistical methods, such as analyzing historical data and the moment information, or some non-parametric statistical estimation [5]. For the RO models, the uncertainty sets are designed to preserve the computational tractability of robust models [2]. Consequently, these sets have well-defined structures such as *box and ellipsoid uncertainty set* [2]. Similarly, for the DRO, the distributional sets contain distributions with similar properties about the uncertainties, and these sets have well defined structures [18]. Among the more applied methods to build uncertainty and distributional sets from data, we cite the statistical hypothesis testing validation [3] and machine learning techniques [12].

**iii. Optimization methods:** The solution approaches for the LSPs usually rely on mixed-integer linear programming, and the choice of the model must be adapted to the production context. First, multiple formulations of the LSPs exist, and the most efficient ones change depending on the context [9]. Second, the incorporation of uncertainty in these models depends on the decision framework [15]. Within a static decision framework, the lot-sizes are fixed for the entire horizon, and so they are frozen. This situation corresponds to a two-stage stochastic optimization model or a classical robust optimization model. Within a dynamic decision framework, decisions can be updated in each period $t$ after that some uncertain parameters are revealed for periods up to $t$. This situation corresponds to a multi-stage stochastic optimization model or an adjustable robust optimization model. Finally, in a non-deterministic context, the models must include a mechanism to dampen uncertainty without escalating the costs. This mechanism may be a service level constraint [14], though an appropriate balance between the lost sales/backordering costs and inventory costs can also improve the quality of the solution.

Solving large-scale LSPs in an uncertain context can require intensive computations. To solve practical size instances with the SP method, the resolution approaches often rely on sampling methods, such as sample average approximation algorithm, or decomposition such as L-shaped, stochastic dual dynamic programming, or Progressive Hedging. The solution strategies for RO and DRO often cover the reformulation per constraint and dualization, and adversarial approaches, such as heuristics, branch-and-bound, or decomposition approaches.

# 3   Comparison of RO, SP, and DRO Methodologies

The choice of an optimization depends on decision makers' preferences, instance structures, available information, and expected trade-off in terms of solution quality and computing time. Although the problem involves three possible decision frameworks, namely static, static-dynamic and dynamic strategies, we focus our study on the static case. Based on the existing literature, we present a brief analysis of the performance of each method in terms of scalability, tractability, conservatism and flexibility of adaptation to unforeseen events.

First, RO can be used when little or no data is available, whereas SP requires large historic data to accurately estimate the probability distribution. The SP formulations often suffer from scalability issues, because the model must properly account for the uncertainty (often by relying on large scenario samples). On the contrary, RO approaches often remain tractable for practical size problems, when the robust model can be formulated as a convex problem.

RO typically optimizes the worst possible realization of the uncertain parameters, which leads to conservative solutions. On the contrary, the SP optimizes the expected costs, but it requires an sufficiently good probability distribution. Therefore SP solutions are poorly flexible to unforeseen or misrepresented events.

The DRO proposes a trade-off between these two approaches since it compensates the conservatism of the RO by taking advantage of partial distributional information obtained from the probability distributions from the SP framework. Hence, DRO emerges as a method sufficiently flexible to unforeseen events, while it remains computationally tractable, and it provides a less conservative solution.

# 4   Discussion

The LSPs have been studied for decades, but it still has room for improvements and further investigation. The SP, RO and DRO methodologies have stood out either for the quality of the solution, or for the ease of calculation within an uncertain environment. Most of the studies on optimization under uncertainties rely on statistical approaches to expound available data, and to reduce the conservatism of solutions. The growth of learning methods and the increase in data availability have motivated recent works to develop some data-based approaches that deal with the uncertain information [6].

Data-driven approaches enhance the quality and the performance of the methodologies for optimization under uncertainties. Among the methodologies presented in this work, a natural application of the data-driven methodology leads to distributionally robust optimization. Here, some data processing and analysis is implemented to extract more quality information from the decision context, and propose better predictions about the information of the uncertain parameters. Therefore, a worst-case perspective can be applied over all gathered distributional data. Thus, an optimization combining the expected value from the SP and the robustness from RO would propose more realistic solutions.

DDO is an emerging field of research, whose techniques, approaches, and applications are still under development. More applications should be analyzed to report the feasibility and tractability of the proposed approaches in real applications. On the other hand, further investigation into the application and feasibility of different data-driven approaches must be carried out to deal with different versions of LSPs, considering not only different versions of the problem but also different uncertain parameters. In addition, a deeper study of data processing, data analysis, and machine learning techniques is envisaged to develop data-driven approaches, and to better understand their challenges, limits, and prospects for improving different optimization methodologies.

# 5   Conclusion

There is a growing interest about data-driven optimization for lot-sizing problems. These methods learn the uncertainty representation from the data, and they incorporate these uncertainties in the lot-sizing models. The DRO can be applied to tackle different types of uncertainties which would derive more benefits from the DDO approach, being more stochastic or robust according to the decision maker's needs and decision environment. Although DRO is a promising method integrating data-driven approaches to the conception of flexible production plans, there is a lack of research on the DDO methods for the LSPs. Further studies should be envisaged to fulfill this knowledge gap.

# References

1. Ben-Tal, A., El Ghaoui, L., Nemirovski, A.: Robust Optimization. Princeton University Press, Princeton (2009)
2. Bertsimas, D., Brown, D.B., Caramanis, C.: Theory and applications of robust optimization. SIAM Rev. **53**(3), 464–501 (2011)
3. Bertsimas, D., Gupta, V., Kallus, N.: Data-driven robust optimization. Math. Program. **167**(2), 235–292 (2017). https://doi.org/10.1007/s10107-017-1125-8
4. Bertsimas, D., Shtern, S., Sturt, B.: A data-driven approach for multi-stage linear optimization. Available at Optimization Online (2018)
5. Birge, J.R., Louveaux, F.: Introduction to Stochastic Programming. Springer, New York (2011). https://doi.org/10.1007/978-1-4614-0237-4
6. Delage, E., Ye, Y.: Distributionally robust optimization under moment uncertainty with application to data-driven problems. Oper. Res. **58**(3), 595–612 (2010)
7. Dolgui, A., Ammar, O.B., Hnaien, F., Louly, M.A., et al.: A state of the art on supply planning and inventory control under lead time uncertainty. Stud. Inform. Control **22**(3), 255–268 (2013)
8. Dolgui, A., Prodhon, C.: Supply planning under uncertainties in MRP environments: a state of the art. Ann. Rev. Control **31**(2), 269–279 (2007)
9. Gruson, M., Bazrafshan, M., Cordeau, J.F., Jans, R.: A comparison of formulations for a three-level lot sizing and replenishment problem with a distribution structure. Comput. Oper. Res. **111**, 297–310 (2019)
10. Jiang, R., Guan, Y.: Data-driven chance constrained stochastic program. Math. Program. **158**(1–2), 291–327 (2016)
11. Jiang, R., Guan, Y.: Risk-averse two-stage stochastic program with distributional ambiguity. Oper. Res. **66**(5), 1390–1405 (2018)
12. Ning, C., You, F.: Data-driven stochastic robust optimization: general computational framework and algorithm leveraging machine learning for optimization under uncertainty in the big data era. Comput. Chem. Eng. **111**, 115–133 (2018)
13. Shang, C., You, F.: Data analytics and machine learning for smart process manufacturing: recent advances and perspectives in the big data era. Engineering **5**(6), 1010–1016 (2019)
14. Tempelmeier, H.: Stochastic lot sizing problems. In: Smith, J., Tan, B. (eds.) Handbook of Stochastic Models and Analysis of Manufacturing System Operations. ISOR, vol. 192, pp. 313–344. Springer, New York (2013). https://doi.org/10.1007/978-1-4614-6777-9_10
15. Thevenin, S., Adulyasak, Y., Cordeau, J.F.: Material requirements planning under demand uncertainty using stochastic optimization. Prod. Oper. Manag. **30**(2), 475–493 (2020)
16. Wagner, H.M., Whitin, T.M.: Dynamic version of the economic lot size model. Manag. Sci. **5**(1), 89–96 (1958)
17. Wiesemann, W., Kuhn, D., Sim, M.: Distributionally robust convex optimization. Oper. Res. **62**(6), 1358–1376 (2014)

18. Zhang, Y., Shen, Z.J.M., Song, S.: Distributionally robust optimization of two-stage lot-sizing problems. Prod. Oper. Manag. **25**(12), 2116–2131 (2016)
19. Zhao, L., Ning, C., You, F.: A data-driven robust optimization approach to operational optimization of industrial steam systems under uncertainty. Comput. Aided Chem. Eng. **46**, 1399–1404 (2019)

# Author Index

Printed in the United States
by Baker & Taylor Publisher Services